BIOCHEMISTRY

BIOCHEMISTRY

Lubert Stryer
YALE UNIVERSITY

W. H. FREEMAN AND COMPANY
San Francisco

Library of Congress Cataloging in Publication Data

Stryer, Lubert.
    Biochemistry.

    Includes bibliographies.
    1. Biological chemistry. I. Title.
[DNLM: 1. Biochemistry. QU4 S928b]
QP514.2.S66   574.1′92   74-23269
ISBN 0-7167-0174-X

Copyright © 1975 by Lubert Stryer

Printed in the United States of America

4   5   6   7   8   9

To my teachers

> *Paul F. Brandwein*
> *Daniel L. Harris*
> *Douglas E. Smith*
> *Elkan R. Blout*
> *Edward M. Purcell*

PREFACE

This book is an outgrowth of my teaching of biochemistry to undergraduates, graduate students, and medical students at Yale and Stanford. My aim is to provide an introduction to the principles of biochemistry that gives the reader a command of its concepts and language. I also seek to give an appreciation of the process of discovery in biochemistry. My exposition of the principles of biochemistry is organized around several major themes:

1. Conformation—exemplified by the relationship between the three-dimensional structure of proteins and their biological activity
2. Generation and storage of metabolic energy
3. Biosynthesis of macromolecular precursors
4. Information—storage, transmission, and expression of genetic information
5. Molecular physiology—interaction of information, conformation, and metabolism in physiological processes

The elucidation of the three-dimensional structure of proteins, nucleic acids, and other biomolecules has contributed much in recent years to our understanding of the molecular basis of life. I have emphasized this aspect of biochemistry by making extensive use of molecular models to give a vivid picture of architecture and dynamics at the molecular level. Another stimulating and heartening aspect of contemporary biochemistry is its increasing interaction with medicine. I have presented many examples of this interplay. Discussions of molecular diseases such as sickle-cell anemia and of the mechanism of action of drugs such as penicillin enrich the teaching of biochemistry. Finally, I have tried to define several challenging areas of inquiry in biochemistry today, such as the molecular basis of excitability.

In writing this book, I have benefitted greatly from the advice, criticism, and encouragement of many colleagues and students. Leroy Hood, Arthur Kornberg, Jeffrey Sklar, and William Wood gave me invaluable counsel on its overall structure. Richard Caprioli, David Cole, Alexander Glazer, Robert Lehman, and Peter Lengyel read much of the manuscript and made many very helpful suggestions. I am indebted to Frederic Richards for sharing his thoughts on macromolecular conformation and for extensive advice on how to depict three-dimensional structures. Deric Bownds, Thomas Broker, Jack Griffith, Hugh Huxley, and George Palade made available to me many striking electron micrographs. I am also very thankful for the advice and criticism that were given at various times in the preparation of this book by Richard Dickerson, David Eisenberg, Moises Eisenberg, Henry Epstein, Joseph Fruton, Michel Goldberg, James Grisolia, Richard Henderson, Harvey Himel, David Hogness, Dale Kaiser, Samuel Latt, Susan Lowey, Vincent Marchesi, Peter Moore, Allan Oseroff, Jordan Pober, Russell Ross, Edward Reich, Mark Smith, James Spudich, Joan Steitz, Thomas Steitz, and Alan Waggoner.

I am grateful to the Commonwealth Fund for a grant that enabled me to initiate the writing of this book. The interest and support of Robert Glaser, Terrance Keenan, and Quigg Newton came at a critical time. One of my aims in writing this book has been to achieve a close integration of word and picture and to illustrate chemical transformations and three-dimensional structures vividly. I am especially grateful to Donna Salmon, John Foster, and Jean Foster for their work on the drawings, diagrams, and graphs. Many individuals at Yale helped to bring this project to fruition. I particularly wish to thank Margaret Banton and Sharen Westin for typing the manuscript, William Pollard for photographing space-filling models, and Martha Scarf for generating the computer drawings of molecular structures on which many of the illustrations in this book are based. John Harrison and his staff at the Kline Science Library helped in many ways.

Much of this book was written in Aspen. I wish to thank the Aspen Center for Physics and the Given Institute of Pathobiology for their kind hospitality during several summers. I have warm memories of many stimulating discussions about biochemistry and molecular aspects of medicine that took place in the lovely garden of the Given Institute and while hiking in the surrounding wilderness areas. The concerts in Aspen were another source of delight, especially after an intensive day of writing.

I am deeply grateful to my wife, Andrea, and to my children, Michael and Daniel, for their encouragement, patience, and good spirit during the writing of this book. They have truly shared in its gestation, which was much longer than expected. Andrea offered advice on style and design and also called my attention to the remark of the thirteenth-century Chinese scholar Tai T'ung (*The Six Scripts: Principles of Chinese Writing*): "Were I to await perfection, my book would never be finished."

I welcome comments and criticisms from readers.

October 1974   *Lubert Stryer*

CONTENTS

CHAPTER   1. Molecules and Life   1

Part I   CONFORMATION: exemplified by the relationship between the three-dimensional structure of proteins and their biological activity   9

CHAPTER   2. Introduction to Protein Structure and Function   11
           3. Oxygen Transporters: Myoglobin and Hemoglobin   46
           4. Hemoglobin: An Allosteric Protein   71
           5. Molecular Disease: Sickle-Cell Anemia   95
           6. Introduction to Enzymes   115
           7. Mechanisms of Enzyme Action: Lysozyme and Carboxypeptidase   153
           8. Zymogen Activation: Digestive Enzymes and Clotting Factors   178
           9. Connective-Tissue Proteins: Collagen and Elastin   206
          10. Introduction to Biological Membranes   227

Part II   GENERATION AND STORAGE OF METABOLIC ENERGY   255

CHAPTER 11. Metabolism: Basic Concepts and Design   257
          12. Glycolysis   277
          13. Citric Acid Cycle   307
          14. Oxidative Phosphorylation   331
          15. Pentose Phosphate Pathway and Gluconeogenesis   356
          16. Glycogen and Disaccharide Metabolism   378
          17. Fatty Acid Metabolism   404
          18. Amino Acid Degradation and the Urea Cycle   432
          19. Photosynthesis   456

Part III   BIOSYNTHESIS OF MACROMOLECULAR PRECURSORS   477

CHAPTER 20. Biosynthesis of Membrane Lipids and Steroid Hormones   479
        21. Biosynthesis of Amino Acids and Heme   503
        22. Biosynthesis of Nucleotides   528

Part IV   INFORMATION: storage, transmission, and expression of genetic
                    information   555

CHAPTER 23. DNA: Genetic Role, Structure, and Replication   557
        24. Messenger RNA and Transcription   594
        25. The Genetic Code and Gene-Protein Relationships   623
        26. Protein Synthesis   645
        27. Control of Gene Expression   679
        28. Eucaryotic Chromosomes   693
        29. Viruses   708

Part V   MOLECULAR PHYSIOLOGY: interaction of information, conformation,
                        and metabolism in physiological
                        processes   729

CHAPTER 30. Immunoglobulins   731
        31. Bacterial Cell Walls   754
        32. Membrane Transport   767
        33. Excitable Membranes   787
        34. Hormone Action   807
        35. Muscle Contraction and Cell Motility   826

        Appendixes   843
        Answers to Problems   847
        Index   855

BIOCHEMISTRY

CHAPTER 1

MOLECULES AND LIFE

Biochemistry is the study of the molecular basis of life. There is much excitement and activity in biochemistry today for several reasons. *First, the chemical basis of some central processes in biology are now understood.* The discovery of the double-helical structure of DNA, the elucidation of the genetic code, the determination of the three-dimensional structure of some protein molecules, and the unraveling of the central metabolic pathways are some of the outstanding achievements of biochemistry. *Second, there are common molecular patterns and principles that underlie the diverse expressions of life.* Organisms as different as the bacterium *Escherichia coli* and man have many common features at the molecular level. They use the same building blocks to construct macromolecules. The flow of genetic information from DNA to RNA to protein is essentially the same in both species. Both use ATP as the currency of energy. *Third, biochemistry is making an increasing impact on medicine.* For example, assays for enzyme activity now play an important role in clinical diagnosis. The level of certain enzymes in the serum is a valuable criterion of whether a patient has recently had a myocardial infarction. Furthermore, biochemistry is beginning to provide a basis for the rational design of drugs. Most significant, the molecular mechanisms of diseases

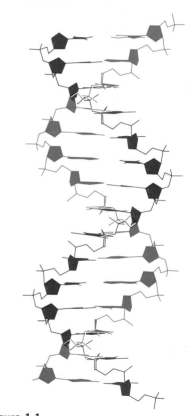

Figure 1-1
A model of the DNA double helix. The diameter of the helix is about 20 Å.

are being elucidated, as exemplified by sickle-cell anemia and the large number of inborn errors of metabolism that have been characterized. *Fourth, the rapid development of biochemistry in recent years has enabled investigators to tackle some of the most challenging and fundamental problems in biology and medicine.* How does a single cell give rise to cells as different as those in muscle, the brain, and the liver? How do cells find each other in forming a complex organ? How is the growth of cells controlled? What is the mechanism of memory? How does light elicit a nerve impulse in the retina? What are the causes of cancer? What are the causes of schizophrenia?

MOLECULAR MODELS

The interplay between the three-dimensional structure of biomolecules and their biological function is the unifying motif of this book. We will use three types of atomic models to depict molecular architecture: space-filling, ball-and-stick, and skeletal. The *space-filling* models are the most realistic. The size and configuration of an atom in a space-filling model are determined by its bonding properties and van der Waals radius (Figure 1-2). The colors of the atoms are set by convention.

Hydrogen: white   Oxygen: red
Carbon: black   Phosphorus: yellow
Nitrogen: blue   Sulfur: yellow

Space-filling models of several simple molecules are shown in Figure 1-3.

Ball-and-stick models will be used frequently. They are not as realistic as space-filling models because the atoms are depicted as spheres of smaller radius than the van der Waals radius. However, the bonding arrangement is easier to see than in space-filling models because the bonds are explicitly represented by sticks. The taper of a stick tells whether the bond is directed in front of or behind the plane of the page. More of a complex structure can be seen in ball-and-stick models than in space-filling models. An even simpler image is achieved with *skeletal models,* which show only the molecular framework. In these models, atoms are not explicitly shown. Rather, their position is implied by the junctions of bonds and their termini. Skeletal models are frequently used to depict

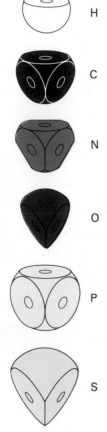

Figure 1-2
Space-filling models of hydrogen, carbon, nitrogen, oxygen, phosphorus, and sulfur atoms.

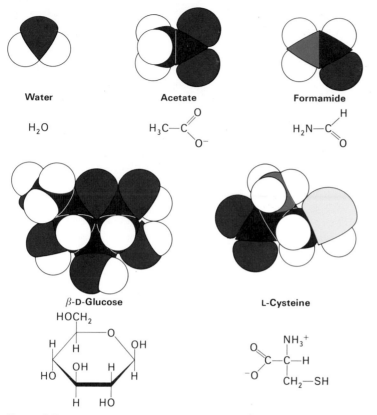

Water

H_2O

Acetate

$$H_3C-\overset{\displaystyle O}{\underset{\displaystyle O^-}{C}}$$

Formamide

$$H_2N-\overset{\displaystyle H}{\underset{\displaystyle O}{C}}$$

β-D-Glucose

L-Cysteine

Figure 1-3
Space-filling models of water, acetate, formamide, glucose, and cysteine.

large biological macromolecules, such as protein molecules having several thousand atoms. Space-filling, ball-and-stick, and skeletal models of ATP are compared in Figure 1-4.

DESIGN OF THIS BOOK

This book has five parts, each centered on a major theme:

      I: Conformation
     II: Generation and Storage of Metabolic Energy
   III: Biosynthesis of Macromolecular Precursors
   IV: Information
    V: Molecular Physiology

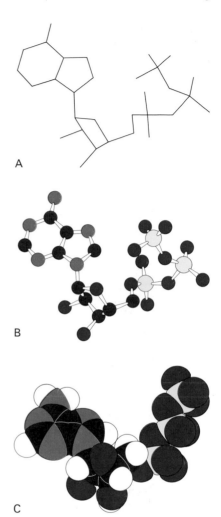

A

B

C

Figure 1-4
A comparison of (A) skeletal, (B) ball-and-stick, and (C) space-filling models of ATP. Hydrogen atoms are not shown in models A and B.

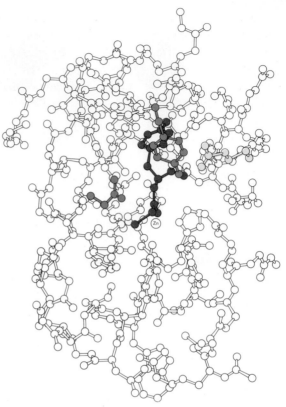

Figure 1-5
Structure of an enzyme-substrate complex.
Glycyltyrosine (shown in red) is bound to
carboxypeptidase A, a digestive enzyme. Only a
quarter of the enzyme is shown. [Based on W. N.
Lipscomb. *Proc. Robert A. Welch Found. Conf. Chem. Res.*
15(1971):141.]

The interplay of three-dimensional structure and biological
activity as exemplified in proteins is the major theme of Part I.
The structure and function of myoglobin and hemoglobin, the
oxygen-carrying proteins in vertebrates, are presented in detail
because these proteins illustrate some general principles. Hemoglobin is especially interesting because its binding of O_2 is regulated
by specific molecules in its environment. The molecular pathology
of hemoglobin, particularly sickle-cell anemia, is also presented. We
then turn to enzymes and consider how they recognize substrates

and enhance reaction rates by factors of a million or more. The enzymes lysozyme, carboxypeptidase A, and chymotrypsin are examined in detail because they provide insight into many general principles of catalysis. Rather different facets of the theme of conformation emerge in the chapter on collagen and elastin, two connective-tissue proteins. The final chapter in Part I is an introduction to biological membranes, which are organized assemblies of lipids and proteins. One of the major functions of membranes in biological systems is to create compartments.

Part II deals with the generation and storage of metabolic energy. First, the overall strategy of metabolism is presented. Cells convert energy from fuel molecules into ATP. In turn, ATP drives most energy-requiring processes in cells. In addition, reducing power in the form of NADPH is generated for use in biosyntheses. The metabolic pathways that carry out these reactions are then presented in detail. For example, the generation of ATP from glucose involves a sequence of three series of reactions—glycolysis, the citric acid cycle, and oxidative phosphorylation. The latter two are also common to the generation of ATP from the oxidation of fats and some amino acids, the other major fuels. We see here an illustration of molecular economy. Two storage forms of fuel molecules, glycogen and triacylglycerols (neutral fats), are also discussed in Part II. The concluding topic of this section is photosynthesis, in which the primary event is the light-activated transfer of an electron from one substance to another against a chemical potential gradient.

Part III deals with the biosynthesis of macromolecular precursors, starting with the synthesis of membrane lipids and steroids. The pathway for the synthesis of cholesterol, a 27-carbon steroid, is of particular interest because all of its carbon atoms come from a 2-carbon precursor. The reactions leading to the synthesis of selected amino acids and the heme group are discussed in the next chapter. The control mechanisms in these pathways are of general significance. The final chapter in this part is concerned with the biosynthesis of nucleotides, the activated precursors of DNA and RNA.

The storage, transmission, and expression of genetic information is the central theme of Part IV. The experiments that revealed that DNA is the genetic material and the discovery of the DNA double helix are discussed first, followed by a presentation of the enzymatic mechanism of DNA replication. We then turn to the expression

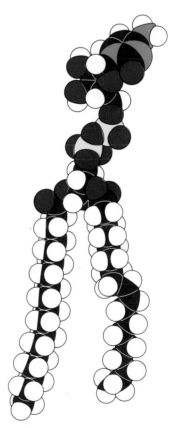

Figure 1-6
Model of CDP-diacylglycerol, an activated intermediate in the synthesis of some membrane lipids.

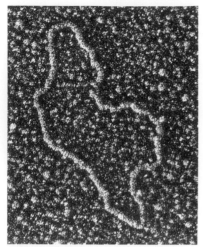

Figure 1-7
Electron micrograph of a DNA
molecule. [Courtesy of Dr. Thomas
Broker.]

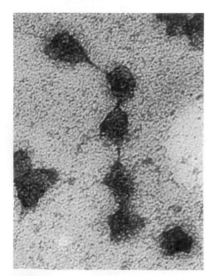

Figure 1-8
Electron micrograph of several
ribosomes bound to a messenger
RNA molecule. [Courtesy of Dr.
Henry Slayter.]

of genetic information in DNA, starting with the evidence that
showed that messenger RNA is the information-carrying intermediate in protein synthesis. The process of transcription, which is
the synthesis of RNA according to instructions given by a DNA
template, is then considered. This brings us to the genetic code,
which is the relationship between the sequence of bases in DNA
(or its messenger RNA transcript) and the sequence of amino acids
in a protein. The code, which is the same in all organisms, is
beautiful in its simplicity. Three bases constitute a codon, the unit
that specifies one amino acid. Codons on messenger RNA are read
sequentially by transfer RNA molecules, which are the adaptors
in protein synthesis. We then examine the mechanism of protein
synthesis, a process called translation because it converts the four
base-pairing letters of nucleic acids into the twenty-letter alphabet
of proteins. Translation is carried out by the coordinated interplay
of more than 100 kinds of macromolecules, centered on ribosomes.
The next chapter deals with the control of gene expression in
bacteria. The focus here is on the lactose operon of *E. coli,* which
is now understood in detail. Recent research on the chromosomes
of higher organisms, which are much larger and more complex
than those of bacteria, is then presented. Part IV concludes with
a discussion of viral multiplication and assembly. The assembly
of viruses exemplifies some general principles of how biological
macromolecules form highly ordered structures. Some viruses that
cause cancers in experimental animals are also considered here.

Part V, entitled Molecular Physiology, is a transition from biochemistry to physiology. Here we use many of the concepts that
were developed in the preceding parts of the book, since physiology
involves the interplay of information, conformation, and metabolism. We start with the molecular basis of the immune response.
How does an organism recognize a foreign substance? Bacterial
cell walls are considered next. Here the mechanism of action of
penicillin is described. Several subsequent chapters deal with
aspects of membrane structure and dynamics. How do cells transport ions such as Na^+, K^+, and Ca^{2+} and molecules across their
membranes? What is the molecular basis of action potentials in
nerve cells? How does acetylcholine trigger receptors in the postsynaptic membrane? How is a retinal rod cell triggered by a single
photon? What are some unifying principles of hormone action?
The final chapter deals with the problem of how chemical bond

energy is transformed into coordinated motion. The focus here is on the mechanism of contraction of striated muscle and its control by a membrane system called the sarcoplasmic reticulum.

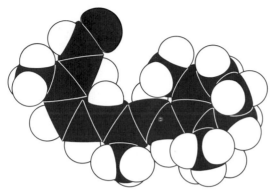

Figure 1-9
A model of 11-*cis*-retinal, the light-absorbing group in rhodopsin. The isomerization of this chromophore by light is the first event in visual excitation.

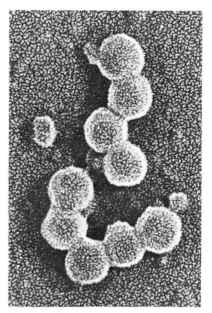

Figure 1-10
Electron micrograph of SV40 virus particles. This virus causes tumors in certain experimental animals. [Courtesy of Dr. Jack Griffith.]

On the facing page: Complex of the enzyme ribonuclease S and a dinucleotide substrate analog. The structure of this complex was elucidated by William Carlson and Harold Wyckoff. [Based on a drawing kindly provided by Mr. John Mouning and Dr. Frederic Richards.]

PART I

CONFORMATION

exemplified by the relationship between the three-dimensional structure of proteins and their biological activity

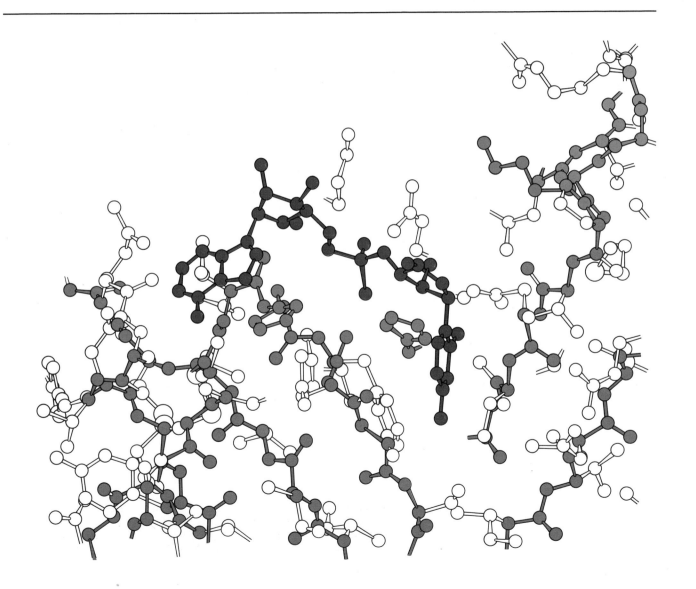

CHAPTER 2

INTRODUCTION TO
PROTEIN STRUCTURE AND FUNCTION

Proteins play crucial roles in virtually all biological processes. The significance and remarkable scope of their functions are exemplified in:

1. *Enzymatic catalysis*. Nearly all chemical reactions in biological systems are catalyzed by specific macromolecules called enzymes. Some of these reactions, such as the hydration of carbon dioxide, are quite simple. Others, such as the replication of an entire chromosome, are highly intricate. Nearly all enzymes exhibit enormous catalytic power. Enzymes usually enhance reaction rates by at least a millionfold. Indeed, chemical transformations rarely occur at perceptible rates in vivo in the absence of enzymes. More than a thousand enzymes have been characterized, and many more have been crystallized. The striking fact is that all known enzymes are proteins. Thus, proteins play the unique role of determining the pattern of chemical transformations in biological systems.

2. *Transport and storage*. Many small molecules and ions are transported by specific proteins. For example, hemoglobin transports oxygen in erythrocytes, whereas myoglobin, a related protein, transports oxygen in muscle. Iron is carried in the plasma of blood by transferrin and is stored in the liver as a complex with ferritin, a different protein.

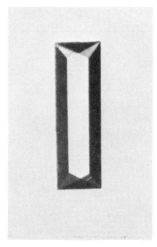

Figure 2-1
Photomicrograph of a crystal of hexokinase, a key enzyme in the utilization of glucose. [Courtesy of Dr. Thomas Steitz and Mr. Mark Yeager.]

Figure 2-2
Electron micrograph of a cross section of insect flight muscle, showing a hexagonal array of two kinds of protein filaments. [Courtesy of Dr. Michael Reedy.]

Figure 2-3
Electron micrograph of a fiber of collagen. [Courtesy of Dr. Jerome Gross and Dr. Romaine Bruns.]

3. *Coordinated motion.* Proteins are the major component of muscle. Muscle contraction is accomplished by the sliding motion of two kinds of protein filaments. On the microscopic scale, such coordinated motions as the movement of chromosomes in mitosis and the propulsion of sperm by their flagellae are also produced by contractile assemblies consisting of proteins.

4. *Mechanical support.* The high tensile strength of skin and bone is due to the presence of collagen, an elongated protein that readily forms fibers.

5. *Immune protection.* Antibodies are highly specific proteins that recognize and combine with such foreign substances as viruses, bacteria, and cells from other organisms. Proteins thus play a vital role in distinguishing between self and nonself.

6. *Generation and transmission of nerve impulses.* The response of nerve cells to specific stimuli is mediated by receptor proteins. For example, rhodopsin is the photoreceptor protein in retinal rod cells. Receptor molecules that can be triggered by specific small molecules, such as acetylcholine, are responsible for the transmission of nerve impulses at synapses—that is, at junctions between nerve cells.

7. *Control of growth and differentiation.* Controlled sequential expression of genetic information is essential for the orderly growth

Figure 2-4
Scanning electron micrograph of the outer segments of several retinal rod cells. Each outer segment contains about 10^9 rhodopsin molecules. Absorption of a single photon by one of these photoreceptor protein molecules can elicit a nerve impulse. [Courtesy of Dr. Deric Bownds.]

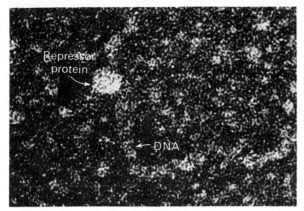

Figure 2-5
Electron micrograph of a repressor protein bound to
a DNA molecule. [Courtesy of Dr. Jack Griffith.]

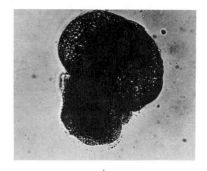

Nerve growth
factor

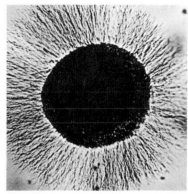

Figure 2-6
Photo micrograph of a ganglion
showing the proliferation of nerves
after addition of nerve growth
factor, a complex of proteins.
[Courtesy of Dr. Eric Shooter.]

and differentiation of cells. Only a small fraction of the genome
of a cell is expressed at any one time. In bacteria, repressor proteins
are important control elements that silence specific segments of the
deoxyribonucleic acid (DNA) of a cell. A quite different way in
which proteins act in differentiation is exemplified by nerve growth
factor, a protein complex that guides the formation of neural
networks in higher organisms.

PROTEINS ARE BUILT FROM AMINO ACIDS

Amino acids are the basic structural units of proteins. An amino
acid consists of an amino group, a carboxyl group, a hydrogen atom,
and a distinctive R group bonded to a carbon atom, which is called
the α-carbon (Figure 2-7). An R group is referred to as a *side chain*

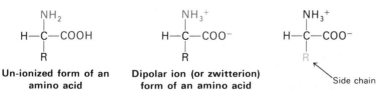

Un-ionized form of an
amino acid

Dipolar ion (or zwitterion)
form of an amino acid

Side chain

Figure 2-7
Structure of the un-ionized and zwitterion forms of an amino acid.

for reasons that will be evident shortly. Amino acids in solution at neutral pH are predominantly *dipolar ions* (or *zwitterions*) rather than un-ionized molecules. In the dipolar form of an amino acid, the amino group is protonated ($—NH_3^+$) and the carboxyl group is dissociated ($—COO^-$). The ionization state of an amino acid varies with pH (Figure 2-8). In acid solution (e.g., pH 1), the

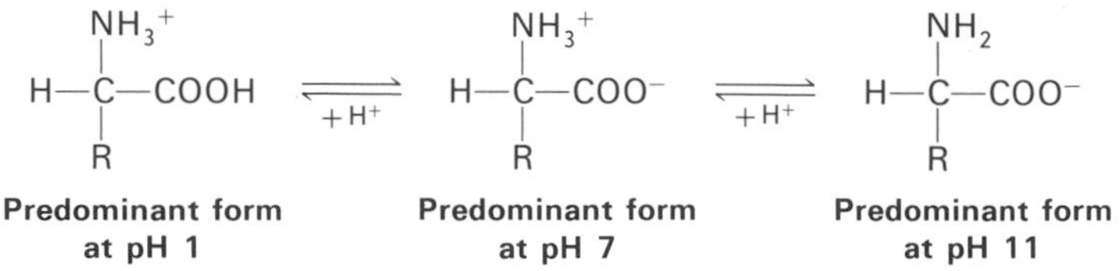

$$NH_3^+ \quad\quad NH_3^+ \quad\quad NH_2$$
$$H—C—COOH \underset{+H^+}{\rightleftharpoons} H—C—COO^- \underset{+H^+}{\rightleftharpoons} H—C—COO^-$$
$$R \quad\quad R \quad\quad R$$

Predominant form   **Predominant form**   **Predominant form**
at pH 1   **at pH 7**   **at pH 11**

Figure 2-8
Ionization states of an amino acid as a function of pH.

carboxyl group is un-ionized ($—COOH$) and the amino group is ionized ($—NH_3^+$). In alkaline solution (e.g., pH 11), the carboxyl group is ionized ($—COO^-$) and the amino group is un-ionized ($—NH_2$). The concept of pH and the acid-base properties of amino acids are discussed further in the Appendix.

The tetrahedral array of four different groups about the α-carbon atom confers optical activity on amino acids. The two mirror-image forms are called the L-isomer and the D-isomer (Figure 2-9). *Only L-amino acids are constituents of proteins.* Hence, we will omit the designation of the optical isomer in our discussion of proteins. Unless otherwise noted, the L-isomer is always implied.

Twenty kinds of side chains varying in *size, shape, charge, hydrogen-bonding capacity, and chemical reactivity* are commonly found in proteins. Indeed, all proteins in all species, from bacteria to man, are constructed from the same set of twenty amino acids. This fundamental alphabet of proteins is at least two billion years old. The remarkable range of functions mediated by proteins results from the diversity and versatility of these twenty kinds of building blocks. In subsequent chapters, we will explore some ways in which this alphabet is used to create the intricate three-dimensional structures that enable proteins to participate in so many biological processes.

Let us look at this repertoire of amino acids. The simplest one is glycine, which contains a hydrogen atom as its side chain (Figure

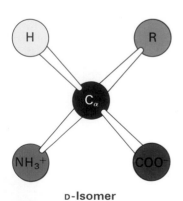

L-Isomer

D-Isomer

Figure 2-9
Absolute configurations of the L- and D-isomers of amino acids.

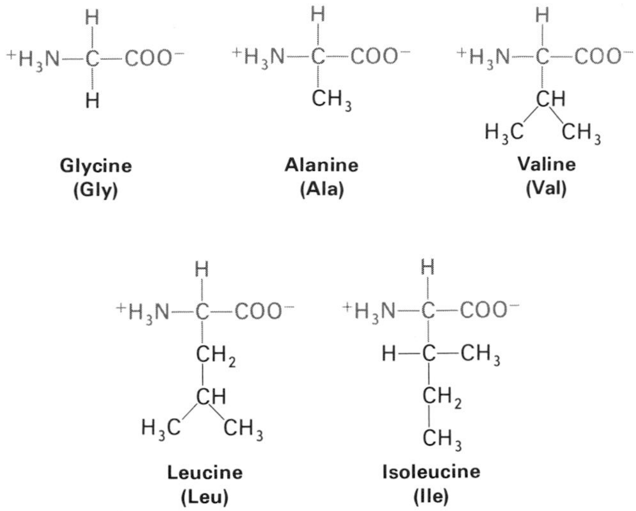

Figure 2-10
Amino acids having aliphatic side chains.

2-10). Alanine has a methyl group as its side chain. The other amino acids that have hydrocarbon side chains are valine, leucine, isoleucine, and proline. However, proline differs from the other amino acids in the basic set of twenty in containing a secondary rather than a primary amino group (Figure 2-11). Strictly speaking, proline is an imino acid rather than an amino acid. The side chain of proline is bonded to both the amino group and the α-carbon, thereby forming a cyclic structure.

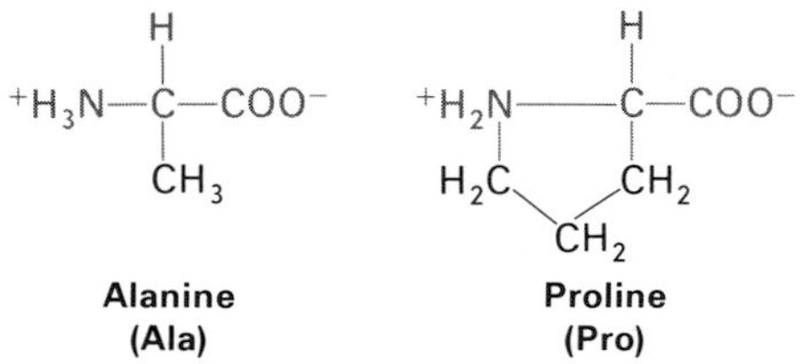

Figure 2-11
Proline differs from the other common amino acids in having a secondary amino group.

Two amino acids, serine and threonine, contain an aliphatic hydroxyl group (Figure 2-12).

There are three common aromatic amino acids: phenylalanine, tyrosine, and tryptophan (Figure 2-13).

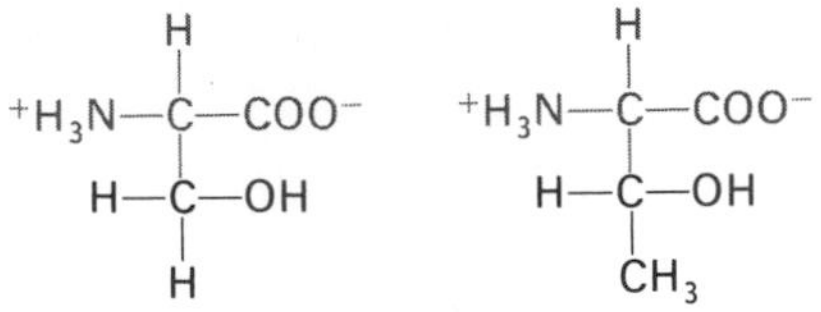

Serine
(Ser)

Threonine
(Thr)

Figure 2-12
Serine and threonine have aliphatic hydroxyl side chains.

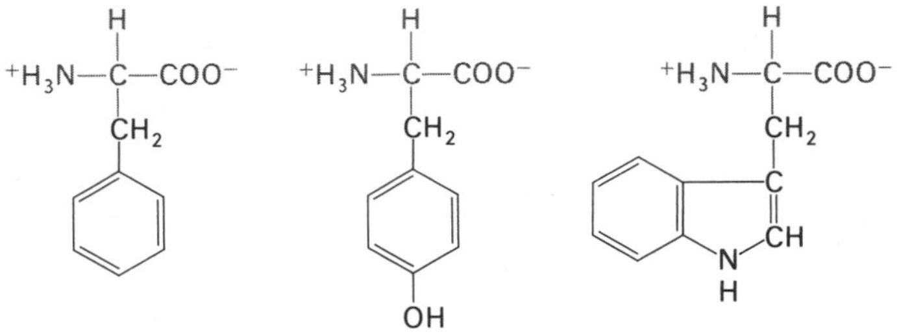

Phenylalanine
(Phe)

Tyrosine
(Tyr)

Tryptophan
(Trp)

Figure 2-13
Phenylalanine, tyrosine, and tryptophan have aromatic side chains.

The side chains of the amino acids mentioned so far are uncharged at physiological pH. We turn now to some charged side chains. Lysine and arginine are positively charged at neutral pH, whereas histidine may be positively charged or neutral, depending on its local environment. These basic amino acids are shown in Figure 2-14. The negatively charged side chains are those of glutamic acid and aspartic acid (Figure 2-15). These amino acids will

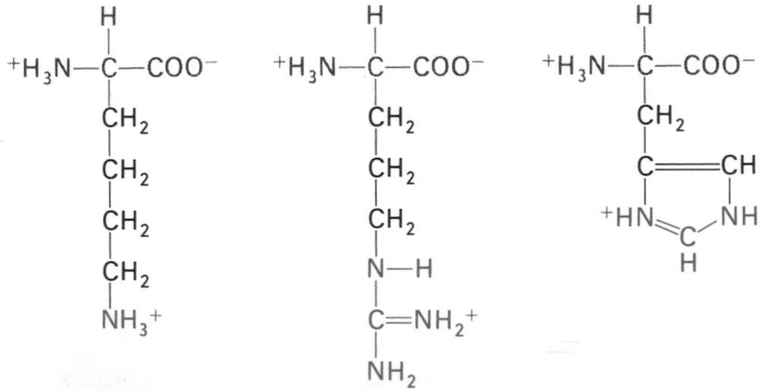

Lysine
(Lys)

Arginine
(Arg)

Histidine
(His)

Figure 2-14
Lysine, arginine, and histidine have basic side chains.

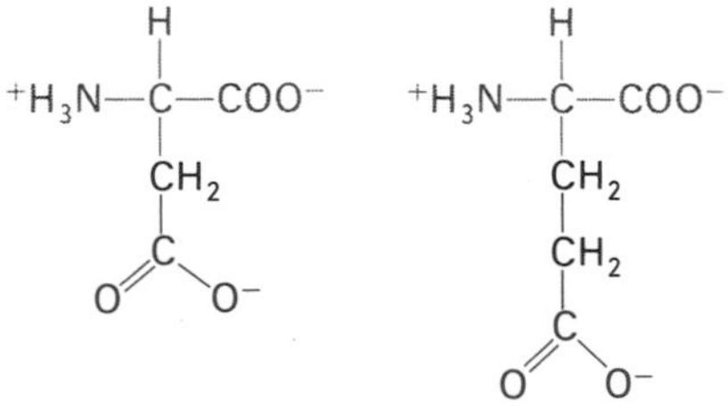

Aspartate
(Asp)

Glutamate
(Glu)

Figure 2-15
Aspartate and glutamate have acidic side chains.

be called glutamate and aspartate, to emphasize the fact that they are negatively charged at physiological pH. The uncharged derivatives of glutamate and aspartate are glutamine and asparagine (Figure 2-16), which contain a terminal amide group rather than a carboxylate. Finally, there are two amino acids whose side chains contain a sulfur atom: methionine and cysteine (Figure 2-17). As will be discussed shortly, cysteine plays a special role in some proteins by forming disulfide cross-links.

Some proteins contain special amino acids that supplement the basic set of twenty. For example, collagen contains hydroxyproline, a hydroxylated derivative of proline. All the special amino acids in proteins are formed by modification of a common amino acid after it has been incorporated into a polypeptide chain.

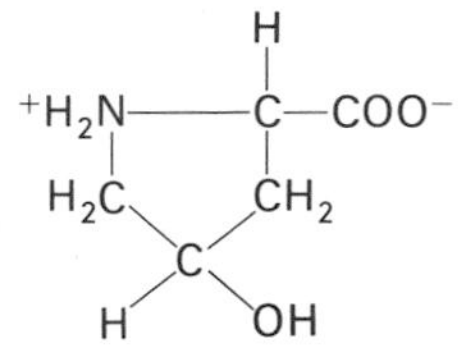

Hydroxyproline

Table 2-1
Abbreviations for amino acids

Amino acid	Three-letter abbreviation	One-letter symbol
Alanine	Ala	A
Arginine	Arg	R
Asparagine	Asn	N
Aspartic acid	Asp	D
Asparagine or aspartic acid	Asx	B
Cysteine	Cys	C
Glutamine	Gln	Q
Glutamic Acid	Glu	E
Glutamine or glutamic acid	Glx	Z
Glycine	Gly	G
Histidine	His	H
Isoleucine	Ile	I
Leucine	Leu	L
Lysine	Lys	K
Methionine	Met	M
Phenylalanine	Phe	F
Proline	Pro	P
Serine	Ser	S
Threonine	Thr	T
Tryptophan	Trp	W
Tyrosine	Tyr	Y
Valine	Val	V

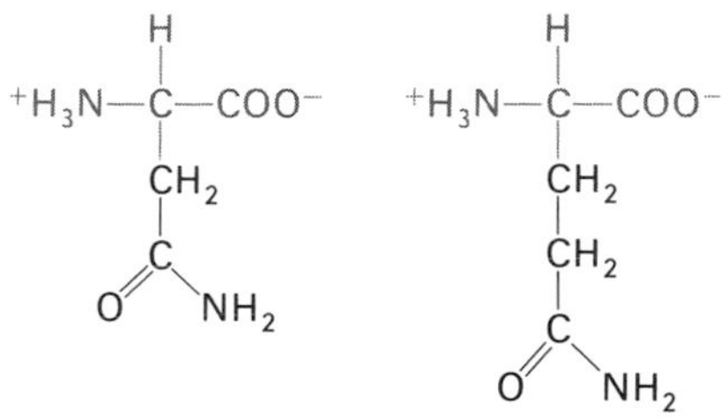

Asparagine (Asn)   **Glutamine (Gln)**

Figure 2-16
Asparagine and glutamine have amide side chains.

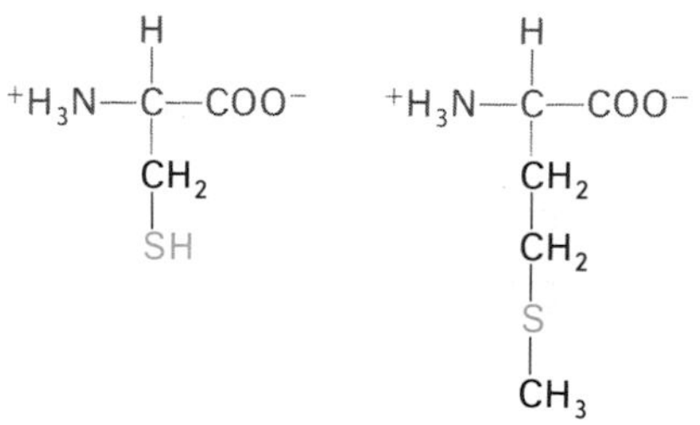

Cysteine (Cys)   **Methionine (Met)**

Figure 2-17
Cysteine and methionine have sulfur-containing side chains.

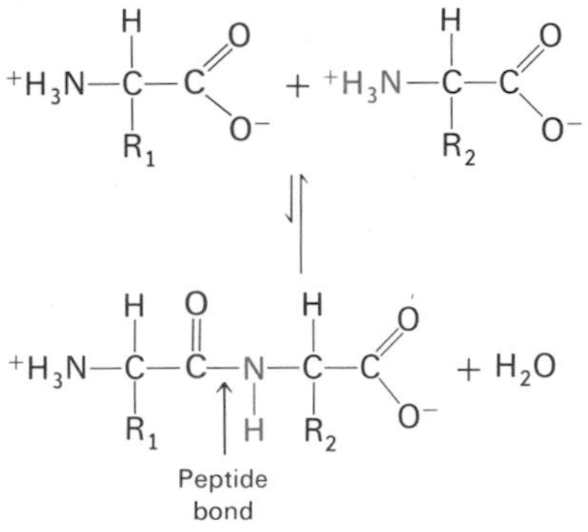

Figure 2-18
Formation of a peptide bond.

AMINO ACIDS ARE LINKED BY PEPTIDE BONDS TO FORM POLYPEPTIDE CHAINS

In proteins, the α-carboxyl group of one amino acid is joined to the α-amino group of another amino acid by a *peptide bond* (also called an amide bond). The formation of a dipeptide from two amino acids by loss of a water molecule is shown in Figure 2-18. The equilibrium of this reaction lies far on the side of hydrolysis, rather than synthesis. Hence, the biosynthesis of peptide bonds requires an input of free energy, whereas their hydrolysis is thermodynamically downhill.

Many amino acids, usually more than 100, are joined by peptide bonds to form a *polypeptide chain*, which is an unbranched structure (Figure 2-19). An amino acid unit in a polypeptide is called a *residue*.

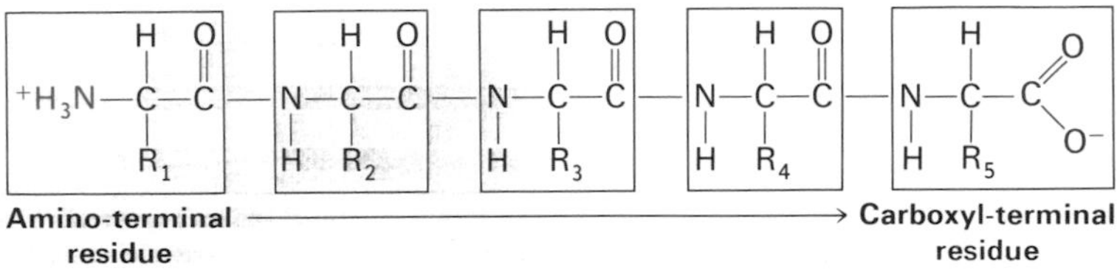

Figure 2-19
A pentapeptide. The constituent amino acid residues are outlined. The chain starts at the amino end.

A polypeptide chain has direction because its building blocks have different ends—namely, the α-amino and the α-carboxyl groups. By convention, *the amino end is taken to be the beginning of a polypeptide chain.* The sequence of amino acids in a polypeptide chain is written starting with the amino-terminal residue. Thus, in the tripeptide alanine-glycine-tryptophan, alanine is the amino-terminal residue and tryptophan is the carboxyl-terminal residue. Note that tryptophan-glycine-alanine is a different tripeptide.

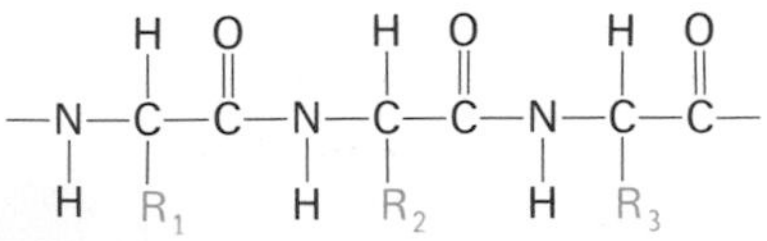

A polypeptide chain consists of a regularly repeating part, called the *main chain,* and a variable part, the distinctive *side chains* (Figure 2-20). The main chain is sometimes termed the backbone.

In some proteins, a few side chains are cross-linked by *disulfide bonds.* These cross-links are formed by the oxidation of cysteine

Figure 2-20
A polypeptide chain is made up of a regularly repeating *backbone* and distinctive *side chains* (R_1, R_2, R_3, shown in green).

residues. The resulting disulfide is called *cystine* (Figure 2-21). There are no other common covalent cross-links in proteins.

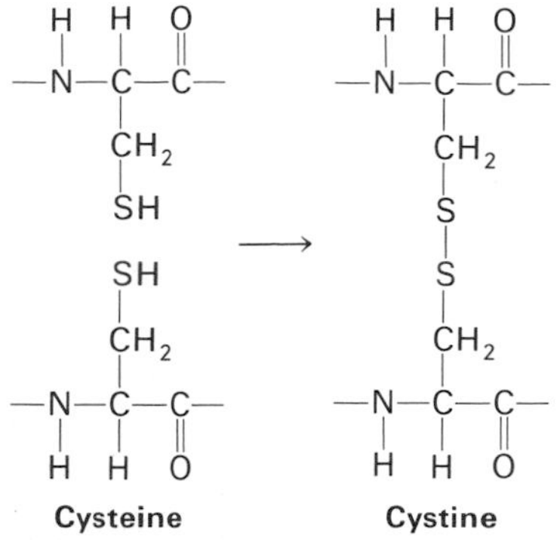

PROTEINS CONSIST OF ONE OR MORE POLYPEPTIDE CHAINS

Many proteins, such as myoglobin, consist of a single polypeptide chain. Others contain two or more chains, which may be identical or different. For example, hemoglobin is made up of two chains of one kind and two of another kind. These four chains are held together by noncovalent forces. Alternatively, the polypeptide chains of some multichain proteins are linked by disulfide bonds. The two chains of insulin, for example, are joined by two disulfide bonds.

Figure 2-21
A disulfide bridge (—S—S—) is formed from the sulfhydryl groups (—SH) of two cysteine residues. The product is a *cystine* residue.

PROTEINS CAN BE PURIFIED BY A VARIETY OF TECHNIQUES

The purification of a protein is an indispensable step toward the elucidation of its mechanism of action. A large number of proteins, more than a thousand, have been isolated in pure form. Proteins can be separated from each other and from other kinds of molecules on the basis of characteristics such as *size, solubility, charge, and specific binding affinity.*

Proteins can be separated from small molecules by dialysis through a semipermeable membrane (Figure 2-22). Molecules with a molecular weight greater than about 15,000 are retained inside a typical dialysis bag, whereas smaller molecules and ions traverse the pores of such a dialysis membrane and emerge in the dialysate outside the bag. Separations on the basis of size also can be achieved by the technique of *gel-filtration chromatography* (Figure 2-23). The sample is applied to the top of a column consisting of an insoluble but highly hydrated carbohydrate polymer in the form of beads, which are typically 0.1 mm in diameter. Sephadex is a commonly used commercial preparation. Small molecules can enter these beads, but large ones cannot. The result is that small molecules are distributed in the aqueous solution inside the beads and also between them, whereas large molecules are located only in the

Dalton—
A unit of mass very nearly equal to that of a hydrogen atom (precisely equal to 1.0000 on the atomic mass scale).
The terms "dalton" and "molecular weight" are used interchangeably; for example, a 20,000-dalton protein has a molecular weight of 20,000.
Named after John Dalton (1766–1844), who developed the atomic theory of matter.

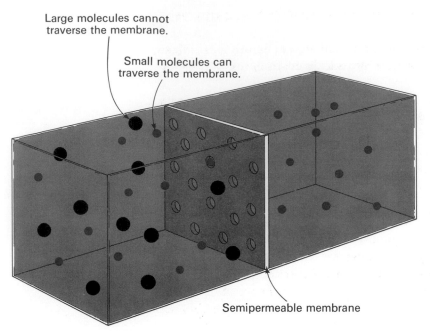

Large molecules cannot
traverse the membrane.

Small molecules can
traverse the membrane.

Semipermeable membrane

Figure 2-22
Separation of molecules on the basis of size by dialysis.

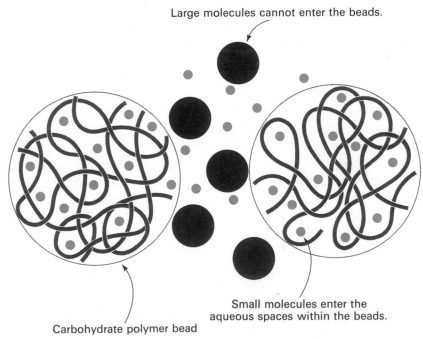

Large molecules cannot enter the beads.

Carbohydrate polymer bead

Small molecules enter the
aqueous spaces within the beads.

Figure 2-23
Separation of molecules on the basis of size by gel-filtration chromatography.

solution between the beads. Large molecules flow more rapidly through this column and emerge first because a smaller volume is accessible to them.

Proteins can also be separated on the basis of their net charge by *ion-exchange chromatography*. If a protein has a net positive charge at pH 7, it will usually bind to an ion-exchange column containing carboxylate groups, whereas a negatively charged protein will not. Such a positively charged protein can be released from the column by adding NaCl or another salt to the eluting buffer. Sodium ions compete with positively charged groups on the protein for binding to the column. Proteins that have a low density of net positive charge will tend to emerge first, followed by those having a higher charge density. Factors other than net charge also influence the behavior of proteins on ion-exchange columns. The net charge on a protein also influences its rate of migration in an electric field. This principle is exploited in *electrophoresis,* a technique that will be discussed further in Chapter 5.

The purification of proteins is as much an art as a science. In seeking to purify a protein, a variety of separation methods are tried and their efficacy is assessed by *assaying* for the protein of interest. The assay for an enzyme, for example, is usually based on its specific catalytic activity. The total amount of protein is also measured. In this way, the degree of purification obtained in a particular step can be determined.

PROTEINS HAVE UNIQUE AMINO ACID SEQUENCES THAT ARE SPECIFIED BY GENES

In 1953, Frederick Sanger determined the amino acid sequence of insulin, a protein hormone (Figure 2-24). *This work is a landmark*

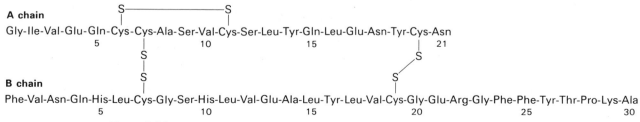

Figure 2-24
Amino acid sequence of bovine insulin.

Protein—
A word coined by Jöns J. Berzelius in 1838 to emphasize the importance of this class of molecules. Derived from the Greek word *proteios,* which means "of the first rank."

in biochemistry because it showed for the first time that a protein has a precisely defined amino acid sequence. This accomplishment also stimulated other laboratories to carry out sequence studies of a wide variety of proteins. Indeed, the complete amino acid sequences of more than 100 proteins are now known. The striking fact is that each protein has a unique, precisely defined amino acid sequence. A series of incisive studies in the late 1950s and early 1960s revealed that the amino acid sequences of proteins are genetically determined. The sequence of nucleotides in deoxyribonucleic acid, the molecule of heredity, specifies a complementary sequence of nucleotides in a ribonucleic acid (RNA), which in turn specifies the amino acid sequence of a protein (see Chapters 23 to 26). Furthermore, proteins are synthesized from their constituent amino acids by a common mechanism.

The importance of determining amino acid sequences of proteins is fourfold. First, the determination of the sequence of a protein is a significant step toward the elucidation of the molecular basis of its biological activity. A sequence is particularly informative if it is considered together with other chemical and physical data. Second, the sequences and detailed three-dimensional structures of numerous proteins need to be known so that the rules governing the folding of polypeptide chains into highly specific three-dimensional forms can be deduced. The amino acid sequence is the link between the genetic message in DNA and the three-dimensional structure that is the basis of a protein's biological function. Third, alterations in amino acid sequence can produce abnormal function and disease. Fatal disease, such as sickle-cell anemia, can result from a change in a single amino acid in a single protein. Sequence determination is thus part of molecular pathology, an emerging area of medicine. Fourth, the amino acid sequence of a protein reveals much about its evolutionary history. Amino acid sequences of unrelated proteins are very different. Proteins resemble one another in their amino acid sequences only if they have a common ancestor. Consequently, molecular events in evolution can be traced from amino acid sequences.

EXPERIMENTAL METHODS FOR THE DETERMINATION OF AMINO ACID SEQUENCE

Let us first consider how the sequence of a short peptide is determined. Suppose the peptide has six amino acid residues in this sequence:

Ala-Gly-Asp-Phe-Arg-Gly

The abbreviations used are the standard ones given in Table 2-1 (page 17). The *amino acid composition* of the peptide is first determined. The peptide is hydrolyzed into its constituent amino acids by heating it in 6 N HCl at 110°C for 24 hours. The amino acids in the hydrolysate are separated by ion-exchange chromatography on a column of sulfonated polystyrene. The separated amino acids are detected by the color produced on heating with ninhydrin: α-amino acids give an intense blue color, whereas imino acids, such as proline, give a yellow color. The technique is very sensitive; even a microgram of an amino acid can be detected, which is about the amount present in a thumbprint. The quantity of amino acid is proportional to the optical absorbance of the solution after heating with ninhydrin. The identity of the amino acid is revealed by its elution volume, which is the volume of buffer used to remove the amino acid from the column (Figure 2-25). A comparison of

Ninhydrin

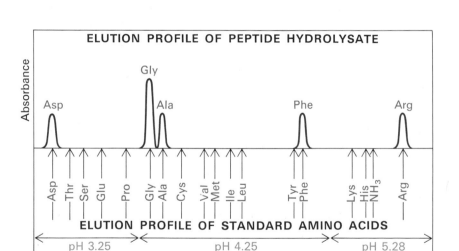

Figure 2-25
Different amino acids in a peptide hydrolysate can be separated by ion-exchange chromatography on a sulfonated polystyrene resin (such as Dowex-50). Buffers of increasing pH are used to elute the amino acids from the column. Aspartate, which has an acidic side chain, is first to emerge, whereas arginine, which has a basic side chain, is the last.

the chromatographic patterns of the hydrolysate with that of a standard mixture of amino acids shows that the amino acid composition of the peptide is:

$$(Ala, Arg, Asp, Gly_2, Phe)$$

The parentheses denote that this is the amino acid composition of the peptide, not its sequence.

The amino-terminal residue of a protein or peptide can be identified by labeling it with a compound with which it forms a stable covalent link (Figure 2-26). Fluorodinitrobenzene (FDNB), first used for this purpose by Sanger, reacts with the uncharged α-NH_2 group to form a yellow dinitrophenyl (DNP) derivative of the peptide. The bond between the DNP and the terminal amino group is stable under conditions that hydrolyze peptide bonds. Hydrolysis of the DNP-peptide in 6 N HCl yields a DNP-amino acid, which is identified as DNP-alanine by its chromatographic properties.

Dansyl chloride is also used to identify amino-terminal residues. It reacts with amino groups to form highly fluorescent sulfonamide derivatives. Indeed, it is feasible to detect as little as 0.01 microgram of an amino-terminal residue.

| Fluorodinitrobenzene | Dansyl chloride | Phenyl isothiocyanate |

Although the DNP and dansyl methods for the determination of the amino-terminal residue are powerful, they cannot be used repetitively on the same peptide because the peptide is degraded in the acid hydrolysis step. Pehr Edman devised a method for labeling the amino-terminal residue and cleaving it from the peptide without disrupting the peptide bonds between the other amino acid residues. The *Edman degradation* sequentially removes one resi-

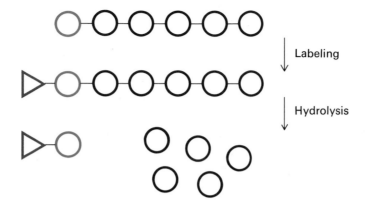

Figure 2-26
Determination of the amino-terminal residue of a peptide. The peptide is reacted with fluorodinitrobenzene (Sanger's reagent), and the labeled peptide is then hydrolyzed. The DNP-amino acid (DNP-alanine in this example) is identified by its chromatographic characteristics.

due at a time from the amino end of a peptide (Figure 2-27). Phenyl isothiocyanate reacts with the uncharged terminal amino group of the peptide to form a phenylthiocarbamoyl derivative. Under mildly acidic conditions, a cyclic derivative of the terminal amino acid is liberated, leaving an intact peptide shortened by one amino acid. The cyclic compound is a phenylthiohydantoin (PTH) amino acid. The PTH-amino acid can be identified by paper chromatography. Furthermore, the amino acid composition of the shortened peptide

$$(\text{Arg, Asp, Gly}_2, \text{Phe})$$

can be compared with that of the original peptide

$$(\text{Ala, Arg, Asp, Gly}_2, \text{Phe})$$

The difference between these analyses is one alanine residue, which shows that alanine is the amino-terminal residue of the original peptide. The Edman procedure can then be repeated on the shortened peptide. The amino acid analysis after the second round of degradation is

$$(\text{Arg, Asp, Gly, Phe})$$

showing that the second residue from the amino end is glycine. This conclusion can be confirmed by chromatographic identification of PTH-glycine obtained in the second round of the Edman degradation. Three more rounds of the Edman degradation will reveal the complete sequence of the original peptide.

The experimental strategy for determining the amino acid sequence of proteins is to divide and conquer. A protein is *specifically cleaved into smaller peptides* (ideally containing about fifteen residues each) that can be readily sequenced by the Edman method. Specific cleavage can be achieved by chemical or enzymatic methods. For example, cyanogen bromide (CNBr) splits the polypeptide chain only on the carboxyl side of methionine residues (Figure 2-28). A protein that has ten methionines will usually yield eleven peptides on cleavage with CNBr. Highly specific cleavage is also obtained with trypsin, a proteolytic enzyme from intestinal juice. Trypsin cleaves polypeptide chains on the carboxyl side of arginine and lysine residues (Figure 2-29). A protein that contains nine lysines and seven arginines will usually yield seventeen peptides on digestion with trypsin. Each of these tryptic peptides, except for the

EDMAN DEGRADATION

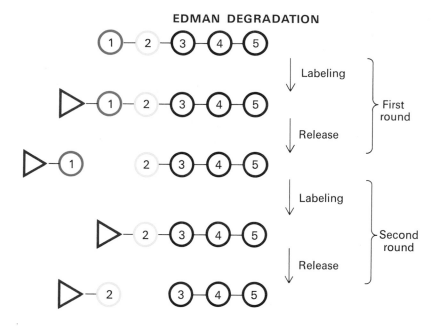

Figure 2-27
The Edman degradation. The labeled amino-terminal residue (PTH-alanine in the first round) can be released without hydrolyzing the rest of the peptide. Hence, the amino-terminal residue of the shortened peptide (Gly-Asp-Phe-Arg-Gly) can be determined in the second round. Three more rounds of the Edman degradation reveal the complete sequence of the original peptide.

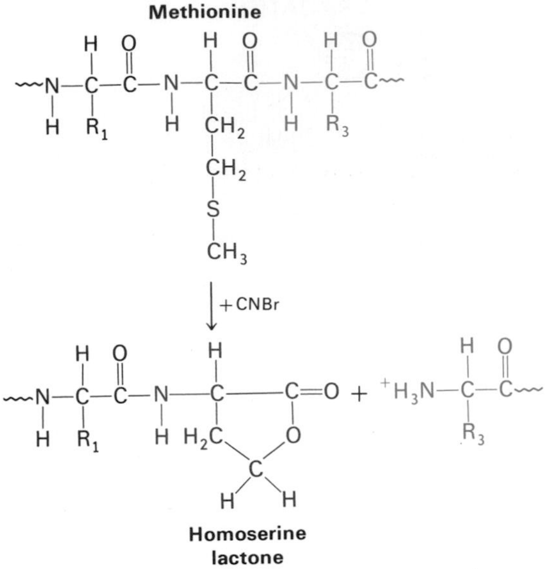

Figure 2-28
Cyanogen bromide cleaves polypeptides on the carboxyl
side of methionine residues.

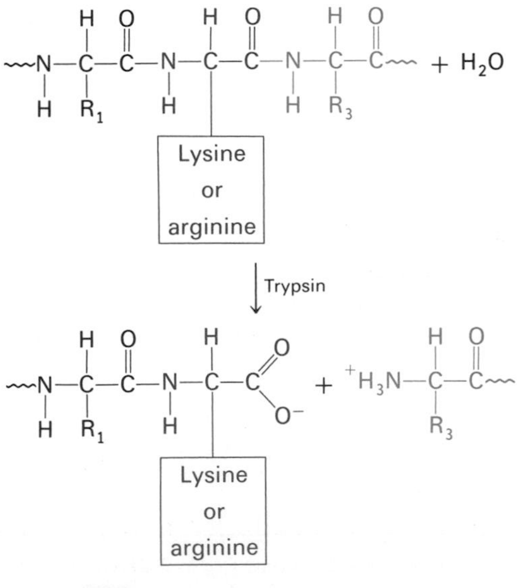

Figure 2-29
Trypsin hydrolyzes polypeptides on the carboxyl
side of arginine and lysine residues.

carboxyl-terminal peptide of the protein, will end with either arginine or lysine.

The peptides obtained by specific chemical or enzymatic cleavage are separated by chromatographic methods. The sequence of each purified peptide is then determined by the Edman method. At this point, the amino acid sequences of segments of the protein are known, but the order of these segments is not yet defined. The necessary additional information is obtained from what are called *overlap peptides* (Figure 2-30). An enzyme different from trypsin is

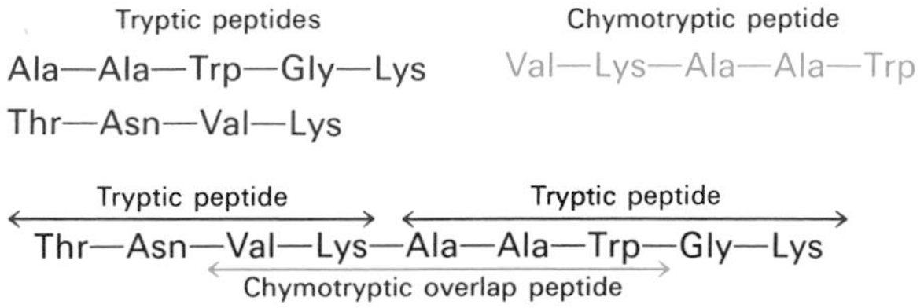

Figure 2-30
The peptide obtained by chymotryptic digestion overlaps two tryptic peptides, thus establishing their order.

used to split the polypeptide chain at different linkages. For example, chymotrypsin cleaves preferentially on the carboxyl side of aromatic and other bulky nonpolar residues. Since these chymotryptic peptides overlap two or more tryptic peptides, they can be used to establish their order. The entire amino acid sequence of the protein is then determined.

These methods apply to proteins that consist of a single polypeptide chain without any disulfide bonds. Additional steps in the elucidation of sequence are necessary if a protein has disulfide bonds or more than one chain. For proteins made up of two or more polypeptide chains held together by noncovalent bonds, denaturing agents, such as guanidine hydrochloride, are used to dissociate the chains. The dissociated chains must be separated before sequence determination can begin. For polypeptide chains that are covalently linked by disulfide bonds, as in insulin, oxidation with performic acid is used to cleave the disulfide bonds, yielding cysteic acid residues (Figure 2-31).

A promising recent development is the *sequenator*, which is an instrument for the automatic determination of amino acid sequence. The Edman degradation is used and the PTH-amino acid is identified by thin-layer or gas-liquid chromatography. Each

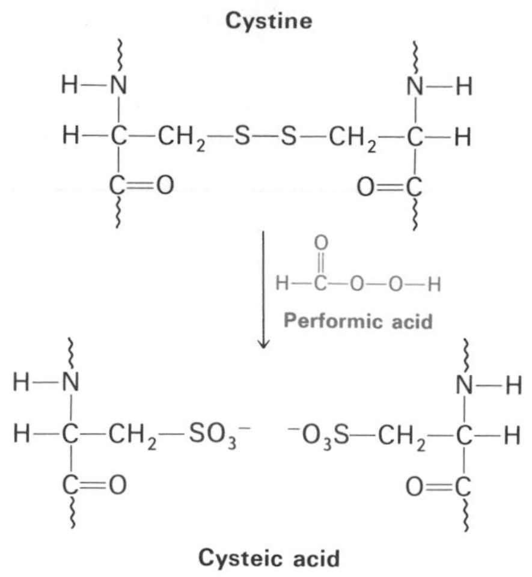

Cystine

H—N
|
H—C—CH₂—S—S—CH₂—C—H
| |
C=O   O=C

$$H—C—O—O—H$$
Performic acid

H—N   N—H
| |
H—C—CH₂—SO₃⁻   ⁻O₃S—CH₂—C—H
| |
C=O   O=C

Cysteic acid

Figure 2-31
Performic acid cleaves disulfides.

round takes less than two hours. For example, the amino acid sequence of the first sixty amino acids of myoglobin has been determined with the sequenator, starting with the intact protein.

CONFORMATION OF POLYPEPTIDE CHAINS

A striking characteristic of proteins is that they have well-defined three-dimensional structures. A stretched-out or randomly arranged polypeptide chain is devoid of biological activity, as will be discussed shortly. Function arises from *conformation,* which is the three-dimensional arrangement of atoms in a structure. Amino acid sequences are important because they specify the conformation of proteins.

In the late 1930s, Linus Pauling and Robert Corey initiated x-ray crystallographic studies of the precise structure of amino acids and peptides. Their aim was to obtain a set of standard bond distances and bond angles for these building blocks and then use this information to predict the conformation of proteins. One of their important findings was that *the peptide group is a rigid planar unit*. The hydrogen of the —NH— group is nearly always *trans* to the oxygen of the carbonyl group (Figure 2-32). There is no freedom of rotation

Angstrom (Å)—
A unit of length equal to 10^{-10} meter.
$1 \text{ Å} = 10^{-10} \text{ m} = 10^{-8} \text{ cm}$
$= 10^{-4} \, \mu\text{m} = 10^{-1} \text{ nm}$
Named after Anders J. Ångström (1814–1874), a spectroscopist.

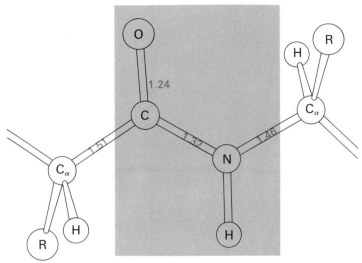

Figure 2-32
The peptide group is a rigid planar unit. Standard bond distances
(in Å) are shown.

about the bond between the carbonyl carbon atom and the nitrogen
atom of the peptide group because this link has partial double-bond
character (Figure 2-33). The length of this bond is 1.32 Å, which
is between that of a C—N single bond (1.49 Å) and a C=N double
bond (1.27 Å). In contrast, the link between the α-carbon atom and
a carbonyl carbon atom is a pure single bond. The bond between
an α-carbon atom and a nitrogen atom is also a pure single bond.
Consequently, *there is a large degree of rotational freedom about these bonds
on either side of the rigid peptide unit* (Figure 2-34). Rotations about

Figure 2-33
The peptide group is planar
because the carbon–nitrogen
bond has partial
double-bond character.

Figure 2-34
There is considerable freedom of rotation
about the bonds joining the peptide groups
to the α-carbon atoms.

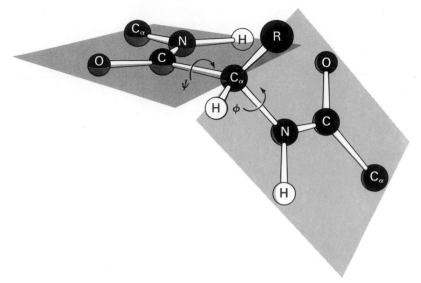

Figure 2-35
Definition of ψ and φ: ψ refers to rotations about the C_α—C single bond; φ refers to rotations about the C_α—N single bond. [Based on C. Levinthal. Molecular model-building by computer. Copyright © 1966 by Scientific American, Inc. All rights reserved.]

these bonds are designated by the angles ψ and ϕ (Figure 2-35). The conformation of the main chain of a polypeptide is fully defined when ψ and ϕ for each amino acid residue are known.

PERIODIC STRUCTURES: THE ALPHA HELIX, BETA PLEATED SHEET, AND COLLAGEN HELIX

Can a polypeptide chain fold into a regularly repeating structure? To answer this question, Pauling and Corey evaluated a variety of potential polypeptide conformations by building precise molecular models of them. They adhered closely to the experimentally observed bond angles and distances for amino acids and small peptides. In 1951, they proposed two periodic polypeptide structures, called the α helix and β pleated sheet.

The α *helix* is a rodlike structure. The tightly coiled polypeptide main chain forms the inner part of the rod, and the side chains extend outward in a helical array (Figure 2-36). The α helix is stabilized by hydrogen bonds between the NH and CO groups of the main chain. The CO group of each amino acid is hydrogen

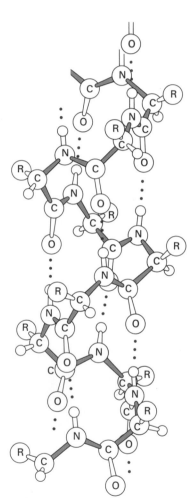

Figure 2-36
Drawing of a right-handed α helix. [From L. Pauling. *The Nature of the Chemical Bond,* 3rd ed. © 1960 by Cornell University. Used by permission of Cornell University Press.]

bonded to the NH group of the amino acid that is situated four residues ahead in the linear sequence (Figure 2-37). Thus, *all the*

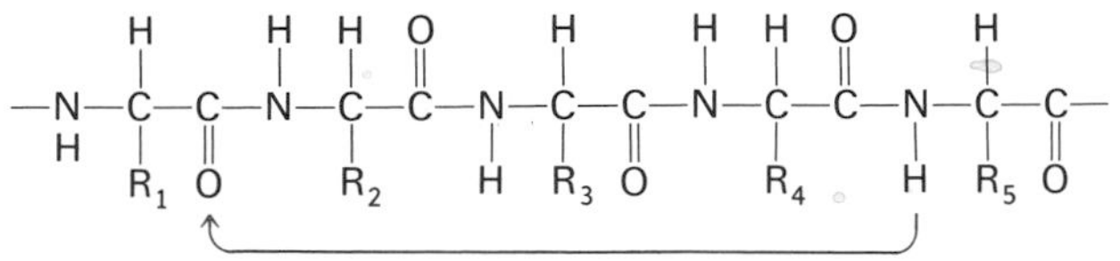

Figure 2-37
In the α helix, the NH group of residue n is hydrogen-bonded to the CO group of residue $(n - 4)$.

main-chain CO and NH groups are hydrogen-bonded. Each residue is related to the next one by a translation of 1.5 Å along the helix axis and a rotation of 100°, which gives 3.6 amino acid residues per turn of helix. Thus, amino acids spaced three and four apart in the linear sequence are spatially quite close to one another in an α helix. In contrast, amino acids two apart in the linear sequence are situated on opposite sides of the helix and so are unlikely to make contact. The pitch of the α helix is 5.4 Å, the product of the translation (1.5 Å) and the number of residues per turn (3.6). The screw-sense of a helix can be right-handed or left-handed; the α helices found in proteins are right-handed.

The α helix content of proteins of known three-dimensional structure is highly variable. In some, such as myoglobin and hemoglobin, the α helix is the major structural motif. Other proteins, such as the digestive enzyme chymotrypsin, are virtually devoid of α helix. The single-stranded α helix discussed above is usually a rather short rod, typically less than 40 Å in length. A variation of the α-helical theme is used to construct much longer rods, extending to 1000 Å or more. Two or more α helices can entwine around each other to form a cable. Such α-*helical coiled coils* are found in several proteins: keratin in hair, myosin and tropomyosin in muscle, epidermin in skin, and fibrin in blood clots. The helical cables in these proteins serve a mechanical role in forming stiff bundles of fibers.

The structure of the α helix was deduced by Pauling and Corey six years before it was actually seen in the x-ray reconstruction of the structure of myoglobin. The elucidation of the structure of the α helix is a landmark in molecular biology because it demonstrated that the conformation of a polypeptide chain can be predicted if the properties of its components are rigorously and precisely known. In the same year, Pauling and Corey discovered another periodic structural motif, which they named the β pleated sheet (β because it was the second structure they elucidated, the α helix being the first).

The *β-pleated sheet* differs markedly from the α helix in that it is a sheet rather than a rod. The polypeptide chain in the β pleated sheet is almost fully extended (Figure 2-38) rather than being tightly

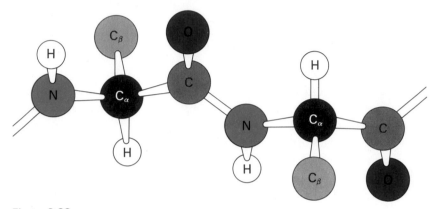

Figure 2-38
Conformation of a dipeptide unit in a β pleated sheet. The polypeptide chain is almost fully stretched out.

coiled as in the α helix. The axial distance between adjacent amino acids is 3.5 Å, in contrast with 1.5 Å for the α helix. Another difference is that the β pleated sheet is stabilized by hydrogen bonds between NH and CO groups in different polypeptide strands, whereas in the α helix the hydrogen bonds are between NH and CO groups in the same polypeptide chain. Extensive regions of the β pleated sheet structure (Figure 2-39) have been found thus far only in silk fibroin. However, short segments of the polypeptide backbone of a number of proteins are arranged in a conformation similar to that of the β pleated sheet.

A third periodic structure is the *collagen helix,* which will be discussed in detail in Chapter 9. This structure is responsible for the high tensile strength of collagen, the major component of skin, bone, and tendon.

In discussing the architecture of proteins, it is convenient to refer to four levels of structure. *Primary structure* is simply the sequence

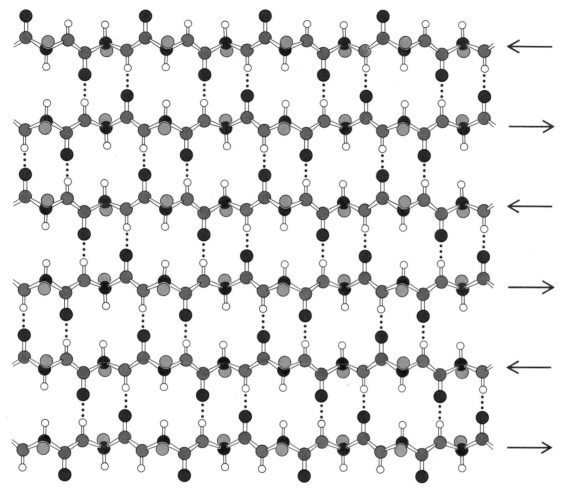

Figure 2-39
Antiparallel β pleated sheet. Adjacent strands run in opposite directions.
Hydrogen bonds between NH and CO groups of adjacent strands stabilize the
structure. The side chains are above and below the plane of the sheet.

of amino acids and location of disulfide bridges, if there are any.
The primary structure is thus a complete description of the covalent
connections of a protein. *Secondary structure* refers to the steric relationship of amino acid residues that are close to one another in
the linear sequence. Some of these steric relationships are of a
regular kind, giving rise to a periodic structure. The α helix, the
β pleated sheet, and the collagen helix are examples of secondary

structure. *Tertiary structure* refers to the steric relationship of amino acid residues that are far apart in the linear sequence. It must be emphasized that the dividing line between secondary and tertiary structure is arbitrary and so these terms can be ambiguous. Proteins that contain more than one polypeptide chain display an additional level of structural organization, namely *quaternary structure,* which refers to the way in which the chains are packed together. Each polypeptide chain in such a protein is called a *subunit*.

AMINO ACID SEQUENCE SPECIFIES THREE-DIMENSIONAL STRUCTURE

Insight into the relationship between the amino acid sequence of a protein and its conformation came from the work of Christian Anfinsen on ribonuclease, an enzyme that hydrolyzes RNA. Ribonuclease is a single polypeptide chain consisting of 124 amino acid residues (Figure 2-40). It contains four disulfide bonds, which can be irreversibly oxidized by periodate to give cysteic acid residues (Figure 2-31). Alternatively, these disulfide bonds can be cleaved reversibly by reducing them with reagents such as β-mercaptoethanol, which forms mixed disulfides with cysteine side chains (Figure 2-41). In the presence of a large excess of β-mercaptoethanol, the mixed disulfides are also reduced, so that the final product is a protein in which the disulfides (cystines) are fully converted to sulfhydryls (cysteines). However, it was found that ribonuclease at 37°C and pH 7 cannot readily be reduced by β-mercaptoethanol unless the protein is partially unfolded by denaturing agents such as urea or guanidine hydrochloride. Although the mechanism of action of these denaturing agents is not fully understood, it is evident that they disrupt noncovalent interactions. Polypeptide chains devoid of cross-links usually assume a *random-coil* conformation in 8 M urea or 6 M guanidine HCl, as evidenced by physical properties such as viscosity and optical rotatory spectra. When ribonuclease was treated with β-mercaptoethanol in 8 M urea, the product was a fully reduced, randomly coiled polypeptide chain *devoid of enzymatic activity*. In other words, ribonuclease was *denatured* by this treatment (Figure 2-42).

Anfinsen then made the critical observation that the denatured ribonuclease, freed of urea and β-mercaptoethanol by dialysis,

HO—CH_2—CH_2—SH
β-Mercaptoethanol

H_2N—C(=O)—NH_2
Urea

H_2N—C(=NH_2^+ Cl^-)—NH_2
Guanidine hydrochloride

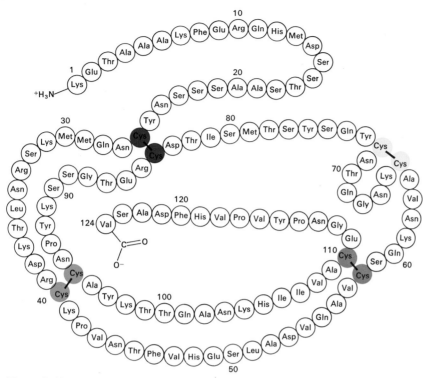

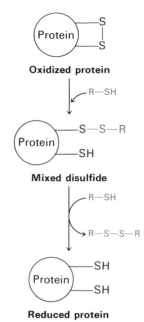

Figure 2-40
Amino acid sequence of bovine ribonuclease. The four disulfide bonds are shown in color. [Based on C. H. W. Hirs, S. Moore, and W. H. Stein. *J. Biol. Chem.* 235(1960):633.]

Figure 2-41
Reduction of the disulfide bonds in a protein by an excess of a sulfhydryl reagent such as β-mercaptoethanol.

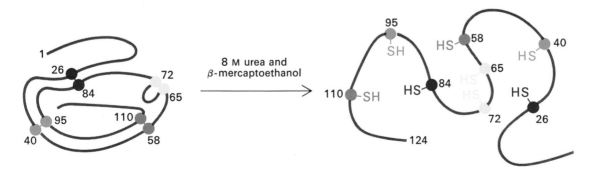

Native ribonuclease

Denatured reduced ribonuclease

Figure 2-42
Reduction and denaturation of ribonuclease.

slowly regained enzymatic activity. He immediately perceived the significance of this chance finding: the sulfhydryls of the denatured enzyme became oxidized by air and the enzyme spontaneously refolded into a catalytically active form. Detailed studies then showed that nearly all of the original enzymatic activity was regained if the sulfhydryls were oxidized under suitable conditions (Figure 2-43). All of the measured physical and chemical properties

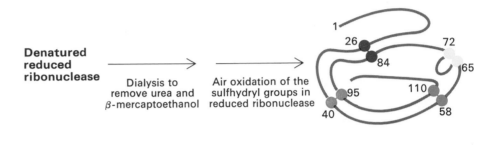

Denatured reduced ribonuclease → Dialysis to remove urea and β-mercaptoethanol → Air oxidation of the sulfhydryl groups in reduced ribonuclease

Native ribonuclease

Figure 2-43
Renaturation of ribonuclease.

of the refolded enzyme were virtually identical to those of the native enzyme. These experiments showed that *the information needed to specify the complex three-dimensional structure of ribonuclease is contained in its amino acid sequence.* Subsequent studies of other proteins have established the generality of this principle, which is a central one in molecular biology: *sequence specifies conformation.*

A quite different result was obtained when reduced ribonuclease was reoxidized while it was still in 8 M urea. This preparation was then dialyzed to remove the urea. Ribonuclease reoxidized in this way had only 1% of the enzymatic activity of the native protein. Why was the outcome of this experiment different from the one in which reduced ribonuclease was reoxidized in a solution free of urea? The reason is that wrong disulfide pairings were formed when the random-coil form of the reduced molecule was reoxidized. There are 105 different ways of pairing eight cysteines to form four disulfides; only one of these combinations is enzymatically active. The 104 wrong pairings have been picturesquely termed "scrambled" ribonuclease. Anfinsen then found that "scrambled" ribo-

nuclease spontaneously converted into fully active, native ribonuclease when trace amounts of β-mercaptoethanol were added to
the aqueous solution of the reoxidized protein (Figure 2-44). The

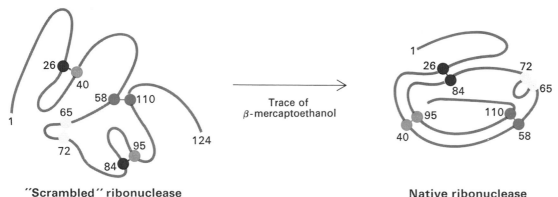

"Scrambled" ribonuclease   Native ribonuclease

Figure 2-44
Formation of native ribonuclease from "scrambled" ribonuclease in the
presence of a trace of β-mercaptoethanol.

added β-mercaptoethanol catalyzed the rearrangement of disulfide
pairings, until the native structure was regained, which took about
ten hours. This process is driven entirely by the free energy change
in going from the "scrambled" conformations to the stable, native
conformation of the enzyme. *Thus, the native form of ribonuclease appears
to be the thermodynamically most stable structure.*
   Anfinsen (1964) wrote:

   It struck me recently that one should really consider the sequence
   of a protein molecule, about to fold into a precise geometric form,
   as a line of melody written in canon form and so designed by Nature
   to fold back upon itself, creating harmonic chords of interaction
   consistent with biological function. One might carry the analogy
   further by suggesting that the kinds of chords formed in a protein
   with scrambled disulfide bridges, such as I mentioned earlier, are
   dissonant, but that, by giving an opportunity for rearrangement by
   the addition of mercaptoethanol, they modulate to give the pleasing
   harmonics of the native molecule. Whether or not some conclusion
   can be drawn about the greater thermodynamic stability of Mozart's
   over Schoenberg's music is something I will leave to the philosophers
   of the audience.

SUMMARY

Proteins play key roles in nearly all biological processes. Enzymes, the catalysts of chemical reactions in biological systems, are always proteins. Hence, proteins determine the pattern of chemical transformations in cells. Proteins mediate a wide range of other functions, such as transport and storage, coordinated motion, mechanical support, immune protection, excitability, and the control of growth and differentiation.

The basic structural units of proteins are amino acids. All proteins in all species from bacteria to man are constructed from the same set of twenty amino acids. The side chains of these building blocks differ in size, shape, charge, hydrogen-bonding capacity, and chemical reactivity. They can be grouped as follows: (a) aliphatic side chains—glycine, alanine, valine, leucine, isoleucine, and proline; (b) hydroxyl aliphatic side chains—serine and threonine; (c) aromatic side chains—phenylalanine, tyrosine, and tryptophan; (d) basic side chains—lysine, arginine, and histidine; (e) acidic side chains—aspartic acid and glutamic acid; (f) amide side chains—asparagine and glutamine; and (g) sulfur side chains—cysteine and methionine.

Many amino acids, usually more than 100, are joined by peptide bonds to form a polypeptide chain. A peptide bond links the α-carboxyl group of one amino acid and the α-amino group of the next one. In some proteins, a few side chains are cross-linked by disulfide bonds, which result from the oxidation of cysteine residues. A protein consists of one or more polypeptide chains. Each kind of protein has a unique amino acid sequence that is genetically determined. The amino acid sequence of a protein is elucidated in the following way. First, its amino acid composition is determined by ion-exchange chromatography of an acid hydrolysate of the protein. The amino-terminal residue is identified by using an end-group reagent such as dansyl chloride. Second, the protein is specifically cleaved into small peptides. For example, trypsin hydrolyzes proteins on the carboxyl side of lysine and arginine residues. Third, the amino acid sequence of these peptides is then determined by the Edman technique, which successively removes the amino-terminal residue. Finally, the order of the peptides is established from the amino acid sequences of overlap peptides.

The critical determinant of the biological function of a protein is its conformation, which is defined as the three-dimensional arrangement of the atoms of a molecule. Three regularly repeating conformations of polypeptide chains are known: the α helix, the β pleated sheet, and the collagen helix. Short segments of the α helix and the β pleated sheet are found in many proteins. An important principle is that the amino acid sequence of a protein specifies its three-dimensional structure, as was first shown for ribonuclease. Reduced, unfolded ribonuclease spontaneously forms the correct disulfide pairings and regains full enzymatic activity when oxidized by air following removal of mercaptoethanol and urea.

SELECTED READINGS

Where to start:

Dickerson, R. E., and Geis, I., 1973. *The Structure and Action of Proteins.* Benjamin.

Moore, S., and Stein, W. H., 1973. Chemical structures of pancreatic ribonuclease and deoxyribonuclease. *Science* 180:458–464.

Anfinsen, C. B., 1973. Principles that govern the folding of protein chains. *Science* 181:223–230.

Determination of amino acid sequence:

Schroeder, W. A., 1968. *The Primary Structure of Proteins.* Harper and Row.

Needleman, S. B., (ed.), 1970. *Protein Sequence Determination.* Springer-Verlag.

Edman, P., and Begg, G., 1967. A protein sequenator. *Eur. J. Biochem.* 1:80–91.

Dayhoff, M. O., (ed.), 1972. *Atlas of Protein Sequence and Structure.* National Biomedical Research Foundation, Silver Springs, Maryland. [A valuable compilation of all known amino acid sequences. This atlas is revised every few years.]

Books on protein chemistry:

Cold Spring Harbor Laboratory, 1972. *Structure and Function of Proteins at the Three-Dimensional Level* (Cold Spring Harbor Symposia on Quantitative Biology, vol. 36). [An outstanding collection of papers on the conformation of proteins, as presented in a symposium in 1972.]

Haschemeyer, R. H., and Haschemeyer, A. E. V., 1973. *Proteins: A Guide to Study by Physical and Chemical Methods.* Wiley-Interscience.

Bailey, J. L., 1967. *Techniques of Protein Chemistry* (2nd ed.). American Elsevier.

Means, G. E., and Feeney, R. E., 1971. *Chemical Modification of Proteins.* Holden-Day.

Neurath, H., (ed.), 1963. *The Proteins: Composition, Structure and Function* (2nd ed.). Academic Press. [A five-volume treatise that contains many fine articles.]

Folding of proteins:

Anfinsen, C. B., 1964. On the possibility of predicting tertiary structure from primary sequence. *In* Sela, M., (ed.), *New Perspectives in Biology,* pp. 42–50. American Elsevier.

Wetlaufer, D. B., and Ristow, S., 1973. Acquisition of three-dimensional structure of proteins. *Annu. Rev. Biochem.* 42:135–158.

Tanford, C., 1970. Protein denaturation. *Advan. Protein Chem.* 24:1–95.

Review articles:

Advances in Protein Chemistry.
Annual Review of Biochemistry.
Annual Review of Biophysics and Bioengineering.

APPENDIX
Acid-Base Concepts

Ionization of Water

Water dissociates into hydronium (H_3O^+) and hydroxyl (OH^-) ions. For simplicity, we refer to the hydronium ion as a hydrogen ion (H^+) and write the equilibrium as

$$H_2O \rightleftharpoons H^+ + OH^-$$

The equilibrium constant K_{eq} from this dissociation is given by

$$K_{eq} = \frac{[H^+][OH^-]}{[H_2O]} \qquad (1)$$

where the terms in brackets denote molar concentrations. Since the concentration of water (55.5 M) is changed little by ionization, the above expression can be simplified to give

$$K_w = [H^+][OH^-] \qquad (2)$$

where K_w is the ion product of water. At 25°C, K_w is 1.0×10^{-14}.

Note that the concentrations of H^+ and OH^- are reciprocally related. If the concentration of H^+ is high, then the concentration of OH^- must be low, and vice versa. For example, if $[H^+] = 10^{-2}$ M, then $[OH^-] = 10^{-12}$ M.

Definition of Acid and Base

An acid is a proton donor. A base is a proton acceptor.

$$\text{Acid} \rightleftharpoons H^+ + \text{Base}$$

$$CH_3{-}COOH \rightleftharpoons H^+ + CH_3{-}COO^-$$
$$\text{Acetic acid} \qquad\qquad\qquad \text{Acetate}$$

$$NH_4^+ \rightleftharpoons H^+ + NH_3$$
$$\text{Ammonium ion} \qquad\qquad \text{Ammonia}$$

The species formed by the ionization of an acid is its conjugate base. Conversely, protonation of a base yields its conjugate acid. Acetic acid and acetate ion are a conjugate acid-base pair.

Definition of pH and pK

The pH of a solution is a measure of its concentration of H^+. The pH is defined as

$$pH = \log_{10} \frac{1}{[H^+]} = -\log_{10}[H^+] \qquad (3)$$

The ionization equilibrium of a weak acid is given by

$$HA \rightleftharpoons H^+ + A^-$$

The apparent equilibrium constant K for this ionization is

$$K = \frac{[H^+][A^-]}{[HA]} \qquad (4)$$

The pK of an acid is defined as

$$pK = -\log K = \log \frac{1}{K} \qquad (5)$$

Inspection of equation (4) shows that *the pK of an acid is the pH at which it is half-dissociated.*

Henderson-Hasselbalch Equation

What is the relationship between pH and the ratio of acid to base? A useful expression can be derived starting with equation (4). Rearrangement of that equation gives

$$\frac{1}{[H^+]} = \frac{1}{K} \frac{[A^-]}{[HA]} \qquad (6)$$

Taking the logarithm of both sides of equation (6) gives

$$\log \frac{1}{[H^+]} = \log \frac{1}{K} + \log \frac{[A^-]}{[HA]} \qquad (7)$$

Substituting pH for $\log 1/[H^+]$ and pK for $\log 1/K$ in equation (7) yields

$$pH = pK + \log \frac{[A^-]}{[HA]} \qquad (8)$$

which is commonly known as the Henderson-Hasselbalch equation.

The pH of a solution can be calculated from equation (8) if the molar proportion of A^- to HA and the pK of HA are known. Consider a solution of 0.1 M acetic acid and 0.2 M acetate ion. The pK of acetic acid is 4.8. Hence, the pH of the solution is given by

$$pH = 4.8 + \log \frac{0.2}{0.1} = 4.8 + \log 2$$

$$= 4.8 + 0.3 = 5.1$$

Conversely, the pK of an acid can be calculated if the molar proportion of A^- to HA and the pH of the solution are known.

Buffering Power

An acid-base conjugate pair (such as acetic acid and acetate ion) has an important property: it resists changes in the pH of a solution. In other words, it acts as a *buffer.* Consider the addition of OH^- to a solution of acetic acid (HA).

$$HA + OH^- \rightleftharpoons A^- + H_2O$$

A plot of the dependence of the pH of this solution on the amount of OH^- added is called a *titration curve.* Note that there is an inflection point in the curve at pH 4.8, which is the pK of acetic acid.

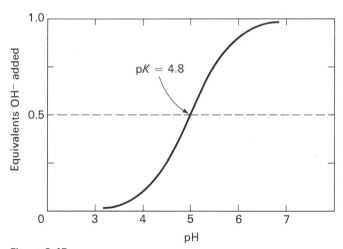

Figure 2-45
Titration curve of acetic acid.

In the vicinity of this pH, a relatively large amount of OH^- produces little change in pH. In general, a weak acid is most effective in buffering against pH changes in the vicinity of its pK value.

pK Values of Amino Acids

An amino acid such as glycine contains two ionizable groups: an α-carboxyl group and a protonated α-amino group. As base is added, these two groups are titrated (Figure 2-46). The pK of the α-COOH group is 2.3, whereas that of the α-NH$_3^+$ group is 9.6. The pK values of these groups in other amino acids are similar. Some amino acids, such as aspartic acid, also contain an ionizable side chain. The pK values of ionizable side chains in amino acids range from 3.9 (aspartic acid) to 12.5 (arginine).

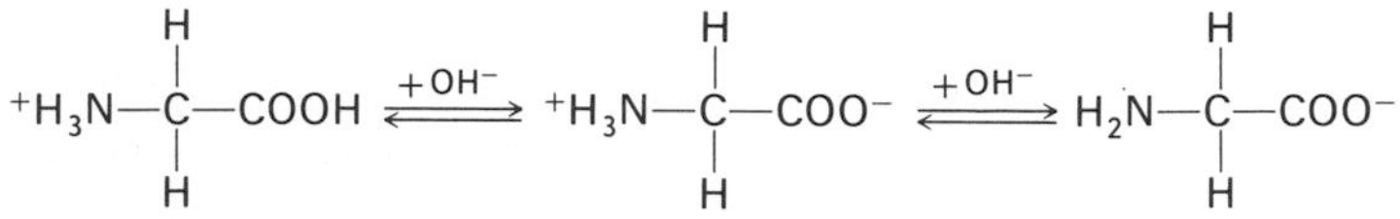

Table 2-2
pK values of some amino acids

Amino acid	α-COOH group	α-NH$_3^+$ group	Side chain
Alanine	2.3	9.9	
Glycine	2.4	9.8	
Phenylalanine	1.8	9.1	
Serine	2.1	9.2	
Valine	2.3	9.6	
Aspartic acid	2.0	10.0	3.9
Glutamic acid	2.2	9.7	4.3
Histidine	1.8	9.2	6.0
Cysteine	1.8	10.8	8.3
Tyrosine	2.2	9.1	10.9
Lysine	2.2	9.2	10.8
Arginine	1.8	9.0	12.5

pK values (25°C)

Based on J. T. Edsall and J. Wyman. *Biophysical Chemistry* (Academic Press, 1958), Ch. 8.

Figure 2-46
Titration of the ionizable groups of an amino acid.

PROBLEMS

1. The following reagents are often used in protein chemistry:

CNBr Dansyl chloride
Urea 6 N HCl
β-Mercaptoethanol Ninhydrin
Trypsin Phenyl isothiocyanate
Performic acid Chymotrypsin

Which one is the best suited for accomplishing each of the following tasks?
   (a) Determination of the amino acid sequence of a small peptide.
   (b) Identification of the amino-terminal residue of a peptide (of which you have less than 10^{-7} gram).
   (c) Reversible denaturation of a protein devoid of disulfide bonds. Which additional reagent would you need if disulfide bonds were present?
   (d) Hydrolysis of peptide bonds on the carboxyl side of aromatic residues.
   (e) Cleavage of peptide bonds on the carboxyl side of methionines.
   (f) Hydrolysis of peptide bonds on the carboxyl side of lysine and arginine residues.

2. What is the pH of each of the following solutions?

   (a) 10^{-3} N HC1.
   (b) 10^{-2} N NaOH.
   (c) 0.1 M acetic acid and 0.03 M sodium acetate.
   (d) 0.1 M glycine and 0.05 M NaOH.
   (e) 0.1 M glycine and 0.05 M HCl.

3. What is the ratio of base to acid at pH 4, 5, 6, 7, and 8 for an acid with a pK of 6?

4. How many hydrogen ions are there in a spherical cell 10 μm in diameter in which the pH is 7?

5. Tropomyosin, a muscle protein, is a two-stranded α-helical coiled coil. The molecular weight of the protein is 70,000. The average residue weight is about 110. What is the length of the molecule?

6. Poly-L-glutamic acid, a synthetic polypeptide, is α-helical at pH 4. However, it is a random coil at pH 7. Why?

7. Anhydrous hydrazine has been used to cleave peptide bonds in proteins. What are the reaction products? How might this technique be used to identify the carboxyl-terminal amino acid?

8. The amino acid sequence of human adrenocorticotrophin, a polypeptide hormone, is:

   Ser-Tyr-Ser-Met-Glu-His-Phe-Arg-Trp-Gly-Lys-Pro-Val-Gly-Lys-Lys-Arg-Arg-Pro-Val-Lys-Val-Tyr-Pro-Asp-Ala-Gly-Glu-Asp-Gln-Ser-Ala-Glu-Ala-Phe-Pro-Leu-Glu-Phe

   (a) What is the approximate net charge of this molecule at pH 7? Assume that its side chains have the pK values given in Table 2-2 (p. 44), and that the pKs of the terminal —NH_3^+ and —COOH groups are 7.8 and 3.6, respectively.
   (b) How many peptides result from the treatment of the hormone with cyanogen bromide?

9. Ethyleneimine reacts with cysteine side chains in proteins to form S-aminoethyl derivatives. The peptide bonds on the carboxyl side of these modified cysteine residues are susceptible to hydrolysis by trypsin. Why?

10. An enzyme that catalyzes disulfide-sulfhydryl exchange reactions has been isolated. Inactive "scrambled" ribonuclease is rapidly converted to enzymatically active ribonuclease by this enzyme. In contrast, insulin is rapidly inactivated by this enzyme. What does this important observation imply about the relationship between the amino acid sequence and three-dimensional structure of insulin?

For additional problems, see W. B. Wood, J. H. Wilson, R. M. Benbow, and L. E. Hood, *Biochemistry: A Problems Approach* (Benjamin, 1974), Chapter 2.

CHAPTER 3

OXYGEN TRANSPORTERS:
MYOGLOBIN AND HEMOGLOBIN

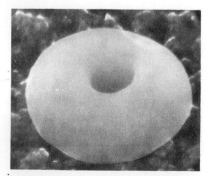

Scanning electron micrograph of an erythrocyte. [Courtesy of Dr. Mark Goldman and Dr. Robert Leif.]

The transition from anaerobic to aerobic life was a major step in evolution because it uncovered a rich reservoir of energy. Eighteen times as much energy is extracted from glucose in the presence of oxygen than in its absence. Vertebrates have evolved two principal mechanisms for providing their cells with a continuous and adequate flow of oxygen. The first is a circulatory system that actively delivers oxygen to the cells. In the absence of a circulatory system, the limit on the size of an aerobic organism would be about a millimeter since the diffusion of oxygen over larger dimensions would be too slow to meet cellular needs. The second major adaptation is the acquisition of *oxygen-carrying molecules* to overcome the limitation imposed by the low solubility of oxygen in water. The oxygen carriers in vertebrates are the proteins *hemoglobin* and *myoglobin*. Hemoglobin, which is contained in red blood cells, serves as the oxygen carrier in blood. The presence of hemoglobin increases the oxygen transporting capacity of a liter of blood from 5 to 250 ml of O_2. Hemoglobin also plays a vital role in the transport of carbon dioxide and hydrogen ion. Myoglobin, which is located in muscle, serves as a reserve supply of oxygen and also facilitates the movement of oxygen within muscle.

OXYGEN BINDS TO A HEME PROSTHETIC GROUP

The capacity of myoglobin and hemoglobin to bind oxygen depends on the presence of a nonpolypeptide unit, namely a *heme group*. The heme also gives myoglobin and hemoglobin their distinctive color. Indeed, many proteins require a tightly bound, specific nonpolypeptide unit for their biological activity. Such a unit is called a *prosthetic group*. A protein without its characteristic prosthetic group is termed an *apoprotein*.

The heme consists of an organic part and an iron atom. The organic part, *protoporphyrin*, is made up of four *pyrrole* groups. The four pyrroles are linked by methene bridges to form a tetrapyrrole ring. Four methyl, two vinyl, and two propionate side chains are attached to the tetrapyrrole ring. These substituents can be arranged in fifteen different ways. Only one of these isomers, called protoporphyrin IX, is present in biological systems.

Protoporphyrin IX

Heme
(Fe-protoporphyrin IX)

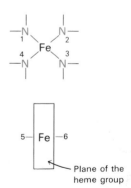

Figure 3-1
The iron atom in heme can form six bonds. Four of these bonds are in the plane of the heme. The fifth bond is on one side of this plane, and the sixth bond on the other side. The locations of the bonded atoms are sometimes called coordination positions.

The iron atom in heme binds to the four nitrogens in the center of the protoporphyrin ring (Figures 3-1 and 3-2). The iron can form two additional bonds, one on either side of the heme plane. These bonding sites are termed the fifth and sixth coordination positions. The iron atom in the heme can be in the ferrous ($+2$) or the ferric ($+3$) oxidation state. The corresponding forms of hemoglobin are

called *ferrohemoglobin* and *ferrihemoglobin*, respectively. Ferrihemoglobin is also called methemoglobin. Only ferrohemoglobin, the $+2$
oxidation state, can bind oxygen. The same nomenclature applies
to myoglobin.

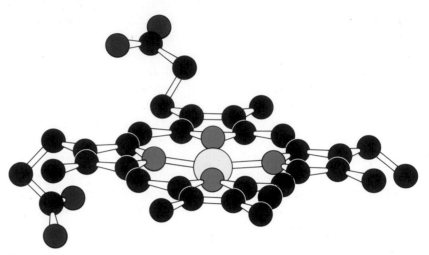

Figure 3-2
Model of the heme group in myoglobin (yellow for Fe,
blue for N, red for O, and black for C).

X-RAY CRYSTALLOGRAPHY REVEALS THREE-DIMENSIONAL STRUCTURE IN ATOMIC DETAIL

The elucidation of the three-dimensional structure of myoglobin
by John Kendrew and of hemoglobin by Max Perutz are landmarks
in molecular biology. These studies came to fruition in the late
1950s and showed that x-ray crystallography can reveal the structure of molecules as large as proteins. The largest structure solved
before 1957 was Vitamin B_{12}, which is an order of magnitude
smaller than myoglobin (molecular weight 17,800) and hemoglobin
(molecular weight 66,000). The determination of the three-dimensional structure of these proteins was a great stimulus to the field
of protein crystallography. The detailed three-dimensional structures of more than twenty proteins are now known, and the elucidation of the structure of many others is in progress. *X-ray crystallography is making a profound contribution to our understanding of protein
structure and function, since it is the only method that can reveal the three-
dimensional position of most of the atoms in a protein.* Electron microscopy

Myoglobin in
dilute buffer

Addition of
$(NH_4)_2SO_4$

Several days

Myoglobin in
$3 \, \text{M} \, (NH_4)_2SO_4$, pH 7

Myoglobin crystals

Figure 3-3
Crystallization
of myoglobin.

can be highly informative about the structure of biological macro-molecules, but the technique is not yet effective in delineating molecular architecture in atomic detail.

Before turning to the three-dimensional structure of myoglobin and hemoglobin, let us consider some basic aspects of the x-ray crystallographic method. First, crystals of the protein are needed. Myoglobin, for example, is crystallized by adding ammonium sulfate to a concentrated solution of the protein (Figure 3-3). Ammonium sulfate at a concentration of 3 M markedly reduces the solubility of myoglobin and leads to its crystallization. The solubility of most proteins is reduced on addition of high concentrations of any salt. This effect is called *salting-out*. Crystals of myoglobin (Figure 3-4) produced in this way may be several millimeters long.

The three components in an x-ray crystallographic experiment are a *source of x-rays*, a *protein crystal*, and a *detector* (Figure 3-5). A beam of x-rays of wavelength 1.54 Å is produced by accelerating electrons against a copper target. A narrow beam of x-rays strikes the protein crystal. Part of the beam goes through the crystal without any change in direction. The rest is *scattered* in a variety

Figure 3-4
Photograph of myoglobin crystals.
[Courtesy of Dr. John Kendrew.]

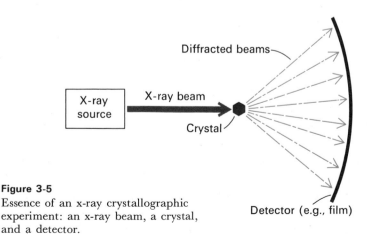

Diffracted beams

X-ray
source

X-ray beam

Crystal

Detector (e.g., film)

Figure 3-5
Essence of an x-ray crystallographic
experiment: an x-ray beam, a crystal,
and a detector.

of different directions. The scattered (or *diffracted*) beams can be detected by photographic film, the blackening of the emulsion being proportional to the intensity of the scattered x-ray beam. The basic physical principles underlying the technique are:

1. Electrons scatter x-rays. The amplitude of the wave scattered by an atom is proportional to its number of electrons. Thus, a carbon atom scatters six times as strongly as a hydrogen atom.

2. The scattered waves recombine. Each atom contributes to each scattered beam. The scattered waves can reinforce or cancel one another, depending on whether they are in phase or out of phase.

3. The way in which the scattered waves recombine depends only on the atomic arrangement.

STAGES IN THE X-RAY STUDY OF MYOGLOBIN

Kendrew chose myoglobin for x-ray analysis for several reasons: it is a rather small protein (molecular weight 17,600), it is easily prepared in quantity, and it is readily crystallized. Furthermore, myoglobin had the additional advantage of being closely related to hemoglobin, which was already being studied by his colleague Perutz. Myoglobin from the skeletal muscle of the sperm whale (Figure 3-6) was chosen because it is stable and forms excellent crystals. The skeletal muscle of diving mammals such as whale, seal, and porpoise is particularly rich in myoglobin, which serves as a store of oxygen that is used during a dive.

Figure 3-6
A sperm whale.

The myoglobin crystal is mounted in a capillary and positioned in a precise orientation with respect to the x-ray beam and the film. A precession motion of the crystal yields an x-ray photograph in which there is a regular array of spots. The x-ray photograph shown in Figure 3-7 is a two-dimensional section through a three-dimensional array of spots. For myoglobin, there are 25,000 of these spots. The intensity of each of these spots is measured. These intensities are the basic experimental data of an x-ray crystallographic analysis.

Figure 3-7
X-ray precession photograph of a myoglobin crystal.

The problem now is to reconstruct an image of myoglobin from these observed intensities. This is accomplished by a mathematical relationship called a Fourier series, which is a sum of sine and cosine terms. However, the scattered intensities of myoglobin crystals provide only part of the information needed to calculate the Fourier series. The essential missing data—namely, the phases of the scattered beams—are obtained from the complete diffraction patterns of myoglobin crystals that contain heavy atoms such as uranium and lead at one or two sites in the molecule. The stage is now set for the calculation of an electron-density map using high-speed computers. Fortunately, the development of fast computers with large memories kept pace with the needs of the myoglobin project. The final Fourier synthesis for myoglobin contained about a billion terms.

The Fourier synthesis gives the density of electrons at a large number of regularly spaced points in the crystal. This three-dimensional electron-density distribution is represented by a series of parallel sections stacked on top of each other. Each section is a transparent plastic sheet on which the electron-density distribution is represented by contour lines (Figure 3-8). This mode of representing electron density is analogous to the use of contour lines in geological survey maps to depict altitude (Figure 3-9). The next step is to interpret the electron-density map. A critical factor is the

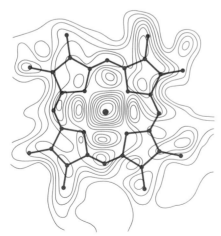

Figure 3-8
Section from the electron-density map of myoglobin showing the heme group. The peak of the center of this section corresponds to the position of the iron atom. [From J. C. Kendrew. The three-dimensional structure of a protein molecule. Copyright © 1961 by Scientific American, Inc. All rights reserved.]

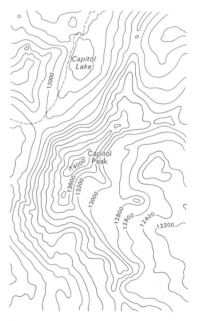

Figure 3-9
Section from a U.S. Geological Survey Map of the Capitol Peak Quadrangle, Colorado.

A

B

C

D

Figure 3-10
Effect of resolution on the quality of a reconstructed image, as shown by an optical analog of x-ray diffraction: (A) the Parthenon; (B) a diffraction pattern of the Parthenon; (C and D) images reconstructed from the pattern in part B. More data were used to obtain D than C, which accounts for the higher quality of the image in D. [Courtesy of Dr. Thomas Steitz (part A) and Dr. David De Rosier (part B).]

resolution of the x-ray analysis, which is determined by the number of scattered intensities used in the Fourier synthesis. The fidelity of the image depends on the resolution of the Fourier synthesis as shown by the optical analogy in Figure 3-10. The analysis of myoglobin was carried out in three stages. In the first, completed in 1957, only the innermost 400 spots of the diffraction pattern were used, corresponding to a resolution of 6 Å. As we shall see shortly, this low-resolution electron-density map revealed the course of the polypeptide chain but few other structural details. The reason is that polypeptide chains pack together so that their centers are 5 to 10 Å apart. Maps at higher resolution are needed to delineate groups of atoms, which lie 2.8 to 4 Å apart, and individual atoms, which are 1 to 1.5 Å apart. Such maps of myoglobin were obtained in 1959 at a resolution of 2 Å (10,000 intensities) and in 1962 at a resolution of 1.4 Å (25,000 intensities). The ultimate resolution of an x-ray analysis is determined by the degree of perfection of the crystal. For proteins, this limiting resolution is usually not much better than 2 Å.

In 1957, Kendrew and his colleagues saw what no one before had ever seen: a three-dimensional picture of a protein molecule in all its complexity. The model derived from their 6 Å Fourier synthesis contained a set of high-density rods of just the dimensions expected for a polypeptide chain (Figure 3-11). The molecule

Figure 3-11
A model of myoglobin at low resolution. [Courtesy of Dr. John Kendrew.]

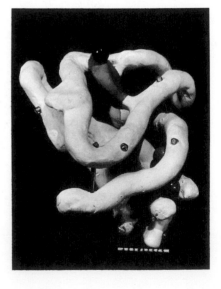

appeared very compact. Closer examination showed that the molecule consisted of a complicated and intertwining set of these rods, going straight for a distance, then turning a corner and going off in a new direction. The location of the iron atom of the heme was also evident because it contains many more electrons than any other atom in the structure. The most striking features of the molecule were its irregularity and its total lack of symmetry.

MYOGLOBIN HAS A COMPACT STRUCTURE AND A HIGH CONTENT OF ALPHA HELICES

The high-resolution electron-density map of myoglobin, obtained two years later, contained a wealth of structural detail. The positions of 1200 of the 1260 nonhydrogen atoms were clearly defined to a precision of better than 0.3 Å. The course of the main chain and the position of the heme group are shown in Figure 3-12. Some important features of myoglobin are:

1. Myoglobin is *extremely compact*. The overall dimensions are about $45 \times 35 \times 25$ Å. There is very little empty space inside.

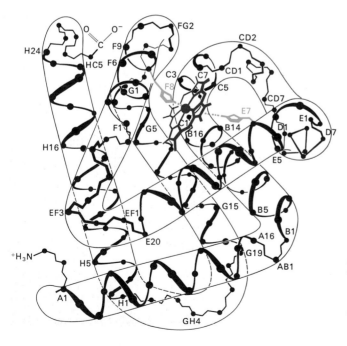

Figure 3-12
A model of myoglobin at high resolution. Only the α-carbon atoms are shown. The heme group is shown in red. [Based on R. E. Dickerson. In *The Proteins*, H. Neurath, ed., 2nd ed., vol. 2 (Academic Press, 1964), p. 634.]

Val-Leu-Ser-Glu-Gly-Glu-Trp-Gln-Leu-Val-
NA1 NA2 A1 A2 A3 A4 A5 A6 A7 A8   10

Leu-His-Val-Trp-Ala-Lys-Val-Glu-Ala-Asp-
A9 A10 A11 A12 A13 A14 A15 A16 AB1 B1   20

Val-Ala-Gly-His-Gly-Gln-Asp-Ile-Leu-Ile-
B2 B3 B4 B5 B6 B7 B8 B9 B10 B11   30

Arg-Leu-Phe-Lys-Ser-His-Pro-Glu-Thr-Leu-
B12 B13 B14 B15 B16 C1 C2 C3 C4 C5   40

Glu-Lys-Phe-Asp-Arg-Phe-Lys-His-Leu-Lys-
C6 C7 CD1 CD2 CD3 CD4 CD5 CD6 CD7 CD8   50

Thr-Glu-Ala-Glu-Met-Lys-Ala-Ser-Glu-Asp-
D1 D2 D3 D4 D5 D6 D7 E1 E2 E3   60

Leu-Lys-Lys-His-Gly-Val-Thr-Val-Leu-Thr-
E4 E5 E6 E7 E8 E9 E10 E11 E12 E13   70

Ala-Leu-Gly-Ala-Ile-Leu-Lys-Lys-Lys-Gly-
E14 E15 E16 E17 E18 E19 E20 EF1 EF2 EF3   80

His-His-Glu-Ala-Glu-Leu-Lys-Pro-Leu-Ala-
EF4 EF5 EF6 EF7 EF8 F1 F2 F3 F4 F5   90

Gln-Ser-His-Ala-Thr-Lys-His-Lys-Ile-Pro-
F6 F7 F8 F9 FG1 FG2 FG3 FG4 FG5 G1   100

Ile-Lys-Tyr-Leu-Glu-Phe-Ile-Ser-Glu-Ala-
G2 G3 G4 G5 G6 G7 G8 G9 G10 G11   110

Ile-Ile-His-Val-Leu-His-Ser-Arg-His-Pro-
G12 G13 G14 G15 G16 G17 G18 G19 GH1 GH2   120

Gly-Asn-Phe-Gly-Ala-Asp-Ala-Gln-Gly-Ala-
GH3 GH4 GH5 GH6 H1 H2 H3 H4 H5 H6   130

Met-Asn-Lys-Ala-Leu-Glu-Leu-Phe-Arg-Lys-
H7 H8 H9 H10 H11 H12 H13 H14 H15 H16   140

Asp-Ile-Ala-Ala-Lys-Tyr-Lys-Glu-Leu-Gly-
H17 H18 H19 H20 H21 H22 H23 H24 HC1 HC2   150

Tyr-Gln-Gly
HC3 HC4 HC5   153

Figure 3-13
Amino acid sequence of sperm whale myoglobin. The labels below each residue in the sequence refer to its position in an α-helical region or a nonhelical region. For example, B4 is the fourth residue in the B helix; EF7 is the seventh residue in the nonhelical region between the E and F helices. [Based on A. E. Edmundson. *Nature* 205(1965):883; and H. C. Watson. *Progr. Stereochem.* 4(1969):299–333.]

2. About 75% of the main chain is folded in an α-*helical conformation*. All of the α helices are right-handed. There are eight major helical segments, referred to as A, B, C, . . . , H. The first residue in helix A is designated as A1, the second as A2, and so forth (Figure 3-13). There are five nonhelical segments between helical regions (named CD, for example, if located between the C and D helices). There are two other nonhelical regions: the two residues at the amino-terminal end (named NA1 and NA2) and the five residues at the carboxyl-terminal end (named HC1 through HC5).

3. The factors responsible for termination of helical segments are not all known. An important one is the occurrence of a proline residue. *Proline cannot be accommodated in an α helix* (except at one of its ends). The five-membered pyrrolidone ring simply does not fit within a straight stretch of α helix. There are four prolines in myoglobin, whereas there are eight terminations of helices. Clearly, other factors are involved. For instance, the interaction of the OH group of serine and threonine with the main-chain carbonyl group sometimes disrupts an α helix.

4. The peptide groups are *planar*. The carbonyl group is *trans* to the main-chain NH. Also, the bond angles and distances are like those in dipeptides and related compounds.

5. The inside and outside are well defined. *The interior consists almost entirely of nonpolar residues* such as leucine, valine, methionine, and phenylalanine. There are no glutamic, aspartic, glutamine, asparagine, lysine, or arginine side chains in the interior of the molecule. Residues that have both a polar and nonpolar part, such as threonine, tyrosine, and tryptophan, are oriented so that their nonpolar portions point inward. The only polar residues inside myoglobin are two histidines, which have a critical function at the active site. The outside of the molecule contains both polar and nonpolar residues.

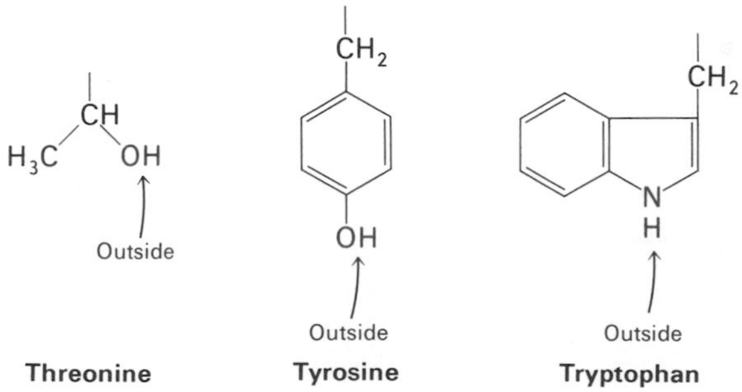

THE OXYGEN-BINDING SITE IN MYOGLOBIN

The heme group is located in a crevice in the myoglobin molecule. The highly polar propionate side chains of the heme are on the surface of the molecule. At physiological pH, these groups are ionized. The rest of the heme is inside the molecule, where it is surrounded by nonpolar residues except for two histidines. The iron atom of the heme is directly bonded to one of these histidines, namely residue F8 (Figures 3-14 and 3-15). This histidine, which occupies the fifth coordination position, is called the proximal histidine. The iron atom is about 0.3 Å out of the plane of the porphyrin, on the side of histidine F8. *The oxygen-binding site is on the other side of the heme plane, at the sixth coordination position.* A second

Figure 3-14
Schematic diagram of the oxygen-binding site in myoglobin: the fifth coordination position is occupied by histidine F8 (the proximal histidine); O_2 is bound at the sixth coordination position; and histidine E7 (the distal histidine) is near the sixth coordination position.

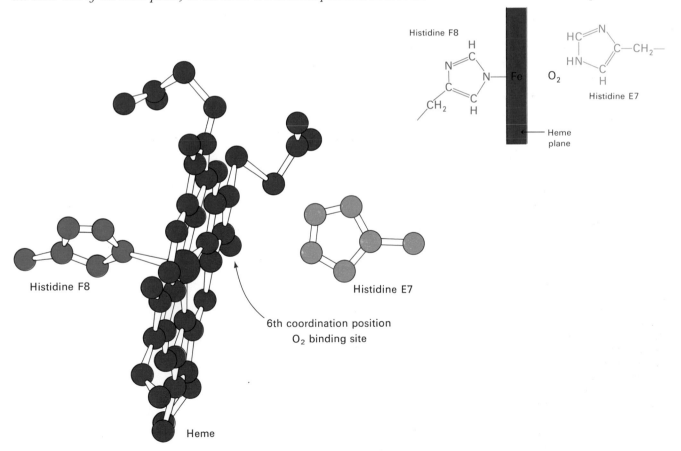

Figure 3-15
Model of the oxygen-binding site in myoglobin. The heme group, the proximal histidine (F8), and the distal histidine (E7) are shown.

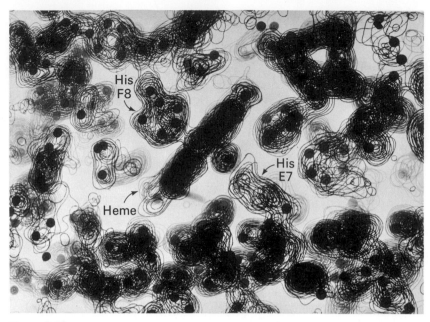

Figure 3-16
Section of the electron-density map of myoglobin near the oxygen-binding site. Electron density extending across the lower part of the map is the E helix. [Courtesy of Dr. John Kendrew.]

histidine residue (E7), termed the distal histidine, is nearby. This E7 distal histidine is not bonded to the heme. A section of the electron-density map near the heme is shown in Figure 3-16.

Three physiologically pertinent forms of myoglobin have been investigated: deoxymyoglobin, oxymyoglobin, and ferrimyoglobin. Their conformations appear to be very similar, except at the sixth coordination position (Table 3-1). In ferrimyoglobin, this site is occupied by water; in deoxymyoglobin, it is empty; in oxymyoglobin, it is occupied by O_2. Thus, *the oxygenation of deoxymyoglobin entails only the insertion of oxygen at the sixth coordination position, where it forms a bond with the iron atom of the heme.*

The residues in contact with the heme come from different segments of the linear sequence of amino acids. Contributions are made by residues ranging from C4 to H14, corresponding to amino acids 39 and 138 in the linear sequence. Thus, *the heme-binding site is very much a three-dimensional entity.*

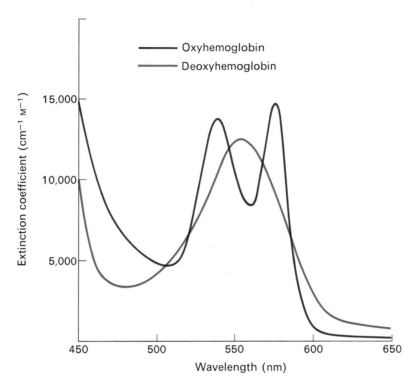

Figure 3-17
The visible absorption spectra of myoglobin and hemoglobin change markedly on binding oxygen. Myoglobin and hemoglobin have very similar visible absorption spectra.

Table 3-1
Heme environment

Form	Oxidation state of Fe	Occupant	
		5th coordination position	6th coordination position
Deoxymyoglobin	+2	His F8	Empty
Oxymyoglobin	+2	His F8	O_2
Ferrimyoglobin	+3	His F8	H_2O

THE NONPOLAR HEME ENVIRONMENT IS ESSENTIAL FOR REVERSIBLE OXYGENATION

The oxygen-binding site comprises only a small fraction of the volume of the myoglobin molecule. Indeed, oxygen is directly bonded only to the iron atom of the heme. Why is the polypeptide portion of myoglobin needed for oxygen transport and storage? The answer comes from a consideration of the oxygen-binding properties of an isolated heme group. In water, an isolated ferrous heme group can bind oxygen, but it does so for only a fleeting moment. The reason is that in water the ferrous heme is very rapidly oxidized to ferric heme, which cannot bind oxygen. In contrast, in myoglobin there is little likelihood of oxidation to the ferric form on binding oxygen. How does myoglobin stabilize the ferrous form of its heme group? Experiments by Jui Wang on a model system have provided insight in this regard. The heme group can undergo reversible oxygenation without oxidation if it is imbedded in a polystyrene matrix that contains a derivative of imidazole. This well-defined model system mimics some features of the heme environment in myoglobin and also in hemoglobin, which has a very similar oxygen-binding site, as will be discussed shortly. The imidazole has a role similar to that of the proximal F8 histidine, whereas the

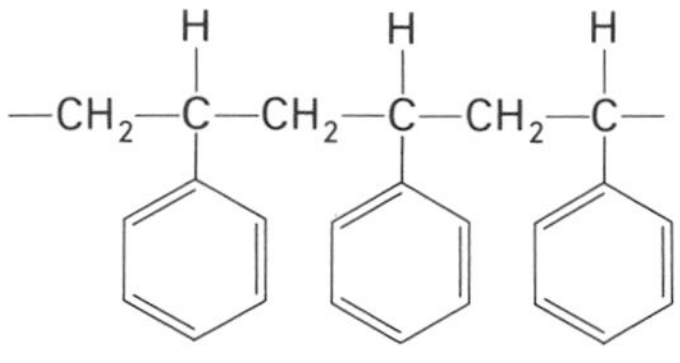

Polystyrene
(Provides a nonpolar matrix)

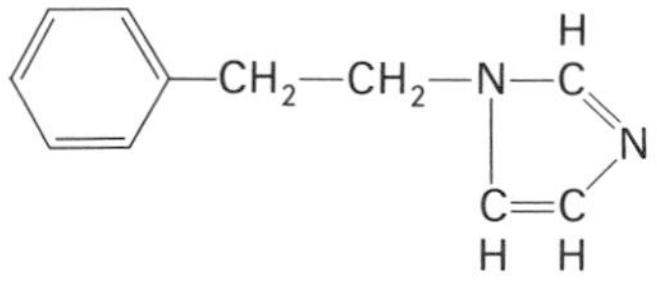

Phenylethyl imidazole
(Analog of histidine F8)

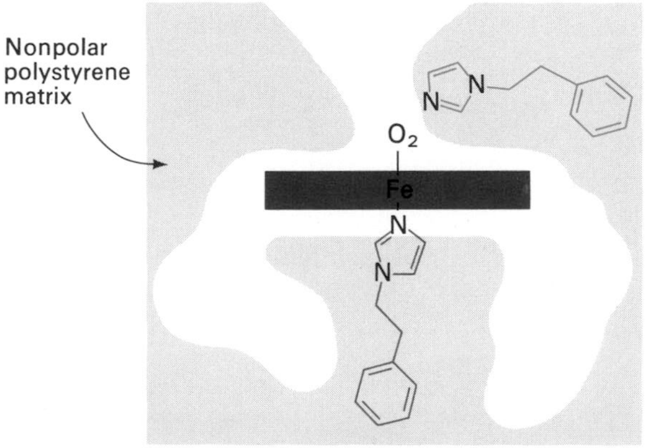

Figure 3-18
A synthetic model system that can bind oxygen reversibly. The heme group is in a nonpolar environment. [Based on J. H. Wang. *Accounts Chem. Res.* 3(1970):97.]

polystyrene provides the heme with a markedly nonpolar environment (Figure 3-18). In a nonpolar milieu, it is much harder to strip an electron away from the ferrous iron than in water. Thus, *the nonpolar heme-binding site in myoglobin protects the ferrous state of the heme from oxidation by excluding water.*

In this way, myoglobin has created a special microenvironment for its prosthetic group, thereby conferring distinctive properties on it. In general, *the function of a prosthetic group is partly dependent on its polypeptide environment.* For example, the same heme group has quite a different function in cytochrome *c*, a protein in the terminal oxidation chain in mitochondria of all aerobic organisms. In cytochrome *c*, the heme is a reversible carrier of electrons rather than of oxygen. The heme has yet another function in the enzyme catalase, where it catalyzes the conversion of hydrogen peroxide to water and oxygen.

STRUCTURES OF MYOGLOBIN IN SOLUTION AND IN THE CRYSTALLINE STATE ARE VERY SIMILAR

The high-resolution x-ray analysis of myoglobin provided a magnificently detailed, though static, picture of the molecule. It is pertinent to ask whether the structure of myoglobin in the crystalline state is like that in solution. The crystalline state of myoglobin differs from the solution state in its high ionic strength, $3 \text{ M } (NH_4)_2SO_4$, and in the existence of interactions between different myoglobin molecules. Do these factors distort the structure of myoglobin so that the picture that emerges from the x-ray analysis is not relevant toward an understanding of its biological function? The answer to this question is a decisive no. There are several lines of evidence that indicate that *the structures in solution and in the crystalline state are very similar.*

1. The crystalline state of myoglobin is functionally active. Crystalline myoglobin can bind oxygen. However, the reaction is not as fast as in solution. The crystalline state is generally more sluggish in its reactivity.

2. The absorption spectrum of the heme group of myoglobin is virtually the same in solution and in the crystal. The absorption

spectrum is a very sensitive index of the detailed environment of the heme group.

3. The α-helix content of a molecule in solution can be estimated from optical rotatory dispersion and circular dichroism measurements. The α-helix content of myoglobin in solution determined in this way agrees well with the α-helix content derived from the electron-density map of the crystal. It seems improbable that a major change in conformation would be unaccompanied by a change in the amount of helix present.

4. An x-ray analysis of seal myoglobin shows that this molecule has a tertiary structure very much like that of sperm whale myoglobin. In contrast, the crystal lattices of these myoglobins are quite different, and so the interactions between myoglobin molecules in the two crystals are dissimilar. Thus, it is unlikely that there are major distortions imposed by the crystal lattice.

NONPOLAR INTERACTIONS ARE IMPORTANT IN STABILIZING THE CONFORMATION OF MYOGLOBIN

The x-ray analysis reveals the structure of myoglobin but does not directly explain why myoglobin folds in that way. Indeed, one of the challenges in protein chemistry is to discover the rules that determine how the amino acid sequence specifies the three-dimensional structure. Rather little can be said now in this regard. The one generalization that can be made about the structure of myoglobin is that the inside consists of tightly packed nonpolar residues. Residues such as valine, leucine, isoleucine, methionine, and phenylalanine are *hydrophobic*. Given a choice between water and a nonpolar environment, they markedly prefer a nonpolar milieu. Furthermore, there are van der Waals attractive forces between these closely packed side chains. Thus, a significant driving force in the folding of myoglobin is the tendency of these residues to flee from water. These hydrophobic side chains are thermodynamically more stable when clustered in the interior of the molecule than when extended into the aqueous surroundings. This stability arises in the folding process because of the gain in entropy when ordered water molecules around the exposed hydrophobic groups are dispersed (see Chapter 6). The polypeptide chain therefore folds spontaneously in aqueous solution so that its hydrophobic side chains are buried and its polar charged chains are on the surface.

UNFOLDED MYOGLOBIN CAN SPONTANEOUSLY REFOLD INTO AN ACTIVE MOLECULE

What is the effect of the heme group on the three-dimensional structure of myoglobin? This question was answered by preparing *apomyoglobin*, which is myoglobin devoid of its heme group. Apomyoglobin was obtained by lowering the pH of a myoglobin solution to 3.5. The heme group binds only weakly to the protein at this acidic pH, and so it can be removed by extraction with an organic solvent. The aqueous phase, which contained the apomyoglobin, was then neutralized. Optical rotatory dispersion measurements showed that the α-helix content of apomyoglobin at neutral pH is 60%, which is appreciably lower than that of myoglobin (75%). Hydrodynamic studies indicated that apomyoglobin is less compact than myoglobin. Furthermore, apomyoglobin is much less stable than myoglobin. Thus, it is evident that *the presence of the heme group has an appreciable effect on the structure of myoglobin.*

Is the folding of apomyoglobin and the subsequent insertion of the heme group a spontaneous process? Apomyoglobin at neutral pH was unfolded by addition of urea or guanidine. The α-helix content under these conditions (e.g., 8 M urea) was nearly zero. The urea was then removed by dialysis. The α-helix content increased to 60%, which showed that apomyoglobin had refolded. The subsequent addition of heme to this solution of apomyoglobin resulted in the formation of a biologically active molecule (Figure 3-19).

Figure 3-19
Formation of myglobin from unfolded apomyoglobin.

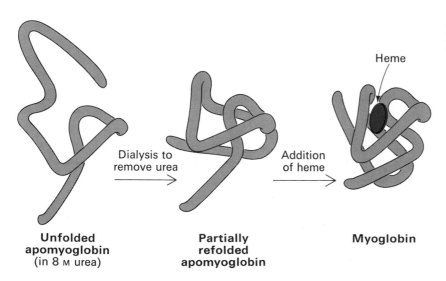

Dialysis to remove urea

Addition of heme

Heme

Unfolded apomyoglobin (in 8 M urea)

Partially refolded apomyoglobin

Myoglobin

On reduction to the ferrous state, the refolded myoglobin was fully effective in reversibly binding oxygen. Thus, *the complex three-dimensional structure of myoglobin is inherent in the amino acid sequence of apomyoglobin*, when it is combined with the heme prosthetic group. This finding reinforced the generality of the principle first discovered in studies of the renaturation of ribonuclease: *amino acid sequence specifies conformation*.

HEMOGLOBIN CONSISTS OF FOUR POLYPEPTIDE CHAINS

We turn now to hemoglobin, a related protein. Hemoglobin is made up of four polypeptide chains, in contrast to myoglobin, which is a single-chain protein. The four chains are held together by non-covalent interactions. There are four binding sites for oxygen on the hemoglobin molecule, since each chain contains one heme group. Hemoglobin A, the principal hemoglobin in adults, consists of two chains of one kind, called α chains, and two of another kind, called β chains. Thus, *the subunit structure of hemoglobin A is $\alpha_2\beta_2$.* There is a minor hemoglobin in adults, called hemoglobin A_2, which constitutes about 2% of the total hemoglobin. Hemoglobin A_2 has the subunit structure $\alpha_2\delta_2$. The fetus has distinctive hemoglobins. An embryonic $\alpha_2\epsilon_2$ hemoglobin has been detected early in fetal life. This is followed by hemoglobin F, which has the subunit structure $\alpha_2\gamma_2$. The biological significance of these different hemoglobins is an intriguing question, which will be considered in the next chapter. The α chain, which is common to these hemoglobins, contains 141 amino acid residues. The β, δ, and γ chains each have 146 amino acids, and rather similar amino acid sequences (Figure 3-20).

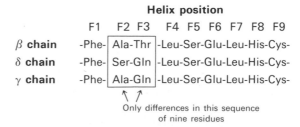

Figure 3-20
The β, γ, and δ chains of human hemoglobins have similar amino acid sequences, as shown here for residues F1 to F9.

X-RAY ANALYSIS OF HEMOGLOBIN

The three-dimensional structure of hemoglobin A has been determined by Max Perutz and his colleagues. This monumental accomplishment had its start in 1936, when Perutz left Austria to
pursue graduate work in Cambridge, England. Perutz joined the
laboratory of John Bernal, who had taken the first x-ray pictures
of protein crystals two years earlier. Bernal and Dorothy Crowfoot
Hodgkin, a graduate student, had obtained excellent diffraction
patterns from the enzyme pepsin, thus showing that proteins in fact
possess well-defined structures. They envisioned in 1934 the high
promise of x-ray crystallography for "arriving at far more detailed
conclusions about protein structure than previous physical or
chemical methods have been able to give." However, more than
20 years passed before this promise was realized. When Perutz chose
hemoglobin as his thesis subject, the largest structure that had been
solved was the dye phthalocyanin, which contains 58 atoms. Thus,
Perutz chose to tackle a molecule one hundred times larger. It was
little wonder that "my fellow students regarded me with a pitying
smile. . . . Fortunately, the examiners of my doctoral thesis did not
insist on a determination of the structure, otherwise I should have
had to remain a graduate student for 23 years." Fortunately,
Lawrence Bragg, who with his father in 1912 was the first to use
x-ray crystallography to solve structures, became director of the
Cavendish Laboratory at this time and supported the project. He
wrote: "I was frank about the outlook. It was like multiplying a
zero probability that success would be achieved by an infinity of
importance if the structure came out; the result of this mathematic
operation was anyone's guess." Success came in 1959, when Perutz
obtained a low-resolution electron-density image of horse oxyhemoglobin. Since then, high-resolution maps have been obtained
for both oxyhemoglobin and deoxyhemoglobin of horse, and a lower
resolution for these hemoglobins of man. Human and horse hemoglobins are very similar.

QUATERNARY STRUCTURE OF HEMOGLOBIN

The hemoglobin molecule is nearly spherical, with a diameter of
55 Å. The four chains are packed together in a tetrahedral array
(Figure 3-21). The heme groups are located in crevices near the

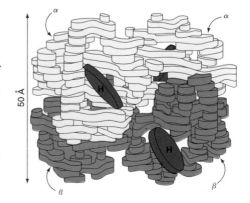

Figure 3-21
A model of hemoglobin at low
resolution. The α chains in this
model are yellow; the β chains,
blue; and the heme groups, red.
[Based on M. F. Perutz. The
hemoglobin molecule. Copyright ©
1964 by Scientific American, Inc.
All rights reserved.]

64

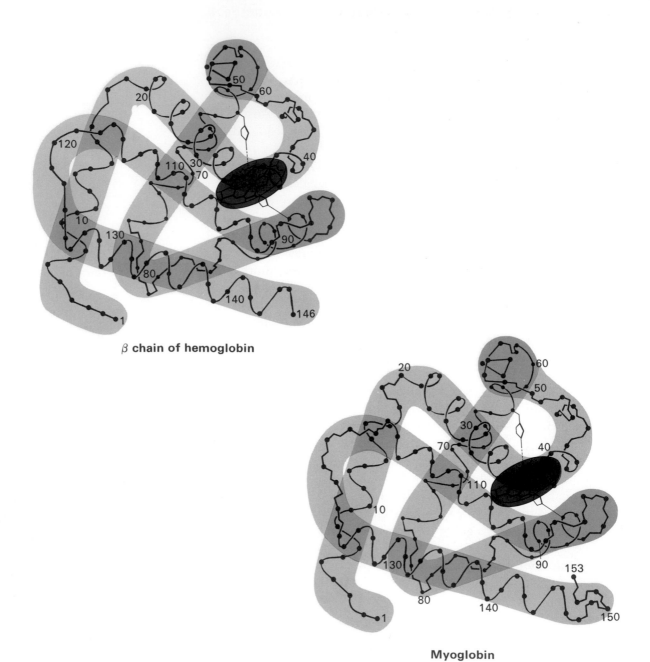

β chain of hemoglobin

Myoglobin

Figure 3-22
Comparison of the conformations of the main chain of myoglobin and of the β chain of hemoglobin. The similarity of their conformations is evident.

exterior of the molecule, one in each subunit. The four oxygen-binding sites are far apart; the distance between the two closest iron atoms is 25 Å. Each α chain is in contact with both β chains. In contrast, there are few interactions between the two α chains or between the two β chains.

THE ALPHA AND BETA CHAINS OF HEMOGLOBIN CLOSELY RESEMBLE MYOGLOBIN

The three-dimensional structures of myoglobin and of the α and β chains of human hemoglobin are strikingly similar (Figure 3-22). The close resemblance in the folding of their main chains was unexpected because there are many differences in the amino acid sequences of these three polypeptide chains. In fact, the sequence is the same for the three chains at only 24 out of 141 positions, showing that quite different amino acid sequences can specify very similar three-dimensional structures (Figure 3-23).

Figure 3-23
Comparison of the amino acid sequences of sperm whale myoglobin and of the α and β chains of human hemoglobin, for residues F1 to F9. The amino acid sequences of these three polypeptide chains are much less alike than are their three-dimensional structures.

It is evident that the three-dimensional form of sperm whale myoglobin and of the α and β chains of human hemoglobin has broad biological significance. In fact, this structure seems to be common to all vertebrate myoglobins and hemoglobins. *The intricate folding of the polypeptide chain first evident in myoglobin is nature's fundamental design for an oxygen carrier: it places the heme in an environment that enables it to carry oxygen reversibly.*

CRITICAL RESIDUES IN THE AMINO ACID SEQUENCE

The amino acid sequences of hemoglobins from more than twenty species (ranging from lamprey to man) are known. A comparison of these sequences shows considerable variability at most positions. However, there are nine positions in the sequence that contain the same amino acid in all or nearly all species studied thus far (Table 3-2). These conserved positions are especially important for the function of the hemoglobin molecule. Several of them are directly involved in the oxygen-binding site. Another conserved residue is tyrosine HC2, which stabilizes the molecule by forming a hydrogen bond, thereby bridging the H and F helices. A glycine (B6) is conserved because of its small size. A side chain larger than a hydrogen atom would not allow the B and E helices to approach each other as closely as they do (Figure 3-24). Proline C2 is important because it terminates the C helix.

The nonpolar residues in the interior of hemoglobin show considerable variability. However, the change is always of one nonpolar residue for another (as from alanine to isoleucine). Thus, *the striking nonpolar character of the interior of the molecule is conserved.* As discussed previously, the reversible oxygenation of the heme group depends on its location in a nonpolar niche where it is protected from water. Also, the nonpolar core of the hemoglobin molecule is important in stabilizing its three-dimensional structure.

Table 3-2
Conserved amino acid residues in hemoglobin

Position	Amino acid	Role
F8	Histidine	Proximal heme-linked histidine
E7	Histidine	Distal histidine near the heme
CD1	Phenylalanine	Heme contact
F4	Leucine	Heme contact
B6	Glycine	Allows the close approach of the B and E helices
C2	Proline	Helix termination
HC2	Tyrosine	Cross-links the H and F helices
C4	Threonine	Uncertain
H10	Lysine	Uncertain

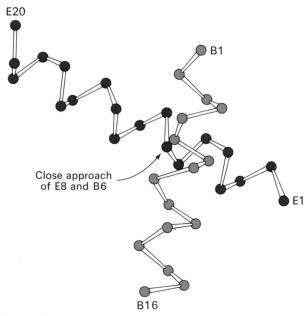

E20

B1

Close approach
of E8 and B6

E1

B16

Figure 3-24
Crossing of the B and E helices in myoglobin. Residue
B6 is almost invariably glycine because there is no space
for a larger side chain.

In contrast, the residues on the surface of the molecule are highly variable. Indeed, few are consistently positively or negatively charged. It might be thought that the prolines would be constant because of their role as helix-breakers. However, this is not so. Only one proline is constant, yet the lengths and directions of the helices in all myoglobins are very similar. Obviously, there must be other ways of terminating or bending α helices.

HEMOGLOBIN: A STEP UP IN EVOLUTION

Thus far, we have stressed the structural similarity of myoglobin and hemoglobin. However, these molecules are functionally quite different. The subunits of hemoglobin have the same structural design as myoglobin. But in coming together to form an $\alpha_2\beta_2$ tetramer, new properties of profound biological importance emerge. These will be discussed in the next chapter.

SUMMARY

Myoglobin and hemoglobin are the oxygen-carrying proteins in vertebrates. Myoglobin facilitates the transport of oxygen in muscle and serves as a reserve store of oxygen in that tissue. Hemoglobin, which is packaged in erythrocytes, is the oxygen carrier in blood. These proteins can bind oxygen because they contain a tightly bound heme prosthetic group. Heme is a substituted porphyrin with a central iron atom. The iron atom in heme can be in the ferrous ($+2$) or ferric ($+3$) state. Only the ferrous form can bind O_2.

Myoglobin, a single polypeptide chain of 153 residues (molecular weight 17,800), has a compact shape. The inside of myoglobin consists almost exclusively of nonpolar residues, whereas the outside contains both polar and nonpolar residues. About 75% of the polypeptide chain is α-helical. There are eight helical segments. The single ferrous heme group is located in a nonpolar niche, which protects it from oxidation to the ferric form. The iron atom of the heme is directly bonded to a nitrogen atom of a histidine side chain. This proximal histidine (F8) occupies the fifth coordination position. The sixth coordination position on the other side of the heme plane is the binding site for O_2. A second histidine, termed the distal histidine (E7), is next to this site.

Functionally active myoglobin can be reformed from unfolded apomyoglobin and heme. This renaturation experiment shows that the conformation of myoglobin is inherent in its amino acid sequence, as was first demonstrated for ribonuclease. The sequestration of nonpolar residues in the interior of myoglobin seems to be an important driving force in the folding of this molecule.

Hemoglobin consists of four polypeptide chains, each with a heme group. Hemoglobin A, the predominant hemoglobin in adults, has the subunit structure $\alpha_2\beta_2$. Hemoglobin A_2, a minor hemoglobin in adults, has the subunit structure $\alpha_2\delta_2$, whereas that of hemoglobin F, a fetal hemoglobin, is $\alpha_2\gamma_2$. The three-dimensional structures of the α and β chains of hemoglobin are strikingly similar to that of myoglobin, although their amino acid sequences are quite different. A comparison of the amino acid sequences of hemoglobins from many species shows that nine positions are virtually invariant. This group of conserved amino acid residues includes several near the heme, such as the proximal and distal histidines. The markedly nonpolar character of the interior of each subunit is also conserved.

SELECTED READINGS

Where to start:

Kendrew, J. C., 1961. The three-dimensional structure of a protein molecule. *Sci. Amer.* 205(6):96–111. [Available as *Sci. Amer.* Offprint 121.]

Perutz, M. F., 1964. The hemoglobin molecule. *Sci. Amer.* 211(5):64–76. [Offprint 196.]

Kendrew, J. C., 1963. Myoglobin and the structure of proteins. *Science* 139:1259–1266.

Structure of myoglobin and hemoglobin:

Watson, H. C., 1969. The stereochemistry of the protein myoglobin. *Progr. Stereochem.* 4:299–333. [A detailed presentation of the three-dimensional structure of sperm whale ferrimyoglobin, including the atomic coordinates of all nonhydrogen atoms in the structure.]

Nobbs, C. L., Watson, H. C., and Kendrew, J. C., 1966. Structure of deoxymyoglobin: A crystallographic study. *Nature* 209:339–341.

Watson, H. C., and Nobbs., C. L., 1968. The structure of oxygenated and deoxygenated myoglobin. *Colloq. Ges. Biol. Chem. (Mosbach/Baden)*, pp. 37–48. Springer-Verlag.

Perutz, M. F., 1969. The haemoglobin molecule. *Proc. Roy. Soc.* (B)173:113–140.

Refolding of myoglobin:

Harrison, S. C., and Blout, E. R., 1965. Reversible conformational changes of myoglobin and apomyoglobin. *J. Biol. Chem.* 240:299–303.

Physiological aspects:

Wittenberg, J. B., 1970. Myoglobin-facilitated oxygen diffusion: role of myoglobin in oxygen entry into muscle. *Physiol. Rev.* 50:559–636.

Roughton, F. J. W., 1964. Transport of oxygen and carbon dioxide. *Handbook of Physiology*, Section 3: *Respiration*, vol. 1, pp. 767–825. American Physiological Society, Washington, D.C.

Molecular evolution:

Zuckerkandl, E., 1965. The evolution of hemoglobin. *Sci. Amer.* 212(5):110–118. [Offprint 1012.]

Dayhoff, M. O., Hunt, L. T., McLaughlin, P. J., and Jones, D. D., 1972. Gene duplications in evolution: the globins. *In* Dayhoff, M. O., (ed.), *Atlas of Protein Sequence and Structure*, vol. 5, pp. 17–30. National Biomedical Research Foundation, Silver Springs, Maryland.

Model systems:

Wang, J. H., 1970. Synthetic biochemical models. *Accounts Chem. Res.* 3:90–97. [A review of several model systems that have provided insight into complex biological processes.]

X-ray crystallography:

Holmes, K. C., and Blow, D. M., 1965. *The Use of X-ray Diffraction in the Study of Protein and Nucleic Acid Structure.* Wiley-Interscience. [An excellent introduction to one of the most important techniques in biochemistry.]

Glusker, J. P., and Trueblood, K. N., 1972. *Crystal Structure Analysis: A Primer.* Oxford University Press. [A lucid and concise introduction to x-ray crystallography in general.]

PROBLEMS

1. The average volume of a red blood cell is 87 cubic microns. The mean concentration of hemoglobin in red cells is 34 g/100 ml.
   - (a) What is the weight of the hemoglobin contained in a red cell?
   - (b) How many hemoglobin molecules are there in a red cell?
   - (c) Could the hemoglobin concentration in red cells be much higher than the observed value? (Hint: Suppose that a red cell contained a crystalline array of hemoglobin molecules 65 Å apart in a cubic lattice).

2. How much iron is there in the hemoglobin of a 70-kg adult? Assume that the blood volume is 70 ml/kg of body weight and that the hemoglobin content of blood is 16 g/100 ml.

3. The myoglobin content in some human muscles is about 8 g/kg of muscle. In sperm whale, the myoglobin content is about 80 g/kg of muscle.
   - (a) How much O_2 is bound to myoglobin in the muscle of man and sperm whale? Assume that the myoglobin is saturated with O_2.
   - (b) The amount of oxygen dissolved in tissue water (in equilibrium with venous blood at 37°) is about 3.5×10^{-5} M. What is the ratio of oxygen bound to myoglobin to that directly dissolved in the water of sperm whale muscle?

4. The amino acid composition of sperm whale myoglobin is:

Ala 5	Gln 5	Leu 18	Ser 6
Arg 4	Glu 14	Lys 19	Thr 5
Asn 2	Gly 11	Met 2	Trp 2
Asp 6	His 12	Phe 6	Tyr 3
Cys 0	Ile 9	Pro 4	Val 8

   - (a) Estimate the net charge of ferromyoglobin at pH 2, 7, and 9.
   - (b) Estimate the isoelectric point of myoglobin. The isoelectric point is defined as the pH at which there is no net charge.

5. A solution of myoglobin is treated with cyanogen bromide.
   - (a) What are the cleavage products? Refer to the amino acid sequence of myoglobin.
   - (b) In aqueous solution, the cleavage products have a low content of α helix. What does this finding reveal about the stability of α helices in aqueous solution?

6. The binding of O_2 to myoglobin can be described by the simple equilibrium

$$Mb + O_2 \rightleftharpoons MbO_2$$

Let [Mb] denote the concentration of deoxymyoglobin; $[MbO_2]$, the concentration of oxymyoglobin; pO_2, the concentration of O_2 (expressed as the partial pressure of O_2); and P_{50}, the pO_2 at which $[Mb] = [MbO_2]$. Derive an equation analogous to the Henderson-Hasselbalch equation that gives the ratio of oxymyoglobin to deoxymyoglobin as a function of pO_2 and P_{50}.

7. The equilibrium constant K for the binding of oxygen to myoglobin is 10^{-6} M, where K is defined as

$$K = [Mb][O_2]/[MbO_2]$$

The rate constant for the combination of O_2 with myoglobin is 2×10^7 M^{-1} sec^{-1}.
   - (a) What is the rate constant for the dissociation of O_2 from oxymyoglobin?
   - (b) What is the mean duration of the oxymyoglobin complex?

CHAPTER 4

HEMOGLOBIN:
AN ALLOSTERIC PROTEIN

New properties emerge in going from monomeric myoglobin to tetrameric hemoglobin. Hemoglobin is a much more intricate molecule than myoglobin. First, hemoglobin transports H^+ and CO_2 in addition to O_2. Second, the binding of oxygen by hemoglobin is regulated by specific molecules in its environment, namely H^+, CO_2, and organic phosphate compounds. These regulatory molecules profoundly alter the oxygen-binding properties of hemoglobin although they are bound to sites on the protein that are

Figure 4-1
A model of hemoglobin at low resolution. The α chains in this model are yellow, the β chains are blue, and the heme groups red. [Based on M. F. Perutz. The hemoglobin molecule. Copyright © 1964 by Scientific American, Inc. All rights reserved.]

far from the hemes. Indeed, interactions between spatially distinct sites, termed *allosteric interactions*, occur in many proteins. Allosteric effects play a critical role in the control and integration of molecular events in biological systems. Hemoglobin is the best understood of the allosteric proteins, and so it is rewarding to examine its structure and function in some detail.

FUNCTIONAL DIFFERENCES BETWEEN MYOGLOBIN AND HEMOGLOBIN

Hemoglobin is an allosteric protein, whereas myoglobin is not. This difference is expressed in three ways:

1. The shape of the oxygen dissociation curve of hemoglobin is sigmoidal, whereas that of myoglobin is hyperbolic. The sigmoidal curve shows that the binding of oxygen to hemoglobin is cooperative.

2. The affinity of hemoglobin for oxygen depends on pH, whereas that of myoglobin is independent of pH. The CO_2 molecule also affects the oxygen-binding characteristics of hemoglobin.

3. The oxygen affinity of hemoglobin is further regulated by organic phosphates such as diphosphoglycerate. The result is that hemoglobin has a lower affinity for oxygen than does myoglobin.

THE BINDING OF OXYGEN TO HEMOGLOBIN IS COOPERATIVE

Torr—

A unit of pressure equal to that exerted by a column of mercury 1 mm high at 0°C and standard gravity (1 mm Hg).
Named after Evangelista Torricelli (1608–1647), the inventor of the mercury barometer.

The saturation Y is defined as the fractional occupancy of the oxygen-binding sites. The value of Y can range from 0 (all sites empty) to 1 (all sites filled). A plot of Y versus pO_2, the partial pressure of oxygen, is called an *oxygen dissociation curve*. The oxygen dissociation curves of myoglobin and hemoglobin differ in two ways (Figures 4-2 and 4-3). For any given pO_2, Y is higher for myoglobin than for hemoglobin. In other words, *myoglobin has a higher affinity for oxygen than does hemoglobin.* The oxygen affinity can be characterized by a quantity called P_{50}, which is the partial pressure of oxygen at which 50% of sites are filled (i.e., at which $Y = 0.5$). For myoglobin, P_{50} is typically 1 torr, whereas for hemoglobin P_{50} is 26 torr.

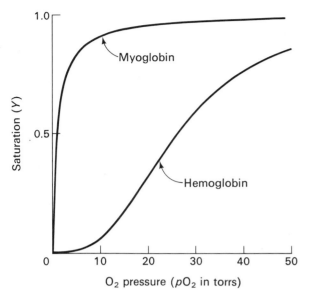

Figure 4-2
Oxygen dissociation curves of myoglobin and
hemoglobin. The saturation of the oxygen-binding
sites is plotted as a function of the partial pressure
of oxygen surrounding the solution.

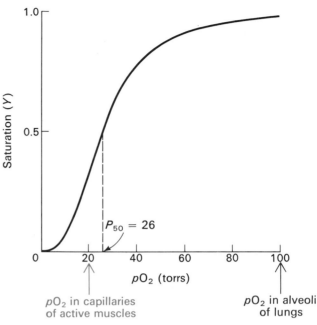

Figure 4-3
Oxygen dissociation curve of hemoglobin. Typical values
for pO_2 in the capillaries of active muscle and in the
alveoli of the lung are marked on the x-axis. Note that P_{50}
for hemoglobin under physiological conditions lies between
these values.

The second difference is that *the oxygen dissociation curve of myoglobin
has a hyperbolic shape, whereas that of hemoglobin is sigmoidal.* As will
be discussed shortly, the sigmoidal shape of the curve for hemoglobin is ideally suited to its physiological role as an oxygen carrier
in blood. In molecular terms, the sigmoidal shape means that the
binding of oxygen to hemoglobin is cooperative—that is, the binding of oxygen at one heme facilitates the binding of oxygen to the
other hemes.

Let us consider the oxygen dissociation curves of these molecules
in quantitative terms, starting with myoglobin because it is simpler.
The binding of oxygen to myoglobin can be described in terms
of a simple equilibrium:

$$MbO_2 \rightleftharpoons Mb + O_2 \qquad (1)$$

The equilibrium constant K for the dissociation of oxymyoglobin is

$$K = \frac{[\text{Mb}][\text{O}_2]}{[\text{MbO}_2]} \tag{2}$$

where $[\text{MbO}_2]$ is the concentration of oxymyoglobin, $[\text{Mb}]$ is the concentration of deoxymyoglobin, and $[\text{O}_2]$ is the concentration of uncombined oxygen, all in moles per liter. The fractional saturation Y is defined as

$$Y = \frac{[\text{MbO}_2]}{[\text{MbO}_2] + [\text{Mb}]} \tag{3}$$

Substitution of equation (2) into equation (3) yields

$$Y = \frac{[\text{O}_2]}{[\text{O}_2] + K} \tag{4}$$

Since oxygen is a gas, it is convenient to express its concentration in terms of $p\text{O}_2$, the partial pressure of oxygen (in torrs) in the atmosphere surrounding the solution. Equation (4) then becomes

$$Y = \frac{p\text{O}_2}{p\text{O}_2 + P_{50}} \tag{5}$$

Equation (5) plots as a hyperbola. In fact, the oxygen dissociation curve calculated from equation (5), taking P_{50} to be 1 torr, closely matches the experimentally observed curve for myoglobin.

It is useful to derive a binding relationship that gives a straight line plot. From equation (5), it follows that

$$\frac{Y}{1 - Y} = \frac{p\text{O}_2}{P_{50}} \tag{6}$$

Taking the logarithm of both sides of equation (6) gives

$$\log \frac{Y}{1 - Y} = \log p\text{O}_2 - \log P_{50} \tag{7}$$

A plot of $\log Y/(1 - Y)$ versus $\log p\text{O}_2$ yields a straight line with a slope of 1. This representation is called a *Hill plot* and its slope (n) is called the Hill coefficient. Myoglobin in fact gives a linear Hill plot with $n = 1$ (Figure 4-4).

The binding data for hemoglobin cannot be expressed in terms of equation (5). The plot of Y versus pO_2 is sigmoid, not hyperbolic. Furthermore, the Hill plot for hemoglobin has a slope of 2.8 rather than 1.0 (Figure 4-4). A Hill coefficient greater than 1 indicates that the oxygen-binding sites are not independent of each other. Rather, *the binding of oxygen in hemoglobin is cooperative.* Binding at one heme facilitates the binding of oxygen at the other heme on the same tetramer. Conversely, the unloading of oxygen at one heme facilitates the unloading of oxygen at the others. In other words, there is communication among the heme groups of a hemoglobin molecule. The cooperative binding of oxygen by hemoglobin is sometimes called *heme-heme interaction.* The mechanism of this interaction will be discussed shortly.

The cooperative binding of oxygen by hemoglobin makes hemoglobin a more efficient oxygen transporter. The oxygen saturation of hemoglobin changes more rapidly with changes in the partial pressure of O_2 than it would if the oxygen-binding sites were independent of each other. Let us consider a specific example. Assume that the alveolar pO_2 is 100 torr, and that the pO_2 in the capillary of an active muscle is 20 torr. Let $P_{50} = 30$ torr, and take $n = 2.8$. Then Y in the alveolar capillaries will be 0.97, and Y in the muscle capillaries will be 0.25. The oxygen delivered will be

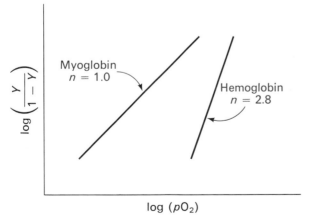

Figure 4-4
Hill plot for the binding of O_2 to myoglobin and hemoglobin. The slope of 2.8 for hemoglobin indicates that it binds oxygen cooperatively, in contrast to myoglobin, which has a slope of 1.0.

proportional to the difference in Y, which is 0.72. Let us now make the same calculation for a hypothetical oxygen carrier for which P_{50} is also 30 torr, but in which the binding of oxygen is not cooperative ($n = 1$). Then $Y_{\text{alveoli}} = 0.77$, and $Y_{\text{muscle}} = 0.41$, so that $\Delta Y = 0.36$. Thus, *the cooperative binding of oxygen by hemoglobin enables it to deliver twice as much oxygen than if the sites were independent.*

H+ AND CO₂ PROMOTE THE RELEASE OF OXYGEN (THE BOHR EFFECT)

Myoglobin shows no change in oxygen binding over a broad range of pH, nor does CO_2 have an appreciable effect. In hemoglobin, acidity enhances the release of oxygen. In the physiological range, a lowering of pH shifts the oxygen dissociation curve to the right, so that the oxygen affinity is reduced (Figure 4-5). Increasing the concentration of CO_2 (at constant pH) also lowers the oxygen affinity. In rapidly metabolizing tissue, such as contracting muscle,

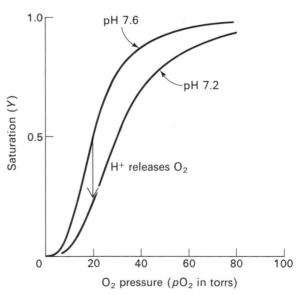

Figure 4-5
Effect of pH on the oxygen affinity of hemoglobin.
A lowering of the pH from 7.6 to 7.2 results in the
release of O_2 from oxyhemoglobin.

much CO_2 and acid are produced. *The presence of higher levels of CO_2 and H^+ in the capillaries of such metabolically active tissue promotes the release of O_2 from oxyhemoglobin.* This is one mechanism for meeting the higher oxygen needs of metabolically active tissue.

The reciprocal effect occurs in the alveolar capillaries of the lungs. The high concentration of H^+ and CO_2 drives off O_2; increasing the concentration of O_2 drives off H^+ and CO_2. The former occurs in the tissues of actively metabolizing tissue, the latter in the alveolar capillaries of the lung. These linkages are known as the Bohr effect (Figure 4-6).

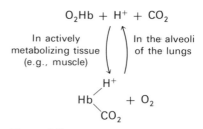

Figure 4-6
A summary of the Bohr effect. The actual mechanism and stoichiometry are more complex than indicated in this diagram.

DPG LOWERS THE OXYGEN AFFINITY

The oxygen affinity of hemoglobin within the red cell is lower than that of hemoglobin in free solution. As early as 1921, Joseph Barcroft wondered, "Is there some third substance present . . . which forms an integral part of the oxygen-hemoglobin complex?" Indeed there is. Reinhold Benesch and Ruth Benesch showed in 1967 that 2,3-diphosphoglycerate (DPG) binds to hemoglobin and has a large effect on its oxygen affinity. DPG is present in the red cells of man and many other species at about the same molar concentration as hemoglobin. In the absence of DPG, the P_{50} of hemoglobin is 1 torr, like that of myoglobin. In its presence, P_{50} becomes 26 torr (Figure 4-7). Thus, *DPG reduces the oxygen affinity of hemoglobin by a*

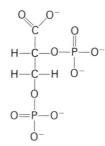

2,3-Diphosphoglycerate (DPG)

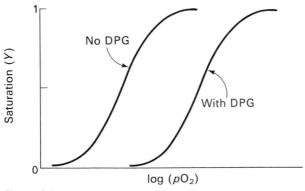

Figure 4-7
Diphosphoglycerate decreases the oxygen affinity of hemoglobin.

factor of 26. DPG is physiologically important. In its absence, hemoglobin would unload little oxygen in passing through tissue capillaries, since the pO_2 there is about 26 torr.

DPG exerts its effect on the oxygen affinity of hemoglobin by binding to deoxyhemoglobin but not to oxyhemoglobin. The binding of oxygen and DPG to hemoglobin is mutually exclusive. To a good approximation, the oxygenation of hemoglobin in the presence of DPG can be represented by the equation

$$\text{Hb}\!-\!\text{DPG} + 4\,O_2 \rightleftharpoons \text{Hb}(O_2)_4 + \text{DPG}$$

CLINICAL SIGNIFICANCE OF DPG

The discovery of the role of DPG in oxygen transport has led to new insights in several areas of clinical medicine. For example, it has been known for some time that blood stored in *acid-citrate-dextrose, a conventional medium, has an increased oxygen affinity:* P_{50} is 16 instead of 26 torr. The reason is now apparent. The oxygen affinity increases during storage because there is a concomitant decrease in the level of DPG, from 4.5 mM to less than 0.5 mM in ten days. The oxygen affinity of stored blood may be of critical clinical importance. If a patient is given a large volume of blood with high oxygen affinity, the amount of oxygen unloaded in his tissues may be compromised. Transfused red cells that are totally depleted of DPG can regain half the normal level within 24 hours, but this restitution may not be rapid enough to be effective in a severely ill patient. Thus, under certain conditions, it may be advantageous to give transfusions of red cells that have a normal oxygen affinity. The DPG level of deficient cells cannot be restored by adding DPG to them, because red cell membranes are not permeable to this highly charged molecule. However, the decrease in DPG level of stored cells can be prevented if the medium in which they are suspended contains *inosine*. This uncharged nucleoside molecule can penetrate the red cell membrane. Inside the red cell, inosine is converted to DPG by a complex series of reactions (see Chapter 15). Inosine is now frequently used in this way to preserve the functional integrity of stored blood.

Adaptive mechanisms in disorders of oxygen delivery to the

Inosine

tissues (termed *hypoxia*) have also been illuminated by the work on DPG. Let us consider a patient who has severe obstructive pulmonary emphysema. In this disease, airflow in the bronchioles is blocked, and consequently the arterial blood is not fully saturated with oxygen. Indeed, the pO_2 of the arterial blood of such a patient may be only 50 torr, half the normal value. There is a compensatory shift in his oxygen dissociation curve, which is due to an increase in the level of DPG from 4.5 mM to 8 mM. At this higher DPG level, P_{50} is 31 torr, rather than 26 torr. With a P_{50} of 31 torr, the arterial saturation (Y_A) is 0.82, whereas the venous saturation (Y_V) is 0.49; hence, the change in oxygen saturation (ΔY) is 0.33. For a normal P_{50} of 26 torr, $Y_A = 0.86$, $Y_V = 0.60$, and $\Delta Y = 0.26$. The shift of the oxygen dissociation curve is advantageous to this patient since it increases ΔY from 0.26 to 0.33. Thus, the increased DPG level leads to a 27% increase in the amount of oxygen delivered to the tissues.

In some *metabolic disorders of the red cell*, the level of DPG is abnormal and there is a concomitant change in the oxygen affinity. Several of these diseases are discussed in the next chapter.

An interesting problem in respiratory physiology is the mechanism of *high-altitude adaptation*. Recent studies suggest that this process may be influenced by adjustments in the DPG level. When a person goes from sea level to 15,000 feet, his oxygen affinity decreases within a day and there is a simultaneous increase in the DPG level in his red cells. Both changes are reversed on returning to sea level. The same effects occur when indigenous highlanders descend to sea level. The half-time of these changes in DPG level and in oxygen affinity is typically six hours.

FETAL HEMOGLOBIN HAS A HIGH OXYGEN AFFINITY

The fetus has its own kind of hemoglobin, called hemoglobin F $(\alpha_2\gamma_2)$, which differs from adult hemoglobin A $(\alpha_2\beta_2)$, as mentioned previously. An important property of hemoglobin F is that it has a higher oxygen affinity under physiological conditions that does hemoglobin A (Figure 4-8). This higher oxygen affinity of hemoglobin F optimizes the transfer of oxygen from the maternal to the fetal circulation. Hemoglobin F is oxygenated at the expense of hemoglobin A on the other side of the placental circulation.

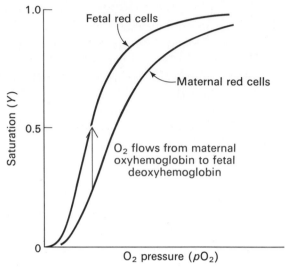

Figure 4-8
Fetal red blood cells have a higher oxygen affinity
than maternal red blood cells. *In the presence of DPG,*
the oxygen affinity of fetal hemoglobin is higher
than that of maternal hemoglobin.

The higher oxygen affinity of fetal blood has been known for
many years, but its basis has only recently been elucidated. *Hemo-*
globin F binds DPG less strongly than hemoglobin A and consequently has
a higher oxygen affinity. Indeed, in the absence of DPG, the oxygen
affinities of the two hemoglobins are in reverse order. This observation was puzzling until it was appreciated that DPG must be
present when the oxygen dissociation curves are measured, since
DPG is present in both fetal and adult red cells.

SUBUNIT INTERACTION IS REQUIRED
FOR ALLOSTERIC EFFECTS

Let us now consider the structural basis of the allosteric effects.
Hemoglobin can be dissociated into its constituent chains. The
properties of the isolated α chain are very much like those of
myoglobin. The α chain by itself has a high oxygen affinity, a
hyperbolic oxygen dissociation curve, and oxygen-binding characteristics that are insensitive to pH, CO_2 concentration, and DPG

level. The isolated β chain readily forms a tetramer, β_4, which is called hemoglobin H. Like the β chain and myoglobin, β_4 entirely lacks the allosteric properties of hemoglobin. In short, *the allosteric properties of hemoglobin arise from subunit interaction. The functional unit of hemoglobin is a tetramer consisting of two kinds of polypeptide chains.*

THE QUATERNARY STRUCTURE OF HEMOGLOBIN CHANGES MARKEDLY ON OXYGENATION

In 1937, Felix Haurowitz observed that crystals of deoxyhemoglobin shatter when they are oxygenated. The crystal shatters because the structure of oxyhemoglobin is appreciably different from that of deoxyhemoglobin. The oxygenated molecules can no longer fit precisely on the crystal lattice of the deoxygenated hemoglobin. In contrast, crystals of myoglobin are unaffected by oxygenation, as are crystals of β_4.

X-ray crystallographic studies have shown that oxy and deoxyhemoglobin differ markedly in quaternary structure (Figure 4-9). The oxygenated molecule is more compact. For instance, the distance between the iron atoms of the β chains decreases from 39.9 to 33.4 Å on oxygenation. The changes in the contacts between the α and β chains are of special interest. There are two kinds of contact regions between the α and β chains, called the $\alpha_1\beta_1$ and the $\alpha_1\beta_2$ contacts

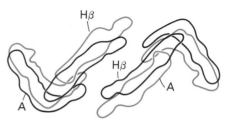

Figure 4-9
Projection of part of the electron-density maps of oxyhemoglobin (shown in red) and deoxyhemoglobin (shown in blue) at a resolution of 5.5 Å. The A and H helices of the two β chains of hemoglobin are shown here. The center of the diagram corresponds to the central cavity of the molecule. One of the conformational changes accompanying oxygenation—a movement of the H helices toward each other—is shown here. [Based on M. F. Perutz. *Nature* 228(1970):738.]

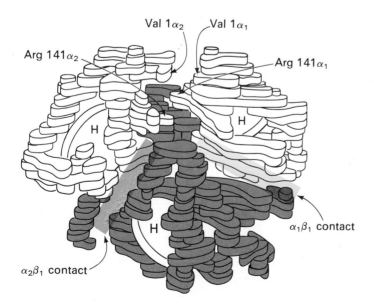

Figure 4-10
Model of oxyhemoglobin at low resolution showing the two kinds of interfaces between α and β chains. The α chains are white, the β chains black. The three hemes that can be seen in this view of the molecule are labeled H. The $\alpha_2\beta_1$ contact region is shown in blue, the $\alpha_1\beta_1$ contact region in yellow. [Based on M. F. Perutz and L. F. TenEyck. *Cold Spring Harbor Symp. Quant. Biol.* 36(1971):296.]

(Figure 4-10). In the transition from oxy to deoxyhemoglobin, movement in the $\alpha_1\beta_1$ contact is slight. The β_1 subunit rotates by about 4° relative to the α_1 subunit. The relative displacement of atoms at the contact is only about 1 Å. In contrast, movement in the $\alpha_1\beta_2$ contact is large. The β_2 subunit rotates by 13.5° relative to the α_1 subunit, and there is a translation of 1.9 Å. This leads to displacements of some atoms by as much as 5.7 Å. The $\alpha_1\beta_2$ contact is seen to be both smaller and smoother than the $\alpha_1\beta_1$ contact, and appears to be constructed to allow the two subunits to slide past each other.

The $\alpha_1\beta_2$ contact is closely connected to the heme groups. Hence, structural changes in this region can be expected to affect the hemes. The importance of the $\alpha_1\beta_2$ contact is reinforced by the finding that most residues in it are unaltered in going from one species to another. Also mutations in the $\alpha_1\beta_2$ contact almost always diminish heme-heme interaction whereas $\alpha_1\beta_1$ contact mutations do not.

In oxyhemoglobin, the carboxyl-terminal residues of all four chains have almost complete freedom of rotation. In contrast, these terminal groups are anchored in deoxyhemoglobin (Figure 4-11).

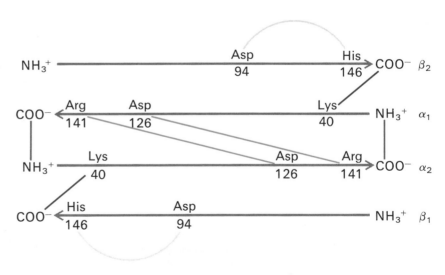

Figure 4-11
Cross-links between different subunits in deoxyhemoglobin. These noncovalent, electrostatic interactions are disrupted on oxygenation.

Asp⁻ . . . His⁺
terminal-COO⁻ . . . Lys⁺
Asp⁻ . . . Arg⁺
terminal-COO⁻ . . . NH₃⁺-terminal

First, the carboxyl terminus of the α_1 chain interacts with the amino terminus of the α_2 chain. Second, the arginine side chain of this carboxyl-terminal residue is linked to an aspartate on the α_2 chain. Third, the carboxyl terminus of the β_1 chain is linked to a lysine side chain on the α_2 chain. Finally, the imidazole side chain of the carboxyl-terminal residue of the β_1 chain interacts with an aspartate side chain on the same chain. These pairs of residues are held together by noncovalent, electrostatic interactions between oppositely charged groups, which are often called *salt links*. Deoxyhemoglobin is a tauter, more constrained molecule than oxyhemoglobin because of these eight salt links (Figure 4-11).

ALLOSTERIC CHANGES IN HEMOGLOBIN ARE MEDIATED BY CHANGES IN TERTIARY AND QUATERNARY STRUCTURE

The trigger for the allosteric changes in hemoglobin is the movement of the iron atom on oxygenation (Figure 4-12). The iron atom in deoxyhemoglobin is 0.75 Å out of the plane of the porphyrin, displaced toward the proximal histidine F8. The reason for the out-of-plane location of the iron atom in deoxyhemoglobin is that the radius of the iron is too large for it to fit into the hole in the center of the porphyrin. In oxyhemoglobin, the radius of the iron atom is smaller and so *the iron atom moves into the plane of the porphyrin on oxygenation, and pulls the proximal histidine F8 with it.* This appears to be the primary structural change underlying all of the allosteric interactions in hemoglobin.

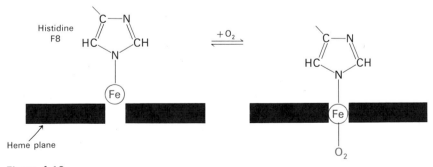

Figure 4-12
The iron atom moves 0.75 Å into the plane of the heme on oxygenation because its diameter becomes smaller. Histidine F8 is pulled along with the iron atom.

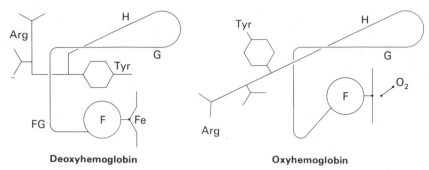

Figure 4-13
Diagrammatic representation of the change in conformation of a subunit in hemoglobin on oxygenation. Movement of the iron atom into the heme plane leads to a movement of helix F toward helix H. Tyrosine HC2 is thereby expelled from the pocket between these helices. [Based on M. F. Perutz. *Nature* 228(1970):231.]

The displacement of histidine F8 produces small changes in the tertiary structure of the same subunit (Figure 4-13). Helix F moves towards helix H, and narrows the pocket between them. The narrowing results in the expulsion of tyrosine HC2, the next to last residue in all four chains. The expelled tyrosine pulls the adjacent carboxyl-terminal residue with it. This motion has important consequences because the carboxyl-terminal residue of a subunit interacts with residues on *other* subunits in deoxyhemoglobin but not in oxyhemoglobin. The expulsion of tyrosine HC2 from the pocket between the F and H helices ruptures some links between subunits, and consequently destabilizes the deoxy quaternary structure. Thus, *a structural change within a subunit (oxygenation) is translated into structural changes at the interfaces between subunits. The binding of oxygen at one heme site is thereby communicated to parts of the molecule that are far away.*

DPG DECREASES OXYGEN AFFINITY
BY CROSS-LINKING DEOXYHEMOGLOBIN

DPG binds specifically to deoxyhemoglobin in the ratio of 1 DPG per hemoglobin tetramer. This is a very interesting stoichiometry. A protein with a subunit structure of $\alpha_2\beta_2$ is usually expected to have at least two binding sites for any small molecule. The finding of one binding site immediately suggested that DPG binds on the symmetry axis of the hemoglobin molecule in the central cavity, where the four subunits are near each other (Figure 4-14). The binding site for DPG is contributed by the following positively charged residues on *both* β chains: the α-amino group, lysine EF6, and histidine H21. These groups can interact with the strongly negatively charged DPG, which carries nearly four negative charges at physiological pH. DPG is stereochemically complementary to a constellation of six positively charged groups in the β chains that face the central cavity of the hemoglobin molecule (Figure 4-15). DPG binds more weakly to fetal hemoglobin than to hemoglobin A because residue H21 in fetal hemoglobin is serine rather than histidine.

On oxygenation, DPG is extruded because the central cavity becomes too small. Specifically, the gap between the H helices of the β chains becomes narrowed. Also, the distance between the α-amino groups increases from 16 Å to 20 Å, so that they can no longer make contact with the phosphates of the DPG.

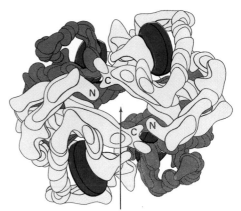

DPG binding site

Figure 4-14
The binding site for diphosphoglycerate is in the central cavity of deoxyhemoglobin. [Based on M. F. Perutz. The hemoglobin molecule. Copyright © 1964 by Scientific American, Inc. All rights reserved.]

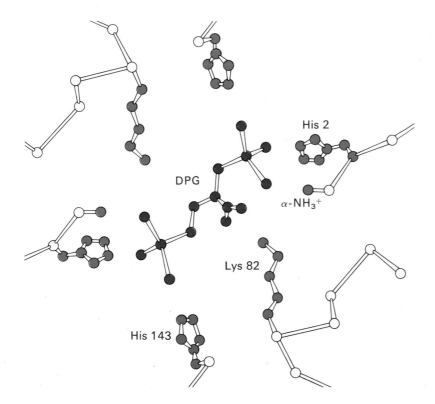

Figure 4-15
Mode of binding of diphosphoglycerate to human deoxyhemoglobin. DPG interacts with three positively charged groups on each β chain [Based on A. Arnone. *Nature* 237(1972):148.]

The basis of the effect of DPG in decreasing oxygen affinity is now evident. *DPG stabilizes the deoxyhemoglobin quaternary structure by cross-linking the β chains.* As mentioned previously, deoxyhemoglobin has eight salt links involving its carboxyl-terminal residues. These must be broken for oxygenation to occur. The binding of DPG contributes additional cross-links that must be broken, and so the oxygen affinity of hemoglobin is reduced by the binding of DPG.

MECHANISM OF THE BOHR EFFECT

How does hemoglobin take up protons in going from the oxy to the deoxy form? The affinity of some sites for H^+ must be increased as a consequence of this transition. Specifically, the pKs of some groups must be *raised* in the transition from oxy to deoxyhemoglobin; an increase in pK means stronger binding of H^+.

Which groups have their pKs raised? Let us consider the possible binding sites for H^+ in hemoglobin (Figure 4-16 and Table 4-1)

Table 4-1
pK values of ionizable groups in proteins

Group	Acid $\rightleftharpoons$ Base + H^+	Typical pK
Terminal carboxyl	$-COOH \rightleftharpoons -COO^- + H^+$	3.1
Aspartic and glutamic acid	$-COOH \rightleftharpoons -COO^- + H^+$	4.4
Histidine		6.5
Terminal amino	$-NH_3^+ \rightleftharpoons -NH_2 + H^+$	8.0
Cysteine	$-SH \rightleftharpoons -S^- + H^+$	8.5
Tyrosine		10.0
Lysine	$-NH_3^+ \rightleftharpoons -NH_2 + H^+$	10.0
Arginine		12.0

and then narrow the choice. The side-chain carboxylates of glutamate and aspartate normally have a pK of about 4. It is unlikely
that their pKs will be shifted to a value between 7 and 8, which
would be necessary for a group that participates in the Bohr effect.
Similarly, pKs of the side chains of tyrosine, lysine, and arginine
are usually above 10, and so it is unlikely that the pK of any of
them would be shifted far enough. Hence, the most plausible
candidates for the groups participating in the Bohr effect are
histidine, cysteine, and the terminal amino group, which normally
have pK values near 7.

The identification of the specific groups responsible for the Bohr
effect was achieved by combining information derived from chemical and x-ray studies. The carboxyl-terminal histidines of the β
chains (histidine 146β) were implicated by x-ray studies. This
suggestion was tested by preparing a hemoglobin devoid of histidine
146 on its β chains. Carboxypeptidase B, a specific proteolytic
enzyme that splits the peptide bond involving carboxyl-terminal
basic amino acids in some polypeptide chains, was used for this
purpose. The modified hemoglobin produced in this way had only
half of the Bohr effect of normal hemoglobin (0.35 H$^+$ bound per
O$_2$, rather than 0.7). Hence, *histidine 146 on each β chain appears to
make a major contribution to the Bohr effect.*

The role of the terminal amino groups of hemoglobin in the Bohr
effect was ascertained by specifically modifying them with cyanate
to form a *carbamoylated* derivative, which can no longer bind H$^+$.
The Bohr effect was unchanged when the terminal amino groups
of only the β chains were carbamoylated. In contrast, the Bohr
effect was significantly reduced when the terminal amino groups
of the α chains were carbamoylated.

$$R{-}NH_2 \ + \ {}^-NCO \ + \ H^+ \ \longrightarrow \ R{-}\overset{\displaystyle H}{\underset{}{N}}{-}\overset{}{\underset{\displaystyle O}{C}}{-}NH_2$$

| Terminal amino group | Cyanate | Carbamoylated derivative |

The x-ray results provide a concrete picture of how these groups
might function in the Bohr effect. In oxyhemoglobin, histidine 146β
rotates freely, whereas, in deoxyhemoglobin, this terminal residue
participates in a number of interactions. Of particular significance
is the interaction of its imidazole ring with the negatively charged

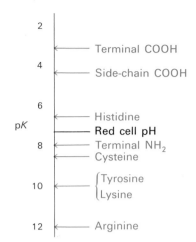

Figure 4-16
Typical pKs of acidic groups in
proteins. The actual pK of a
particular group can be higher or
lower than the value given here,
depending on its environment.

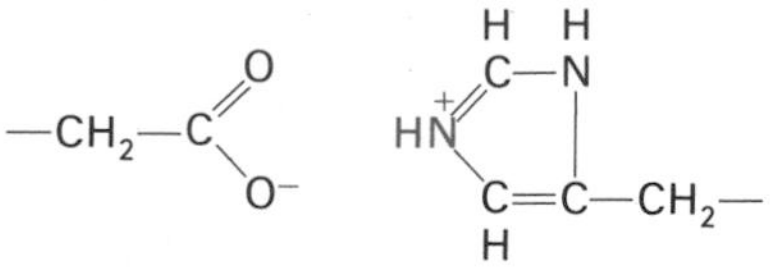

Figure 4-17
Aspartate 194 raises the
pK of histidine 146 in
deoxyhemoglobin but not in
oxyhemoglobin. The proximity
of the negative charge on
aspartate 94 favors protonation
of histidine 146 in
deoxyhemoglobin.

aspartate 94 on the same β chain. The close proximity of this negatively charged group enhances the likelihood that the imidazole group will bind a proton (Figure 4-17). In other words, the proximity of aspartate 94 raises the pK of histidine 146. Thus, *in going from oxy to deoxyhemoglobin, histidine 146 acquires a greater affinity for H^+ because its local charge environment is altered.*

The environment of the α-NH_2 groups of the α chains changes in a similar way on deoxygenation. In oxyhemoglobin, these groups are free. In deoxyhemoglobin, the α-NH_2 group of one α chain interacts with the carboxyl-terminal group of the other α chain. *The proximity of a negatively charged carboxylate residue in deoxyhemoglobin increases the affinity of this terminal amino group for H^+.* X-ray results have implicated a third group in the Bohr effect, namely histidine 122 of the α chains.

In summary, three pairs of proton-binding groups (a terminal amino group and two histidines) have different environments in oxy and deoxyhemoglobin. Their immediate environments are more negatively charged in deoxyhemoglobin. Consequently, these groups take up H^+ when oxygen is released.

CO_2 BINDS TO THE TERMINAL AMINO GROUPS OF ALL FOUR CHAINS

In aerobic metabolism, about one molecule of CO_2 is produced per molecule of O_2 consumed. Hemoglobin participates directly in the transport of CO_2 from the tissue to the lungs. The direct transport of CO_2 by hemoglobin accounts for about 60% of the total CO_2 exchanged by the red cell during respiration. Human oxyhemoglobin binds 0.15 moles CO_2/mole heme, whereas the deoxy form binds 0.40. The bound CO_2 is combined with the α-amino group of all four chains in the form of carbamate. The CO_2 molecule reacts only with uncharged amino groups, and so the amount of carbamate formed depends on the pK of these groups.

$$R-NH_3^+ \rightleftharpoons R-NH_2 + H^+$$

$$R-NH_2 + CO_2 \rightleftharpoons R-NHCOO^- + H^+$$

The pKs of α-amino groups are typically between 7 and 8, which means that an appreciable fraction of these groups is unionized at pH 7.3, and therefore capable of forming carbamino compounds. Recent studies have shown that *all four α-amino groups react with CO_2. Carbamoylation of any of these groups blocks CO_2 binding.*

MECHANISM OF THE COOPERATIVE BINDING OF OXYGEN

The sigmoid oxygen-binding curve of hemoglobin means that oxygenation becomes progressively easier as hemoglobin becomes saturated with oxygen. The mechanism of this allosteric property of hemoglobin is an intriguing problem, which has not yet been completely solved. One way of looking at the problem is depicted in Figure 4-18. Deoxyhemoglobin is a taut molecule, constrained by its eight salt links between the four subunits. Oxygenation cannot occur unless some of these salt links are broken so that the iron atom can move into the plane of the heme group. The number of salt links that need to be broken for the binding of an O_2 molecule depends on whether it is the first, second, third, or fourth to be bound. More salt links must be broken to permit the entry of the first O_2 than of subsequent ones. Since energy is required to break salt links, the binding of the first O_2 is energetically less favorable than of subsequent oxygen molecules. In this scheme, the binding affinity of the second and third O_2 molecules is intermediate between that of the first and last (Figure 4-19). This sequential increase in oxygen affinity would give the sigmoid oxygen-binding curve that is experimentally observed.

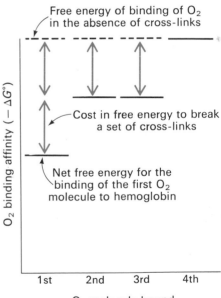

Figure 4-19
Free energy of binding of successive molecules of O_2 for the model depicted in Figure 4-18. The binding affinity $(-\Delta G°)$ increases because fewer cross-links must be broken to bind the fourth O_2 molecule than the first one.

Figure 4-18
A postage stamp analogy of the oxygenation of hemoglobin. Two sets of perforations must be torn to remove the first stamp. Only one perforated edge must be torn to remove the second stamp, and one edge again to remove the third stamp. The fourth stamp is then free.

COMMUNICATION WITHIN A PROTEIN MOLECULE

We have seen that the binding of O_2, H^+, CO_2, and DPG are linked. These molecules are bound to spatially distinct sites that communicate with each other by means of conformational changes within the protein. The sites for these molecules are separate because they are structurally very different and thus cannot bind to the same site. *The interplay between the different sites is mediated by changes in quaternary structure.* In fact, all known allosteric proteins consist of two or more polypeptide chains. The contact region between two chains can amplify and transmit conformational changes from one subunit to another. An allosteric protein does not have fixed properties. Rather, its functional characteristics are responsive to specific molecules in its environment. Consequently, allosteric interactions have immense importance in cellular function. *In the step from myoglobin to hemoglobin, we see the emergence of a macromolecule capable of perceiving information from its environment.*

SUMMARY

New properties emerge in going from monomeric myoglobin to tetrameric hemoglobin. Hemoglobin transports H^+ and CO_2 in addition to O_2. Furthermore, the binding of these molecules is regulated by allosteric interactions, which are defined as interactions between spatially distinct sites that are transmitted by conformational changes in the protein. Indeed, hemoglobin is the best understood allosteric protein. Hemoglobin exhibits three kinds of allosteric effects. First, the oxygen-binding curve of hemoglobin is sigmoidal, which means that the binding of oxygen is cooperative. Binding of oxygen to one heme facilitates the binding of oxygen to the other hemes on the same molecule. Second, H^+ and CO_2 promote the release of O_2 from hemoglobin, an effect that is physiologically important in enhancing the release of O_2 in metabolically active tissues such as muscle. Conversely, O_2 promotes the release of H^+ and CO_2 in the alveolar capillaries of the lungs. These allosteric linkages between the binding of H^+, CO_2, and O_2 are known as the Bohr effect. Third, the affinity of hemoglobin for O_2 is further regulated by 2,3-diphosphoglycerate (DPG), a small molecule with a high density of negative charge. DPG can bind to deoxyhemoglobin but not to oxyhemoglobin. Hence, DPG lowers the oxygen affinity of hemoglobin. DPG plays a role in adaptation to high altitude and hypoxia. Fetal hemoglobin has a higher oxygen affinity than adult hemoglobin because it binds less DPG.

The allosteric properties of hemoglobin arise from interactions among its α and β subunits. Structural changes within a subunit produced by the binding of O_2 (or H^+ or CO_2) are amplified into structural changes at the interfaces between subunits. In turn, these alterations affect the binding properties of all four subunits. First, oxyhemoglobin and deoxyhemoglobin have markedly different quaternary structures. Deoxyhemoglobin contains four pairs of salt links not present in oxyhemoglobin. Oxygenation requires the breakage of these salt links, which is triggered by the movement of the iron atom into the plane of the heme. Second, DPG binds to positively charged groups on the two β chains surrounding the central cavity of hemoglobin. These cross-links stabilize the deoxy-hemoglobin quaternary structure and thereby reduce the oxygen affinity of hemoglobin. Third, the hydrogen ions involved in the Bohr effect are bound to three pairs of sites. The environments of a pair of terminal amino groups and two pairs of histidine side chains become more negatively charged in deoxyhemoglobin, causing them to take up H^+ when O_2 is released. Fourth, CO_2 binds to the terminal amino groups of all four chains by forming a readily reversible carbamate linkage.

SELECTED READINGS

Where to start:

Perutz, M. F., 1964. The hemoglobin molecule. *Sci. Amer.* 211(11):2–14. [Available as *Sci. Amer.* Offprint 196. A delightful account of the x-ray crystallography effort on hemoglobin up to 1964.]

Perutz, M. F., 1970. Stereochemistry of cooperative effects of haemoglobin. *Nature* 228:726–739. [An important and very interesting article in which the harvest of a thirty-year labor is presented. Perutz compares the atomic models of deoxy and ferrihemoglobin and proposes a structural hypothesis for the mechanism of the allosteric properties of hemoglobin.]

Structure, reactivity, and regulation of hemoglobin:

Perutz, M. F., 1969. The haemoglobin molecule. *Proc. Roy. Soc.* (B)173:113–140. [A detailed and well-illustrated account of the change in quaternary structure accompanying oxygenation.]

Antonini, E., and Brunori, M., 1971. *Hemoglobin and Myoglobin in their Reactions with Ligands.* North-Holland.

Roughton, F. J. W., 1964. Transport of oxygen and carbon dioxide. *Handbook of Physiology*, Section 3: *Respiration*, vol. 1, pp. 767–825. American Physiological Society, Washington, D.C. [Contains extensive data on equilibria and kinetics of binding.]

McConnell, H. M., 1971. Spin-label studies of cooperative oxygen binding to hemoglobin. *Annu. Rev. Biochem.* 40:227–236. [A review of the application of paramagnetic probe molecules to elucidate the mechanism of cooperative oxygen binding.]

Wyman, J., 1968. Regulation in macromolecules as illustrated by haemoglobin. *Quart. Rev. Biophys.* 1:35–80. [A thermodynamic analysis of the allosteric properties of hemoglobin based on the binding potential, a concept introduced by Wyman.]

Diphosphoglycerate:

Benesch, R., and Benesch, R. E., 1969. Intracellular organic phosphates as regulators of oxygen release by haemoglobin. *Nature* 221:618–622.

Lenfant, C., Torrance, J. D., Woodson, R. D., Jacobs, P., and Finch, C. A., 1970. Role of organic phosphates in the adaptation of man to hypoxia. *Fed. Proc.* 29:1115–1117.

Tyuma, I., and Shimizu, K., 1970. Effect of organic phosphates on the difference in oxygen affinity between fetal and adult human hemoglobin. *Fed. Proc.* 29:1112–1114.

Arnone, A., 1972. X-ray diffraction study of binding of 2,3-diphosphoglycerate to human deoxyhaemoglobin. *Nature* 237:146–149.

Bohr effect:

Kilmartin, J. V., and Rossi-Bernardi, L., 1973. Interaction of hemoglobin with hydrogen ions, carbon dioxide, and organic phosphates. *Physiol. Rev.* 53:836–890. [A review of recent studies of the allosteric properties of hemoglobin.]

Kilmartin, J. V., and Rossi-Bernardi, L., 1969. Inhibition of CO_2 combination and reduction of the Bohr effect in haemoglobin chemically modified at its α-amino groups. *Nature* 222:1243–1246.

Perutz, M. F., Muirhead, H., Mazzarella, L., Crowther, R. A., Greer, J., and Kilmartin, J. V., 1969. Identification of residues responsible for the alkaline Bohr effect in haemoglobin. *Nature* 222:1240–1243.

PROBLEMS

1. What is the effect of each of the following treatments on the oxygen affinity of hemoglobin A in vitro?
   (a) Increase in pH from 7.2 to 7.4.
   (b) Increase in pCO_2 from 10 to 40 torr.
   (c) Increase in [DPG] from 2×10^{-4} to 8×10^{-4} M.
   (d) Dissociation of $\alpha_2\beta_2$ into monomer subunits.

2. What is the effect of each of the following treatments on the number of H^+ bound to hemoglobin A in vitro?
   (a) Increase in pO_2 from 20 to 100 torr (at constant pH and pCO_2).
   (b) Reaction of hemoglobin with excess cyanate (at constant pH).

3. The erythrocytes of birds and turtles contain a regulatory molecule different from DPG. This substance is also effective in reducing the oxygen affinity of human hemoglobin stripped of DPG. Which of the following substances would you predict to be most effective in this regard?
   (a) Glucose 6-phosphate.
   (b) Inositol hexaphosphate.
   (c) HPO_4^{2-}.
   (d) Malonate.
   (e) Arginine.
   (f) Lactate.

4. The oxygen dissociation curve of hemoglobin can be described to a good approximation by the empirical equation

$$Y = \frac{(pO_2)^n}{(pO_2)^n + (P_{50})^n}$$

where $n = 2.8$. Note that this is a modification of equation (5) for noncooperative binding, where $n = 1$.

(a) The amount of oxygen transported is proportional to ΔY, the change in saturation. Calculate ΔY in going from the lung to active muscle. Let $P_{50} = 26$ torr, $P_{lung} = 100$ torr, and $P_{muscle} = 20$ torr.

(b) Suppose that the binding of oxygen to hemoglobin were noncooperative but that P_{50} remained at 26 torr. What would ΔY be for such a noncooperative oxygen binder?

5. The pK of an acid depends in part on its environment. Predict the effect of these environmental changes on the pK of a glutamic acid side chain.

(a) A lysine side chain is brought into close proximity.

(b) The terminal carboxyl group of the protein is brought into close proximity.

(c) The glutamic acid side chain is shifted from the outside of the protein to a non-polar site inside.

6. DPG plays a role in high-altitude adaptation. What is the effect of DPG on the amount of oxygen transported to muscle in a person living at 10,000 feet? Assume that the increased DPG level has shifted P_{50} from 26 torr to 35 torr, that the alveolar P at 10,000 feet is 67 torr, and that P_{muscle} is 20 torr.

7. The concept of linkage is crucial for the understanding of many biochemical processes. Consider a protein molecule P that can bind A or B or both:

$$P \underset{K_B}{\overset{K_A}{\rightleftharpoons}} PA$$

$$PB \underset{K_{BA}}{\rightleftharpoons} PAB$$

The dissociation constants for these equilibria are defined as:

$$K_A = \frac{[P][A]}{[PA]} \qquad K_B = \frac{[P][B]}{[PB]}$$

$$K_{BA} = \frac{[PB][A]}{[PAB]} \qquad K_{AB} = \frac{[PA][B]}{[PAB]}$$

(a) Suppose $K_A = 5 \times 10^{-4}$ M, $K_B = 10^{-3}$ M, and $K_{BA} = 10^{-5}$ M. Is the value of the fourth dissociation constant K_{AB} defined? If so, what is it?

(b) What is the effect of A on the binding of B? What is the effect of B on the binding of A?

8. Carbon monoxide combines with hemoglobin to form CO-hemoglobin. Crystals of CO-hemoglobin are isomorphous with those of oxyhemoglobin. Each heme in hemoglobin can bind one carbon monoxide molecule but O_2 and CO cannot simultaneously bind to the same heme. The binding affinity for CO is about 200 times greater than for oxygen. Exposure for 1 hour to a CO concentration of 0.1% in inspired air leads to the occupancy by CO of about half of the heme sites in hemoglobin, which is frequently fatal.

An interesting problem was posed (and partially solved) by J. S. Haldane and J. G. Priestley in 1935:

If the action of CO were simply to diminish the oxygen-carrying power of the hemoglobin, without other modification of its properties, the symptoms of CO poisoning would be very difficult to understand in the light of other knowledge. Thus, a person whose blood is half-saturated with CO is practically helpless, as we have just seen; but a person whose hemoglobin percentage is simply diminished to half by anemia may be going about his work as usual.

What is the key to this seeming paradox?

9. A protein molecule P reversibly binds a small molecule L. The dissociation constant K for the equilibrium

$$P + L \rightleftharpoons PL$$

is defined as

$$K = \frac{[P][L]}{[PL]}$$

The protein transports the small molecule from a region of high concentration $[L_A]$ to one of low concentration $[L_B]$. Assume that the concentrations of unbound small molecules remain constant. The protein goes back and forth between regions A and B.

(a) Suppose $[L_A] = 10^{-4}$ M and $[L_B] = 10^{-6}$ M. What value of K yields maximal transport? One way of solving this problem is to write an expression for ΔY, the change in saturation of the ligand-binding site in going from region A to B. Then, take the derivative of ΔY with respect to K.

(b) Treat oxygen transport by hemoglobin in a similar way. What value of P_{50} would give a maximal ΔY? Assume the P in the lungs is 100 torr, whereas P in the tissue capillaries is 20 torr. Compare your calculated value of P_{50} with the physiological value of 26 torr.

CHAPTER 5

MOLECULAR DISEASE: SICKLE-CELL ANEMIA

In 1904, James Herrick, a Chicago physician, examined a twenty-year-old black college student who was admitted to the hospital because of a cough and fever. The patient felt weak and dizzy and had a headache. For about a year he had been experiencing palpitation and shortness of breath. Also, for several years he had participated less in physical activities. Three years earlier, there had been a discharge of pus from one ear, lasting six months. Since childhood, he had frequently had sores on the legs, which had been slow to heal.

On physical examination, the patient appeared rather well developed physically and was intelligent. There was a tinge of yellow in the whites of his eyes, and the visible mucous membranes were pale. His lymph nodes were enlarged. There were about twenty scars on his legs and thighs. The heart was distinctly abnormal, in that it was enlarged and a murmur was detected. Herrick noted that "the heart's action reminded one of a heart under strong stimulation, though no history of ingestion of a stimulant of any kind was obtainable."

The laboratory examination included careful scrutiny of the stools to determine whether any parasites were present, a strong possibility in a patient who had grown up in the tropics. None were found. The sputum showed no tubercle bacilli. The urine contained

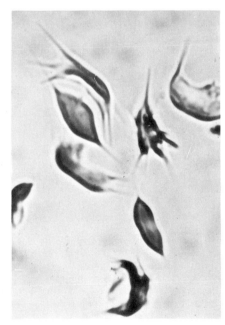

Figure 5-1
Sickle cells from the blood of a patient with sickle-cell anemia, as viewed under a light microscope. [Courtesy of Dr. Frank Bunn.]

cellular debris indicative of diseased kidneys. The blood was highly abnormal:

	Observed	Normal range
Red cell count	2.6×10^6/ml	$4.6–6.2 \times 10^6$/ml
Hemoglobin content	8 g/100 ml	14–18 g/100 ml
White cell count	15,250/ml	4,000–10,000/ml

The patient was definitely anemic, his hemoglobin content being half of the normal value. The red cells varied much in size, many abnormally small ones being evident. There were many nucleated red cells present. These are the precursors of normal red cells, which do not contain nuclei. Herrick described the unusual red cells in these terms: "The shape of the red cells was very irregular, but *what especially attracted attention was the large number of thin, elongated, sickle-shaped and crescent-shaped forms.* These were seen in fresh specimens, no matter in what way the blood was spread on the slide. . . . They were not seen in specimens of blood taken at the same time from other individuals and prepared under exactly similar conditions. They were surely not artifacts, nor were they any form of parasite."

The treatment was supportive, consisting of rest and nourishing food. The patient left the hospital four weeks later, less anemic and feeling much better. However, his blood still exhibited a "tendency to the peculiar crescent-shape in the red corpuscles though this was by no means as noticeable as before."

Herrick was puzzled by the clinical picture and laboratory findings. Indeed, he waited six years before publishing the case history, and then candidly asserted that "not even a definite diagnosis can be made." He noted the chronic nature of the disease, and the diversity of abnormal physical and laboratory findings: cardiac enlargement, a generalized swelling of lymph nodes, jaundice, anemia, and evidence of kidney damage. He concluded that the disease could not be explained on the basis of an organic lesion in any one organ. He singled out the abnormal blood picture as the key finding and entitled his case report *Peculiar Elongated and Sickle-Shaped Red Blood Corpuscles in a Case of Severe Anemia.* Herrick suggested that "some unrecognized change in the composition of the corpuscle itself may be the determining factor."

SICKLE-CELL ANEMIA IS A CHRONIC, HEMOLYTIC DISEASE THAT IS GENETICALLY TRANSMITTED

Other cases of this disease, called sickle-cell anemia, were found soon after the publication of Herrick's description. Indeed, sickle-cell anemia is not a rare disease. It is a significant public health problem wherever there is a substantial black population. The incidence of sickle-cell anemia in blacks is about four per thousand. In the past, sickle-cell anemia has usually been a fatal disease, often before age thirty, as a result of infection, renal failure, cardiac failure, or thrombosis.

The sickled red cells become trapped in the small blood vessels. Circulation is impaired, resulting in the damage of multiple organs, particularly bone and kidney. The sickled cells are more fragile than normal cells. They hemolyze readily and consequently have a shorter life than normal cells. The resulting anemia is usually severe. The chronic course of the disease is punctuated by crises in which the proportion of sickled cells is especially high. During such a crisis, the patient may go into shock.

Sickle-cell anemia is genetically transmitted. Patients with *sickle-cell anemia are homozygous for an abnormal gene* located on one of the autosomal chromosomes. Individuals who receive the abnormal gene from one parent but its normal allele from the other have *sickle-cell trait*. Such *heterozygous* individuals are usually not symptomatic. Only 1% of the red cells in their venous circulation are sickled, in contrast with about 50% in the homozygote. However, sickle-cell trait, which occurs in about one of ten blacks, is not entirely benign. Vigorous physical activity at high altitude, air travel in unpressurized planes, and anesthesia may be potentially hazardous to individuals with sickle-cell trait. The reasons will be evident shortly.

THE SOLUBILITY OF DEOXYGENATED SICKLE HEMOGLOBIN IS ABNORMALLY LOW

Herrick correctly surmised the location of the defect in sickle-cell anemia. Red cells from a patient with this disease will sickle on a microscope slide in vitro if the concentration of oxygen is reduced. In fact, hemoglobin in these cells is itself defective. Deoxygenated sickle-cell hemoglobin has an abnormally low solubility, which is

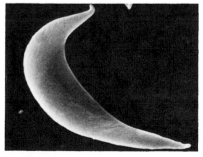

Scanning electron micrograph of an erythrocyte from a patient with sickle-cell anemia. [Courtesy of Dr. Jerry Thornthwaite and Dr. Robert Leif.]

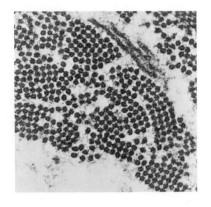

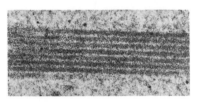

$$\xleftrightarrow{\text{2000 Å}}$$

Figure 5-2
Electron micrograph of fibers of
deoxygenated sickle-cell
hemoglobin. The upper figure is a
cross section and the lower a
longitudinal section. The fibers are
170 Å thick. [From J. T. Finch,
M. F. Perutz, J. F. Bertles, and
J. Döbler. *Proc. Nat. Acad. Sci.*
70(1973):719.]

about twenty-five times less than that of normal deoxygenated hemoglobin. A fibrous precipitate is formed if a concentrated solution of sickle-cell hemoglobin is deoxygenated (Figure 5-2). This precipitate deforms red cells and gives them their sickle shape. We will refer to sickle-cell hemoglobin as hemoglobin S (Hb S) to distinguish it from hemoglobin A (Hb A), the normal adult hemoglobin.

HEMOGLOBIN S HAS AN ABNORMAL ELECTROPHORETIC MOBILITY

In 1949, Pauling and his associates examined the physical-chemical properties of hemoglobin from normal individuals and from patients with sickle-cell trait and sickle-cell anemia. Their experimental approach was to search for differences in these hemoglobins by measuring their mobilities in an electrical field. This technique is called *electrophoresis*. The velocity of migration (V) of a protein (or of any other molecule) in an electric field depends on the strength of the electric field (E), the net electrical charge on the protein (Z), and the frictional resistance (f). The frictional resistance is a function of the size and shape of the protein. The migration velocity is related to these variables by

$$V = \frac{EZ}{f} \tag{1}$$

The *isoelectric point* is the pH at which there is no net electric charge on a protein. At this pH, the electrophoretic mobility is zero because Z in equation (1) is zero. Below the isoelectric pH, the molecule is positively charged; above the isoelectric pH, it is negatively charged. The isoelectric points for normal and sickle-cell hemoglobin are:

	Normal	Sickle-cell anemia	Difference
Oxyhemoglobin	6.87	7.09	0.22
Deoxyhemoglobin	6.68	6.91	0.23

Thus, the electrophoretic difference between normal and sickle-cell hemoglobin is the same in both the oxygenated and deoxygenated forms.

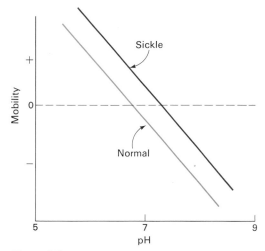

Figure 5-3
Electrophoretic mobility of sickle-cell hemoglobin and
of normal hemoglobin as a function of pH.

The different electrophoretic velocities of hemoglobin A and S
could arise either from a change in the net charge Z or in the
frictional coefficient f. A purely frictional effect (produced by a
change in shape) would cause one species to move more slowly than
the other throughout the entire pH range. This is not the case, since
the slope of the plot of the electrophoretic velocity versus pH is
the same for hemoglobin A and S. Furthermore, other physical-
chemical experiments, namely velocity sedimentation and free
diffusion measurements, showed that the frictional coefficient is the
same for the oxygenated forms of these hemoglobins.

These observations suggested that *there is a difference in the number
or kind of ionizable groups in the two hemoglobins.* How many are in
each? An estimate was made from the acid-base titration curve of
hemoglobin. In the neighborhood of pH 7, this curve is nearly
linear. A change of one pH unit in the hemoglobin solution is
associated with a change of about thirteen charges. The difference
in isoelectric pH of 0.23 therefore corresponds to about three charges
per hemoglobin molecule. It was concluded that *sickle-cell hemoglobin
has between two and four more net positive charges per molecule than normal
hemoglobin.*

Is the difference in the polypeptide chain or in the heme group?
The porphyrin isolated from sickle-cell hemoglobin was shown to
be normal on the basis of its melting point and x-ray crystal-

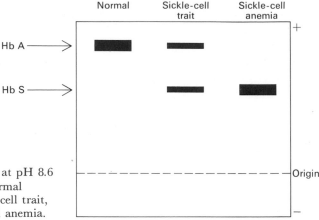

Figure 5-4
Starch-gel electrophoresis pattern at pH 8.6 of hemoglobin isolated from a normal person, from a person with sickle-cell trait, and from a person with sickle-cell anemia.

lographic pattern. Hence, it was deduced that *the difference between the two hemoglobins must be in their polypeptide chains.*

Individuals with sickle-cell anemia (who are homozygous for the sickle gene) have hemoglobin S but no hemoglobin A. In contrast, individuals with sickle-cell trait (who are heterozygous for the sickle gene) have both kinds of hemoglobin in approximately equal amounts (Figure 5-4). Thus, Pauling's study revealed "a clear case of a change produced in a protein molecule by an allelic change in a single gene" involved in its synthesis. This was the first demonstration of a *molecular disease.*

FINGERPRINTING: DETECTION OF THE ALTERED AMINO ACID IN SICKLE HEMOGLOBIN

The electrophoretic analysis showed that hemoglobin S has two to four more positive charges than hemoglobin A. A difference of this kind could arise in several ways:

Hb A	Hb S
Neutral amino acid $\longrightarrow$	Positively charged amino acid
Negatively charged amino acid $\longrightarrow$	Positively charged amino acid
Negatively charged amino acid $\longrightarrow$	Neutral amino acid

The breakthrough in the elucidation of the precise change came in 1954, when Vernon Ingram devised a new technique for detecting amino acid substitutions in proteins. The hemoglobin molecule was split into smaller units for analysis since it was anticipated that it would be easier to detect an altered amino acid in a peptide containing about 20 residues than in a protein ten times larger. Trypsin was used to specifically cleave hemoglobin on the carboxyl side of its lysine and arginine residues. Since there are a total of 27 lysine and arginine residues in the αβ half of hemoglobin, 28 different peptides were formed by *tryptic digestion*. The next step was to separate these peptides. This was accomplished by a *two-dimensional procedure* (Figure 5-5). The mixture of peptides was placed in a spot at one corner of a large sheet of filter paper. *Electrophoresis* was first carried out in one direction, separating the peptides according to their net charge. However, electrophoresis did not suffice to separate the peptides completely. There was much overlapping. A second separation technique, namely paper chromatography, was carried out at right angles to the direction of electrophoresis. The end of the filter paper closest to the peptides was immersed in a mixture of organic solvents and water at the bottom of a sealed glass jar. The solvent ascends into the paper. Each peptide now has a choice: it can either migrate with the solvent, which is quite nonpolar, or cling to the hydrated cellulose support, which is highly polar. This separatory technique is called *partition chromatography*. A highly nonpolar peptide will partition with the rising solvent, and consequently will move with the solvent front to the top of the paper, whereas a markedly polar peptide will remain near the bottom of the paper. Paper chromatography and electrophoresis are complementary techniques because they separate peptides on the basis of different characteristics—namely, their degree of polarity and net charge, respectively. This sequence of steps—selective

Chromatography—
A term introduced by Mikhail Tswett in 1906 to describe the separation of a mixture of leaf pigments on a calcium carbonate column, like "light rays in the spectrum."
Derived from the Greek words *chroma*, color, and *graphein*, to draw or write.

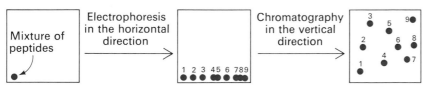

Figure 5-5
A mixture of peptides is resolved by electrophoresis in the horizontal direction, followed by partition chromatography in the vertical direction.

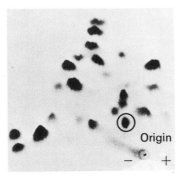

Hemoglobin A

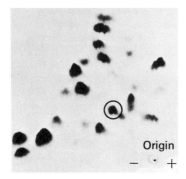

Hemoglobin S

Figure 5-6
Comparison of the
ninhydrin-stained fingerprints
of hemoglobin A and
hemoglobin S. The position of
the peptide that is different in
these hemoglobins is encircled
in red. [Courtesy of Dr.
Corrado Baglioni.]

cleavage of a protein into small peptides, followed by their separation in two dimensions—is called *fingerprinting*.

The resulting fingerprint was highly revealing. The peptide spots were made visible by staining the fingerprint with ninhydrin. A comparison of the maps for hemoglobin A and S showed that *all but one of the 28 peptide spots matched*. The one spot that was different was eluted from each fingerprint and then shown to be a single peptide consisting of eight amino acids. Amino acid analysis indicated that this peptide from hemoglobin S differed from the one in hemoglobin A by a single amino acid.

A SINGLE AMINO ACID IN THE BETA CHAIN IS ALTERED

The α and β chains were separated by ion-exchange chromatography. Fingerprints of the separate chains showed that the change in hemoglobin S is in its β chains. In fact, the difference is located in the amino-terminal peptide of the β chain. Ingram determined the sequence of this peptide and showed that *hemoglobin S contains valine instead of glutamate at position 6 of the β chain*.

Hemoglobin A	Val-His-Leu-Thr-Pro-Glu-Glu-Lys-
Hemoglobin S	Val-His-Leu-Thr-Pro-Val-Glu-Lys-
	$\beta 1$ 2 3 4 5 6 7 8

SICKLE HEMOGLOBIN HAS STICKY PATCHES ON ITS SURFACE

The side chain of valine is distinctly nonpolar, whereas that of glutamate is highly polar. The substitution of valine for glutamate at position 6 of the β chains places a nonpolar residue on the outside of hemoglobin S (Figure 5-7). *This alteration markedly reduces the solubility of deoxygenated hemoglobin S, but has little effect on the solubility of oxygenated hemoglobin S.* This fact is crucial to an understanding of the clinical picture in sickle-cell anemia and sickle-cell trait.

The molecular basis of sickling can be visualized as follows:

1. The substitution of valine for glutamic acid gives hemoglobin S a sticky patch on the outside of each of its β chains (Figure 5-8).

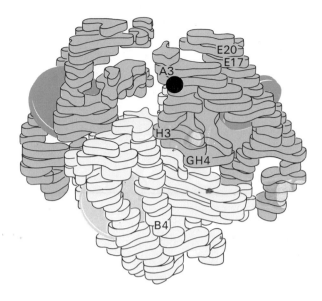

Figure 5-7
Model of deoxyhemoglobin A at low resolution.
The α chains are shown in yellow, the β
chains in blue. The position of the amino
acid change in hemoglobin S is marked in
red. Note that this site is at the surface of
the molecule. [Based on J. T. Finch, M. F.
Perutz, J. F. Bertles, and J. Döbler. *Proc.
Nat. Acad. Sci.* 70(1973):721.]

This sticky patch is present on both oxy and deoxyhemoglobin S
and is missing in hemoglobin A.

2. There is a site on deoxyhemoglobin S that is complementary
to the sticky patch (Figure 5-8). Its location is not yet known. This
complementary site on one deoxyhemoglobin S molecule interacts
with the sticky patch on another deoxyhemoglobin S molecule,
leading to the formation of long aggregates that distort the red cell.

3. In oxyhemoglobin S, the complementary site is masked. The
sticky patch is present, but it cannot bind to another hemoglobin
S because the complementary site is unavailable.

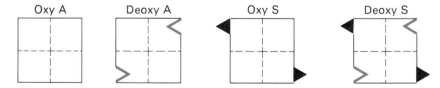

Figure 5-8
The red triangle represents the sticky patch that is present on both oxy and
deoxyhemoglobin S, but not on either form of hemoglobin A. The
complementary site is represented by an indentation that can accommodate
the triangle. This complementary site is present in deoxyhemoglobin S and
probably also in deoxyhemoglobin A.

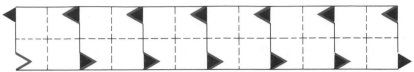

Figure 5-9
A schematic representation of the interaction of a sticky patch on deoxyhemoglobin S with the complementary site on another deoxyhemoglobin S molecule to form long aggregates. One strand of the helical fiber is shown here.

4. Thus, *sickling occurs when there is a high concentration of the deoxygenated form of hemoglobin S* (Figure 5-9).

These facts account for several clinical characteristics of sickle-cell anemia. A vicious cycle is set up when sickling occurs in a small blood vessel. The blockage of the vessel creates a local region of low oxygen concentration. Hence, more hemoglobin goes into the deoxy form and so more sickling occurs. The individual with sickle-cell trait usually is asymptomatic because not more than half of his hemoglobin is hemoglobin S. This is too low a concentration for extensive sickling at normal oxygen levels. However, if the oxygen level is unusually low (as at high altitude), sickling can occur in an individual with sickle-cell trait.

DEOXYHEMOGLOBIN S FORMS LONG, TUBULAR FIBERS

Deoxyhemoglobin S forms fibrous precipitates. The long tubular fibers consist of a helical array of six filaments of hemoglobin molecules (Figure 5-10). Each hemoglobin molecule makes contact with four neighbors in this tubular helix, which has a diameter of about 170 Å.

HIGH INCIDENCE OF THE SICKLE GENE IS DUE TO THE PROTECTION CONFERRED AGAINST MALARIA

The frequency of the sickle gene is as high as 40% in certain parts of Africa. Since homozygotes have usually died before reaching adulthood, it is evident that there must be strong selective pressures

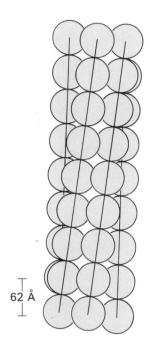

62 Å

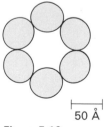

50 Å

Figure 5-10
Proposed helical structure of a fiber of aggregated deoxyhemoglobin S. [Based on J. T. Finch, M. F. Perutz, J. F. Bertles, and J. Döbler. *Proc. Nat. Acad. Sci.*70(1973):721.]

to maintain the high incidence of the gene. James Neel proposed that the heterozygote enjoys advantages not shared by either the normal homozygote or the sickle-cell homozygote. In fact, Anthony Allison found that individuals with sickle-cell trait are protected against the most lethal form of malaria. The incidence of malaria and the frequency of the sickle gene in Africa are definitely correlated (Figure 5-11). This is a clear-cut example of *balanced polymorphism*—the heterozygote is protected against malaria and does not suffer from sickle-cell disease, whereas the normal homozygote is vulnerable to malaria.

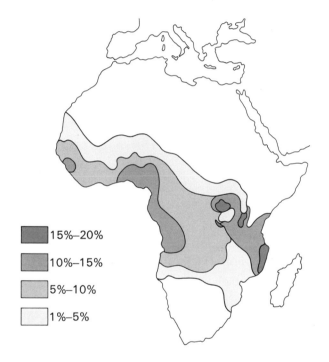

15%–20%

10%–15%

5%–10%

1%–5%

Figure 5-11
Frequency of the sickle-cell gene in Africa. High frequencies are restricted to regions where malaria is a major cause of death. [Based on A. C. Allison. Sickle cells and evolution. Copyright © 1956 by Scientific American, Inc. All rights reserved.]

PROSPECTS FOR THE THERAPY OF SICKLE-CELL ANEMIA

Several years ago, it was reported that the administration of large amounts of urea is of value in treating the crisis phase of sickle-cell anemia. The rationale for this therapy was the hope that urea would disrupt noncovalent interactions between different molecules of aggregated hemoglobin S. Other workers questioned whether the reported clinical improvement was due to urea itself, since high concentrations of urea in the blood are not readily attained because urea is rapidly removed by the kidneys. They thought that *cyanate* might be the substance that actually interferes with sickling. It has been known for a long time that cyanate is in equilibrium with urea in solution:

$$NH_2-\overset{\overset{O}{\|}}{C}-NH_2 \rightleftharpoons NH_4^+ + \cdot NCO^-$$

Urea   **Ammonium**   **Cyanate**
                    ion   **ion**

At 37° and pH 7.4, a 1% solution of urea contains 5×10^{-3} M cyanate ion, at equilibrium.

Indeed, *relatively low concentrations of potassium cyanate* (3×10^{-2} M) *inhibit the sickling of sickle-cell erythrocytes* (Figure 5-12). This inhibition of sickling is irreversible, as evidenced by the finding that cyanate-treated cells that are washed or dialyzed still do not sickle. Hemoglobin S is carbamoylated by cyanate:

$$R-NH_2 + H-N=C=O \longrightarrow R-\overset{\overset{H}{|}}{N}-\overset{\underset{\|}{C}}{\underset{O}{}}-NH_2$$

The reactive species is the un-ionized form, which is called isocyanic acid. The sites of carbamoylation are the α-amino groups of hemoglobin. The incorporation of about one cyanate per hemoglobin S tetramer suffices to prevent most of the cells from sickling. The high specificity of the reaction is due to the fact that $HN=C=O$ (isocyanic acid) is a reactive analog of $O=C=O$. Recall that the terminal amino groups of hemoglobin participate in the *reversible* binding of CO_2.

$$R-NH_2 + O=C=O \longrightarrow R-\overset{\overset{H}{|}}{N}-\overset{\underset{\|}{C}}{\underset{O}{}}-O^- + H^+$$

In contrast, the carbamoylation reaction at this site is *irreversible*.

These in vitro studies are promising and suggest that potassium cyanate might be used to treat sickle-cell anemia. It will first be necessary to determine whether potassium cyanate has any toxic effects at the level required to prevent sickling. Information will

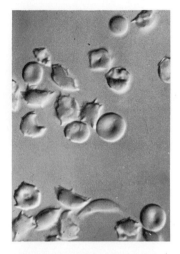
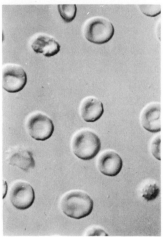

Figure 5-12
Electron micrographs of deoxygenated sickle-cell erythrocytes (upper figure) and deoxygenated sickle-cell erythrocytes that have been treated in the oxygenated state with 0.03 M potassium cyanate (lower figure). [From A. Cerami and J. M. Manning. *Proc. Nat. Acad. Sci.* 68(1971):1181.]

also be needed on the effect of cyanate treatment on oxygen and carbon dioxide transport. It will be very interesting to learn how sickling is prevented by this highly specific carbamoylation reaction.

MOLECULAR PATHOLOGY OF HEMOGLOBIN

More than 100 abnormal hemoglobins have been discovered by examination of patients with clinical symptoms and by electro-phoretic surveys of normal populations. In northern Europe, one of three hundred persons is heterozygous for a variant of hemo-globin A. The frequency of any one mutant allele is less than 10^{-4}, which is lower by several orders of magnitude than the frequency of the sickle allele in regions where malaria is endemic. In other words, most abnormal hemoglobins do not confer a selective ad-vantage on the individual. They are almost always neutral or harmful.

The abnormal hemoglobins are of several types:

1. *Altered exterior.* Nearly all substitutions on the surface of the hemoglobin molecule are harmless. Hemoglobin S is a striking exception.

2. *Altered active site.* The defective subunit cannot bind oxygen because of a structural change near the heme that directly affects oxygen binding.

3. *Altered tertiary structure.* The polypeptide chain is prevented by the amino acid substitution from folding into its normal confor-mation. These hemoglobins are usually unstable.

4. *Altered quaternary structure.* Some mutations at subunit interfaces result in the loss of allosteric properties without impairing the binding of oxygen.

HEMOGLOBIN M: ACTIVE-SITE MUTATION

Substitution of tyrosine for the proximal or distal heme histidine results in the stabilization of the heme in the ferric form, which can no longer bind oxygen (Figure 5-13). The tyrosine side chain is ionized in this complex with the ferric ion of the heme. This change can occur in either the α or β chain. Indeed, all four kinds of mutants have been observed. Mutant hemoglobins characterized by a permanent ferric state of two of the hemes are called *hemoglobin M*. The letter M signifies that the altered chains are in the *methemoglobin* (ferrihemoglobin) form. The patients are usually cyanotic. The disease has been seen only in recessive form, since homozygosity would almost certainly be lethal.

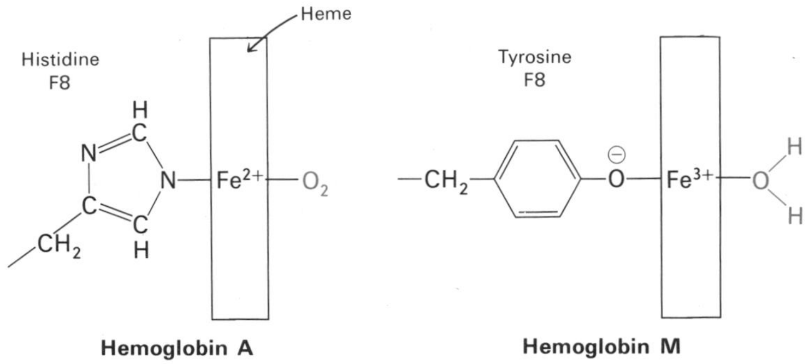

Figure 5-13
Substitution of tyrosine for the proximal histidine (F8) results in the formation of a hemoglobin M. The negatively charged oxygen atom of tyrosine is coordinated to the iron atom, which is in the ferric state. Water, rather than O_2, is bound at the sixth coordination position.

POLAR GROUPS IN THE HEME POCKET IMPAIR THE BINDING OF HEME

A different kind of active site mutation occurs when a polar group is substituted for a nonpolar one in the heme crevice. There are sixty interatomic contacts between the polypeptide chain and the heme group. These contacts are nonpolar in nature. The conservation of these interactions in normal hemoglobins of many species suggests that most of them are essential for the function of the hemoglobin molecule. Indeed, mutations at these heme-binding sites almost always have harmful consequences. For example, in Hemoglobin Hammersmith, the phenylalanine residue at CD1 (Figure 5-14) is replaced by serine. This change impairs the binding of the heme, probably because the polar serine residue enables water to enter the crevice in which the heme is normally bound.

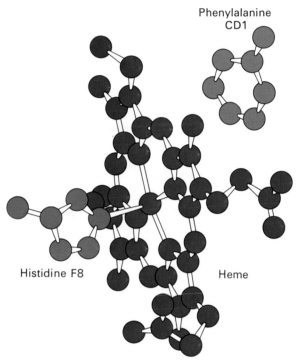

Figure 5-14
Location of phenylalanine CD1, one of the conserved residues of hemoglobin. The aromatic ring of this residue is in contact with the heme.

SOME MUTATIONS RESULT IN A DISTORTED TERTIARY STRUCTURE

Amino acid substitutions at sites far from the heme can prevent the molecule from folding into its normal conformation, thereby severely disrupting its function. An interesting example is Hemoglobin Riverdale-Bronx, which has arginine at B6, in place of glycine. The arginine is too large to fit into the narrow region where the B and E helices usually cross (Figure 5-15). In fact, B6 is glycine

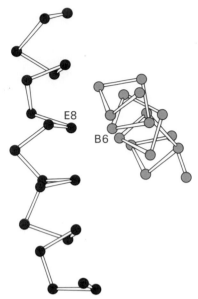

Figure 5-15
A hemoglobin molecule can have a normal structure only if the residue at B6 is glycine. There is no space for a larger side chain because this is where the B and E helices cross.

in all known normal myoglobins and hemoglobins. The effect of this substitution is to make Hemoglobin Riverdale unstable.

Hemoglobin Gun Hill is one of the few known abnormal hemoglobins in which a mutation affects more than one amino acid residue in a chain. In this hemoglobin, five amino acids in each of the β chains are deleted. These residues correspond to the last amino acid in the F helix and the next four in the corner between the F and G helices. This deletion mutation removes a section of the polypeptide chain that forms essential contacts with the heme group. As a result there is no heme in the β chains. The clinical picture is a hemolytic anemia.

ALLOSTERIC PROPERTIES ARE LOST IN SOME SUBUNIT INTERFACE MUTANTS

Effective oxygen transport requires that the allosteric properties of hemoglobin be fully operative. The integrity of the heme-binding site and the normal folding of a hemoglobin subunit are necessary but not sufficient for optimal function. Residues at contacts between subunits also are important because some of them transmit information between different subunits. Contacts between like chains in hemoglobin are tenuous and polar, whereas those between unlike chains are extensive and predominantly nonpolar. There are two different kinds of interfaces between the α and β chains, named $\alpha_1\beta_1$ and $\alpha_1\beta_2$. As noted previously, the $\alpha_1\beta_2$ *contacts* appear to play a key role in transmitting allosteric interactions, since there is considerable movement at these interfaces on oxygenation. This inference is reinforced by the finding that all five known abnormal hemoglobins with replacements at the $\alpha_1\beta_2$ contact have *diminished cooperativity*. The slope of their Hill plots for the binding of oxygen is close to 1.

IMPACT OF THE DISCOVERY OF MUTANT HEMOGLOBINS

Mutations affecting oxygen transport have had a major impact on molecular biology, medicine, genetics, and anthropology. Their significance is threefold:

1. *They provide insight into relationships between the structure and function of normal hemoglobin.* Mutations of a single amino acid residue

are highly specific chemical modifications provided by nature. They show which parts of the molecule are critical for function.

2. *The discovery of mutant hemoglobins has revealed that disease can arise from a change of a single amino acid in one kind of polypeptide chain.* The concept of molecular disease, which is now an integral part of medicine, had its origins in the incisive studies of sickle-cell hemoglobin.

3. *The finding of mutant hemoglobins has enhanced our understanding of evolutionary processes.* Mutations are the raw materials of evolution; the studies of sickle-cell anemia have shown that a mutation may be simultaneously beneficial and harmful. The disease of an individual may be a concomitant of the evolutionary process, as has been picturesquely described by Zuckerkandl and Pauling (1962):

> Subjectively, to evolve must most often have amounted to suffering from a disease. And these diseases were of course molecular. The appearance of the concept of good and evil, interpreted by man as his painful expulsion from Paradise, was probably a molecular disease that turned out to be evolution.

SUMMARY

The change of a single amino acid in a single protein caused by gene mutation can produce disease. The best understood molecular disease is sickle-cell anemia. Hemoglobin S, the abnormal hemoglobin in patients with this disease, consists of two normal α chains and two mutant β chains. The amino acid change in the β chains of hemoglobin S is the replacement of glutamate at position 6 by valine. This substitution of a nonpolar side chain for a polar one drastically reduces the solubility of deoxyhemoglobin S. In contrast, the solubility of oxyhemoglobin S is quite normal. Deoxygenated hemoglobin S forms fibrous precipitates that deform the red cell and give it a sickle shape. These red cells are thereby destroyed and so the clinical picture is a chronic hemolytic anemia. Sickle-cell anemia occurs when an individual is homozygous for the mutant sickle gene for the β chain. Heterozygotes synthesize a mixture of hemoglobin A and hemoglobin S. The heterozygous condition, called sickle-cell trait, is relatively asymptomatic. About one of ten blacks is heterozygous for the sickle gene. This high incidence of a gene that is harmful in homozygotes is due to the fact that it is beneficial in heterozygotes, a state known as balanced polymorphism. Individuals with sickle-cell trait are protected against the most lethal form of malaria.

More than 100 mutant hemoglobins have been discovered by examination of the hemoglobin of patients with hematologic symptoms and by surveys of normal populations. Hemoglobins with electrophoretic mobilities different from that of hemoglobin A are fingerprinted and the amino acid sequence of the new peptide in the fingerprint is then determined. Several classes of mutant hemoglobins are known. First, nearly all substitutions on the surface of the hemoglobin molecule are harmless. Hemoglobin S is the striking exception. Second, the oxygen-binding site is often impaired by substitutions near the heme. For example, substitution of tyrosine for the proximal or distal histidine results in the formation of a hemoglobin M, which is locked in the ferric (met) state and therefore unable to bind oxygen. Third, alterations in the interior of the molecule often distort the tertiary structure and result in an unstable molecule, as in the hemoglobin that contains arginine in place of glycine at a site where glycine is required because of its small size. Fourth, alterations at subunit interfaces often lead to the loss of allosteric properties without impairing the binding of oxygen.

SELECTED READINGS

Where to start:

Allison, A. C., 1956. Sickle cells and evolution. *Sci. Amer.* 195(2):87–94. [Available as *Sci. Amer.* Offprint 1065.]

Perutz, M. F., and Lehmann, H., 1968. Molecular pathology of human haemoglobin. *Nature* 219:902–909. [A landmark analysis of more than fifty mutant hemoglobins. Clinical symptoms and the inferred alterations in structure are correlated.]

Sickle-cell anemia:

Herrick, J. B., 1910. Peculiar elongated and sickle-shaped red blood corpuscles in a case of severe anemia. *Arch. Intern. Med.* 6:517–521.

Neel, J. V., 1949. The inheritance of sickle-cell anemia. *Science* 110:64–66.

Pauling, L., Itano, H. A., Singer, S. J., and Wells, I. C., 1949. Sickle cell anemia, a molecular disease. *Science* 110:543–548.

Ingram, V. M., 1957. Gene mutation in human haemoglobin: the chemical difference between normal and sickle cell haemoglobin. *Nature* 180:326–328.

Cerami, A., and Manning, J. M., 1971. Potassium cyanate as an inhibitor of the sickling of erythrocytes in vitro. *Proc. Nat. Acad. Sci.* 68:1180–1183.

Gillette, P. N., Peterson, C. M., Lu, Y. S., and Cerami, A., 1974. Sodium cyanate as a potential treatment for sickle-cell disease. *N. Engl. J. Med.* 290:654–660.

Finch, J. T., Perutz, M. F., Bertles, J. F., and Dobler, J., 1973. Structure of sickled erythrocytes and of sickle-cell hemoglobin. *Proc. Nat. Acad. Sci.* 70:718–722.

Other mutant hemoglobins:

Lehmann, H., and Huntsman, R. G., 1972. The hemoglobinopathies. *In* Stanburg, J. B., Wyngaarden, J. B., and Fredrickson, D. S., (eds.), *The Metabolic Basis of Inherited Disease* (3rd ed.), pp. 1398–1431. McGraw-Hill.

Hunt, L. T., Sochard, M. R., and Dayhoff, M. O., 1972. Mutations in human genes: abnormal hemoglobins and myoglobins. *In* Dayhoff, M. O., (ed.), *Atlas of Protein Sequence and Structure*, vol. 5, pp. 67–87. National Biomedical Research Foundation, Silver Springs, Maryland. [A valuable compilation and discussion of 145 mutant hemoglobins and myoglobins.]

Ranney, H. M., 1970. Clinically important variants of human hemoglobin. *N. Engl. J. Med.* 282:144–152.

Morimoto, H., Lehmann, H., and Perutz, M. F., 1971. Molecular pathology of human haemoglobin: stereochemical interpretation of abnormal oxygen affinities. *Nature* 232:408–413.

Greer, J. N., 1971. Three-dimensional structure of abnormal mutant hemoglobins. *Cold Spring Harbor Symp. Quant. Biol.* 36:315–323.

Delivoria-Papadopoulos, M., Oski, F. A., and Gottlieb, A. J., 1969. Oxygen-hemoglobin dissociation curves: effect of inherited enzyme defects of the red cell. *Science* 165:601–602.

Disease and evolution:

Zuckerkandl, E., and Pauling, L., 1962. Molecular disease, evolution, and genic heterogeneity. *In* Kasha, M., and Pullman, B., (eds.), *Horizons in Biochemistry*, pp. 189–225. Academic Press.

Cavalli-Sforza, L. L., and Bodmer, W. F., 1971. *The Genetics of Human Populations.* W. H. Freeman and Company. [An excellent discussion of genetic polymorphism is given in chapters 4 and 5.]

PROBLEMS

1. A hemoglobin with an abnormal electrophoretic mobility is detected in a screening program. Fingerprinting after tryptic digestion reveals that the amino acid substitution is in the β chain. The normal amino-terminal tryptic peptide (Val-His-Leu-Thr-Pro-Glu-Glu-Lys) is missing. A new tryptic peptide consisting of six amino acid residues is found. Valine is the amino terminal residue of this peptide.

   (a) Which amino acid substitutions are consistent with these data?
   (b) What is the electrophoretic mobility of this hemoglobin likely to be in relation to Hb A and Hb S at pH 8?

2. The electrophoretic mobilities at pH 8.6 of a series of mutant Hb A molecules is shown below.

   Electrode $\overline{}$ $\overset{+}{}$ Electrode
    ↑ ↑ ↑ ↑ ↑
    a b Normal c d

   Match the positions a, b, c, and d with these mutant hemoglobins:

Hb D (α68)	Lysine instead of asparagine
Hb J (β69)	Aspartate instead of glycine
Hb N (β95)	Glutamate instead of lysine
Hb C (β6)	Lysine instead of glutamate

3. Some mutations in a hemoglobin gene affect all three of the hemoglobins A, A_2, and F, whereas others affect only one of them. Why?

4. The starch-gel electrophoresis pattern of hemoglobin from an individual with sickle-cell trait contains two major bands. One of them is Hb A ($\alpha_2\beta_2$) and the other is Hb S ($\alpha_2\beta_2^S$). The absence of a third band with an intermediate electrophoretic mobility initially suggested that the hybrid molecule $\alpha_2\beta\beta^S$ is not present. However, it was subsequently found that this hybrid hemoglobin is in fact present in solution. Why is it absent from the gel pattern? (Hint: Consider the effect of an electric field on the equilibrium $2(\alpha_2\beta\beta^S) \rightleftharpoons \alpha_2\beta_2 + \alpha_2\beta_2^S$).

5. Hb A inhibits the formation of long fibers of Hb S and the subsequent sickling of the red cell upon deoxygenation. Why does Hb A have this effect?

CHAPTER 6

INTRODUCTION TO ENZYMES

Chemical reactions in biological systems rarely occur in the absence of a catalyst. These catalysts are specific proteins called enzymes. The striking characteristics of all enzymes are their catalytic power and specificity. Furthermore, the activity of many enzymes is regulated. In addition, some enzymes are intimately involved in the transformation of different forms of energy. Let us examine these highly distinctive and biologically crucial properties of enzymes.

ENZYMES HAVE ENORMOUS CATALYTIC POWER

Enzymes accelerate reactions by factors of at least a million. Indeed, most reactions in biological systems do not occur at perceptible rates in the absence of enzymes. Even a reaction as simple as the hydration of carbon dioxide is catalyzed by an enzyme.

$$CO_2 + H_2O \rightleftharpoons H_2CO_3$$

Otherwise, the transfer of CO_2 from the tissues into the blood and then to the alveolar air would be incomplete. Carbonic anhydrase, the enzyme that catalyzes this reaction, is one of the fastest known. Each enzyme molecule can hydrate 10^5 molecules of CO_2 in one second. This catalyzed reaction is 10^7 times faster than the uncatalyzed reaction.

ENZYMES ARE HIGHLY SPECIFIC

Enzymes are highly specific both in the reaction catalyzed and in their choice of reactants, which are called *substrates*. An enzyme usually catalyzes a single chemical reaction or a set of closely related reactions. The degree of specificity for substrate is usually high and sometimes virtually absolute.

Let us consider *proteolytic enzymes* as an example. The reaction catalyzed by these enzymes is the hydrolysis of a peptide bond.

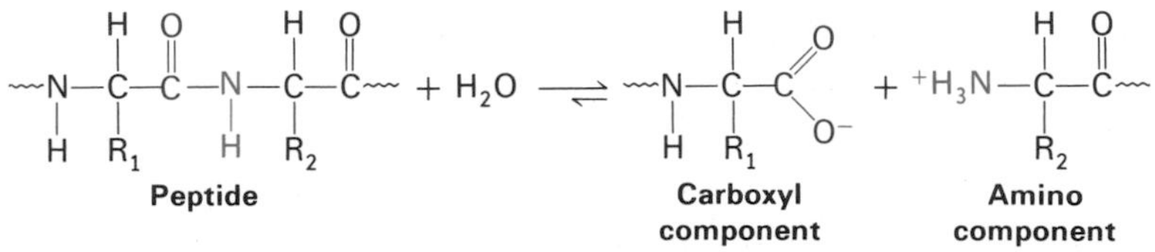

$$\sim N-\underset{\underset{H}{|}}{\overset{\overset{H}{|}}{C}}-\underset{}{\overset{\overset{O}{\|}}{C}}-N-\underset{\underset{R_2}{|}}{\overset{\overset{H}{|}}{C}}-\overset{\overset{O}{\|}}{C}\sim + H_2O \rightleftharpoons \sim N-\underset{\underset{H}{|}}{\overset{\overset{H}{|}}{C}}-\underset{}{\overset{\overset{O}{}}{C}}_{O^-} + {}^+H_3N-\underset{\underset{R_2}{|}}{\overset{\overset{H}{|}}{C}}-\overset{\overset{O}{\|}}{C}\sim$$

Peptide **Carboxyl component** **Amino component**

Most proteolytic enzymes also catalyze a different but related reaction, namely the hydrolysis of an ester bond.

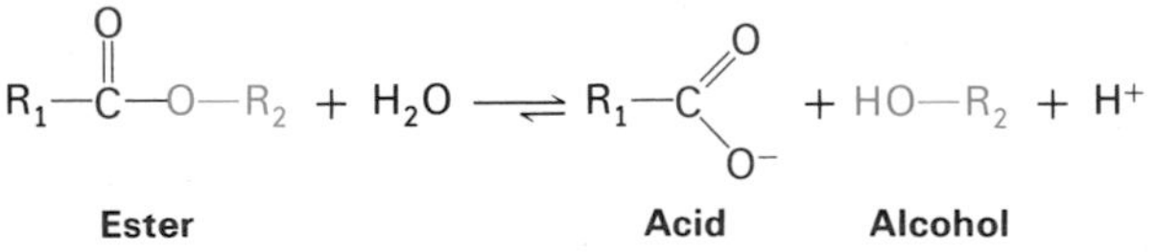

$$R_1-\overset{\overset{O}{\|}}{C}-O-R_2 + H_2O \rightleftharpoons R_1-\overset{\overset{O}{}}{C}_{O^-} + HO-R_2 + H^+$$

Ester **Acid** **Alcohol**

Proteolytic enzymes vary markedly in their degree of substrate specificity. Subtilisin, which comes from certain bacteria, is quite undiscriminating about the nature of the side chains adjacent to the peptide bond to be cleaved. Trypsin, as was mentioned in Chapter 2, is quite specific, in that it splits peptide bonds on the carboxyl side of lysine and arginine residues only (Figure 6-1). Thrombin, an enzyme that is involved in blood clotting, is even more specific than trypsin. The side chain on the carboxyl side of the susceptible peptide bond must be arginine, whereas the one on the amino side must be glycine (Figure 6-2).

Another example of the high degree of specificity of enzymes is provided by DNA polymerase I. This enzyme synthesizes DNA by linking together four kinds of nucleotide building blocks. The sequence of nucleotides in the DNA strand that is being synthesized is determined by the sequence of nucleotides in another DNA strand that serves as a template (Figure 6-3). DNA polymerase I

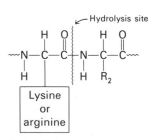

Figure 6-1
Specificity of trypsin.

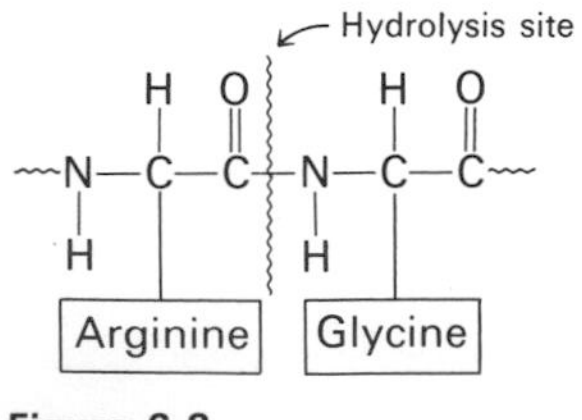

Figure 6-2
Specificity of thrombin, a clotting factor.

is remarkably precise in carrying out the instructions given by the template. The wrong nucleotide is inserted into the new DNA strand less than once in a million times.

THE ACTIVITY OF SOME ENZYMES IS REGULATED

Some enzymes are synthesized in an *inactive precursor form*, and are activated at a physiologically appropriate time and place. The digestive enzymes exemplify this kind of control. For example, trypsinogen is synthesized in the pancreas and is activated by peptide-bond cleavage in the small intestine to form the active enzyme trypsin (Figure 6-4). This type of control is also repeatedly used in the sequence of enzymatic reactions leading to the clotting of blood. The enzymatically inactive precursors of proteolytic enzymes are called *zymogens*.

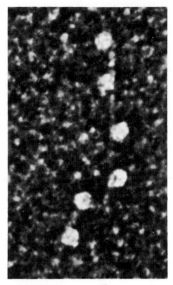

Figure 6-3
Electron micrograph of DNA polymerase I molecules (white spheres) bound to a threadlike synthetic DNA template. [Courtesy of Dr. Jack Griffith.]

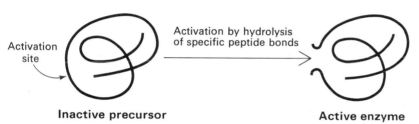

Inactive precursor

Active enzyme

Figure 6-4
Zymogen activation by hydrolysis of specific peptide bonds.

Another mechanism that controls activity is the covalent insertion of a small group on an enzyme. This control mechanism is called *covalent modification*. For example, the activities of the enzymes that synthesize and degrade glycogen are regulated by the attachment of a phosphoryl group to a specific serine residue on these enzymes (see Chapter 16). This modification can be reversed by hydrolysis. Specific enzymes catalyze the insertion and removal of phosphoryl and other modifying groups.

$$-CH_2-O-\overset{\overset{\displaystyle O}{\|}}{\underset{\underset{\displaystyle O^-}{|}}{P}}-O^-$$

Phosphorylated derivative of a serine residue

A different kind of regulatory mechanism affects many reaction sequences resulting in the synthesis of small molecules such as amino acids. The enzyme that catalyzes the first step in such a biosynthetic pathway is inhibited by the ultimate product (Figure 6-5). The biosynthesis of isoleucine illustrates this type of control, which is called *feedback inhibition*. Threonine is converted to isoleucine in five steps, the first of which is catalyzed by threonine deaminase. This enzyme is inhibited when the concentration of isoleucine reaches a sufficiently high level. Isoleucine binds to a different site from threonine. The inhibition of threonine deaminase is mediated by an *allosteric interaction*, which is reversible. When the level of isoleucine drops sufficiently, threonine deaminase becomes active again, and consequently isoleucine is again synthesized.

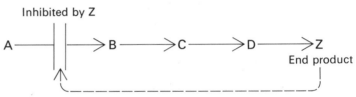

Figure 6-5
Feedback inhibition of the first enzyme in a pathway by reversible binding of the final product.

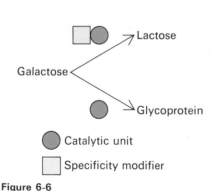

Galactose unit   **Glucose unit**

LACTOSE

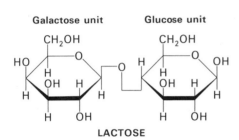

○ Catalytic unit

□ Specificity modifier

Figure 6-6
Lactose, a sugar consisting of a galactose and a glucose residue, is synthesized by an enzyme that contains a catalytic subunit and a specificity-modifier subunit. A different reaction is catalyzed by the catalytic subunit alone.

The specificity of some enzymes is under physiological control. The synthesis of lactose by the mammary gland is a particularly striking example (Figure 6-6). Lactose synthetase, the enzyme that catalyzes the synthesis of lactose, consists of a catalytic subunit and a modifier subunit. The catalytic subunit by itself cannot synthesize lactose. It has a different role, which is to catalyze the attachment of galactose to proteins that contain a covalently linked carbohydrate chain. *The modifier subunit alters the specificity of the catalytic subunit so that it links galactose to glucose to form lactose.* The level of the modifier subunit is under hormonal control. During pregnancy, the catalytic subunit is formed in the mammary gland, but little modifier subunit is formed. At the time of birth, hormonal levels change drastically, and the modifier subunit is synthesized in large amounts. The modifier subunit then binds to the catalytic subunit to form an active lactose synthetase complex that produces large amounts of lactose. This system clearly shows that hormones can exert their physiological effects by altering the specificity of enzymes.

ENZYMES TRANSFORM DIFFERENT KINDS OF ENERGY

In many biochemical reactions, *the energy of the reactants is converted to a different form with high efficiency*. For example, in photosynthesis, light energy is converted into chemical bond energy. In mitochondria, the free energy contained in small molecules derived from foods is converted into a different currency, that of adenosine triphosphate (ATP). The chemical-bond energy of ATP is then utilized in many different ways. In muscular contraction, the energy of ATP is converted into mechanical energy. Cells and organelles have pumps that utilize ATP to transport molecules and ions against chemical and electrical gradients. These transformations of energy are carried out by enzyme molecules that are integral parts of highly organized assemblies.

ENZYMES DO NOT ALTER REACTION EQUILIBRIA

An enzyme is a catalyst and consequently it cannot alter the equilibrium of a chemical reaction. This means that an enzyme accelerates the forward and reverse reaction by precisely the same factor. Consider the interconversion of A and B. Suppose that in the absence of enzyme the forward rate (k_F) is 10^{-4} sec^{-1} and the reverse rate (k_R) is 10^{-6} sec^{-1}. The equilibrium constant K is given by the ratio of these rates.

$$A \underset{10^{-6} \text{ sec}^{-1}}{\overset{10^{-4} \text{ sec}^{-1}}{\rightleftharpoons}} B$$

$$K = \frac{[B]}{[A]} = \frac{k_F}{k_R} = \frac{10^{-4}}{10^{-6}} = 100$$

The equilibrium concentration of B is 100 times that of A, whether or not enzyme is present. However, it would take several hours to approach this equilibrium without enzyme, whereas equilibrium would be attained within a second when enzyme is present. Thus, enzymes accelerate the attainment of equilibria but do not shift their positions.

ENZYMES REDUCE THE ACTIVATION ENERGIES
OF REACTIONS CATALYZED BY THEM

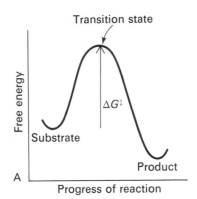

A chemical reaction A $\rightleftharpoons$ B goes through a *transition state* that has a higher energy than either A or B. The rate of the forward reaction depends on the temperature and on the difference in free energy between that of A and the transition state, which is called the *free energy of activation* and symbolized by $\Delta G^{\ddagger}$ (Figure 6-7A).

$$\Delta G^{\ddagger} = G_{\text{transition state}} - G_{\text{substrate}}$$

The reaction rate is proportional to the fraction of molecules that have a free energy equal to or greater than $\Delta G^{\ddagger}$. The proportion of molecules that have an energy equal to or greater than $\Delta G^{\ddagger}$ increases with temperature.

Enzymes accelerate reactions by reducing $\Delta G^{\ddagger}$, the activation barrier. The combination of substrate with enzyme creates a new reaction pathway that has a transition state of lower energy than in the absence of enzyme (Figure 6-7B).

FORMATION OF AN ENZYME-SUBSTRATE COMPLEX
IS THE FIRST STEP IN ENZYMATIC CATALYSIS

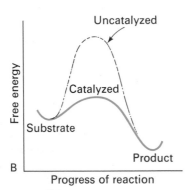

Figure 6-7
A. Definition of $\Delta G^{\ddagger}$, the free energy of activation. B. Enzymes accelerate catalysis by reducing $\Delta G^{\ddagger}$.

The making and breaking of chemical bonds by an enzyme are preceded by the formation of an *enzyme-substrate* (ES) complex. The substrate is bound to a specific region of the enzyme called the *active site*. Most enzymes are highly selective in their binding of substrates. Indeed, the catalytic specificity of enzymes depends in large part on the specificity of the binding process. Furthermore, the control of enzymatic activity may also take place at this stage.

The existence of ES complexes has been shown in a variety of ways:

1. ES complexes have been *directly visualized by electron microscopy and x-ray crystallography*. Complexes of nucleic acids and their polymerase enzymes are evident in electron micrographs (Figure 6-3). Detailed information concerning the location and interactions of glycyl-L-tyrosine, a substrate of carboxypeptidase A, has been obtained from x-ray studies of that ES complex.

2. The *physical properties* of an enzyme, such as its solubility or heat stability, frequently change upon formation of an ES complex.

3. The *spectroscopic characteristics* of many enzymes and substrates change upon formation of an ES complex just as the absorption spectrum of deoxyhemoglobin changes markedly when it binds oxygen or when it is oxidized to the ferric state, as described previously (Figure 3-17). These changes are particularly striking if the enzyme contains a colored prosthetic group. Tryptophan synthetase, a bacterial enzyme that contains a pyridoxal phosphate prosthetic group, affords a nice illustration. This enzyme catalyzes the synthesis of L-tryptophan from L-serine and indole. The addition of L-serine to the enzyme produces a marked increase in the fluorescence of the pyridoxal phosphate group (Figure 6-8). The subsequent addition of indole, the second substrate, quenches this fluorescence to a level lower than that of the enzyme alone. Thus, fluorescence spectroscopy reveals the existence of an enzyme-serine complex and of an enzyme-serine-indole complex. Other spectroscopic techniques such as nuclear and electron magnetic resonance are also highly informative about ES interactions.

4. A high degree of *stereospecificity* is displayed in the formation of ES complexes. For example, D-serine is not a substrate of tryptophan synthetase. Indeed, the D-isomer does not even bind to the enzyme. This implies that the substrate-binding site has a very well defined shape.

5. ES complexes can sometimes be *isolated in pure form*. For an enzyme that catalyzes the reaction A + B ⇌ C, it is sometimes possible to isolate an EA complex. This can be done if the enzyme has a sufficiently high affinity for A and if B is absent from the mixture.

6. At a constant concentration of enzyme, the reaction rate increases with increasing substrate concentration until a maximal velocity is reached (Figure 6-9). In contrast, uncatalyzed reactions do not show this saturation effect. In 1913, Leonor Michaelis interpreted the *maximal velocity of an enzyme-catalyzed reaction* in terms of the formation of a discrete ES complex. At a sufficiently high substrate concentration, the catalytic sites are filled and so the reaction rate reaches a maximum. This is the oldest and most general evidence for the existence of ES complexes.

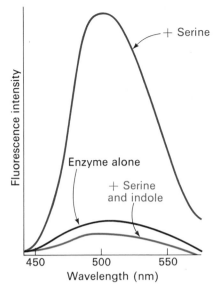

Figure 6-8
The fluorescence intensity of the pyridoxal phosphate group at the active site of tryptophan synthetase changes upon addition of serine and indole, the substrates.

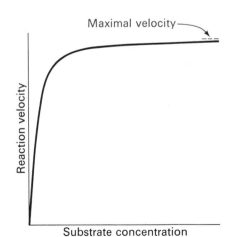

Figure 6-9
Velocity of an enzyme-catalyzed reaction as a function of the substrate concentration.

SOME FEATURES OF ACTIVE SITES

The active site of an enzyme is the region that binds the substrates (and the prosthetic group, if any) and contributes the residues that directly participate in the making and breaking of bonds. These residues are called the *catalytic groups*. Although enzymes differ widely in structure, specificity, and mode of catalysis, a number of generalizations concerning their active sites can be stated:

1. *The active site takes up a relatively small portion of the total volume of an enzyme.* Most of the amino acid residues in an enzyme are not in contact with the substrate. This raises the intriguing question of why enzymes are so big. Nearly all enzymes are made up of more than 100 amino acid residues, giving them a molecular weight greater than 10,000 and a diameter of more than 25 Å.

2. *The active site is a three-dimensional entity.* The active site of an enzyme is not a point, a line, or even a plane. It is an intricate three-dimensional form made up of groups that come from different parts of the linear amino acid sequence—indeed, residues far apart in the linear sequence may interact more strongly than adjacent residues in the amino acid sequence, as we have already seen for myoglobin and hemoglobin. In lysozyme, an enzyme that will be discussed in more detail in the next chapter, the important groups in the active site are contributed by residues numbered 35, 52, 62, 63, and 101 in the linear sequence of 129 amino acids.

3. *The specificity of binding depends on the precisely defined arrangement of atoms in an active site.* A substrate must have a matching shape to fit into the site. Emil Fischer's metaphor of the lock and key (Figure 6-10), stated in 1890, has proven to be an essentially correct and highly fruitful way of looking at the stereospecificity of catalysis. However, recent work suggests that the active sites of some enzymes are not rigid. In these enzymes, the shape of the active site is modified by the binding of substrate. The active site has a shape complementary to that of the substrate only *after* the substrate is bound. This process of dynamic recognition is called *induced fit* (Figure 6-11). Furthermore, some enzymes (such as lysozyme, which will be discussed in Chapter 7) preferentially bind a strained form of the substrate corresponding to the transition state. Thus, the

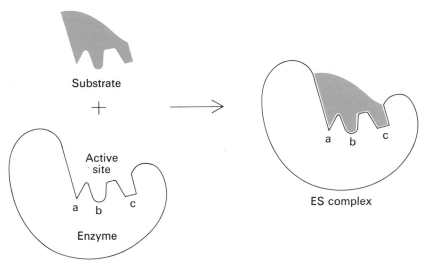

Figure 6-10
Lock-and-key model of the interaction of substrates and enzymes. The active site of the enzyme by itself is complementary in shape to that of the substrate.

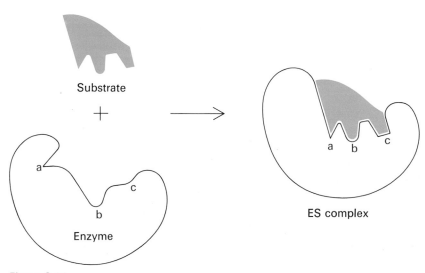

Figure 6-11
Induced-fit model of the interaction of substrates and enzymes. The enzyme changes shape upon binding substrate. The active site has a shape complementary to that of the substrate only *after* the substrate is bound.

flexibility of both enzymes and substrates may contribute to catalysis.

4. *Substrates are bound to enzymes by relatively weak forces.* ES complexes usually have equilibrium constants that range from 10^{-2} to 10^{-8} M, corresponding to free energies of interaction of -3 to -12 kcal/mol. These values should be compared with the strengths of covalent bonds, which are between -50 and -110 kcal/mol.

5. *Active sites are clefts or crevices.* In all enzymes of known structure, substrate molecules are bound to a cleft or crevice from which water is largely excluded. The cleft also contains several polar residues that are essential for binding and catalysis. The nonpolar character of the cleft enhances the binding of substrate. In addition, the cleft creates a microenvironment in which certain polar residues acquire special properties essential for their catalytic role.

THE MICHAELIS-MENTEN MODEL ACCOUNTS FOR THE KINETIC PROPERTIES OF MANY ENZYMES

For many enzymes, the rate of catalysis, V, varies with the substrate concentration, [S], in a manner shown in Figure 6-12. At a fixed concentration of enzyme, V is nearly linearly proportional to [S] when [S] is small. At high [S], V is nearly independent of [S]. In 1913, Leonor Michaelis and Maud Menten proposed a simple model to account for these kinetic characteristics. The critical feature in their treatment is that a specific ES complex is a necessary intermediate in catalysis. The model proposed, which is the simplest one that accounts for the kinetic properties of many enzymes, is:

$$E + S \underset{k_2}{\overset{k_1}{\rightleftharpoons}} ES \xrightarrow{k_3} E + P \qquad (1)$$

The enzyme E combines with S to form an ES complex, with a rate constant k_1. The ES complex has two possible fates. It can dissociate to E and S with a rate constant k_2, or it can proceed to form product P with a rate constant k_3. It is assumed that none of the product reverts to the initial substrate.

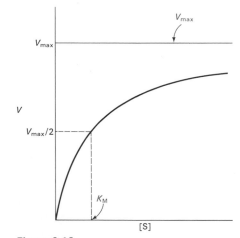

Figure 6-12
A plot of the reaction velocity, V, as a function of the substrate concentration, [S], for an enzyme that obeys Michaelis-Menten kinetics (V_{max} is the maximal velocity and K_M is the Michaelis constant).

We want an expression that relates the rate of catalysis to the concentrations of substrate and enzyme and the rates of the individual steps. The starting point is that the catalytic rate is equal to the product of the concentration of the ES complex and k_3:

$$V = k_3[\text{ES}] \tag{2}$$

Now we need to express ES in terms of known quantities. The rates of formation and breakdown of ES are given by

$$\text{Rate of formation of ES} = k_1[\text{E}][\text{S}] \tag{3}$$

$$\text{Rate of breakdown of ES} = (k_2 + k_3)[\text{ES}] \tag{4}$$

We are interested in the catalytic rate under steady-state conditions. A *steady-state* situation is one in which the concentrations of intermediates stay the same while the concentrations of starting materials and products are changing. This occurs when the rates of formation and breakdown of the ES complex are equal. On setting the right-hand sides of equations (3) and (4) equal,

$$k_1[\text{E}][\text{S}] = (k_2 + k_3)[\text{ES}] \tag{5}$$

By rearranging equation (5),

$$[\text{ES}] = \frac{[\text{E}][\text{S}]}{(k_2 + k_3)/k_1} \tag{6}$$

Equation (6) can be simplified by defining a new constant K_M called the *Michaelis constant*:

$$K_M = \frac{k_2 + k_3}{k_1} \tag{7}$$

and substituting it into equation (6), which then becomes

$$[\text{ES}] = \frac{[\text{E}][\text{S}]}{K_M} \tag{8}$$

Now let us examine the numerator of equation (8). The concentration of uncombined substrate, [S], is very nearly equal to the total substrate concentration, provided that the concentration of enzyme is much lower than that of the substrate. The concentration of uncombined enzyme, [E], is equal to the total enzyme concentration E_T minus the concentration of the ES complex.

$$[E] = [E_T] - [ES] \tag{9}$$

On substituting this expression for [E] in equation (8),

$$[ES] = ([E_T] - [ES])[S]/K_M \tag{10}$$

On rearranging equation (10),

$$[ES] = [E_T]\frac{[S]/K_M}{1 + [S]/K_M} \tag{11}$$

or

$$[ES] = [E_T]\frac{[S]}{[S] + K_M} \tag{12}$$

By substituting this expression for [ES] into equation (2), we get

$$V = k_3[E_T]\frac{[S]}{[S] + K_M} \tag{13}$$

The maximal rate V_{max} is attained when the enzyme sites are saturated with substrate—that is, when [S] is much greater than K_M—so that $[S]/([S] + K_M)$ approaches 1. Thus

$$V_{max} = k_3[E_T] \tag{14}$$

Substituting equation (14) into equation (13) yields the Michaelis-Menten equation.

$$V = V_{max}\frac{[S]}{[S] + K_M} \tag{15}$$

This equation accounts for the kinetic data given in Figure 6-12. At low substrate concentration, when [S] is much less than K_M, $V = [S]V_{max}/K_M$; that is, the rate is directly proportional to the substrate concentration. At high substrate concentration, when [S] is much greater than K_M, $V = V_{max}$; that is, the rate is maximal, independent of substrate concentration.

The meaning of K_M is evident from equation (15). When $[S] = K_M$, then $V = V_{max}/2$. Thus, K_M *is equal to the substrate concentration at which the reaction rate is half of its maximal value.*

V_{MAX} AND K_M CAN BE DETERMINED BY VARYING THE SUBSTRATE CONCENTRATION

The Michaelis constant K_M and maximal rate V_{max} can readily be derived from rates of catalysis at different substrate concentrations, if an enzyme operates according to the simple scheme given in equation (1). It is convenient to transform the Michaelis-Menten equation into one that gives a straight line plot. This can be done by taking the reciprocal of both sides of equation (15) to give

$$\frac{1}{V} = \frac{1}{V_{max}} + \frac{K_M}{V_{max}} \cdot \frac{1}{[S]} \qquad (16)$$

A plot of $1/V$ versus $1/[S]$ yields a straight line with an intercept of $1/V_{max}$ and a slope of K_M/V_{max} (Figure 6-13).

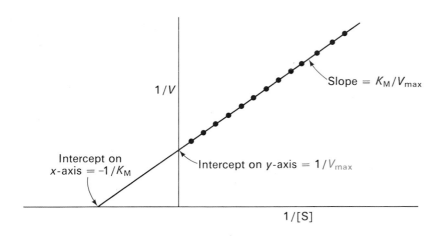

Figure 6-13
A double-reciprocal plot of enzyme kinetics: $1/V$ is plotted as a function of $1/[S]$. The slope is K_M/V_{max}, the intercept on the y-axis is $1/V_{max}$, and the intercept on the x-axis is $-1/K_M$.

SIGNIFICANCE OF K_M AND V_{MAX} VALUES

The K_M values of enzymes range widely (Table 6-1). For most enzymes, K_M lies between 10^{-1} and 10^{-6} M. The K_M value for an enzyme depends on the particular substrate and also on environmental conditions such as the temperature and ionic strength. The Michaelis constant, K_M, has two meanings. First, K_M is the concentration of substrate at which half the active sites are filled. Once

Table 6-1

K_M values of some enzymes

Enzyme	Substrate	K_M
Chymotrypsin	Acetyl-L-tryptophanamide	5×10^{-3} M
Lysozyme	Hexa-N-acetylglucosamine	6×10^{-6} M
β-Galactosidase	Lactose	4×10^{-3} M
Threonine deaminase	Threonine	5×10^{-3} M
Carbonic anhydrase	CO_2	8×10^{-3} M
Penicillinase	Benzylpenicillin	5×10^{-5} M
Pyruvate carboxylase	Pyruvate	4×10^{-4} M
	HCO_3^-	1×10^{-3} M
	ATP	6×10^{-5} M
Arginine-tRNA synthetase	Arginine	3×10^{-6} M
	tRNA	4×10^{-7} M
	ATP	3×10^{-4} M

the K_M is known, the fraction of sites filled, f_{ES}, at any substrate concentration can be calculated from

$$f_{ES} = \frac{V}{V_{max}} = \frac{[S]}{[S] + K_M} \tag{17}$$

Second, K_M is related to the rate constants of the individual steps in the catalytic scheme given in equation (1). In equation (7), K_M is defined as $(k_2 + k_3)/k_1$. Consider a limiting case in which k_2 is much greater than k_3. This means that dissociation of the ES complex to E and S is much more rapid than formation of E and product. Under these conditions,

$$K_M = \frac{k_2}{k_1} \tag{18}$$

when $k_2 \gg k_3$. The dissociation constant of the ES complex is given by

$$K_{ES} = \frac{[E][S]}{[ES]} = \frac{k_2}{k_1} \tag{19}$$

In other words, K_M *is equal to the dissociation constant of the ES complex if* k_3 *is much smaller than* k_2. When this condition is met, K_M is a measure of the strength of the ES complex: a high K_M indicates weak binding, a low K_M indicates strong binding. It must be stressed that K_M indicates the affinity of the ES complex only when k_2 is much greater than k_3. Experimental data that this is in fact the case are needed before K_M can be equated with K_{ES}.

The maximal rate, V_{max}, reveals the turnover number of an enzyme, if the concentration of active sites $[E_T]$ is known, since

$$V_{max} = k_3[E_T] \qquad (20)$$

For example, a 10^{-6} M solution of carbonic anhydrase catalyzes the formation of 0.6 M H_2CO_3 per second when it is fully saturated with substrate. Hence, k_3 is 6×10^5 sec^{-1}. The kinetic constant k_3 is called the *turnover number*. The turnover number of an enzyme is *the number of substrate molecules converted into product per unit time, when the enzyme is fully saturated with substrate*. The turnover number of 600,000 sec^{-1} for carbonic anhydrase is one of the largest known. Each round of catalysis occurs in a time equal to $1/k_3$, which is 1.7 microseconds for carbonic anhydrase. The turnover numbers of most enzymes for their physiological substrates fall in the range of 1 to 10^4 per second (Table 6-2).

ENZYMES CAN BE INHIBITED BY SPECIFIC MOLECULES

The inhibition of enzymatic activity by specific small molecules and ions is important because it serves as a major control mechanism in biological systems. Also, many drugs and toxic agents act by inhibiting enzymes. Furthermore, enzyme inhibition can provide insight into the mechanism of enzyme action. Enzyme inhibition can be either a reversible or an irreversible process. In *irreversible inhibition*, the inhibitor is covalently linked to the enzyme or bound so tightly that its dissociation from the enzyme is very slow. The action of nerve gas poisons on acetylcholinesterase, an enzyme that plays an important role in the transmission of nerve impulses, is an example of irreversible inhibition. Diisopropylphosphofluoridate (DIPF), one of these agents, reacts with a critical serine residue at the active site on the enzyme to form an inactive diisopropylphosphoryl enzyme (Figure 6-14). Alkylating reagents, such as iodoacetamide, may irreversibly inhibit enzymatic activity by modifying cysteine and other side chains (Figure 6-15).

Table 6-2

Maximum turnover numbers of some enzymes

Enzyme	Turnover number (per second)
Carbonic anhydrase	600,000
Acetylcholinesterase	25,000
Penicillinase	2,000
Lactate dehydrogenase	1,000
Chymotrypsin	100
DNA polymerase I	15
Tryptophan synthetase	2
Lysozyme	0.5

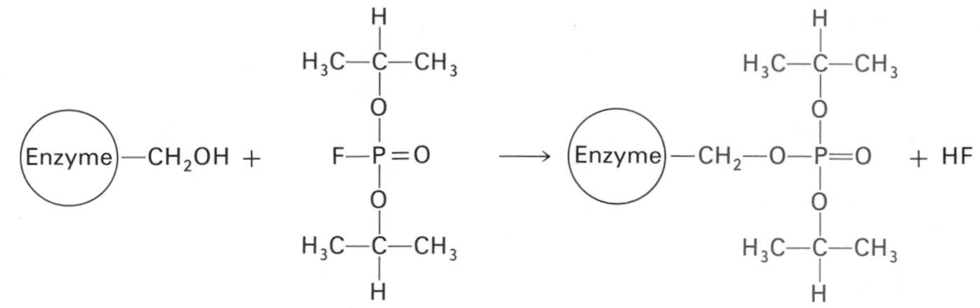

Diisopropylphospho-fluoridate (DIPF)

Figure 6-14
Inactivation of chymotrypsin by diisopropylphosphofluoridate (DIPF).

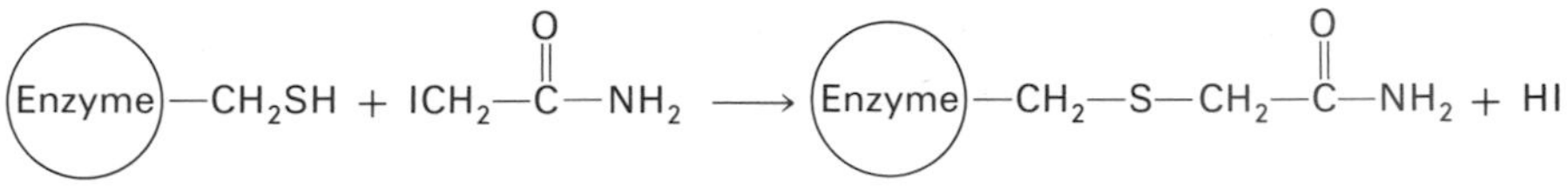

Iodoacetamide

Figure 6-15
Inactivation of an enzyme with a critical cysteine residue by iodoacetamide.

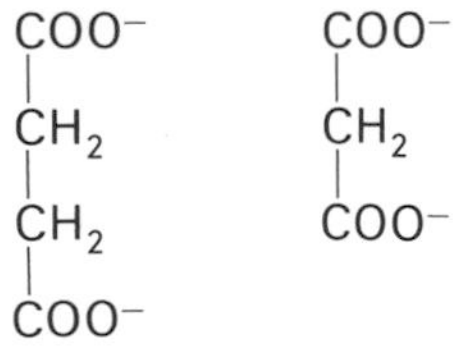

Succinate **Malonate**

In contrast, *reversible inhibition* is characterized by a rapid equilibrium of the inhibitor and enzyme. The simplest type of reversible inhibition is competitive inhibition. A *competitive inhibitor* resembles the substrate and binds to the active site of the enzyme (Figure 6-16). The substrate is then prevented from binding to the same active site. In other words, the binding of substrate and a competitive inhibitor are mutually exclusive events. *A competitive inhibitor diminishes the rate of catalysis by reducing the proportion of enzyme molecules that have a bound substrate.* A classic example of competitive inhibition is the action of malonate on succinate dehydrogenase, an enzyme that dehydrogenates succinate. Malonate differs from succinate in having one rather than two methylene groups. A physiologically important example of competitive inhibition is found in the formation of 2,3-diphosphoglycerate from 1,3-diphosphoglycerate. Di-

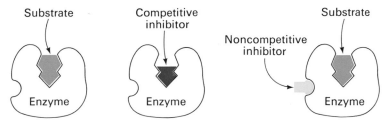

Figure 6-16
Distinction between a *competitive* inhibitor and a *noncompetitive* inhibitor: (left) enzyme-substrate complex; (middle) a competitive inhibitor prevents the substrate from binding; and (right) a noncompetitive inhibitor does not prevent the substrate from binding.

phosphoglycerate mutase, the enzyme catalyzing this isomerization, is competitively inhibited by low levels of 2,3-diphosphoglycerate. In fact, it is not uncommon for the product of an enzymatic reaction to be a competitive inhibitor of the substrate because of their structural resemblance.

$$
\begin{array}{ccc}
\begin{array}{c}
\text{O} \\
\| \\
\text{C}-\text{OPO}_3{}^{2-} \\
| \\
\text{H}-\text{C}-\text{OH} \\
| \\
\text{H}_2\text{C}-\text{OPO}_3{}^{2-}
\end{array}
&
\rightleftharpoons
&
\begin{array}{c}
\text{O} \\
\| \\
\text{C}-\text{O}^- \\
| \\
\text{H}-\text{C}-\text{OPO}_3{}^{2-} \\
| \\
\text{H}_2\text{C}-\text{OPO}_3{}^{2-}
\end{array}
\\
\textbf{1,3-Diphosphoglycerate} & & \textbf{2,3-Diphosphoglycerate}
\end{array}
$$

In *noncompetitive inhibition*, which is also reversible, the inhibitor and substrate can bind simultaneously to an enzyme molecule. This means that their binding sites do not overlap. A noncompetitive inhibitor acts by decreasing the turnover number of an enzyme rather than by diminishing the proportion of enzyme molecules that have a bound substrate. More complex patterns of inhibition occur when the inhibitor affects both the binding of substrate and the turnover number of the enzyme.

Enzyme activity may also be inhibited by interactions between sites on different subunits of an oligomeric enzyme. This kind of inhibition, which is called *allosteric inhibition*, is very important physiologically. It will be discussed shortly.

COMPETITIVE AND NONCOMPETITIVE INHIBITION
ARE KINETICALLY DISTINGUISHABLE

Measurements of the rates of catalysis at different concentrations of substrate and inhibitor serve to distinguish between competitive and noncompetitive inhibition. In *competitive inhibition*, the intercept of the plot of $1/V$ versus $1/[S]$ is the same in the presence and absence of inhibitor, although the slope is different (Figure 6-17).

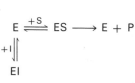

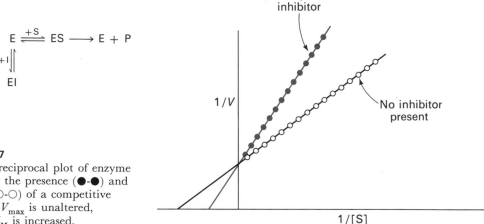

Figure 6-17
A double-reciprocal plot of enzyme kinetics in the presence (●-●) and absence (○-○) of a competitive inhibitor; V_{max} is unaltered, whereas K_M is increased.

This reflects the fact that V_{max} is not altered by a competitive inhibitor. *The hallmark of competitive inhibition is that the inhibition can be overcome at a sufficiently high substrate concentration.* Substrate and inhibitor compete for the same site. At a sufficiently high substrate concentration, virtually all the active sites are filled by substrate, and the enzyme is fully operative. The increase in the slope of the $1/V$ versus $1/[S]$ plot indicates the strength of binding of competitive inhibitor. In the presence of a competitive inhibitor, equation (16) is replaced by

$$\frac{1}{V} = \frac{1}{V_{max}} + \frac{K_M}{V_{max}}\left(1 + \frac{[I]}{K_i}\right)\left(\frac{1}{S}\right) \qquad (21)$$

where [I] is the concentration of inhibitor and K_i is the dissociation constant of the enzyme-inhibitor complex

$$E + I \rightleftharpoons EI$$

$$K_i = \frac{[E][I]}{[EI]} \tag{22}$$

In other words, the slope of the plot is increased by the factor $(1 + [I]/K_i)$ in the presence of a competitive inhibitor. Consider an enzyme with a K_M of 10^{-4} M. In the absence of inhibitor, $V = V_{max}/2$ when $[S] = 10^{-4}$ M. In the presence of 2×10^{-3} M competitive inhibitor that is bound to the enzyme with a K_i of 10^{-3} M, the apparent K_M will be 3×10^{-4} M. Consequently, $V = V_{max}/4$.

In *noncompetitive inhibition* (Figure 6-18), V_{max} is reduced and so the intercept on the *y*-axis is increased. The slope, which is equal to K_M/V^I_{max}, is increased by the same factor. In contrast to V_{max}, K_M is not affected by this kind of inhibition. *Noncompetitive inhibition cannot be overcome by increasing the substrate concentration.* The maximal velocity in the presence of a noncompetitive inhibitor, V^I_{max}, is given by

$$V^I_{max} = V_{max}\Big/\left(1 + \frac{[I]}{K_i}\right) \tag{23}$$

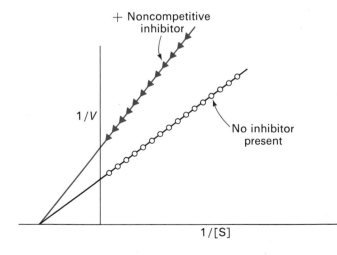

Figure 6-18
A double-reciprocal plot of enzyme kinetics in the presence (▲-▲-▲) and absence (○-○-○) of a noncompetitive inhibitor; K_M is unaltered by the noncompetitive inhibitor, whereas V_{max} is reduced.

TREATMENT OF ETHYLENE GLYCOL POISONING
BY COMPETITIVE INHIBITION

About fifty deaths occur annually from the ingestion of ethylene glycol, a constituent of permanent-type automobile antifreeze. Ethylene glycol itself is not lethally toxic. Rather, the harm is done by oxalic acid, an oxidation product of ethylene glycol. The first step in this conversion is the oxidation of ethylene glycol by alcohol dehydrogenase (Figure 6-19). This reaction can be effectively inhibited by the administration of a nearly intoxicating dose of ethanol. The basis for this effect is that *ethanol acts as a competitive inhibitor of oxidation of ethylene glycol to aldehyde products*. The ethylene glycol is then excreted harmlessly. The same rationale is used in the therapy of methanol poisoning.

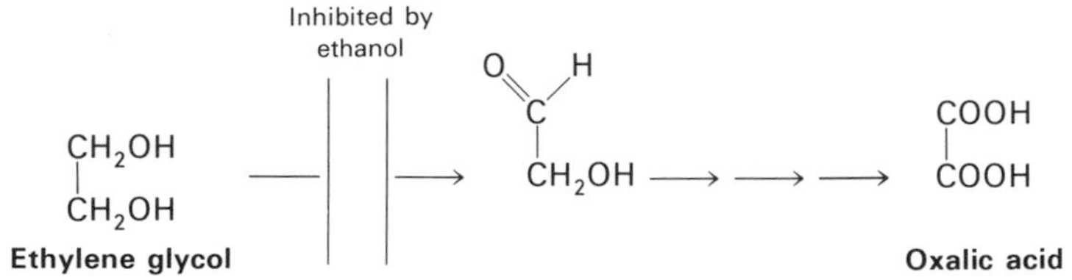

Figure 6-19
Formation of oxalic acid from ethylene glycol.

ALLOSTERIC ENZYMES DO NOT OBEY
MICHAELIS-MENTEN KINETICS

The Michaelis-Menten model has had a great impact on the development of enzyme chemistry. Its virtues are its simplicity and broad applicability. However, the kinetic properties of many enzymes cannot be accounted for by the Michaelis-Menten model. An important group consists of the *allosteric enzymes*, which often display sigmoidal plots of the reaction velocity, V, versus substrate concentration, [S], rather than the hyperbolic plots predicted by the Michaelis-Menten equation [eq. (15)]. Recall that the oxygen-binding curve of myoglobin is hyperbolic, whereas that of hemoglobin is sigmoidal. The situation with enzymes is analogous. In allosteric enzymes, one active site in an enzyme molecule can affect

another active site in the same enzyme molecule. A possible outcome of this interaction across subunits is that the binding of
substrate becomes cooperative, which would give a sigmoidal plot
of V versus S. In addition, the activity of allosteric enzymes may
be altered by regulatory molecules that are bound to sites other
than the catalytic sites, just as oxygen binding in hemoglobin is
affected by DPG, H^+, and CO_2.

THE CONCERTED MODEL FOR ALLOSTERIC INTERACTIONS

An elegant and incisive model for allosteric enzymes was proposed
in 1965 by Jacques Monod, Jeffries Wyman, and Jean-Pierre
Changeux. Let us apply their approach to an allosteric enzyme
made up of two identical subunits, each with one active site.
Suppose that a subunit can exist in either of two conformations,
called R and T, and that R has a high affinity for substrate, whereas
T has a low affinity (Figure 6-20). Forms R and T are interconvertible. *An important assumption of this model is that both subunits must be
in the same conformational state, so that the symmetry of the dimer is
conserved.* Thus, RR and TT are allowed conformations, but RT
is not permitted. In the absence of substrate, the two allowed states
are symbolized as T_0 and R_0, and L is the ratio of their concentrations.

T form
(Low affinity for substrate)

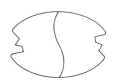

R form
(High affinity for substrate)

Figure 6-20
Schematic representation of
the R and T forms of
an allosteric enzyme.

$$R_0 \rightleftharpoons T_0 \tag{24}$$

$$L = T_0/R_0 \tag{25}$$

For simplicity, let us assume that the substrate does not bind to
the T state. The R state of the dimer can bind one or two substrate
molecules; these species are denoted by R_1 and R_2, respectively.

$$R_0 + S \rightleftharpoons R_1 \tag{26}$$

$$R_1 + S \rightleftharpoons R_2 \tag{27}$$

$$K_R = \frac{2[R_0][S]}{[R_1]} = \frac{[R_1][S]}{2[R_2]} \tag{28}$$

In this model, the microscopic dissociation constant K_R is the same
for the binding of the first and second substrate molecules to the
R form of the dimeric enzyme molecule.

We want an expression that gives us the *fractional saturation Y,* which is the fraction of active sites that have a bound substrate, as a function of the substrate concentration.

$$Y = \frac{[\text{occupied sites}]}{[\text{total sites}]} = \frac{[R_1] + 2[R_2]}{2([T_0] + [R_0] + [R_1] + [R_2])} \quad (29)$$

Substituting equations (24)–(28) into equation (29) gives the derived expression for Y.

$$Y = \left(\frac{[S]}{K_R}\right)\frac{1 + [S]/K_R}{L + (1 + [S]/K_R)^2} \quad (30)$$

Let us plot equation (30) with $K_R = 10^{-5}$ M and $L = 10^4$. Such a plot of Y versus [S] is sigmoidal rather than hyperbolic (see Figure 6-23 on the facing page). In other words, *the binding of substrate is cooperative.* If the turnover number per active site is the same for ES complexes in R_1 and R_2, then the plot of reaction velocity versus substrate concentration will also be sigmoidal, since

$$V = YV_{max} \quad (31)$$

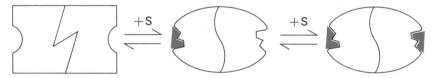

Figure 6-21
Concerted model for the cooperative binding of substrate in an allosteric enzyme. The low affinity form, TT, switches to the high affinity form, RR, upon binding the first substrate molecule.

Let us look at this binding process (Figure 6-21). In the absence of substrate, nearly all the enzyme molecules are in the T form. Specifically, there is only one molecule in the R form for every 10,000 in the T form, in the above example. The addition of substrate shifts this conformational equilibrium in the direction of the R form, since substrate binds only to the R form. When substrate binds to one site, the other site on the same enzyme molecule must also be in the R form, according to the basic postulate of this model. In other words, the transition from T to R or vice versa is *concerted*. Hence, *the proportion of enzyme molecules in the R form increases progressively as more substrate is added, and so the binding of substrate is cooperative.* When the active sites are fully saturated, all of the enzyme molecules are in the R form.

The effects of allosteric activators and inhibitors can readily be accounted for by this concerted model. An allosteric inhibitor binds preferentially to the T form, whereas an allosteric activator binds preferentially to the R form (Figure 6-22). Consequently, *an allosteric inhibitor shifts the* $R \rightleftharpoons T$ *conformational equilibrium toward T, whereas an allosteric activator shifts it toward R.* These effects can be expressed quantitatively by a change in the allosteric equilibrium constant L, which is a variable in equation (30). An allosteric inhibitor increases L, whereas an allosteric activator decreases L. These effects are shown in Figure 6-23, in which Y is plotted versus S for these values of L: 10^3 (activator present), 10^4 (no activator, no inhibitor), and 10^5 (inhibitor present). The fractional saturation Y at all values of S is reduced by the presence of the inhibitor and increased by the presence of activator.

Allosteric
inhibitor

Allosteric
activator

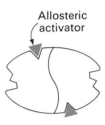

Figure 6-22
In the concerted model, an allosteric inhibitor (represented by a hexagon) stabilizes the T state, whereas an allosteric activator (represented by a triangle) stabilizes the R state.

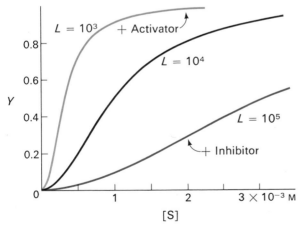

Figure 6-23
Saturation, Y, as a function of substrate concentration, [S], according to the concerted model [equation (30)]. The effects of an allosteric activator and inhibitor are also shown.

It is useful to define two terms at this point. *Homotropic* effects refer to allosteric interactions between identical ligands, whereas *heterotropic* effects refer to interactions between different ligands. In the example cited above, the cooperative binding of substrate to the enzyme is a homotropic effect. In contrast, the effect of an activator or inhibitor on the binding of substrate is heterotropic because it involves interactions between different kinds of molecules. In this concerted model for allosteric interactions, homotropic effects are necessarily positive (cooperative), whereas heterotropic effects can be either positive or negative.

THE SEQUENTIAL MODEL FOR ALLOSTERIC INTERACTIONS

Allosteric interactions can also be accounted for by the *sequential model*, which has been developed by Daniel Koshland, Jr. The simplest form of this model makes three assumptions:

1. There are only two conformational states (R and T) accessible to any one subunit.

2. The binding of substrate changes the shape of the subunit to which it is bound. However, the conformation of the other subunits in this enzyme molecule are not appreciably altered.

3. The conformational change elicited by the binding of substrate in one subunit can increase or decrease the substrate binding affinity of the other subunits in the same enzyme molecule.

The sequential model pictures the binding process in an allosteric enzyme to occur as shown in Figure 6-24. The binding is cooperative if the affinity of RT for substrate is greater than that of TT.

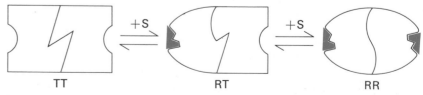

Figure 6-24
Sequential model for the cooperative binding of substrate in an allosteric enzyme. The empty active site in RT has a higher affinity for substrate than the sites in TT.

The simple sequential model differs from the concerted model in several ways. First, an equilibrium between the R and T forms in the absence of substrate is not assumed. Rather, the conformational transition from T to R is *induced* by the binding of substrate. Second, the conformational change from T to R in different subunits of an enzyme molecule is sequential, not concerted. The hybrid species RT is prominent in the sequential model but excluded in the concerted model. The concerted model supposes that symmetry is essential for the interaction of subunits in oligomeric proteins and therefore requires that it be conserved in allosteric transitions. In contrast, the sequential model assumes that subunits can interact even if they are in different conformational states. Finally, these models differ in that homotropic interactions are necessarily positive in the concerted model but can be either positive or negative in the sequential model. In the sequential model, the second substrate molecule can be bound more or less tightly than the first, depending on the nature of the distortion induced by the binding of the first substrate molecule.

Which model is correct? For some allosteric proteins, the concerted model fits well, whereas for others the sequential model appears applicable. However, neither model is satisfactory for yet another group of allosteric proteins. For these proteins, the assumption that there are only two significant conformational states (R and T) appears to be too restrictive. More complex models are needed to account for the allosteric properties of these proteins.

ELECTROSTATIC, HYDROGEN, AND VAN DER WAALS BONDS IN ENZYME-SUBSTRATE COMPLEXES

Reversible molecular interactions in biological systems are mediated by three different kinds of forces. The folding of macromolecules, the binding of substrates to enzymes, and the interactions of cells—indeed, all molecular interactions in biological systems—involve the interplay of *electrostatic bonds, hydrogen bonds,* and *van der Waals bonds.* These three fundamental noncovalent bonds differ in their geometrical requirements, strength, and specificity. Furthermore, they are affected in different ways by the presence of water, which has a profound influence. Let us consider the characteristics of each of these three basic bonds.

CHARGED SUBSTRATES CAN BIND TO
OPPOSITELY CHARGED GROUPS ON ENZYMES

A charged group on a substrate can interact with an oppositely charged group on an enzyme. The force of such an *electrostatic interaction* is given by Coulomb's law:

$$F = \frac{q_1 q_2}{r^2 D}$$

where q_1 and q_2 are the charges of the two groups, r is the distance between them, and D is the dielectic constant of the medium. An electrostatic interaction is strongest in a vacuum (where D is 1) and is weakest in a medium such as water (where D is 80).

The binding of glycyl-L-tyrosine to carboxypeptidase A, a proteolytic enzyme that cleaves carboxyl-terminal residues, provides an example of an electrostatic interaction. The negatively charged terminal carboxylate group of the dipeptide substrate interacts with the positively charged guanidinium group of an arginine residue of the enzyme. The distance between these oppositely charged groups is 2.8 Å.

Substrate   **Arginine side chain
of enzyme**

This kind of interaction is also called an ionic bond, salt linkage, salt bridge, or ion pair. These terms all have the same meaning: an electrostatic interaction between oppositely charged groups. A negatively charged substrate can form an electrostatic bond with the positively charged side chain of a lysine or arginine residue. The imidazole group of a histidine residue and the terminal amino group are also potential binding sites for a negatively charged substrate, if their pKs render them positively charged at the pH of the medium. For a positively charged substrate, the potential binding sites on the enzyme are the negatively charged carboxylate groups of aspartate and glutamate and the terminal carboxylate of the polypeptide chain.

SUBSTRATES BIND TO ENZYMES BY PRECISELY DIRECTED HYDROGEN BONDS

Many substrates are uncharged, yet they bind to enzymes with high affinity and specificity. The significant interactions for these substrates and indeed also for most charged substrates are hydrogen bonds. *In a hydrogen bond, a hydrogen atom is shared by two other atoms.* The atom to which the hydrogen is more tightly linked is called the hydrogen donor, whereas the other atom is the hydrogen acceptor. In fact, a hydrogen bond can be considered an intermediate in the transfer of a proton from an acid to a base. The acceptor atom has a partial negative charge that attracts the hydrogen atom. In short, the hydrogen bond is reminiscent of a ménage à trois.

Table 6-3
Typical hydrogen-bond distances

Bond	Distance (Å)
O—H···O	2.70
O—H···O⁻	2.63
O—H···N	2.88
N—H···O	3.04
N⁺—H···O	2.93
N—H···N	3.10

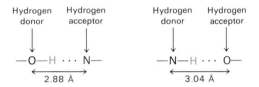

The donor atom in a hydrogen bond in biological systems is an oxygen or nitrogen atom that has a covalently attached hydrogen atom. The acceptor atom is either oxygen or nitrogen. The kinds of hydrogen bonds formed and their bond lengths are given in Table 6-3. The bond energies range from about 3 to 7 kcal/mol. Hydrogen bonds are stronger than van der Waals bonds but much weaker than covalent bonds. The length of a hydrogen bond is intermediate between that of a covalent bond and a van der Waals bond. *An important feature of hydrogen bonds is that they are highly directional.* The strongest hydrogen bond is one in which the donor, hydrogen, and acceptor atoms are colinear. If the acceptor atom is at an angle to the line joining the donor and hydrogen atoms, the bond is much weaker.

We have already encountered hydrogen bonds in our discussion of the structure of myoglobin and hemoglobin. In the α helix, the peptide —NH and —CO groups are hydrogen-bonded to each other. The nitrogen atom is the hydrogen donor, whereas the oxygen atom is the hydrogen acceptor. The distance between the nitrogen and oxygen atoms is 2.9 Å. The hydrogen atom is closer to the nitrogen than to the oxygen atom by 0.9 Å.

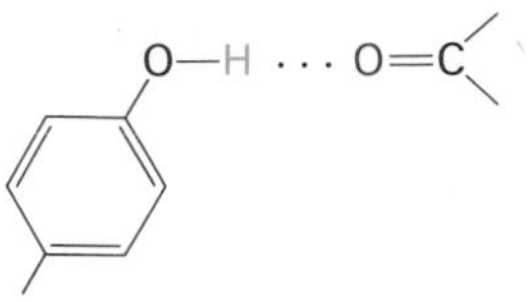

Another example of a hydrogen bond in myoglobin and hemoglobin is the link between the hydroxyl group of tyrosine HC2 and the peptide carbonyl of FG4. The oxygen atom of the hydroxyl group of tyrosine is the hydrogen donor, whereas the oxygen atom of the peptide carbonyl is the hydrogen acceptor.

The role of hydrogen bonding in the interaction of substrates with enzymes is nicely illustrated by the binding of the uridine portion of the substrate to pancreatic ribonuclease, an enzyme that cleaves ribonucleic acid (Figure 6-25). Three hydrogen bonds are involved:

1. One of the C=O groups of the uridine ring is hydrogen-bonded to a peptide N—H.

2. The N—H group of the uridine ring is hydrogen-bonded to the —OH of a threonine residue.

3. The other ring C=O is hydrogen-bonded to the —OH group of a serine residue.

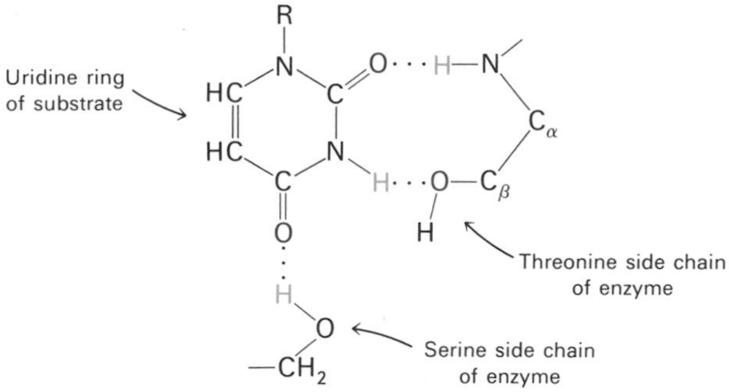

Figure 6-25
Hydrogen-bond interactions in the binding of a substrate to ribonuclease. [Based on F. M. Richards, H. W. Wyckoff, and N. Allewell. In *The Neurosciences: Second Study Program,* F. O. Schmitt, ed. (Rockfeller University Press, 1970), p. 970.]

PROTEINS ARE RICH IN HYDROGEN-BONDING POTENTIALITY

Amino acid side chains and the peptide main chain can form a variety of different kinds of hydrogen bonds. In fact, eleven of the twenty fundamental amino acids can form hydrogen bonds through their side chains. It is convenient to group these residues according to the kinds of hydrogen bonds they can form.

1. The side chains of tryptophan and arginine can serve as *hydrogen-bond donors only*.

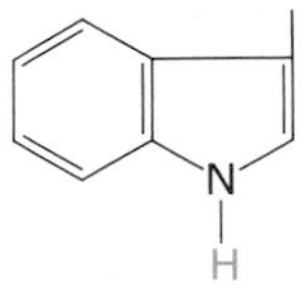

**Hydrogen donor group
of tryptophan**

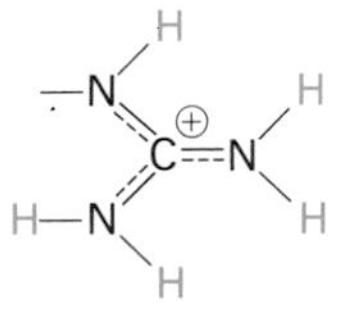

**Hydrogen donor groups
of arginine**

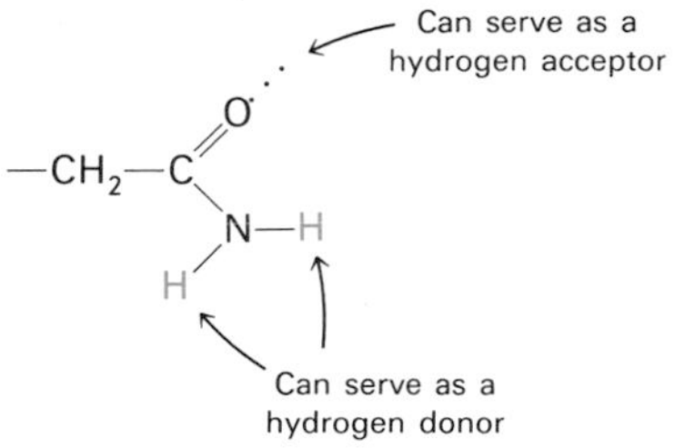

2. The side chains of asparagine, glutamine, serine, and threonine can serve as both *hydrogen-bond donors and acceptors*, as can the peptide group.

3. The hydrogen-bonding capabilities of lysine (and the terminal amino group), aspartic and glutamic acid (and the terminal carboxyl group), tyrosine, and histidine vary with pH. These groups can serve as both hydrogen-bond acceptors and donors over a certain range of pH, and as acceptors or donors (but not both) at other pH values, as shown for aspartate and glutamate in Figure 6-26. *The hydrogen-bonding modes of these ionizable residues are pH dependent.*

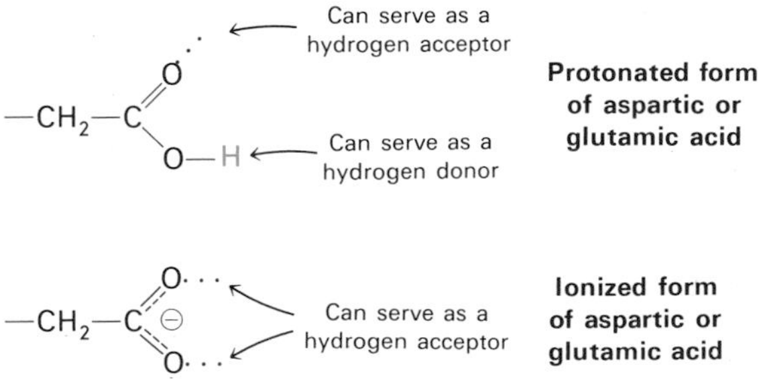

Can serve as a
hydrogen acceptor

Can serve as a
hydrogen donor

Asparagine or glutamine

Can serve as a
hydrogen acceptor

Can serve as a
hydrogen donor

**Protonated form
of aspartic or
glutamic acid**

Can serve as a
hydrogen acceptor

**Ionized form
of aspartic or
glutamic acid**

Figure 6-26
The hydrogen-bonding potentialities of aspartate and glutamate.

VAN DER WAALS INTERACTIONS ARE IMPORTANT WHEN THERE IS STERIC COMPLEMENTARITY

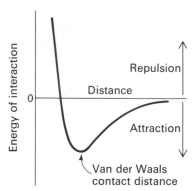

Figure 6-27
Energy of a van der Waals interaction as a function of the distance between two atoms.

Table 6-4
Van der Waals contact radii of atoms

Atom	Radius (Å)
H	1.2
C	2.0
N	1.5
O	1.4
S	1.85
P	1.9

There is a nonspecific attractive force between any two atoms when they are about 3 to 4 Å apart. This interaction, called a van der Waals bond, is weaker and less specific than electrostatic and hydrogen bonds but no less important in biology. The basis of the van der Waals bond is that the distribution of electronic charge around an atom changes with time. At any instant, the charge distribution is not perfectly symmetric. This transient asymmetry in the electronic charge around an atom will alter the electron distribution around its neighboring atoms. The attraction between the pair of atoms increases as they come closer, until they are separated by the van der Waals *contact distance* (Figure 6-27). At a shorter distance, very strong repulsive forces become dominant because the outer electron clouds overlap. The contact distance between an oxygen and carbon atom, for example, is 3.4 Å, which is obtained by adding 1.4 and 2.0 Å, the contact radii of the O and C atoms, respectively.

The van der Waals bond energy for a pair of atoms is about 1 kcal/mol. This is considerably weaker than a hydrogen or electrostatic bond, which is in the 3 to 7 kcal/mol range. Thus, a single van der Waals bond counts for very little. Its strength is only a little more than the average thermal energy of molecules at room temperature (0.6 kcal/mol). Van der Waals forces become significant in binding only when numerous substrate atoms can simultaneously come close to numerous enzyme atoms. The van der Waals force fades rapidly when the distance between a pair of atoms is even 1 Å greater than their contact distance. Numerous atoms of a substrate can interact with many atoms of an enzyme only if their shapes match. In other words, an effective van der Waals interaction between a substrate and an enzyme can occur only if they are *sterically complementary*. Thus, though there is virtually no specificity in a single van der Waals interaction, *specificity arises when there is an opportunity to make a large number of van der Waals bonds simultaneously.* Repulsions between atoms closer than the van der Waals contact distance are as important as the attractive forces in the generation of specificity.

THE BIOLOGICALLY IMPORTANT PROPERTIES OF WATER ARE ITS POLARITY AND COHESIVENESS

Thus far, we have not considered the effect of water on the three basic kinds of bonds. In fact, water is an active participant in molecular interactions in biology. Two properties of water are especially important in this regard:

1. *Water is a polar molecule.* The shape of the molecule is triangular, not linear, and so there is an asymmetrical distribution of charge. The oxygen nucleus draws electrons away from the hydrogen nuclei, thus leaving the region around them with a net positive charge. If a tetrahedron is described about the oxygen atom, with the hydrogen nuclei at two of the corners, then the other two corners are electronegative. The water molecule is thus an electrically polar structure.

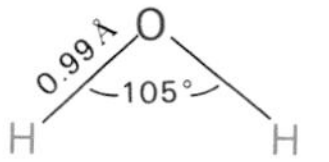

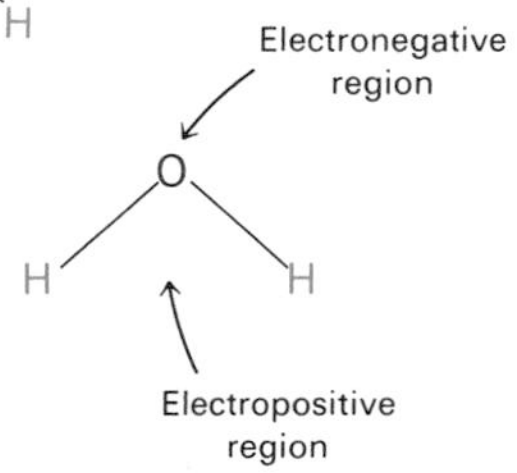

2. *Water molecules have a high affinity for each other.* In a group of water molecules clustered together, a positively charged region in one molecule tends to orient itself toward a negatively charged region in one of its neighbors. Each of its two electronegative regions attracts a proton of a neighbor molecule. Each of its own protons attracts the oxygen end of a neighbor. Thus, each oxygen molecule is the center of a tetrahedron of other oxygens, the O—O distance being 2.76 Å.

The structure of a form of ice is shown in Figure 6-28. When ice melts, its highly regular crystalline structure breaks in many places. However, only about 15% of the bonds are broken. In liquid water, each H_2O is hydrogen-bonded to 3.4 neighbors on the average. Thus, liquid water has a partially ordered structure. The hydrogen-bonded aggregates are constantly forming and breaking up. The striking feature of water is that it can *bond with itself.* Indeed, the number of protons that form the positive ends of the hydrogen bonds around any given oxygen atom is equal to the number of unshared electron pairs that form the negative ends. The tetrahedral arrangement of these bonds gives rise to extensive three-dimensional structures. *Water is highly cohesive.*

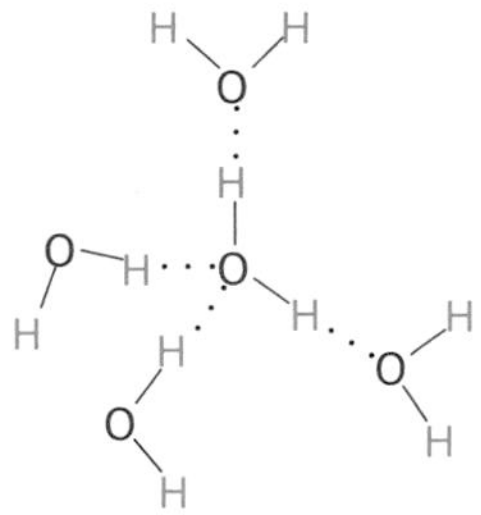

Figure 6-28
Structure of a form of ice.

In a nonpolar environment

In water

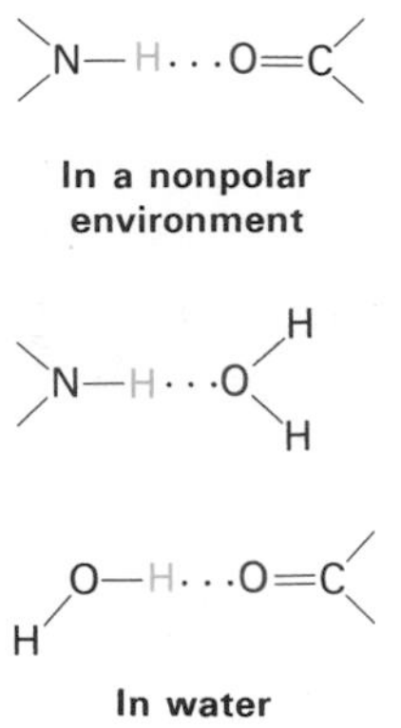

Figure 6-29
Water competes for hydrogen bonds.

Table 6-5
Dielectric constants of some solvents

Substance	Dielectric constant ($20°C$)
Hexane	1.9
Benzene	2.3
Diethyl ether	4.3
Chloroform	5.1
Acetone	21.4
Ethanol	24
Methanol	33
Water	80
Hydrogen cyanide	116

WATER WEAKENS POLAR INTERACTIONS

The polarity and hydrogen-bonding capabilites of water make it a highly interacting molecule. The consequence is that water weakens electrostatic and hydrogen-bond interactions between other molecules. Water is a very effective competitor in these polar interactions. Consider the action of water on the hydrogen bonding of a carbonyl and amide group (Figure 6-29). Water can replace the NH group as a hydrogen-bond donor. Furthermore, the oxygen atom of water can replace the carbonyl oxygen in the other hydrogen bond. Thus, a strong hydrogen-bond interaction between a CO and NH group occurs only if water is excluded.

Water diminishes the strength of electrostatic interactions by a factor of 80, the dielectric constant of water, compared with the same interactions in a vacuum. The high dielectric constant of water is an expression of its polarity and its capacity to form an oriented solvent shell around an ion. This attenuates the electrostatic interaction of one ion with another (Figure 6-30). These oriented solvent shells produce electric fields of their own, which oppose the fields produced by the ions. The consequence is that electrostatic interactions of these ions are markedly weakened. Water has an unusually high dielectric constant.

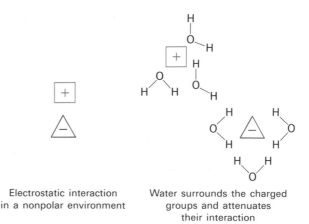

Electrostatic interaction
in a nonpolar environment

Water surrounds the charged
groups and attenuates
their interaction

Figure 6-30
Water attenuates electrostatic interactions between charged groups.

HYDROPHOBIC INTERACTIONS: NONPOLAR GROUPS TEND TO ASSOCIATE IN WATER

The sight of dispersed oil droplets coming together in water to form a single large oil drop is a familiar one. An analogous process occurs at the atomic level: *nonpolar molecules or groups tend to cluster together in water.* These associations are called *hydrophobic interactions.* In a figurative sense, water tends to squeeze nonpolar molecules together.

Let us examine the basis of hydrophobic interactions, which are a major driving force in the folding of macromolecules, the binding of substrates to enzymes, and most other molecular interactions in biology. Consider the solution of a single nonpolar molecule, such as hexane, in water. A hexane molecule will occupy a cavity in the water. The formation of this cavity is energetically unfavorable because it requires the separation of water molecules. We already have noted that water is a highly cohesive substance, primarily because it forms a large number of hydrogen bonds with itself. The water molecules around the molecule of hexane will tend to reorient themselves to form a maximum number of new hydrogen bonds. This is accomplished but at a price: the number of configurations that give rise to favorable hydrogen bonding is much fewer than in the absence of hexane. The water molecules around the hexane molecule are more ordered than elsewhere in the solution. In other words, the entropy of the solution has decreased. Now let us consider the arrangement of two hexane molecules in water. In part A of Figure 6-31 each sits in a separate cavity. In part B, they occupy a single larger cavity. Which is more favorable? Since it takes less work to form one large cavity than two small ones, the two hexane molecules will associate and occupy a single large cavity. Some of the oriented water molecules around the separated hexane molecules will be freed as the hexanes associate. The entropy of the solution will increase because of the enhanced freedom of these water molecules. Thus, a hydrophobic interaction is characterized by an increase in entropy. *Nonpolar solute molecules are driven together in water not primarily because they have a high affinity for each other but rather because water bonds strongly to itself.*

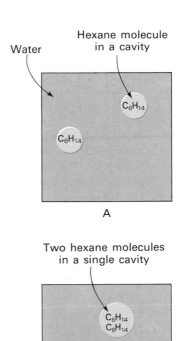

Figure 6-31
A schematic representation of two molecules of hexane in a small volume of water: (A) the hexane molecules occupy different cavities in the water structure and (B) they occupy the same cavity, which is energetically more favorable.

SUMMARY

The catalysts in biological systems are enzymes, which invariably are proteins. Enzymes are highly specific and have great catalytic power. They usually enhance reaction rates by a factor of at least 10^7. Enzymes do not alter reaction equilibria. Rather, they serve as catalysts by reducing the activation energy of chemical reactions. The Michaelis-Menten model accounts for the kinetic properties of some enzymes. In this model, the enzyme (E) combines with a substrate (S) to form an enzyme-substrate (ES) complex, which can proceed to form product (P) or dissociate into E and S.

$$\text{E} + \text{S} \underset{k_2}{\overset{k_1}{\rightleftharpoons}} \text{ES} \overset{k_3}{\longrightarrow} \text{E} + \text{P}$$

The rate (V) of formation of product is given by the Michaelis-Menten equation:

$$V = V_{\max}\frac{[\text{S}]}{[\text{S}] + K_{\text{M}}}$$

where $V_{\max}$ is the rate when the enzyme is fully saturated with substrate, and K_{M}, the Michaelis constant, is the substrate concentration at which the reaction rate is half maximal. The maximal rate, $V_{\max}$, is equal to the product of k_3 and the total concentration of enzyme. The kinetic constant k_3, called the turnover number, is the number of substrate molecules converted into product per unit time by a single catalytic site when the enzyme is fully saturated with substrate. Turnover numbers for most enzymes are between 1 and 10^4 per second.

Enzymes can be inhibited by specific small molecules or ions. In irreversible inhibition, the inhibitor is covalently linked to the enzyme or bound so tightly that its dissociation from the enzyme is very slow. In contrast, reversible inhibition is characterized by a rapid equilibrium between enzyme and inhibitor. A competitive inhibitor prevents the substrate from binding to the active site. It reduces the reaction velocity by diminishing the proportion of enzyme molecules that have a bound substrate. In noncompetitive inhibition, the inhibitor decreases the turnover number. Competitive inhibition can be distinguished from noncompetitive inhibition by determining whether the inhibition can be overcome by raising the substrate concentration.

The catalytic activity of many enzymes is regulated in vivo. Allosteric interactions, which are defined as interactions between spatially distinct sites, are particularly important in this regard. All known allosteric enzymes consist of two or more subunits. Allosteric interactions are mediated by conformational changes that are transmitted from one subunit to another. Allosteric enzymes usually exhibit sigmoidal rather than hyperbolic plots of V versus [S]. Two limiting models—the concerted model and the sequential model—have been postulated to account for some properties of allosteric enzymes.

Reversible molecular interactions in biological systems are due to electrostatic bonding, hydrogen bonding, and van der Waals bonding. These interactions are profoundly affected by the presence of water. Some important characteristics of water are its polarity, capacity to serve as both a hydrogen donor and acceptor, and cohesiveness. Consequently, water diminishes the strength of electrostatic and hydrogen bonding between other molecules and ions. In contrast, water accentuates the interaction of nonpolar molecules. Substrates are bound to enzymes at active-site clefts from which water is largely excluded when the substrate is bound. The exclusion of water strengthens electrostatic and hydrogen bonds between enzyme and substrate. A significant part of the binding energy comes from the association of nonpolar parts of the substrate with those of the active site. The specificity of the enzyme-substrate interaction arises from hydrogen bonding, which is highly directional, and the shape of the active site, which rejects molecules that do not have a complementary shape. The recognition of substrate by many enzymes is a dynamic process accompanied by conformational changes at the active site of the enzyme.

SELECTED READINGS

Where to start:

Koshland, D. E., Jr., 1973. Protein shape and biological control. *Sci. Amer.* 229(4):52–64. [Available as *Sci. Amer.* Offprint 1280. An excellent introduction to the importance of conformational flexibility for the specificity and regulation of enzyme action.]

Books on enzymes:

Boyer, P. D., (ed.), 1970. *The Enzymes* (3rd ed.). Academic Press. [This multivolume treatise on enzymes contains a wealth of information. Volumes 1 and 2 (available in a paperback edition) deal with general aspects of enzyme structure, mechanism, and regulation. Volume 3 and subsequent ones contain detailed and authoritative articles on individual enzymes.]

Cold Spring Harbor Laboratory, 1972. *Structure and Function of Proteins at the Three-Dimensional Level* (Cold Spring Harbor Symposia on Quantitative Biology, vol. 36). [Contains many important articles on enzyme structure and function.]

Jencks, W. P., 1969. *Catalysis in Chemistry and Enzymology.* McGraw-Hill. [A rigorous account of catalytic mechanisms and intermolecular forces.]

Bernhard, S. A., 1968. *The Structure and Function of Enzymes.* Benjamin. [An introduction to enzyme structure, kinetics, and catalytic mechanisms.]

Gutfreund, H., 1972. *Enzymes: Physical Principles.* Wiley-Interscience.

Bender, M. L., 1971. *Mechanisms of Homogeneous Catalysis from Protons to Proteins.* Wiley.

Allosteric interactions:

Monod, J., Changeux, J-P., and Jacob, F., 1963. Allosteric proteins and cellular control systems. *J. Mol. Biol.* 6:306–329. [A classic paper that introduced the concept of allosteric interactions.]

Monod, J., Wyman, J., and Changeux, J-P., 1965. On the nature of allosteric transitions. *J. Mol. Biol.* 12:88–118. [Presentation of the concerted model for allosteric transitions.]

Koshland, D. E., Jr., Nemethy, G., and Filmer, D., 1966. Comparison of experimental binding data and theoretical models in proteins containing subunits. *Biochemistry* 5:365–385. [Presentation of the sequential model for allosteric transitions.]

Matthews, B. W., and Bernhard, S. A., 1973. Structure and symmetry of oligomeric enzymes. *Annu. Rev. Biophys. Bioeng.* 2:257–317.

Review articles:

Kirsch, J. F., 1973. Mechanism of enzyme action. *Annu. Rev. Biochem.* 42:205–234.

Koshland, D. E., Jr., and Neet, K. E., 1968. The catalytic and regulatory properties of enzymes. *Annu. Rev. Biochem.* 37:359–410.

Richards, F. M., Wyckoff, H. W., and Allewell, N., 1970. The origin of specificity in binding: a detailed example in a protein–nucleic acid interaction. *In* Schmitt, F. O., (ed.), *The Neurosciences: Second Study Program,* pp. 901–912. Rockefeller University Press.

Davidson, N., 1967. Weak interactions and the structure of biological macromolecules. *In* Quarton, G. C., Melnechuk, T., and Schmitt, F. O., (eds.), *The Neurosciences: A Study Program,* pp. 46–56. Rockefeller University Press.

Eigen, M., 1968. New looks and outlooks on physical enzymology. *Quart. Rev. Biophys.* 1:3–33. [A review of relaxation spectrometry of enzymes.]

PROBLEMS

1. The hydrolysis of pyrophosphate to orthophosphate is important in driving forward biosynthetic reactions such as the synthesis of DNA. This hydrolytic reaction is catalyzed in *E. coli* by a pyrophosphatase that has a molecular weight of 120,000 daltons and consists of six identical subunits. Purified enzyme has a V_{max} of 2800 units per milligram of enzyme. For this enzyme, a unit of activity is defined as the amount of enzyme that hydrolyzes 10 μmol of pyrophosphate in 15 minutes at 37° under standard assay conditions.

   (a) How many moles of substrate are hydrolyzed per second per milligram of enzyme?
   (b) How many moles of active site are there in 1 mg of enzyme? Assume that each subunit has one active site.
   (c) What is the turnover number of the enzyme? Compare this value with others mentioned in this chapter.

2. Penicillin is hydrolyzed and thereby rendered inactive by penicillinase, an enzyme present in some resistant bacteria. The molecular weight of this enzyme in *Staphylococcus aureus* is 29,600.

[Penicillin]	Amount hydrolyzed (moles)
0.1 × 10⁻⁵ M	0.11 × 10⁻⁹
0.3 × 10⁻⁵ M	0.25 × 10⁻⁹
0.5 × 10⁻⁵ M	0.34 × 10⁻⁹
1.0 × 10⁻⁵ M	0.45 × 10⁻⁹
3.0 × 10⁻⁵ M	0.58 × 10⁻⁹
5.0 × 10⁻⁵ M	0.61 × 10⁻⁹

The amount of penicillin hydrolyzed in 1 minute in a 10-ml solution containing 10^{-9} g of purified penicillinase was measured as a function of the concentration of penicillin. Assume that the concentration of penicillin does not change appreciably during the assay.

   (a) Make a $1/V$ versus $1/[S]$ plot of these data. Does penicillinase appear to obey Michaelis-Menten kinetics? If so, what is the value of K_M?
   (b) What is the value of V_{max}?
   (c) What is the turnover number of penicillinase under these experimental conditions? Assume one active site per enzyme molecule.

3. The kinetics of an enzyme are measured as a function of substrate concentration in the presence and absence of 2×10^{-3} M inhibitor (I).

[S]	Velocity ($\mu mol/min$)	
	No inhibitor	Inhibitor
0.3 × 10⁻⁵ M	10.4	4.1
0.5 × 10⁻⁵ M	14.5	6.4
1.0 × 10⁻⁵ M	22.5	11.3
3.0 × 10⁻⁵ M	33.8	22.6
9.0 × 10⁻⁵ M	40.5	33.8

   (a) What are the values of V_{max} and K_M in the absence of inhibitor? in its presence?
   (b) What type of inhibition is this?
   (c) What is the binding constant of this inhibitor?
   (d) When $[S] = 1 \times 10^{-5}$ M and $[I] = 2 \times 10^{-3}$ M, what fraction of the enzyme molecules have a bound substrate? a bound inhibitor?

(e) When $[S] = 3 \times 10^{-5}$ M, what fraction of the enzyme molecules have a bound substrate in the presence and absence of 2×10^{-3} M inhibitor? Compare this ratio with the ratio of the reaction velocities under the same conditions.

4. The kinetics of the enzyme discussed in problem 3 are measured in the presence of a different inhibitor. The concentration of this inhibitor is 10^{-4} M.

[S]	Velocity ($\mu mol/min$)	
	No inhibitor	Inhibitor
0.3×10^{-5} M	10.4	2.1
0.5×10^{-5} M	14.5	2.9
1.0×10^{-5} M	22.5	4.5
3.0×10^{-5} M	33.8	6.8
9.0×10^{-5} M	40.5	8.1

(a) What are the values of V_{max} and K_M in the presence of this inhibitor? Compare them with those obtained in problem 3.
(b) What type of inhibition is this?
(c) What is the binding constant of this inhibitor?
(d) When $[S] = 3 \times 10^{-5}$ M, what fraction of the enzyme molecules have a bound substrate in the presence and absence of 10^{-4} M inhibitor?

5. The plot of $1/V$ versus $1/[S]$ is sometimes called a Lineweaver-Burk plot. Another way of expressing the kinetic data is to plot V versus $V/[S]$, which is known as an Eadie-Hofstee plot.
(a) Rearrange the Michaelis-Menten equation to give V as a function of $V/[S]$.
(b) What is the significance of the slope, the y-intercept, and the x-intercept in a plot of V versus $V/[S]$?
(c) Make a sketch of a plot of V versus $V/[S]$ in the absence of an inhibitor, in the presence of a competitive inhibitor, and in the presence of a noncompetitive inhibitor.

6. For allosteric enzymes, low concentrations of competitive inhibitors frequently act as activators. Why? (Hint: Consider the analogy of CO-hemoglobin.)

7. The hormone progesterone contains two ketone groups. Little is known about the properties of the receptor protein that recognizes progesterone. At pH 7, which amino acid side chains might form hydrogen bonds with progesterone? (Assume that the side chains in the receptor protein have the same pKs as in the amino acids in aqueous solution).

For additional problems, see W. B. Wood, J. H. Wilson, R. M. Benbow, and L. E. Hood, *Biochemistry: A Problems Approach* (Benjamin, 1974), Chapters 6 and 7.

CHAPTER 7

MECHANISMS OF ENZYME ACTION: LYSOZYME AND CARBOXYPEPTIDASE

In 1922, Alexander Fleming, a bacteriologist in London, had a cold. He was not one to waste a moment, and consequently used his cold as an opportunity to do an experiment. He allowed a few drops of his nasal mucus to fall on a culture plate containing bacteria. He was excited to find some time later that the bacteria near the mucus had been dissolved away and thought that the mucus might contain the universal antibiotic he was seeking. Fleming showed that the antibacterial substance was an enzyme, which he named lysozyme—*lyso* because of its capacity to lyse bacteria and *zyme* because it was an enzyme. He also discovered a small round bacterium that was particularly susceptible to lysozyme, which he named *Micrococcus lysodeikticus* since it was a displayer of lysis ("deiktikos" means able to show). Fleming found that tears are a rich source of lysozyme. Volunteers provided tears after they suffered a few squirts of lemon—an "ordeal by lemon." The St. Mary's Hospital Gazette published a cartoon showing children coming for a few pennies to Fleming's laboratory, where one attendant administered the beatings, while another collected their tears! Fleming was disappointed to find that lysozyme was not effective against the most harmful bacteria. But seven years later, he did discover a highly effective antibiotic: penicillin—a striking illustration of Pasteur's comment that chance favors the prepared mind.

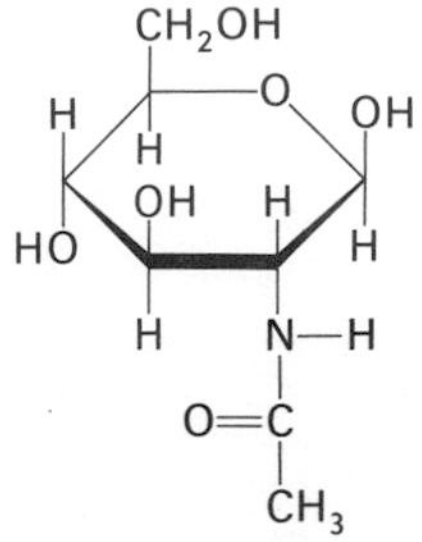

N-Acetylglucosamine
(NAG)

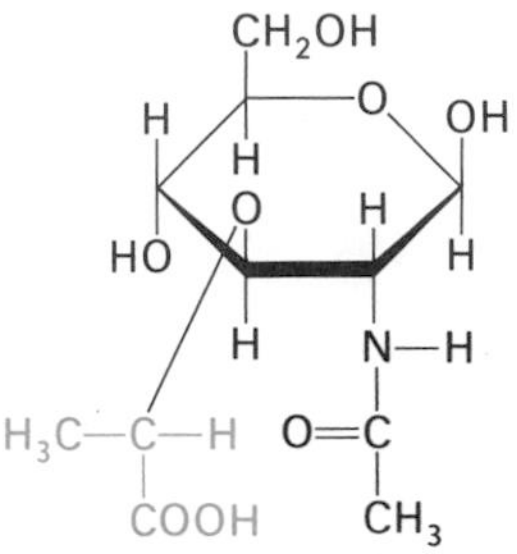

N-Acetylmuramic acid
(NAM)

Figure 7-1
Sugar residues in the polysaccharide
of bacterial cell walls.

LYSOZYME CLEAVES BACTERIAL CELL WALLS

Lysozyme dissolves certain bacteria by cleaving the polysaccharide component of their cell walls. The function of the cell wall in bacteria is to confer mechanical support. A bacterial cell devoid of a cell wall usually bursts because of the high osmotic pressure inside the cell. The detailed structure of bacterial cell walls will be discussed in a later chapter. Let us consider here the structure of the polysaccharide portion.

The cell-wall polysaccharide is made up of two kinds of sugars: N-*acetylglucosamine* (NAG) and N-*acetylmuramic acid* (NAM). NAM and NAG are derivatives of glucosamine in which the amino group is acetylated (Figure 7-1). In NAM, a lactyl side chain is attached to C-3 of the sugar ring by an ether bond. In bacterial cell walls, NAM and NAG are joined by *glycosidic linkages* between C-1 of one sugar and C-4 of the other. The oxygen atom in a glycosidic bond can be located either above or below the plane of the sugar ring. In the α configuration, the oxygen is below the plane of the sugar; in the β configuration, it is above (see Chapter 12 for a more detailed discussion of the properties and nomenclature of sugars). All of the glycosidic bonds of the cell-wall polysaccharide have a β *configuration* (Figure 7-2). NAM and NAG alternate in sequence. Thus, the cell-wall polysaccharide is an alternating polymer of NAM and NAG residues joined by $\beta(1 \longrightarrow 4)$ glycosidic linkages. Different polysaccharide chains are cross-linked by short peptides that are attached to some of the NAM residues.

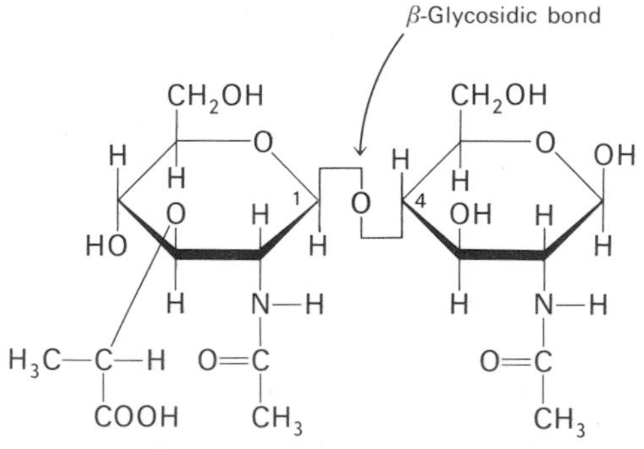

Figure 7-2
NAM is linked to NAG by a $\beta(1 \longrightarrow 4)$ glycosidic bond.

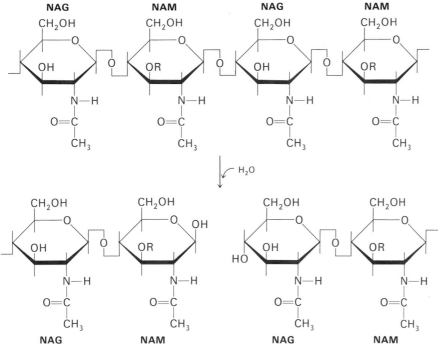

Figure 7-3
Lysozyme hydrolyzes the glycosidic bond between NAM and NAG (R refers to the lactyl group of NAM).

Lysozyme hydrolyzes the glycosidic bond between C-1 of NAM and C-4 of NAG (Figure 7-3). The other glycosidic bond, between C-1 of NAG and C-4 of NAM, is not cleaved. *Chitin*, a polysaccharide found in the shell of crustacea, is also a substrate for lysozyme. Chitin consists only of NAG residues joined by $\beta(1 \longrightarrow 4)$ glycosidic links.

THREE-DIMENSIONAL STRUCTURE OF LYSOZYME

Lysozyme is a relatively small enzyme. The enzyme from hen egg white, a rich source, is made up of a single polypeptide chain of 129 amino acids, and has a molecular weight of 14,600. The enzyme is cross-linked by four disulfide bridges, which contribute to its high

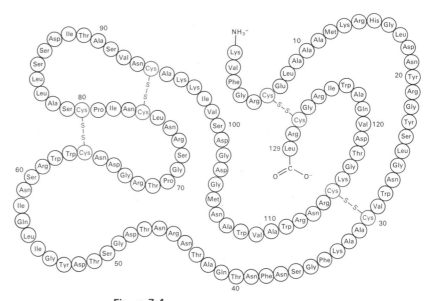

Figure 7-4
Amino acid sequence of hen egg-white lysozyme. Residues that are part of the active site are circled in red. [Based on R. E. Canfield, and A. K. Liu. *J. Biol. Chem.* 240(1965): 2000; and D. C. Phillips. *Sci. Amer.* (5)215(1966):79.]

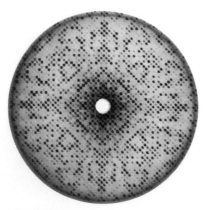

Figure 7-5
X-ray precession photograph of a lysozyme crystal. [Courtesy of Dr. David Phillips.]

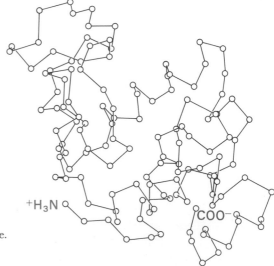

Figure 7-6
Three-dimensional structure of lysozyme. Only the α-carbon atoms are shown. [Courtesy of Dr. David Phillips.]

stability. The amino acid sequence of lysozyme is shown in Figure 7-4.

In 1965, David Phillips and his colleagues determined the three-dimensional structure of lysozyme. Their high-resolution electron-density map was the first for an enzyme molecule. Lysozyme is a compact molecule, roughly ellipsoidal in shape, with dimensions $45 \times 30 \times 30$ Å. The folding of this molecule is complex (Figure 7-6). There is much less α helix than in myoglobin and hemoglobin. In a number of regions, the polypeptide chain is in an extended conformation. In one of these sections, the chain doubles back on itself, and the two strands are hydrogen-bonded through their peptide groups. This hairpin section is similar to a regularly repeating secondary structure mentioned earlier—namely, the *antiparallel β pleated sheet* seen in silk protein. The interior of lysozyme, like that of myoglobin and hemoglobin, is almost entirely nonpolar. Hydrophobic interactions evidently play an important role in the folding of lysozyme, as they do for most proteins.

FINDING THE ACTIVE SITE IN LYSOZYME

Knowledge of the detailed three-dimensional structure of lysozyme did not immediately reveal the catalytic mechanism. In fact, the location of the active site was not obvious on looking at the electron-density map. Lysozyme does not contain a prosthetic group, and thus lacks a built-in marker at its active site, in contrast to some proteins, such as myoglobin and hemoglobin. The essential information needed to identify the active site, specify the mode of binding of substrate, and elucidate the enzymatic mechanism came from an x-ray crystallographic study of the interaction of lysozyme with inhibitors. When the three-dimensional structure of a protein is known, the mode of binding of small molecules can often be determined quite readily by x-ray crystallographic methods. These experiments are feasible because protein crystals are quite porous. Rather large inhibitor molecules can diffuse in the channels between different protein molecules and find their way to specific binding sites. The electron density corresponding to the additional molecule can be calculated directly from the intensities of the x-ray reflections (using the phases already determined for the native protein crystal) if the crystal structure is not markedly altered. This technique is called the *difference Fourier method.*

Ideally, one would like to use the difference Fourier method to elucidate the structure of an enzyme-substrate (ES) complex undergoing catalysis. Unfortunately, this is not yet technically feasible. However, much information can be derived from a study of a complex of an enzyme with an unreactive (or very slowly reactive) analog of its substrate. For lysozyme, this was achieved with the trimer of N-acetylglucosamine (tri-NAG or NAG_3) (Figure 7-7). Oligomers of N-acetylglucosamine consisting of fewer than five residues are hydrolyzed very slowly or not at all. However, they do bind to the active site of the enzyme. Indeed, tri-NAG is a potent competitive inhibitor of lysozyme.

**Tri-*N*-acetylglucosamine
(Tri-NAG)**

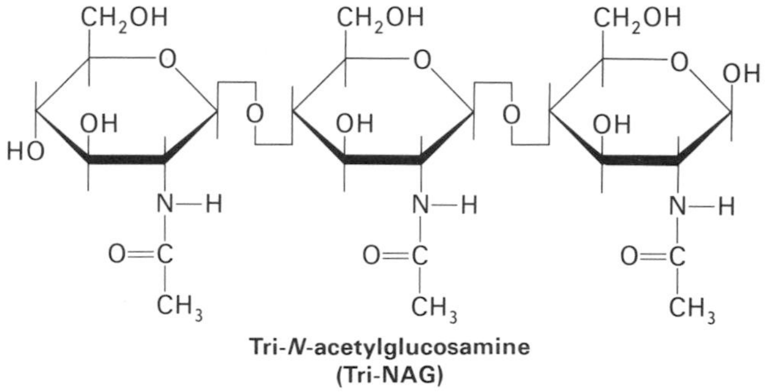

Figure 7-7
Formula of tri-N-acetylglucosamine (tri-NAG or NAG_3), a competitive inhibitor of lysozyme.

MODE OF BINDING OF A COMPETITIVE INHIBITOR

The x-ray study of the tri-NAG lysozyme complex showed the location of the active site, revealed the interactions responsible for the specific binding of substrate, and led to the proposal of a detailed enzymatic mechanism. Tri-NAG binds to lysozyme in a cleft at the surface of the enzyme, and occupies about half of the cleft. Tri-NAG is bound to lysozyme by hydrogen-bond and van der Waals interactions. Electrostatic interactions cannot occur because tri-NAG lacks ionic groups.

The *hydrogen-bond interactions* between tri-NAG and lysozyme are shown in Figure 7-8. The carboxylate group of aspartate 101 is hydrogen-bonded to residues A and B. The most specific and extensive hydrogen bonds are made between the enzyme and sugar residue C. There are four hydrogen bonds. The NH of the indole ring of tryptophan 62 is hydrogen-bonded to the oxygen attached to C-6. The adjacent amino acid residue, tryptophan 63, is similarly hydrogen-bonded to the oxygen attached to C-3. The ring of tryptophan 62 moves by 0.75 Å when tri-NAG binds to the enzyme. There are very good hydrogen bonds between the acetamido side-chain CO and NH groups of sugar residue C and the main-chain NH and CO groups of amino acid residues 59 and 107, respectively.

There are a large number of *van der Waals contacts* between tri-NAG and the enzyme. Sugar residue B is involved in few polar contacts with the enzyme but is closely associated with the indole ring of tryptophan 62. Residue A has rather tenuous contacts with the enzyme.

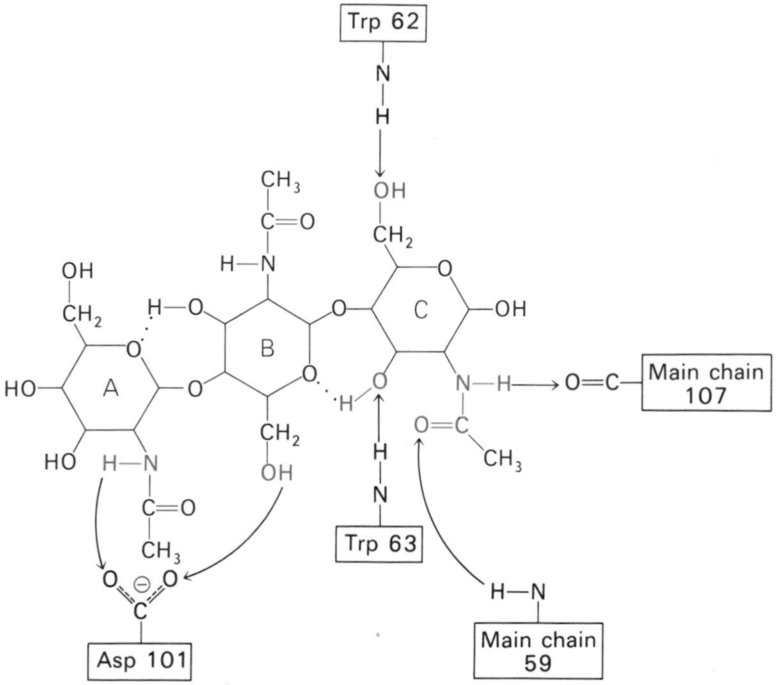

Figure 7-8
Hydrogen bonds between tri-NAG and lysozyme. The groups on tri-NAG involved in hydrogen bonding are shown in blue; those on lysozyme, in red.

FROM STRUCTURE TO ENZYMATIC MECHANISM

1. *How would a substrate bind?* As noted previously, the mode of binding of an effective substrate cannot be established directly by the x-ray crystallographic method. However, the structure of an enzyme-competitive inhibitor complex may provide key clues to how a substrate binds to the enzyme. The finding that tri-NAG fills only half of the cleft in lysozyme was a very suggestive starting point. It was assumed that the observed binding of tri-NAG as an inhibitor involves interactions that are also utilized in the binding of substrate. It seemed likely that additional sugar residues, which would fill the rest of the cleft, are required for the formation of a reactive ES complex. There was space for three additional sugar residues. This was encouraging, since it was known that the hexamer of *N*-acetylglucosamine (hexa-NAG) is rapidly hydrolyzed by the enzyme.

Three additional sugar residues, named D, E, and F, were fitted into the cleft by careful model building (Figure 7-9). Residues E and F went in nicely, making a number of good hydrogen bonds and van der Waals contacts. However, residue D would not fit unless it was distorted. Its C-6 and O-6 atoms came too close to several groups on the enzyme unless the ring was distorted from its normal conformation, which has the appearance of a chair. The cleft simply does not have sufficient space for residue D in the chair configuration customary for sugars of this kind. This distortion was at first disquieting to Phillips and his colleagues, but soon proved to be very exciting because it is a crucial aspect of the catalytic mechanism.

2. *Which bond is cleaved?* The rate of hydrolysis of oligomers of *N*-acetylglucosamine increases strikingly in going from $(NAG)_4$ to $(NAG)_5$ (Table 7-1). There is a further increase in the cleavage rate in going to $(NAG)_6$, but no change as the number of residues is increased to 8. This finding is consistent with the crystallographic results, which show that the active-site cleft would be fully occupied by six sugar residues.

Which bond of hexa-NAG is cleaved? Since tri-NAG is stable, the A—B bond (that is, the glycosidic bond between the A and B sugar residues) cannot be the site of cleavage. Similarly, the B—C bond cannot be the site of cleavage. A second crucial piece of

Table 7-1
Effectiveness of oligomers of *N*-acetylglucosamine as substrates

Substrate	Relative rate of hydrolysis
$(NAG)_2$	0
$(NAG)_3$	1
$(NAG)_4$	8
$(NAG)_5$	4,000
$(NAG)_6$	30,000
$(NAG)_8$	30,000

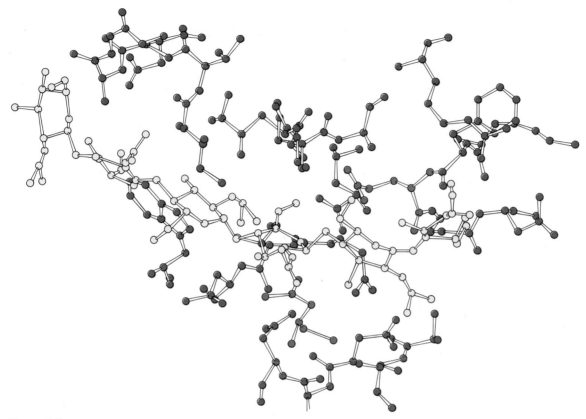

Figure 7-9
Mode of binding of hexa-NAG (shown in yellow) to
lysozyme. The locations of sugar residues A, B, and C are
those observed in the tri-NAG-lysozyme complex,
whereas those of residues D, E, and F are inferred by
model building.

evidence is that site C cannot be occupied by NAM. NAG fits nicely
into site C, but NAM is too large because of its lactyl side chain.
Since the bond cleaved in bacterial cell walls is the NAM–NAG
link, it follows that the C—D bond cannot be the site of cleavage
if the bacterial cell-wall polysaccharide binds to the enzyme in the
same manner as hexa-NAG. The inability of NAM to fit into site
C excludes yet another cleavage site: the E—F bond. Since the
cell-wall polysaccharide is an alternating polymer of NAM and
NAG, it follows that NAM cannot occupy site E if it cannot occupy
site C.

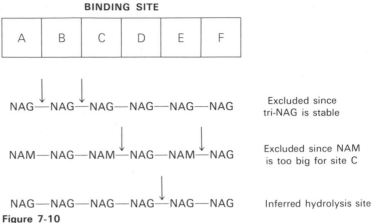

BINDING SITE

| A | B | C | D | E | F |

NAG—NAG—NAG—NAG—NAG—NAG   Excluded since tri-NAG is stable

NAM—NAG—NAM—NAG—NAM—NAG   Excluded since NAM is too big for site C

NAG—NAG—NAG—NAG—NAG—NAG   Inferred hydrolysis site

Figure 7-10
Steps in deducing that the glycosidic bond between sugar residues D and E is the one cleaved by lysozyme.

These observations eliminated the A—B, B—C, C—D, and E—F bonds as possible sites of enzymatic cleavage of the hexamer substrate. Hence, *the D—E bond was the only remaining candidate for the cleavage site* (Figure 7-10).

3. *Which groups on the enzyme participate in catalysis?* The inference that it is the D—E bond that is split was an important step toward defining the groups on the enzyme that carry out the hydrolysis reaction. However, it was necessary to localize the site of cleavage even more precisely: on which side of the glycosidic oxygen atom is the bond cleaved? The question was answered by carrying out the enzymatic hydrolysis in the presence of water enriched with ^{18}O, the stable heavy isotope of oxygen (Figure 7-11). The sugars isolated contained ^{18}O attached to C-1 of the D sugar. In contrast, the hydroxyl group attached to C-4 of the E sugar contained the ordinary isotope of oxygen. *Hence, the bond split is the one between C-1 of residue D and the oxygen of the glycosidic linkage to residue E.* This experiment illustrates the use of isotopes in elucidating enzymatic mechanisms. In the absence of an isotopic marker, it would have been difficult, if not impossible, to establish the precise site of cleavage.

A search was then made for possible catalytic groups close to the glycosidic bond that is cleaved. A *catalytic group* is one that

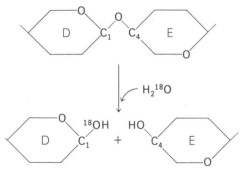

Figure 7-11
Hydrolysis in ^{18}O water showed that lysozyme cleaves the C_1—O bond rather than the O—C_4 bond. (Only the skeletons of the D and E residues are shown here.)

directly participates in the formation or breakage of covalent bonds. The most plausible candidates are groups that can serve as *hydrogen-bond donors or acceptors*. The donation or abstraction of a hydrogen ion is a critical step in most enzymatic reactions. The only plausible catalytic residues near the glycosidic bond that is cleaved are aspartic 52 and glutamic 35. The aspartic acid residue is on one side of the glycosidic linkage, whereas the glutamic acid residue is on the other. These two acidic side chains have markedly different environments. Aspartic 52 is in a distinctly polar environment, where it serves as a hydrogen-bond acceptor in a complex network of hydrogen bonds. In contrast, glutamic 35 lies in a nonpolar region. Thus, it seemed likely that at pH 5, which is the pH optimum for the hydrolysis of chitin by lysozyme, *aspartic 52 is in the ionized COO^- form, whereas glutamic 35 is in the un-ionized COOH form*. The nearest oxygen of each of these acid groups is located about 3 Å away from the glycosidic linkage (Figure 7-12).

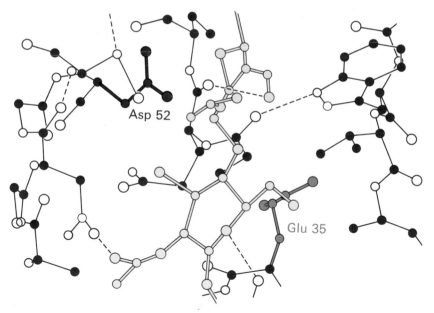

Figure 7-12
Structure of part of the active site of lysozyme. The D and E rings of the hexa-NAG substrate are shown in yellow. The side chains of aspartate 52 (red) and glutamic 35 (green) are in close proximity. [From W. N. Lipscomb. *Proc. Robert A. Welch Found. Conf. Chem. Res.* 15(1971):150.]

DISTORTION OF THE SUBSTRATE IS IMPORTANT IN THE PROPOSED CATALYTIC MECHANISM FOR LYSOZYME

Phillips and his colleagues have proposed a detailed catalytic mechanism for lysozyme based on the preceding structural data. The essential steps are:

1. The —COOH group of glutamic 35 donates an H^+ to the bond between C-1 of the D ring and the glycosidic oxygen atom, thereby cleaving this bond (Figure 7-13).

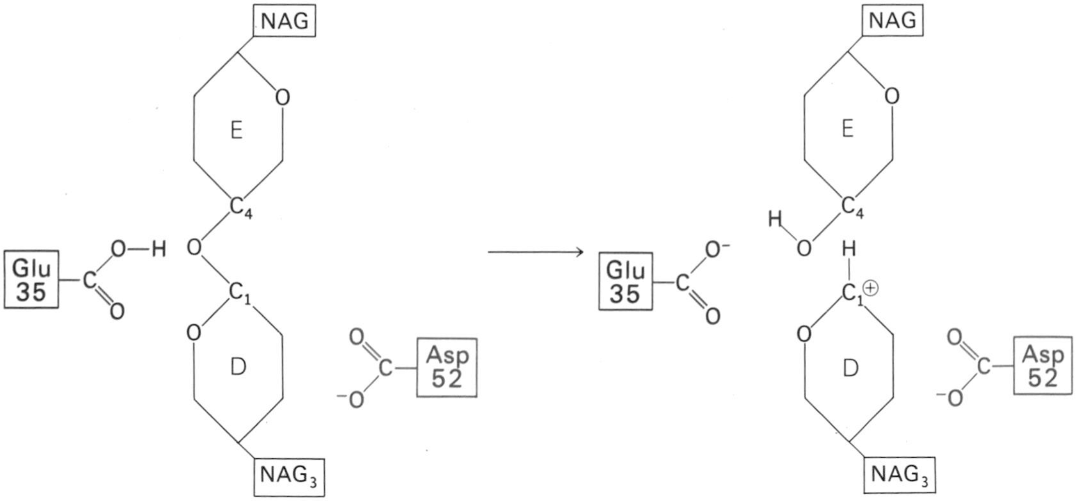

Figure 7-13
The first step in the proposed catalytic mechanism for lysozyme is the transfer of an H^+ from Glu 35 to the oxygen atom of the glycosidic bond. The glycosidic bond is thereby cleaved, and a carbonium ion intermediate is formed.

2. This creates a positive charge on C-1 of the D ring. This transient species is called a *carbonium ion* because it contains a positively charged carbon atom.

3. The dimer of NAG consisting of residues E-F diffuses away from the enzyme.

4. The carbonium ion intermediate then reacts with OH^- from the solvent (Figure 7-14). Tetra-NAG, consisting of residues A-B-C-D, diffuses away from the enzyme.

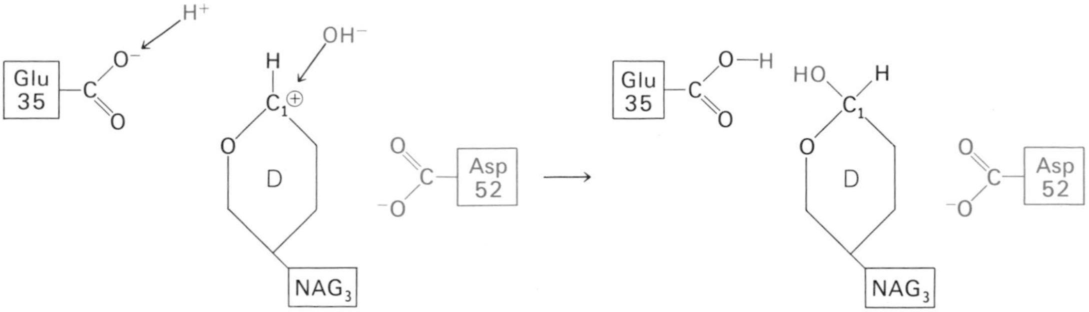

Figure 7-14
The hydrolysis reaction is completed by the addition of OH⁻ to the carbonium ion intermediate and of H⁺ to the side chain of Glu 35.

5. Glutamic 35 becomes protonated, and the enzyme is ready for another round of catalysis.

The critical elements of this catalytic scheme are:

1. *General acid catalysis.* A proton is transferred from glutamic 35, which is un-ionized and optimally located 3 Å away from the glycosidic oxygen atom.

2. *Promotion of the formation of the carbonium ion intermediate.* The enzymatic reaction is markedly facilitated by two different factors that stabilize the carbonium ion intermediate:

   a. The electrostatic factor is the presence of a negatively charged group 3 Å away from the carbonium ion intermediate. Aspartic 52, which is in the negatively charged carboxylate form, interacts electrostatically with the positive charge on C_1 of ring D.

   b. The geometrical factor is the distortion of ring D (Figure 7-15). Recall that hexa-NAG fits into the active-site cleft only if sugar residue D is distorted out of its customary chair conformation into a half-chair form. This distortion is critical to catalysis because the half-chair geometry markedly promotes carbonium ion formation. The planarity of carbon atoms 1, 2, and 5 and oxygen atom 5 in the half-chair form enables the positive charge to be shared between C-1 and the ring oxygen atom. Thus, *in the process of binding, the enzyme forces the substrate to assume the geometry of the transition state,* which is a carbonium ion.

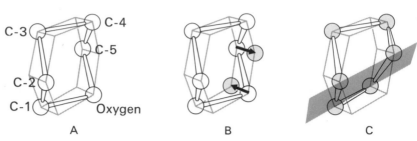

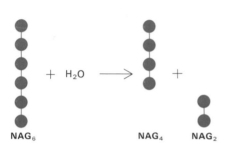

Figure 7-16
Hexa-NAG is hydrolyzed to
tetra-NAG and di-NAG.

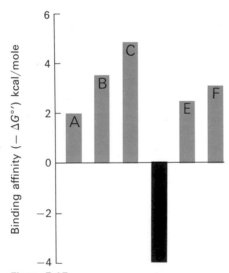

Figure 7-17
Contributions made by each of the six
sugars of hexa-NAG to the standard
free energy of binding of this substrate.
The binding of residue D *costs* free
energy because it must be distorted to
fit into the active site.

EXPERIMENTAL SUPPORT FOR THE PROPOSED MECHANISM

The mode of binding of substrate and the mechanism of catalysis
proposed on the basis of the crystallographic studies have been
tested in a variety of chemical experiments. All of the experimental
results thus far support the crystallographic hypothesis. A number
of experimental findings are especially pertinent in this regard:

1. *Cleavage pattern.* Hexa-NAG is split into tetra-NAG and
di-NAG, which confirms the crystallographic hypothesis that
cleavage occurs between the fourth and fifth residues of a hexamer
(Figure 7-16).

2. *Binding affinity.* The contributions of each of the six sugar
residues to the total free energy of binding to a hexameric substrate
have been determined from measurements of binding equilibria
(Figure 7-17). The striking finding is that sugar residue D makes
a negative contribution to the binding affinity. The cost of binding
residue D is about 4 kcal/mol. This result confirms the crystallographic inference that residue D is bound in a distorted form. A
price must be paid for the distortion of residue D from its customary
chair form into a half-chair form. Also of interest is the finding
that residue C makes the largest positive contribution to the binding
affinity. The crystallographic model shows that residue C makes
a large number of hydrogen bonds and van der Waals interactions.

3. *Transition-state analogs.* The distortion of sugar residue D into a half-chair form is a critical aspect of the postulated enzymatic mechanism since the half-chair form is assumed to be the conformation of the transition state. As noted above, this idea is supported by the finding that the binding of residue D costs energy, the price of distortion. Further evidence comes from a study of a transition-state analog of the substrate—that is, a compound that has the geometry of the transition state in catalysis *before* it is bound to the enzyme, as well as when it is bound. The D ring of the lactone analog of tetra-NAG has a half-chair conformation (Figure 7-18). Hence, it need not be distorted on binding to lysozyme if ring D is indeed bound in a half-chair form. In fact, this lactone analog of tetra-NAG binds to the A-D subsites of lysozyme 3600 times more strongly than does tetra-NAG. This finding suggests that *the distortion of the D ring of a normal substrate accelerates catalysis by a factor of the order of 3600.*

This factor in catalysis was clearly foreseen by Pauling in a lecture that he gave in 1948:

> I think that enzymes are molecules that are complementary in structure to the activated complexes of the reactions that they catalyze, that is, to the molecular configuration that is intermediate between the reacting substances and the products of reaction for these catalyzed processes. The attraction of the enzyme molecule for the activated complex would thus lead to a decrease in its energy and hence to a decrease in the energy of activation of the reaction and to an increase in the rate of reaction.

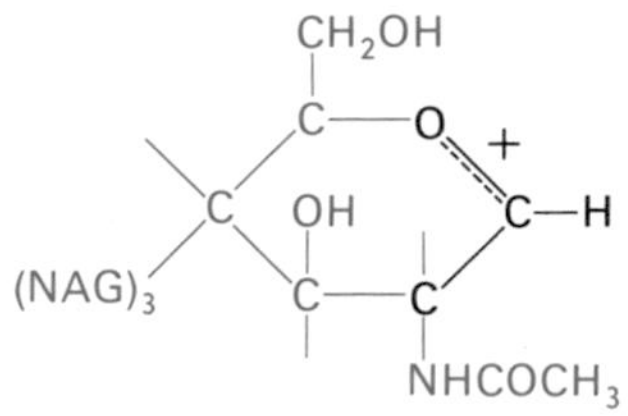

4. *pH dependence of the catalytic rate.* The rate of hydrolysis of chitin is most rapid at pH 5 (Figure 7-19). The enzymatic activity drops sharply on either side of this optimal pH. The decrease on the alkaline side is due to the ionization of glutamic 35, whereas the decrease in rate on the acid side reflects the protonation of aspartate 52. Lysozyme is active only when glutamic 35 is un-ionized and aspartate 53 is ionized.

5. *Selective chemical modification.* Lysozyme remains active when all of its carboxyl groups except those of glutamic 35 and aspartate 52 are modified. Glutamic 35 and aspartate 52 are unmodified if this reaction is carried out in the presence of substrate. When the substrate is removed, aspartate 52 also becomes modified (whereas glutamic 35 remains unaltered). The modification of aspartate 52

Carbonium ion derivative of tetra-NAG

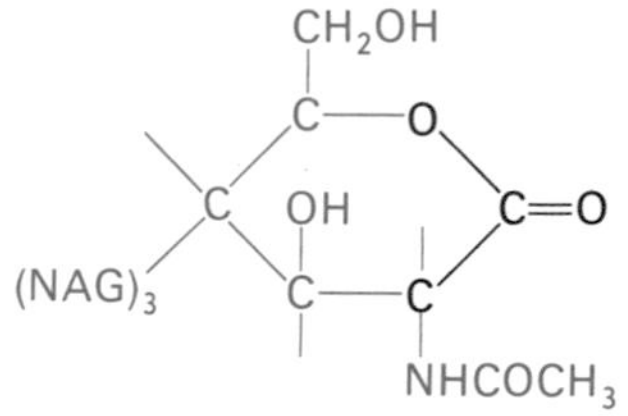

Lactone analog of tetra-NAG

Figure 7-18
The lactone analog of tetra-NAG resembles the transition-state intermediate in the reaction catalyzed by lysozyme because its D ring has a half-chair form.

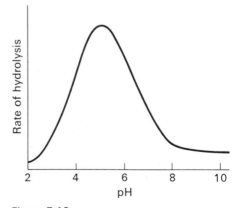

Figure 7-19
The rate of hydrolysis of chitin (poly-NAG) by lysozyme as a function of pH.

produces totally inactive enzyme, which supports the proposal that the precise position of the carboxylate group of aspartate 52 is important in stabilizing the carbonium ion intermediate.

6. *Transglycosylation.* Hexa-NAG and di-NAG are formed at a slow rate when tetra-NAG is added to lysozyme (Figure 7-20). This reaction, called a transglycosylation, supports an important feature of the proposed enzymatic mechanism, namely, the occurrence of a *glycosyl-enzyme intermediate.* In the usual hydrolytic reaction, this glycosyl-enzyme intermediate reacts with OH^-. In the transglycosylation reaction, the reaction is with another sugar: ROH. The transglycosylation reaction exhibits specificity since the acceptor is bound to sites E and F of the active-site cleft. Moreover, the glycosidic bond formed has the β configuration, the same as the substrate. The proposed geometry of the catalytic intermediate is supported by this finding.

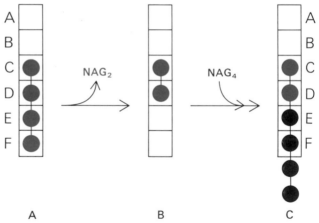

Figure 7-20
The occurrence of a *glycosyl-enzyme intermediate* is supported by the finding that lysozyme catalyzes a slow transglycosylation reaction. NAG_4 (shown in red) adds to the glycosyl-enzyme intermediate (shown in blue in part B) to yield NAG_6.

CARBOXYPEPTIDASE A: A ZINC-CONTAINING PROTEOLYTIC ENZYME

Let us now turn to carboxypeptidase A, a digestive enzyme that hydrolyzes the carboxyl-terminal peptide bond in polypeptide chains. Hydrolysis occurs most readily if the carboxyl-terminal residue has an aromatic or a bulky aliphatic side chain (Figure 7-21). The catalytic mechanism of this enzyme is especially interesting because it is very different from that of lysozyme. Two aspects of the catalytic mechanism of carboxypeptidase A, which will be discussed shortly, are particularly noteworthy:

1. *Induced fit*. The binding of substrate is accompanied by quite large alterations in the structure of the enzyme.

2. *Electronic strain*. The enzyme contains a zinc atom and other groups at its active site that induce rearrangements in the electronic distribution of the substrate, thereby rendering it more susceptible to hydrolysis.

The three-dimensional structure of carboxypeptidase A at a resolution of 2 Å was solved by William Lipscomb in 1967 (Figure 7-22). This enzyme is a single polypeptide chain of 307 amino acid residues. Carboxypeptidase A has a compact shape, which can be approximated by an ellipsoid of dimensions $50 \times 42 \times 38$ Å. The

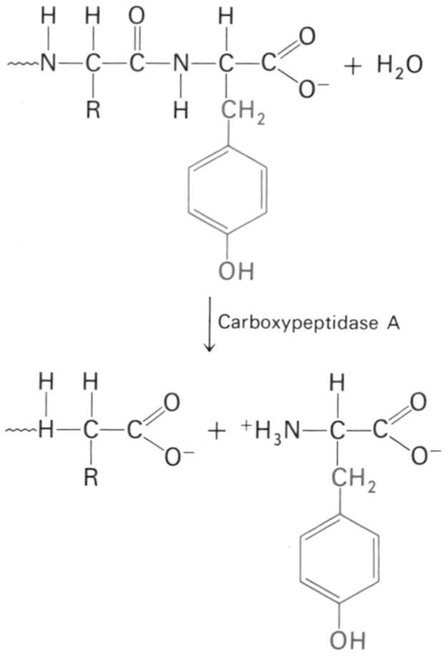

Figure 7-21
Reaction catalyzed by carboxypeptidase A.

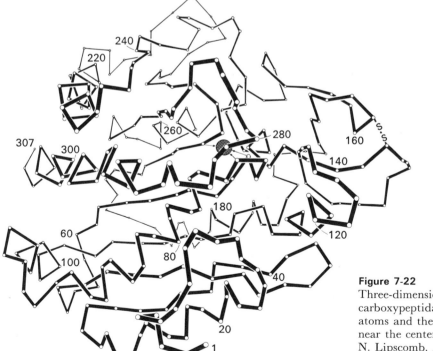

Figure 7-22
Three-dimensional structure of carboxypeptidase A. Only the α-carbon atoms and the zinc ion (shaded circle near the center) are shown. [From W. N. Lipscomb. *Proc. Robert A. Welch Found. Conf. Chem. Res.* 15(1971):134.]

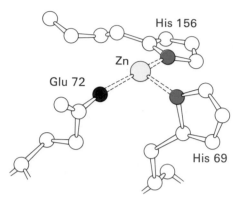

Figure 7-23
A zinc ion is coordinated to two histidine side chains and a glutamate side chain at the active site of carboxypeptidase A. A water molecule coordinated to the zinc is not shown here. [Based on D. M. Blow and T. A. Steitz. X-ray diffraction studies of enzymes, *Annu. Rev. Biochem.* 39:78 Copyright © 1970 by Annual Reviews, Inc. All rights reserved.]

enzyme contains regions of α helix (38%) and of β pleated sheet (17%). There is a tightly bound zinc ion, which is essential for enzymatic activity. This zinc ion is located in a groove near the surface of the molecule, where it is coordinated to a tetrahedral array of two histidine side chains, a glutamate side chain, and a water molecule (Figure 7-23). There is a large pocket near the zinc ion, which accommodates the side chain of the terminal residue of the peptide substrate.

BINDING OF SUBSTRATE INDUCES LARGE STRUCTURAL CHANGES AT THE ACTIVE SITE OF CARBOXYPEPTIDASE A

The mode of binding of substrates to carboxypeptidase A has been deduced from the structure of a complex of glycyltyrosine and this enzyme. Glycyltyrosine is a slowly hydrolyzed substrate, and thus its interaction with the enzyme probably resembles that of a good substrate. The binding of glycyltyrosine (Figures 7-24 and 7-25) can be described in terms of five interactions.

1. The negatively charged terminal carboxylate of glycyltyrosine interacts electrostatically with the positively charged side chain of arginine 145.

2. The tyrosine side chain of the substrate binds to a nonpolar pocket in the enzyme.

3. The NH hydrogen of the peptide bond to be cleaved is hydrogen bonded to the OH group of the aromatic side chain of tyrosine 248.

4. The carbonyl oxygen of the peptide bond to be cleaved is coordinated to the zinc ion.

5. The terminal amino group of the substrate is hydrogen-bonded through an intervening water molecule to the side chain of glutamate 270. This interaction probably does not occur in productive ES complexes, and in fact it may account for the very slow hydrolysis of glycyltyrosine.

The binding of glycyltyrosine is accompanied by a structural rearrangement of the active site (Figure 7-26). In fact, *the catalytic groups of the enzyme are brought into the correct orientation by the binding*

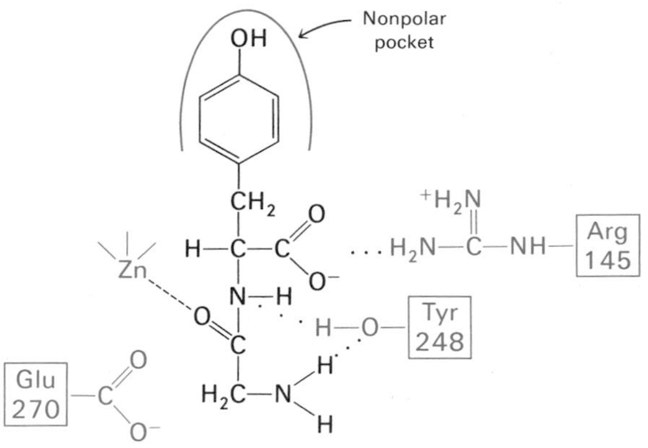

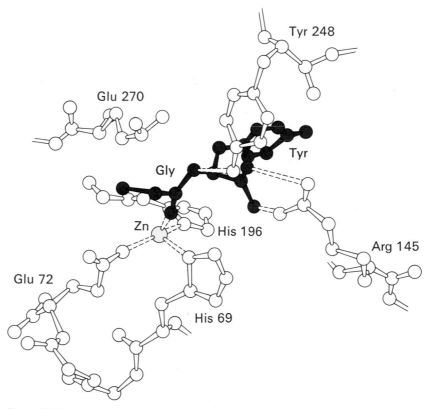

Nonpolar
pocket
OH
CH2
O
H—C—C
O−
Zn
N—H
O
C
H—O—
Tyr
248
+H2N
...H2N—C—NH—
Arg
145
H
H2C—N
H
Glu
270
C
O
O−

Figure 7-24
Schematic representation of the binding of glycyltyrosine to the active site of carboxypeptidase A. The proposed catalytically active complex is shown here.

Tyr 248
Glu 270
Tyr
Gly
Zn
His 196
Arg 145
Glu 72
His 69

Figure 7-25
Three-dimensional structure of glycyltyrosine at the active site of carboxypeptidase A. Glycyltyrosine, the substrate, is shown in red. [Redrawn from D. M. Blow and T. A. Steitz. X-ray diffraction studies of enzymes, *Annu. Rev. Biochem.* 39:79. Copyright © 1970 by Annual Reviews, Inc. All rights reserved.]

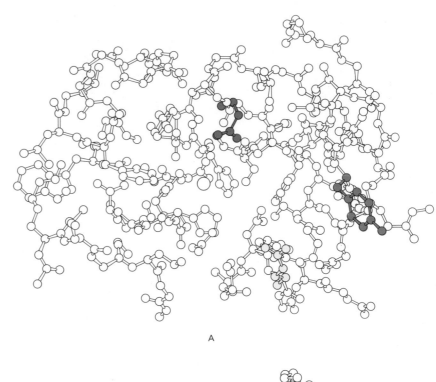

A

Figure 7-26
The structure of carboxypeptidase A changes upon binding substrate: (A) enzyme alone (Arg 145 is shown in yellow, Glu 270 in green, and Tyr 248 in blue); and (B) enzyme-substrate complex (glycyltyrosine, the substrate, is shown in red). [From W. N. Lipscomb. *Proc Robert A. Welch Found. Conf. Chem. Res.* 15(1971): 140–141.]

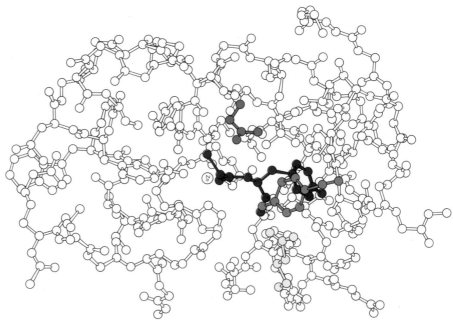

B

of substrate, as originally proposed by Koshland in his *induced-fit model* of enzyme action. The guanidinium group of arginine 145 moves 2 Å, as does the carboxylate group of glutamate 270. The binding of the carbonyl group of the substrate to the zinc ion displaces a bound water molecule. At least four other water molecules are displaced from the nonpolar pocket when the tyrosine side chain of the substrate binds them. The largest conformational change is the movement of the phenolic hydroxyl group of tyrosine 248 by 12 Å, a distance equal to about a quarter of the diameter of the protein. This motion is accomplished primarily by a facile rotation about a single carbon–carbon bond. The hydroxyl group of tyrosine 248 moves from the surface of the molecule to the vicinity of the peptide bond of the substrate. An important consequence of this motion is that it closes the active-site cavity and completes its conversion from a water-filled to a hydrophobic region. These structural changes may be initiated by the binding of arginine 145 to the terminal carboxylate group of the substrate.

ELECTRONIC STRAIN ACCELERATES CATALYSIS BY CARBOXYPEPTIDASE A

Lipscomb has proposed a catalytic mechanism for carboxypeptidase A based on x-ray crystallographic studies. The productive ES complex is postulated to have the structure shown in Figure 7-27. In this proposed mechanism, the hydroxyl group of tyrosine 248 donates a proton to the NH of the peptide bond to be split. The

Figure 7-27
A proposed catalytic mechanism for carboxypeptidase A in which Glu 270 directly attacks the carbonyl carbon atom of the susceptible peptide bond and Tyr 248 donates a proton to the NH group of this peptide. The resulting anhydride is then hydrolyzed.

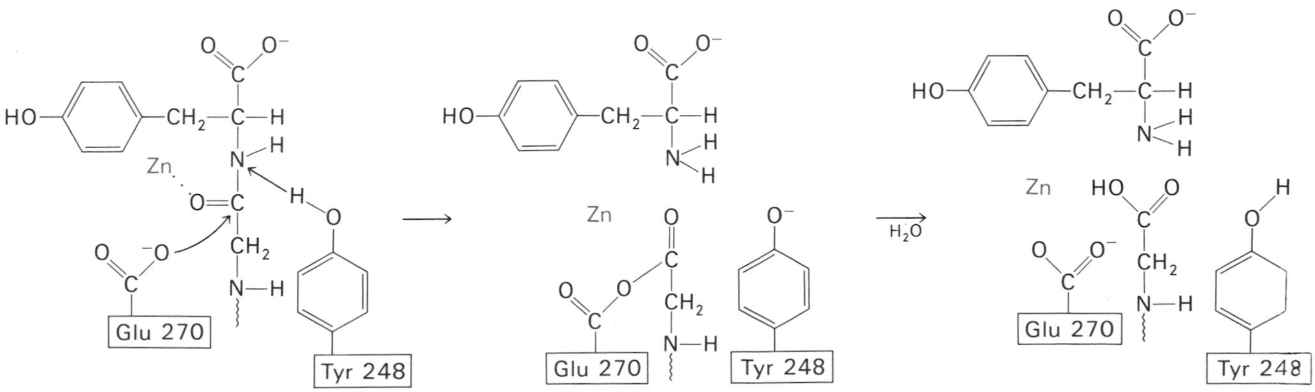

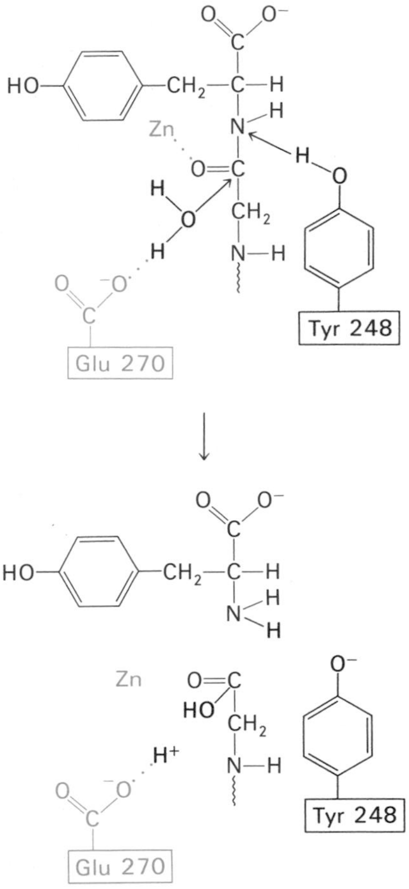

Figure 7-28
An alternative catalytic mechanism for carboxypeptidase A. Tyr 248 has the same role as in the mechanism shown in Figure 7-27. In contrast, Glu 270 activates a water molecule, which attacks the carbonyl carbon atom of the susceptible peptide bond. Hydrolysis is direct; an anhydride is not formed.

carbonyl carbon atom of this peptide bond is attacked by the carboxylate group of glutamate 270, which acts as a nucleophile. The resulting anhydride of glutamate 270 and the acid component of the substrate is hydrolyzed in a subsequent step.

An alternative mechanism that is also compatible with the x-ray data is shown in Figure 7-28. In this scheme, glutamate 270 activates a water molecule. The resulting OH^- directly attacks the carbonyl carbon atom of the susceptible peptide bond. At the same time, tyrosine 248 donates a proton to the NH group of the susceptible peptide bond, which results in its hydrolysis. This mechanism differs from the one in Figure 7-27 in that the susceptible peptide bond of the substrate, rather than an anhydride intermediate, is hydrolyzed by water.

What is the role of zinc in this catalytic scheme? The carbonyl group of the susceptible peptide bond is pointed toward the zinc ion so that the C=O bond is polarized more than usual, thus rendering the carbonyl carbon atom more vulnerable to nucleophilic attack. The induction of a dipole is enhanced by the nonpolar environment of the zinc ion, which increases its effective charge. The proximity of the negative charge on glutamate 270 also contributes to the induction of a large dipole in the carbonyl group. Thus, *carboxypeptidase A induces electronic strain in its substrate to accelerate catalysis.*

The need for substrate-induced structural changes in the active site of carboxypeptidase A can now be appreciated. The bound substrate is surrounded on all sides by catalytic groups of the enzyme. This arrangement promotes catalysis for the reasons cited above. *It is evident that a substrate could not enter such an array of catalytic groups (nor could a product leave) unless the enzyme were flexible.* In general, a flexible enzyme has a much larger repertoire of potential conformations that can be exploited for catalysis and selected in the course of evolution than a rigid enzyme. Furthermore, an induced fit may contribute to the specificity of an enzyme. A substrate for carboxypeptidase A must have a terminal carboxylate group. The enzyme tests for the presence of this group in the following way. If present, the terminal carboxylate forms a salt link with arginine 145, which triggers the movement of tyrosine 248 into a catalytically active position. If there is no terminal carboxylate, this movement fails to occur, and the enzyme is inactive. In other words, *induced fit can serve as a dynamic recognition process.*

SUMMARY

Lysozyme is a small enzyme that cleaves the polysaccharide component of bacterial cell walls. This polysaccharide is an alternating polymer of *N*-acetylglucosamine (NAG) and *N*-acetylmuramic acid (NAM) residues joined by $\beta(1 \longrightarrow 4)$ glycosidic linkages. Lysozyme hydrolyzes the glycosidic bond between C-1 of NAM and C-4 of NAG. Oligomers of *N*-acetylglucosamine are also hydrolyzed by lysozyme. Hexa-NAG and higher oligomers are good substrates, whereas tri-NAG and di-NAG are hydrolyzed at very slow rates. Tri-NAG is an effective competitive inhibitor. The three-dimensional structure of lysozyme and of the tri-NAG-lysozyme complex are known at atomic resolution. Tri-NAG occupies half of a cleft that runs across the enzyme, to which it is bound by many hydrogen bonds and van der Waals interactions. The mode of binding of hexa-NAG, a good substrate, was deduced by model building starting with the structure of the tri-NAG complex.

A plausible catalytic mechanism has been proposed for lysozyme. First, the critical groups in catalysis are the un-ionized carboxyl of glutamic 35 and the ionized carboxylate of aspartate 52. Both are about 3 Å from the glycosidic linkage that is hydrolyzed, namely, the one between residues D and E of a hexameric substrate. Second, glutamate 35 donates an H^+ to the bond between C-1 of the D ring and the glycosidic oxygen atom, thereby cleaving this bond. Carbon-1 of the D ring becomes positively charged; this transient species is called a *carbonium* ion. Third, this carbonium ion reacts with OH^- from the solvent, and glutamic 35 becomes protonated again. Lysozyme is ready for another round of catalysis after the products diffuse away. Fourth, catalysis is markedly enhanced by two factors that promote the formation of the carbonium ion intermediate. The electrostatic factor is the proximity of the negatively charged side chain of aspartate 52. The geometrical factor is the distortion of ring D into a half-chair form, which enables the positive charge in the carbonium ion to be shared between C-1 and the ring oxygen atom. Lysozyme clearly demonstrates the importance of *geometrical strain* in catalysis.

Carboxypeptidase A, a digestive enzyme that hydrolyzes the carboxyl-terminal peptide in polypeptides, exemplifies some different principles of catalysis. The structure of this enzyme and its complex with glycyltyrosine, a substrate analog, are known at atomic resolution. The binding of glycyltyrosine induces large structural changes at the active site, which convert it from a water-filled to a hydrophobic region. Carboxypeptidase A exemplifies the role of *induced fit* in catalysis. Another noteworthy feature of this enzyme is that it contains a zinc ion at its active site, which is essential for catalysis. The carbonyl atom of the susceptible peptide bond is polarized by the zinc so that it becomes more vulnerable to nucleophilic attack. This is an example of the induction of *electronic strain* in a substrate. In a proposed catalytic mechanism for carboxypeptidase A, the phenolic hydroxyl of tyrosine 248 donates a H^+ to the NH group of the susceptible peptide bond. The CO group of this bond is attacked by the carboxylate of glutamate 270 to form an anhydride, which is then hydrolyzed. Alternatively, a water molecule activated by glutamate 270 may directly attack the carbonyl group of the susceptible peptide bond.

SELECTED READINGS

Where to start:

Phillips, D. C., 1966. The three-dimensional structure of an enzyme molecule. *Sci. Amer.* 215(5):78–90. [Available as *Sci. Amer.* Offprint 1055. A superb article on the three-dimensional structure and catalytic mechanism of lysozyme.]

Lipscomb, W. N., 1971. Structures and mechanisms of enzymes. *Proc. Robert A. Welch Found. Conf. Chem. Res.* 15:131–156. [A beautifully illustrated and incisive discussion of the structure and mechanism of carboxypeptidase A and several other enzymes.]

Structure and enzymatic mechanism of lysozyme:

Osserman, E. F., Canfield, R. E., and Beychok, S., (eds.), 1974. *Lysozyme.* Academic Press.

Chipman, D. M., and Sharon, N., 1969. Mechanism of lysozyme action. *Science* 165:454–465.

Imoto, T., Johnson, L. N., North, A. C. T., Phillips, D. C., and Rupley, J. A., 1972. Vertebrate lysozymes. *In* Boyer, P. D., (ed.), *The Enzymes* (3rd ed.), vol. 7, pp. 666–868. Academic Press.

Dahlquist, F. W., Rand-Meir, T., and Raftery, M. A., 1968. Demonstration of carbonium ion intermediate during lysozyme catalysis. *Proc. Nat. Acad. Sci.* 61:1194–1198.

Rupley, J. A., 1967. The binding and cleavage by lysozyme of N-acetylglucosamine oligosaccharides. *Proc. Roy. Soc.* (B)167:416–428. [Presents results on hydrolysis in ^{18}O water, $\Delta G°$ for the binding of oligosaccharides, and rates of hydrolysis by lysozyme.]

Lin, T. Y., and Koshland, D. E., Jr., 1969. Carboxyl group modification and the activity of lysozyme. *J. Biol. Chem.* 244:505–508. [Presents evidence for the catalytic role of aspartate 52.]

Role of strain in catalysis:

Pauling, L., 1948. Nature of forces between large molecules of biological interest. *Nature* 161:707–709. [Includes a visionary statement of the importance of strain in enzymatic catalysis.]

Wolfenden, R., 1972. Analog approaches to the structure of the transition state in enzyme reactions. *Accounts Chem. Res.* 5:10–18.

Lienhard, G. E., 1973. Enzymatic catalysis and transition-state theory. *Science* 180:149–154.

Secemski, I. I., and Lienhard, G. E., 1971. The role of strain in catalysis by lysozyme. *J. Amer. Chem. Soc.* 93:3549–3550.

Structure and enzymatic mechanism of carboxypeptidase A:

Quiocho, F. A., and Lipscomb, W. N., 1971. Carboxypeptidase A: a protein and an enzyme. *Advan. Protein Chem.* 25:1–78. [A review of the crystallographic studies of this enzyme. Includes atomic coordinates.]

Johansen, J. T., and Vallee, B. L., 1973. Conformations of arsanilazotyrosine-248 carboxypeptidase $A_{\alpha,\beta,\gamma}$. Comparison of crystals and solution. *Proc. Nat. Acad. Sci.* 70:2006–2010. [This paper questions the occurrence of an inward movement of tyrosine 248 toward the zinc atom as a necessary part of the catalytic mechanism.]

Lipscomb, W. N., 1973. Enzymatic activities of carboxypeptidase A's in solution and in crystals. *Proc. Nat. Acad. Sci.* 70:3797–3801.

Neurath, H., Bradshaw, R. A., Pétra, P. H., and Walsh, K. A., 1970. Bovine carboxypeptidase A: activation, chemical structure and molecular heterogeneity. *Phil. Trans. Roy. Soc. London* (B)257:159–176.

Vallee, B. L., Riordan, J. F., Auld, D. S., and Latt, S. A., 1970. Chemical approaches to the mode of action of carboxypeptidase A. *Phil. Trans. Roy. Soc. London* (B)257:215–230.

PROBLEMS

1. Predict the relative rates of hydrolysis by lysozyme of these oligosaccharides (G stands for an *N*-acetylglucosamine residue, and M for *N*-acetylmuramic acid):
    (a) M-M-M-M-M-M
    (b) G-M-G-M-G-M
    (c) M-G-M-G-M-G

2. Predict on the basis of the data given in Figure 7-17 which of the sugar binding sites A to F on lysozyme will be occupied in the major complex with each of these oligosaccharides (same abbreviations as in Problem 1):
    (a) G-G
    (b) G-M   (c) G-G-G-G

3. Suppose that hexa-NAG is synthesized so that the glycosidic oxygen between its D and E sugar residues is labeled with ^{18}O. Where will this isotope appear in the products formed by hydrolysis with lysozyme?

4. Compare the coordination of the zinc atom in carboxypeptidase A with that of the iron atom in oxymyoglobin and oxyhemoglobin.
    (a) Which atoms are directly bonded to these metal ions?
    (b) Which side chains contribute these metal-binding groups?
    (c) Which other side chains in proteins are potential metal-binding groups?

ZYMOGEN ACTIVATION: DIGESTIVE ENZYMES AND CLOTTING FACTORS

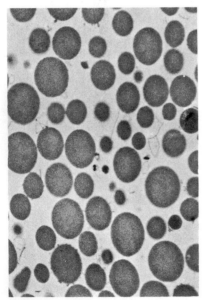

Figure 8-1
Electron micrograph of zymogen granules in an acinar cell of the pancreas. [Courtesy of Dr. George Palade.]

Lysozyme acquires full enzymatic activity as it spontaneously folds into its characteristic three-dimensional form. In contrast, many other proteins are synthesized as inactive precursors that subsequently are activated by cleavage of one or a few specific peptide bonds. If the active protein is an enzyme, the inactive precursor is called a *zymogen* or a *proenzyme*.

Activation of proteins by specific proteolysis recurs frequently in biological systems. Several examples follow.

1. The *digestive enzymes* that hydrolyze proteins are synthesized as zymogens in the stomach and pancreas (Table 8-1).

Table 8-1
Gastric and pancreatic zymogens

Site of synthesis	Zymogen	Active enzyme
Stomach	Pepsinogen	Pepsin
Pancreas	Chymotrypsinogen	Chymotrypsin
Pancreas	Trypsinogen	Trypsin
Pancreas	Procarboxypeptidase	Carboxypeptidase
Pancreas	Proelastase	Elastase

2. *Blood clotting* is mediated by a cascade of proteolytic activations that assure a rapid and amplified response to trauma.

3. Some protein hormones are synthesized as inactive precursors. For example, *insulin* is derived from *proinsulin* by proteolytic removal of a peptide.

4. The fibrous protein *collagen,* which occurs in skin and bone, is derived from *procollagen,* a soluble precursor.

CHYMOTRYPSINOGEN IS ACTIVATED BY SPECIFIC CLEAVAGE OF A SINGLE PEPTIDE BOND

Chymotrypsin is a digestive enzyme that hydrolyzes proteins in the small intestine. The inactive precursor *chymotrypsinogen* is synthesized in the pancreas, as are several other zymogens and digestive enzymes. Indeed, the pancreas is one of the most active organs in synthesizing proteins. The enzymes and zymogens are synthesized in the acinar cells of the pancreas (Figure 8-2). The proteins travel from the endoplasmic reticulum to the Golgi apparatus, where they are covered with a membrane made up of lipid and protein. These *zymogen granules* appear in electron micrographs as very dense bodies because they have a high concentration of protein (Figure 8-1). The zymogen granules accumulate at the apex of the acinar cell and are secreted into a duct leading into the duodenum when stimulated by a hormonal or nerve impulse signal.

Chymotrypsinogen is a single polypeptide chain consisting of 245 amino acid residues. It is cross-linked by five disulfide bonds. Chymotrypsinogen is virtually devoid of enzymatic activity. It is converted into a fully active enzyme when the peptide bond joining arginine 15 and isoleucine 16 is cleaved by trypsin (Figure 8-3). The resulting active enzyme, called π-chymotrypsin, then acts on other π-chymotrypsin molecules. Two peptides are removed to yield α-chymotrypsin, the stable form of the enzyme. The additional cleavages that are made in going from π- to α-chymotrypsin are superfluous, since π-chymotrypsin is already fully active. The striking feature of this activation process is that *cleavage of a single specific peptide bond transforms the protein from a catalytically inactive form to one that is fully active.*

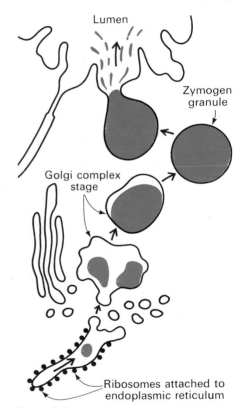

Figure 8-2
Diagrammatic representation of the secretion of zymogens by an acinar cell of the pancreas. [Based on a drawing kindly provided by Dr. George Palade.]

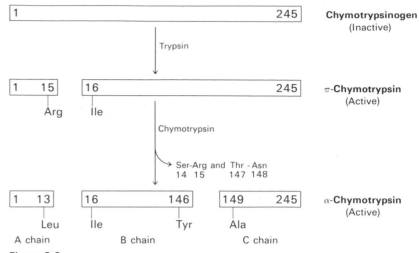

Figure 8-3
Activation of chymotrypsinogen.

THREE-DIMENSIONAL STRUCTURE OF CHYMOTRYPSIN

An understanding of this remarkable activation process depends on a detailed knowledge of the structure and catalytic mechanism of chymotrypsin. Fortunately, much is known about this enzyme from chemical and x-ray crystallographic studies. In fact, chymotrypsin is one of the most thoroughly studied of all enzymes, and so it is rewarding to look at it in some detail.

Alpha-chymotrypsin consists of three polypeptide chains connected by two interchain disulfide bonds (Figure 8-4). The molecular weight of the enzyme is about 25,000. The three-dimensional structure of the enzyme at 2 Å resolution (Figure 8-5) is known from the x-ray crystallographic studies of David Blow. The molecule is

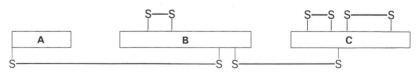

Figure 8-4
Alpha-chymotrypsin contains two interchain disulfide bonds and three intrachain disulfide bonds.

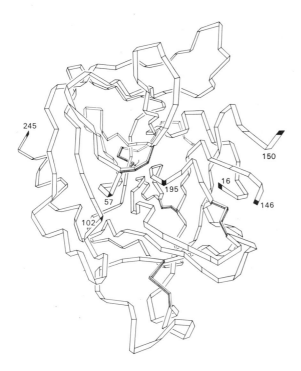

Figure 8-5
Three-dimensional structure of
α-chymotrypsin. Only the
α-carbon atoms are shown.
[Courtesy of Dr. David Blow.]

a compact ellipsoid of dimensions 51 × 40 × 40 Å. All charged
groups are on the surface of the molecule except for three that play
a critical role in catalysis. The folding of the molecule is complex.
Chymotrypsin contains very little α helix, in contrast to myoglobin
and hemoglobin. The chains tend to be fully extended and often
run parallel to each other, separated by about 5 Å. There is extensive hydrogen bonding between the peptide groups of adjacent
strands. Parts of the molecule have a secondary structure that
resembles an antiparallel pleated sheet, as was also found in
lysozyme.

CHYMOTRYPSIN IS SPECIFIC FOR AROMATIC
AND BULKY NONPOLAR SIDE CHAINS

The biological role of chymotrypsin is to catalyze the hydrolysis
of proteins in the small intestine (Figure 8-6). The equilibrium of
this reaction overwhelmingly favors hydrolysis (>99%). Chymotrypsin does not cleave all peptide bonds at a significant rate.

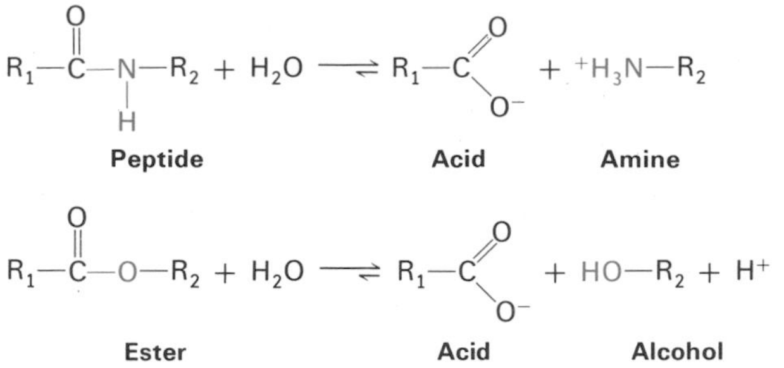

Figure 8-6
Chymotrypsin catalyzes the hydrolysis of peptide and ester bonds.

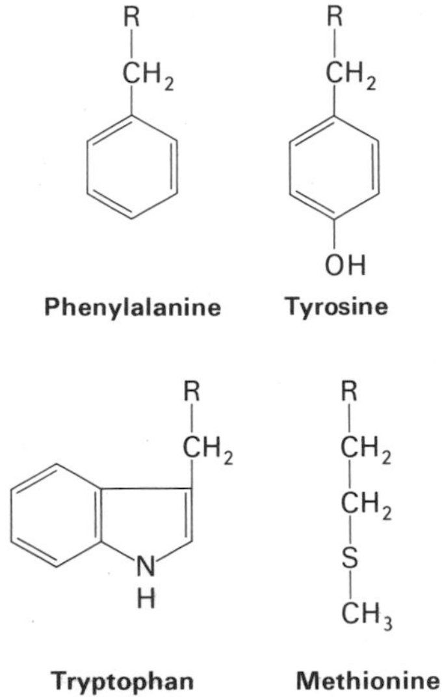

Figure 8-7
Chymotrypsin preferentially hydrolyzes peptide bonds on the carboxyl side of aromatic and bulky nonpolar side chains.

Rather, it is selective for peptide bonds on the *carboxyl side* of the *aromatic side chains* tyrosine, tryptophan, and phenylalanine, and of large *hydrophobic residues such as methionine* (Figure 8-7).

Chymotrypsin also hydrolyzes *ester bonds*. Though this reaction is not important physiologically, it is of interest because of its close relationship to peptide-bond hydrolysis (Figure 8-6). Indeed, much of our knowledge of the catalytic mechanism of chymotrypsin comes from studies of the hydrolysis of simple esters.

PART OF THE SUBSTRATE IS COVALENTLY BOUND TO CHYMOTRYPSIN DURING CATALYSIS

Chymotrypsin catalyzes the hydrolysis of peptide or ester bonds in two distinct stages. This was first revealed by studies of the kinetics of hydrolysis of *p*-nitrophenyl acetate. The liberation of *p*-nitrophenol, one of the products, is clearly biphasic when large amounts of enzyme are used (Figure 8-8). There is an initial *rapid burst* of *p*-nitrophenol product, followed by its formation at a much *slower steady-state rate*.

The first step is the combination of *p*-nitrophenyl acetate with chymotrypsin to form an enzyme-substrate (ES) complex (Figure 8-9). The ester bond of this substrate is cleaved. One of the products, *p*-nitrophenol, is then released from the enzyme, whereas the acetyl group of the substrate becomes covalently attached to the enzyme. Water then attacks the acetyl-enzyme complex to yield acetate ion and regenerate the enzyme (Figure 8-10). The initial rapid burst of *p*-nitrophenol production corresponds to the formation of the acetyl-enzyme complex. This step is called *acylation.* The slower steady-state production of *p*-nitrophenol corresponds to the hydrolysis of the acetyl-enzyme complex to regenerate the free enzyme. This second step, called *deacylation,* is rate-limiting in the hydrolysis of esters by chymotrypsin. In fact, the acetyl-enzyme complex is sufficiently stable to be isolated under appropriate conditions. The catalytic mechanism of chymotrypsin can thus be represented by the following scheme, in which P_1 is the amine (or alcohol) component of the substrate, $E{-}P_2$ is the covalent intermediate, and P_2 is the acid component of the substrate.

$$E + S \rightleftharpoons ES \xrightarrow[\textstyle\downarrow P_1]{} E{-}P_2 \xrightarrow[\textstyle\downarrow P_2]{} E$$

A distinctive feature of this mechanism is the occurrence of a covalent intermediate. In the reaction discussed above, an acetyl

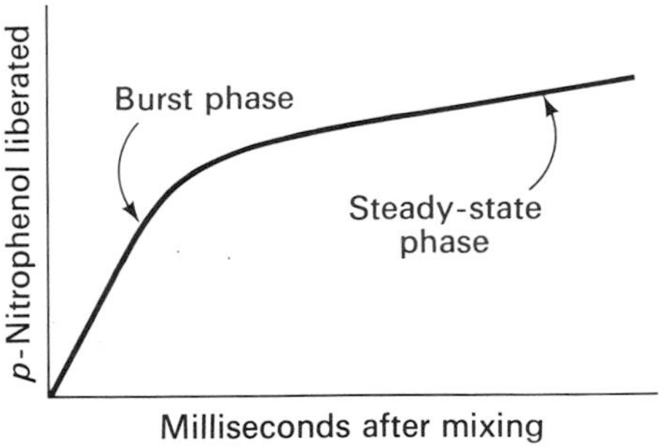

Figure 8-8
Two phases in the formation of *p*-nitrophenol are evident following the mixing of chymotrypsin and *p*-nitrophenyl acetate.

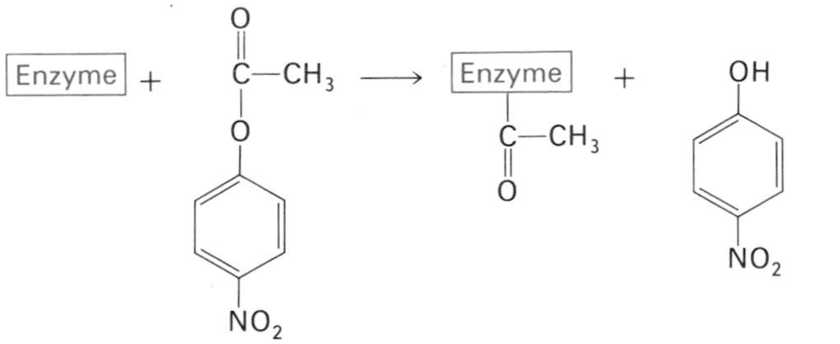

***p*-Nitrophenyl acetate** **Acetyl-enzyme intermediate** ***p*-Nitrophenol**

Figure 8-9
Acylation: formation of the acetyl-enzyme intermediate.

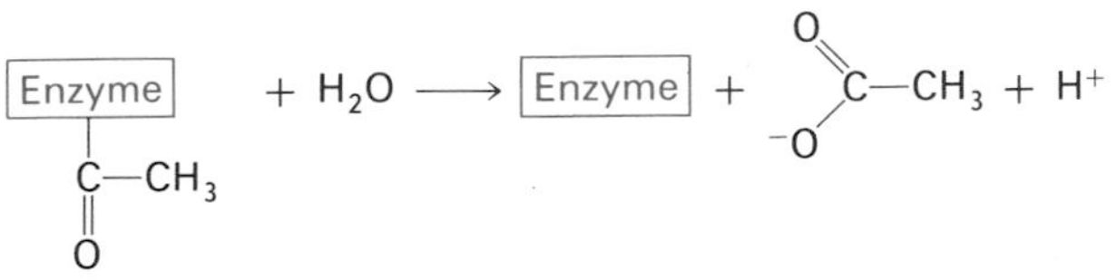

Acetyl-enzyme intermediate **Acetate**

Figure 8-10
Deacylation: hydrolysis of the acetyl-enzyme intermediate.

group is covalently bonded to the enzyme. In general, the group attached to chymotrypsin at the E—P$_2$ stage is an acyl group. Thus, E—P$_2$ is an *acyl-enzyme intermediate*.

THE ACYL GROUP IS ATTACHED TO AN UNUSUALLY REACTIVE SERINE RESIDUE ON THE ENZYME

The site of attachment of the acyl group was identified following the isolation of E—P$_2$, which is quite stable at pH 3. The acyl group is linked to the oxygen atom of a specific serine residue, namely serine 195.

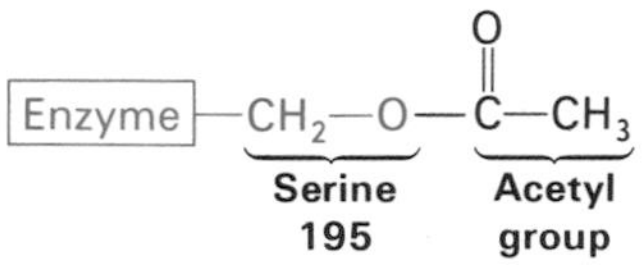

$$\boxed{\text{Enzyme}}\!-\!CH_2\!-\!O\!-\!\overset{\displaystyle O}{\overset{\|}{C}}\!-\!CH_3$$

Serine   Acetyl
195   group

Serine 195 is unusually reactive. It can be specifically labeled with *organic fluorophosphates* such as diisopropylphosphofluoridate (DIPF). DIPF reacts only with serine 195 to form an inactive *diisopropyl-phosphoryl-enzyme complex,* which is indefinitely stable (Figure 8-11). The remarkable reactivity of serine 195 is highlighted by the fact that the other 27 serine residues in chymotrypsin are untouched by DIPF.

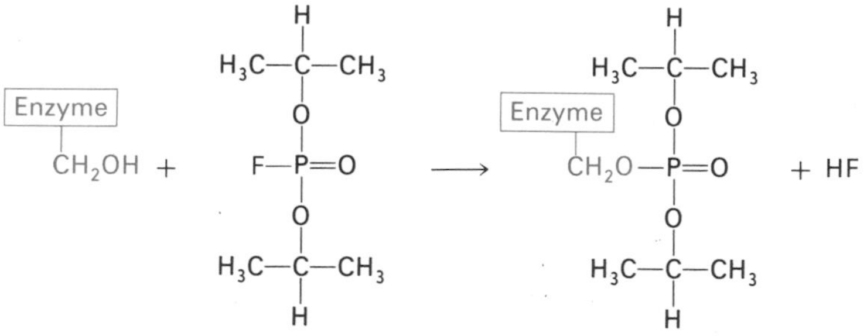

Figure 8-11
Diisopropylphosphofluoridate (DIPF) inactivates chymotrypsin by forming a diisopropylphosphoryl derivative of serine 195.

Chymotrypsin is not the only enzyme to be inactivated by DIPF. Numerous other proteolytic enzymes, such as trypsin, elastase, thrombin, and subtilisin, react specifically with DIPF and are thereby inactivated. The reaction occurs at a unique serine residue, as in chymotrypsin. Hence, these enzymes are called the *serine proteases*. DIPF also reacts with the serine residue of *acetylcholinesterase*, an enzyme crucial for the transmission of nerve impulses at certain synapses. In fact, the inactivation of acetylcholinesterase by DIPF is the basis for its use in *insecticides* and *nerve gases*.

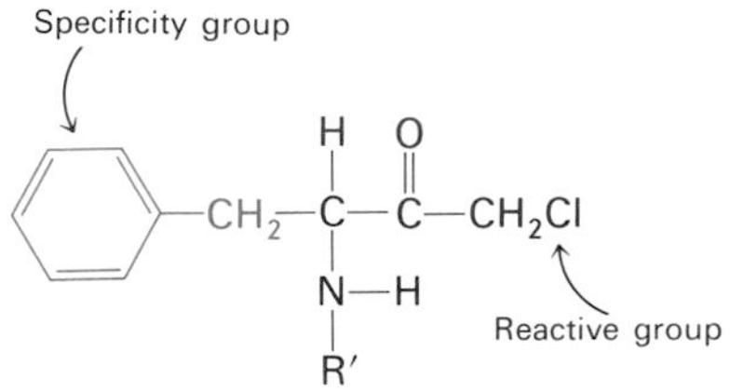

DEMONSTRATION OF THE CATALYTIC ROLE OF HISTIDINE 57 BY AFFINITY LABELING

The importance of a second residue in catalysis was shown by *affinity-labeling* studies. The strategy was to react chymotrypsin with a molecule that (*a*) specifically binds to the active site because it resembles a substrate, and then (*b*) forms a stable covalent bond with a group on the enzyme that is in close proximity. These criteria are met by tosyl-L-phenylalanine chloromethyl ketone (TPCK) (Figure 8-12). TPCK binds specifically to chymotrypsin because it contains a phenylalanine side chain. The reactive group in TPCK is the chloromethyl ketone function. TPCK attacks chymotrypsin only at histidine 57, which is alkylated at one of its ring nitrogens (Figure 8-13). The TPCK derivative of chymotrypsin is enzymatically inactive. Three lines of evidence indicated that histidine 57 is part of the active site. First, the affinity-labeling reaction was highly stereospecific; the D-isomer of TPCK was totally ineffective. Second, the reaction was inhibited when a competitive inhibitor of chymotrypsin, β-phenylpropionate, was present. Third, the rate of inactivation by TPCK varied with pH in nearly the same way as the rate of catalysis.

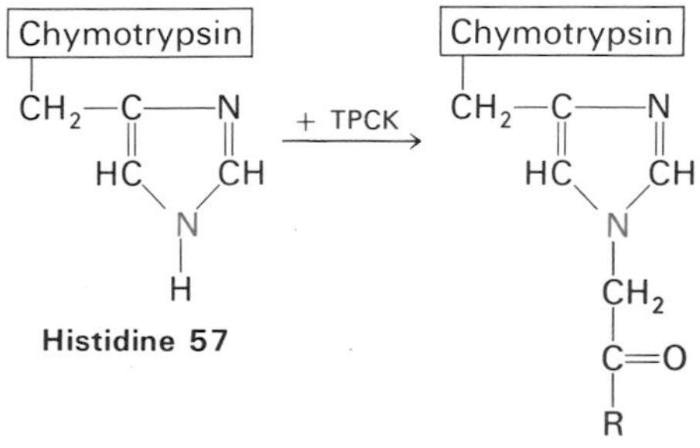

Figure 8-12
Structure of tosyl-L-phenylalanine chloromethyl ketone (TPCK), an affinity labeling reagent for chymotrypsin (R′ represents a tosyl group).

Figure 8-13
Alkylation of histidine 57 in chymotrypsin by TPCK.

A CHARGE RELAY NETWORK MAKES THE ACTIVE-SITE SERINE A POWERFUL NUCLEOPHILE

The catalytic activity of chymotrypsin depends on the unusual reactivity of serine 195. A —CH$_2$OH group is usually quite unreactive under physiological conditions. What makes it so reactive

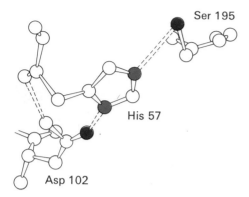

Figure 8-14
Conformation of the charge relay system
in chymotrypsin. [Based on D. M. Blow
and T. A. Steitz. X-ray diffraction studies
of enzymes, *Annu. Rev. Biochem.* 39(1970):
86. Copyright © 1970 by Annual
Reviews Inc. All rights reserved.]

in the active site of chymotrypsin? A plausible explanation has
emerged from the x-ray studies of the three-dimensional structure
of the enzyme. As could be expected from the affinity labeling work,
histidine 57 is adjacent to serine 195. The carboxyl side chain of
aspartic 102 is also nearby (Figure 8-14).

In fact, these three residues interact to enhance the catalytic
power of chymotrypsin. Aspartic 102 is hydrogen-bonded to histidine 57, which is in turn hydrogen-bonded to serine 195. These
three residues are said to form a *charge relay network* (Figure 8-15).
In the hydrogen-bond arrangement shown in Figure 8-15A, aspartic
102 is in the ionized —COO⁻ form and serine 195 is in the un-
ionized —CH₂OH form. However, aspartic 102 is located in the
interior of the enzyme, where it is shielded from the solvent by
histidine 57 and by several nonpolar side chains. The pK_A of
aspartic 102 is therefore expected to be higher than normal. Hence,
a small but significant fraction of the enzyme molecules probably
have the hydrogen-bond array shown in Figure 8-15B, in which
aspartic 102 draws a proton away from serine 195, through histidine
57. *The operation of this charge relay network makes serine 195 a more
powerful nucleophile.* Recent studies have indicated that the rate of
catalysis is enhanced by a factor of the order of 10^3 by this charge
relay network.

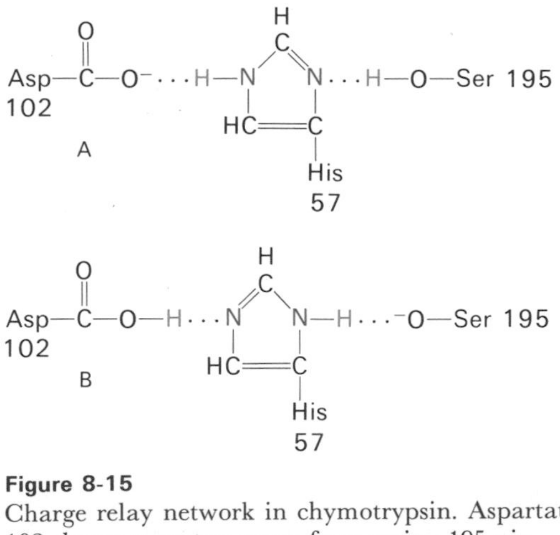

Figure 8-15
Charge relay network in chymotrypsin. Aspartate
102 draws a proton away from serine 195 via
histidine 57.

MODE OF BINDING OF A SUBSTRATE ANALOG TO CHYMOTRYPSIN

Crystallographic studies of complexes of chymotrypsin with substrate analogs have shown the location of the specificity site and the likely orientation of the susceptible peptide bond of a good substrate. Formyl-L-tryptophan binds to chymotrypsin with its indole side chain fitted neatly into a pocket near serine 195 (Figure 8-16). The hydrogen of its NH group is hydrogen-bonded to the main-chain carbonyl group of residue 214. The carboxylate group of this substrate analog is in close proximity to histidine 57 and serine 195.

PROPOSED CATALYTIC MECHANISM OF CHYMOTRYPSIN

A plausible catalytic mechanism for chymotrypsin has been proposed on the basis of x-ray crystallographic and chemical data (Figures 8-17 and 8-18). In this mechanism, *histidine 57 and serine 195 participate directly in the cleavage of the susceptible peptide bond of the substrate.* Peptide-bond hydrolysis starts with an attack of the oxygen atom of the hydroxyl group of serine 195 on the carbonyl carbon atom of the susceptible peptide bond (Figure 8-17). *A transient tetrahedral intermediate is formed.* This reaction is facilitated

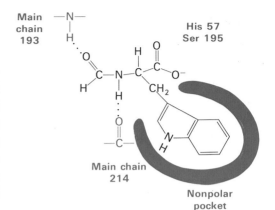

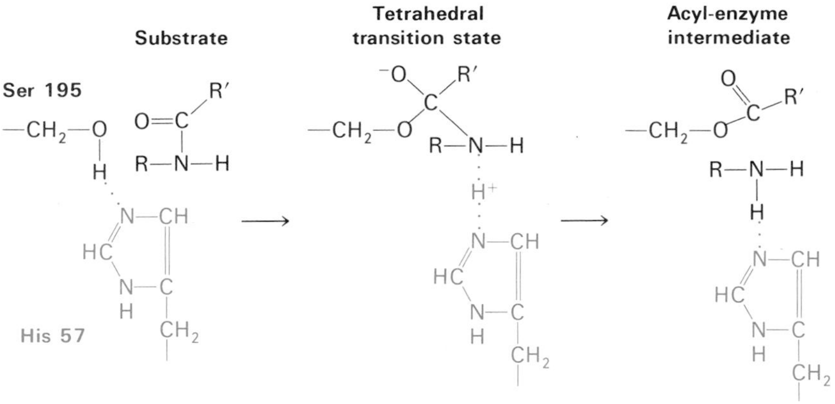

Figure 8-16
A schematic representation of the binding of formyl-L-tryptophan, a substrate analog, to chymotrypsin.

Figure 8-17
First stage in the hydrolysis of a peptide by chymotrypsin: *acylation.* A tetrahedral transition state is postulated in this proposed mechanism.

by the charge relay network described above, which serves to draw a proton away from the hydroxyl group of serine 195, thereby making it a powerful nucleophile. Histidine 57 then donates a proton to the nitrogen atom of the susceptible peptide bond. As a result, the susceptible peptide bond is cleaved. At this stage, the amine component is hydrogen-bonded to histidine 57, whereas the acid component of the substrate is esterified to serine 195. The *acylation stage* of the hydrolytic reaction is now completed.

The next stage is *deacylation* (Figure 8-18). The amine component of the substrate diffuses away, and a water molecule takes its place at the active site. In essence, *deacylation is the reverse of acylation, with H_2O substituting for the amine component*. First, the charge relay network draws a proton away from water. The resulting OH^- ion simultaneously attacks the carbonyl carbon atom of the acyl group that is attached to serine 195. As in acylation, a transient tetrahedral intermediate is formed. Histidine 57 then donates a proton to the oxygen atom of serine 195, which results in the release of the acid component of the substrate. The acid component of the substrate diffuses away and the enzyme is ready for another round of catalysis.

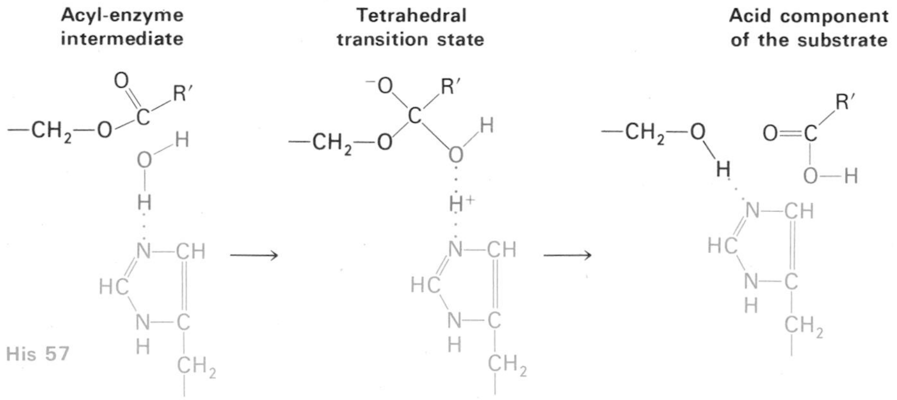

Figure 8-18
Second stage in the hydrolysis of a peptide of chymotrypsin: *deacylation*. The acyl-enzyme intermediate is hydrolyzed by water. Note that deacylation is essentially the reverse of acylation, with water replacing the amine component of the substrate.

MECHANISM OF ZYMOGEN ACTIVATION

We now return to the question of how cleavage of a single peptide bond in chymotrypsinogen converts it into an active enzyme. The

three-dimensional structure of chymotrypsinogen has been elucidated by Joseph Kraut and some of the conformational changes in
activation have been identified:

1. Hydrolysis of the peptide bond between arginine 15 and
isoleucine 16 creates new carboxyl- and amino-terminal groups.

2. The newly formed *amino-terminal group of isoleucine 16 turns
inward and interacts with aspartate 194* in the interior of the chymotrypsin molecule (Figure 8-19). This amino group must be protonated for chymotrypsin to be active, as shown by the dependence
of enzyme activity on pH.

3. This electrostatic interaction between a positively charged
amino group and a negatively charged carboxylate ion in a nonpolar region triggers a number of conformational changes. Methionine 192 moves from a deeply buried position in the zymogen to
the surface of the molecule in the enzyme, and residues 187 and
193 become more extended. These changes result in the formation
of the *substrate specificity site* for aromatic and bulky nonpolar groups.
One side of this site is made up of residues 189–192. *This cavity
for part of the substrate does not exist in the zymogen.*

4. The tetrahedral transition-state intermediate in acylation and
deacylation is stabilized by a hydrogen bond that can be formed
in chymotrypsin but not in chymotrypsinogen. The carbonyl oxygen atom of the substrate acyl group appears to be hydrogen-
bonded to the main-chain NH groups of glycine 193 and serine
195 (Figure 8-20). The hydrogen bond to glycine 193 is not sterically
feasible in chymotrypsinogen.

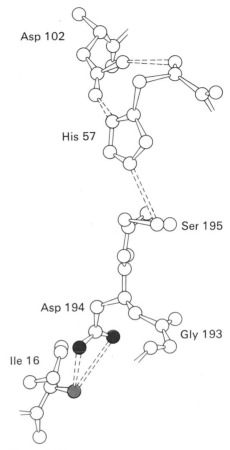

Figure 8-19
Environment of asparate 194 and
isoleucine 16 in chymotrypsin. The
electrostatic interaction between the
carboxylate of Asp 194 (red) and the
α-NH₂ group of Ile 16 (blue) is essential
for the activity of chymotrypsin. These
groups are adjacent to the charge relay
network. [Based on D. M. Blow and T. A.
Steitz. X-ray diffraction studies of
enzymes, *Annu. Rev. Biochem.* 39 (1970):
86. Copyright © 1970 by Annual Reviews
Inc. All rights reserved.]

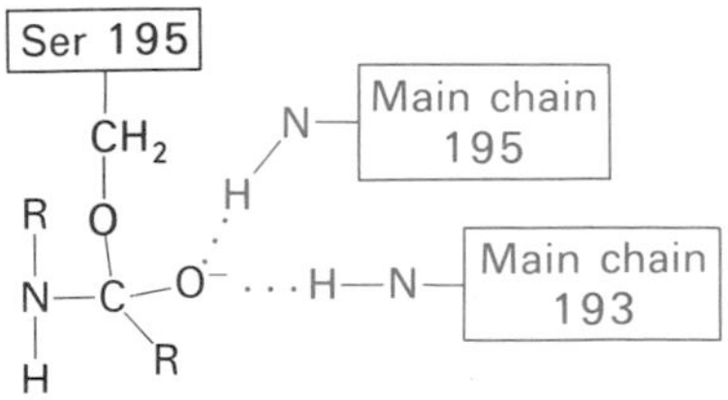

Figure 8-20
The proposed tetrahedral transition-state intermediate
in the acylation and deacylation reactions of
chymotrypsin is probably stabilized by hydrogen
bonding to two NH groups in the main chain.

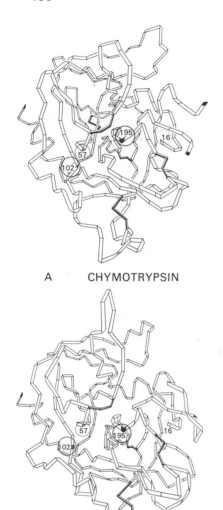

A   CHYMOTRYPSIN

B   ELASTASE

Figure 8-21
Comparison of the conformation of the
main chain of (A) chymotrypsin and (B)
elastase. The locations of the charge
relay network (residues 102, 57, and 195)
and of the α-amino group of residue 16
are shown in color to emphasize the
similarity of these enzymes. [Based on
B. S. Hartley and D. M. Shotton. In
The Enzymes, P. D. Boyer, ed., 3rd ed.,
vol. 3 (Academic Press, 1971), p. 362.]

5. The conformational changes elsewhere in the molecule are
very small. Thus, *the switching on of enzymatic activity in a protein can
be accomplished by discrete, highly localized conformational changes that are
triggered by the hydrolysis of a single peptide bond.*

TRYPSIN AND ELASTASE: VARIATIONS ON A THEME

Trypsin and elastase are like chymotrypsin in a number of respects:

1. They are secreted by the pancreas as zymogens and have
similar activation mechanisms.

2. They are inhibited by fluorophosphates such as DIPF. Trypsin
and elastase also have an active-site serine residue. In fact, the
amino acid sequence around this serine is the same in all three
enzymes: Gly-Asp-Ser-Gly-Gly-Pro.

3. Furthermore, they are alike in amino acid sequence in other
regions of the molecule also. About 40% of their sequences are
identical. The degree of identity is even higher for amino acid
residues located in the interior of these enzymes.

4. X-ray studies have shown that the tertiary structures of these
enzymes are very similar (Figure 8-21). Elastase and trypsin have
a charge relay network just like the one in chymotrypsin. The
catalytic mechanism of these three enzymes is probably nearly the
same.

Although similar in structure and mechanism, these enzymes
differ strikingly in specificity. Chymotrypsin requires an aromatic
or bulky nonpolar side chain. Trypsin requires a lysine or arginine
residue. Elastase cannot cleave either of these kinds of substrates.
The specificity of elastase is directed toward the smaller, uncharged
side chains. X-ray studies have shown that these different substrate
specificities arise from quite small structural changes in the binding
site (Figure 8-22). In chymotrypsin, a nonpolar pocket serves as a
niche for the aromatic or bulky nonpolar side chain. In trypsin,
one residue in this pocket is changed relative to chymotrypsin: a
serine is replaced by an aspartate. This aspartate in the nonpolar
pocket of trypsin can form a strong electrostatic bond with a
positively charged lysine or arginine side chain of a substrate. In

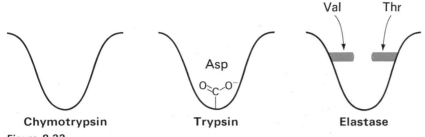

Figure 8-22
A highly simplified representation of part of the substrate binding site in chymotrypsin, trypsin, and elastase.

elastase, the pocket no longer exists because the two glycine residues lining it in chymotrypsin are replaced by the much bulkier valine and threonine.

DIVERGENT AND CONVERGENT EVOLUTION OF THE SERINE PROTEASES

Chymotrypsin, elastase, and trypsin have similar overall amino acid sequences because they evolved from a common ancestor. The gene for this common ancestral enzyme probably duplicated several times. Mutations in these genes eventually gave rise to the present-day proteins. The acquisition of different specificities by mutations of genes descended from a common ancestor is called *divergent evolution*.

A different kind of evolutionary process is revealed by comparing chymotrypsin with subtilisin, a bacterial enzyme. Subtilisin is also a serine protease. However, the amino acid sequences of chymotrypsin and subtilisin are very different, indicating that they arose independently in evolution. For example, chymotrypsin contains five disulfide bonds, whereas subtilisin has none. The sequence around the active-site serine is:

-Gly-Thr-Ser*-Met-Ala-Ser   (in subtilisin)

-Gly-Asp-Ser*-Gly-Gly-Pro   (in chymotrypsin)

The three-dimensional structures of these enzymes are entirely different. It was therefore surprising to find that *subtilisin has a charge relay network* that resembles the one in chymotrypsin in consisting of an aspartic, a histidine, and a serine residue. However, the similarity ends there, since the orientations of these three side chains are quite different in the two enzymes. Furthermore, the position of participating residues in the amino acid sequences of the two proteins is different:

Aspartic 32   Histidine 64   Serine 221   (in subtilisin)

Aspartic 102   Histidine 57   Serine 195   (in chymotrypsin)

The occurrence of charge relay networks in both subtilisin and chymotrypsin is a nice example of *independent convergent evolution* at the level of an enzyme molecule. There may be only a few ways, perhaps just one, of markedly enhancing the nucleophilic character of a serine residue in an enzyme. The vertebrate pancreatic proteases and the bacterial subtilisin protease appear to have attained the same solution by independent, parallel evolutionary processes.

THE ACTIVATIONS OF THE PANCREATIC ZYMOGENS ARE COORDINATED

The digestion of proteins in the duodenum requires the concurrent action of several proteolytic enzymes, since each is specific for a limited number of side chains. Thus the zymogens must be switched on at the same time. Coordinated control is achieved by the action of *trypsin as the common activator of all the pancreatic zymogens*—trypsinogen, chymotrypsinogen, proelastase, and procarboxypeptidase. How is enough trypsin produced to initiate these activation processes? The cells that line the duodenum produce an enzyme *enterokinase* that hydrolyzes a unique lysine–isoleucine peptide bond in trypsinogen, as it enters the duodenum. The small amount of trypsin produced in this way activates trypsinogen and the other zymogens. Thus, *the formation of trypsin by enterokinase is the master activation step.*

PREMATURE ACTIVATION OF ZYMOGENS CAN BE LETHAL, AS IN PANCREATITIS

The pancreas is protected in several ways from the proteolytic action of the enzymes it synthesizes. All of the pancreatic enzymes that hydrolyze proteins are synthesized as inactive zymogens. Furthermore, they are packaged in granules, surrounded by membranes made of protein and lipid. As a further safeguard, the pancreatic secretion contains an inhibitor of trypsin. This protein inhibitor complexes and renders inactive any small amounts of trypsin that may be present in the pancreas.

The pancreas also synthesizes lipases. The activation of these fat-splitting enzymes in the duodenum occurs by a mechanism entirely different from proteolytic activation. The lipases are active only in the presence of bile salts. These steroid derivatives solubilize lipids, and thereby render them susceptible to enzymatic cleavage by the lipases. Bile is synthesized in the liver, stored in the gall bladder, and released into the duodenum, where it acts in concert with the pancreatic lipases.

Acute pancreatitis is a serious and sometimes lethal disease that is characterized by the *premature activation of the proteolytic and lipolytic enzymes of the pancreas*. In pancreatitis, these enzymes are unleashed while they are still inside the pancreas. The effect is to destroy the pancreas itself and its blood vessels. Trauma to the acinar tissue can cause acute pancreatitis.

CLOTTING OCCURS BY A CASCADE OF ZYMOGEN ACTIVATIONS

The activation of a precursor protein by peptide-bond cleavage is a basic means of control that recurs in many different biochemical systems. We now turn to the role of zymogen activations in clot formation, which is one of three mechanisms of hemostasis. The other two are a rapid constriction of the injured vessel and the aggregation of platelets to form a plug on the injured surface of the blood vessel.

A clot is formed by a series of transformations involving more than ten different proteins. A striking feature of the process is the occurrence of a *series of zymogen activations*. In this enzymatic *cascade,* the activated form of a factor catalyzes the activation of the next factor. Very small amounts of the initial factors are needed because of the catalytic nature of the activation process. The numerous steps yield a large amplification, assuring a rapid response to trauma.

CLOT FORMATION INVOLVES THE INTERPLAY
OF TWO KINDS OF ENZYMATIC PATHWAYS

In 1863, Joseph Lister showed that blood stayed fluid in the excised jugular vein of an ox, but that it rapidly clotted when it was transferred to a glass vessel. The abnormal surface triggered components already present in the blood. Consequently, this pathway of clotting is called the *intrinsic* one. Clotting can also be triggered by the addition of substances that are normally not present in the blood. Extracts from many tissues, particularly brain, rapidly cause clot formation when added to plasma. This pathway of clotting is called the *extrinsic* one.

Clotting involves the interplay of the intrinsic and extrinsic pathways (Figure 8-23). Both are needed for proper clotting, as evidenced by various clotting disorders in which there is a deficiency of a single protein in one of the pathways. Furthermore, the intrinsic and extrinsic pathways converge on a final *common pathway* that results in the production of a *fibrin clot.*

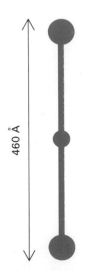

460 Å

Figure 8-24
Diagram of a fibrinogen molecule, based on electron micrographs.

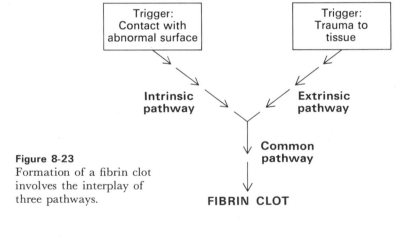

Figure 8-23
Formation of a fibrin clot involves the interplay of three pathways.

FIBRINOGEN IS CONVERTED BY THROMBIN
INTO A FIBRIN CLOT

The best-characterized part of the clotting process is the conversion of fibrinogen to fibrin by thrombin, a proteolytic enzyme. Fibrinogen is a much larger and more elongated protein than the ones discussed thus far (e.g., lysozyme and chymotrypsin). Electron micrographs show that fibrinogen is made up of three nodules connected by two rods (Figure 8-24). Fibrinogen is 460 Å long and has a molecular weight of 340,000, which makes it about ten times as large as chymotrypsin. It consists of six polypeptide chains. There are pairs of three kinds of chains: one of them is called Aα, the second Bβ, and the third γ.

Fibrinogen, a highly soluble molecule in the plasma, is converted into insoluble fibrin monomer by the proteolytic action of thrombin. Thrombin cleaves four *arginine–glycine peptide bonds* in fibrinogen. Four peptides are released: an A peptide of 18 residues from each of the two α chains and a B peptide of 20 residues from each of the two β chains. These A and B peptides are called *fibrinopeptides*. A fibrinogen molecule devoid of these fibrinopeptides is called *fibrin monomer*. It has the subunit structure $(\alpha\beta\gamma)_2$, and about 97% of the amino acid residues in fibrinogen.

FIBRIN MONOMERS SPONTANEOUSLY FORM FIBERS

Fibrin monomers have a much lower solubility than the parent fibrinogen molecules. They spontaneously associate to form *fibrin*, which has the form of long, insoluble fibers. Electron micrographs and low-angle x-ray patterns show that fibrin has a periodic structure that repeats every 230 Å (Figure 8-25). Since fibrinogen is about 460 Å long, it seems likely that fibrin monomers come together to form a half-staggered array (Figure 8-26).

Figure 8-25
Electron micrograph of fibrin. The 230-Å period along the fiber axis is half the length of a fibrinogen molecule. [Courtesy of Dr. Henry Slayter.]

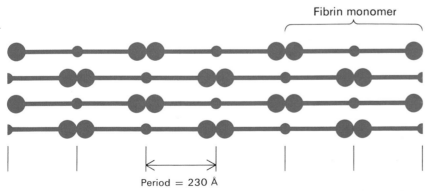

Figure 8-26
Proposed arrangement of fibrin monomers in a fibrin clot. This half-staggered array of fibrin monomers would yield the observed 230-Å period.

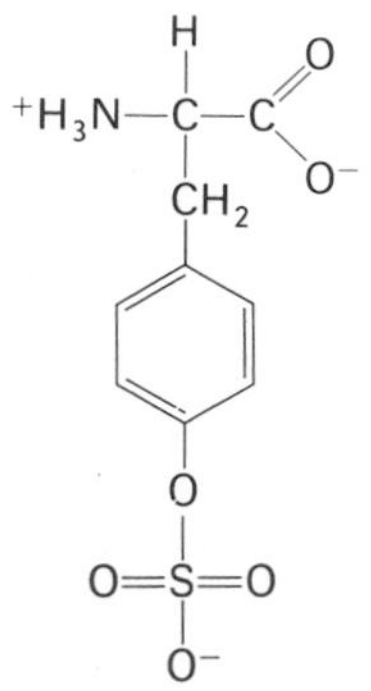

Tyrosine-*O*-sulfate

Why do fibrin monomers aggregate whereas their parent fibrinogen molecules stay in solution? A definitive answer to this question should come from detailed structural studies now in progress. The fibrinopeptides of all vertebrate species studied thus far have a *large net negative charge.* Aspartate and glutamate residues are found in abundance. An unusual negatively charged derivative of tyrosine, namely *tyrosine-O-sulfate,* is found in fibrinopeptide B. The presence of these and other negatively charged groups in the fibrinopeptides probably keeps fibrinogen molecules apart. *Their release by thrombin gives fibrin monomers a different surface-charge pattern, leading to their specific aggregation.* Recall that a change of a single charged group, glutamic to valine, causes the aggregation of deoxyhemoglobin molecules in sickle-cell anemia.

THE FIBRIN CLOT IS STRENGTHENED BY COVALENT CROSS-LINKS

The clot produced by the spontaneous aggregation of fibrin monomer is quite fragile. It is subsequently stabilized by the formation of covalent cross-links between the side chains of different molecules in the fibrin fiber. In fact, *peptide bonds are formed between specific glutamine and lysine side chains in a transamidation reaction* (Figure 8-27). This cross-linking reaction is catalyzed by a transamidase enzyme. Cross-links of this type are rarely found in proteins. The importance of this unusual bond in fibrin is highlighted by the finding that patients with a deficiency of this transamidase enzyme have a pronounced tendency to bleed.

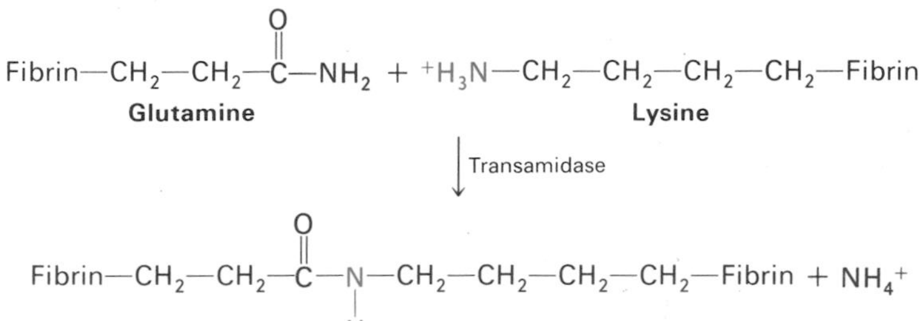

Figure 8-27
Fibrin is cross-linked by transamidation.

THROMBIN IS HOMOLOGOUS TO TRYPSIN

The specificity of thrombin for arginine–glycine bonds suggests that thrombin might resemble trypsin. Indeed it does, as shown by amino acid sequence studies. Thrombin has a molecular weight of 33,700 and consists of two chains. The A chain of 49 residues exhibits no detectable homology to the pancreatic enzymes. The B chain, however, is quite similar in sequence to trypsin, chymotrypsin, and elastase. The sequence around its active-site serine is Gly-Asp-Ser-Gly-Gly-Pro, the same as that in the pancreatic serine proteases. Moreover, thrombin also contains a charge relay network. Its three-dimensional structure is not yet known, but one important feature of its specificity site is already apparent. Thrombin, like trypsin, contains an aspartate residue at the bottom of its substrate binding cleft. This negatively charged group undoubtedly forms an electrostatic bond with the positively charged arginine side chain.

Thrombin is much more specific than trypsin. Thrombin cleaves certain arginine–glycine bonds, whereas trypsin cleaves most peptide bonds following arginine or lysine residues. Thrombin, like the pancreatic serine proteases, is synthesized as a zymogen called *prothrombin*, which has a molecular weight of about 70,000. The activation mechanism appears to be more complex than in the homologous pancreatic enzymes. However, at least one feature of the activation process is common to these enzymes. An ion pair like the one formed between the positively charged amino group of isoleucine 16 and the negatively charged aspartate 194 in chymotrypsin also occurs in thrombin.

Similarities in amino acid sequence indicate that *thrombin is evolutionarily related to the pancreatic serine proteases*. This is reinforced by the occurrence of a charge relay system and the likelihood of a similar activation mechanism. It is of interest to note that thrombin is formed in the liver, which has a common embryological origin with the pancreas.

PROTHROMBIN IS ACTIVATED BY FACTOR X_a

The cascade nature of the clotting process is evident in Figure 8-28. Prothrombin is activated by a proteolytic enzyme which is called Factor X_a. Clotting factors are assigned a Roman numeral for ease

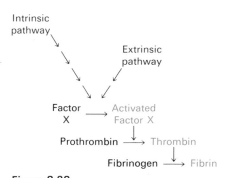

Figure 8-28
Final steps in the formation of a fibrin clot. These three reactions constitute the common pathway. Factor X is activated by products of the intrinsic and extrinsic pathways.

of discussion. The subscript "a" means that the factor is in the active form. Prothrombin is converted to thrombin by the proteolytic action of Factor X_a. This activation is accelerated by Factor V, which is not itself an enzyme. Factor V can be regarded as a *modifier protein*. In addition, Ca^{2+} ion and a phospholipid surface promote the activation of prothrombin. In fact, *Ca^{2+} and phospholipid are required in many of the steps in clotting*.

How is Factor X activated? This activation step is where the intrinsic and extrinsic pathways meet. The products of both the intrinsic and extrinsic pathways are proteolytic enzymes that activate Factor X.

HEMOPHILIA AND OTHER BLEEDING DISORDERS HAVE REVEALED SOME OF THE EARLY STEPS IN CLOTTING

It becomes progressively more difficult to carry out biochemical studies of the clotting factors in going from later to earlier stages. The amount of fibrinogen in 1 ml of blood is 3 mg, but the amount of Factor X is only 0.01 mg. The concentrations of some of the earlier clotting factors are even lower. Furthermore, these proteins are highly labile. Some of the important breakthroughs in the elucidation of the pathways of clotting have therefore come from studies of patients with bleeding disorders.

The best known of these diseases is *hemophilia*. This clotting disorder is genetically transmitted as a sex-linked recessive characteristic. Heterozygous females are asymptomatic carriers. A famous carrier of this disease was Queen Victoria, who transmitted it to the royal families of Prussia, Spain, and Russia (Figure 8-29).

In 1904, the tsarevich Alexis was born. He was the first male heir born to a reigning Russian tsar since the seventeenth century, which was taken as an omen of hope. Four healthy daughters had previously been born to Tsar Nicholas II and Empress Alexandra, a granddaughter of Queen Victoria. The mood changed six weeks after the birth of Alexis, as reflected in his father's diary: "A hemorrhage began this morning without the slightest cause from the navel of our small Alexis. It lasted with but a few interruptions until evening." The evidence grew stronger and more ominous as Alexis started to crawl and toddle, which caused large, blue swellings on his legs and arms. The hemorrhages became more serious

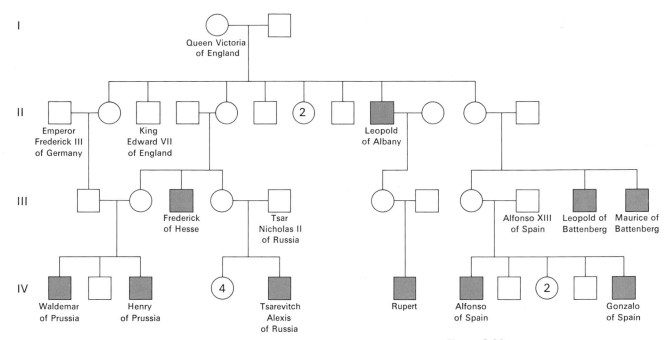

Figure 8-29
Pedigree of hemophilia in the royal families of Europe. All of Queen Victoria's children, but not all individuals in later generations, are included in this diagram. Females are symbolized by circles, normal males by white squares, and hemophilic males by red squares. [From C. Stern. *Principles of Human Genetics,* 3rd ed. (W. H. Freeman and Company). Copyright © 1973.]

and there was little that physicians could do to alleviate the pain. The anguished empress then turned to Rasputin, who was reputed to be a miracle man. No man knew less about molecular disease, no man ever profited more from it. Rasputin held a position of great power in the Russian court for many years because the empress placed great faith in his healing abilities.

Why did Nicholas and Alexandra marry when it was already known that her brother, nephews, and uncle had hemophilia? In 1873, Grandidier, a French physician, counseled that "all members of bleeder families should be advised against marriage." Indeed, the hereditary nature of the disease was fully appreciated as early as 1803, by John Otto:

About seventy or eighty years ago, a woman by the name of Smith settled in the vicinity of Plymouth, New Hampshire, and transmitted the following idiosyncrasy to her decendants. . . . It is a surprising circumstance that the males only are subject to this strange affection, and that all of them are not liable to it. . . . Although the females are exempt, they are still capable of transmitting it to their male children.

J. B. S. Haldane has suggested that "kings are carefully protected against disagreeable realities. . . . The hemophilia of the Tsarevich was a symptom of the divorce between royalty and reality."

INTRINSIC PATHWAY OF CLOTTING

The defect in hemophilia is in the intrinsic pathway (Figure 8-30). A protein called *antihemophilic factor* is missing or has a markedly reduced activity. Antihemophilic factor (VIII) acts in concert with Factor IX_a, a proteolytic enzyme, to activate Factor X. Antihemophilic factor is not itself an enzyme; rather, it is a modifier protein.

Factor IX was discovered in an interesting way. A boy named Stephen Christmas had a clotting disorder that was clinically indistinguishable from classical hemophilia in its symptoms and mode of inheritance. The clotting time of his blood measured in a glass tube was prolonged, as in hemophilia. His blood was then mixed with the blood from a hemophiliac. The striking result was that the mixed blood had a nearly normal clotting time. Thus, it was evident that Christmas could not have the same molecular defect as the hemophiliac patient, despite the virtually identical clinical picture. The clotting factors missing in these patients must have been different, since their plasmas complemented each other in clotting. In this way, a new factor termed IX or *Christmas factor* was discovered. In general, a *complementation test is a powerful means of determining whether two inactive systems lack the same component.* The power of the complementation approach is that the components of a system can be enumerated before they are isolated and purified.

Figure 8-30
Intrinsic pathway of clotting.
Inactive forms of factors are
shown in red, active ones in
green. Activated factors
catalyze the activation of
other factors.

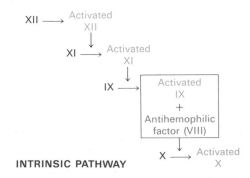

Complementation tests are extensively used in bacterial and viral genetics.

Christmas factor is two steps removed from the initial trigger in the intrinsic pathway. This reaction sequence is initiated by the contact of Factor XII with an abnormal surface. Activated Factor XII_a then converts Factor XI into its active form. In turn, Christmas factor (IX) is activated by XI_a. These activation steps are probably conversions of zymogens to active enzymes.

EXTRINSIC PATHWAY OF CLOTTING

As in the intrinsic pathway, the end result is the activation of Factor X. The extrinsic pathway (Figure 8-31) appears relatively simple. Trauma to the blood vessel releases a lipoprotein called *tissue factor*. A complex of tissue factor and Factor VII then catalyzes the activation of X. Tissue factor is a modifier protein, since Factor VII is inactive in the absence of tissue factor. Thus, there are at least three modifier proteins in the clotting process: Factor V, Factor VIII, and tissue factor.

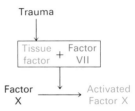

EXTRINSIC PATHWAY
Figure 8-31
Extrinsic pathway of clotting.

Table 8-2
Blood coagulation factors

Factor	Pathway	Function of active form
Hageman factor (XII)	Intrinsic	Activates XI
Plasma thromboplastin antecedent (PTA) (XI)	Intrinsic	Activates IX
Christmas factor (IX)	Intrinsic	Both required to activate X
Antihemophilic factor (VIII)	Intrinsic	
Tissue factor	Extrinsic	Both required to activate X
Proconvertin (VII)	Extrinsic	
Stuart factor (X)	Common	Activates II
Accelerin (V)	Common	Stimulates activation of II
Prothrombin (II)	Common	Activates fibrinogen (I)
Fibrinogen (I)	Common	FORMS FIBRIN CLOT
Fibrin-stabilizing factor (XIII)	Common	Stabilizes clot by cross-linking fibrin

CONTROL OF CLOTTING: AN UNSOLVED PROBLEM

It is evident that the clotting process must be precisely regulated. There is a fine line between hemorrhage and thrombosis. Clotting must occur rapidly yet remain confined to the area of injury. We are far from understanding the critical elements in the control of this remarkable process. Why are there so many stages? Why are there distinct intrinsic and extrinsic pathways and how are they related? What are the mechanisms that normally limit clot formation to the site of injury? These intriguing questions are of interest for their own sake and also because of their pertinence to preventing and treating cardiovascular diseases, which constitute a major health problem.

SUMMARY

The activation of a protein by proteolytic cleavage of one or a few peptide bonds is a recurring control mechanism in biochemistry. If the active protein is an enzyme, the inactive precursor is called a zymogen or a proenzyme. The best-understood zymogen activation is the conversion of chymotrypsinogen into chymotrypsin. Trypsin hydrolyzes the peptide bond between residues 15 and 16 in chymotrypsinogen, and thereby converts it into a fully active chymotrypsin. The newly formed amino-terminal group of isoleucine 16 turns inward and forms an electrostatic bond with aspartate 194 in the interior. This interaction triggers a series of localized conformational changes that result in the creation of a pocket for the binding of the aromatic (or bulky nonpolar) side chain of the substrate. In addition, chymotrypsin can form a hydrogen bond with the substrate that cannot occur in chymotrypsinogen.

A highly reactive serine residue (serine 195) plays a critical role in the catalytic mechanism of chymotrypsin. The first stage in the hydrolysis of a peptide substrate is the formation of a covalent *acyl-enzyme intermediate,* in which the carboxyl component of the substrate is esterified to the hydroxyl group of serine 195. This acyl-enzyme intermediate is hydrolyzed in the second stage of catalysis. Histidine 57 also plays a key role in both the acylation and deacylation reactions. In fact, serine 195 is made a strong nucleophile by a series of hydrogen bonds. Serine 195 is hydrogen-bonded to histidine 57, which is in turn hydrogen-bonded to aspartate 102 in the inside of the enzyme. These three residues constitute a *charge relay network,* which accelerates catalysis by a factor of about 10^3. Such a charge relay network also occurs in trypsin, elastase, and thrombin. Indeed, these enzymes are like chymotrypsin in amino acid sequence, conformation, and catalytic mechanism, but differ in specificity. These variations on a theme probably arose by divergent evolution from a common ancestor.

Zymogen activations also play a major role in the control of blood clotting. A striking feature of the clotting process is that it occurs by a cascade of zymogen conversions, in which the activated form of a clotting factor catalyzes the activation of the next precursor. Clotting involves the interplay of two reaction sequences, called the intrinsic and extrinsic pathways. Both are needed for normal clotting. They converge on a common pathway that results in the formation of a fibrin clot. Fibrinogen, a highly soluble molecule in the plasma, is converted into fibrin by the hydrolysis of four arginyl–glycine bonds. This reaction is catalyzed by thrombin, an enzyme that is similar to trypsin. Two A and two B peptides comprising 3% of the fibrinogen are released. The resulting fibrin monomer then spontaneously forms long, insoluble fibers called fibrin. The fibrin clot is subsequently strengthened by the formation of covalent cross-links as a result of transamidation reactions between specific glutamine and lysine side chains.

SELECTED READINGS

Where to start:

Stroud, R. M., 1974. A family of protein-cutting proteins. *Sci. Amer.* 231(1):24–88. [Available as *Sci. Amer.* Offprint 1301. An excellent introduction to the structure, catalytic properties, and evolution of serine proteases.]

Blow, D. M., and Steitz, T. A., 1970. X-ray diffraction studies of enzymes. *Annu. Rev. Biochem.* 39:63–100.

McKusick, V. A., 1965. The royal hemophilia. *Sci. Amer.* 213(2):88–95.

Chymotrypsin and elastase:

Henderson, R., Wright, C. S., Hess, G. P., and Blow, D. M., 1971. α-Chymotrypsin: what can we learn about catalysis from x-ray diffraction? *Cold Spring Harbor Symp. Quant. Biol.* 36:63–70.

Blow, D. M., 1971. The structure of chymotrypsin. *In* Boyer, P. D., (ed.), *The Enzymes* (3rd ed.), vol. 3, pp. 185–212. Academic Press.

Hess, G. P., 1971. Chymotrypsin: chemical properties and catalysis. *In* Boyer, P. D., (ed.), *The Enzymes* (3rd ed.), vol. 3, pp. 213–248.

Henderson, R., and Wang, J. H., 1972. Catalytic configurations. *Annu. Rev. Biophys. Bioeng.* 1:1–26. [A review that includes a critical discussion of the catalytic mechanism of chymotrypsin.]

Hartley, B. S., and Shotton, D. M., 1971. Pancreatic elastase. *In* Boyer, P. D., (ed.), *The Enzymes* (3rd ed.), vol. 3, pp. 323–373.

Activation of chymotrypsinogen:

Kassell, B., and Kay, J., 1973. Zymogens of proteolytic enzymes. *Science* 180:1022–1027. [Recent studies show that some zymogens have a low level of inherent proteolytic activity.]

Kraut, J., 1971. Chymotrypsinogen: x-ray structure. *In* Boyer, P. D., (ed.), *The Enzymes* (3rd ed.), vol. 3, pp. 165–183.

Wright, H. T., 1973. Activation of chymotrypsinogen-A: an hypothesis based upon comparison of the crystal structures of chymotrypsinogen-A and α-chymotrypsin. *J. Mol. Biol.* 79:13–23.

Clotting factors:

Ratnoff, O. D., 1972. Hereditary disorders of hemostasis. *In* Stanbury, J. B., Wyngaarden, J. B., and Fredrickson, D. S., (eds.), *The Metabolic Basis of Inherited Disease* (3rd ed.), pp. 1671–1709. McGraw-Hill.

Ratnoff, O. D., and Bennett, B., 1973. The genetics of hereditary disorders of blood coagulation. *Science* 179:1291–1298.

Doolittle, R. F., 1973. Structural aspects of the fibrinogen to fibrin conversion. *Advan. Protein Chem.* 27:1–109.

Massie, R. K., 1967. *Nicholas and Alexandra.* Dell. [A fascinating account of the interplay of hemophilia and Russian history.]

PROBLEMS

1. Let us compare *lysozyme, carboxypeptidase A,* and *chymotrypsin.*
   (a) Which of these enzymes requires a metal ion for catalytic activity?
   (b) Which of these enzymes is a single polypeptide chain?
   (c) Which of these enzymes is rapidly inactivated by DIPF?
   (d) Which of these enzymes is formed by proteolytic cleavage of a precursor that has been identified?

2. The transfer of a proton from an enzyme to its substrate is often a key step in catalysis.
   (a) Does this occur in the proposed catalytic mechanisms for chymotrypsin, lysozyme, and carboxypeptidase A?
   (b) If so, identify the proton donor in each case.

3. A substrate is frequently attacked by a nucleophilic group of an enzyme. What is the nucleophile in the proposed mechanism for chymotrypsin and for carboxypeptidase A?

4. Cite the factors that contribute to the catalytic power of enzymes.

5. TPCK is an affinity-labeling reagent for chymotrypsin. It inactivates chymotrypsin by alkylating histidine 57.
   (a) Design an affinity-labeling reagent for trypsin that resembles TPCK.
   (b) How would you test its specificity?
   (c) Which other serine protease might also be inactivated by this affinity-labeling reagent for trypsin?

6. Elastase is specifically inhibited by an aldehyde derivative of one of its substrates:

$$N\text{-Acetyl}-Pro-Ala-Pro-N-\overset{\overset{\displaystyle H}{|}}{C}-\overset{\overset{\displaystyle H}{}}{\underset{\underset{\displaystyle CH_3}{|}}{C}}\overset{O}{\diagup}_{H}$$

   In fact, this aldehyde is an analog of the transition state for catalysis by elastase.
   (a) Which residue at the active site of elastase is most likely to form a covalent bond with this aldehyde?
   (b) What type of covalent link would be formed?

7. Boron acids are another kind of transition-state analog for enzymes that form acyl-enzyme intermediates. Acetylcholinesterase is an enzyme that catalyzes the hydrolysis of the ester bond in acetylcholine:

$$CH_3-\overset{\overset{\displaystyle O}{\|}}{C}-O-CH_2-CH_2-\overset{+}{N}(CH_3)_3 + H_2O$$

Acetylcholine

$$\downarrow$$

$$CH_3-\overset{\overset{\displaystyle O}{\|}}{C}-O^- + HO-CH_2-CH_2-\overset{+}{N}(CH_3)_3 + H^+$$

Acetate   **Choline**

   Acetylcholinesterase is specifically inhibited by this boron analog of acetylcholine:

$$CH_3-\overset{\overset{\displaystyle OH}{|}}{B}-CH_2-CH_2-CH_2-\overset{+}{N}(CH_3)_3$$

Boron analog of acetylcholine

   How might this boron analog be covalently attached to the active site of acetylcholinesterase?

CHAPTER 9

CONNECTIVE-TISSUE PROTEINS:
COLLAGEN AND ELASTIN

Collagen is a fibrous protein found in all multicellular organisms. Indeed, it is the most abundant protein in mammals, constituting a quarter of the total. Collagen is the major fibrous element of skin, bone, tendon, cartilage, and teeth. It is present to some extent in nearly all organs and serves to hold cells together in discrete units. In addition to its structural role in mature tissue, collagen has a directive role in developing tissue. *The distinctive property of collagen is that it forms insoluble fibers that have a high tensile strength* (Figure 9-1). Furthermore, the basic structural motif of collagen can be modified to meet the specialized needs of particular tissues.

TROPOCOLLAGEN IS THE BASIC STRUCTURAL UNIT OF COLLAGEN

The insolubility of collagen fibers was for many years an impasse in the chemical characterization of collagen. The breakthrough came when it was found that collagen from the tissues of young animals can be extracted in soluble form because it is not yet extensively cross-linked. The absence of covalent cross-links in immature collagen makes it feasible to extract the basic structural unit, called tropocollagen, in intact form.

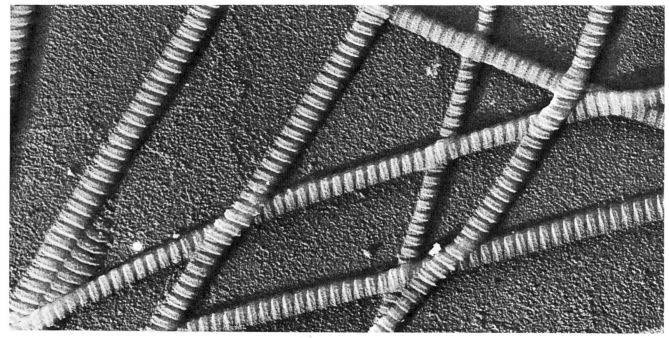

Figure 9-1
Electron micrograph of intact collagen fibrils obtained from skin. The
preparation was shadowed with chromium. The period along the fiber
axis is 640 Å. [Courtesy of Dr. Jerome Gross.]

Tropocollagen has a molecular weight of 285,000 and consists of
three polypeptide chains of the same size. Two of the chains, called
$\alpha 1$, are identical, whereas the other one, called $\alpha 2$, is similar. Each
chain has about a thousand amino acid residues. Thus, the basic
structural unit of collagen is very large—indeed, more than ten
times the size of chymotrypsin.

COLLAGEN HAS AN UNUSUAL AMINO ACID COMPOSITION AND SEQUENCE

The proportion of *glycine* residues in all collagen molecules is nearly
one-third, which is unusually high for a protein. In hemoglobin,
for example, the glycine content is 5%. Also, *proline* is present to
a much greater extent in collagen than in most other proteins.

Furthermore, collagen contains two amino acids that are present in very few other proteins, namely *hydroxyproline* and *hydroxylysine*.

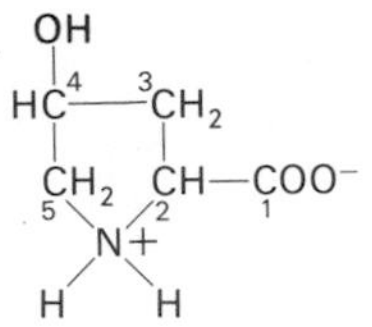

4-Hydroxyproline
(Hyp)

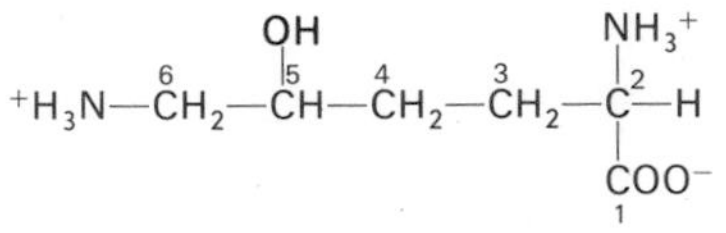

5-Hydroxylysine
(Hyl)

13
-Gly-Pro-Met-Gly-Pro-Ser-Gly-Pro-Arg-
22
-Gly-Leu-**Hyp**-Gly-Pro-**Hyp**-Gly-Ala-**Hyp**-
31
-Gly-Pro-Gln-Gly-Phe-Gln-Gly-Pro-**Hyp**-

Figure 9-2
Amino acid sequence of part of the α1 chain of collagen. Every third residue is glycine in this region.

The determination of the amino acid sequence of collagen, a formidable task, has only recently been started. However, it is already evident that the sequence is remarkably regular: *nearly every third residue is glycine*. Moreover, the sequence glycine-proline-hydroxyproline recurs frequently (Figure 9-2). In contrast, globular proteins do not exhibit regularities in their amino acid sequences. The only other proteins known to have some regularly repeating sequences are silk fibroin and elastin.

SOME PROLINE AND LYSINE RESIDUES BECOME HYDROXYLATED

Hydroxyproline and hydroxylysine are not incorporated in the polypeptide chains of collagen by the usual mechanisms of protein synthesis. If ^{14}C-labeled hydroxyproline is administered to a rat, none of the radioactivity appears in the collagen that is synthesized. In contrast, the hydroxyproline in collagen is radioactive if ^{14}C-proline is given. Thus, proline is a precursor of the hydroxyproline residues in collagen, whereas exogenous hydroxyproline is not.

Proline is converted into hydroxyproline by an enzyme called *protocollagen hydroxylase*. Molecular oxygen, ferrous ion, and α-keto-glutarate are required for activity. A reducing agent such as ascorbic acid is also needed for the hydroxylation reaction.

The hydroxylation reaction is highly specific (Figure 9-3). Proline as a free amino acid is not a substrate. Rather, hydroxylation takes place at specific sites on relatively large polypeptide chains. *A proline*

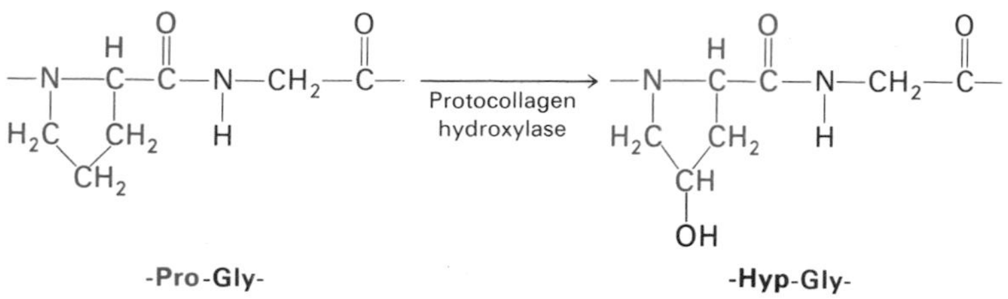

Figure 9-3
Hydroxylation of a proline residue in collagen
by protocollagen hydroxylase.

can be hydroxylated only if it is situated on the amino side of a glycine residue. There may be additional restrictions since not all prolines in that position are hydroxylated. The conversion of lysine to hydroxylysine is catalyzed by a different enzyme.

SUGARS ARE ATTACHED TO HYDROXYLYSINE RESIDUES

Collagen contains carbohydrate units covalently attached to its hydroxylysine residues. A disaccharide of glucose and galactose is commonly found (Figure 9-4). The number of carbohydrate units per tropocollagen depends on the tissue. For example, tendon collagen contains 6, whereas the lens capsule has 110.

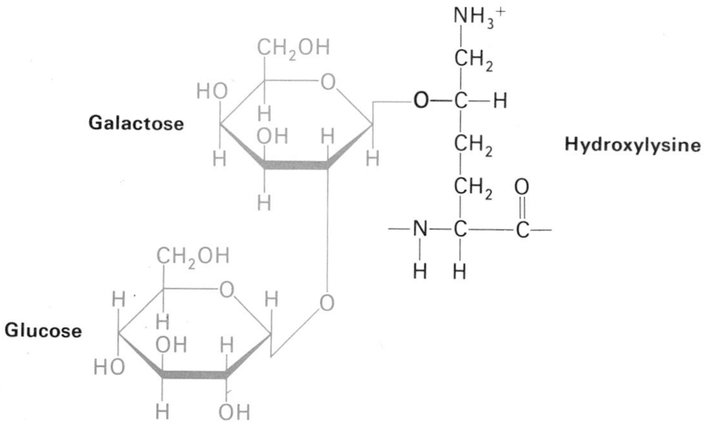

Figure 9-4
A carbohydrate unit in collagen.

DEFECTIVE HYDROXYLATION IS ONE OF THE BIOCHEMICAL LESIONS IN SCURVY

The biological roles of the carbohydrate units and of hydroxyproline residues in collagen are uncertain. There is no doubt, however, about the importance of the hydroxylation of collagen, as evidenced by the disease *scurvy*. A vivid description of scurvy was given by Jacques Cartier in 1536, when it afflicted his men as they were exploring the Saint Lawrence River:

> Some did lose all their strength, and could not stand on their feet. . . . Others also had all their skins spotted with spots of blood of a purple colour: then did it ascend up to their ankles, knees, thighs, shoulders, arms, and necks. Their mouths became stinking, their gums so rotten, that all the flesh did fall off, even to the roots of the teeth, which did also almost all fall out.

The means of preventing scurvy was succinctly stated by James Lind, an English physician, in 1753:

> Experience indeed sufficiently shows that as greens or fresh vegetables, with ripe fruits, are the best remedies for it, so they prove the most effectual preservatives against it.

Lind urged the inclusion of lemon juice in the diet of sailors. His advice was adopted by the British Navy some forty years later.

Scurvy is caused by a dietary deficiency of ascorbic acid (Vitamin C).

$$\underset{\text{O}}{\overset{}{\text{C}}}=\underset{\text{OH}}{\overset{\text{O}}{\text{C}}}=\underset{\text{OH}}{\text{C}}-\underset{\text{H}}{\text{C}}-\underset{\text{OH}}{\overset{\text{H}}{\text{C}}}-\text{CH}_2\text{OH}$$

Ascorbic acid

Primates and guinea pigs are unable to synthesize their own ascorbic acid. In scurvy, hydroxylation of collagen is impaired possibly because ascorbic acid is needed as a reducing agent in the reaction catalyzed by protocollagen hydroxylase. The collagen synthesized in the absence of ascorbic acid cannot properly form fibers, thereby resulting in the skin lesions and blood vessel fragility that are so prominent in scurvy.

TROPOCOLLAGEN IS A TRIPLE-STRANDED HELICAL ROD

We turn to the conformation of the basic structural unit of the collagen fiber. Electron microscopic and hydrodynamic studies have shown that *tropocollagen has the shape of a rod 3000 Å long and 15 Å in diameter*. It is the longest known protein. Its elongated character can be appreciated by noting that tropocollagen is sixty times longer than the diameter of chymotrypsin, but its diameter is less than half that of chymotrypsin. Each of the three polypeptide chains is in a helical conformation (Figure 9-5). Furthermore, the three helical strands wind around each other to form a stiff cable. Indeed, the strength of a collagen fiber is remarkable: a load of at least 10 kg is needed to break a fiber 1 mm in diameter.

The helical motif of the individual chains of the triple-stranded collagen cable is nicely illustrated by a model compound, *poly*-L-*proline*. This synthetic polypeptide has a helical form (Figure 9-6) that is entirely different from the α helix. There are no hydrogen bonds in the poly-L-proline helix. Instead, *this helix is stabilized by steric repulsion of the pyrrolidone rings of the proline residues*. The pyrrolidone rings keep out of each other's way when the polypeptide chain assumes this helical form (known as the type II *trans* helix), which is much more open than the tightly coiled α helix. The rise per residue along the helix axis of poly-L-proline is 3.12 Å, whereas it is 1.5 Å in the α helix. There are three residues per turn in the poly-L-proline helix.

Let us now look at the conformation of a single strand of the triple-stranded collagen helix (Figure 9-7). Each strand has a helical

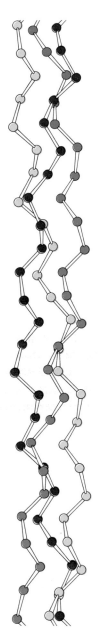

Figure 9-5
Model of the triple-stranded collagen helix. Only the α-carbon atoms are shown.

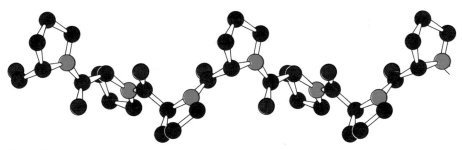

Figure 9-6
Model of a helical form of poly-L-proline (type II *trans* helix). This kind of helix is the basic structural motif of each of the three strands of tropocollagen.

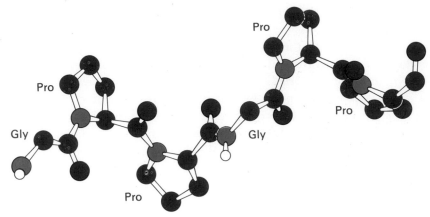

Figure 9-7
Conformation of a single strand of the collagen triple helix.
The sequence shown here is -Gly-Pro-Pro-Gly-Pro-Pro-.

conformation like that shown for poly-L-proline. The three strands
wind around each other to form a *superhelical cable* (Figure 9-8). The
rise per residue in this superhelix is 2.9 Å and the number of
residues per turn is nearly 3.3. *The three strands are hydrogen-bonded
to each other*. The hydrogen donors are the peptide NH groups of
glycine residues, and the hydrogen acceptors are the peptide CO
groups of residues on the other chains. The direction of the hydrogen bond is transverse to the long axis of the tropocollagen rod.
A space-filling model of the triple-stranded collagen helix is shown
in Figure 9-10.

GLYCINE IS CRITICAL BECAUSE IT IS SMALL

We are now in a position to see why glycine occupies every third
position in the amino acid sequence of tropocollagen. The inside
of the triple-stranded helical cable is very crowded (Figure 9-9).
In fact, *the only residue that can fit in an interior position is glycine*. Since
there are three residues per turn of the helix, it follows that every
third residue on each strand must be glycine. The two amino acid
residues on either side of glycine are located on the outside of the
cable, where the bulky rings of proline and hydroxyproline residues
can be readily accommodated.
We might think of glycine as an insignificant amino acid, the
least important of its kind because its side chain consists only of

Figure 9-8
Skeletal model of the
triple-stranded collagen
helix. The repeating
sequence shown here is
-Gly-Pro-Pro-.

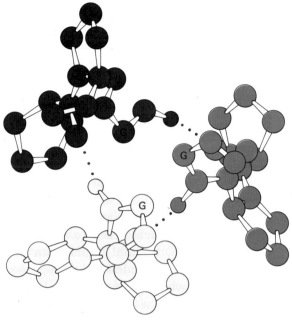

Figure 9-9
Cross section of a model of collagen. Each strand is
hydrogen-bonded to the other two strands (. . . denotes a hydrogen
bond). The α-carbon atom of a glycine residue in each strand is
labeled G. Every third residue must be glycine because there is no
space near the helix axis (center) for a larger amino acid residue.
Note that the pyrrolidone rings are on the outside. Each strand of
the triple helix is shown in a different color.

Figure 9-10
A space-filling model of the collagen
triple helix. [Courtesy of Dr. Alexander
Rich.]

a hydrogen atom. In protein structure, however, simplicity is sometimes advantageous. Glycine is a very important amino acid precisely because it occupies very little space and thereby allows
different polypeptide strands to come together. We have already
seen this in myoglobin and hemoglobin, where glycine B6 is invariant because it allows the close approach of the B and E helices.
In chymotrypsin, the substrate-binding cavity can accommodate
a large aromatic group because two of the residues lining the cavity
are glycine. In the evolution of cytochrome *c*, an electron carrier,
the amino acid residue that is most conserved is glycine. Indeed,

Table 9-1
Dependence of thermal stability on imino acid content

Source	Proline plus hydroxyproline (per 1000 residues)	Thermal stability (°C)		Body temperature (°C)
		T_s	T_m	
Calf skin	232	65	39	37
Shark skin	191	53	29	24–28
Cod skin	155	40	16	10–14

the content of the imino acids (proline plus hydroxyproline) in the collagen. *The higher the imino acid content, the more stable the helix.* The imino acid content of collagens increased in the evolution of warm-blooded species from cold-blooded ones.

In short, the stability of the helical form of a single strand of tropocollagen depends on the locking effect of proline and hydroxyproline residues. The triple helix is further stabilized by transverse hydrogen bonding and van der Waals interactions between residues on different strands. The superhelix is sterically allowed because glycine occupies every third position in the amino acid sequence.

PROCOLLAGEN IS THE BIOSYNTHETIC PRECURSOR OF COLLAGEN

The triple-stranded collagen helix is rapidly assembled in vivo. In contrast, the process takes days in vitro from a solution of the constituent α1 and α2 chains. Furthermore, the yield is low. Why this disparity between in vitro and in vivo assembly of the tropocollagen helix? The answer is suggested by comparing the refolding of denatured chymotrypsinogen and chymotrypsin. The unfolded zymogen spontaneously refolds into the correct three-dimensional structure, whereas the enzyme does not. The reason is that part of the structure needed to specify the three-dimensional form is missing in chymotrypsin; namely, the two dipeptides that were cleaved in the activation process.

It was surmised that the α1 and α2 chain do not spontaneously form the correct tropocollagen structure because part of the necessary information is missing. This is indeed so. *The constituent chains of collagen are synthesized in the form of larger precursors.* The precursor of the α1 chain, called *pro-α1* has a molecular weight of 120,000, in contrast with 95,000 for α1. The additional peptide is located at the amino-terminal end of the α1 chain. Similarly, the α2 chain is synthesized as *pro-α2*, which also has a molecular weight of 120,000. The additional peptide segments in pro-α1 and pro-α2 have an amino acid composition that is quite different from the rest of the chain. They are not rich in glycine, hydroxyproline, and proline. The additional segments contain interchain disulfide cross-links, which are not found in α1 or α2.

THE ADDITIONAL PEPTIDE REGIONS OF THE PRECURSOR CHAINS ARE ENZYMATICALLY EXCISED

The pro-α1 and pro-α2 chains are synthesized by fibroblast cells and secreted as such into the extracellular space of connective tissues. The additional amino-terminal peptide regions on these precursor chains are cleaved by a specific proteolytic enzyme called *procollagen peptidase*.

Defects in the conversion of precursor chains to α1 and α2 can lead to generalized disorders of connective tissue. Patients with the *Ehlers-Danlos syndrome* have stretchable skin, hypermobile joints, and short stature. Patients with one form of this disease have significant amounts of procollagen in extracts of their skin and tendon (Figure 9-12). Furthermore, the activity of procollagen peptidase is reduced in cultures of their fibroblasts. In addition, there is a connective-tissue disease in cattle that is caused by the absence of procollagen peptidase. *Dermatosparaxis* is a genetically transmitted (recessive) disease characterized by extreme fragility of the skin. The dermis of these afflicted animals contains disorganized collagen bundles. A significant proportion of the collagen of these animals consists of pro-α1 and pro-α2 chains. Thus, it is evident that the removal of the additional peptide segments is essential for the orderly formation of collagen fibers.

THE COLLAGEN FIBER IS A STAGGERED ARRAY OF TROPOCOLLAGEN MOLECULES

Tropocollagen fibers spontaneously associate in a specific way to form collagen fibers in the extracellular space. The overall geometry

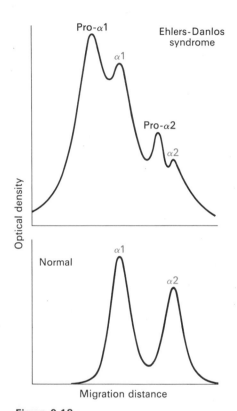

Figure 9-12
Acrylamide gel electrophoresis patterns of skin extracts. Appreciable amounts of procollagen (pro-α1 and pro-α2) are present in acid extracts of skin from patients with Ehlers-Danlos syndrome, whereas none is evident in controls from normal individuals. [Based on J. R. Lichtenstein, G. R. Martin, L. D. Kohn, P. H. Byers, and V. A. McKusick. *Science* 182(1973):299. Copyright 1973 by the American Association for the Advancement of Science.]

of the fiber has been deduced from x-ray and electron microscopic studies. Collagen fibers exhibit cross striations every 680 Å (Figure 9-1). In contrast, the length of a tropocollagen molecule is 3000 Å. A fiber period several times smaller than the length of the constituent molecules indicates that tropocollagen molecules in adjacent rows cannot be in register. Instead, adjacent rows of tropocollagens are displaced by approximately one-fourth of the length of the basic unit. *The fundamental structural design of the collagen fiber is a quarter-staggered array of tropocollagen molecules* (Figure 9-13). This arrangement is reminiscent of a fugue in music (Figure 9-14).

An interesting structural feature is that tropocollagen molecules along a row are not linked end-to-end. There is a gap of about 400 Å between the end of one tropocollagen molecule and the start of another (Figure 9-13). This gap may play a role in *bone formation.* Bone consists of an organic phase, which is nearly entirely collagen, and an inorganic phase, which is calcium phosphate. Specifically, the structure is like that of hydroxylapatite, which has the composition $Ca_{10}(PO_4)_6(OH)_2$. Collagen is required for the deposition of calcium phosphate crystals to form bone. In fact, the initial crystals are found at intervals of about 680 Å, which is the period of the collagen fibers. It is possible that the gaps between the tropocollagen molecules along a row are the *nucleation sites* for the mineral phase of bone.

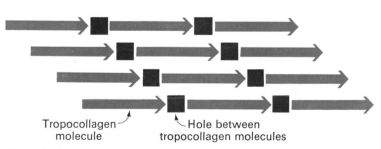

Tropocollagen molecule

Hole between tropocollagen molecules

Figure 9-13
A schematic representation of the basic structural design of a collagen fiber. Tropocollagen molecules (shown as blue arrows) form a quarter-staggered array. The gaps between tropocollagen molecules along a row (represented by red squares) may be nucleation sites in bone formation.

Figure 9-14
A passage from the Fugue in D Major, Well-tempered Clavier, by J. S. Bach. [From R. Erickson *The Structure of Music: A Listener's Guide* (Noonday Press, 1955), p. 130.]

PROCOLLAGEN PEPTIDASE CONTROLS FIBER FORMATION

Procollagen molecules (which contain two pro-α1 and one pro-α2 chains) do not spontaneously associate into fibers as do tropocollagen molecules. Excision of the amino-terminal peptides appears to be essential for the formation of normal collagen fibers, as evidenced by the previously discussed defect in dermatosparaxis. Thus, *the formation of collagen fibers is analogous to the formation of fibrin fibers.* Procollagen is akin to fibrinogen, tropocollagen to fibrin monomer, and procollagen peptidase to thrombin. In both systems, specific proteolytic cleavage is required for fiber formation.

Procollagen, not tropocollagen, is secreted by the fibroblast cell (Figure 9-15). Collagen fiber formation occurs in the extracellular fluid rather than inside the fibroblast because of the location of procollagen peptidase. The additional peptide regions of pro-α1 and pro-α2 prevent the premature formation of fibers. They may also guide the transport of the molecule across the permeability barrier of the fibroblast. Finally, the additional peptide regions may facilitate the specific alignment of the three chains and promote formation of the triple helix.

Figure 9-15
Steps in the formation of mature collagen fibers.

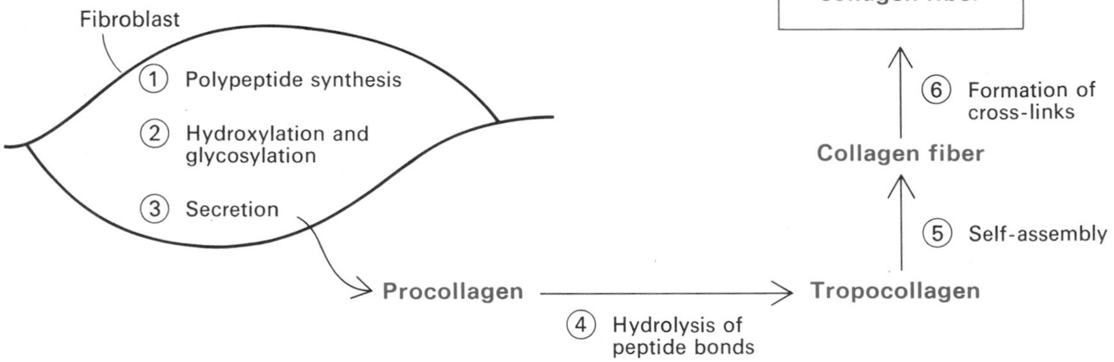

COLLAGEN FIBERS ARE STRENGTHENED BY CROSS-LINKS

Collagen, like fibrin, is stabilized by the formation of covalent cross-links. Two kinds of cross-links are formed in the collagen fiber: intramolecular (within a tropocollagen unit) and intermolecular

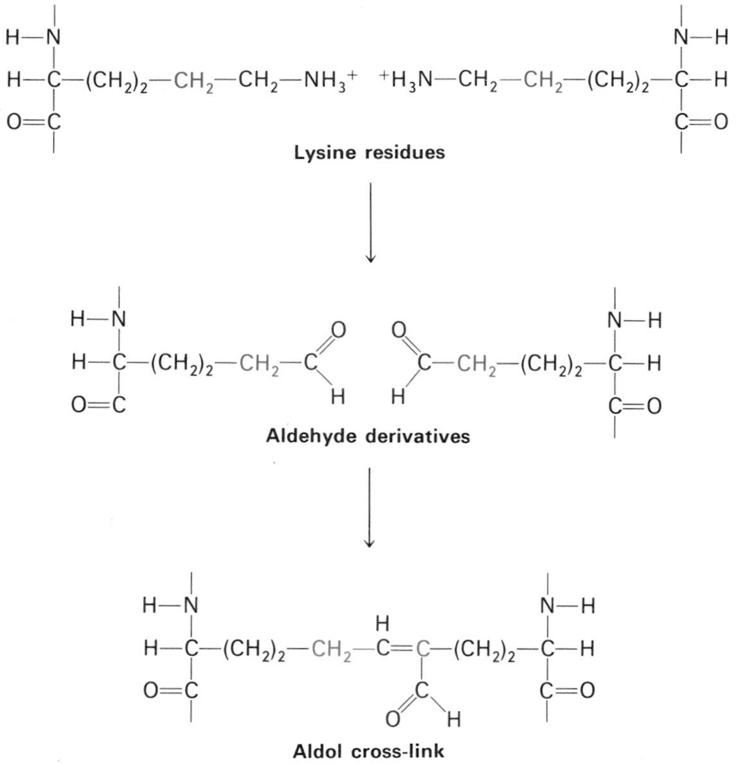

Figure 9-16
Formation of an aldol cross-link from two lysine side chains.

(between different tropocollagen units). The bonds formed appear to be unique to collagen and elastin, a related protein.

The intramolecular cross-links in collagen are derived from lysine side chains. The first step is the conversion of the ϵ-amino terminus of lysine to an aldehyde, a reaction catalyzed by lysyl oxidase. Two of these aldehydes then undergo an *aldol condensation* (Figure 9-16). In this condensation reaction, an enolate ion derived from one aldehyde adds to the carbonyl group of the other aldehyde. The location of this intramolecular cross-link is interesting. The amino acid sequence of the amino-terminal segment of the $\alpha 1$ chain of collagen shows that every third residue is glycine starting at position 13. Thus, lysine 5 is in a nonhelical region of the polypeptide chain, near the start of the triple-stranded helix. The flexibility of this nonhelical segment may facilitate the formation of this cross-link.

This aldol cross-link can react with a histidine side chain to form a histidine–aldol cross-link (Figure 9-17). The aldehyde group in the aldol–histidine cross-link can form a Schiff base with yet another side chain, such as that of hydroxylysine. In this way, four side chains can be covalently bonded to each other. Some of these cross-links can be between side chains on different tropocollagen molecules (Figure 9-18).

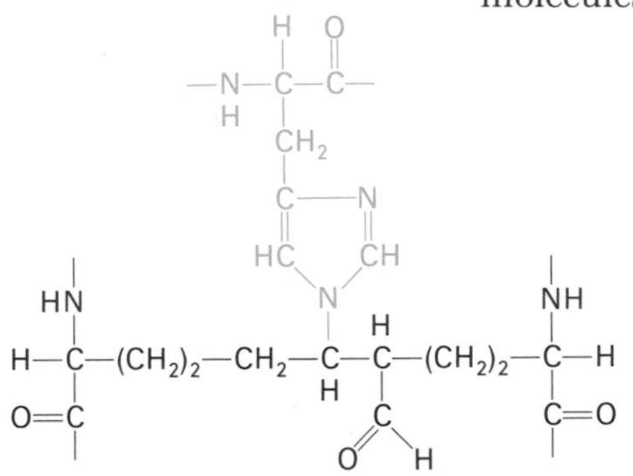

Histidine–aldol cross-link

Figure 9-17
A histidine side chain can add across the carbon–carbon double bond in an aldol cross-link to yield a histidine–aldol cross-link.

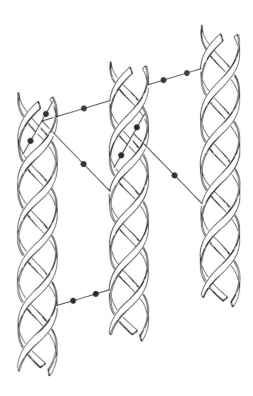

Figure 9-18
Schematic diagram of intramolecular and intermolecular cross-links in collagen. [Based on M. L. Tanzer. *Science* 180 (1973):562. Copyright 1973 by the American Association for the Advancement of Science.]

The extent and type of cross-linking varies with the physiological function and age of the tissue. The collagen in the Achilles' tendon of mature rats is highly cross-linked, whereas that of the flexible tail tendon is much less so. The importance of cross-linking in conferring mechanical strength on collagen fibers is evident in *lathyrism*, an experimental disease in animals produced by the administration of nitriles. These agents act by inhibiting the transformation of lysyl side chains to aldehydes. The result is that the collagen in these animals is extremely fragile.

COLLAGENASES ARE ENZYMES THAT SPECIFICALLY DEGRADE COLLAGEN

A collagenase is an enzyme that cleaves peptide bonds located in the characteristic helical regions of collagen. Collagen molecules are otherwise very resistant to enzymatic attack. Two types of collagenases are known:

1. One type is formed by certain microorganisms. *Clostridium histolyticum,* which causes *gas gangrene,* secretes a collagenase that splits each polypeptide chain of collagen at more than two hundred sites. The bond cleaved by this bacterial enzyme is:

$$-X\overset{\downarrow}{-}Gly-Pro-Y-$$

This enzyme contributes to the invasiveness of this highly pathogenic clostridium bacterium by *destroying the connective-tissue barriers of the host.* The bacteria are unaffected by the enzyme because they contain no collagen.

2. *Tissue collagenases,* the other type, have been found in amphibian and mammalian tissues undergoing growth or remodeling. This enzyme activity was first demonstrated in the tail fin of tadpoles undergoing metamorphosis (Figure 9-19). Large amounts of collagen are resorbed in that tissue in the course of a few days. Collagenolytic activity was assayed by culturing small pieces of metamorphosing tail fin on a layer of fibrous collagen, which is opaque. The formation of a clear zone within several hours revealed the presence of a high level of collagenase (Figure 9-20). High enzyme activity was then found in mammalian tissue undergoing rapid

Figure 9-19
Collagen is rapidly removed from the tail fin of a tadpole during metamorphosis. [From C. M. Lapiere and J. Gross. In *Mechanisms of Hard Tissue Destruction,* R. F. Sognnaes, ed., p. 665. Copyright 1963 by the American Association for the Advancement of Science.]

Figure 9-20
Explants from tadpole fin were placed on the surface of a reconstituted collagen gel (left). After 24-hour incubation at 37°C, there was a clear zone around each explant, indicative of collagenase activity (right). [From C. M. Lapiere and J. Gross. In *Mechanisms of Hard Tissue Destruction,* R. F. Sognnaes, ed., p. 681. Copyright 1963 by the American Association for the Advancement of Science.]

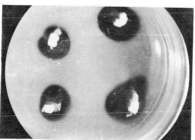

reorganization, as in the uterus following pregnancy. It is evident that the activity of tissue collagenases must be precisely controlled in timing and magnitude.

The specificity of tadpole collagenase is remarkable. It cleaves tropocollagen across its three chains at a unique site near amino acid residue 750 in the sequence of 1000 residues. The one-quarter and three-quarter length fragments spontaneously unfold at body temperature and are then cleaved by other proteolytic enzymes.

ELASTIN IS A RUBBERLIKE PROTEIN IN ELASTIC FIBERS

Elastin is found in most connective tissues in conjunction with collagen and polysaccharides. *It is the major component of elastic fibers, which can stretch to several times their length and then rapidly return to their original size and shape when the tension is released.* Large amounts of elastin are found in the walls of blood vessels, particularly in the arch of the aorta near the heart, and in ligaments. The elastic ligaments prominent in the necks of grazing animals are an especially rich source of elastin. There is relatively little elastin in skin, tendon, and loose connective tissue.

The amino acid composition of elastin is highly distinctive. As in collagen, one-third of the residues are glycine. Also, elastin is rich in proline. In contrast to collagen, elastin contains very little hydroxyproline, no hydroxylysine, and few polar amino acids. Elastin is very rich in nonpolar aliphatic residues, such as alanine, valine, leucine, and isoleucine.

Mature elastin contains many cross-links, which render it highly insoluble and therefore difficult to analyze. However, *a soluble precursor of elastin* has been isolated from *copper-deficient* pigs. Copper deficiency blocks the formation of aldehydes, which are essential for cross-linking, as in collagen. Amino acid sequence studies of this soluble precursor, which has a molecular weight of about 70,000, are in progress. Some interesting results have been obtained thus far. There are several regions that are highly rich in alanine and lysine residues. In particular, each of the sequences

-Lys-Ala-Ala-Lys-

and

-Lys-Ala-Ala-Ala-Lys-

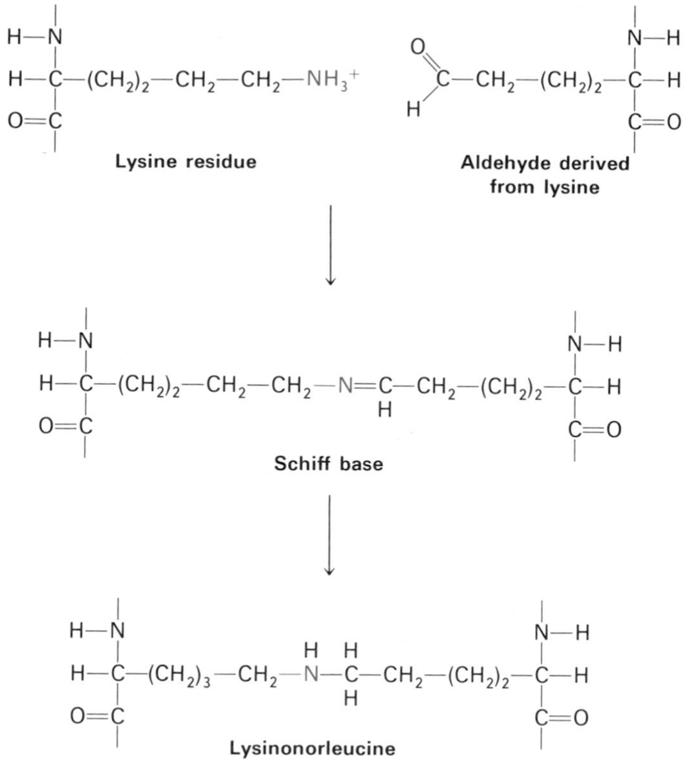

Figure 9-21
Formation of a lysinonorleucine cross-link in collagen or elastin.

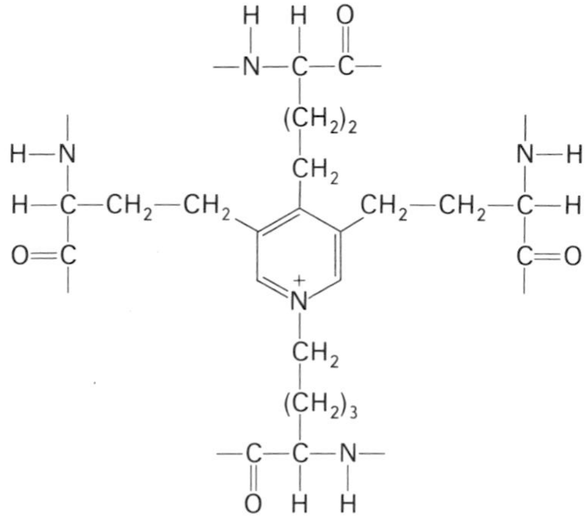

Figure 9-22
Desmosine, a cross-link in elastin, is derived from four lysine side chains.

occurs several times. Cross-links are formed in these regions. The aldol cross-link found in collagen (Figure 9-16) is also present in elastin. Another type of cross-link derived from lysine residues is *lysinonorleucine* (Figure 9-21), which also occurs in both collagen and elastin. *Desmosine,* found only in elastin, is derived from four lysine side chains (Figure 9-22). *These cross-links may be important in enabling elastin fibers to return to their original size and shape after stretching.*

Regions of elastin between cross-links are rich in glycine, proline, and valine. The amino acid sequence of some of these regions displays regularities (Figure 9-23).

The elucidation of the conformation of these regular regions and their relationship to the remarkable elasticity of elastin are challenging areas of inquiry. Detailed information concerning the molecular basis of the function of this protein is likely to be important for the elucidation of some cardiovascular diseases in view of the role of elastin in arterial dynamics.

-Pro-Gly-Val-Gly-Val-Pro-Gly-Val-Gly-Val-

-Pro-Gly-Val-Gly-Val-Pro-Gly-Val-Ser-Val-

-Pro-Gly-Val-Gly-Val-Pro-Gly-Val-Gly-Val-

Figure 9-23
Part of the amino acid sequence of a soluble precursor of elastin. The repeating nature of this sequence is evident.

SUMMARY

Collagen is a fibrous protein that has a very high tensile strength. It is the major fibrous component of skin, bone, tendon, cartilage, and teeth. The basic structural unit of collagen is tropocollagen, which consists of two $\alpha 1$ chains and an $\alpha 2$ chain, each about 1000 residues long. Collagen is unusually rich in glycine and proline. Furthermore, collagen contains hydroxyproline and hydroxylysine, which occur in very few other proteins. The amino acid sequence of collagen is highly distinctive: nearly every third residue is glycine. Tropocollagen is a triple-stranded helical rod, 3000 Å long and 15 Å in diameter. The helical motif of each of its chains is entirely different from that of an α helix. There are hydrogen bonds between NH groups of glycine residues on each strand and CO groups on the other two strands. The stability of the collagen helix also depends on the locking effect of its proline and hydroxyproline residues. The inside of this triple-stranded helix is crowded, and so every third residue in the amino acid sequence must be glycine.

Proteolytic activation plays an important role in the biosynthesis of collagen. The three chains of tropocollagen are synthesized as larger precursors called pro-$\alpha 1$ and pro-$\alpha 2$. Some proline residues in these precursor chains are converted to hydroxyproline by protocollagen hydroxylase, an enzyme that requires O_2, Fe^{2+}, and α-ketoglutarate for activity; a reducing agent such as ascorbic acid is also needed. Lysine is hydroxylated by a different enzyme. Sugars are then attached to hydroxylysine residues on these precursor chains. These hydroxylation and glycosylation reactions occur inside the fibroblast. The modified precursor chains are secreted into the extracellular medium. The additional amino-terminal peptide regions on these precursor chains are cleaved by procollagen peptidase to yield tropocollagen, which spontaneously associates into fibers. The basic design of a collagen fiber is a quarter-staggered array of tropocollagen molecules. Finally, collagen fibers are cross-linked to give them additional strength. For example, some lysine side chains in collagen are oxidized to aldehydes, which condense to give aldol cross-links.

Elastin is an insoluble, rubberlike protein in the elastic fibers of connective tissues, which can be reversibly stretched to several times their initial length. Connective tissues such as ligaments and the arch of the aorta contain much elastin. Elastin, like collagen, is rich in proline and glycine. In contrast to collagen, elastin contains very little hydroxyproline and no hydroxylysine. Its amino acid composition is highly nonpolar. The amino acid sequence of elastin displays certain regularities. For example, the sequence -Pro-Gly-Val-Gly-Val- recurs frequently. Elastin is synthesized as a soluble precursor, which then becomes cross-linked in a variety of ways. Desmosine, one of these cross-links, is derived from four lysine residues. Aldehyde intermediates participate in the formation of cross-links in elastin, as they do in collagen.

SELECTED READINGS

Where to start:

Gross, J., 1961. Collagen. *Sci. Amer.* 204(5):120–130. [Available as *Sci. Amer.* Offprint 88.]

Ross, R., and Bornstein, P., 1971. Elastic fibers in the body. *Sci. Amer.* 224(6):44–52. [Offprint 1225.]

Books and review articles:

Balazs, E. A., (ed.), 1970. *Chemistry and Molecular Biology of the Intercellular Matrix*. Academic Press.

Bailey, A. J., 1968. The nature of collagen. *Compr. Biochem.* 26B:297–423.

Traub, W., and Piez, K. A., 1971. The chemistry and structure of collagen. *Advan. Protein Chem.* 25:243–352.

Gallop, P. M., Blumenfeld, O. O., and Seifter, S., 1972. Structure and metabolism of connective tissue proteins. *Annu. Rev. Biochem.* 41:617–672.

Biosynthesis of collagen:

Bornstein, P., 1974. The biosynthesis of collagen. *Annu. Rev. Biochem.* 43:567–604.

Bellamy, G., and Bornstein, P., 1971. Evidence for procollagen, a biosynthetic precursor of collagen. *Proc. Nat. Acad. Sci.* 68:1138–1142.

Dehm, P., Jimenez, S. A., Olsen, B. R., and Prockop, D. J., 1972. A transport form of collagen from embryonic tendon: electron microscopic demonstration of an NH_2-terminal extension and evidence suggesting the presence of cystine in the molecule. *Proc. Nat. Acad. Sci.* 69:60–64.

Tanzer, M. L., 1973. Cross-linking of collagen. *Science* 180:561–566.

Collagenases:

Gross, J., and Lapiere, C., 1962. Collagenolytic activity in amphibian tissues: a tissue culture assay. *Proc. Nat. Acad. Sci.* 48:1014–1022.

Gross, J., and Nagai, Y., 1965. Specific degradation of the collagen molecule by tadpole collagenolytic enzyme. *Proc. Nat. Acad. Sci.* 54:1197–1204.

Seifter, S., and Harper, E., 1971. The collagenases. *In* Boyer, P. D. (ed.), *The Enzymes* (3rd ed.), vol. 3, pp. 649–697. Academic Press.

Elastin:

> Franzblau, C., and Lent, R. W., 1969. Studies on the chemistry of elastin. *Brookhaven Symp. Biol.* 21:358–377.
>
> Gray, W. R., Sandberg, L. B., and Foster, J. A., 1973. Molecular model for elastin structure and function. *Nature* 246:461–466.

Connective tissue diseases:

> Bornstein, P., 1969. Disorders of connective tissue. *In* Bondy, P. K., (ed.), *Duncan's Diseases of Metabolism* (6th ed.), pp. 654–710. Saunders.
>
> Krane, S. M., Pinnell, S. R., and Erbe, R. W., 1972. Lysyl-protocollagen hydroxylase deficiency in fibroblasts from siblings with hydroxylysine-deficient collagen. *Proc. Nat.*

Acad. Sci. 69:2899–2903.

> Lichtenstein, J. R., Martin, G. R., Kohn, L. D., Byers, P. H., and McKusick, V. A., 1973. Defect in conversion of procollagen to collagen in a form of Ehlers-Danlos syndrome. *Science* 182:298–299.
>
> Lapiere, C. M., Lenaers, A., and Kohn, L. D., 1971. Procollagen peptidase: an enzyme excising the coordination peptides of procollagen. *Proc. Nat. Acad. Sci.* 68:3054–3058. [Demonstration of the molecular defect in dermatosparaxis, an inherited connective-tissue disorder in cattle.]
>
> Major, R. H., (ed.), 1945. *Classic Descriptions of Disease* (3rd ed.). Thomas. [Cartier's description of scurvy is given on page 587 of this volume.]

PROBLEMS

1. Poly-L-proline, a synthetic polypeptide, can adopt a helical conformation like that of a single strand in the collagen triple helix.

    (a) Poly-L-proline cannot form a triple helix. Why?

    (b) Poly(Gly-Pro-Pro) can form a triple helix like that of collagen. Predict the thermal stability of the triple helix of poly(Gly-Pro-Gly) relative to that of poly(Gly-Pro-Pro).

    (c) Would you expect poly(Gly-Pro-Gly-Pro) to form a triple helix like that of collagen?

2. Consider the amino acid sequence

    -Gly-Leu-Pro-Gly-Pro-Pro-Gly-Ala-Pro-Gly-

    (a) Which residues are susceptible to hydroxylation by protocollagen hydroxylase?

    (b) Which peptide bonds are most susceptible to hydrolysis by the collagenase from *Clostridium histolyticum?*

3. Several kinds of covalent cross-links in proteins have been discussed in the preceding chapters. Identify the covalent cross-links, if there are any, in the following proteins.

    (a) Ribonuclease.

    (b) Hemoglobin.

    (c) Fibrin.

    (d) Collagen.

    (e) Elastin.

CHAPTER **10**

INTRODUCTION TO
BIOLOGICAL MEMBRANES

We now turn to biological membranes, which are organized assemblies consisting mainly of proteins and lipids. The functions carried out by membranes are indispensable for life. Membranes give cells their individuality by separating them from their environment. *Membranes are highly selective permeability barriers* rather than impervious walls because they contain specific molecular *pumps* and *gates*. These transport systems regulate the molecular and ionic composition of the intracellular medium. Eucaryotic cells also contain internal membranes that form the boundaries of organelles such as mitochondria, chloroplasts, and lysosomes. Functional specialization in the course of evolution has been closely linked to the formation of these compartments.

Membranes also control the flow of information between cells and their environment. They contain *specific receptors for external stimuli*. The movement of bacteria toward food, the response of target cells to hormones such as insulin, and the perception of light are examples of processes in which the primary event is the detection of a signal by a specific receptor in a membrane. In turn, *some*

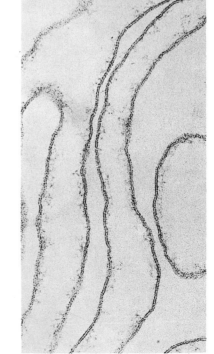

Figure 10-1
Electron micrograph of a preparation of plasma membranes from red blood cells. [Courtesy of Dr. Vincent Marchesi.]

membranes generate signals, which may be chemical or electrical. Thus, membranes play a central role in biological communication.

The two most important *energy conversion processes* in biological systems are carried out by membrane systems that contain highly ordered arrays of enzymes and other proteins. *Photosynthesis,* in which light is converted into chemical-bond energy, occurs in the inner membranes of chloroplasts, whereas *oxidative phosphorylation,* in which adenosine triphosphate (ATP) is formed by the oxidation of fuel molecules, takes place in the inner membranes of mitochondria. These and other membrane processes will be discussed in detail in later chapters. This chapter deals with some essential features that are common to most biological membranes.

COMMON FEATURES OF BIOLOGICAL MEMBRANES

Membranes are as diverse in structure as they are in function. However, they do have in common a number of important attributes:

1. Membranes are *sheetlike structures,* only a few molecules thick, that form closed boundaries between compartments of different composition. Membranes are usually from 60 to 100 Å thick.

2. Membranes consist mainly of *lipids* and *proteins.* The weight ratio of protein to lipid in most biological membranes ranges from 1:4 to 4:1.

3. *Membrane lipids are relatively small molecules* that have both a hydrophilic and a hydrophobic moiety. These lipids spontaneously form bimolecular sheets in aqueous media. These *lipid bilayer regions* are barriers to the flow of polar molecules.

4. *Specific proteins mediate distinctive functions of membranes.* Proteins serve as pumps, gates, receptors, energy transducers, and enzymes. Membrane lipids create a suitable environment for the action of these proteins.

5. Membranes are *noncovalent assemblies.* The constituent protein and lipid molecules are held together by many noncovalent interactions, which are cooperative in character.

6. Membranes are *asymmetric.* The inside and outside faces of membranes are usually different.

PHOSPHOLIPIDS ARE THE MAJOR CLASS OF MEMBRANE LIPIDS

Lipids are a group of biomolecules that are strikingly different from amino acids and proteins. By definition, lipids are water-insoluble biomolecules that have high solubility in organic solvents such as chloroform. Lipids have a variety of biological roles: they serve as fuel molecules, as highly concentrated energy stores, and as components of membranes. The first two roles of lipids will be discussed in Chapter 17. Here, we are concerned with lipids as membrane constituents. The three major kinds of membrane lipids are *phospholipids, glycolipids,* and *cholesterol.*

Let us start with phospholipids since they are abundant in all biological membranes. Phospholipids are derived from either *glycerol,* a three-carbon alcohol, or *sphingosine,* a more complex alcohol. Phospholipids derived from glycerol are called *phosphoglycerides.* A phosphoglyceride consists of a glycerol backbone, two fatty acid chains, and a phosphorylated alcohol.

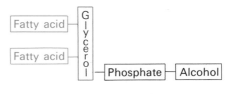

Components of a phosphoglyceride

The *fatty acid chains* in phospholipids and glycolipids usually contain an even number of carbon atoms, typically between 14 and 24. The 16- and 18-carbon fatty acids are the most common ones. In animals, the hydrocarbon chain in fatty acids is unbranched. Fatty acids may be saturated or unsaturated. The configuration of double bonds in unsaturated fatty acids is nearly always *cis.* As will

Figure 10-2
Space-filling models of (A) palmitate (C_{16}, saturated) and (B) oleate (C_{18}, unsaturated). The *cis* double bond in oleate produces a bend in the hydrocarbon chain.

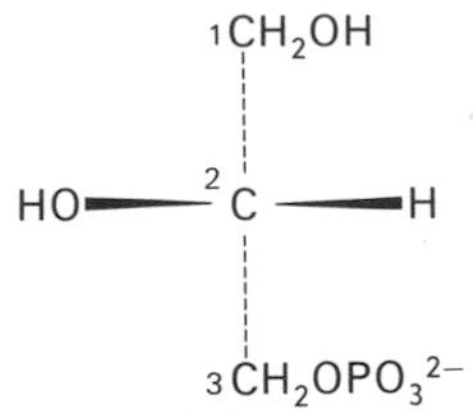

$_1CH_2OH$

$HO\!\!=\!\!\!^2C\!\!=\!\!H$

$_3CH_2OPO_3^{2-}$

Figure 10-3
Absolute configuration of the glycerol 3-phosphate moiety of membrane lipids. (H and OH, attached to C-2, are in front of the page, whereas C-1 and C-3 are behind the page.)

be discussed shortly, the length and degree of unsaturation of fatty acid chains in membrane lipids have a profound effect on membrane fluidity. The structures of the ionized form of two common fatty acids—palmitic acid (C_{16}, saturated) and oleic acid (C_{18}, one double bond)—are shown below. We will refer to them as palmitate and oleate to emphasize the fact that they are ionized under physiological conditions. The nomenclature of fatty acids is discussed in Chapter 17.

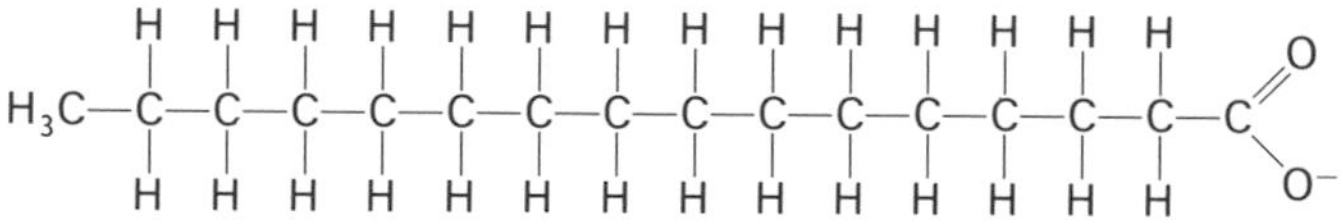

Palmitate
(Ionized form of palmitic acid)

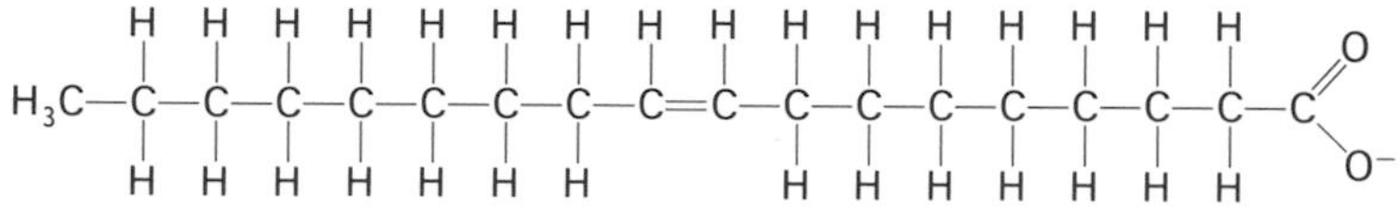

Oleate
(Ionized form of oleic acid)

In phosphoglycerides, the hydroxyl groups at C-1 and C-2 of glycerol are esterified to the carboxyl groups of two fatty acid chains. The C-3 hydroxyl group of the glycerol backbone is esterified to phosphoric acid. The resulting compound, called *phosphatidate* (or diacylglycerol-3-phosphate), is the simplest phosphoglyceride. Only small amounts of phosphatidate are present in membranes. However, it is a key intermediate in the biosynthesis of the other phosphoglycerides.

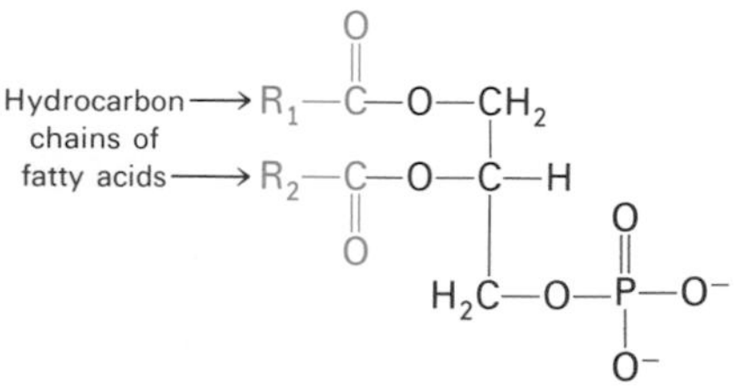

The major phosphoglycerides are derivatives of phosphatidate. The phosphate group of phosphatidate becomes esterified to the hydroxyl group of one of several alcohols. The common alcohol moieties of phosphoglycerides are serine, ethanolamine, choline, glycerol, and inositol.

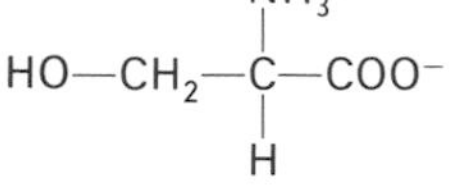

Hydrocarbon chains of fatty acids

$R_1\!-\!\overset{O}{\overset{\|}{C}}\!-\!O\!-\!CH_2$

$R_2\!-\!\overset{O}{\overset{\|}{C}}\!-\!O\!-\!\overset{|}{C}\!-\!H$

$H_2C\!-\!O\!-\!\overset{O}{\overset{\|}{\underset{|}{P}}}\!-\!O^-$
$\qquad\qquad O^-$

Phosphatidate
(Diacylglycerol 3-phosphate)

$HO\!-\!CH_2\!-\!\overset{NH_3^+}{\underset{H}{\overset{|}{C}}}\!-\!COO^-$

Serine

$HO\!-\!CH_2\!-\!CH_2\!-\!NH_3^+$

Ethanolamine

$HO\!-\!CH_2\!-\!CH_2\!-\!\overset{+}{N}(CH_3)_3$

Choline

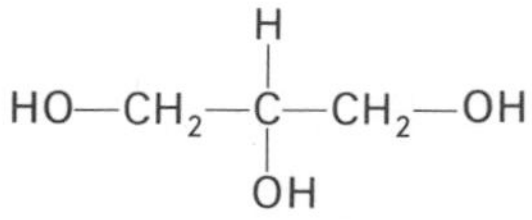

$HO\!-\!CH_2\!-\!\overset{H}{\underset{OH}{\overset{|}{C}}}\!-\!CH_2\!-\!OH$

Glycerol

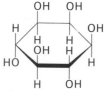

Inositol

Now let us link some of these components to form phosphatidyl choline, a phosphoglyceride found in most membranes of higher organisms.

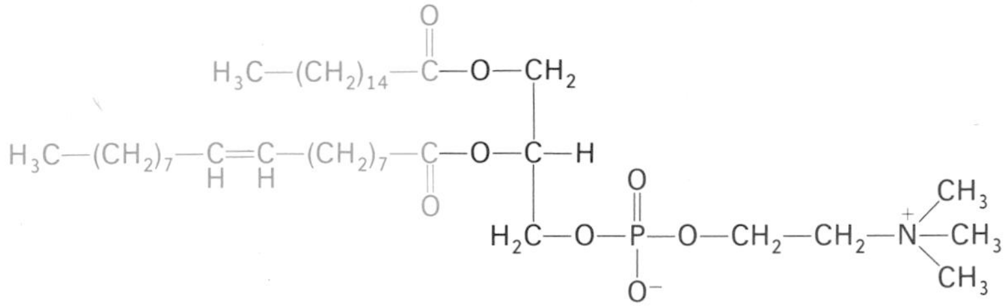

A phosphatidyl choline
(1-Palmitoyl-2-oleoyl-phosphatidyl choline)

The structural formulas of the other principal phosphoglycerides—namely, phosphatidyl ethanolamine, phosphatidyl serine, phosphatidyl inositol, and diphosphatidyl glycerol—are given in Figure 10-4.

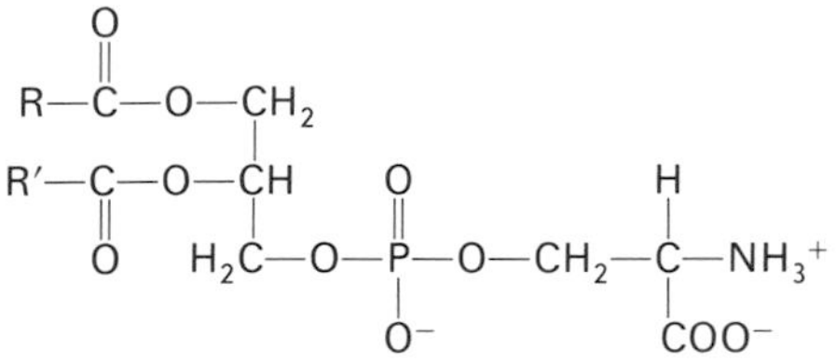

Phosphatidyl serine

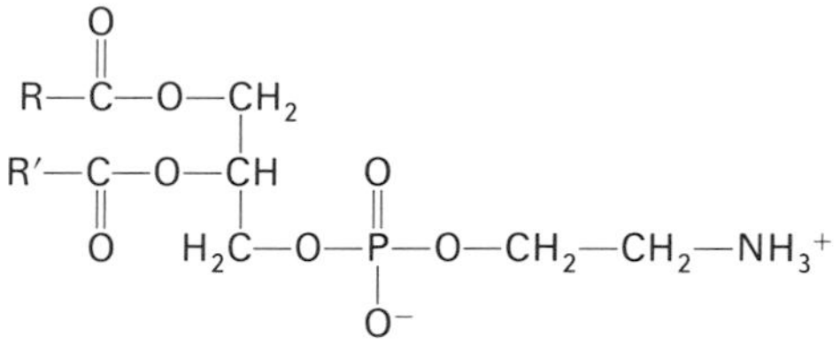

Phosphatidyl ethanolamine

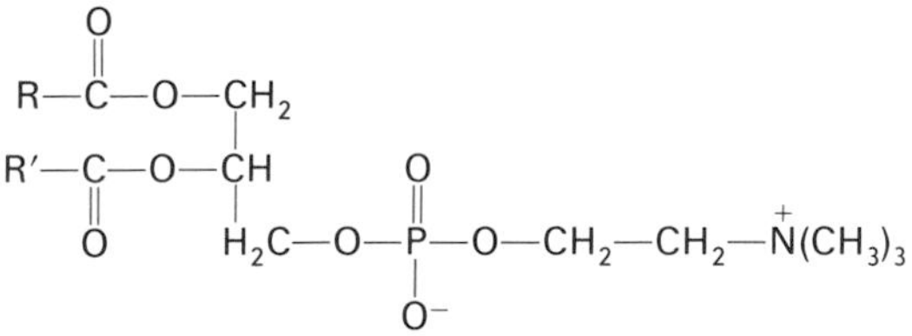

Phosphatidyl choline

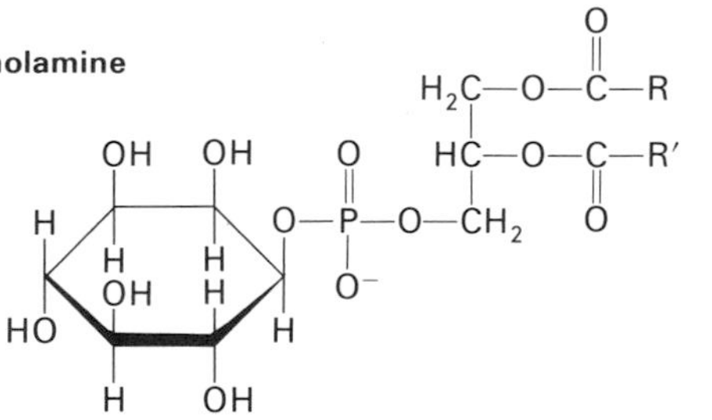

Phosphatidyl inositol

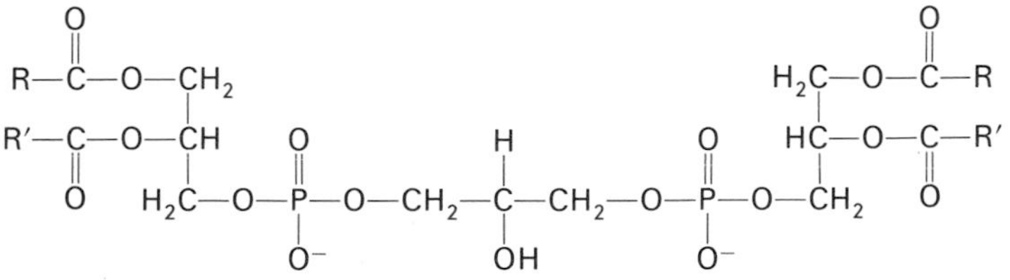

Diphosphatidyl glycerol

Figure 10-4
Formulas of some phosphoglycerides.

Sphingomyelin is the only phospholipid in membranes that is not derived from glycerol. Instead, the backbone in sphingomyelin is *sphingosine,* an amino alcohol that contains a long, unsaturated hydrocarbon chain.

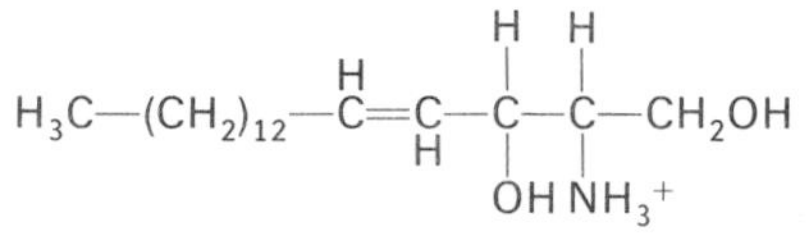

Sphingosine

In sphingomyelin, the amino group of the sphingosine backbone is linked to a fatty acid by an amide bond. In addition, the primary hydroxyl group of sphingosine is esterified to phosphoryl choline. As will be shown shortly, the conformation of sphingomyelin resembles that of phosphatidyl choline.

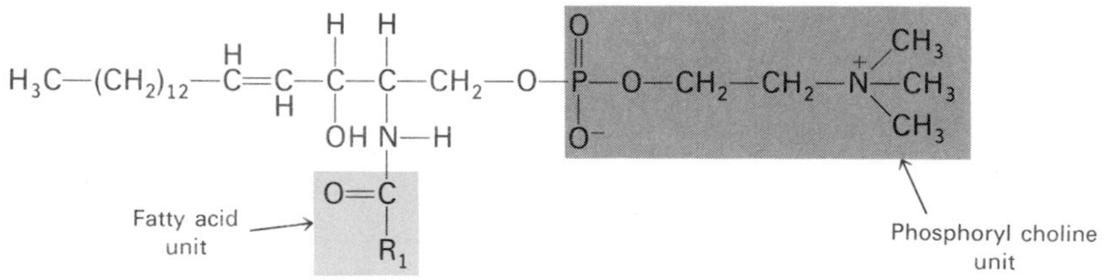

Fatty acid unit

Phosphoryl choline unit

Sphingomyelin

MANY MEMBRANES ALSO CONTAIN GLYCOLIPIDS AND CHOLESTEROL

Glycolipids, as their name implies, are *sugar-containing lipids.* Glycolipids, like sphingomyelin, are derived from sphingosine. The amino group of the sphingosine backbone is acylated by a fatty acid, as in sphingomyelin. Glycolipids differ from sphingomyelin in the nature of the unit that is linked to the primary hydroxyl group of the sphingosine backbone. In glycolipids, one or more sugars (rather than phosphoryl choline) are attached to this group. The simplest glycolipid is *cerebroside,* in which there is only one sugar residue, either glucose or galactose. More complex glycolipids, such as *gangliosides,* contain a branched chain of as many as seven sugar residues.

H₃C—(CH₂)₁₂—C=C—C—C—CH₂—O—[Glucose or galactose] ← Sugar unit

Cerebroside
(A glycolipid)

Fatty acid unit → O=C R₁

Another important lipid in some membranes is *cholesterol*. This steroid is present in eucaryotes but not in procaryotes. The plasma membranes of cells such as erythrocytes, liver cells, and myelinated nerve cells are rich in cholesterol.

Cholesterol

PHOSPHOLIPIDS AND GLYCOLIPIDS READILY FORM BILAYERS

The repertoire of membrane lipids is extensive, perhaps even bewildering at first sight. However, they possess a critical common structural theme: *membrane lipids are amphipathic molecules*. They contain both a *hydrophilic* and a *hydrophobic* moiety (Table 10-1).

Let us look at a space-filling model of a phosphoglyceride, such as phosphatidyl choline (Figure 10-5). Its overall shape is roughly rectangular. The two fatty acid chains are approximately parallel to one another, whereas the phosphoryl choline moiety points in

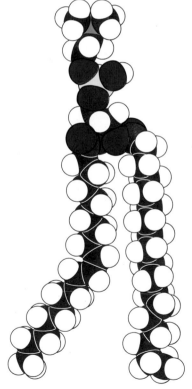

Figure 10-5
Space-filling model of a phosphatidyl choline molecule.

the opposite direction. Sphingomyelin has a similar conformation (Figure 10-6). The sugar group of a glycolipid occupies nearly the same position as the phosphoryl choline unit of sphingomyelin. We can therefore adopt the following shorthand representation for these membrane lipids. Their hydrophilic unit, also called the *polar head group,* is represented by a circle, whereas their hydrocarbon tails are depicted by straight or wavy lines (Figure 10-7).

Now let us consider the arrangement of phospholipids and glycolipids in an aqueous medium. It is evident that their polar head groups will have affinity for water, whereas their hydrocarbon tails

Figure 10-6
Space-filling model of a sphingomyelin molecule.

Table 10-1
Hydrophobic and hydrophilic units of membrane lipids

Membrane lipid	Hydrophobic unit	Hydrophilic unit
Phosphoglycerides	Fatty acid chains	Phosphorylated alcohol
Sphingomyelin	Fatty acid chain and hydrocarbon chain of sphingosine	Phosphoryl choline
Glycolipid	Fatty acid chain and hydrocarbon chain of sphingosine	One or more sugar residues
Cholesterol	Entire molecule except for OH group	OH group at C-3

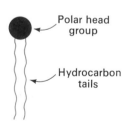

Figure 10-7
Symbol for a phospholipid or glycolipid molecule.

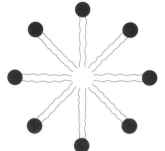

Figure 10-8
Diagram of a section of a micelle formed from phospholipid molecules.

will avoid water. This can be accomplished by forming a *micelle*, in which the polar head groups are on the surface and the hydrocarbon tails are sequestered inside (Figure 10-8).

Another arrangement that satisfies both the hydrophilic and hydrophobic preferences of these membrane lipids is a *bimolecular sheet,* which is also called a *lipid bilayer* (Figure 10-9). *In fact, the favored structure for most phospholipids and glycolipids in aqueous media is a bimolecular sheet rather than a micelle.* The preference for a bilayer structure is of critical biological importance. A micelle is a limited structure, usually less than 200 Å in diameter. In contrast, a bi-

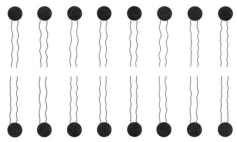

Figure 10-9
Diagram of a section of a bilayer membrane formed from phospholipid molecules.

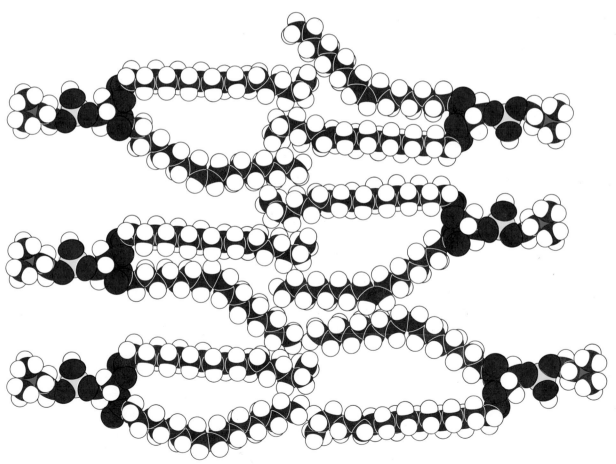

Figure 10-10
A space-filling model of a section of a highly fluid phospholipid bilayer membrane.

molecular sheet can have macroscopic dimensions such as a milli-meter (10^7 Å). Phospholipids and glycolipids are key membrane constituents because they readily form extensive bimolecular sheets. Furthermore, these sheets serve as a permeability barrier, yet they are quite fluid.

The formation of lipid bilayers is a *self-assembly process*. In other words, the structure of a bimolecular sheet is inherent in the structure of the constituent lipid molecules, specifically in their amphipathic character. The formation of lipid bilayers from glycolipids and phospholipids is a rapid and spontaneous process in water. *Hydrophobic interactions are the major driving force for the formation of lipid bilayers*. Recall that hydrophobic interactions also play a dominant role in the folding of proteins in aqueous solution. Water molecules are released from the hydrocarbon tails of membrane lipids as these tails become sequestered in the nonpolar interior of the bilayer. This release of water results in a large increase in entropy. Furthermore, there are *van der Waals attractive forces* between the hydrocarbon tails. These van der Waals forces favor close packing of the hydrocarbon tails. Finally, there are favorable *electrostatic* and *hydrogen-bonding* interactions between the polar head groups and water molecules. Thus, lipid bilayers are stabilized by the full array of forces that mediate molecular interactions in biological systems.

LIPID BILAYERS ARE NONCOVALENT, COOPERATIVE STRUCTURES

Another important feature of lipid bilayers is that they are *cooperative structures*. They are held together by many *reinforcing, noncovalent interactions*. A pertinent analogy is the clustering together of sheep in the cold to minimize the area of exposed body surface. Phospholipids and glycolipids cluster together in water to minimize the number of exposed hydrocarbon chains. Clustering is also favored by the van der Waals attractive forces between adjacent hydrocarbon chains. These energetic factors have three significant biological consequences: (1) lipid bilayers have an inherent tendency to be *extensive;* (2) lipid bilayers will tend to *close on themselves* so that there are no ends with exposed hydrocarbon chains, which results in the formation of a compartment; and (3) lipid bilayers are *self-sealing* since a hole in a bilayer is energetically unfavorable.

LIPID BILAYERS ARE HIGHLY IMPERMEABLE TO IONS AND MOST POLAR MOLECULES

The permeability properties of lipid bilayers have been measured in two well-defined *model systems: liposomes* and *planar bilayer membranes*. These model systems have provided insight into a major function of biological membranes—namely, their role as permeability barriers. The key finding is that this property of biological membranes is an inherent characteristic of their lipid bilayer regions.

Liposomes (also called *lipid vesicles*) are aqueous compartments enclosed by a lipid bilayer. They are spherical or slightly elongated in shape and have a diameter of several hundred angstroms (Figure 10-11). X-ray diffraction measurements have shown that the bilayers are typically 50 Å thick. Liposomes can be formed by suspending a suitable lipid in an aqueous medium. This mixture is then sonicated to yield a dispersion of liposomes that are quite uniform in size. There are about 3000 lipid molecules in the bilayer of each liposome. Alternatively, liposomes can be made by rapidly mixing a solution of lipid in ethanol with water. This can be accomplished by injection of the lipid through a fine needle.

Ions or molecules can be trapped in the aqueous compartment of liposomes by forming the liposomes in the presence of these

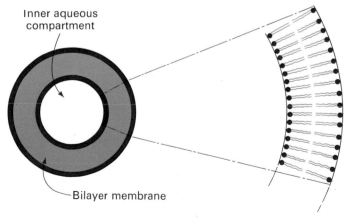

Figure 10-11
Diagram of a liposome.

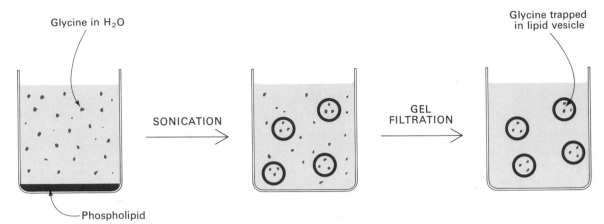

Figure 10-12
Preparation of a suspension of liposomes containing glycine molecules.

substances (Figure 10-12). Suppose that liposomes are formed in a 0.1 M glycine solution. Then, the aqueous compartment of each liposome will contain about thirty glycine molecules. These glycine-containing liposomes can be separated from the surrounding solution of glycine by dialysis or by gel-filtration chromatography. The permeability of the bilayer membrane to glycine can then be determined by measuring the rate of efflux of glycine from the inner compartment of the liposomes to the ambient solution.

A *planar bilayer membrane* more than a millimeter wide can be formed across a hole in a partition between two aqueous compartments. Such a membrane is very well suited for electrical studies, which cannot be performed on liposomes because of their small size. Paul Mueller and Donald Rudin showed that a large bilayer membrane can readily be formed in the following way. A fine paint brush is dipped into a membrane-forming solution such as phosphatidyl choline in decane. The tip of the brush is then stroked across a hole (1 mm in diameter) in a partition between two aqueous media. The lipid film across the hole thins spontaneously; the excess lipid forms a torus at the edge of the hole. A bilayer membrane consisting primarily of phosphatidyl choline is formed within a few minutes. The thinning process can be followed by looking at light reflected by the lipid film. A series of interference colors are seen as the lipid film thins from 5000 to 1000 Å. It then turns silvery at about 1000 Å, and finally appears black (denoting the loss of reflectivity) when it is less than 300 Å thick. These membranes are usually called *black lipid films* or *Mueller-Rudin membranes*. Other physical measurements have shown that a completely thinned phosphatidyl choline membrane is a bilayer about 50 Å wide. The electrical conduction properties of this macroscopic bilayer membrane can be readily studied by inserting electrodes into each aqueous compartment (Figure 10-13). The permeability of this membrane to ions can then be determined by measuring the current across the membrane as a function of the applied voltage.

Permeability studies of liposomes and electrical conductance measurements of planar bilayers have shown that *lipid bilayer membranes have a very low permeability for ions and most polar molecules*. Water is a conspicuous exception to this generalization; it readily traverses such membranes. The range of measured permeability coefficients is very wide (Figure 10-14). For example, Na^+ and K^+ traverse these membranes 10^9 times more slowly than does H_2O. Tryptophan, a zwitterion at pH 7, crosses the membrane 10^3 times more slowly than indole, a structurally related molecule that lacks ionic groups.

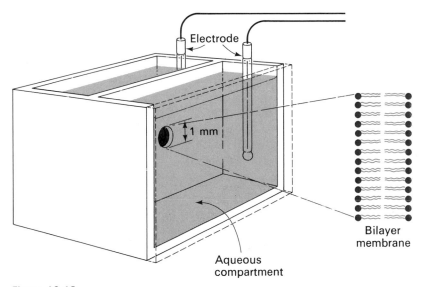

Electrode

1 mm

Bilayer
membrane

Aqueous
compartment

Figure 10-13
Experimental arrangement for the study of planar bilayer membranes. A bilayer membrane is formed across a 1-mm hole in a septum that separates two aqueous compartments.

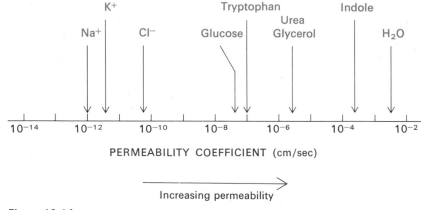

Figure 10-14
Permeability coefficients of some ions and molecules
in lipid bilayer membranes.

The permeability coefficients of small molecules are correlated with their solubility in a nonpolar solvent relative to their solubility in water. This relationship suggests that a small molecule might traverse a lipid bilayer membrane in the following way: first, it sheds its solvation shell of water; then, it becomes dissolved in the hydrocarbon core of the membrane; finally, it diffuses through this core to the other side of the membrane, where it becomes resolvated by water.

PROTEINS CARRY OUT MOST MEMBRANE PROCESSES

We now turn to membrane proteins, which are responsible for most of the dynamic processes carried out by membranes. Membrane lipids form a permeability barrier and thereby establish compartments, whereas *specific proteins mediate distinctive membrane functions,* such as transport, communication, and energy transduction. Membrane lipids create an appropriate environment for the action of such proteins.

Membranes differ in their protein content. Myelin, a membrane that serves as an insulator around certain nerve fibers, has a low content of protein (18%). Lipid, the major molecular species in myelin, is well suited for insulation. In contrast, the plasma membranes of most other cells are much more active. They contain many pumps, gates, receptors, and enzymes. The protein content of these plasma membranes is typically 50%. Membranes involved in energy transduction, such as the internal membranes of mitochondria and chloroplasts, have the highest content of protein, typically 75%.

The major proteins in a membrane can readily be visualized by *SDS-acrylamide gel electrophoresis.* In this technique, the membrane to be analyzed is first solubilized in a 1% solution of sodium dodecyl sulfate (SDS). This detergent disrupts most protein-protein and protein-lipid interactions. This solution is layered on top of an acrylamide gel containing SDS, which is then subjected to an electric field for a few hours. The electrophoretic mobility of many proteins in this gel depends on their molecular weight rather than on their net charge in the absence of SDS. The negative charge contributed by SDS molecules bound to the protein is much larger than the net charge of the protein itself. A pattern of bands appears when the gel is stained with a dye such as coomassie blue. A few micrograms of a protein can be visualized in this way. The gel electrophoresis patterns of three membranes—the plasma membrane of erythrocytes, the photoreceptor membrane of retinal rod cells, and the sarcoplasmic reticulum membrane of muscle—are shown in Figure 10-15. The gel patterns reveal that these three membranes have very different protein compositions. Furthermore, they differ in the number of kinds of polypeptide chains and in their molecular-weight distribution. In short, *membranes that perform different functions have different proteins.*

Numerous membrane proteins have been solubilized and purified (see Chapters 32 to 34). Some of them are active in detergent solution. For example, rhodopsin, a photoreceptor protein, has the same 500-nm absorption band in detergent solution as in the retinal disc membrane. Furthermore, the prosthetic group of this protein undergoes the same structural change on illumination in both environments. Calsequestrin, a calcium-binding protein from the sarcoplasmic reticulum of muscle, retains this ion-binding property when it is removed from its membrane environment. The calcium pump (called the Ca^{2+} ATPase) from this membrane system has also been isolated. In fact, functionally active vesicles can be formed from a mixture of phospholipid and the pump protein. These vesicles accumulate Ca^{2+} if ATP, the energy source for the pump, is provided. *The reconstitution of functionally active membrane systems from purified components is a powerful experimental approach in the elucidation of membrane processes.*

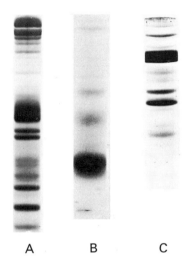

A   B   C

Figure 10-15
SDS-acrylamide gel patterns of (A) the plasma membrane of erythrocytes, (B) the disc membranes of retinal rod cells, and (C) the sarcoplasmic reticulum membrane of muscle cells. [Courtesy of Dr. Theodore Steck (part A) and Dr. David MacLennan (part C).]

SOME MEMBRANE PROTEINS ARE DEEPLY IMBEDDED
IN THE LIPID BILAYER

Some membrane proteins can be dissociated from the rest of the membrane by relatively mild means, such as extraction by a solution of high ionic strength (e.g., 1 M NaCl). In contrast, other membrane proteins are bound much more tenaciously. They can be separated only by using a detergent or an organic solvent. This difference in dissociability has led some investigators to classify membrane proteins as being *peripheral* or *integral* (Figure 10-16). Since peripheral proteins can be dissociated by the addition of salts, it is inferred that they are bound to the surface of membranes by electrostatic and hydrogen-bond interactions. In contrast, it is presumed that integral proteins interact to a large extent with the hydrocarbon chains of membrane lipids because they are solubilized by detergents, which would compete for these nonpolar interactions.

Freeze-fracture electron microscopy is a valuable technique for ascertaining whether proteins are located in the interior of biological membranes. Cells or membrane fragments are rapidly frozen at liquid-nitrogen temperatures. The frozen membrane is then fractured by the impact of a microtome knife. Cleavage usually occurs along a plane in the middle of the bilayer (Figure 10-17). Hence, extensive regions *within* the lipid bilayer are exposed. These exposed regions can then be shadowed with carbon and platinum, which yields a replica of the interior of the bilayer. The external surfaces of membranes can also be viewed by combining freeze-

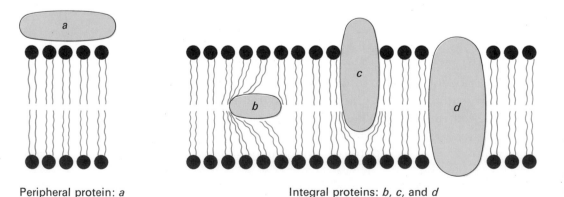

Peripheral protein: *a*   Integral proteins: *b, c,* and *d*

Figure 10-16
Postulated location of peripheral and integral membrane proteins.

fracture and deep-etching techniques. First, the interior of a frozen membrane is exposed by fracturing. The ice that covers one of the adjacent membrane surfaces is then sublimed away; this process is termed deep-etching. The combined technique, called *freeze-etching electron microscopy,* provides a view of the interior of a membrane and of both its surfaces. An attractive feature of the freeze-etching technique is that fixatives and dehydrating agents are not required.

Freeze-etching studies have provided direct evidence for the presence of integral proteins in many biological membranes. Erythrocyte membrane, for example, contains a high density of globular particles, approximately 75 Å in diameter, in its interior (Figure 10-18). The inside of the sarcoplasmic reticulum membrane is also rich in globular particles. In contrast, synthetic bilayers formed from phosphatidyl choline yield smooth fracture faces. Also, the fracture faces of myelin membranes are smooth, as might be expected for a relatively inert membrane that serves primarily as an insulator.

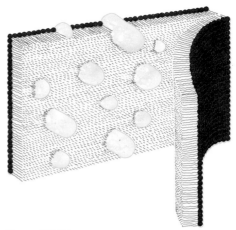

Figure 10-17
Technique of freeze-fracture electron microscopy. The cleavage plane passes through the middle of the bilayer membrane. [Based on S. J. Singer. *Hosp. Pract.* 8(1973):81.]

Interior of the bilayer (Cleaved surface) External surface of the membrane

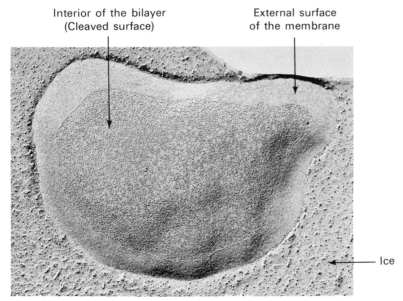

Ice

Figure 10-18
Freeze-etch electron micrograph of the plasma membrane of red blood cells. [Courtesy of Dr. Vincent Marchesi.] The interior of the membrane, which has been exposed by fracture of the membrane, is rich in globular particles that have a diameter of about 75 Å. These globular particles are thought to be integral membrane proteins.

MEMBRANES ARE ASYMMETRIC

Membranes are structurally and functionally asymmetric. *The two surfaces of a membrane have different components and different enzymatic activities.* A clear-cut example is provided by the pump that regulates the concentration of Na^+ and K^+ ions in cells. This transport system is located in the plasma membrane of nearly all cells in higher organisms. The Na^+-K^+ pump assembly is oriented in the plasma membrane so that it pumps Na^+ out of and K^+ into the cell (Figure 10-19). Furthermore, ATP must be on the inside of the cell to drive the pump. Ouabain, a specific inhibitor of the pump, is effective only if it is located outside.

The distribution of sugar residues in membranes is also highly distinctive. The membranes of eucaryotic cells usually contain between 2% and 10% carbohydrate, in the form of *glycolipids* and *glycoproteins*. As mentioned earlier (p. 232), glycolipids are derivatives of sphingosine that contain one or more sugar residues. In membrane glycoproteins, one or more chains of sugar residues are attached to serine, threonine, or asparagine side chains of the protein, usually through *N*-acetylglucosamine or *N*-acetylgalactosamine. The location of these carbohydrate groups in membranes can be determined by specific labeling techniques. There are many plant proteins, called lectins, that have high affinity for specific sugar residues. For example, *concanavalin A* binds to α-mannosyl residues at the nonreducing ends of carbohydrate chains, whereas *wheat-germ agglutinin* binds to *N*-acetylglucosamine residues. These lectins can be readily seen in electron micrographs if they are conjugated to ferritin, a protein with a very electron-dense core of

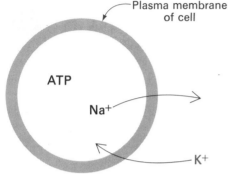

Figure 10-19
Asymmetry of the Na^+-K^+ transport system in plasma membranes.

N-Acetylglucosamine
linked to an
asparagine residue

N-Acetylgalactosamine
linked to a
serine residue

iron. The ferritin conjugate of concanavalin A binds specifically to the outer surface of the erythrocyte membranes, and not to the inner cytoplasmic surface. The same asymmetry has been observed in the binding of other lectins. Indeed, *the sugar residues in the plasma membranes of all mammalian cells studied thus far are located exclusively on the external surface* (Figure 10-20).

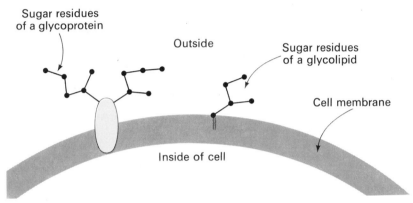

Figure 10-20
Sugar residues of glycoproteins and glycolipids are usually located on the outside surface of mammalian plasma membranes.

SUGAR RESIDUES MAY ORIENT GLYCOPROTEINS IN MEMBRANES

Why do some membrane proteins contain covalently attached sugar residues? The answer is not yet known. One possibility is that *carbohydrate groups may orient glycoproteins in membranes*. Since sugars are highly hydrophilic, the sugar residues of a glycoprotein or of a glycolipid will tend to be located at a membrane surface, rather than in the hydrocarbon core. The cost in free energy of inserting an oligosaccharide chain into the hydrocarbon core of a membrane is very high. Consequently, there is a high barrier to the rotation of a glycoprotein from one side of a membrane to the other. The carbohydrate moieties of a membrane glycoprotein probably help to establish and maintain the asymmetric character of biological membranes.

Carbohydrates on cell surfaces may also play an important role in *intercellular recognition*. The interaction of different cells to form a tissue and the detection of foreign cells by the immune system of a higher organism are examples of processes that depend on the recognition of one cell surface by another. Carbohydrates have the potential for great structural diversity. An enormous number of patterns of surface sugars is possible because (a) monosaccharides can be joined to each other through any of several hydroxyl groups, (b) the C-1 linkage can have either an α or β configuration, and (c) extensive branching is possible. Indeed, many more different oligosaccharides can be formed from four sugars than oligopeptides from four amino acids.

MANY PROTEINS AND LIPIDS CAN DIFFUSE RAPIDLY IN MEMBRANES

Biological membranes are not rigid structures. On the contrary, many membrane proteins and lipids are constantly in motion. The rapid movement of membrane proteins has been visualized by means of fluorescence microscopy. Human cells and mouse cells in culture can be induced to fuse with each other. The resulting hybrid cell is called a *heterokaryon*. Part of the plasma membrane of this heterokaryon comes from a mouse cell, the rest from a human cell. Do the membrane proteins derived from the mouse and human cells stay segregated in the heterokaryon or do they intermingle? This question was answered by using fluorescent-labeled antibodies as markers that could be followed by light microscopy. An antibody specific for mouse membrane proteins was labeled to show a green fluorescence, and an antibody specific for human membrane proteins was labeled to show a red fluorescence (Figure 10-21). In a newly formed heterokaryon, half of the surface displayed green fluorescing patches, the other half red. However, in less than an hour (at 37°C), the red and green fluorescing patches became completely intermixed. This experiment and studies of other plasma membranes have revealed that *membrane proteins can diffuse a distance of several microns in approximately one minute.*

Phospholipid molecules diffuse even more rapidly in membranes than do proteins because they are smaller. Magnetic resonance studies have shown that the diffusion constant D for a phospholipid

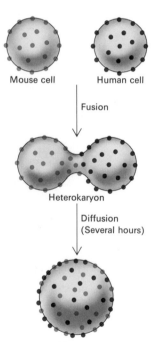

Figure 10-21
Diagram showing the fusion of a mouse cell and a human cell, followed by diffusion of membrane components in the plane of the plasma membrane. The green and red fluorescing markers are completely intermingled after several hours.

molecule in a variety of membranes is about 10^{-8} cm^2 sec. In two-dimensional diffusion, the average distance s (in cm) traversed by a molecule in time t (in sec) is given by:

$$s = (4Dt)^{1/2}$$

Thus, a phospholipid molecule diffuses an average distance of 2×10^{-4} cm or $2\ \mu$m in 1 sec. This means that a *lipid molecule can travel from one end of a bacterium to the other in a second.* The magnitude of the observed diffusion coefficient indicates that the viscosity of the membrane is about 100 times that of water, rather like that of olive oil.

MOST PROTEINS AND LIPIDS DO NOT READILY ROTATE ACROSS MEMBRANES

The rotation of most proteins and lipids from one side of a membrane to the other is a very slow process, in contrast to their movement parallel to the plane of the bilayer. The transition of a molecule from one membrane surface to the other is called *transverse diffusion* or *flip-flop,* whereas diffusion in the plane of a membrane is termed *lateral diffusion.*

As mentioned previously, the carbohydrate units of membrane proteins are usually located on only one of the membrane surfaces. This asymmetric distribution implies that glycoproteins either do not flip-flop or do so at a very slow rate. The flip-flop of phospholipid molecules in phosphatidyl choline vesicles has been directly measured by electron spin resonance techniques, which showed that *a phospholipid molecule flip-flops once in several hours* (see problem 5 on page 253 for the experimental design). Thus, a phospholipid molecule takes about 10^9 times as long to flip-flop across a 50 Å membrane as it takes to diffuse a distance of 50 Å in the lateral direction.

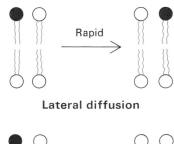

Lateral diffusion

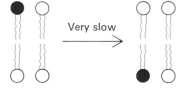

Transverse diffusion
(Flip-flop)

MEMBRANE FLUIDITY IS CONTROLLED BY FATTY ACID COMPOSITION

The fatty acyl chains of lipid molecules in bilayer membranes can exist in an ordered, rigid state or in a relatively disordered, fluid state. The transition from the rigid to the fluid state occurs as the temperature is raised above a "melting" temperature. This transition temperature depends on the length of the fatty acyl chains and on their degree of unsaturation. The preferred conformation of a saturated hydrocarbon is a straight chain. The rigid state is favored by the presence of saturated fatty acyl residues because their straight hydrocarbon chains interact very favorably with each other. On the other hand, *a cis double bond produces a bend in the hydrocarbon chain. This bend interferes with the optimal packing of fatty acyl chains* (Figure 10-22).

Furthermore, a *cis* double bond enhances rotation about the carbon–carbon single bonds on either side. Thus, *an unsaturated fatty acyl chain is intrinsically more flexible than a saturated one.* This effect is enhanced by the presence of more than one double bond.

The length of the fatty acyl chain also affects membrane fluidity. Long hydrocarbon chains interact more strongly than do short ones.

Rotation about these single bonds is enhanced by the proximity of the double bond.

Figure 10-22
The packing of fatty acid chains is disrupted by the presence of *cis* double bonds. Space-filling model showing the packing of (A) three molecules of stearate (C_{18}, saturated) and (B) a molecule of oleate (C_{18}, unsaturated) between two molecules of stearate.

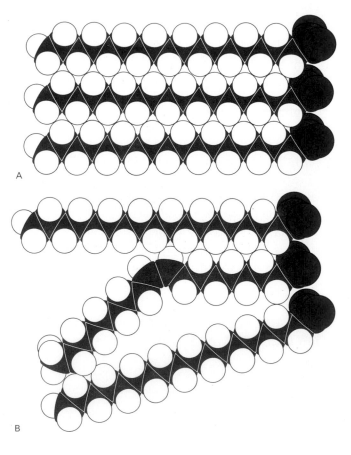

Specifically, each additional —CH$_2$— group makes a favorable contribution of about -0.5 kcal/mol to the free energy of interaction of two adjacent hydrocarbon chains.

In short, *membrane fluidity is enhanced by the presence of double bonds in fatty acyl chains and by their short chain length.* There is experimental evidence that the fluidity of the membrane of *E. coli* is regulated by the variation of these factors. For example, the ratio of saturated to unsaturated fatty acyl chains in *E. coli* membranes decreases from 1.6 to 1.0 as the temperature decreases from 42°C to 27°C. This decrease in the proportion of saturated residues prevents the membrane from becoming too rigid at the lower temperature. In eucaryotes, cholesterol influences membrane fluidity. It seems likely that the activity of many membrane proteins depends on a fluid membrane.

FLUID MOSAIC MODEL OF BIOLOGICAL MEMBRANES

A *fluid mosaic model* depicting the gross organization of biological membranes has been proposed by S. Jonathan Singer and Garth Nicolson (Figure 10-23). The essence of their model is that membranes are *two-dimensional solutions of oriented globular proteins and lipids.*

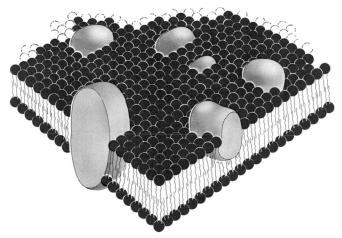

Figure 10-23
Fluid mosaic model. [Based on S. J. Singer and G. L. Nicolson. *Science* 175(1972):723. Copyright 1972 by the American Association for the Advancement of Science.]

The major features of this model are:

1. Most of the membrane phospholipid and glycolipid molecules are in bilayer form. This lipid bilayer has a dual role: it is a *solvent* for integral membrane protein and it is also a *permeability barrier*.

2. A small proportion of membrane lipids interact specifically with particular membrane proteins and may be essential for their function.

3. Membrane proteins are free to diffuse laterally in the lipid matrix unless restricted by special interactions, whereas they are not free to rotate from one side of a membrane to the other.

SUMMARY

Biological membranes are sheetlike structures, typically 75 Å wide, that are composed of protein and lipid molecules held together by noncovalent interactions. Membranes are highly selective permeability barriers. They create compartments, which may be an entire cell or an organelle within a cell. Pumps and gates in membranes regulate the molecular and ionic composition of these compartments. Membranes also control the flow of information between cells. For example, some membranes contain receptors for hormones such as insulin. Furthermore, membranes are intimately involved in energy conversion processes such as photosynthesis and oxidative phosphorylation.

The major classes of membrane lipids are phospholipids, glycolipids, and cholesterol. Phosphoglycerides, a type of phospholipid, consist of a glycerol backbone, two fatty acid chains, and a phosphorylated alcohol. The fatty acid chains usually contain between 14 and 24 carbon atoms; they may be saturated or unsaturated. Phosphatidyl choline, phosphatidyl serine, and phosphatidyl ethanolamine are major phosphoglycerides. Sphingomyelin, a different type of phospholipid, contains a sphingosine backbone instead of glycerol. Glycolipids are sugar-containing lipids derived from sphingosine. A common feature of these membrane lipids is that they are amphipathic molecules. They spontaneously form extensive bimolecular sheets in aqueous solutions because they contain both a hydrophilic and a hydrophobic moiety. These lipid bilayers are highly impermeable to ions and most polar molecules, yet they are quite fluid, which enables them to act as a solvent for membrane proteins.

Distinctive membrane functions such as transport, communication, and energy transduction are mediated by specific proteins. Some membrane proteins are deeply imbedded in the hydrocarbon region of the lipid bilayer. Membranes are structurally and functionally asymmetric, as exemplified by the directionality of ion transport systems in them and the localization of sugar residues on the external surface of mammalian plasma membranes. Membranes are dynamic structures in which proteins and lipids diffuse rapidly in the plane of the membrane (lateral diffusion), unless restricted by special interactions. In contrast, the rotation of proteins and lipids from one side of a membrane to the other (transverse diffusion or flip-flop) is usually very slow. The degree of fluidity of membranes partly depends on the chain length and extent of unsaturation of their constituent fatty acids.

SELECTED READINGS

Where to start:

Singer, S. J., and Nicolson, G. L., 1972. The fluid mosaic model of the structure of cell membranes. *Science* 175:720–731.

Fox, C. F., 1972. The structure of cell membranes. *Sci. Amer.* 226(2):30–38. [Available as *Sci. Amer.* Offprint 1241.]

Reviews:

Singer, S. J., 1974. The molecular organization of membranes. *Annu. Rev. Biochem.* 43:805–830.

Guidotti, G., 1972. Membrane proteins. *Annu. Rev. Biochem.* 41:731–752.

Oseroff, A. R., Robbins, P. W., and Burger, M. W., 1972. The cell surface membrane: biochemical aspects and biophysical probes. *Annu. Rev. Biochem.* 42:647–680.

Bretscher, M. S., 1973. Membrane structure: some general principles. *Science* 181:622–629.

Tooze, J., (ed.), 1973. *The Molecular Biology of Tumor Viruses*. Cold Spring Harbor Laboratory. [Chapter 3 reviews the changes in cell surfaces that accompany transformation to a cancerous state.]

Chavin, S. I., 1971. Isolation and study of functional membrane proteins. *FEBS (Fed. Eur. Biochem. Soc.) Lett.* 14:269–282.

Books:

Fox, C. F., and Keith, A., (eds.), 1972. *Membrane Molecular Biology*. Sinauer Associates. [A collection of articles on membrane composition and isolation, physical properties of membranes, some membrane functions, and membrane assembly.]

Tanford, C., 1973. *The Hydrophobic Effect: Formation of Micelles and Biological Membranes.* Wiley-Interscience.

Rothfield, L. I., (ed.), 1971. *Structure and Function of Biological Membranes.* Academic Press. [A collection of articles on membrane spectroscopy, membrane-bound enzymes, membrane models, active transport, mitochondrial membranes, and mammalian cell surfaces.]

Freeze-etching electron microscopy of membranes:

Branton, D., 1966. Fracture faces of frozen membranes. *Proc. Nat. Acad. Sci.* 55:1048–1056.

Pinto da Silva, P., and Branton, D., 1970. Membrane splitting in freeze-etching: covalently bound ferritin as a membrane marker. *J. Cell Biol.* 45:598–605.

Membrane mobility:

Frye, C. D., and Edidin, M., 1970. The rapid intermixing of cell surface antigens after formation of mouse-human heterokaryons. *J. Cell Sci.* 7:319–335.

Kornberg, R. D., and McConnell, H. M., 1971. Inside-outside transitions of phospholipids in vesicle membranes. *Biochemistry* 10:1111–1120.

Cone, R. A., 1972. Rotational diffusion of rhodopsin in the visual receptor membrane. *Nature New Biol.* 236:39–43.

Poo, M., and Cone, R. A., 1974. Lateral diffusion of rhodopsin in the photoreceptor membrane. *Nature* 247:438–441.

Edidin, M., 1974. Rotational and translational diffusion in membranes. *Annu. Rev. Biophys. Bioeng.* 3:179–201.

Membrane glycoproteins:

Ginsburg, V., and Kobata, A., 1971. Structure and function of surface components of mammalian cells. *In* Rothfield, L. I., (ed.), *Structure and Function of Biological Membranes,* pp. 439–459.

Marchesi, V. T., Tillack, T. W., Jackson, R. L., Segrest, J. P., and Scott, R. E., 1972. Chemical characterization and surface orientation of the major glycoprotein of the human erythrocyte membrane. *Proc. Nat. Acad. Sci.* 69:1445–1449.

Model membrane systems:

Mueller, P., and Rudin, D. O., 1969. Translocators in bimolecular lipid membranes. *Curr. Top. Bioenergetics* 3:157–249.

Jain, J. K., 1972. *The Bimolecular Lipid Membrane.* Van Nostrand Reinhold.

Thompson, T. E., and Henn, F. A., 1970. Experimental phospholipid model membrane. *In* Racker, E., (ed.), *Membranes of Mitochondria and Chloroplasts,* pp. 1–52. Van Nostrand Reinhold.

PROBLEMS

1. How many phospholipid molecules are there in a 1-μm^2 region of a phospholipid bilayer membrane? Assume that a phospholipid molecule occupies 70 Å^2 of the surface area.

2. Bacterial phospholipids contain cyclopropane fatty acid residues.

$$\text{H}_3\text{C}-(\text{CH}_2)_5-\overset{\displaystyle \text{CH}_2}{\underset{\displaystyle \text{H}}{\text{C}}}-\overset{}{\underset{\displaystyle \text{H}}{\text{C}}}-(\text{CH}_2)_9-\text{C}\overset{\displaystyle \text{O}}{\underset{\displaystyle \text{O}^-}{}}$$

Predict the effect of the cyclopropane ring on the packing of hydrocarbon chains in the interior of a bilayer. Would these cyclopropane fatty acid groups tend to make a membrane more or less fluid?

3. What is the average distance traversed by a membrane lipid in 1 μsec, 1 msec, and 1 sec? Assume a diffusion coefficient of 10^{-8} cm^2/sec.

4. The diffusion coefficient D of a rigid spherical molecule is given by

$$D = kT/(6\pi\eta r)$$

where η is the viscosity of the solvent, r is the radius of the sphere, k is the Boltzman constant (1.38×10^{-16} erg/deg), and T is the absolute temperature. What is the diffusion coefficient at 37°C of a 100,000-dalton protein in a membrane that has an effective viscosity of 1 poise (1 poise = 1 erg sec/cm^3)? What is the average distance traversed by this protein in 1 μsec, 1 msec, and 1 sec? Assume that this protein is an unhydrated, rigid sphere of density 1.35 g/cm^3.

5. R. D. Kornberg and H. M. McConnell (1971) investigated the transverse diffusion (flip-flop) of phospholipids in a bilayer membrane by using a paramagnetic analog of phosphatidyl choline, called *spin-labeled phosphatidyl choline*.

The nitroxide (NO) group in spin-labeled phosphatidyl choline gives a distinctive paramagnetic resonance spectrum. This spectrum disappears when nitroxides are converted to amines by reducing agents such as ascorbate.

Liposomes containing phosphatidyl choline (95%) and the spin-labeled analog (5%) were prepared by sonication and purified by gel-filtration chromatography. The outside diameter of these liposomes was about 250 Å. The amplitude of the paramagnetic resonance spectrum decreased to 35% of its initial value within a few minutes of the addition of ascorbate. There was no detectable change in the spectrum within a few minutes after the addition of a second aliquot of ascorbate. However, the amplitude of the residual spectrum decayed exponentially with a half-time of 6.5 hr. How would you interpret these changes in the amplitude of the paramagnetic spectrum?

On the facing page: Model of acetyl coenzyme A, a key intermediate in the generation of metabolic energy.

PART II

GENERATION AND STORAGE
OF METABOLIC ENERGY

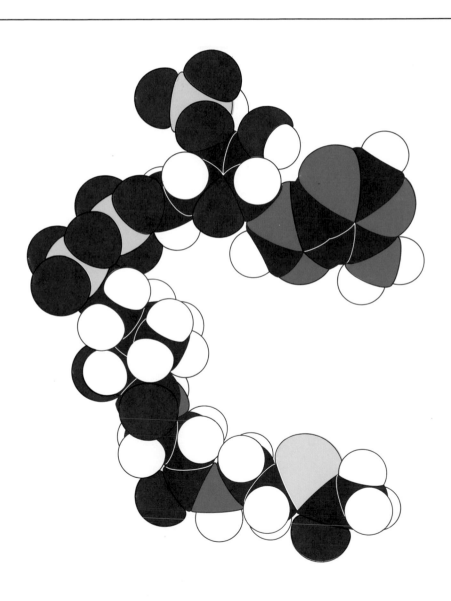

CHAPTER 11

METABOLISM:
BASIC CONCEPTS AND DESIGN

In Part I, we considered the theme of conformation. The concepts developed there—especially those dealing with specificity and catalytic power of enzymes, the regulation of their catalytic activity, and the formation of compartments by membranes—enable us to turn now to two major questions of biochemistry:

1. *How do cells extract energy from their environments?*

2. *How do cells synthesize the building blocks of their macromolecules?*

These processes are carried out by a highly integrated network of chemical reactions, which are collectively known as *metabolism*.

There are at least a thousand chemical reactions in even a simple organism such as *E. coli*. This array of reactions may seem overwhelming at first glance. However, closer scrutiny reveals that metabolism has a *coherent design containing many common motifs*. The number of reactions in metabolism is large, but the number of *kinds* of reactions is relatively small. Furthermore, a group of about a hundred molecules plays a central role. Moreover, metabolic pathways are regulated in common ways. The purpose of this chapter is to introduce some general principles and motifs of metabolism.

FREE ENERGY IS THE MOST USEFUL THERMODYNAMIC FUNCTION IN BIOCHEMISTRY

Let us start by reviewing some thermodynamic relationships that are essential for an understanding of metabolism. In thermodynamics, a *system* is the matter within a defined region. The matter in the rest of the universe is called the *surroundings*. *The first law of thermodynamics states that the total energy of a system and its surroundings is a constant.* In other words, energy is conserved. The mathematical expression of the first law is

$$\Delta E = E_{\text{B}} - E_{\text{A}} = Q - W \tag{1}$$

where E_{A} is the energy of a system at the start of a process and E_{B} at the end of the process, Q is the heat absorbed by the system, and W is the work done by the system. An important feature of equation (1) is that *the change in energy of a system depends only on the initial and final states and not on the path of the transformation.*

The first law of thermodynamics cannot be used to predict whether a reaction can occur spontaneously. Some reactions do occur spontaneously although ΔE is positive. In such cases, the system absorbs heat from its surroundings so that the sum of the energies of the system and its surroundings remains the same. It is evident that a function different from ΔE is required. One such function is the *entropy* (S), which is a measure of the *degree of randomness or disorder of a system.* The entropy of a system increases (ΔS is positive) when it becomes more disordered. *The second law of thermodynamics states that a process can occur spontaneously only if the sum of the entropies of the system and its surroundings increases.*

$$(\Delta S_{\text{system}} + \Delta S_{\text{surroundings}}) > 0 \text{ for a spontaneous process} \tag{2}$$

Note that the entropy of a system may decrease during a spontaneous process, provided that the entropy of the surroundings increases so that their sum is positive. For example, the formation of a highly ordered biological structure is thermodynamically feasible because the decrease in entropy in such a system is more than offset by an increase in entropy in its surroundings.

One difficulty in using entropy as a criterion of whether a biochemical process can occur spontaneously is that entropy changes of chemical reactions are not readily measured. Furthermore, the

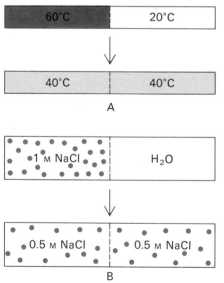

Figure 11-1
Examples of processes that are driven by an increase in the entropy of a system: (A) diffusion of heat and (B) diffusion of a solute.

criterion of spontaneity given in equation (2) requires a knowledge of the entropy change of the surroundings as well as of the system of interest. These difficulties are obviated by using a different thermodynamic function called the *free energy*, which is denoted by the symbol G (or F, in the older literature). In 1878, Josiah Willard Gibbs created the free-energy function by combining the first and second laws of thermodynamics. The basic equation is

$$\Delta G = \Delta H - T\Delta S \qquad (3)$$

where ΔG is the change in free energy of a system undergoing a transformation at constant pressure (P) and temperature (T), ΔH is the change in enthalpy of this system, and ΔS is the change in entropy of this system. Note that the properties of the surroundings do not enter into this equation. The enthalpy change is given by

$$\Delta H = \Delta E + P\Delta V \qquad (4)$$

Since the volume change ΔV is small for nearly all biochemical reactions, ΔH is nearly equal to ΔE. Hence,

$$\Delta G \cong \Delta E - T\Delta S \qquad (5)$$

Thus, the ΔG of a reaction depends on both the change in internal energy and the change in entropy of the system.

The change in free energy (ΔG) of a reaction, in contrast to the change in internal energy (ΔE) of a reaction, is a valuable criterion of whether it can occur spontaneously.

1. *A reaction can occur spontaneously only if* ΔG *is negative.*

2. A system is at equilibrium and no net change can take place if ΔG is zero.

3. A reaction cannot occur spontaneously if ΔG is positive. An input of free energy is required to drive such a reaction.

Two additional points need to be emphasized here. First, the ΔG of a reaction depends only on the free energy of the products (the final state) minus that of the reactants (the initial state). *The ΔG of a reaction is independent of the path of the transformation.* The mechanism of a reaction has no effect on ΔG. For example, the ΔG for the oxidation of glucose to CO_2 and O_2 is the same whether it occurs by combustion in vitro or by a series of many enzyme-catalyzed steps in a cell. Second, ΔG *provides no information about the rate of a reaction.* A negative ΔG indicates that a reaction can occur spontaneously but it does not signify that it will occur at a perceptible rate. As previously discussed (p. 120), the rate of a reaction depends on the *free energy of activation* $(\Delta G^{\ddagger})$, which is unrelated to ΔG.

STANDARD FREE-ENERGY CHANGE OF A REACTION AND ITS RELATIONSHIP TO THE EQUILIBRIUM CONSTANT

Consider the reaction

$$A + B \rightleftharpoons C + D$$

The ΔG of this reaction is given by

$$\Delta G = \Delta G^{\circ} + RT \log_e \frac{[C][D]}{[A][B]} \tag{6}$$

where ΔG° is the *standard free-energy change,* R is the gas constant, T is the absolute temperature, and [A], [B], [C], and [D] are the molar concentrations (more rigorously, the activities of the reactants). ΔG° is the free-energy change for this reaction under standard conditions—that is, where A, B, C, and D are each present at a concentration of 1.0 M. Thus, the ΔG of a reaction depends on the *nature* of the reactants [expressed in the ΔG° term of equation (6)] and on their *concentrations* [expressed in the logarithmic term in equation (6)].

A convention has been adopted to simplify free-energy calculations for biochemical reactions. The standard state is defined as having a pH of 7. Consequently, the activity of H^+ corresponding to a pH of 7 has a value of 1 in equations (6) and (9). Also, the activity of water is taken to be 1 in these equations. The *standard free-energy change at pH 7,* denoted by the symbol $\Delta G^{\circ\prime}$, will be used throughout this book.

The relationship between the standard free energy and the equilibrium constant of a reaction is readily derived. At equilibrium,

$\Delta G = 0$. Equation (6) then becomes

$$0 = \Delta G^{\circ\prime} + RT \log_e \frac{[C][D]}{[A][B]} \tag{7}$$

and so

$$\Delta G^{\circ\prime} = -RT \log_e \frac{[C][D]}{[A][B]} \tag{8}$$

The equilibrium constant under standard conditions, K'_{eq}, is defined as

$$K'_{eq} = \frac{[C][D]}{[A][B]} \tag{9}$$

Substituting equation (9) into (8) gives

$$\Delta G^{\circ\prime} = -RT \log_e K'_{eq} \tag{10}$$

$$\Delta G^{\circ\prime} = -2.303 \, RT \log_{10} K'_{eq} \tag{11}$$

Thus, the standard free energy and the equilibrium constant of a reaction are related by a simple expression. For example, an equilibrium constant of 10 corresponds to a standard free-energy change of -1.36 kcal/mol at 25°C.

Let us calculate $\Delta G^{\circ\prime}$ and ΔG for the isomerization of dihydroxyacetone phosphate to glyceraldehyde 3-phosphate as an example. This reaction occurs in glycolysis (p. 286).

Dihydroxyacetone phosphate ⇌ Glyceraldehyde 3-phosphate

At equilibrium, the ratio of glyceraldehyde 3-phosphate to dihydroxyacetone phosphate is .0475 at 25°C (298°K) at pH 7. Hence, $K'_{eq} = .0475$. The standard free-energy change for this reaction is then calculated from equation (11).

$$\begin{aligned}
\Delta G^{\circ\prime} &= -2.303 \, RT \log_{10} K'_{eq} \\
&= -2.303 \times 1.98 \times 298 \times \log_{10}(.0475) \\
&= +1800 \text{ cal/mol} = +1.8 \text{ kcal/mol}
\end{aligned}$$

Table 11-1
Relationship between $\Delta G^{\circ\prime}$ and K'_{eq} (at 25°C)

K'_{eq}	$\Delta G^{\circ\prime}$ (kcal/mol)
10^{-5}	6.82
10^{-4}	5.46
10^{-3}	4.09
10^{-2}	2.73
10^{-1}	1.36
1	0
10	-1.36
10^2	-2.73
10^3	-4.09
10^4	-5.46
10^5	-6.82

Now let us calculate ΔG for this reaction when the initial concentration of dihydroxyacetone phosphate is 2×10^{-4} M and the initial concentration of glyceraldehyde 3-phosphate is 3×10^{-6} M. Substituting these values into equation (6) gives

$$\Delta G = 1.8 \text{ kcal/mol} + 2.303 \, RT \log_{10} \frac{3 \times 10^{-6} \text{ M}}{2 \times 10^{-4} \text{ M}}$$
$$= 1.8 \text{ kcal/mol} - 2.5 \text{ kcal/mol}$$
$$= -0.7 \text{ kcal/mol}$$

This negative value for the ΔG indicates that the isomerization of dihydroxyacetone phosphate to glyceraldehyde 3-phosphate can occur spontaneously when these species are present at the concentrations stated above. Note that ΔG for this reaction is negative although $\Delta G°$ is positive. *It is important to stress that the ΔG for a reaction can be larger, smaller, or the same as $\Delta G°$, depending on the concentrations of the reactants. The criterion of spontaneity for a reaction is ΔG, not $\Delta G°$.*

A THERMODYNAMICALLY UNFAVORABLE REACTION CAN BE DRIVEN BY A FAVORABLE ONE

An important thermodynamic fact is that *the overall free-energy change for a series of reactions is equal to the sum of the free-energy changes of the individual steps.* Consider the reactions

A $\rightleftharpoons$ B + C	$\Delta G°' = +5 \text{ kcal/mol}$
B $\rightleftharpoons$ D	$\Delta G°' = -8 \text{ kcal/mol}$
A $\rightleftharpoons$ C + D	$\Delta G°' = -3 \text{ kcal/mol}$

Under standard conditions, A cannot be spontaneously converted to B and C since ΔG is positive. However, the conversion of B to D under standard conditions is thermodynamically feasible. Because free-energy changes are additive, the conversion of A to C and D has a $\Delta G°'$ of -3 kcal/mol, which means that it can occur spontaneously under standard conditions. Thus, *a thermodynamically unfavorable reaction can be driven by a thermodynamically favorable one.* These reactions are *coupled* by B, the common intermediate. We will encounter many examples of energy coupling in metabolism.

ATP IS THE UNIVERSAL CURRENCY OF FREE ENERGY IN BIOLOGICAL SYSTEMS

Living things require a continual input of free energy for three major purposes: the performance of mechanical work in muscle contraction and other cellular movements, the active transport of molecules and ions, and the synthesis of macromolecules and other biomolecules from simple precursors. The free energy used in these processes, which maintain an organism in a state that is far from equilibrium, is derived from the environment. *Chemotrophs* obtain this energy by the oxidation of foodstuffs, whereas *phototrophs* obtain it by trapping light energy. The free energy derived from the oxidation of foodstuffs and from light is partly transformed into a special form before it is used for motion, active transport, and biosyntheses. This special carrier of free energy is *adenosine triphosphate* (ATP). The central role of ATP in energy exchanges in biological systems was perceived by Fritz Lipmann and by Herman Kalckar in 1941.

ATP is a nucleotide consisting of an adenine, a ribose, and a triphosphate unit (Figure 11-2; see p. 529 for a discussion of nucleotide nomenclature). The active form of ATP is usually a complex of ATP with Mg^{2+} or Mn^{2+}. In considering the role of ATP as an energy carrier, we can focus on its triphosphate moiety. *ATP is an energy-rich molecule because its triphosphate unit contains two phosphoanhydride bonds*. A large amount of free energy is liberated when ATP is hydrolyzed to adenosine diphosphate (ADP) and orthophosphate (P_i), or when ATP is hydrolyzed to adenosine monophosphate (AMP) and pyrophosphate (PP_i). The $\Delta G°'$ for these reactions depends on the ionic strength of the medium and on the concentrations of Mg^{2+} and Ca^{2+}. We will use a value of

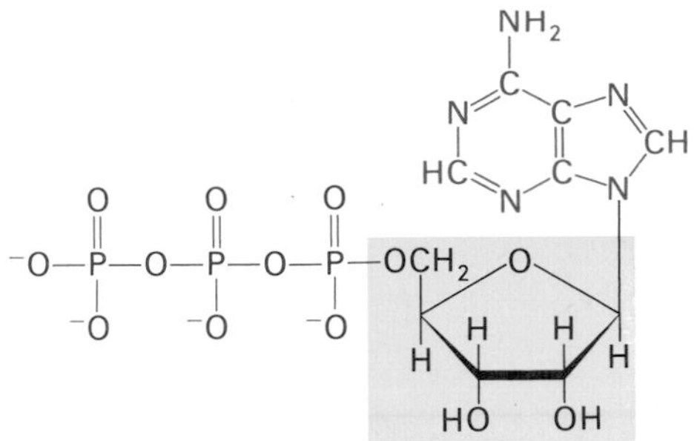

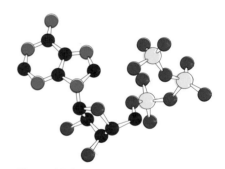

Adenosine triphosphate (ATP)

Figure 11-2
Adenosine triphosphate (ATP) consists of an adenine (blue), a ribose (yellow), and a triphosphate (red) unit.

Figure 11-3
A model of adenosine triphosphate (ATP).

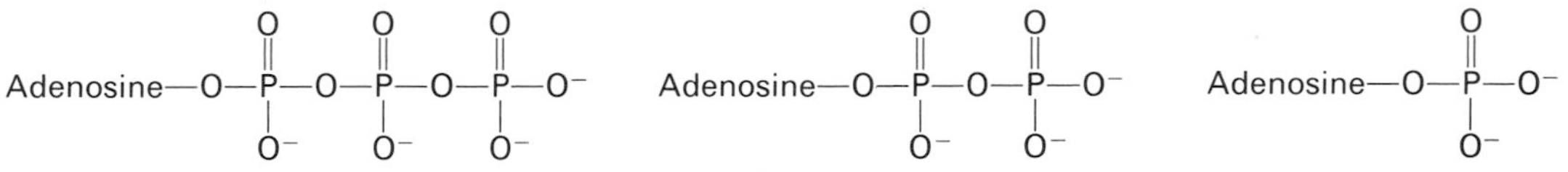

Adenosine triphosphate (ATP)

Adenosine diphosphate (ADP)

Adenosine monophosphate (AMP)

Figure 11-4
Structure of ATP, ADP, and AMP. (Adenosine consists of adenine linked to ribose.)

-7.3 kcal/mol. Under typical cellular conditions, the actual ΔG for these hydrolyses is approximately -12 kcal/mol.

$$\text{ATP} + \text{H}_2\text{O} \rightleftharpoons \text{ADP} + \text{P}_i + \text{H}^+ \qquad \Delta G^{\circ\prime} = -7.3 \text{ kcal/mol}$$

$$\text{ATP} + \text{H}_2\text{O} \rightleftharpoons \text{AMP} + \text{PP}_i + \text{H}^+ \qquad \Delta G^{\circ\prime} = -7.3 \text{ kcal/mol}$$

ATP, AMP, and ADP are interconvertible. The enzyme adenylate kinase (also called myokinase) catalyzes the reaction

$$\text{ATP} + \text{AMP} \rightleftharpoons \text{ADP} + \text{ADP}$$

The free energy liberated in the hydrolysis of an anhydride bond of ATP is used to drive reactions that require an input of free energy, such as muscle contraction. In turn, ATP is formed from ADP and P_i when fuel molecules are oxidized in chemotrophs or when light is trapped by phototrophs. *This ATP-ADP cycle is the fundamental mode of energy exchange in biological systems.*

Some biosynthetic reactions are driven by nucleotides that are analogous to ATP, namely guanosine triphosphate (GTP), uridine triphosphate (UTP), and cytidine triphosphate (CTP). The diphosphate forms of these nucleotides are denoted by GDP, UDP, and CDP, respectively. Enzymes catalyze the transfer of the terminal phosphoryl group from one nucleotide to another, as in the reactions

$$\text{ATP} + \text{GDP} \rightleftharpoons \text{ADP} + \text{GTP}$$

$$\text{ATP} + \text{GMP} \rightleftharpoons \text{ADP} + \text{GDP}$$

STRUCTURAL BASIS OF THE HIGH GROUP-TRANSFER POTENTIAL OF ATP

Let us compare the standard free energy of hydrolysis of ATP with that of a phosphate ester, such as glycerol 3-phosphate.

$$\text{ATP} + \text{H}_2\text{O} \rightleftharpoons \text{ADP} + \text{P}_i + \text{H}^+ \qquad \Delta G^{\circ\prime} = -7.3 \text{ kcal/mol}$$

$$\text{Glycerol 3-phosphate} + \text{H}_2\text{O} \rightleftharpoons \text{glycerol} + \text{P}_i$$
$$\Delta G^{\circ\prime} = -2.2 \text{ kcal/mol}$$

The magnitude of $\Delta G^{\circ\prime}$ for the hydrolysis of glycerol 3-phosphate

CH$_2$OH
H—C—OH
CH$_2$O—P—O$^-$

Glycerol 3-phosphate

is much smaller than that of ATP. This means that ATP has a stronger tendency to transfer its terminal phosphoryl group to water than does glycerol 3-phosphate. In other words, ATP has a higher *phosphate group-transfer potential* than glycerol 3-phosphate.

What is the structural basis of the high phosphate group-transfer potential of ATP? The structures of ATP and its hydrolysis products, ADP and P_i, must be examined to answer this question since $\Delta G°'$ depends on the *difference* in free energies of the products and reactants. Two factors prove to be important in this regard: *electrostatic repulsion* and *resonance stabilization*. At pH 7, the triphosphate unit of ATP carries about four negative charges. These charges repel each other strongly because they are in close proximity. *The electrostatic repulsion between these negatively charged groups is reduced when ATP is hydrolyzed. The other factor contributing to the high group-transfer potential of ATP is that ADP and P_i enjoy greater resonance stabilization than does ATP.* For example, orthophosphate has a number of resonance forms of similar energy (Figure 11-5). In contrast, the terminal portion of ATP has fewer significant resonance forms per phosphate group. Forms of the type shown in Figure 11-6 are unlikely to occur because the two phosphorus atoms compete for electron pairs on oxygen. Furthermore, there is a positive charge on an oxygen atom adjacent to a positively charged phosphorus atom, which is electrostatically unfavorable.

There are a variety of other compounds in biological systems that have a high phosphate group-transfer potential. In fact, some of them, such as phosphoenolpyruvate, acetyl phosphate, and creatine phosphate (Figure 11-7), have a higher group-transfer potential than does ATP. This means that phosphoenolpyruvate can transfer its phosphoryl group to ADP to form ATP. In fact, this is one of the ways in which ATP is generated in the breakdown of sugars. It is significant that ATP has a group-transfer potential that is intermediate among the biologically important phosphorylated

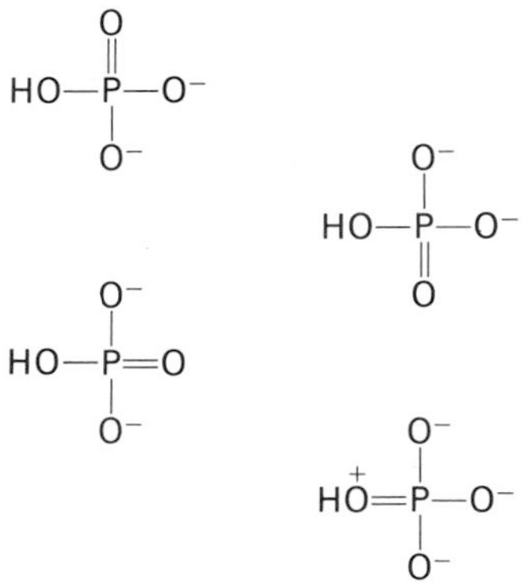

Figure 11-5
Significant resonance forms of orthophosphate.

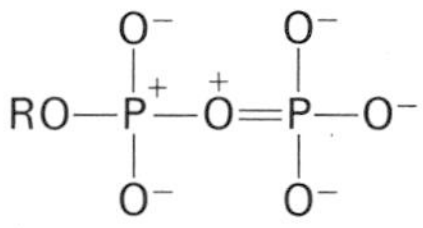

Figure 11-6
An improbable resonance form of the terminal portion of ATP.

Phosphoenolpyruvate   **Acetyl phosphate**   **Creatine phosphate**

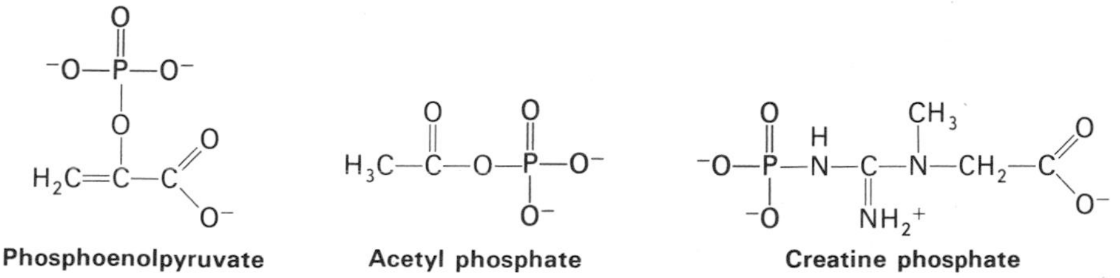

Figure 11-7
Compounds with a higher phosphate group-transfer potential than that of ATP.

Table 11-2
Free energies of hydrolysis of some phosphorylated compounds

Compound	$\Delta G°'$ (kcal/mol)
Phosphoenolpyruvate	−14.8
Carbamoyl phosphate	−12.3
Acetyl phosphate	−10.3
Creatine phosphate	−10.3
Pyrophosphate	−8.0
ATP (to ADP)	−7.3
Glucose 1-phosphate	−5.0
Glucose 6-phosphate	−3.3
Glycerol 3-phosphate	−2.2

Note: Phosphoenolpyruvate has the highest phosphate group-transfer potential of the compounds listed here.

molecules (Table 11-2). This intermediate position enables ATP to function efficiently as a carrier of phosphoryl groups.

ATP is often called a high-energy phosphate compound and its phosphoanhydride bonds are referred to as high-energy bonds. It should be noted that there is nothing special about the bonds themselves. *They are high-energy bonds in the sense that much free energy is released when they are hydrolyzed,* for the reasons given above. Lipmann's term "high-energy bond" and his symbol ~P for a compound having a high phosphate group-transfer potential are vivid, concise, and useful notations. In fact, Lipmann's squiggle did much to stimulate interest in bioenergetics.

NADH, NADPH, AND $FADH_2$ ARE THE MAJOR ELECTRON CARRIERS

Chemotrophs derive free energy from the oxidation of fuel molecules such as glucose and fatty acids. In aerobic organisms, the ultimate electron acceptor is O_2. However, electrons are not transferred directly from fuel molecules and their breakdown products to O_2. Instead, these substrates transfer electrons to special carriers, which are either *pyridine nucleotides* or *flavins*. The reduced forms of these carriers then transfer their high-potential electrons to O_2 via an electron-transport chain located in the inner membrane of mitochondria. ATP is formed from ADP and P_i as a result of this flow of electrons. This process, called *oxidative phosphorylation* (Chapter 14), is the major source of ATP in aerobic organisms. Alternatively, the high-potential electrons derived from the oxidation of fuel molecules can be used in biosyntheses that require *reducing power* in addition to ATP.

Nicotinamide adenine dinucleotide (NAD^+) is a major electron acceptor in the oxidation of fuel molecules (Figure 11-8). The reactive part of NAD^+ is its nicotinamide ring. *In the oxidation of a substrate, the nicotinamide ring of NAD^+ accepts a hydrogen ion and two*

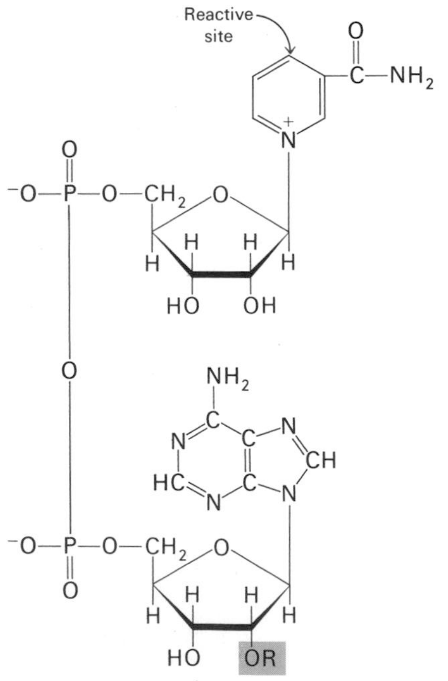

Figure 11-8
Structure of the oxidized form of nicotinamide adenine dinucleotide (NAD^+) and of nicotinamide adenine dinucleotide phosphate ($NADP^+$). In NAD^+, R = H; in $NADP^+$, R = PO_3^{2-}.

electrons, which is equivalent to a hydride ion. The reduced form of this carrier is called NADH.

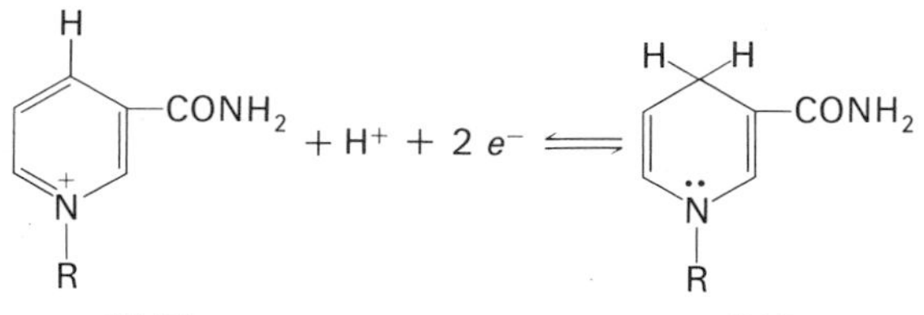

NAD$^+$ is the electron acceptor in many reactions of the type

$$NAD^+ + R-\underset{\underset{OH}{|}}{\overset{\overset{H}{|}}{C}}-R' \rightleftharpoons NADH + R-\overset{\overset{O}{\|}}{C}-R' + H^+$$

In this dehydrogenation, one hydrogen atom of the substrate is directly transferred to NAD$^+$, whereas the other appears in the solvent. Both electrons that are lost by the substrate are transferred to the nicotinamide ring.

The other major electron carrier in the oxidation of fuel molecules is *flavin adenine dinucleotide* (Figure 11-9). The abbreviations for the oxidized and reduced forms of this carrier are FAD and FADH$_2$, respectively. FAD is the electron acceptor in reactions of the type

$$FAD + R-\underset{\underset{H}{|}}{\overset{\overset{H}{|}}{C}}-\underset{\underset{H}{|}}{\overset{\overset{H}{|}}{C}}-R' \rightleftharpoons FADH_2 + R-\underset{\underset{H}{|}}{\overset{\overset{H}{|}}{C}}=\underset{\underset{H}{|}}{\overset{\overset{H}{|}}{C}}-R'$$

The reactive part of FAD is its isoalloxazine ring (Figure 11-10). FAD, like NAD$^+$, is a two-electron acceptor. In contrast to NAD$^+$, FAD accepts both hydrogen atoms that are lost by the substrate. The electron-transfer potentials of NADH and FADH$_2$ and the thermodynamics of oxidation-reduction reactions will be discussed in Chapter 14.

The electron donor in most reductive biosyntheses is the reduced form of nicotinamide adenine dinucleotide phosphate (NADPH)

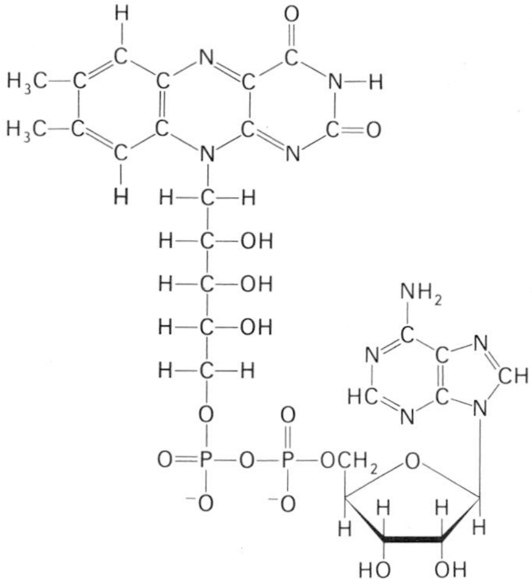

Figure 11-9
Structure of the oxidized form of flavin adenine dinucleotide (FAD).

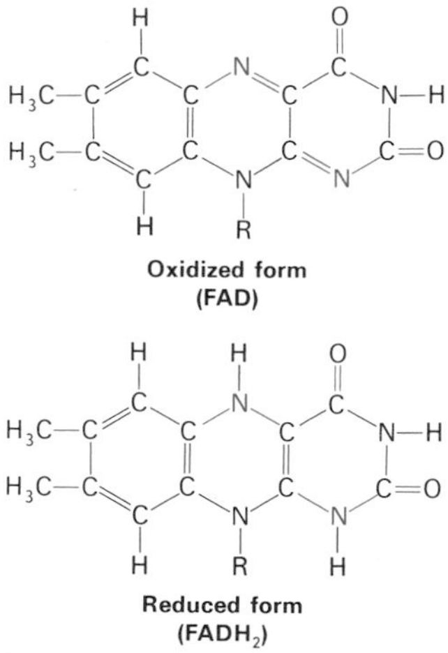

Oxidized form
(FAD)

Reduced form
(FADH$_2$)

Figure 11-10
Structures of the reactive parts of FAD and FADH$_2$.

(Figure 11-8). NADPH differs from NADH in that the 2'-hydroxyl group of its adenosine moiety is esterified with phosphate. The oxidized form of NADPH is denoted as NADP⁺. NADPH carries electrons in the same way as NADH. However, *NADPH is used almost exclusively for the reductive biosyntheses, whereas NADH is used primarily for the generation of ATP.* The extra phosphate group on NADPH is a tag that reveals the designated purpose of the molecule to discerning enzymes. The biological significance of the distinction between NADPH and NADH will be discussed later (p. 356).

It is important to note that NADH, NADPH, and $FADH_2$ react very slowly with O_2 in the absence of catalysts. Likewise, ATP is hydrolyzed at a slow rate in the absence of a catalyst. These molecules are kinetically quite stable, although there is a large thermodynamic driving force for the reaction of these electron carriers with O_2 and of ATP with water. *The stability of these molecules in the absence of specific catalysts is essential for their biological function since it enables enzymes to control the flow of free energy and reductive power.*

COENZYME A IS A UNIVERSAL CARRIER OF ACYL GROUPS

Coenzyme A is another central molecule in metabolism. In 1945, Lipmann found that a heat-stable cofactor was required in many enzyme-catalyzed acetylations. This cofactor was named *coenzyme A* (CoA), the A standing for *acetylation*. It was isolated, and its structure was determined several years later (Figure 11-11). The terminal sulfhydryl group in CoA is the reactive site. Acyl groups are linked to CoA by a thioester bond. The resulting derivative is called an *acyl CoA.* An acyl group often linked to CoA is the acetyl unit; this derivative is called *acetyl CoA* (acetyl coenzyme A). The $\Delta G°'$ for the hydrolysis of acetyl CoA has a large negative value.

$$\text{Acetyl CoA} + H_2O \rightleftharpoons \text{acetate} + \text{CoA}$$

$$\Delta G°' = -7.5 \text{ kcal/mol}$$

In other words, *acetyl CoA has a high acetyl group-transfer potential.* CoA is a carrier of activated acetyl or other acyl groups, just as ATP is a carrier of activated phosphoryl groups.

Acyl CoA

Acetyl CoA

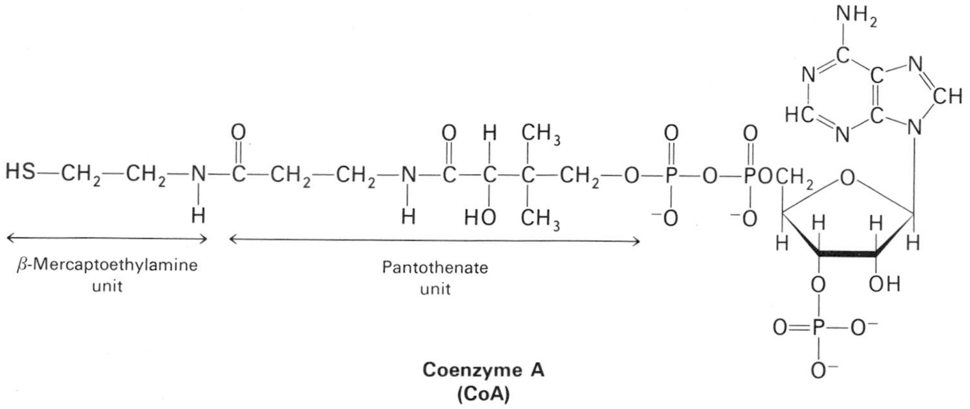

Figure 11-11
Structure of coenzyme A (CoA).

We will encounter other carriers of activated groups in our consideration of metabolism. Several of them are listed in Table 11-3. These carriers mediate the interchange of activated groups in a wide variety of biochemical reactions. Indeed, they have very similar roles in all forms of life. Their universal occurrence is one of the unifying motifs of biochemistry.

Table 11-3
Some activated carriers in metabolism

Carrier molecule	Group carried in activated form
ATP	Phosphoryl
NADH and NADPH	Electrons
FADH$_2$	Electrons
Coenzyme A	Acyl
Lipoamide	Acyl
Thiamine pyrophosphate	Aldehyde
Biotin	CO$_2$
Tetrahydrofolate	One-carbon units
S-Adenosylmethionine	Methyl
Uridine diphosphate glucose	Glucose
Cytidine diphosphate diacylglycerol	Phosphatidate

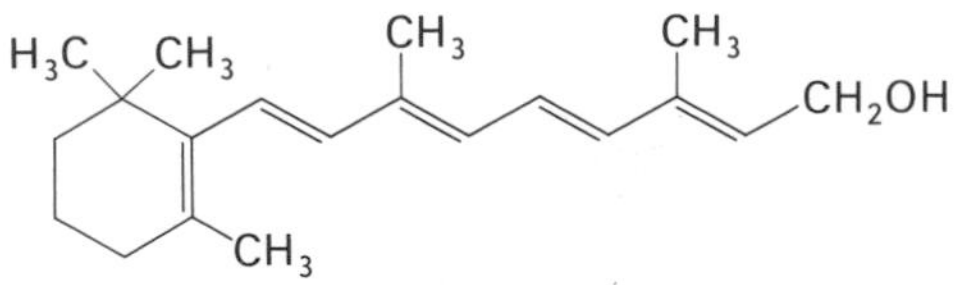

Retinol (Vitamin A)

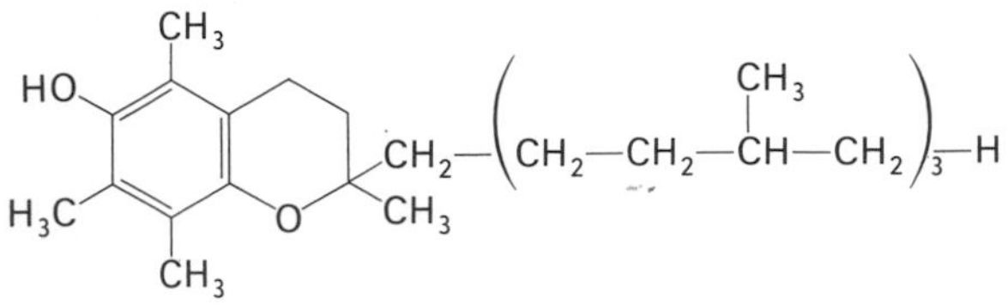

α-Tocopherol (Vitamin E)

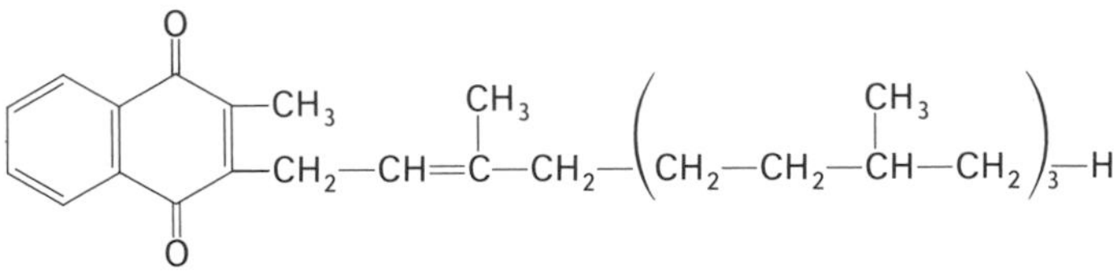

Vitamin K_1

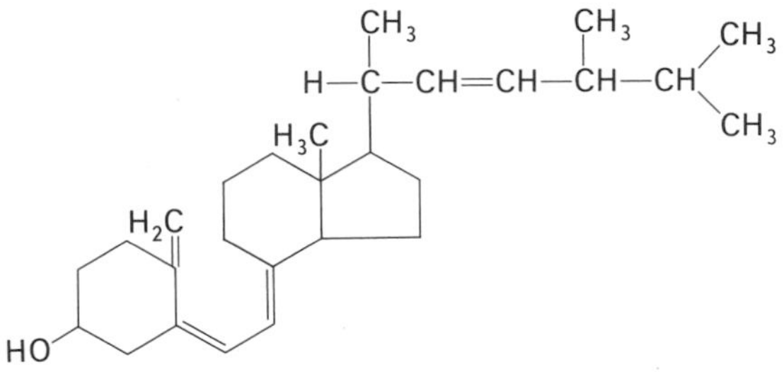

Calciferol (Vitamin D_2)

Figure 11-12
Structures of some fat-soluble vitamins.

Table 11-4
Coenzyme derivatives of some water-soluble vitamins

Vitamin	*Coenzyme derivative*
Thiamine (Vitamin B_1)	Thiamine pyrophosphate
Riboflavin (Vitamin B_2)	Flavin adenine dinucleotide and flavin mononucleotide
Nicotinate (niacin)	Nicotinamide adenine dinucleotide
Pyridoxine, pyridoxal, and pyridoxamine (Vitamin B_6)	Pyridoxal phosphate
Pantothenate	Coenzyme A
Biotin	Covalently attached to carboxylases
Folate	Tetrahydrofolate
Cobalamin (Vitamin B_{12})	Cobamide coenzymes

MOST WATER-SOLUBLE VITAMINS ARE COMPONENTS OF COENZYMES

Lipmann has commented that "doctors like to prescribe vitamins and millions of people take them, but it requires a good deal of biochemical sophistication to understand why they are needed and how the organism uses them." Vitamins are organic molecules that are needed in small amounts in the diet of higher animals. These molecules serve nearly the same roles in all forms of life, but higher animals have lost the capacity to synthesize them. There are two groups of vitamins: the fat-soluble ones, designated by the letters A, D, E, and K, and the water-soluble ones, which are referred to as the vitamin B complex. Most of the biochemical roles of the water-soluble vitamins are known (Table 11-4). In fact, most of them are components of coenzymes. For example, riboflavin (Vitamin B_2) is a precursor of FAD, and pantothenate is a component of coenzyme A.

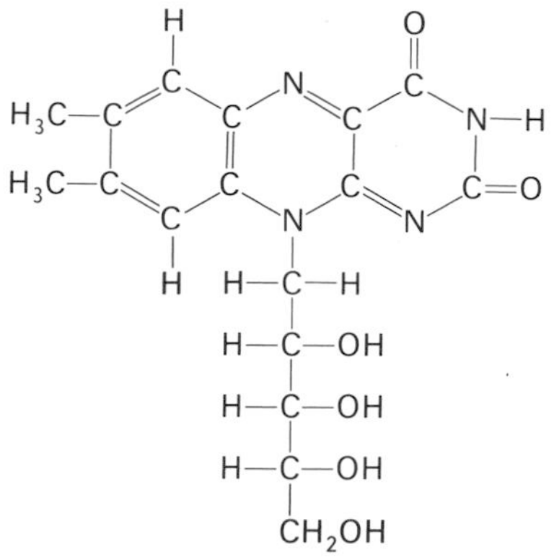

Riboflavin
(Vitamin B_2)

Nicotinate
(Niacin)

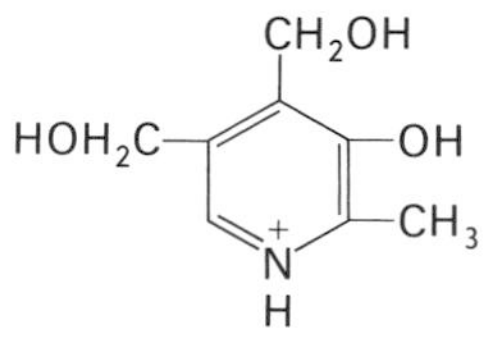

STAGES IN THE EXTRACTION OF ENERGY FROM FOODSTUFFS

Let us take an overview of the process of energy generation before considering these reactions in detail in subsequent chapters. Hans Krebs has described three stages in the generation of energy from the oxidation of foodstuffs. *In the first stage, large molecules in food are broken down to smaller units.* Proteins are hydrolyzed to their twenty kinds of constituent amino acids, polysaccharides are hydrolyzed to simple sugars such as glucose, and fats are hydrolyzed to glycerol and fatty acids. No useful energy is generated in this phase. *In the second stage, these numerous small molecules are degraded to a few simple units that play a central role in metabolism.* In fact, most of them—sugars, fatty acids, glycerol, and several amino acids—are converted into the acetyl unit of acetyl CoA. *The third stage consists of the citric acid cycle and oxidative phosphorylation,* which are the final common pathways in the oxidation of fuel molecules. Acetyl CoA brings acetyl units into this cycle, where they are completely oxidized to CO_2. Four pairs of electrons are transferred to NAD^+ and FAD for each acetyl group that is oxidized. Then, ATP is generated as

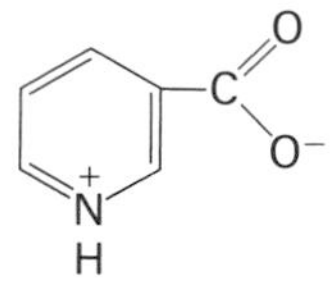

Pyridoxine
(A form of Vitamin B_6)

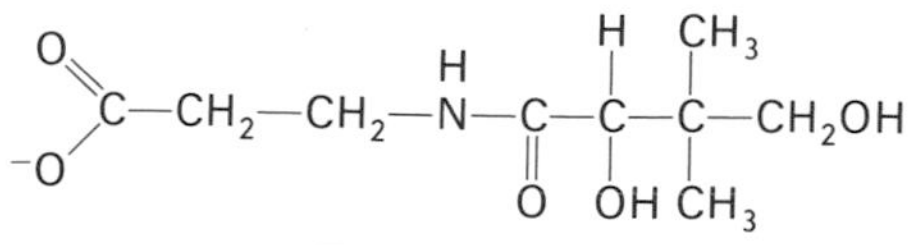

Pantothenate

Figure 11-13
Structures of some water-soluble vitamins.

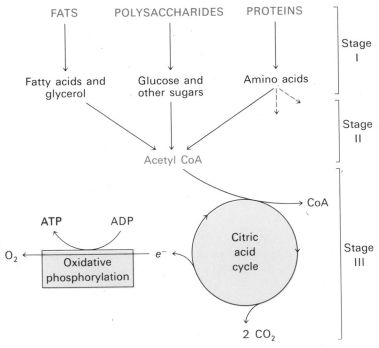

Figure 11-14
Stages in the extraction of
energy from foodstuffs.

electrons flow from the reduced forms of these carriers to O_2, a process called oxidative phosphorylation. Most of the ATP generated by the degradation of foodstuffs is formed in this third stage.

METABOLIC PROCESSES ARE REGULATED BY A VARIETY OF MECHANISMS

Even the simplest bacterial cell has the capacity for carrying out more than a thousand interdependent reactions. It is evident that this complex network must be rigorously regulated. Furthermore, metabolic control must be flexible, since the external environment of cells is not constant. Studies of a wide range of organisms have shown that there are a number of mechanisms for the control of metabolism. It must be emphasized that the central metabolic pathways have been almost fully elucidated, but knowledge concerning their regulation is still in its infancy. Few problems in biochemistry today are as intellectually challenging and important.

A major mechanism of metabolic regulation is the control of the *amounts* of certain enzymes. This mode of regulation has been extensively studied in bacteria. The control of the *rate of synthesis* of β-galactosidase and other proteins needed for the utilization of lactose is a classic example, which will be discussed in detail in Chapter 27. Metabolic regulation is also achieved by control of the *catalytic activities* of certain enzymes. One general and important mechanism is *reversible allosteric control*. For example, the first reaction in many biosynthetic pathways is allosterically inhibited by the ultimate product of the pathway, an interaction called *feedback inhibition*. The activity of some enzymes is also modulated by *covalent modifications* such as the phosphorylation of a specific serine residue.

An important general principle of metabolism is that *biosynthetic and degradative pathways are almost always distinct*. This separation is necessary for energetic reasons, as will be evident in subsequent chapters. It also facilitates the control of metabolism. In eukaryotes, metabolic regulation and flexibility are also enhanced by *compartmentation*. For example, fatty acid oxidation occurs in mitochondria, whereas fatty acid synthesis occurs in the cytosol (the soluble part of the cytoplasm). Compartmentation segregates these opposed reactions.

Many reactions in metabolism are controlled in part by the *energy charge*, which is a measure of the high-energy phosphate state of a cell. The energy stored in the ATP-ADP system is proportional to the mole fraction of ATP plus half the mole fraction of ADP, since ATP contains two anhydride bonds, whereas ADP contains one. Hence, the energy charge is defined as

$$\text{Energy charge} = \frac{[\text{ATP}] + \frac{1}{2}[\text{ADP}]}{[\text{ATP}] + [\text{ADP}] + [\text{AMP}]}$$

The energy charge can have a value ranging from 0 (all AMP) to 1 (all ATP). Daniel Atkinson has shown that *ATP-generating pathways are inhibited by a high energy charge, whereas ATP-utilizing pathways are stimulated by a high energy charge*. In plots of the reaction rates of such pathways versus the energy charge, the curves are steepest near an energy charge of 0.9, where they usually intersect (Figure 11-15). It is evident that the control of these pathways is designed to maintain the energy charge within rather narrow limits. In other words, *the energy charge, like the pH of a cell, is buffered*. The energy charge of most cells is in the range of 0.8 to 0.95.

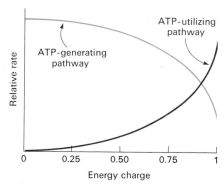

Figure 11-15
Effect of the energy charge on the relative rates of a typical ATP-generating (catabolic) pathway and a typical ATP-utilizing (anabolic) pathway.

SUMMARY

Cells extract energy from their environments and convert foodstuffs into cell components by a highly integrated network of chemical reactions called metabolism. The most valuable thermodynamic concept for the understanding of the energetics of metabolism is free energy, which is a measure of the capacity of a system to do useful work at constant pressure and temperature. A reaction can occur spontaneously only if the change in free energy (ΔG) is negative. The ΔG for a reaction is independent of path, and depends only on the nature of the reactants and their activities (which can sometimes be approximated by their concentrations). The free-energy change of a reaction occurring when reactants and products are at unit activity is called the standard free-energy change ($\Delta G°$). Biochemists usually use $\Delta G°'$, which is the standard free-energy change at pH 7. ATP, the universal currency of energy in biological systems, is an energy-rich molecule because it contains two anhydride bonds. The electrostatic repulsion between these negatively charged groups is reduced when ATP is hydrolyzed. Also, ADP and P_i are stabilized by resonance more than is ATP.

The basic strategy of metabolism is to form ATP, NADPH, and macromolecular precursors. ATP is consumed in muscle contraction and other motions of cells, active transport, and biosyntheses. NADPH, which carries two electrons at a high potential, provides reducing power in the biosynthesis of cell components from more oxidized precursors. There are three stages in the extraction of energy from foodstuffs in aerobic organisms. In the first stage, large molecules are broken down to smaller ones, such as amino acids, sugars, and fatty acids. In the second stage, these small molecules are degraded to a few single units that have a pervasive role in metabolism. One of them is the acetyl unit of acetyl CoA, a carrier of activated acyl groups. The third stage of metabolism is the citric acid cycle and oxidative phosphorylation, in which fuels are completely oxidized to CO_2 and ATP is generated as electrons flow to O_2, the ultimate electron acceptor.

Metabolism is regulated in a variety of ways. The amounts of some critical enzymes are controlled by regulation of the rate of protein synthesis. In addition, the catalytic activities of some enzymes are regulated by allosteric interactions (as in feedback inhibition) and by covalent modification. Compartmentation and distinct pathways for biosynthesis and degradation also contribute to metabolic regulation. The energy charge, which depends on the relative amounts of ATP, ADP, and AMP, plays a role in metabolic regulation. A high energy charge inhibits ATP-generating (catabolic) pathways, whereas it stimulates ATP-utilizing (anabolic) pathways.

SELECTED READINGS

Overviews of metabolism:

Krebs, H. A., and Kornberg, H. L., 1957. *Energy Transformations in Living Matter*. Springer-Verlag. [Includes a valuable appendix by K. Burton containing thermodynamic data.]

Wood, W. B., 1974. *The Molecular Basis of Metabolism*. Unit III in Biocore. McGraw-Hill.

Thermodynamics:

Klotz, I. M., 1967. *Energy Changes in Biochemical Reactions*. Academic Press. [A concise introduction, full of insight.]

Bray, H. G., and White, K., 1966. *Kinetics and Thermodynamics in Biochemistry* (2nd ed.). Academic Press.

Ingraham, L. L., and Pardee, A. B., 1967. Free energy and entropy in metabolism. *In* Greenberg, D. M., (ed.), *Metabolic Pathways* (3rd ed.), vol. 1, pp. 1–46. Academic Press.

Alberty, R. A., 1968. Effect of pH and metal ion concentration on the equilibrium hydrolysis of adenosine triphosphate to adenosine diphosphate. *J. Biol. Chem.* 243:1337–1343.

Jencks, W. P., 1970. Free energies of hydrolysis and decarboxylation. *In* Sober, H. A., *Handbook of Biochemistry* (2nd ed.), pp. J181–J186. Chemical Rubber Co.

Regulation of metabolism:

Newsholme, E. A., and Start, C., 1973. *Regulation in Metabolism*. Wiley.

Stadtman, E. R., 1970. Mechanisms of enzyme regulation in metabolism. *In* Boyer, P. D., (ed.), *The Enzymes* (3rd ed.), vol. 1, pp. 397–459. Academic Press.

Atkinson, D. E., 1968. The energy charge of the adenylate pool as a regulatory parameter. Interaction with feedback modifiers. *Biochemistry* 7:4030–4034.

Historical aspects:

Kalckar, H. M., (ed.), 1969. *Biological Phosphorylations*. Prentice-Hall. [A valuable collection of many classical papers on bioenergetics.]

Fruton, J. S., 1972. *Molecules and Life*. Wiley-Interscience. [Perceptive and scholarly essays on the interplay of chemistry and biology since 1800. Metabolism and bioenergetics are among the topics treated in detail.]

Lipmann, F., 1971. *Wanderings of a Biochemist*. Wiley-Interscience. [Contains reprints of some of the author's classic papers and several delightful essays.]

PROBLEMS

1. What is the direction of each of the following reactions when the reactants are initially present in equimolar amounts? Use the data given in Table 11-2.

   (a) ATP + creatine $\rightleftharpoons$

   creatine phosphate + ADP

   (b) ATP + glycerol $\rightleftharpoons$

   glycerol 3-phosphate + ADP

   (c) ATP + pyruvate $\rightleftharpoons$

   phosphoenolpyruvate + ADP

   (d) ATP + glucose $\rightleftharpoons$

   glucose 6-phosphate + ADP

2. What information do the $\Delta G°'$ data given in Table 11-2 provide about the relative rates of hydrolysis of pyrophosphate and acetyl phosphate?

3. Consider the reaction

   ATP + pyruvate $\rightleftharpoons$

   phosphoenolpyruvate + ADP

   (a) Calculate $\Delta G°'$ and K'_{eq} at 25°C for this reaction, using the data given in Table 11-2.
   (b) What is the equilibrium ratio of pyruvate to phosphoenolpyruvate if the ratio of ATP to ADP is 10?

4. Calculate $\Delta G°'$ for the isomerization of glucose 6-phosphate to glucose 1-phosphate. What is the equilibrium ratio of glucose 6-phosphate to glucose 1-phosphate at 25°C?

5. The formation of acetyl CoA from acetate is an ATP-driven reaction.

   Acetate + ATP + CoA $\rightleftharpoons$

   acetyl CoA + AMP + PP$_i$

   (a) Calculate $\Delta G°'$ for this reaction, using data given in this chapter.
   (b) The PP$_i$ formed in the above reaction is rapidly hydrolyzed in vivo because of the ubiquity of inorganic pyrophosphatase. The $\Delta G°'$ for the hydrolysis of PP$_i$ is -8 kcal/mol. Calculate the $\Delta G°'$ for the overall reaction. What effect does the hydrolysis of PP$_i$ have on the formation of acetyl CoA?

6. The pK of an acid is a measure of its proton group-transfer potential.

   (a) Derive a relationship between $\Delta G°$ and pK.
   (b) What is the $\Delta G°$ for the ionization of acetic acid, which has a pK of 4.8?

7. What is the common structural feature of ATP, FAD, NAD$^+$, and CoA?

8. Fibrinogen contains tyrosine-O-sulfate. Propose an activated form of sulfate that could react in vivo with the aromatic hydroxyl group of a tyrosine residue in a protein to form tyrosine-O-sulfate.

For additional problems, see W. B. Wood, J. H. Wilson, R. M. Benbow, and L. E. Hood, *Biochemistry: A Problems Approach* (Benjamin, 1974), Chapter 8, and I. M. Klotz, *Energy Changes in Biochemical Reactions* (Academic Press, 1967).

Chapter 12

GLYCOLYSIS

We begin our consideration of the generation of metabolic energy with glycolysis, a nearly universal pathway in biological systems. *Glycolysis is the sequence of reactions that converts glucose to pyruvate with the concomitant production of ATP.* In aerobic organisms, glycolysis is the prelude to the citric acid cycle and the electron-transport chain, which together harvest most of the energy contained in glucose. Under aerobic conditions, pyruvate enters mitochondria, where it is completely oxidized to CO_2 and H_2O. If the supply of oxygen is insufficient, as in actively contracting muscle, pyruvate is converted to lactate. In some anaerobic organisms, such as yeast, pyruvate is transformed into ethanol. The formation of ethanol or lactate from glucose are examples of fermentations.

> *Glycolysis*—
> Derived from the Greek words *glycos,* sugar (sweet), and *lysis,* dissolution.

The elucidation of glycolysis has a rich history. Indeed, the development of biochemistry and the delineation of this central pathway went hand-in-hand. A key discovery was made by Hans Buchner and Eduard Buchner in 1897, quite by accident. They were interested in manufacturing cell-free extracts of yeast for possible therapeutic use. These extracts had to be preserved without using antiseptics such as phenol, so they decided to try sucrose, a commonly used preservative in kitchen chemistry. They obtained a startling result: sucrose was rapidly fermented to alcohol by the

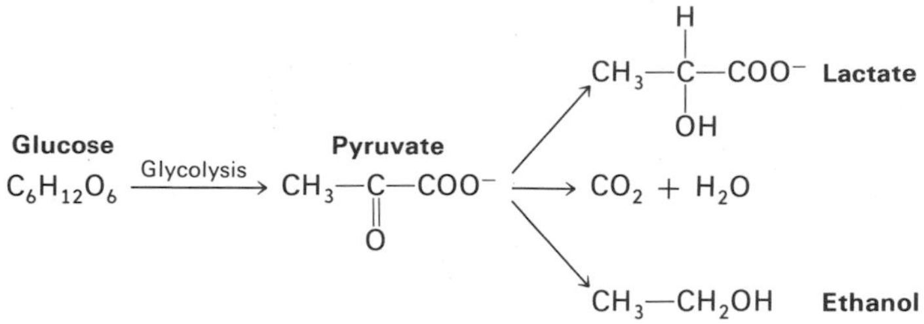

$$\text{Glucose} \atop C_6H_{12}O_6 \xrightarrow{\text{Glycolysis}} \underset{\underset{O}{\overset{\|}{}}{\text{Pyruvate}}}{CH_3-C-COO^-} \nearrow \underset{OH}{\overset{H}{CH_3-\overset{|}{\underset{|}{C}}-COO^-}} \text{Lactate}$$

$$\longrightarrow CO_2 + H_2O$$

$$\searrow CH_3-CH_2OH \quad \text{Ethanol}$$

Figure 12-1
Some fates of glucose.

yeast juice. The significance of this finding was immense. *The Buchners demonstrated for the first time that fermentation could occur outside living cells.* The accepted view of their day, asserted by Louis Pasteur in 1860, was that fermentation is inextricably tied to living cells. The chance discovery of the Buchners refuted this vitalistic dogma and opened the door to modern biochemistry. Metabolism became chemistry.

The next important contribution was made by Arthur Harden and William Young in 1905. They added yeast juice to a solution of glucose and found that fermentation started almost immediately. However, the rate of fermentation soon decreased markedly unless inorganic phosphate was added. Furthermore, they found that the added inorganic phosphate disappeared in the course of fermentation, and so they inferred that *inorganic phosphate was incorporated into a sugar phosphate.* Harden and Young isolated a hexose diphosphate, which was later shown to be fructose 1,6-diphosphate. They also discovered that yeast juice contains two kinds of substances necessary for fermentation: "zymase" and "cozymase." They found that yeast juice lost its activity if it were dialyzed or heated to 50°C. However, the inactive dialyzed juice became active when it was mixed with inactive heated juice. Thus, activity depended on the presence of two kinds of substances: a heat-labile, nondialyzable component (called *zymase*) and a heat-stable, dialyzable fraction (called *cozymase*). We now know that "zymase" consists of a number of enzymes, whereas "cozymase" consists of metal ions, adenosine triphosphate (ATP), adenosine diphosphate (ADP), and coenzymes such as nicotinamide adenine dinucleotide (NAD$^+$).

Studies of muscle extracts carried out several years later showed that many of the reactions of lactic fermentation were the same as those of alcoholic fermentation. *This was an exciting discovery because it revealed an underlying unity in biochemistry.* The complete glycolytic pathway was elucidated by 1940, largely because of the contributions made by Gustav Embden, Otto Meyerhof, Carl Neuberg, Jacob Parnus, Otto Warburg, Gerty Cori, and Carl Cori. Glycolysis is sometimes called the Embden-Meyerhof pathway.

NOMENCLATURE AND CONFORMATION OF MONOSACCHARIDES

Let us consider the nomenclature and structure of monosaccharides, the simplest carbohydrates, before turning to the reactions of glycolysis. Monosaccharides are aldehydes or ketones that have two or more hydroxyl groups; their empirical formula is $(CH_2O)_n$. The simplest ones, for which $n = 3$, are glyceraldehyde and dihydroxyacetone. They are *trioses*. Glyceraldehyde is called an *aldose* because it contains an aldehyde group, whereas dihydroxyacetone is a *ketose* because it contains a keto group.

Glyceraldehyde has a single asymmetric carbon. Thus, there are two stereoisomers of this three-carbon aldose, named D-glyceraldehyde and L-glyceraldehyde. The prefixes D and L designate the absolute configuration.

Sugars with 4, 5, 6, and 7 carbon atoms are called *tetroses, pentoses, hexoses,* and *heptoses,* respectively. Their formulas and stereochemical relationships are given in the Appendix (p. 304). Two common hexoses are D-*glucose* and D-*fructose.* The prefixed symbol D means that the absolute configuration at the asymmetric carbon furthest from the aldehyde or keto group, namely C-5, is the same as in D-glyceraldehyde. Note that glucose is an aldose, whereas fructose is a ketose.

The predominant forms of glucose and fructose in solution are not the open-chain structures shown above. Rather, the open-chain forms of glucose and fructose can cyclize to form rings. In general, an aldehyde can react with an alcohol to form a *hemiacetal.*

D-Glyceraldehyde
(An aldose)

Dihydroxyacetone
(A ketose)

Figure 12-2
Absolute configuration of D-glyceraldehyde.

D-Glucose
(An aldose)

D-Fructose
(A ketose)

Aldehyde   Alcohol   Hemiacetal

The C-1 aldehyde in the open-chain form of glucose reacts with the C-5 hydroxyl group to form an *intramolecular hemiacetal*. The resulting six-membered sugar ring is called *pyranose* because of its similarity to pyran.

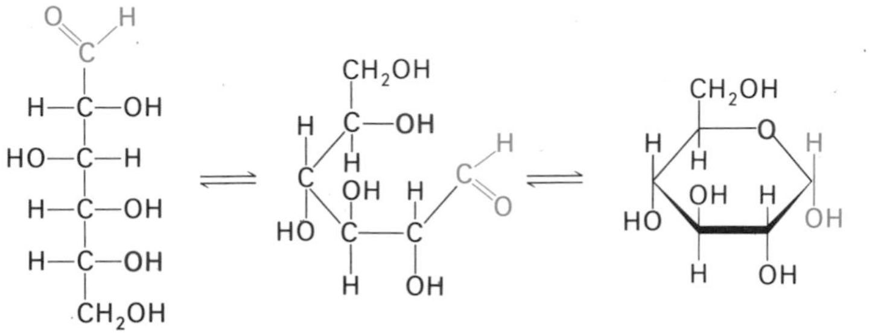

D-Glucose
(Open-chain form)

α-D-Glucopyranose
(A ring form of glucose)

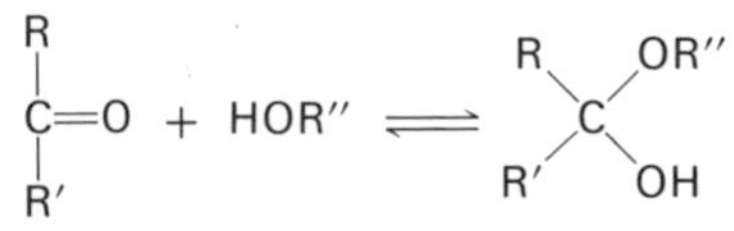

Similarly, a ketone can react with an alcohol to form a *hemiketal*. The C-2 keto group in the open-chain form of fructose reacts with the C-5 hydroxyl group to form an *intramolecular hemiketal*. This five-membered sugar ring is called *furanose* because of its similarity to furan.

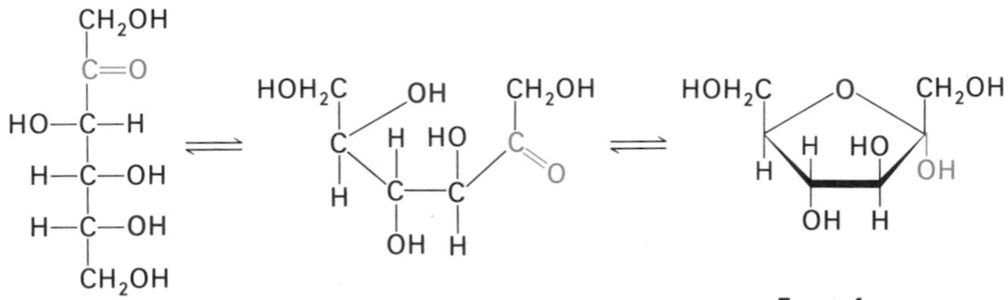

Ketone   Alcohol   Hemiketal

D-Fructose

α-D-Fructofuranose
(A ring form of fructose)

The structural formulas of glucopyranose and fructofuranose shown above are Haworth projections. The carbon atoms in the ring are not explicitly shown. The approximate plane of the ring is perpendicular to the plane of the paper, with the heavy line on the ring closest to the reader.

An additional asymmetric center is created when glucose cyclizes. Carbon-1, the carbonyl carbon atom in the open-chain form, becomes an asymmetric center in the ring form. Two ring structures can be formed: α-D-glucopyranose and β-D-glucopyranose (Figure

12-3). *The designation α means that the hydroxyl group attached to C-1 is below the plane of the ring; β means that it is above the plane of the ring.* The C-1 carbon is called the anomeric carbon atom, and so the α and β forms are *anomers*. The same nomenclature applies to the ring form of fructose, except that α and β refer to the hydroxyl groups attached to C-2, the anomeric carbon atom in ketoses.

α-D-Glucose   β-D-Glucose

α-D-Fructose   β-D-Fructose

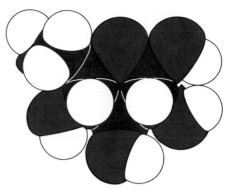

Figure 12-3
Space-filling model of β-D-glucose.

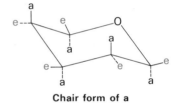

Chair form of a pyranose
(e = equatorial substituent;
a = axial substituent)

β-D-Glucose

Alpha-D-glucopyranose, β-D-glucopyranose, and the open-chain form of glucose interconvert rapidly, as do the furanose and open-chain forms of fructose. The designation glucose (or fructose) will be used to refer to the equilibrium mixture of the open-chain and ring forms of this sugar.

The six-membered pyranose ring is not planar. Rather, the preferred conformation is a *chair form*. The substituents are of two types: *axial* and *equatorial*. In β-D-glucopyranose, all of the hydroxyl groups are equatorial.

AN OVERVIEW OF KEY STRUCTURES AND REACTIONS

The learning of a metabolic pathway is facilitated by having a firm grasp of the structures of the reactants and an understanding of the types of reactions taking place. The intermediates in glycolysis have either six carbons or three carbons. The *six-carbon units* are

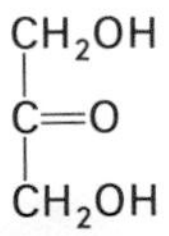

CH₂OH
|
C=O
|
CH₂OH

Dihydroxyacetone

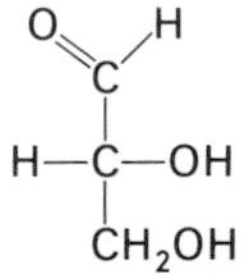

(aldehyde structure)

Glyceraldehyde

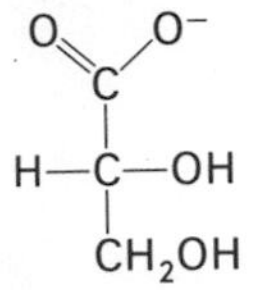

Glycerate

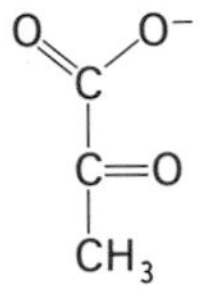

Pyruvate

Ester

Anhydride

derivatives of *glucose* and *fructose.* The *three-carbon units* in glycolysis are derivatives of *dihydroxyacetone, glyceraldehyde, glycerate,* and *pyruvate.*

All intermediates in glycolysis between glucose and pyruvate are *phosphorylated.* The phosphoryl groups in these compounds are in either *ester* or *anhydride* linkage.

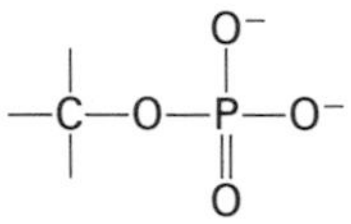

Now let us look at some of the kinds of reactions that occur in glycolysis:

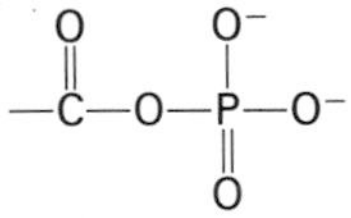

1. *Phosphoryl transfer.* A phosphoryl group is transferred from ATP to a glycolytic intermediate, or vice versa.

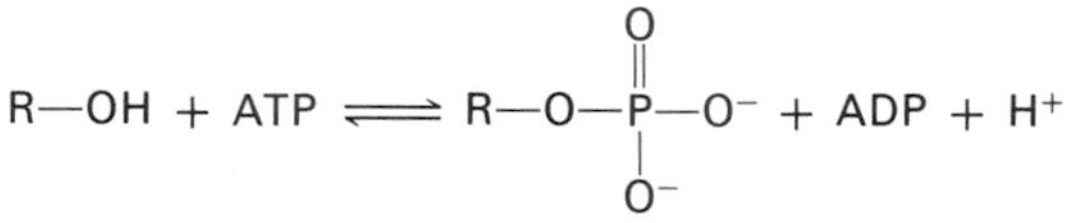

$$R—OH + ATP \rightleftharpoons R—O—\overset{\overset{O}{\parallel}}{\underset{\underset{O^-}{|}}{P}}—O^- + ADP + H^+$$

2. *Phosphoryl shift.* A phosphoryl group is shifted within a molecule from one oxygen atom to another.

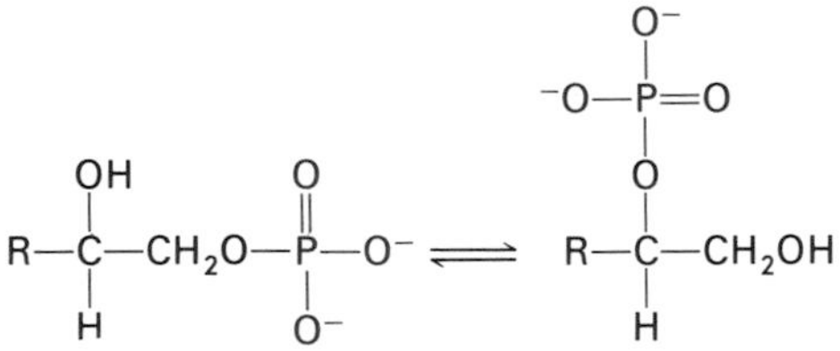

(structure of phosphoryl shift reaction)

3. *Isomerization.* A ketose is converted into an aldose, or vice versa.

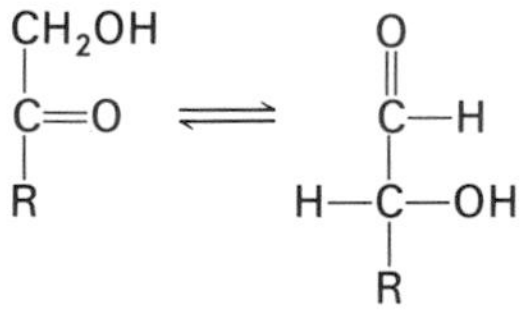

(structure)

Ketose   **Aldose**

4. *Dehydration.* A molecule of water is eliminated.

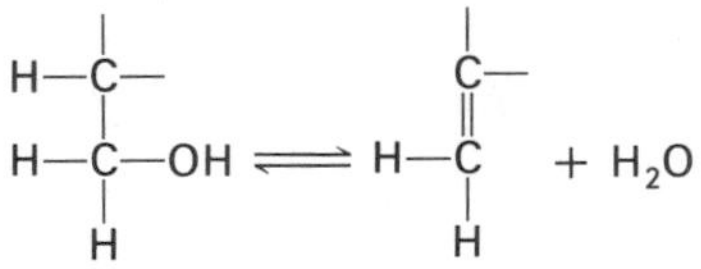

(structure) + H₂O

5. *Aldol cleavage.* A carbon–carbon bond is split in a reversal of an aldol condensation.

FORMATION OF FRUCTOSE 1,6-DIPHOSPHATE FROM GLUCOSE

We now start our journey down the glycolytic pathway. These reactions occur in the cell cytosol. The first stage, which is the conversion of glucose to fructose 1,6-diphosphate, consists of three steps: a phosphorylation, an isomerization, and a second phosphorylation reaction. *The strategy of these initial steps in glycolysis is to form a compound that can be readily cleaved into phosphorylated three-carbon units.* Energy is subsequently extracted from the three-carbon units.

Glucose   Fructose 1,6-diphosphate

The first reaction is the *phosphorylation of glucose by ATP to form glucose 6-phosphate.* The transfer of the phosphoryl group from ATP to the hydroxyl group on C-6 of glucose is catalyzed by *hexokinase.*

Glucose   Glucose 6-phosphate

Phosphoryl transfer is one of the basic reactions in biochemistry. We will encounter it repeatedly. An enzyme that transfers a phosphoryl group from ATP to an acceptor is called a *kinase*. Hexokinase, then, is an enzyme that transfers a phosphoryl group from ATP to a variety of six-carbon sugars (*hexoses*). Hexokinase, like all other kinases, requires Mg^{2+} (or another divalent metal ion such as Mn^{2+}) for activity. The divalent metal ion forms a complex with ATP. The structures of two possible Mg^{2+}-ATP complexes are shown below.

Figure 12-4
X-ray precession photo of a hexokinase crystal. [Courtesy of Dr. Thomas Steitz.]

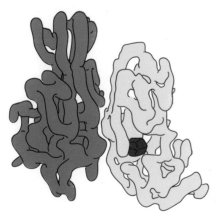

Figure 12-5
Model of hexokinase at low resolution. One of the subunits is shown in blue, the other in yellow. One of the binding sites for glucose is marked in red. [Based on a drawing kindly provided by Dr. Thomas Steitz.]

Another enzyme that catalyzes the synthesis of glucose 6-phosphate from glucose and ATP is *glucokinase*. As their names imply, glucokinase is specific for glucose, whereas hexokinase phosphorylates a variety of hexoses. Glucokinase and hexokinase have distinct roles in metabolism because their activities are regulated differently. Hexokinase participates in glycolysis, whereas glucokinase is involved in the conversion of glucose to glycogen, a storage form of this sugar.

The next step in glycolysis is the *isomerization of glucose 6-phosphate to fructose 6-phosphate*. The *six-membered pyranose ring* of glucose 6-phosphate is converted into the *five-membered furanose ring* of fructose 6-phosphate. Recall that the open-chain form of glucose has an aldehyde group on C-1, whereas the open-chain form of fructose has a keto group on C-2. The aldehyde on C-1 reacts with the hydroxyl group on C-5 to form the pyranose ring, whereas the keto group on C-2 reacts with the C-5 hydroxyl to form the furanose ring. Thus, the isomerization of glucose 6-phosphate to fructose 6-phosphate is a *conversion of an aldose into a ketose.*

$$\text{Glucose 6-phosphate} \quad \underset{\text{isomerase}}{\overset{\text{Phosphoglucose}}{\rightleftarrows}} \quad \text{Fructose 6-phosphate}$$

Glucose 6-phosphate   **Fructose 6-phosphate**

The open-chain representations of these sugars show the essence of this reaction.

$$\text{Glucose 6-phosphate} \quad \underset{\text{isomerase}}{\overset{\text{Phosphoglucose}}{\rightleftarrows}} \quad \text{Fructose 6-phosphate}$$

Glucose 6-phosphate   **Fructose 6-phosphate**
(An aldose)   (A ketose)

Figure 12-6
Photomicrograph of crystals of phosphoglucose isomerase. [Courtesy of Dr. Herman Watson.]

A second phosphorylation reaction follows the isomerization step. *Fructose 6-phosphate is phosphorylated by ATP to fructose 1,6-diphosphate.*

$$\text{Fructose 6-phosphate} + ATP \quad \overset{\text{Phosphofructokinase}}{\longrightarrow} \quad \text{Fructose 1,6-diphosphate} + ADP + H^+$$

Fructose 6-phosphate   **Fructose 1,6-diphosphate**

This reaction is catalyzed by *phosphofructokinase,* an allosteric enzyme. The pace of glycolysis is critically dependent on the level of activity of this enzyme. The catalytic activity of phosphofructokinase is allosterically controlled by ATP and several other metabolites (p. 299).

FORMATION OF GLYCERALDEHYDE 3-PHOSPHATE BY CLEAVAGE AND ISOMERIZATION

The second stage of glycolysis consists of four steps, starting with the splitting of fructose 1,6-diphosphate to yield *glyceraldehyde 3-phosphate* and *dihydroxyacetone phosphate*. The remaining steps in glycolysis involve three-carbon units rather than six-carbon units.

Figure 12-7
Photomicrograph of crystals of aldolase. [Courtesy of Dr. David Eisenberg.]

$$
\begin{array}{l}
CH_2OPO_3{}^{2-} \\
| \\
C{=}O \\
| \\
HO{-}C{-}H \\
| \\
H{-}C{-}OH \\
| \\
H{-}C{-}OH \\
| \\
CH_2OPO_3{}^{2-}
\end{array}
\quad \xrightleftharpoons{\text{Aldolase}} \quad
\begin{array}{l}
CH_2OPO_3{}^{2-} \\
| \\
C{=}O \\
| \\
HO{-}C{-}H \\
| \\
H
\end{array}
\quad + \quad
\begin{array}{l}
H\diagdown\;_{C}\diagup O \\
| \\
H{-}C{-}OH \\
| \\
CH_2OPO_3{}^{2-}
\end{array}
$$

Fructose 1,6-diphosphate   **Dihydroxyacetone phosphate**   **Glyceraldehyde 3-phosphate**

This reaction is catalyzed by *aldolase*. This enzyme derives its name from the nature of the reverse reaction, which is an aldol condensation.

Glyceraldehyde 3-phosphate is on the direct pathway of glycolysis, but dihydroxyacetone phosphate is not. However, dihydroxyacetone phosphate can be readily converted into glyceraldehyde 3-phosphate. These compounds are isomers: dihydroxyacetone phosphate is a ketose, whereas glyceraldehyde 3-phosphate is an aldose. The isomerization of these three-carbon phosphorylated sugars is catalyzed by *triose phosphate isomerase*. This reaction is very fast and reversible. At equilibrium, 96% of the triose phosphate is dihydroxyacetone phosphate. However, the reaction proceeds readily from dihydroxyacetone phosphate to glyceraldehyde 3-phosphate because of efficient removal of this product.

$$
\begin{array}{l}
O\diagdown\;_{C}\diagup H \\
| \\
H{-}C{-}OH \\
| \\
CH_2OPO_3{}^{2-}
\end{array}
\quad \xrightleftharpoons[\text{isomerase}]{\text{Triose phosphate}} \quad
\begin{array}{l}
CH_2OH \\
| \\
C{=}O \\
| \\
CH_2OPO_3{}^{2-}
\end{array}
$$

Glyceraldehyde 3-phosphate   **Dihydroxyacetone phosphate**
(An aldose)   (A ketose)

Thus, two molecules of glyceraldehyde 3-phosphate are formed from one molecule of fructose 1,6-diphosphate by the sequential action of aldolase and triose phosphate isomerase.

Fructose 1,6-diphosphate — Aldolase —

Glyceraldehyde 3-phosphate

Triose phosphate isomerase

Dihydroxyacetone phosphate

ENERGY CONSERVATION: PHOSPHORYLATION COUPLED TO THE OXIDATION OF GLYCERALDEHYDE 3-PHOSPHATE

The preceding steps in glycolysis have transformed one molecule of glucose into two molecules of glyceraldehyde 3-phosphate. No energy has yet been extracted. On the contrary, two molecules of ATP have been invested thus far. We come now to a series of steps that harvest some of the energy contained in glyceraldehyde 3-phosphate. The initial reaction in this sequence is the *conversion of glyceraldehyde 3-phosphate to 1,3-diphosphoglycerate* (1,3-DPG), a reaction catalyzed by *glyceraldehyde 3-phosphate dehydrogenase.*

Glyceraldehyde 3-phosphate + NAD$^+$ + P$_i$ $\rightleftharpoons$
$$1,3\text{-DPG} + \text{NADH} + \text{H}^+$$

A *high-energy phosphate compound* is generated in this oxidation-reduction reaction. The aldehyde group at C_1 is converted into an *acyl phosphate,* which is a *mixed anhydride* of phosphoric acid and a carboxylic acid.

The energy for the formation of the anhydride bond comes from the oxidation of the aldehyde group. The acyl phosphate in 1,3-DPG is at the oxidation level of a carboxylic acid. *NAD$^+$, the tightly bound coenzyme, is concomitantly reduced to NADH.* The mechanism of this complex reaction, which couples oxidation and phosphorylation, will be discussed later.

FORMATION OF ATP FROM 1,3-DPG

In the next step in glycolysis, the high-energy bond in 1,3-DPG is used to generate ATP. Indeed, this is the first ATP-generating reaction that we have encountered thus far. *Phosphoglycerate kinase*

Glyceraldehyde 3-phosphate

1,3-Diphosphoglycerate (1,3-DPG)

Acyl phosphate

catalyzes the transfer of the phosphoryl group from the acyl phosphate of 1,3-DPG to ADP. ATP and 3-phosphoglycerate are the products.

1,3-Diphosphoglycerate   **3-Phosphoglycerate**

Thus, the net outcome of the reactions catalyzed by glyceraldehyde 3-phosphate dehydrogenase and phosphoglycerate kinase is:

1. Glyceraldehyde 3-phosphate, an aldehyde, has been oxidized to 3-phosphoglycerate, a carboxylic acid.

2. NAD^+ has been reduced to NADH.

3. ATP has been formed from P_i and ADP.

FORMATION OF PYRUVATE AND THE GENERATION OF A SECOND ATP

We come now to the last stage of glycolysis. In these three steps, 3-phosphoglycerate is converted to pyruvate and a second molecule of ATP is formed.

3-Phosphoglycerate   **2-Phosphoglycerate**   **Phosphoenolpyruvate**   **Pyruvate**

The first of these reactions is an intramolecular rearrangement. The position of the phosphoryl group shifts in going from *3-phosphoglycerate to 2-phosphoglycerate*, a reaction catalyzed by *phosphoglyceromutase*. In general, a mutase is an enzyme that catalyzes an intramolecular shift of a chemical group, such as a phosphoryl group.

$$\text{3-Phosphoglycerate} \xrightleftharpoons[\text{Phosphoglyceromutase}]{} \text{2-Phosphoglycerate}$$

3-Phosphoglycerate   **2-Phosphoglycerate**

In the second reaction, an *enol* is formed by the dehydration of 2-phosphoglycerate. *Enolase* catalyzes the formation of *phosphoenol-pyruvate*. This dehydration reaction markedly elevates the group transfer potential of the phosphoryl group. An *enol phosphate is a high-energy bond,* whereas the phosphate ester of an ordinary alcohol is a low-energy bond. The reasons for this difference in group transfer potential will be discussed later.

$$\text{2-Phosphoglycerate} \xrightleftharpoons[\text{Enolase}]{} \text{Phosphoenolpyruvate} + H_2O$$

2-Phosphoglycerate   **Phosphoenolpyruvate**

The last reaction is the *formation of pyruvate* with the concomitant generation of an *ATP*. The transfer of a phosphoryl group from phosphoenolpyruvate to ADP is catalyzed by *pyruvate kinase*. This phosphorylation reaction is nonoxidative, in contrast to the one catalyzed by glyceraldehyde 3-phosphate dehydrogenase.

$$\text{Phosphoenolpyruvate} + ADP \xrightleftharpoons[\text{kinase}]{\text{Pyruvate}} \text{Pyruvate} + ATP$$

Phosphoenolpyruvate   **Pyruvate**

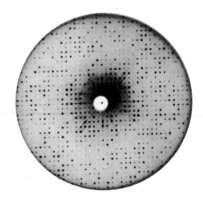
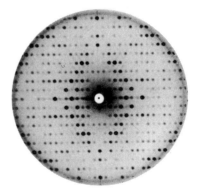

Figure 12-8
X-ray precession photographs of crystals of phosphoglycerate kinase (upper) and phosphoglyceromutase (lower). [Courtesy of Dr. Herman Watson.]

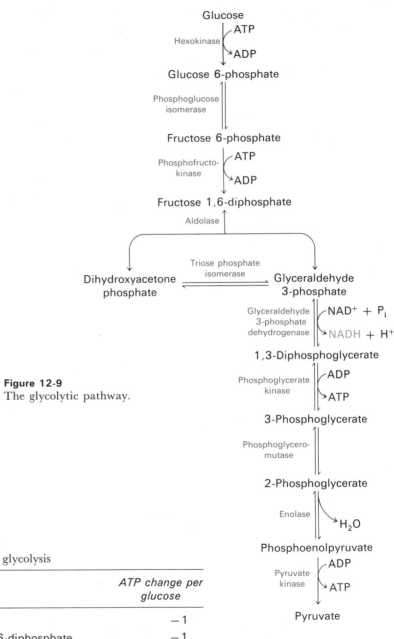

Figure 12-9
The glycolytic pathway.

Table 12-1
Consumption and generation of ATP in glycolysis

Reaction	ATP change per glucose
Glucose ⟶ glucose 6-phosphate	−1
Fructose 6-phosphate ⟶ fructose 1,6-diphosphate	−1
2 1,3-Diphosphoglycerate ⟶ 2 3-phosphoglycerate	+2
2 Phosphoenolpyruvate ⟶ 2 pyruvate	+2
Net	+2

ENERGY YIELD IN GOING FROM GLUCOSE TO PYRUVATE

The net reaction in the transformation of glucose to pyruvate is:

$$\text{Glucose} + 2\,P_i + 2\,\text{ADP} + 2\,\text{NAD}^+ \longrightarrow$$
$$2\text{ pyruvate} + 2\,\text{ATP} + 2\,\text{NADH} + 2\,\text{H}^+$$

Thus, *two molecules of ATP are generated in the conversion of glucose to pyruvate*. A summary of the steps in which ATP is consumed or formed is given in Table 12-1. Recall that two three-carbon units are formed from fructose 1,6-diphosphate. The reactions of glycolysis are summarized in Table 12-2 and Figure 12-9.

Table 12-2
Reactions of glycolysis

Step	Reaction	Enzyme	Type*	$\Delta G^{\circ\prime}$
1	Glucose + ATP $\longrightarrow$ glucose 6-phosphate + ADP + H$^+$	Hexokinase	a	-4.0
2	Glucose 6-phosphate $\rightleftharpoons$ fructose 6-phosphate	Phosphoglucose isomerase	c	$+0.4$
3	Fructose 6-phosphate + ATP $\longrightarrow$ fructose 1,6-diphosphate + ADP + H$^+$	Phosphofructokinase	a	-3.4
4	Fructose 1,6-diphosphate $\rightleftharpoons$ dihydroxyacetone phosphate + glyceraldehyde 3-phosphate	Aldolase	e	$+5.7$
5	Dihydroxyacetone phosphate $\rightleftharpoons$ glyceraldehyde 3-phosphate	Triose phosphate isomerase	c	$+1.8$
6	Glyceraldehyde 3-phosphate + P$_i$ + NAD$^+$ $\rightleftharpoons$ 1,3-diphosphoglycerate + NADH + H$^+$	Glyceraldehyde 3-phosphate dehydrogenase	f	$+1.5$
7	1,3-Diphosphoglycerate + ADP $\rightleftharpoons$ 3-phosphoglycerate + ATP	Phosphoglycerate kinase	a	-4.5
8	3-Phosphoglycerate $\rightleftharpoons$ 2-phosphoglycerate	Phosphoglyceromutase	b	$+1.1$
9	2-Phosphoglycerate $\rightleftharpoons$ phosphoenolpyruvate + H$_2$O	Enolase	d	$+0.4$
10	Phosphoenolpyruvate + ADP + H$^+$ $\longrightarrow$ pyruvate + ATP	Pyruvate kinase	a	-7.5

*Reaction type: (a) Phosphoryl transfer   (d) Dehydration
                (b) Phosphoryl shift   (e) Aldol cleavage
                (c) Isomerization   (f) Phosphorylation coupled to oxidation

PYRUVATE CAN BE CONVERTED TO ETHANOL, LACTATE, OR ACETYL COENZYME A

The sequence of reactions from glucose to pyruvate are very similar in all organisms and in all kinds of cells. In contrast, the fate of pyruvate in the generation of metabolic energy is variable. Three reactions of pyruvate are considered here.

1. *Ethanol* is formed from pyruvate in yeast and several other microorganisms. The first step is the decarboxylation of pyruvate:

$$\text{Pyruvate} + \text{H}^+ \longrightarrow \text{acetaldehyde} + \text{CO}_2$$

This reaction is catalyzed by pyruvate decarboxylase, which contains thiamine pyrophosphate as a coenzyme. Thiamine pyrophosphate is the coenzyme in a variety of decarboxylases (see p. 317) for a discussion of its mechanism of action).

The second step is the reduction of acetaldehyde to ethanol by NADH. *Alcohol dehydrogenase* catalyzes this oxidation-reduction reaction.

$$\text{Acetaldehyde} + \text{NADH} + \text{H}^+ \rightleftharpoons \text{ethanol} + \text{NAD}^+$$

The conversion of glucose to ethanol is called *alcoholic fermentation.* The net reaction of this anaerobic process is:

$$\text{Glucose} + 2\,\text{P}_\text{i} + 2\,\text{ADP} \longrightarrow 2\,\text{ethanol} + 2\,\text{CO}_2 + 2\,\text{ATP}$$

It is important to note that NAD^+ and NADH do not appear in this equation. Acetaldehyde is reduced to ethanol so that NAD^+ is regenerated for use in the reaction catalyzed by glyceraldehyde 3-phosphate dehydrogenase. Thus, there is no net oxidation-reduction in the conversion of glucose to ethanol.

2. *Lactate* is normally formed from pyruvate in a variety of microorganisms. The reaction also occurs in the cells of higher organisms when the amount of oxygen is limiting, as in muscle

during intense activity. The reduction of pyruvate by NADH to form lactate is catalyzed by *lactate dehydrogenase:*

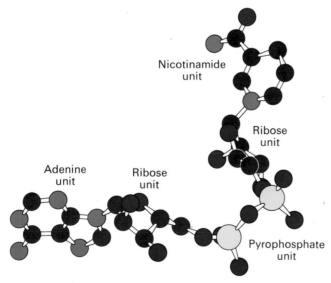

The three-dimensional structure of lactate dehydrogenase, a tetramer with a molecular weight of 140,000, has recently been solved. The structure of its bound NAD^+ is shown in Figure 12-10.

The overall reaction in the conversion of glucose to lactate is:

$$\text{Glucose} + 2\,P_i + 2\,ADP \longrightarrow 2\,\text{lactate} + 2\,ATP$$

As in alcoholic fermentation, there is no net oxidation-reduction.

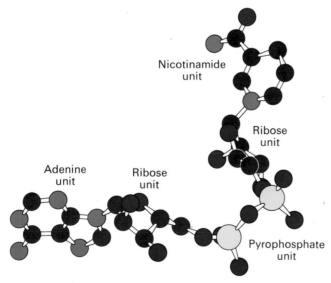

Figure 12-10
Model of NAD^+. The conformation shown here is the one found in the complex of NAD^+ and lactate dehydrogenase.

The NADH formed in the oxidation of glyceraldehyde 3-phosphate is consumed in the reduction of pyruvate. *The regeneration of NAD$^+$ in the reduction of pyruvate to lactate or ethanol sustains the continued operation of glycolysis under anaerobic conditions.* If NAD$^+$ were not regenerated, glycolysis could not proceed beyond glyceraldehyde 3-phosphate, which means that no ATP would be generated. In effect, the formation of lactate buys time, as will be discussed in Chapter 15.

3. Only a small fraction of the energy of glucose is released in the anaerobic conversion to lactate (or ethanol). Much more energy can be extracted aerobically via the citric acid cycle and the electron transport chain. The entry point to this oxidative pathway is *acetyl coenzyme A* (acetyl CoA), which is formed inside mitochondria by the oxidative decarboxylation of pyruvate:

$$\text{Pyruvate} + \text{NAD}^+ + \text{CoA} \longrightarrow$$
$$\text{acetyl CoA} + \text{CO}_2 + \text{NADH} + \text{H}^+$$

This reaction, which is catalyzed by the pyruvate dehydrogenase complex, will be discussed in detail in the next chapter. The NAD$^+$ required for this reaction and for the oxidation of glyceraldehyde 3-phosphate is regenerated when NADH ultimately transfers its electrons to O$_2$ through the electron transport chain in mitochondria.

ALDOLASE FORMS A SCHIFF BASE WITH DIHYDROXYACETONE PHOSPHATE

Now let us turn to aldolase, which catalyzes the condensation of dihydroxyacetone phosphate and glyceraldehyde 3-phosphate to form fructose 1,6-diphosphate. First, dihydroxyacetone phosphate forms a protonated Schiff base with a specific lysine residue in the active site of aldolase.

Protonated
Schiff base

This protonated Schiff base plays a critical role in catalysis because it promotes the formation of the enolate anion of dihydroxyacetone phosphate.

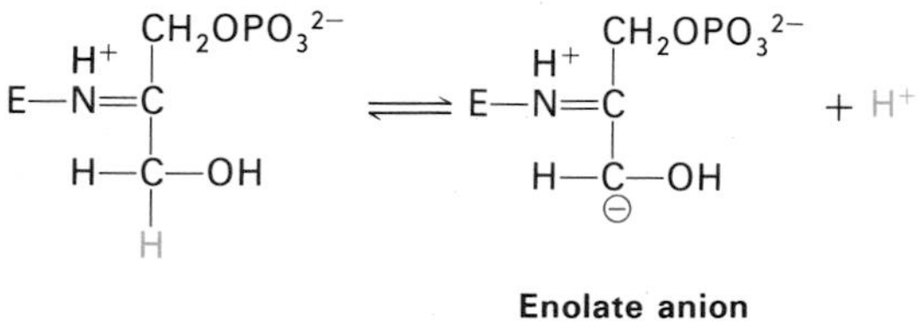

$$
\begin{array}{ccc}
& \overset{\displaystyle CH_2OPO_3{}^{2-}}{\underset{H^+}{\quad}} & \\
E-N=C & & \\
H-C-OH & & \\
H & &
\end{array}
\rightleftharpoons
\begin{array}{c}
\overset{\displaystyle CH_2OPO_3{}^{2-}}{\underset{H^+}{\quad}} \\
E-N=C \\
H-\overset{\ominus}{C}-OH
\end{array}
\quad + H^+
$$

Enolate anion

Glyceraldehyde 3-phosphate then adds to this enolate anion intermediate, yielding the protonated Schiff base of fructose 1,6-diphosphate.

$$
\begin{array}{c}
\overset{\displaystyle CH_2OPO_3{}^{2-}}{\underset{H^+}{\quad}} \\
E-N=C \\
HO-\overset{\ominus}{C}-H
\end{array}
\qquad
\begin{array}{c}
H \\
{}^{\delta+}C=O^{\delta-} \\
H-C-\overset{\cdot}{O}H \\
CH_2OPO_3{}^{2-}
\end{array}
$$

Enolate anion   **Glyceraldehyde 3-phosphate**

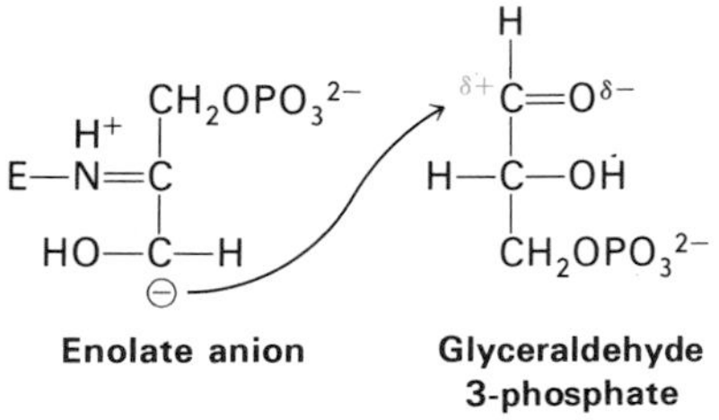

This Schiff base is deprotonated and hydrolyzed to yield fructose 1,6-diphosphate and the regenerated enzyme.

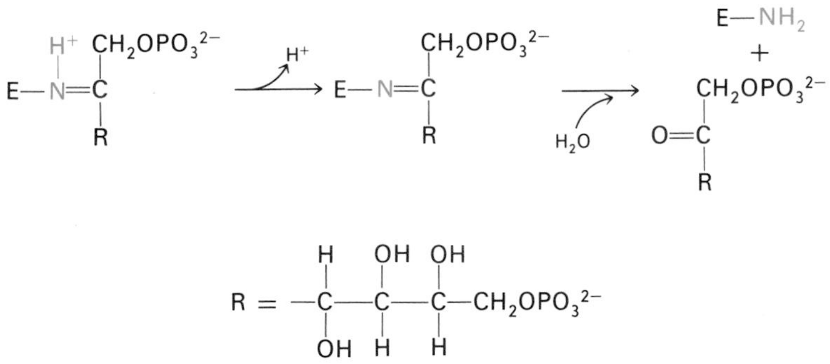

$$
\begin{array}{c}
\overset{\displaystyle CH_2OPO_3{}^{2-}}{\underset{H^+}{\quad}} \\
E-N=C \\
R
\end{array}
\xrightarrow{\ H^+\ }
\begin{array}{c}
CH_2OPO_3{}^{2-} \\
E-N=C \\
R
\end{array}
\xrightarrow[H_2O]{}
\begin{array}{c}
E-NH_2 \\
+ \\
CH_2OPO_3{}^{2-} \\
O=C \\
R
\end{array}
$$

$$
R = \begin{array}{c}
H \quad OH \ OH \\
-C-C-C-CH_2OPO_3{}^{2-} \\
OH \ H \ \ H
\end{array}
$$

The pathway for the cleavage of fructose 1,6-diphosphate is simply the reverse of the one given for its formation.

A THIOESTER IS FORMED IN THE OXIDATION
OF GLYCERALDEHYDE 3-PHOSPHATE

A different kind of covalent enzyme-substrate intermediate is formed in glyceraldehyde 3-phosphate dehydrogenase. This enzyme catalyzes the oxidative phosphorylation of its aldehyde substrate.

$$\text{Glyceraldehyde 3-phosphate} + P_i + NAD^+ \longrightarrow$$
$$1,3\text{-DPG} + NADH + H^+$$

There are two stages in the conversion of an aldehyde to an acyl phosphate:

$$
\underset{\text{Aldehyde}}{R-\overset{\displaystyle O}{\overset{\|}{C}}-H}
\xrightarrow{\text{Oxidation}}
\underset{\text{Intermediate}}{R-\overset{\displaystyle O}{\overset{\|}{C}}-X}
\xrightarrow{\text{Phosphorylation}}
\underset{\text{Acyl phosphate}}{R-\overset{\displaystyle O}{\overset{\|}{C}}-O-\overset{\displaystyle O}{\underset{\displaystyle O^-}{\overset{\|}{P}}}-O^-}
$$

First, the aldehyde is oxidized. This requires the *removal of a hydride ion* ($:H^-$), which is a hydrogen nucleus and two electrons. There is a large barrier to the removal of a hydride ion from an aldehyde because of the dipolar character of the carbonyl group. The carbon atom of the carbonyl group already has a partial positive charge.

$$
\underset{\delta-\quad\delta+}{O=\overset{\displaystyle R}{\overset{|}{C}}-H}
$$

The removal of the hydride ion is facilitated by making the carbon atom less positively charged. This is accomplished by the *addition of a nucleophile,* represented by X^- in this equation:

$$
\underset{\delta-\quad\delta+}{O=\overset{\displaystyle R}{\overset{|}{C}}-H} + X^-
\longrightarrow
{}^-O-\underset{\displaystyle X}{\overset{\displaystyle R}{\overset{|}{\underset{|}{C}}}}-H
\longrightarrow
O=\underset{\displaystyle X}{\overset{\displaystyle R}{\overset{|}{\underset{|}{C}}}} + \underset{\substack{\text{Hydride}\\\text{ion}}}{:H^-}
$$

The hydride ion readily leaves the addition compound because its carbon atom no longer carries a large positive charge. Furthermore, some of the free energy of the oxidation is preserved in the acyl intermediate. Addition of orthophosphate to this acyl intermediate yields an acyl phosphate, which has a high group transfer potential. This sequence of reactions is called a *substrate-level oxidative phosphorylation.*

$$R-\overset{\overset{\displaystyle O}{\|}}{C}-X \;+\; {}^-O-\overset{\overset{\displaystyle O}{\|}}{\underset{\underset{\displaystyle OH}{|}}{P}}-O^- \longrightarrow R-\overset{\overset{\displaystyle O}{\|}}{C}-O-\overset{\overset{\displaystyle O}{\|}}{\underset{\underset{\displaystyle O^-}{|}}{P}}-O^- \;+\; XH$$

Now let us see how glyceraldehyde 3-phosphate dehydrogenase carries out these steps (Figure 12-11). *The nucleophile X^- is the sulfhydryl group of a cysteine residue at the active site.* The aldehyde substrate reacts with the ionized form of this sulfhydryl group to form a hemithioacetal. The next step is the transfer of a hydride ion. *The acceptor for the hydride ion is a molecule of NAD^+ that is tightly bound to the enzyme.* The products of this reaction are the reduced coenzyme NADH and a thioester. *This thioester is an energy-rich intermediate,* corresponding to the acyl intermediate mentioned above. Orthophosphate then attacks the thioester to form 1,3-DPG, an energy-rich phosphate. The active enzyme is regenerated when the bound NADH is oxidized by a free NAD^+ in solution. A crucial aspect of the formation of 1,3-DPG from 3-phosphoglyceraldehyde is that a thermodynamically unfavorable reaction, the formation of an acyl phosphate from a carboxylate, is driven by a thermodynamically favorable reaction, the oxidation of an aldehyde.

$$R-\overset{\overset{\displaystyle O}{\|}}{C}-H \;+\; NAD^+ \;+\; H_2O \rightleftharpoons R-\overset{\overset{\displaystyle O}{\|}}{C}-O^- \;+\; NADH \;+\; 2\,H^+$$

These two reactions are *coupled by the thioester intermediate,* which preserves much of the free energy released in the oxidation reaction. We see here the *use of a covalent enzyme-bound intermediate as a mechanism of energy coupling.*

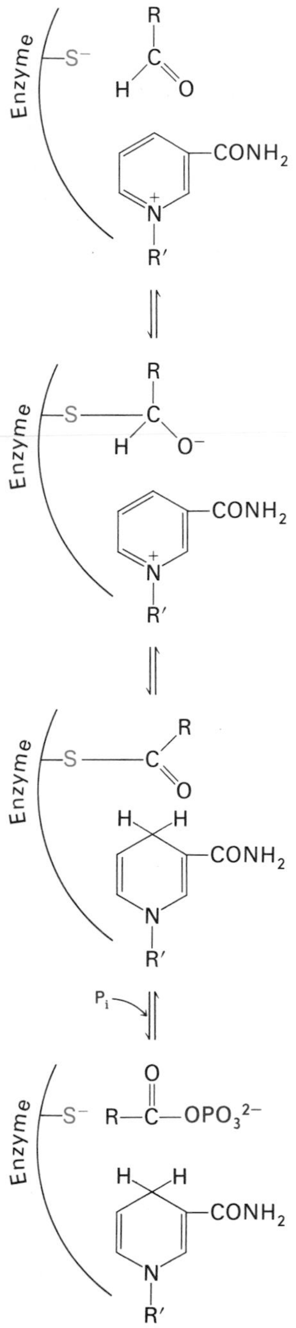

Figure 12-11
Proposed catalytic mechanism of glyceraldehyde 3-phosphate dehydrogenase.

Phosphoenolpyruvate

ENOL PHOSPHATES HAVE A HIGH GROUP-TRANSFER POTENTIAL

Because it is an *acyl phosphate*, 1,3-DPG has a high group-transfer potential. A different kind of high-energy phosphate compound is formed several steps later in glycolysis. Phosphoenolpyruvate, an *enol phosphate*, is formed by the dehydration of 2-phosphoglycerate.

The $\Delta G^{\circ\prime}$ of hydrolysis of a phosphate ester of an ordinary alcohol is -3 kcal/mol, whereas that of phosphoenolpyruvate is -13.3 kcal/mol. Why does phosphoenolpyruvate have such a high phosphate group-transfer potential? The answer is that the reaction does not stop with the enol formed upon transfer of the phosphoryl group. The enol undergoes a conversion to a ketone—namely, pyruvate.

Phosphoenolpyruvate   Enolpyruvate   Pyruvate

The $\Delta G^{\circ\prime}$ of the enol to ketone conversion is very large, of the order of -10 kcal/mol. This should be compared with the $\Delta G^{\circ\prime}$ of hydrolysis of phosphoenolpyruvate to enolpyruvate, which is about -3 kcal/mol. Thus, the *high phosphate group-transfer potential of phosphoenolpyruvate arises from the large driving force of the subsequent enol to ketone conversion.*

ESSENTIALLY IRREVERSIBLE REACTIONS CAN SERVE AS CONTROL SITES

Most of the reactions of glycolysis are readily reversible. Three of them, however, do not proceed appreciably in the reverse direction under physiological conditions. In effect, these reactions are irreversible.

$$\text{Glucose} + \text{ATP} \xrightarrow{\text{hexokinase}} \text{glucose 6-phosphate} + \text{ADP} \quad (1)$$

$$\text{Fructose 6-phosphate} + \text{ATP} \xrightarrow{\text{phosphofructokinase}}$$

$$\text{fructose 1,6-diphosphate} + \text{ADP} \quad (2)$$

$$\text{Phosphoenolpyruvate} + \text{ADP} \xrightarrow{\text{pyruvate kinase}}$$

$$\text{pyruvate} + \text{ATP} \quad (3)$$

The irreversibility of these steps has important consequences. First, it means that they must be bypassed if glucose is to be synthesized from pyruvate. Gluconeogenesis, the formation of glucose from compounds other than sugars, does occur under certain cellular conditions. The necessary bypass steps will be presented in Chapter 15. Second, irreversible reactions are optimal sites for the control of the rate of a pathway. In fact, *the rate of glycolysis is primarily controlled by the level of activity of phosphofructokinase*. This allosteric enzyme is stimulated by ADP and AMP and is inhibited by ATP and citrate. In other words, phosphofructokinase is most active when the energy charge of a cell is low.

METABOLISM OF 2,3-DPG, A REGULATOR OF OXYGEN TRANSPORT

Recall that 2,3-diphosphoglycerate (2,3-DPG) is a regulator of oxygen transport in erythrocytes. It decreases the oxygen affinity of hemoglobin by stabilizing the deoxygenated form of hemoglobin (Chapter 4). Red blood cells have a high concentration of 2,3-DPG, typically 4 mM, in contrast to most other cells, which have only trace amounts. 2,3-DPG has a general role as a cofactor in the conversion of 3-phosphoglycerate to 2-phosphoglycerate by phosphoglycerate mutase.

The synthesis and degradation of 2,3-DPG are detours from the glycolytic pathway (Figure 12-12).

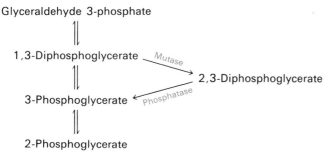

Figure 12-12
Pathway for the synthesis and degradation of 2,3-diphosphoglycerate.

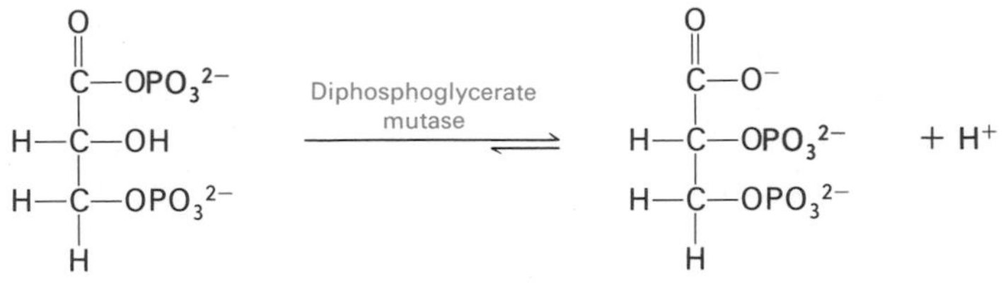

Diphosphoglycerate mutase converts 1,3-DPG to 2,3-DPG.

1,3-Diphosphoglycerate   2,3-Diphosphoglycerate

2,3-DPG is hydrolyzed to 3-phosphoglycerate by *2,3-diphosphoglycerate phosphatase*. A phosphatase is an enzyme that catalyzes the hydrolysis of a phosphate ester.

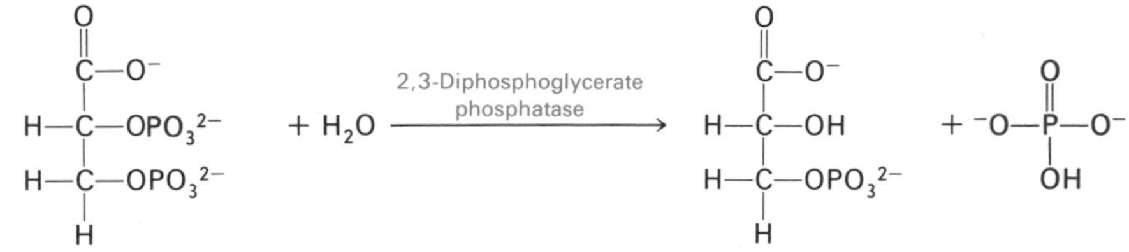

2,3-Diphosphoglycerate   3-Phosphoglycerate

This mutase reaction has an interesting mechanism. *3-Phosphoglycerate* is an obligatory participant although it does not appear in the overall stoichiometry. The mutase binds 1,3-DPG and 3-phosphoglycerate simultaneously. In this ternary complex, a phosphoryl group is transferred from the 1-position of 1,3-DPG to the 2-position of 3-phosphoglycerate (Figure 12-13).

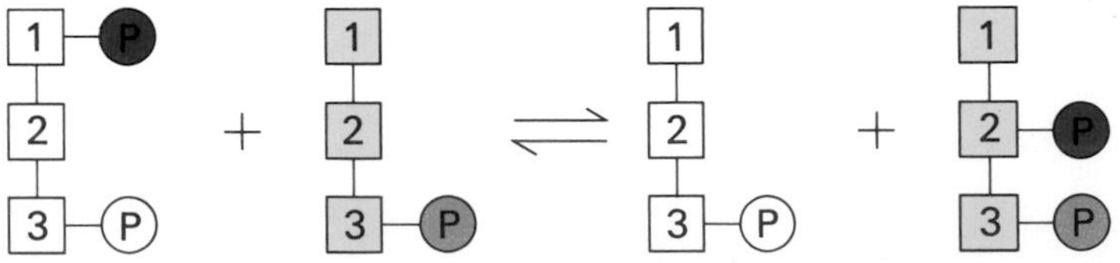

Figure 12-13
Schematic representation of the participation of 3-phosphoglycerate in the conversion of 1,3-diphosphoglycerate to 2,3-diphosphoglycerate.

In this mutase reaction, 2,3-DPG is a potent competitive inhibitor of 1,3-DPG. Thus, the rate of synthesis of 2,3-DPG will depend in part on its own concentration. Another controlling factor is the concentration of 1,3-DPG, since the enzyme is not usually saturated with 1,3-DPG. In contrast, the level of 3-phosphoglycerate in the red cell is nearly adequate to saturate the mutase. Thus, *the rate of synthesis of 2,3-DPG is controlled by the concentration of unbound 1,3-DPG and 2,3-DPG.*

DEFECTS IN RED CELL GLYCOLYSIS ALTER OXYGEN TRANSPORT

Glycolysis in red cells and oxygen transport are linked by 2,3-DPG. This means that defects in glycolysis can affect oxygen transport. Indeed, the oxygen dissociation curves of blood from some patients with inherited defects of glycolysis in red blood cells are altered (Figure 12-14). In *hexokinase deficiency,* the concentration of glycolytic intermediates is low because the phosphorylation of glucose, the first step, is impaired. Hence, the concentration of 2,3-DPG is reduced in these red cells, and the hemoglobin has an *abnormally high oxygen affinity.* Just the opposite is found in *pyruvate kinase deficiency.* The glycolytic intermediates are at an abnormally high concentration because the terminal step is blocked. Consequently, the level of 2,3-DPG is about twice the normal one, which leads to a *low oxygen affinity.* The abnormal oxygen dissociation curves in hexokinase and pyruvate kinase deficiency were puzzling until it was discovered that 2,3-DPG is a regulator of oxygen transport.

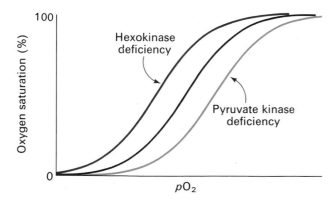

Figure 12-14
Oxygen dissociation curves of normal red cells (black line), red cells from a patient with hexokinase deficiency (red line), and red cells from a patient with pyruvate kinase deficiency (green line).

SUMMARY

Glycolysis is the reaction pathway that converts glucose to pyruvate. In aerobic organisms, glycolysis is the prelude to the citric acid cycle and the electron-transport chain, where most of the free energy in glucose is harvested. The ten reactions of glycolysis occur in the cytosol. In the first stage, glucose is converted to fructose 1,6-diphosphate by a phosphorylation, an isomerization, and a second phosphorylation reaction. Two ATP are consumed per glucose in these reactions, which prepare for the net synthesis of ATP. In the second stage, fructose 1,6-diphosphate is cleaved by aldolase to dihydroxyacetone phosphate and glyceraldehyde 3-phosphate, which are readily interconvertible. Glyceraldehyde 3-phosphate is then oxidized and phosphorylated to form 1,3-DPG, an acyl phosphate with a high phosphate transfer potential. 3-Phosphoglycerate is formed as an ATP is generated. In the last stage of glycolysis, phosphoenolpyruvate, a second intermediate with a high phosphate transfer potential, is formed by a phosphoryl shift and a dehydration. An ATP is generated as phosphoenolpyruvate is converted to pyruvate. There is a net gain of two ATP in the formation of two molecules of pyruvate from one molecule of glucose.

The electron acceptor in the oxidation of glyceraldehyde 3-phosphate is NAD^+, which must be regenerated for glycolysis to proceed. In aerobic organisms, the NADH formed in glycolysis transfers its electrons to O_2 through the electron-transport chain, thereby regenerating NAD^+. Under anaerobic conditions, NAD^+ is regenerated by the reduction of pyruvate to lactate. In some microorganisms, NAD^+ is normally regenerated by the synthesis of lactate or ethanol from pyruvate. These two processes are called fermentations.

The reactions of glycolysis are readily reversible under physiological conditions except for the ones catalyzed by hexokinase, phosphofructokinase, and pyruvate kinase. The rate of glycolysis is primarily controlled by the level of catalytic activity of phosphofructokinase. This allosteric enzyme is most active when the energy charge of a cell is low.

SELECTED READINGS

Books and general reviews:

Davison, E. A., 1967. *Carbohydrate Chemistry.* Holt, Rinehart, and Winston.

Pigman, W. W., and Horton, D., (eds.), 1972. *The Carbohydrates: Chemistry and Biochemistry.* Academic Press.

Dickens, F., Randle, P. J., and Whelan, W. J., 1968. *Carbohydrate Metabolism and Its Disorders,* vols. 1 and 2. Academic Press.

Glycolytic enzymes:

Boyer, P. D., (ed.), 1972. *The Enzymes* (3rd ed.). Academic Press. [Volumes 5 through 9 contain authoritative reviews on each of the glycolytic enzymes.]

Rossman, M. G., Adams, M. J., Buchner, M., Ford, G. C., Hackert, M. C., Lentz, P. J., Jr., McPherson, A., Jr., Schevitz, R. W., and Smiley, I. E., 1971. Structural constraints on possible mechanisms of lactate dehydrogenase as shown by high resolution studies of the apoenzyme and a variety of enzyme complexes. *Cold Spring Harbor Symp. Quant. Biol.* 36:179–191.

Büchner, M., Ford, G. C., Moras, D., Olsen, K. W., and Rossmann, M. G., 1973. Glyceraldehyde 3-phosphate dehydrogenase: three-dimensional structure and evolutionary significance. *Proc. Nat. Acad. Sci.* 70:3052–3054. [The binding sites for the NAD^+ coenzyme in five dehydrogenases are very similar. The authors suggest that the NAD^+ binding site might be one of the earliest and most universal architectural units of proteins.]

Blake, C. C. F., Evans, P. R., and Scopes, R. K., 1972. Structure of horse-muscle phosphoglycerate kinase at 6 Å resolution. *Nature New Biol.* 235:195–198.

Bryant, T. N., Watson, H. C., and Wendell, P. L., 1974. Structure of yeast phosphoglycerate kinase. *Nature* 247:14–17.

Anderson, W. F., Fletterick, R. J., and Steitz, T. A., 1974. Structure of yeast hexokinase III. *J. Mol. Biol.* 86:261–269.

Brändén, C-I., Eklund, H., Nordström, B., Boiwe, T., Söderlund, G., Zeppezauer, E., Ohlsson, I., and Åkeson, Å., 1973. Structure of liver alcohol dehydrogenase at 2.9-Å resolution. *Proc. Nat. Acad. Sci.* 70:2439–2442.

Catalytic mechanisms:

Jencks, W. P., 1969. *Catalysis in Chemistry and Enzymology.* McGraw-Hill. [Chapter 2 on covalent catalysis is especially pertinent to the enzymatic mechanisms discussed in this chapter.]

Rose, I. A., 1970. Enzymology of proton abstraction and transfer reactions. *In* Boyer, P. D., (ed.), *The Enzymes* (3rd ed.), vol. 2, pp. 281–320.

Knowles, J. R., Leadlay, P. F., and Maister, S. G., 1971. Triosephosphate isomerase: isotope studies on the mechanistic pathway. *Cold Spring Harbor Symp. Quant. Biol.* 36:157–164.

2,3-Diphosphoglycerate:

Benesch, R. E., and Benesch, R., 1970. The reaction between diphosphoglycerate and hemoglobin. *Fed. Proc.* 29:1101–1104.

Rose, Z. B., 1970. Enzymes controlling 2,3-diphosphoglycerate in human erythrocytes. *Fed. Proc.* 29:1105–1111.

Delivoria-Papadopoulos, M., Oski, F. A., and Gottlieb, A. J., 1969. Oxygen-hemoglobin dissociation curves: effect of inherited enzyme defects of the red cell. *Science* 165:601–602.

Historical aspects:

Kalckar, H. M., (ed.)., 1969. *Biological Phosphorylations: Development of Concepts.* Prentice-Hall. [Contains many of the classical papers on glycolysis.]

Fruton, J. S., 1972. *Molecules and Life: Historical*

Essays on the Interplay of Chemistry and Biology. Wiley-Interscience. [Includes a meticulously documented account of the elucidation of the nature of fermentation and how it led to enzyme chemistry.]

APPENDIX
Stereochemical Relationships of Some Sugars

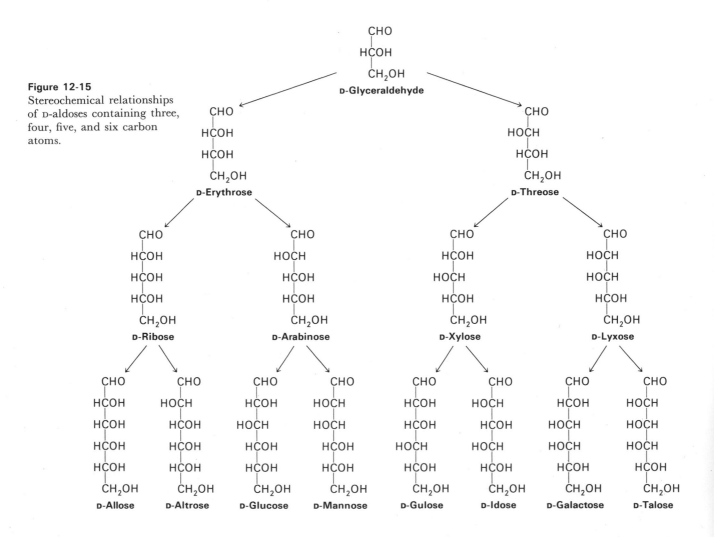

Figure 12-15
Stereochemical relationships of D-aldoses containing three, four, five, and six carbon atoms.

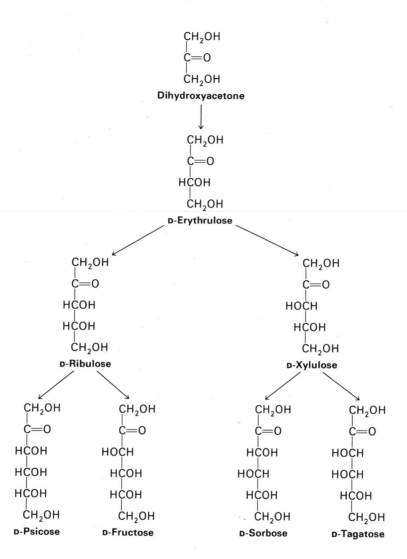

Figure 12-16
Stereochemical relationships of D-ketoses containing three, four, five, and six carbon atoms.

PROBLEMS

1. Carbohydrates that differ only in the configuration about their carbonyl carbon atoms are *anomers*. *Epimers* differ in configuration about a specific carbon atom other than the carbonyl carbon atom. Indicate whether each of the following pairs of sugars are anomers, epimers, or an aldose-ketose pair:

   (a) D-glyceraldehyde and dihydroxyacetone.
   (b) D-glucose and D-mannose.
   (c) D-glucose and D-fructose.
   (d) α-D-glucose and β-D-glucose.
   (e) D-ribose and D-ribulose.
   (f) D-galactose and D-glucose.

2. Glucose labeled with ^{14}C at C-1 is incubated with the glycolytic enzymes and necessary cofactors. What is the distribution of ^{14}C in the pyruvate that is formed? (Assume that the interconversion of glyceraldehyde 3-phosphate and dihydroxyacetone phosphate is very rapid compared with the subsequent step.)

3. Glyceraldehyde 3-phosphate dehydrogenase can use arsenate instead of phosphate as a substrate. The resulting product is 1-arseno-3-phosphoglycerate, a highly unstable compound that is rapidly hydrolyzed to arsenate and 3-phosphoglycerate. Write a balanced equation for glycolysis when arsenate replaces phosphate. What is the effect of this substitution on the yield of ATP in glycolysis?

4. Write a balanced equation for the conversion of glucose to lactate.

   (a) Calculate the standard free-energy change of this reaction using the data given in Table 12-2 (page 291) and the fact that $\Delta G°'$ is -6 kcal for the reaction

   $$\text{Pyruvate} + \text{NADH} + \text{H}^+ \rightleftharpoons \text{lactate} + \text{NAD}^+$$

   (b) What is the free-energy change ($\Delta G'$, not $\Delta G°'$) of this reaction when the concentrations of reactants are: glucose, 5 mM; lactate, .05 mM; ATP, 2 mM; ADP, 0.2 mM; and P_i, 1 mM.

5. What is the equilibrium ratio of phosphoenolpyruvate to pyruvate under standard conditions when [ATP]/[ADP] = 10?

6. What are the equilibrium concentrations of fructose 1,6-diphosphate, dihydroxyacetone phosphate, and glyceraldehyde 3-phosphate when 1 mM fructose 1,6-diphosphate is incubated with aldolase under standard conditions?

7. 3-Phosphoglycerate labeled uniformly with ^{14}C is incubated with 1,3-DPG labeled with ^{32}P at C-1. What is the radioisotope distribution of the 2,3-DPG that is formed on addition of DPG mutase?

For additional problems, see W. B. Wood, J. H. Wilson, R. M. Benbow, and L. E. Hood. *Biochemistry: A Problems Approach* (Benjamin, 1974), Chapter 9.

CHAPTER 13

CITRIC ACID CYCLE

In the previous chapter, we considered the glycolytic pathway, in which glucose is converted to pyruvate. Under aerobic conditions, the next step in the generation of energy from glucose is the oxidative decarboxylation of pyruvate to form acetyl coenzyme A (acetyl CoA). This activated acetyl unit is then completely oxidized to CO_2 by the *citric acid cycle,* a series of reactions that is also known as the *tricarboxylic acid cycle* or the *Krebs cycle.* The citric acid cycle is the *final common pathway for the oxidation of fuel molecules*—amino acids, fatty acids, and carbohydrates. Most fuel molecules enter the cycle as acetyl CoA. The cycle also provides intermediates for biosyntheses. The reactions of the citric acid cycle occur inside mitochondria, in contrast to those of glycolysis, which occur in the cytosol.

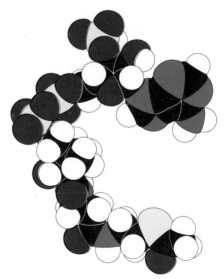

Figure 13-1
Model of acetyl CoA.

FORMATION OF ACETYL COENZYME A FROM PYRUVATE

The oxidative decarboxylation of pyruvate to form acetyl CoA, which occurs in the mitochondrial matrix, is the link between glycolysis and the citric acid cycle:

Pyruvate + CoA + NAD^+ $\longrightarrow$ acetyl CoA + CO_2 + NADH

This irreversible funneling of the product of glycolysis into the citric acid cycle is catalyzed by the *pyruvate dehydrogenase complex*. This very large multienzyme complex is a highly integrated array of three kinds of enzymes, which will be discussed in detail later in this chapter (p. 316):

AN OVERVIEW OF THE CITRIC ACID CYCLE

The overall pattern of the citric acid cycle is shown in Figure 13-2. A four-carbon compound (oxaloacetate) condenses with a two-carbon acetyl unit to yield a six-carbon tricarboxylic acid (citrate). An isomer of citrate is then oxidatively decarboxylated. The resulting five-carbon compound (α-ketoglutarate) is oxidatively decarboxylated to yield a four-carbon compound (succinate). Oxaloacetate is then regenerated from succinate. Two carbon atoms enter the cycle as an acetyl unit and two carbon atoms leave the cycle in the form of two molecules of CO_2. Since CO_2 is more oxidized than an acetyl group, it is evident that there must be some oxidation-reduction reactions in the citric acid cycle. In fact, there are four such reactions. Three hydride ions (hence, six electrons) are transferred to NAD^+, whereas one pair of hydrogen atoms (hence, two electrons) is transferred to flavin adenine dinucleotide (FAD). These electron carriers yield 11 molecules of adenosine triphosphate (ATP) when they are oxidized by O_2 in the electron transport chain. In addition, one high-energy phosphate bond is formed in each round of the citric acid cycle itself.

Figure 13-2
An overview of the citric acid cycle.

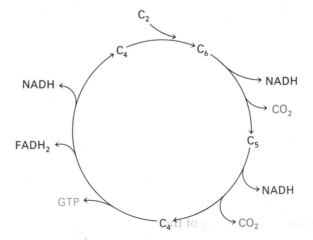

OXALOACETATE CONDENSES WITH ACETYL COENZYME A TO FORM CITRATE

The cycle starts with the joining of a four-carbon unit, oxaloacetate, and a two-carbon unit, the acetyl group of acetyl CoA. Oxaloacetate reacts with acetyl CoA and H_2O to yield citrate and CoA.

$$O=\overset{\displaystyle |}{\underset{\displaystyle |}{C}}-COO^- \; + \; \overset{\displaystyle O}{\underset{\displaystyle S-CoA}{\overset{\|}{C}}}-CH_3 \; + H_2O \longrightarrow HO-\overset{\displaystyle CH_2-COO^-}{\underset{\displaystyle CH_2-COO^-}{C}}-COO^- \; + HS-CoA + H^+$$

| Oxaloacetate | Acetyl CoA | | Citrate |

This reaction, which is an aldol condensation followed by a hydrolysis, is catalyzed by *citrate synthetase* (originally called condensing enzyme). Oxaloacetate first condenses with acetyl CoA to form *citryl CoA*, which is then hydrolyzed to citrate and CoA. Hydrolysis of citryl CoA pulls the overall reaction far in the direction of the synthesis of citrate.

$$\begin{array}{c} O \\ \| \\ CH_2-C-S-CoA \\ | \\ HO-C-COO^- \\ | \\ CH_2-COO^- \end{array}$$

Citryl CoA

CITRATE IS ISOMERIZED TO ISOCITRATE

Citrate must be isomerized to isocitrate to enable the six-carbon unit to undergo oxidative decarboxylation. The isomerization of citrate is accomplished by a *dehydration* step followed by a *hydration* step. The result is an interchange of an H and OH. The enzyme catalyzing both steps is called *aconitase* because the presumed intermediate is cis-*aconitate*.

$$\begin{array}{ccc}
\begin{array}{c} COO^- \\ | \\ CH_2 \\ | \\ HO-C-COO^- \\ | \\ H-C-H \\ | \\ COO^- \end{array}
& \underset{\Large\rightleftharpoons}{\overset{H_2O}{}} &
\begin{array}{c} COO^- \\ | \\ CH_2 \\ \| \\ C-COO^- \\ \\ H-C \\ | \\ COO^- \end{array}
& \underset{\Large\rightleftharpoons}{\overset{H_2O}{}} &
\begin{array}{c} COO^- \\ | \\ CH_2 \\ | \\ H-C-COO^- \\ | \\ H-C-OH \\ | \\ COO^- \end{array}
\end{array}$$

| Citrate | cis-Aconitate | Isocitrate |

ISOCITRATE IS OXIDIZED AND DECARBOXYLATED TO ALPHA-KETOGLUTARATE

We come now to the first of four oxidation-reduction reactions in the citric acid cycle. The oxidative decarboxylation of isocitrate is catalyzed by *isocitrate dehydrogenase:*

$$\text{Isocitrate} + \text{NAD}^+ \rightleftharpoons \alpha\text{-ketoglutarate} + CO_2 + \text{NADH}$$

The intermediate in this reaction is oxalosuccinate, which rapidly loses CO_2 while bound to the enzyme to give α-ketoglutarate.

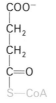

| Isocitrate | Oxalosuccinate | α-Ketoglutarate |

The rate of formation of α-ketoglutarate is important in determining the overall rate of the cycle, as will be discussed later (p. 325). Also noteworthy is that there are two kinds of isocitrate dehydrogenases. One is specific for NAD^+, the other for nicotinamide adenine dinucleotide phosphate (NADP^+). The NAD^+-specific enzyme, which is located in mitochondria, is the important one for the citric acid cycle. The NADP^+-specific enzyme, which is present both in mitochondria and in the cytoplasm, has a different metabolic role.

SUCCINYL COENZYME A IS FORMED BY THE OXIDATIVE DECARBOXYLATION OF ALPHA-KETOGLUTARATE

The conversion of isocitrate to α-ketoglutarate is followed by a second oxidative decarboxylation reaction, the formation of succinyl CoA from α-ketoglutarate:

$$\alpha\text{-Ketoglutarate} + \text{NAD}^+ + \text{CoA} \rightleftharpoons$$
$$\text{succinyl CoA} + CO_2 + \text{NADH}$$

Succinyl CoA

This reaction is catalyzed by the *α-ketoglutarate dehydrogenase complex,* an organized assembly consisting of three kinds of enzymes. *The mechanism of this reaction is very similar to that of the conversion of pyruvate to acetyl CoA.* The same cofactors are used: NAD^+, CoA, thiamine pyrophosphate, lipoamide, and FAD. Indeed, the pyruvate dehydrogenase complex and the α-ketoglutarate dehydrogenase complex have common structural features (p. 320).

A HIGH-ENERGY PHOSPHATE BOND IS GENERATED FROM SUCCINYL COENZYME A

The succinyl thioester of CoA is an energy-rich bond. The $\Delta G°'$ for hydrolysis of succinyl CoA is about -8 kcal/mol, which is comparable to that of ATP (-7.3 kcal/mol). *The cleavage of the thioester bond of succinyl CoA is coupled to the phosphorylation of guanosine diphosphate* (GDP):

$$\text{Succinyl CoA} + P_i + \text{GDP} \rightleftharpoons \text{succinate} + \text{GTP} + \text{CoA}$$

This readily reversible reaction ($\Delta G°' = -0.8$ kcal/mol) is catalyzed by *succinyl CoA synthetase.* The phosphoryl group in guanosine triphosphate (GTP) is readily transferred to adenosine diphosphate (ADP) to form ATP, in a reaction catalyzed by *nucleoside diphosphokinase.*

$$\text{GTP} + \text{ADP} \rightleftharpoons \text{GDP} + \text{ATP}$$

$$
\begin{array}{l}
\text{COO}^- \\
|\\
\text{CH}_2 \\
|\\
\text{CH}_2 \\
|\\
\text{COO}^-
\end{array}
$$
Succinate

The generation of a high-energy phosphate bond from succinyl CoA is an example of a *substrate-level phosphorylation.* In fact, this is the only reaction in the citric acid cycle that directly yields a high-energy phosphate bond. The contrasting process is *respiratory chain phosphorylation* (also called *oxidative phosphorylation*), which is the formation of ATP coupled to the oxidation of NADH and $FADH_2$ by O_2. We had previously encountered substrate-level phosphorylation in two reactions of glycolysis: in the oxidation of glyceraldehyde 3-phosphate and in the conversion of phosphoenol pyruvate to pyruvate. Oxidative phosphorylation will be discussed in the next chapter.

REGENERATION OF OXALOACETATE BY OXIDATION OF SUCCINATE

Reactions of four-carbon compounds constitute the final stage of the citric acid cycle (Figure 13-3). Succinate is converted to oxaloacetate in three steps: an oxidation, a hydration, and a second oxidation reaction. Oxaloacetate is thereby regenerated for another round of the cycle, concomitant with the trapping of energy in the form of $FADH_2$ and NADH.

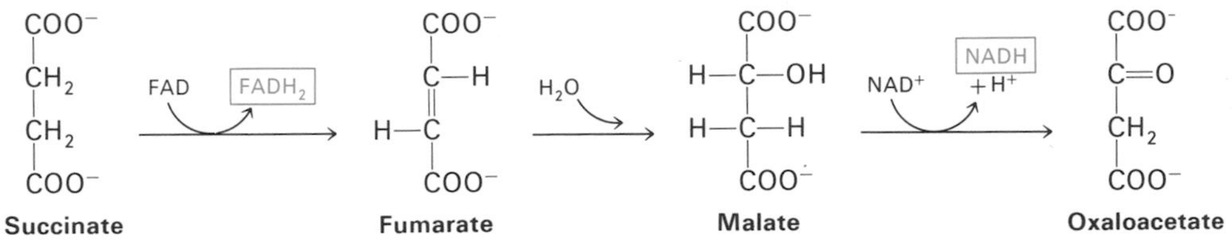

Figure 13-3
Final stage of the citric acid cycle: from succinate to oxaloacetate.

Succinate is oxidized to fumarate by *succinate dehydrogenase*. The hydrogen acceptor is FAD, rather than NAD^+, which is used in the other three oxidation reactions in the cycle. FAD is the hydrogen acceptor in this reaction because the free energy change is insufficient to reduce NAD^+. FAD is covalently attached to a histidine side chain of the enzyme (denoted E-FAD).

$$\text{Succinate} + \text{E-FAD} \rightleftharpoons \text{fumarate} + \text{E-FADH}_2$$

Succinate dehydrogenase contains four iron atoms and four inorganic sulfides in addition to the flavin. There is no heme in this enzyme. Rather, the irons are probably bonded to the inorganic sulfides. Succinate dehydrogenase is an example of a *nonheme iron protein*. This enzyme differs from the others in the citric acid cycle in being an integral part of the inner mitochondrial membrane. In fact, *succinate dehydrogenase is directly linked to the electron-transport chain.* The $FADH_2$ produced by the oxidation of succinate does not dissociate from the enzyme, in contrast to NADH. Rather, two electrons from $FADH_2$ are transferred directly to the Fe^{3+} atoms of the enzyme. The ultimate acceptor of these electrons is molecular oxygen, as will be discussed in the next chapter.

The next step in the cycle is the hydration of fumarate to form L-malate. *Fumarase* catalyzes a stereospecific *trans* addition of H and OH, as shown by deuterium-labeling studies. The OH group adds to only one side of the double bond of fumarate; hence, only the L-isomer of malate is formed.

Finally, malate is oxidized to oxaloacetate. This reaction is catalyzed by *malate dehydrogenase* and NAD^+ is again the hydrogen acceptor.

$$\text{Malate} + NAD^+ \rightleftharpoons \text{oxaloacetate} + NADH + H^+$$

The oxidation of malate is a highly endergonic reaction ($\Delta G^{\circ\prime} = +7$ kcal/mol). However, the reaction goes readily in the forward direction under physiological conditions because the steady-state concentrations of the products are very low. Oxaloacetate and NADH are rapidly consumed in highly exergonic reactions.

STOICHIOMETRY OF THE CITRIC ACID CYCLE

The net reaction of the citric acid cycle is:

$$\text{Acetyl CoA} + 3\,NAD^+ + FAD + GDP + P_i + 2\,H_2O \longrightarrow$$
$$2\,CO_2 + 3\,NADH + FADH_2 + GTP + 2\,H^+ + CoA$$

Let us recapitulate the reactions that give this stoichiometry (Figure 13-4 and Table 13-1):

1. Two carbon atoms enter the cycle in the condensation of an acetyl unit (from acetyl CoA) with oxaloacetate. Two carbon atoms leave the cycle in the form of CO_2 in the successive decarboxylations catalyzed by isocitrate dehydrogenase and α-ketoglutarate dehydrogenase. As will be discussed shortly, the two carbon atoms that leave the cycle are different from the ones that entered in that round.

2. Four pairs of hydrogen atoms leave the cycle in the four oxidation reactions. Two NAD^+ are reduced in the oxidative decarboxylations of isocitrate and α-ketoglutarate, one FAD is reduced

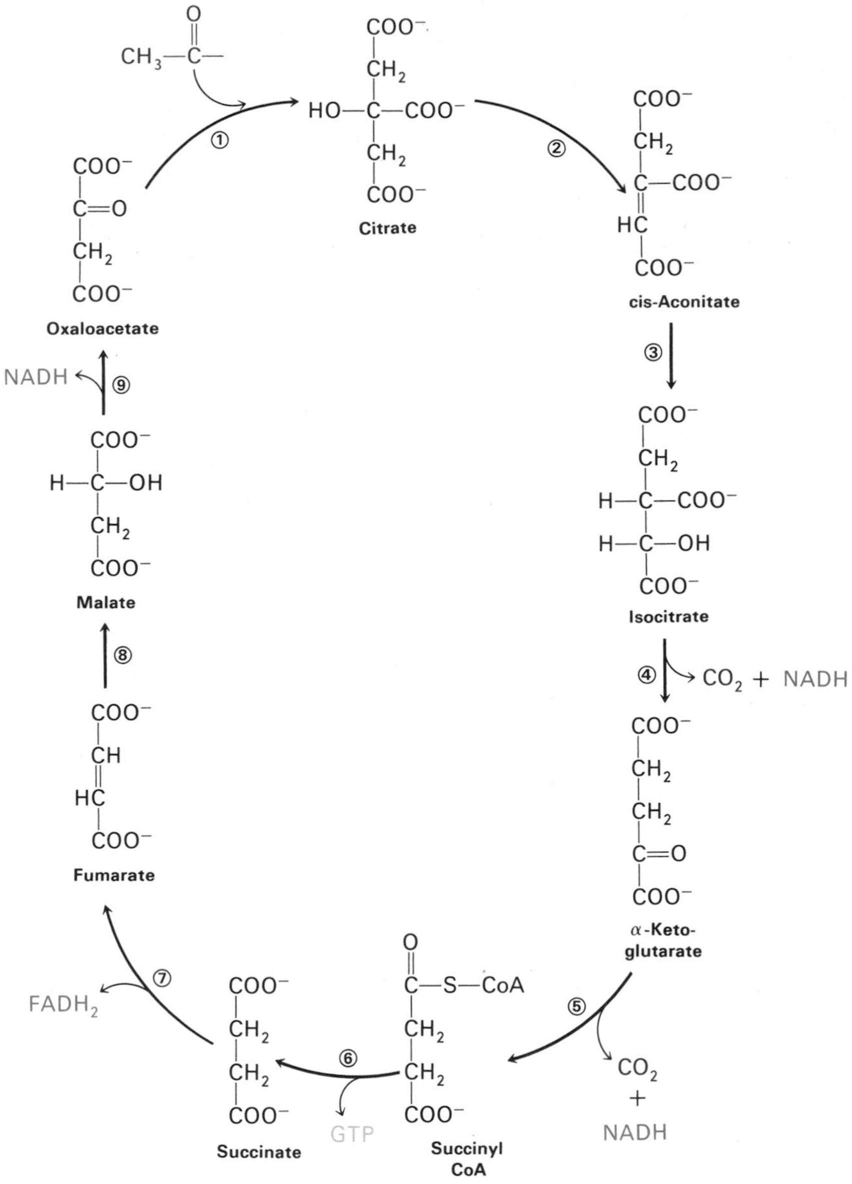

314

Figure 13-4
Citric acid cycle.

in the oxidation of succinate, and one NAD^+ is reduced in the oxidation of malate.

3. One high-energy phosphate bond (in the form of GTP) is generated from the energy-rich thioester linkage in succinyl CoA.

4. Two water molecules are consumed: one in the synthesis of citrate, the other in the hydration of fumarate.

Looking ahead, the NADH and $FADH_2$ formed in the citric acid cycle are oxidized by the electron-transport chain (Chapter 14). ATP is generated as electrons are transferred from these carriers to O_2, the ultimate acceptor. Three ATP are formed for each NADH in the mitochondrion, whereas two ATP are generated per $FADH_2$. Note that only one high-energy phosphate bond per acetyl unit is directly formed in the citric acid cycle. Eleven more high-energy phosphate bonds are generated when three NADH and one $FADH_2$ are oxidized by the electron-transport chain.

Table 13-1
Citric acid cycle

Step	Reaction	Enzyme	Cofactor	Type*	$\Delta G^{\circ\prime}$
1	Acetyl CoA + oxaloacetate + H_2O $\longrightarrow$ citrate + CoA + H^+	Citrate synthetase	CoA	a	-7.5
2	Citrate $\rightleftharpoons$ cis-aconitate + H_2O	Aconitase	Fe^{2+}	b	$+2.0$
3	cis-Aconitate + H_2O $\rightleftharpoons$ isocitrate	Aconitase	Fe^{2+}	c	-0.5
4	Isocitrate + NAD^+ $\rightleftharpoons$ α-ketoglutarate + CO_2 + NADH	Isocitrate dehydrogenase	NAD^+	d + e	-2.0
5	α-Ketoglutarate + NAD^+ + CoA $\rightleftharpoons$ succinyl CoA + CO_2 + NADH	α-Ketoglutarate dehydrogenase complex	NAD^+ CoA TPP Lipoic acid FAD	d + e	-7.2
6	Succinyl CoA + P_i + GDP $\rightleftharpoons$ succinate + GTP + CoA	Succinyl CoA synthetase	CoA	f	-0.8
7	Succinate + FAD (enzyme-bound) $\rightleftharpoons$ fumarate + $FADH_2$ (enzyme-bound)	Succinate dehydrogenase	FAD	e	~ 0
8	Fumarate + H_2O $\rightleftharpoons$ malate	Fumarase	None	c	-0.9
9	L-Malate + NAD^+ $\rightleftharpoons$ oxaloacetate + NADH + H^+	Malate dehydrogenase	NAD^+	e	$+7.1$

*Reaction type: (a) Condensation   (d) Decarboxylation
(b) Dehydration   (e) Oxidation
(c) Hydration   (f) Substrate-level phosphorylation

Molecular oxygen does not participate directly in the citric acid cycle. However, the cycle operates only under aerobic conditions because NAD^+ and FAD can only be regenerated in the mitochondrion by electron transfer to molecular oxygen. Glycolysis has both an aerobic and an anaerobic mode, whereas the citric acid cycle is strictly aerobic. Recall that glycolysis can proceed under anaerobic conditions because NAD^+ is regenerated in the conversion of pyruvate to lactate.

PYRUVATE DEHYDROGENASE COMPLEX: AN ORGANIZED ENZYME ASSEMBLY

We now turn to some reaction mechanisms. The oxidative decarboxylation of pyruvate to acetyl CoA is catalyzed by the *pyruvate dehydrogenase complex,* an organized assembly of three kinds of enzymes (Table 13-2). The net reaction catalyzed by this complex is

$$\text{Pyruvate} + \text{CoA} + NAD^+ \longrightarrow \text{acetyl CoA} + CO_2 + \text{NADH}$$

The mechanism of this reaction is more complex than might be suggested by its stoichiometry. *Thiamine pyrophosphate* (TPP), *lipoamide,* and *FAD* serve as catalytic cofactors, in addition to CoA and NAD^+, the stoichiometric cofactors.

There are four steps in the conversion of pyruvate to acetyl CoA. First, pyruvate is *decarboxylated* after it combines with TPP. This

Table 13-2
Pyruvate dehydrogenase complex of *E. coli*

Enzyme	Number of chains	Prosthetic group	Reaction catalyzed
Pyruvate dehydrogenase component (A)	24	TPP	Decarboxylation of pyruvate
Dihydrolipoyl transacetylase (B)	24	Lipoamide	Oxidation of the C_2 unit and transfer to CoA
Dihydrolipoyl dehydrogenase (C)	12	FAD	Regeneration of the oxidized form of lipoamide

reaction is catalyzed by the *pyruvate dehydrogenase component* of the multienzyme complex.

$$\text{Pyruvate} + \text{TPP} \longrightarrow \text{hydroxyethyl-TPP} + CO_2$$

Thiamine pyrophosphate
(TPP)

A key feature of TPP, the prosthetic group of the pyruvate dehydrogenase component, is that the carbon atom between the nitrogen and sulfur atoms in the thiazole ring is highly acidic. It ionizes to form a *carbanion*, which readily adds to the carbonyl group of pyruvate.

Pyruvate   Carbanion of TPP   Addition compound

The positively charged ring nitrogen of TPP then acts as an electron sink to stabilize the formation of a negative charge, which is necessary for decarboxylation. Protonation then gives *hydroxyethyl thiamine pyrophosphate.*

Addition compound   Resonance forms of ionized
                           hydroxyethyl-TPP

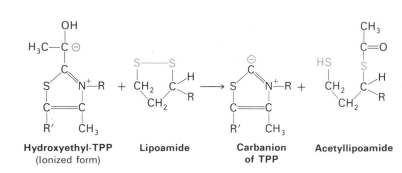

Lipoic acid
(Ionized form)

Reactive
disulfide

Lipoamide

Lysine side chain

Figure 13-5
Structure of lipoic acid and lipoamide. Lipoic acid is covalently attached to a specific lysine side chain of dihydrolipoyl transacetylase. Note that this prosthetic group is at the end of a long, flexible chain that enables it to rotate from one active site to another in the enzyme complex.

Second, the hydroxyethyl group attached to TPP is *oxidized* to an acetyl group and concomitantly *transferred to lipoamide*. This reaction, catalyzed by the *dihydrolipoyl transacetylase* part of the complex, yields *acetyllipoamide*.

Hydroxyethyl-TPP
(Ionized form)   **Lipoamide**   **Carbanion of TPP**   **Acetyllipoamide**

Third, *the acetyl group is transferred from acetyllipoamide to CoA to form acetyl CoA*. Dihydrolipoyl transacetylase also catalyzes this reaction. The energy-rich thioester bond is preserved as the acetyl group is transferred to CoA.

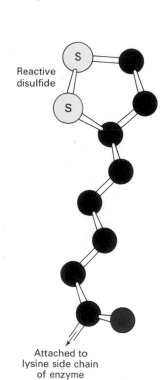

Reactive
disulfide

Attached to
lysine side chain
of enzyme

Figure 13-6
Model of the lipoyl part of lipoamide.

Acetyllipoamide   + HS—CoA ⟶   **Dihydrolipoamide**   **Acetyl CoA**

Fourth, *the oxidized form of lipoamide is regenerated* to complete the reaction. NAD^+ is the oxidant in this reaction, which is catalyzed by the *dihydrolipoyl dehydrogenase* part of the complex. FAD is the prosthetic group of this enzyme.

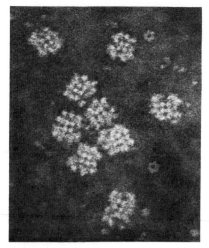

Figure 13-7
Electron micrograph of the pyruvate dehydrogenase complex from *E. coli*. [Courtesy of Dr. Lester Reed.]

The studies of Lester Reed have provided insight into the structure and assembly of the pyruvate dehydrogenase complex. The enzyme complex from *E. coli* has been studied intensively. It has a molecular weight of about 4,600,000 and consists of 60 polypeptide chains. A polyhedral structure with a diameter of about 300 Å is evident in electron micrographs (Figure 13-7). *The transacetylase polypeptide chains are the core of the pyruvate dehydrogenase complex.* The pyruvate dehydrogenase units and lipoyl dehydrogenase units appear to be located on the outside of the transacetylase core (Figure 13-8).

The constituent polypeptide chains of the complex are held together by noncovalent forces. At alkaline pH, the complex dissociates into the pyruvate dehydrogenase component and a subcomplex of the other two enzymes. The transacetylase can then be separated from the dehydrogenase at neutral pH in the presence of urea. *These three enzymes spontaneously associate to form the pyruvate dehydrogenase complex* when they are mixed at neutral pH in the absence of urea, suggesting that the native enzyme complex may be formed by a self-assembly process.

The structural integration of three kinds of enzymes makes possible the coordinated catalysis of a complex reaction. All of the intermediates in the oxidative decarboxylation of pyruvate are tightly bound to the complex. The close proximity of one enzyme to another *increases the overall reaction rate* and *minimizes side reactions.* The activated intermediates are transferred from one active site to another by the lipoamide prosthetic group of the transacetylase. The attachment of the lipoyl group to the ε-amino group of a lysine residue on the transacetylase provides a flexible arm for the reactive ring. An attractive hypothesis is that *this 14-Å molecular string enables the lipoyl*

Figure 13-8
Model of the pyruvate dehydrogenase complex from *E. coli*. [Based on a drawing kindly provided by Dr. Lester Reed.]

moiety of the transacetylase to make contact with the prosthetic groups of the adjacent enzymes. The net charge on the lipoyl moiety during its cycle of transformation is 0, -1, or -2, if the sulfhydryl groups are fully ionized. This change in net charge may provide the driving force for the directed movement of the lipoyl group.

VARIATION ON A MULTIENZYME THEME: THE ALPHA-KETOGLUTARATE DEHYDROGENASE COMPLEX

The oxidative decarboxylation of α-ketoglutarate closely resembles that of pyruvate:

$$\alpha\text{-Ketoglutarate} + CoA + NAD^+ \longrightarrow$$
$$\text{succinyl CoA} + CO_2 + NADH$$

$$\text{Pyruvate} + CoA + NAD^+ \longrightarrow$$
$$\text{acetyl CoA} + CO_2 + NADH$$

The same cofactors are involved: TPP, lipoamide, CoA, FAD, and NAD^+. In fact, *the oxidative decarboxylation of α-ketoglutarate is catalyzed by an enzyme complex that is structurally similar to the pyruvate dehydrogenase complex.* There are three kinds of enzymes in the α-ketoglutarate dehydrogenase complex: an α-ketoglutarate dehydrogenase component (A'), a transsuccinylase (B'), and a dihydrolipoyl dehydrogenase (C'). Again, A' binds to B', and B' binds to C', but A' does not bind directly to C'. Thus, *transsuccinylase* (like transacetylase) *is the core of the complex.*

The α-ketoglutarate dehydrogenase component (A') and transsuccinylase (B') are different from the corresponding enzymes (A and B) in the pyruvate dehydrogenase complex. However, *the dihydrolipoyl dehydrogenase parts (C and C') of the two complexes are similar.* Reconstitution experiments have shown that a complex formed from A, B, and C' is as active in the oxidative decarboxylation of pyruvate as one formed from A, B, and C. Similarly, C and C' are interchangeable in forming a synthetic complex that oxidatively decarboxylates α-ketoglutarate.

We previously noted that chymotrypsin, trypsin, thrombin, and elastase are homologous enzymes. Here we see that the pyruvate and α-ketoglutarate dehydrogenase complexes are *homologous enzyme assemblies.* The structural and mechanistic motifs giving coordinated catalysis in an entrée to the citric acid cycle are used again later in the cycle.

BERIBERI IS CAUSED BY A DEFICIENCY OF THIAMINE

"A certain very troublesome affliction, which attacks men, is called by the inhabitants Beri-beri (which means sheep). I believe those, whom this same disease attacks, with their knees shaking and the legs raised up, walk like sheep. It is a kind of paralysis, or rather Tremor: for it penetrates the motion and sensation of the hands and feet indeed sometimes of the whole body. . . ." Jacobus Bonitus, a Dutch physician, gave this description of beriberi in 1630, while working in Java.

Beriberi is caused by a dietary deficiency of thiamine (also called Vitamin B_1). The disease has been and continues to be a serious health problem in the Orient because rice, the major food, has a rather low content of thiamine. The problem is exacerbated if the rice is polished, since only the outer layer contains appreciable amounts of thiamine. Beriberi is also occasionally seen in alcoholics who are severely malnourished. The disease is characterized by neurologic and cardiac symptoms. Damage to the peripheral nervous system is expressed in terms of pain in the limbs, weakness of the musculature, and distorted skin sensation. The heart may be enlarged and the cardiac output may be inadequate.

How are these symptoms elicited by a deficiency of thiamine? TPP is the prosthetic group of three important enzymes: pyruvate dehydrogenase, α-ketoglutarate dehydrogenase, and transketolase. Transketolase transfers 2-carbon units from one sugar to another; its role in the pentose phosphate pathway will be discussed in Chapter 15. The common feature of enzymatic reactions utilizing TPP is the transfer of an activated aldehyde unit. In beriberi, the level of pyruvate and α-ketoglutarate in the blood is higher than normal. The increase in the level of pyruvate in the blood is especially pronounced after ingestion of glucose. It appears that the enzymatic activities of the pyruvate and α-ketoglutarate dehydrogenase complexes in vivo are abnormally low in beriberi. Furthermore, the transketolase activity of red cells is low in beriberi. This enzymatic assay is useful in the diagnosis of the disease. It is not yet known whether the reduced level of activity of these three enzymes is responsible for the clinical symptoms.

SYMMETRIC MOLECULES MAY REACT ASYMMETRICALLY

Table 13-3
Commonly used radioisotopes

Isotope	Half-life
^3H	12.26 years
^{14}C	5730 years
^{22}Na	2.62 years
^{32}P	14.28 days
^{35}S	87.9 days
^{42}K	12.36 hours
^{45}Ca	163 days
^{59}Fe	45.6 days
^{125}I	60.2 days
^{203}Hg	46.9 days

Let us follow the fate of a particular carbon atom in the citric acid cycle. Suppose oxaloacetate is labeled with ^{14}C in the carboxyl carbon furthest from the keto group. Analysis of the α-ketoglutarate that is formed shows that none of the radioactive label has been lost. Decarboxylation of α-ketoglutarate then yields succinate devoid of radioactivity. All of the label is in the released CO_2.

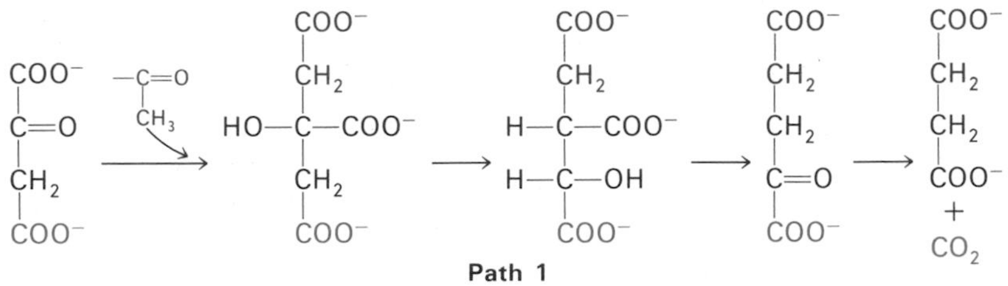

Path 1

The finding that *all* of the label emerged in the CO_2 came as a surprise. Citrate is a symmetric molecule. Consequently, it was assumed that the two $-CH_2COO^-$ groups in it would react identically. Thus, for every citrate undergoing the reactions shown in Path 1, it was thought that another citrate molecule would react as shown in Path 2. If so, then only *half* of the label should have emerged in the CO_2.

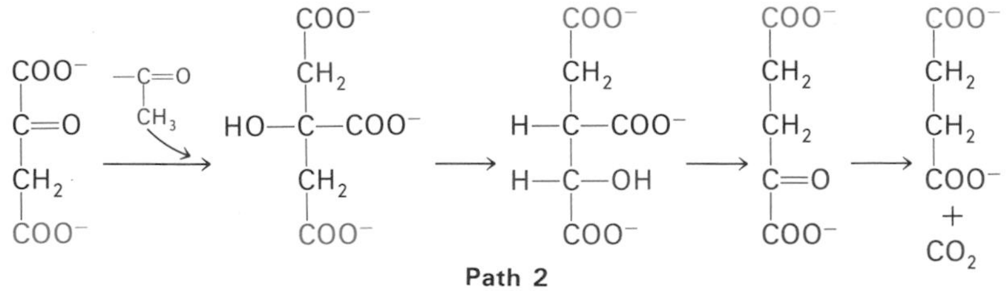

Path 2
(Does not occur)

The interpretation of these experiments, which were carried out in 1941, was that citrate (or any other symmetric compound) could not be an intermediate in the formation of α-ketoglutarate because of the asymmetric fate of the label. This interpretation seemed compelling until Alexander Ogston incisively pointed out in 1948 that it is a fallacy to assume that the two identical groups of a symmetric molecule cannot be distinguished: "On the contrary, it is possible that *an asymmetric enzyme which attacks a symmetrical compound can distinguish between its identical groups* . . . the asymmetrical occurrence of isotope in a product cannot be taken as conclusive evidence against its arising from a symmetrical precursor."

Let us examine Ogston's assertion. For simplicity, consider a molecule in which two hydrogen atoms, a group X, and a different group Y are bonded to a tetrahedral carbon atom. Let us label one hydrogen A, the other B. Now suppose an enzyme binds three groups of this substrate: X, Y, and H. Can H_A be distinguished from H_B? Figure 13-9 shows X, Y, and H_A bound to three points on the enzyme. Note that X, Y, and H_B cannot be bound to this active site; two of these three groups can be bound, but not all three. Thus, *H_A and H_B will have different fates.*

It should be noted that H_A and H_B are sterically not equivalent even though the molecule $CXYH_2$ is optically inactive. Similarly, the $—CH_2COO^-$ groups in citrate are sterically not equivalent even though citrate is optically inactive. *The symmetry rules that determine whether a compound has indistinguishable substituents are different from those that determine whether it is optically inactive:* (1) a molecule is optically inactive if it can be superimposed on its mirror image; (2) a molecule has indistinguishable substituents only if these groups can be brought into coincidence by a rotation that leaves the rest of the structure invariant.

Sterically nonequivalent groups such as H_A and H_B will almost always be distinguished in enzymatic reactions. The essence of the differentiation of these groups is that the enzyme holds the substrate in a specific orientation. Attachment at three points, as depicted in Figure 13-9, is a readily visualized way of achieving a particular orientation of the substrate, but it is not the only means of doing so.

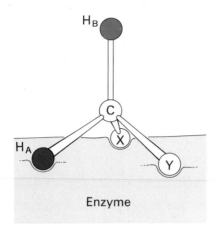

Figure 13-9
H_A and H_B are sterically not equivalent if the substrate $CXYH_2$ is bound to the enzyme at three points.

STEREOSPECIFIC TRANSFER OF HYDROGEN BY NAD⁺-DEHYDROGENASES

In the 1950s, Birgit Vennesland, Frank Westheimer, and their associates carried out some elegant experiments on the stereospecificity of hydrogen transfer by NAD⁺-dehydrogenases. Ethanol labeled with two deuterium atoms at C-1 was the substrate in a reaction catalyzed by alcohol dehydrogenase. They found that the reduced coenzyme contained one atom of deuterium per molecule, whereas the other deuterium was in acetaldehyde. Thus, none of the deuterium was lost to the solvent, showing that deuterium was directly transferred from the substrate to NAD$^+$.

$$H_3C—\overset{\overset{\displaystyle D}{|}}{\underset{\underset{\displaystyle D}{|}}{C}}—OH + NAD^+ \longrightarrow CH_3—C\overset{\displaystyle O}{\underset{\displaystyle D}{\diagdown}} + \begin{array}{c}\text{Reduced NAD}\\ \text{containing}\\ \text{1 deuterium}\end{array} + H^+$$

The deuterated reduced coenzyme formed in this reaction was then used to reduce acetaldehyde. The striking result was that *all of the deuterium was transferred from the coenzyme to the substrate*. None remained in NAD$^+$.

$$CH_3—C\overset{\displaystyle O}{\underset{\displaystyle H}{\diagdown}} + H^+ + \begin{array}{c}\text{Reduced NAD}\\ \text{containing}\\ \text{1 deuterium}\end{array} \longrightarrow \underset{\begin{array}{c}\text{(Contains}\\ \text{1 deuterium)}\end{array}}{CH_3CHDOH} + \underset{\begin{array}{c}\text{(Contains}\\ \text{0 deuterium)}\end{array}}{NAD^+}$$

NADH

These reactions revealed that the transfer catalyzed by alcohol dehydrogenase is stereospecific. *The positions occupied by the two hydrogen atoms at C-4 in NADH are not equivalent.* One of them (H_A) is in front of the nicotinamide plane, the other (H_B) is behind. Alcohol dehydrogenase distinguishes between positions A and B at C-4. Deuterium is transferred from deuterated ethanol to position A only:

$$H_3C—\overset{\overset{\displaystyle D}{|}}{\underset{\underset{\displaystyle D}{|}}{C}}—OH + \;\text{(ring)}—CONH_2 \rightleftharpoons H_3C—C\overset{\displaystyle O}{\underset{\displaystyle D}{\diagdown}} + \;\text{(ring)}—CONH_2 + H^+$$

In the reverse reaction, the deuterium atom at position A is removed and directly transferred to acetaldehyde.

Some dehydrogenases, such as glyceraldehyde phosphate dehydrogenase, transfer hydrogen to position B. Thus, *there are two classes of NAD$^+$ (and NADP$^+$) dehydrogenases: A-stereospecific and B-stereospecific.*

The stereospecificities of a large number of dehydrogenases have been determined. First, all of them are stereospecific. Second, the stereospecificity of a particular reaction is the same for enzymes obtained from all species studied thus far (e.g., yeast and horse alcohol dehydrogenase have the same stereospecificity). Third, the stereospecificity of a particular reaction is the same for NAD^+ and $NADP^+$ in those enzymes that can use either coenzyme. Fourth, when an enzyme reacts with a range of substrates, the stereospecificity of hydrogen transfer is the same for all.

CONTROL OF THE PYRUVATE DEHYDROGENASE COMPLEX AND CITRIC ACID CYCLE

The formation of acetyl CoA from pyruvate is a key irreversible step in metabolism. As expected, the activity of the pyruvate dehydrogenase complex is regulated. In fact, it is controlled in three ways:

1. *Product inhibition.* Acetyl CoA and NADH, the products of the oxidation of pyruvate, inhibit the enzyme complex. Acetyl CoA inhibits the transacetylase component, whereas NADH inhibits the dihydrolipoyl dehydrogenase component. These inhibitory effects are reversed by CoA and NAD^+, respectively.

2. *Feedback regulation by nucleotides.* The activity of the enzyme complex is controlled by the *energy charge* (p. 273). Specifically, the pyruvate dehydrogenase component is inhibited by GTP and activated by adenosine monophosphate (AMP). Again, the activity of the complex is reduced when the cell is rich in immediately available energy.

3. *Regulation by covalent modification.* The enzyme complex becomes inactive when a specific serine residue on the pyruvate dehydrogenase component is phosphorylated by ATP. This reaction is inhibited by pyruvate and ADP. The enzyme complex is inactive until the phosphoryl group is removed by a specific phosphatase.

Covalent modification is an important mechanism for controlling enzymatic activity. We will encounter phosphorylation and dephosphorylation as control reactions again in our consideration of glycogen synthesis and degradation.

The rate of the citric acid cycle is also precisely adjusted to meet the cell's needs for ATP. *The synthesis of citrate from oxaloacetate and acetyl CoA is an important control point in the cycle.* ATP is an allosteric inhibitor of citrate synthetase. The effect of ATP is to increase the K_M for acetyl CoA. Thus, as the level of ATP increases, less of this enzyme is saturated with acetyl CoA and so less citrate is formed.

A second control point is isocitrate dehydrogenase. This enzyme is allosterically stimulated by ADP, which enhances its affinity for substrates. The binding of isocitrate, NAD^+, Mg^{2+}, and ADP is mutually cooperative. In contrast, NADH inhibits isocitrate dehydrogenase by directly displacing NAD^+.

A third control site in the citric acid cycle is *α-ketoglutarate dehydrogenase.* Some aspects of its control are like those of the pyruvate dehydrogenase complex, as might be expected from their structural homology. Alpha-ketoglutarate dehydrogenase is inhibited by succinyl CoA and NADH, the products of the reaction it catalyzes. Also, α-ketoglutarate dehydrogenase is inhibited by a high energy charge. *In short, the funneling of two-carbon fragments into the citric acid cycle and the rate of the cycle are reduced when the cell has a high level of ATP.* This control is achieved by a variety of complementary mechanisms at several sites (Figure 13-10).

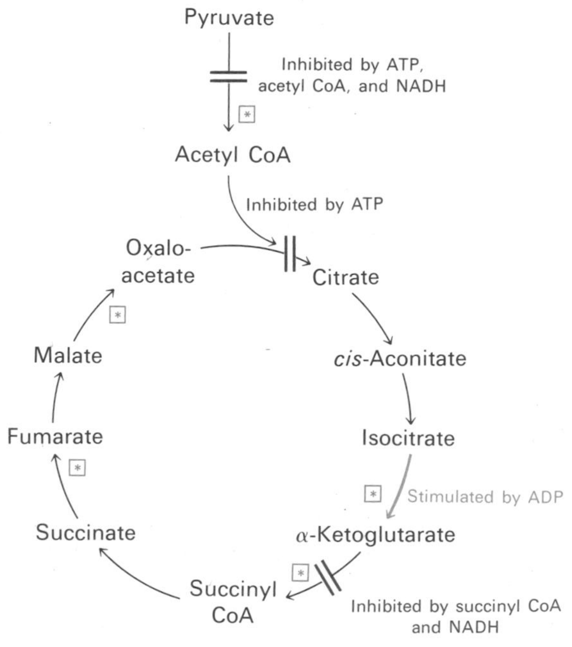

Figure 13-10
Control of the citric acid cycle and the oxidative decarboxylation of pyruvate. [*] indicates steps that require an electron acceptor (NAD^+ or FAD) that is regenerated by the respiratory chain.

KREBS' DISCOVERY OF THE CYCLE

"I have often been asked how the work on the tricarboxylic acid cycle arose and developed. Was the concept perhaps due to a sudden inspiration and vision?" Hans Krebs' reply is that "it was of course nothing of the kind, but a very slow evolutionary process, extending over some five years beginning (as far as I am involved) in 1932." Krebs first studied the rate of oxidation of a variety of compounds by using kidney and liver slices. The substances chosen were ones that might be expected to be intermediates in the oxidation of foodstuffs. The idea behind these experiments was that such intermediates would be rapidly oxidized and thereby identified. The important finding was that *citrate,* succinate, fumarate, and acetate were very readily oxidized in various tissues.

A critical contribution was made by Albert Szent-Györgyi in 1935. He studied the oxidation of various substances by using suspensions of minced pigeon-breast muscle. This very active flight muscle has an exceptionally high rate of oxidation, which facilitated the experimental studies. Szent-Györgyi found that the addition of certain C_4-dicarboxylic acids increased the uptake of O_2 far more than could be caused by their direct oxidation. In other words, they produced a catalytic (rather than a stoichiometric) enhancement of O_2 uptake. *This catalytic stimulation of respiration was obtained with succinate, fumarate, and malate.*

The next breakthrough was the elucidation of the biological pathway of oxidation of citrate by Carl Martius and Franz Knoop in 1937. They showed that citrate is isomerized to isocitrate via *cis*-aconitate and that isocitrate is oxidatively decarboxylated to α-ketoglutarate. It was already known that α-ketoglutarate can be oxidized to succinate. *Thus, their discovery revealed the pathway from citrate to succinate.* This finding came at a propitious moment for it enabled Krebs to interpret the recent observation that *citrate catalytically enhanced* the respiration of minced pigeon-breast muscle.

Additional critical information came from the use of *malonate, a specific inhibitor of succinate dehydrogenase.* Malonate is a competitive inhibitor of this enzyme because it closely resembles succinate. It was known for some time that *malonate poisons respiration.* Krebs reasoned that succinate dehydrogenase must therefore play a key role in respiration. This was reinforced by the finding that succinate accumulated when citrate was added to malonate-poisoned muscle. Furthermore, succinate also accumulated when fumarate was added

```
       COO⁻
        |
       CH₂
        |
       CH₂
        |
       COO⁻
     Succinate


       COO⁻
        |
       CH₂
        |
       COO⁻
      Malonate
```

to malonate-poisoned muscle. The first of these experiments showed that the pathway from citrate to succinate is physiologically significant. The second experiment revealed that there is a pathway from fumarate to succinate distinct from the reaction catalyzed by succinate dehydrogenase.

Krebs then discovered that citrate was readily formed by a muscle suspension if oxaloacetate was added. *The discovery of the synthesis of citrate from oxaloacetate and a substance derived from pyruvate or acetate enabled Krebs to formulate a complete scheme.* His postulated tricarboxylic acid cycle suddenly provided a coherent picture of how carbohydrates are oxidized. Many experimental facts, such as the catalytic enhancement of respiration by succinate and other intermediates, fell neatly into place. It is noteworthy that the citric acid cycle was not the only metabolic cycle, nor the first, to be discovered by Krebs. Six years earlier, he had shown that urea is synthesized by a cyclic metabolic pathway called the ornithine cycle (Chapter 18). Thus, the concept of a cyclic metabolic pathway had already been fully recognized by Krebs when he pondered the data and designed the experiments that led to the proposal of the citric acid cycle.

SUMMARY

The citric acid cycle is the final common pathway for the oxidation of fuel molecules. Most fuel molecules enter the cycle as acetyl CoA. The link between glycolysis and the citric acid cycle is the oxidative decarboxylation of pyruvate to form acetyl CoA. This reaction and those of the cycle occur inside mitochondria, in contrast to glycolysis, which occurs in the cytosol. The cycle starts with the condensation of oxaloacetate (C_4) and acetyl CoA (C_2) to give citrate (C_6), which is isomerized to isocitrate (C_6). Oxidative decarboxylation of this intermediate gives α-ketoglutarate (C_5). The second molecule of CO_2 comes off in the next reaction, in which α-ketoglutarate is oxidatively decarboxylated to succinyl CoA (C_4). The thioester bond of succinyl CoA is cleaved by P_i to yield succinate, and a high-energy phosphate bond in the form of GTP is concomitantly generated. Succinate is oxidized to fumarate (C_4), which is then hydrated to form malate (C_4). Finally, malate is oxidized to regenerate oxaloacetate (C_4). Thus, two carbon atoms from acetyl CoA enter the cycle, and two carbon atoms leave the cycle as CO_2 in the successive decarboxylations catalyzed by isocitrate dehydrogenase and α-ketoglutarate dehydrogenase. In the four oxidation-reductions in the cycle, three pairs of electrons are transferred to NAD^+ and one pair to FAD. These reduced electron carriers are subsequently oxidized by the electron-transport chain to generate eleven ATP. In addition, one high-energy phosphate bond is directly formed in the citric acid cycle. Hence, a total of twelve high-energy phosphate bonds are generated for each two-carbon fragment that is completely oxidized to H_2O and CO_2.

The citric acid cycle operates only under aerobic conditions because it requires a supply of NAD^+ and FAD. These electron acceptors are regenerated when NADH and $FADH_2$ transfer their electrons to O_2 through the electron-transport chain, with the concomitant production of ATP. Consequently, the rate of the citric acid cycle depends on the need for ATP. The regulation of three enzymes in the cycle is also important for control. A high energy charge diminishes the activity of citrate synthetase, isocitrate dehydrogenase, and α-ketoglutarate dehydrogenase. The irreversible formation of acetyl CoA from pyruvate is another important regulatory site. The activity of the pyruvate dehydrogenase complex is controlled by product inhibition, feedback regulation by nucleotides, and covalent modification. These mechanisms complement each other in reducing the rate of formation of acetyl CoA when the energy charge of the cell is high.

SELECTED READINGS

Books and general reviews:

  Lowenstein, J. M., (ed.), 1969. *Citric Acid Cycle: Control and Compartmentation.* Marcel Dekker.

  Goodwin, T. W., (ed.), 1968. *The Metabolic Roles of Citrate.* Academic Press.

  Lowenstein, J. M., 1971. The pyruvate dehydrogenase complex and the citric acid cycle. *Compr. Biochem.* 18S:1–55.

Reaction mechanisms and stereospecificity:

  Popják, G., 1970. Stereospecificity of enzymic reactions. *In* Boyer, P. D., (ed.), *The Enzymes* (3rd ed.), vol. 2, pp. 115–215. Academic Press. [An excellent review containing a discussion of the stereochemistry of the citric acid cycle.]

Ogston, A. G., 1948. Interpretation of experiments on metabolic processes using isotopic tracer elements. *Nature* 162:963.

Hirschmann, H., 1960. The nature of substrate asymmetry in stereoselective reactions. *J. Biol. Chem.* 235:2762–2767.

Pyruvate and α-ketoglutarate dehydrogenase complexes:

Reed, L. J., 1974. Multienzyme complexes. *Accounts Chem. Res.* 7:40–46.

Reed, L. J., and Oliver, R. M., 1968. The multienzyme α-keto acid dehydrogenase complexes. *Brookhaven Symp. Biol.* 21:397–412.

DeRosier, D. J., and Oliver, R. M., 1971. A low resolution electron density map of lipoyl transsuccinylase, the core of the α-ketoglutarate dehydrogenase complex. *Cold Spring Harbor Symp. Quant. Biol.* 36:199–203.

Discovery of the citric acid cycle:

Krebs, H. A., and Johnson, W. A., 1937. The role of citric acid in intermediate metabolism in animal tissues. *Enzymologia* 4:148–156.

Krebs, H. A., 1970. The history of the tricarboxylic acid cycle. *Perspect. Biol. Med.* 14:154–170.

PROBLEMS

1. What is the fate of the radioactive label when each of the following compounds is added to a cell extract containing the enzymes and cofactors of the glycolytic pathway, the citric acid cycle, and the pyruvate dehydrogenase complex? (The ^{14}C label is denoted by an asterisk.)

   (a) $\overset{*}{H_3C}$—C(=O)—COO$^-$

   (b) H_3C—C(=O)—$\overset{*}{C}OO^-$

   (c) H_3C—C(=O)—$\overset{*}{C}OO^-$

   (d) $\overset{*}{H_3C}$—C(=O)—S—CoA

   (e) Glucose 6-phosphate labeled at C-1.

2. Is it possible to get *net synthesis* of oxaloacetate by adding acetyl CoA to an extract that contains the enzymes and cofactors of the citric acid cycle?

3. What are the relative concentrations of citrate, isocitrate, and *cis*-aconitate at equilibrium? (Use the data given in Table 13-1.)

4. What is the $\Delta G^{\circ\prime}$ for the oxidation of the acetyl unit of acetyl CoA by the citric acid cycle?

5. A sample of deuterated reduced NAD was prepared by incubating H_3C—CD_2—OH and NAD^+ with alcohol dehydrogenase. This reduced coenzyme was added to a solution of 1,3-DPG and glyceraldehyde 3-phosphate dehydrogenase. The NAD^+ formed by this second reaction contained one atom of deuterium, whereas glyceraldehyde 3-phosphate, the other product, contained none. What does this experiment reveal about the stereospecificity of glyceraldehyde 3-phosphate dehydrogenase?

For additional problems, see W. B. Wood, J. H. Wilson, R. M. Benbow, and L. E. Hood, *Biochemistry: A Problems Approach* (Benjamin, 1974), Chapter 10.

CHAPTER 14

OXIDATIVE PHOSPHORYLATION

The NADH and FADH$_2$ formed in glycolysis, fatty acid oxidation, and the citric acid cycle are energy-rich molecules because they contain a pair of electrons that have a high transfer potential. When these electrons are transferred to molecular oxygen, a large amount of energy is liberated. This released energy can be used to generate ATP. *Oxidative phosphorylation is the process in which ATP is formed as electrons are transferred from NADH or FADH$_2$ to O$_2$ by a series of electron carriers.* This is the major source of ATP in aerobic organisms. For example, oxidative phosphorylation generates 32 of the 36 ATP that are formed when glucose is completely oxidized to CO$_2$ and H$_2$O. Some salient features of this process are:

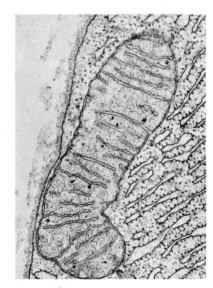

Figure 14-1
Electron micrograph of a mitochondrion. [Courtesy of Dr. George Palade.]

1. Oxidative phosphorylation is carried out by *respiratory assemblies* that are located in the *inner membrane of mitochondria.* The citric acid cycle and the pathway of fatty acid oxidation, which supply most of the NADH and FADH$_2$, take place in the adjacent mitochondrial matrix.

2. The oxidation of NADH yields 3 ATP, whereas the oxidation of FADH$_2$ yields 2 ATP. Oxidation and phosphorylation are coupled processes.

3. The respiratory assembly contains numerous electron carriers such as the cytochromes. The stepwise transfer of electrons from

NADH or FADH$_2$ to O$_2$ through these carriers divides the free-energy change of this highly exergonic reaction, thus making it possible to generate more than one ATP. In fact, there are *three distinct phosphorylation sites* in the respiratory assembly. The mechanism for the coupling of ATP formation to electron transport is one of the most challenging problems in biochemistry today.

OXIDATIVE PHOSPHORYLATION OCCURS IN MITOCHONDRIA

Mitochondria are oval-shaped organelles, typically about 2 μm in length and 0.5 μm in diameter. Techniques for the isolation of mitochondria were devised in the late 1940s. Eugene Kennedy and Albert Lehninger then discovered that *mitochondria contain the respiratory assembly, the enzymes of the citric acid cycle, and the enzymes of fatty acid oxidation.* Electron microscopic studies by George Palade and Fritjof Sjöstrand revealed that mitochondria have two membrane systems: an *outer membrane* and an extensive, highly folded *inner membrane*. The inner membrane is folded into a series of internal ridges called *cristae*. Hence, there are two compartments in mitochondria: the *intermembrane space* between the outer and inner membranes, and the *matrix,* which is bounded by the inner membrane (Figure 14-2). *The respiratory assembly is an integral part of the inner mitochondrial membrane,* whereas most of the reactions of the citric acid cycle and fatty acid oxidation occur in the matrix.

The outer membrane is quite permeable to most small molecules and ions. In contrast, the inner membrane is intrinsically impermeable to many small molecules and ions. There are specific protein

Figure 14-2
Diagram of a mitochondrion. [Based on S. L. Wolfe. *Biology of the Cell,* p. 104. © 1972 by Wadsworth Publishing Company, Inc. Adapted by permission of the publisher.]

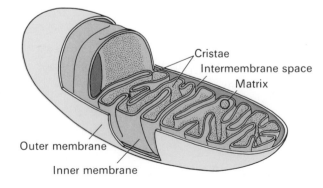

carriers that transport molecules such as ADP and long-chain fatty acids across the inner mitochondrial membrane.

REDOX POTENTIALS AND FREE-ENERGY CHANGES

In oxidative phosphorylation, the *electron transfer potential* of NADH or FADH$_2$ is converted into the *phosphate transfer potential* of ATP. We need quantitative expressions for these forms of free energy. The measure of phosphate transfer potential is already familiar to us: it is given by $\Delta G^{\circ\prime}$ for hydrolysis of the phosphate compound. The corresponding expression for the electron transfer potential is E_0', the redox potential (also called the oxidation-reduction potential).

The redox potential is an electrochemical concept. Consider a substance that can exist in an oxidized form X and a reduced form X$^-$. Such a pair is called a *redox couple* (Figure 14-3). The redox potential of this couple can be determined by measuring the electromotive force generated by a sample *half-cell* with respect to a *standard reference half-cell*. The sample half-cell consists of an electrode immersed in a solution of 1 M oxidant (X) and 1 M reductant (X$^-$). The standard reference half-cell consists of an electrode immersed in a 1 M H$^+$ solution that is in equilibrium with H$_2$ gas at 1 atmosphere pressure. The electrodes are connected to a voltmeter, and

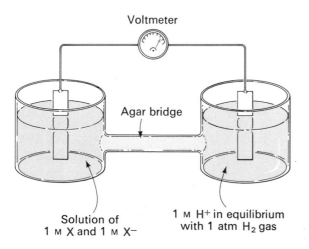

Voltmeter

Agar bridge

Solution of
1 M X and 1 M X$^-$

1 M H$^+$ in equilibrium
with 1 atm H$_2$ gas

Figure 14-3
Arrangement for the measurement of the standard oxidation-reduction potential of a redox couple.

electrical continuity between the half-cells is established by an agar bridge. Electrons then flow from one half-cell to the other. If the reaction proceeds in the direction

$$X^- + H^+ \longrightarrow X + \tfrac{1}{2}H_2$$

the reactions in the half-cells are

$$X^- \longrightarrow X + e^-$$

$$H^+ + e^- \longrightarrow \tfrac{1}{2}H_2$$

Thus, electrons flow from the sample half-cell to the standard reference half-cell, and consequently the sample-cell electrode is negative with respect to the standard-cell electrode. *The redox potential of the $X:X^-$ couple is the observed voltage at the start of the experiment* (when X, X^-, and H^+ are 1 M). *The redox potential of the $H^+:H_2$ couple is defined to be 0 volts.*

The meaning of the redox potential is now evident. A negative redox potential means that a substance has lower affinity for electrons than does H_2, as in the above example. A positive redox potential means that a substance has higher affinity for electrons than does H_2. These comparisons refer to standard conditions, namely 1 M oxidant, 1 M reductant, 1 M H^+, and 1 atmosphere H_2. Thus, *a strong reducing agent (such as NADH) has a negative redox potential, whereas a strong oxidizing agent (such as O_2) has a positive redox potential.*

The redox potentials of many biologically important redox couples are known (Table 14-1). The free-energy change of an oxidation-reduction reaction can readily be calculated from the difference in redox potentials of the reactants. For example, consider the reduction of pyruvate by NADH:

(a) Pyruvate + NADH + H^+ $\rightleftharpoons$ lactate + NAD^+

The redox potential of the $NAD^+:NADH$ couple is -0.32 volt, whereas that of the pyruvate:lactate couple is -0.19 volt. By convention, redox potentials refer to partial reactions written as: oxidant + $e^- \longrightarrow$ reductant. Hence,

(b) Pyruvate + 2 H^+ + 2 $e^- \longrightarrow$ lactate $E_0' = -0.19$ volt

(c) NAD^+ + H^+ + 2 $e^- \longrightarrow$ NADH $E_0' = -0.32$ volt

Table 14-1
Standard oxidation-reduction potentials of some reactions

Reductant	Oxidant	n	E_0' (volts)
Succinate + CO_2	α-Ketoglutarate	2	−0.67
Acetaldehyde	Acetate	2	−0.60
Ferredoxin (reduced)	Ferredoxin (oxidized)	1	−0.43
H_2	$2H^+$	2	−0.42
NADH + H^+	NAD^+	2	−0.32
NADPH + H^+	$NADP^+$	2	−0.32
Lipoate (reduced)	Lipoate (oxidized)	2	−0.29
Glutathione (reduced)	Glutathione (oxidized)	2	−0.23
Ethanol	Acetaldehyde	2	−0.20
Lactate	Pyruvate	2	−0.19
Succinate	Fumarate	2	0.03
Cytochrome b (+2)	Cytochrome b (+3)	1	0.07
Ascorbate	Dehydroascorbate	2	0.08
Ubiquinone (reduced)	Ubiquinone (oxidized)	2	0.10
Cytochrome c (+2)	Cytochrome c (+3)	1	0.22
Ferrocyanide	Ferricyanide	1	0.36
H_2O	$\frac{1}{2}O_2$ + $2H^+$	2	0.82

Note: E_0' is the standard oxidation-reduction potential (pH 7, 25°C) and n is the number of electrons transferred.

Subtracting (c) from (b) yields the desired reaction (a) and a $\Delta E_0'$ of +0.13 volt. Now we can calculate the $\Delta G^{\circ\prime}$ for the reduction of pyruvate by NADH. The standard free-energy change $\Delta G^{\circ\prime}$ is related to the change in redox potential $\Delta E_0'$ by

$$\Delta G^{\circ\prime} = -n\mathrm{F}\Delta E_0'$$

where n is the number of electrons transferred, F is the caloric equivalent of the faraday (23.062 kcal volt^{-1} mol^{-1}), $\Delta E_0'$ is in volts, and $\Delta G^{\circ\prime}$ is in kilocalories per mole. For the reduction of pyruvate, $n = 2$ and so

$$\Delta G^{\circ\prime} = -2 \times 23.062 \times 0.13$$

$$= -6 \text{ kcal/mol}$$

Note that a *positive $\Delta E_0'$ signifies an exergonic reaction.*

THE SPAN OF THE RESPIRATORY CHAIN IS 1.14 VOLTS, WHICH CORRESPONDS TO 53 KCAL

The driving force of oxidative phosphorylation is the electron transfer potential of NADH or $FADH_2$. Let us calculate the $\Delta E_0'$ and $\Delta G^{\circ\prime}$ associated with the oxidation of NADH by O_2. The pertinent partial reactions are:

(a) $\frac{1}{2} O_2 + 2\,H^+ + 2\,e^- \rightleftharpoons H_2O$ $E_0' = +0.82$ volt

(b) $NAD^+ + H^+ + 2\,e^- \rightleftharpoons NADH$ $E_0' = -0.32$ volt

Subtracting (b) from (a) yields

(c) $\frac{1}{2} O_2 + NADH + H^+ \rightleftharpoons H_2O + NAD^+$
$$\Delta E_0' = +1.14 \text{ volt}$$

The free energy of oxidation of this reaction is then given by

$$\Delta G^{\circ\prime} = -n\mathrm{F}\Delta E_0' = -2 \times 23.062 \times 1.14$$
$$= -52.6 \text{ kcal/mol}$$

FLAVIN, QUINONE, AND HEME GROUPS CARRY ELECTRONS FROM NADH TO O_2

Electrons are transferred from NADH to O_2 through a series of electron carriers: flavins, nonheme iron compounds, quinones, and hemes (Figure 14-4). Some of these carriers are prosthetic groups of proteins. The first reaction is the oxidation of NADH by *NADH dehydrogenase*, a flavoprotein that contains tightly bound *flavin mononucleotide* (FMN) as its prosthetic group. Two electrons are trans-

Figure 14-4
Sequence of electron carriers in the respiratory assembly. ATP is generated at three sites.

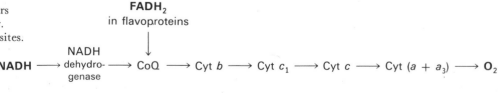

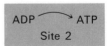

ferred from NADH to FMN to give the reduced form, FMNH$_2$.

$$\text{NADH} + \text{H}^+ + \text{FMN} \longrightarrow \text{FMNH}_2 + \text{NAD}^+$$

Flavin mononucleotide (FMN)

Reduced flavin mononucleotide (FMNH$_2$)

NADH dehydrogenase also contains iron, which probably plays a role in electron transfer. The iron is not part of a heme group. NADH dehydrogenase, like succinate dehydrogenase in the citric acid cycle, is a *nonheme iron protein*.

Electrons are then transferred from the FMNH$_2$ moiety of NADH dehydrogenase to *coenzyme Q* (CoQ).

$$\text{FMNH}_2 + \text{oxidized CoQ} \longrightarrow \text{FMN} + \text{reduced CoQ}$$

Coenzyme Q is a quinone derivative with a long isoprenoid tail. It is also called *ubiquinone* because it is ubiquitous in biological systems. The number of isoprene units in CoQ depends on the species. The most common form in mammals contains ten isoprene units, and so it is designated CoQ$_{10}$.

Oxidized form of coenzyme Q$_{10}$ (Oxidized CoQ$_{10}$)

Reduced form of coenzyme Q$_{10}$ (Reduced CoQ$_{10}$)

The isoprenoid tail makes CoQ highly nonpolar, which enables it to diffuse rapidly in the hydrocarbon phase of the mitochondrial inner membrane. In fact, *CoQ serves as a highly mobile carrier of electrons between the flavoproteins and the cytochromes of the electron-transport chain.* Recall that $FADH_2$ was formed in the citric acid cycle in the oxidation of succinate to fumarate by succinate dehydrogenase. The high-potential electrons of $FADH_2$ are then transferred to CoQ for entry into the electron-transport chain. Likewise, glycerol phosphate dehydrogenase (p. 342) and fatty acyl CoA dehydrogenase (p. 411) transfer their high-potential electrons to CoQ.

The electron carriers between CoQ and O_2 are *cytochromes* (except perhaps for one nonheme iron protein). The central role of the cytochromes in respiration was discovered in 1925 by David Keilin. *A cytochrome is an electron-transporting protein that contains a heme prosthetic group.* The iron atom in cytochromes alternates between a reduced ferrous ($+2$) state and an oxidized ferric ($+3$) state during electron transport. *The heme group is a one-electron carrier,* in contrast with NADH, flavins, and CoQ, which are two-electron carriers, Thus a molecule of reduced CoQ must transfer its two high-potential electrons to two molecules of cytochrome b, the next member of the electron-transport chain.

There are five cytochromes between CoQ and O_2 in the electron-transport chain. Their redox potentials increase sequentially.

$$CoQ \longrightarrow Cyt\ b \longrightarrow Cyt\ c_1 \longrightarrow$$
$$Cyt\ c \longrightarrow Cyt\ (a + a_3) \longrightarrow O_2$$

These cytochromes have distinctive structures and properties. The prosthetic group of cytochromes b, c_1, and c is iron-protoporphyrin IX, commonly called *heme,* which is the same prosthetic group as in myoglobin and hemoglobin. In cytochrome b, the heme is not covalently bonded to the protein, whereas in cytochromes c and c_1, the heme is covalently attached to the protein by thioether linkages (Figure 14-5). These linkages are formed by the addition of the sulfhydryl groups of two cysteine residues to the vinyl groups of the heme.

$$R-CH=CH_2 \ + \ HS-CH_2-R' \ \longrightarrow \ R-\overset{\overset{\displaystyle CH_3}{|}}{C}H-S-CH_2-R'$$

Vinyl group   **Cysteine residue**   **Thioether linkage**
of the heme   **of the protein**

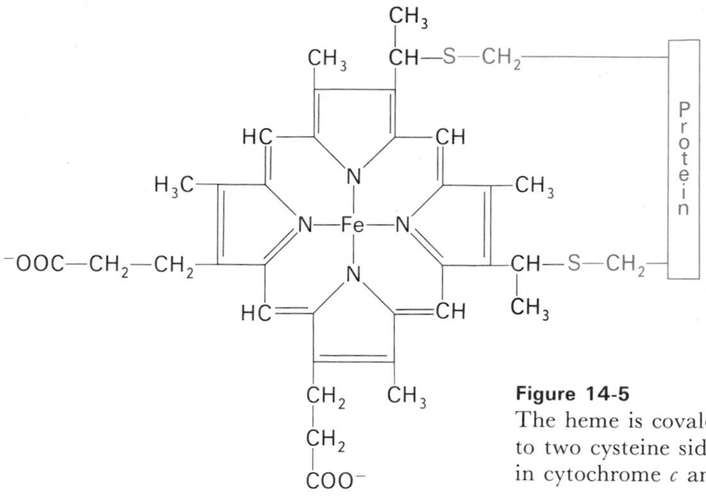

Figure 14-5
The heme is covalently attached to two cysteine side chains in cytochrome c and c_1.

Cytochromes a and a_3 have a different iron-porphyrin prosthetic group, called *heme A*. It differs from the heme in cytochromes c and c_1 in that a formyl group replaces one of the methyl groups, and a hydrocarbon chain replaces one of the vinyl groups.

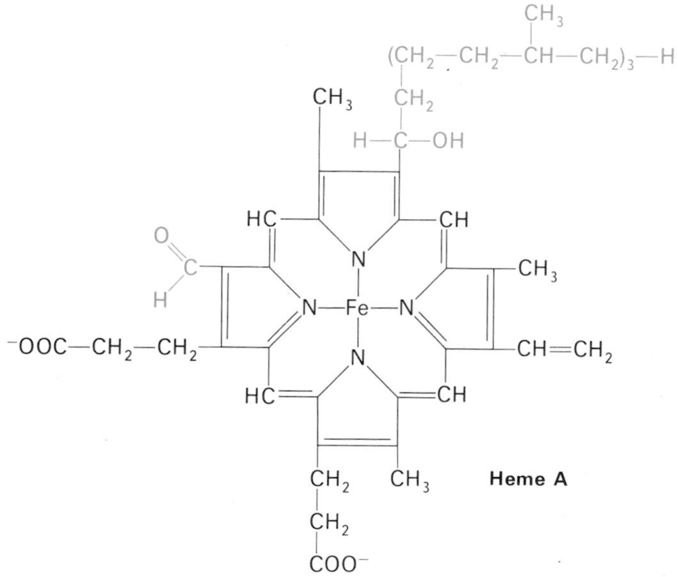

Heme A

Cytochromes a and a_3 are the terminal members of the respiratory chain. They exist as a complex, which is sometimes called *cytochrome oxidase.* Electrons are transferred to the cytochrome a moiety of the complex, and then to cytochrome a_3, which contains copper. This copper atom alternates between a $+2$ oxidized form and a $+1$ reduced form as it transfers electrons from cytochrome a_3 to molecular oxygen. The formation of water is a four-electron process, whereas heme groups are one-electron carriers. It is not known how four electrons converge to reduce a molecule of O_2.

$$O_2 + 4\,H^+ + 4\,e^- \longrightarrow 2\,H_2O$$

ATP IS GENERATED AT THREE SITES

ATP is formed at three sites as electrons flow through the electron-transport chain from NADH to O_2 (Figure 14-4). *Site 1* is between NADH and CoQ; *site 2* is between cytochrome b and cytochrome c; and *site 3* is between cytochrome c and O_2. These sites have been identified by several experimental approaches:

1. *Comparison of the ATP yield from the oxidation of several substrates.* The oxidation of NADH yields 3 ATP, whereas the oxidation of succinate yields 2 ATP. The electrons from $FADH_2$ enter the electron-transport chain at CoQ, which is at a lower energy level than phosphorylation site 1. Only 1 ATP is formed when ascorbate, an artificial substrate, is oxidized, since its electrons enter at cytochrome c, which is at a lower energy level than phosphorylation site 2. The *P:O ratio*, defined as the number of moles of inorganic phosphate incorporated into organic form per atom of oxygen consumed, is a frequently used index of oxidative phosphorylation. The P:O ratio for the oxidation of NADH, succinate, and ascorbate is 3, 2, and 1, respectively.

2. *Thermodynamic estimates.* The $\Delta G°'$ for electron transfer from NADH to the flavin moiety of NADH dehydrogenase is -12 kcal/mol; from cytochrome b to cytochrome c_1, -10 kcal/mol; and from cytochrome $(a + a_3)$ to O_2, -24 kcal/mol. These oxidation-reduction reactions are sufficiently exergonic to drive the synthesis of ATP under standard conditions (which requires 7.3 kcal/mol). The $\Delta G°'$ of the other electron transfer reactions appears to be too small to sustain ATP synthesis under standard conditions.

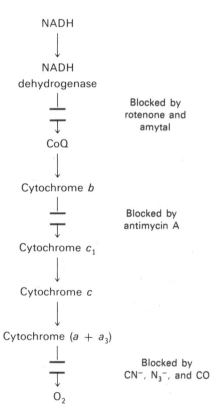

Figure 14-6
Sites of action of some inhibitors of electron transport.

3. *Specific inhibition of electron flow.* *Rotenone* specifically inhibits electron transfer from NADH dehydrogenase to CoQ, which prevents ATP synthesis at site 1. In contrast, rotenone does not inhibit the oxidation of succinate since the electrons of this substrate enter the electron-transport chain at CoQ beyond the block. *Antimycin A* inhibits electron flow between cytochrome b and c_1, which prevents ATP synthesis at site 2. This block can be bypassed by the addition of ascorbate, which directly reduces cytochrome c. Electrons then flow from cytochrome c to O_2, with the concomitant synthesis of ATP at site 3. Finally, electron flow can be blocked between cytochrome $(a + a_3)$ and O_2 by CN^-, N_3^-, and CO. Cyanide and azide react with the ferric form of this carrier, whereas CO inhibits the ferrous form. Since electron flow is blocked, phosphorylation does not occur at site 3.

The sites of action of these inhibitors were revealed by the *crossover technique.* Britton Chance devised some elegant spectroscopic methods for determining the proportions of the oxidized and reduced forms of each carrier. This is feasible because each carrier has a characteristic absorption spectrum for its oxidized and reduced forms. The addition of an electron-transport inhibitor leads to a change in the proportion of the oxidized and reduced forms of each carrier. For example, addition of antimycin A causes the carriers between NADH and cytochrome b to become more reduced. In contrast, the carriers between cytochrome c and O_2 become more oxidized. Hence, it can be concluded that antimycin A inhibits the conversion of cytochrome b to cytochrome c_1, since this step is the *crossover point.*

4. *Reconstitution of particles that contain only one phosphorylation site.* This experimental approach will be discussed shortly (p. 348).

ELECTRONS FROM CYTOPLASMIC NADH ENTER MITOCHONDRIA BY THE GLYCEROL PHOSPHATE SHUTTLE

Intact mitochondria are impermeable to NADH and NAD^+. How then does cytoplasmic NADH get oxidized by the respiratory chain? NADH is formed in glycolysis in the oxidation of glyceraldehyde 3-phosphate. NAD^+ must be regenerated for glycolysis to continue. The solution is that *electrons from NADH,* rather than NADH itself, are carried across the mitochondrial membrane. One carrier is *glycerol 3-phosphate,* which readily traverses the mitochondrial membrane. The first step in this shuttle (Figure 14-7) is the transfer of electrons from NADH to dihydroxyacetone phosphate to form glycerol 3-phosphate. This reaction, catalyzed by glycerol 3-phosphate dehydrogenase, occurs in the cytosol. Glycerol 3-phosphate then enters mitochondria, where it is reoxidized to dihydroxyacetone phosphate by the FAD prosthetic group of a dehydrogenase. This mitochondrial FAD-linked glycerol dehydrogenase is different from the NAD^+-linked glycerol dehydrogenase in the cytosol.

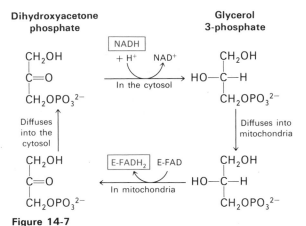

Figure 14-7
Glycerol phosphate shuttle.

The dihydroxyacetone phosphate formed in this reaction then diffuses out of the mitochondria into the cytosol to complete the shuttle. The net reaction is:

$$NADH + H^+ + \text{E-FAD} \longrightarrow NAD^+ + \text{E-FADH}_2$$

| Cyto-plasmic | Mito-chondrial | Cyto-plasmic | Mito-chondrial |

The reduced flavin inside the mitochondria transfers its electrons to the respiratory chain at the level of CoQ. *Consequently, two rather than three ATP are formed when cytoplasmic NADH is oxidized by the respiratory chain.* At first glance, this shuttle may appear to waste one ATP per cycle. This lower yield arises because FAD rather than

NAD^+ is the electron acceptor in mitochondrial glycerol 3-phosphate dehydrogenase. This feature of the shuttle is a necessity, not a luxury, because the concentration of NADH inside the mitochondrion is higher than in the cytosol. Hence, electrons would flow the wrong way if both the mitochondrial and the cytoplasmic glycerol 3-phosphate dehydrogenases were NAD^+-linked. The use of FAD enables electrons from cytoplasmic NADH to be transported into mitochondria against an NADH concentration gradient. The price of this transport is one ATP per two electrons.

THE ENTRY OF ADP INTO MITOCHONDRIA REQUIRES THE EXIT OF ATP

ATP and ADP do not diffuse freely across the inner mitochondrial membrane. Rather, there is a specific carrier that enables these highly charged molecules to traverse the permeability barrier. An interesting property of the carrier is that the flows of ATP and ADP are coupled. *ADP enters the mitochondrial matrix only if ATP exits, and vice versa.* The flows of ATP and ADP are down their concentration gradients. No energy is spent in this transport process, in contrast to the glycerol phosphate shuttle. The coupled flow of ATP and ADP mediated by a carrier is called *facilitated exchange diffusion.* This carrier can be specifically inhibited by *atractyloside,* a plant poison. Oxidative phosphorylation stops soon after the addition of atractyloside because the supply of ADP inside the mitochondrion cannot be replenished.

THE COMPLETE OXIDATION OF GLUCOSE YIELDS 36 ATP

We can now calculate how many ATP are formed when glucose is completely oxidized (Table 14-2). The overall reaction is:

$$Glucose + 36\,ADP + 36\,P_i + 36\,H^+ + 6\,O_2 \longrightarrow$$
$$6\,CO_2 + 36\,ATP + 42\,H_2O$$

The *P:O ratio is* 3 since 36 ATP are formed and 12 atoms of oxygen are consumed. The vast majority of the ATP, 32 out of 36, are generated by oxidative phosphorylation.

344

Table 14-2
ATP yield from the complete oxidation of glucose

Reaction sequence	ATP yield per glucose
Glycolysis: glucose to pyruvate (in the cytosol)	
Phosphorylation of glucose	−1
Phosphorylation of fructose 6-phosphate	−1
Dephosphorylation of 2 molecules of 1,3-DPG	+2
Dephosphorylation of 2 molecules of phosphoenolpyruvate	+2
2 NADH are formed in the oxidation of 2 molecules of glyceraldehyde 3-phosphate	
Conversion of pyruvate to acetyl CoA (inside mitochondria)	
2 NADH are formed	
Citric acid cycle (inside mitochondria)	
Formation of 2 molecules of guanosine triphosphate from 2 molecules of succinyl CoA	+2
6 NADH are formed in the oxidation of 2 molecules of isocitrate, α-ketoglutarate, and malate	
2 FADH$_2$ are formed in the oxidation of 2 molecules of succinate	
Oxidative phosphorylation (inside mitochondria)	
2 NADH formed in glycolysis; each yields 2 ATP (not 3 ATP each, because of the cost of the shuttle)	+4
2 NADH formed in the oxidative decarboxylation of pyruvate; each yields 3 ATP	+6
2 FADH$_2$ formed in the citric acid cycle; each yields 2 ATP	+4
6 NADH formed in the citric acid cycle; each yields 3 ATP	+18
NET YIELD PER GLUCOSE	+36

The overall efficiency of ATP generation is high. The oxidation of glucose yields 686 kcal under standard conditions:

$$\text{Glucose} + 6\,O_2 \longrightarrow 6\,CO_2 + 6\,H_2O \qquad \Delta G^{\circ\prime} = -686 \text{ kcal}$$

The free energy stored in 36 ATP is 263 kcal, since $\Delta G^{\circ\prime}$ for hydrolysis of ATP is -7.3 kcal. Hence, the thermodynamic efficiency of ATP formation from glucose is 263/686 or 38%, under standard conditions.

The *respiratory quotient* (RQ), a frequently used index in metabolic studies of whole organisms, is defined as

$$RQ = \frac{\text{moles of } CO_2 \text{ produced}}{\text{moles of } O_2 \text{ consumed}}$$

The RQ is 1.0 for the complete oxidation of carbohydrate. For fats and proteins, it is about 0.71 and 0.80, respectively. Thus, the RQ can be used as an index of the relative utilization of carbohydrate, fat, and protein by an organism.

THE RATE OF OXIDATIVE PHOSPHORYLATION IS DETERMINED BY THE NEED FOR ATP

Under physiological conditions, electron transport is tightly coupled to phosphorylation. *Electrons do not normally flow through the electron-transport chain to O_2 unless ADP is simultaneously phosphorylated to ATP.* Oxidative phosphorylation requires a supply of NADH (or other source of electrons at high potential), O_2, ADP, and P_i. The most important factor in determining the rate of oxidative phosphorylation is the *level of ADP*. The rate of oxygen consumption of a tissue homogenate increases markedly when ADP is added and then returns to its initial value when the added ADP has been converted to ATP (Figure 14-8).

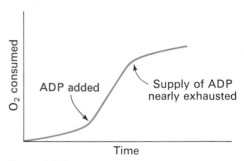

Figure 14-8
Respiratory control. Electrons are transferred to O_2 only if ADP is phosphorylated to ATP.

The regulation of the rate of oxidative phosphorylation by the ADP level is called *respiratory control*. The physiological significance of this regulatory mechanism is evident. The ADP level increases when ATP is consumed, and so oxidative phosphorylation is coupled to the utilization of ATP. *Electrons do not flow from fuel molecules to O_2 unless ATP needs to be synthesized.*

DINITROPHENOL UNCOUPLES OXIDATIVE PHOSPHORYLATION

The tight coupling of electron transport and phosphorylation is disrupted by 2,4-dinitrophenol (DNP) and some other acidic aromatic compounds. In the presence of these uncouplers, electron transport from NADH to O_2 proceeds normally but ATP is not formed by the respiratory assembly. The loss of respiratory control

**2,4-Dinitrophenol
(DNP)**

leads to increased oxygen consumption and oxidation of NADH. In contrast, DNP has no effect on substrate-level phosphorylations. DNP and other uncouplers are very useful tools in metabolic studies because of their specific effect on the respiratory chain.

The uncoupling of oxidative phosphorylation can be biologically useful. *It is a means of generating heat to maintain body temperature in hibernating animals, some newborn animals, and in mammals adapted to cold.* Brown adipose tissue, which is very rich in mitochondria, is specialized for this process of *thermogenesis.* Fatty acids act as uncouplers in brown adipose tissue. In turn, norepinephrine controls the release of fatty acids. Thus, the degree of uncoupling of oxidative phosphorylation in brown adipose tissue is under hormonal control. The mitochondria in this tissue can serve as ATP generators or as miniature furnaces.

There is an interesting case report of a 38-year-old woman who was incapable of performing prolonged physical work. Her basal metabolic rate was more than twice normal, but her thyroid function was normal. A muscle biopsy showed that her mitochondria were highly variable and atypical in structure. Biochemical studies then revealed that these mitochondria were not subject to respiratory control. NADH was oxidized irrespective of whether ADP was present. In other words, *oxidation and phosphorylation were not tightly coupled in these mitochondria.* Furthermore, their P:O ratio was below normal. Thus, in this patient, much of the energy of fuel molecules was converted into heat rather than ATP. The molecular defect in these mitochondria has not yet been elucidated.

PROPOSED MECHANISMS FOR OXIDATIVE PHOSPHORYLATION

The mechanism of oxidative phosphorylation is not yet known. Three kinds of mechanisms have been proposed:

1. In the *chemical-coupling hypothesis,* electron transfer leads to the formation of a covalent *high-energy intermediate* that serves as a precursor of ATP. The simplest scheme is one in which a high-energy compound denoted as A $\sim$ C is formed when the reduced form of A transfers its electrons to the oxidized form of B (A and B are electron carriers in the respiratory chain, whereas C is some

other component). Then, the high-energy bond in A $\sim$ C is split as ADP is phosphorylated to ATP.

$$A_{red} + B_{ox} + C \longrightarrow A_{ox} \sim C + B_{red}$$
$$A_{ox} \sim C + ADP + P_i \longrightarrow A_{ox} + C + ATP$$
$$\text{Sum:} \quad A_{red} + B_{ox} + ADP + P_i \longrightarrow A_{ox} + B_{red} + ATP$$

This model is based on the mechanism of substrate-level phosphorylation, as in the glyceraldehyde 3-phosphate dehydrogenase reaction, in which 1,3-DPG is $A_{ox} \sim C$.

2. *The conformational hypothesis* is a variant of the chemical-coupling hypothesis. The covalent high-energy intermediate $A_{ox} \sim C$ of the chemical-coupling hypothesis is replaced by a *conformationally activated state* A_{ox}^* of an electron carrier. The simplest version of this hypothesis proposes that the energy released in the electron-transfer step is trapped in an activated protein conformation A_{ox}^*, which then drives the synthesis of ATP.

$$A_{red} + B_{ox} \longrightarrow A_{ox}^* + B_{red}$$
$$A_{ox}^* + ADP + P_i \longrightarrow A_{ox} + ATP$$
$$\text{Sum:} \quad A_{red} + B_{ox} + ADP + P_i \longrightarrow A_{ox} + B_{red} + ATP$$

3. The *chemiosmotic hypothesis* proposes that the primary energy conserving event is the *movement of protons across the mitochondrial inner membrane*. The resulting proton concentration gradient and membrane potential drive the synthesis of ATP. One form of this hypothesis, which has been developed by Peter Mitchell, postulates the following molecular organization and reaction sequence:

(a) The electron carriers in the respiratory chain are *vectorially organized*—that is, they are oriented with respect to the two faces of the inner mitochondrial membrane. Electron flow through the respiratory chain results in the *unidirectional release of H^+*. Protons accumulate on the outer face of the membrane, which consequently acquires a positive charge (Figure 14-9).

(b) *A covalent high-energy compound $X \sim I$ is formed at the outer face of the membrane. Components X^- and IO^- diffuse to the positively charged outer face and form $X \sim I$ in a dehydration reaction that utilizes the H^+ ions that were extruded during electron transport.

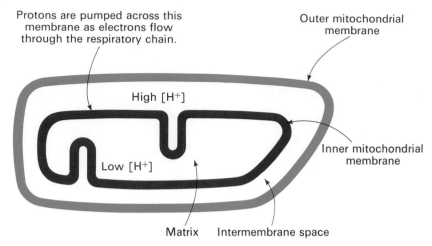

Figure 14-9
The formation of a proton gradient across the inner mitochondrial membrane is the primary energy-conserving event in the chemiosmotic hypothesis.

(c) The energy-rich $X \sim I$ intermediate then diffuses to the matrix side of the membrane. *ATP is formed as $X \sim I$ is split.* The enzyme catalyzing this reaction is oriented so that protons are released on the matrix side of the membrane.

$$X \sim I + ADP + P_i \longrightarrow ATP + X^- + IO^- + 2\,H^+$$

The distinguishing features of the chemiosmotic hypothesis are the proposed *directed movement of protons and the occurrence of a proton and electrical gradient across the inner mitochondrial membrane.* The postulated reactions involving $X \sim I$ are similar to those of the chemical-coupling hypothesis.

SUBMITOCHONDRIAL PARTICLES AND RECONSTITUTED ASSEMBLIES HAVE BEEN HIGHLY INFORMATIVE

Studies of submitochondrial particles and reconstituted membrane assemblies by Efraim Racker have provided insight into the mechanism of oxidative phosphorylation. Mitochondria subjected to sonic oscillation became fragmented into *submitochondrial particles,* which are vesicles formed from the inner mitochondrial membrane.

These particles can carry out oxidative phosphorylation. The large spherical projections on the external surface of these particles, called F_1, can be removed by treatment with urea. The modified sub-mitochondrial particles without F_1 can transfer electrons through their electron-transport chain, but they cannot form ATP. The subsequent addition of F_1 restores the capacity of these particles to form ATP. Hence, F_1 is called a *coupling factor*. Isolated F_1 molecules have ATPase activity, which suggests that *the reaction catalyzed by F_1 in the intact system is the synthesis of ATP.* Several other coupling factors (called F_2, F_3, F_5, F_6, and CF_0) have also been isolated. In contrast to F_1, they do not exhibit any catalytic activity.

The reconstitution of two phosphorylation sites has recently been achieved. Cytochromes *c, a,* and a_3 were purified and then incorporated with F_1 and the other coupling factors into synthetic phospholipid vesicles. This reconstituted system carried out site-3 phosphorylation. The site-1 oxidative phosphorylation system has also been reconstituted from its components.

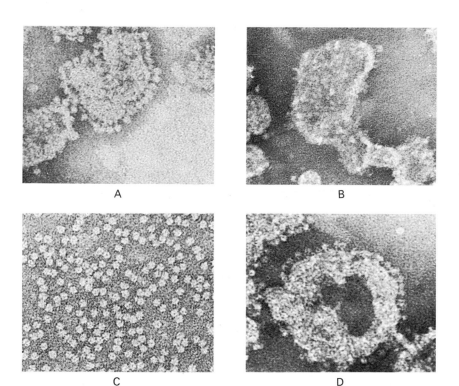

A

B

C

D

Figure 14-10
Electron micrographs of (A) a submitochondrial particle showing the F_1 projections on its surface, (B) a submitochondrial particle treated with urea, which has removed the F_1 projections, (C) isolated F_1 units, and (D) a reconstituted submitochondrial particle formed by adding F_1 (a coupling factor) to stripped membranes. The particle shown in part B can transfer electrons to O_2 but it cannot form ATP. The reconstituted particle shown in part D carries out oxidative phosphorylation. [Courtesy of Dr. Efraim Racker.]

THREE-DIMENSIONAL STRUCTURE OF CYTOCHROME *C*

The elucidation of the mechanism of oxidative phosphorylation will probably require detailed information concerning the structure of the respiratory assembly. At present, rather little is known about the conformation of the electron carriers and coupling factors. Most of the proteins are tenaciously bound to the inner mitochondrial membrane. Their insolubility in aqueous media has made them difficult to purify and crystallize. The exception is cytochrome *c*, which is water-soluble and easily separable from the inner membrane. Cytochrome *c* consists of a single polypeptide chain of 104 amino acid residues and a covalently attached heme group. The three-dimensional structures of the ferrous and ferric forms of cytochrome *c* have been elucidated at nearly atomic resolution by Richard Dickerson (Figure 14-11). The protein is roughly spherical, with a diameter of 34 Å. The heme group is surrounded by many tightly packed hydrophobic side chains. The iron atom is bonded to the sulfur atom of a methionine residue and to the nitrogen atom of a histidine residue (Figure 14-12).

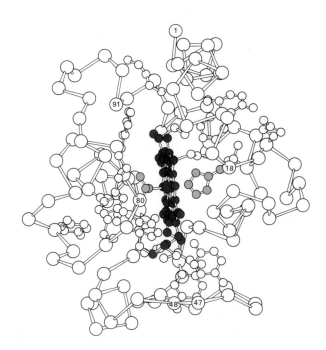

Figure 14-11
Three-dimensional structure of reduced cytochrome *c* from tuna. The heme group and the *a*-carbon atoms are shown. [Based on T. Takano, O. B. Kallai, R. Swanson, and R. E. Dickerson. *J. Biol. Chem.* 248 (1973):5244.]

The overall structure of the molecule can be characterized as a shell one residue thick surrounding the heme. Hydrophobic side chains make up the innermost part of the shell. The main chain comes next, followed by charged side chains on the surface. There is very little α helix and no β pleated sheet. In essence, the polypeptide chain is wrapped around the heme. Residues 1 to 47 are on the histidine 18 side of the heme (called the right side), whereas 48 to 91 are on the methionine 80 side (called the left side). Residues 92 to 104 come back across the heme to the right side.

The distribution of charged residues on the surface of the molecule is striking. These groups are clustered into two positively charged patches with an intervening negatively charged patch. This clustering of charged residues does not occur in any other protein of known structure. The reason for this difference can be surmised. The surfaces of proteins such as myoglobin, lysozyme, and chymotrypsin make them soluble in water, but are otherwise relatively unimportant. In contrast, the detailed nature of the surface of cytochrome c is significant because this molecule interacts with at least two other components of the electron-transport chain—cytochrome reductase (i.e., cytochromes b and c_1) and cytochrome oxidase (i.e., cytochromes a and a_3). *The distribution of charges on the surface of cytochrome* c *almost certainly plays a role in the recognition and binding of these complexes.* For example, the interaction of cytochrome c with cytochrome oxidase is impaired if lysine 13 is modified. This finding suggests that the positively charged cluster on the right side of the molecule may participate in the binding of cytochrome oxidase. Detailed comparisons of the structures of the ferrous and ferric forms, now in progress, should provide valuable clues concerning the mechanisms of electron transfer by this carrier.

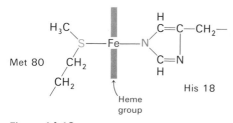

Figure 14-12
The iron atom of the heme group in cytochrome c is bonded to a methionine and a histidine side chain.

THE CONFORMATION OF CYTOCHROME *C* HAS REMAINED ESSENTIALLY CONSTANT FOR A BILLION YEARS

Cytochrome c is present in all organisms that contain a mitochondrial respiratory chain: plants, animals, and eucaryotic microorganisms. This electron carrier evolved more than 1.5 billion years ago, before the divergence of plants and animals. The function of this protein has been conserved throughout this period as evidenced by the fact that *the cytochrome* c *of any eucaryotic species reacts in vitro with the cytochrome oxidase of any other species tested thus far.* For example, wheat-germ cytochrome *c* reacts with human cytochrome oxidase. A second criterion of the conservation of function is that the redox potentials of all cytochrome *c* molecules studied are close to $+0.25$ volt. Third, the absorption spectra of cytochrome *c* molecules of many species are virtually indistinguishable.

The amino acid sequences of cytochrome *c* from more than thirty widely ranging species have been determined by Emil Smith, Emanuel Margoliash, and others. The striking finding is that *35 of 104 residues have been invariant for more than a billion and a half years of evolution.* The reasons for the absolute constancy of many of these residues are evident now that the three-dimensional structure of the molecule is known. As might be expected, the heme ligands methionine 80 and histidine 18 are invariant, as are the two cysteines that are covalently bonded to the heme. A sequence of 11 residues from residue 70 to 80 is the same in all cytochrome *c* molecules. These residues may participate in the conformational change that accompanies electron transfer. Several aromatic residues have remained constant; some of them may participate directly in electron transfer. Many hydrophobic residues in contact with the heme are invariant. Furthermore, most of the glycine residues in cytochrome *c* have been preserved. As discussed previously (p. 66), glycine is important because it is small. The compact folding of the polypeptide chain requires the presence of glycine at certain sites. Several invariant lysine and arginine residues are located in the positively charged clusters on the surface of the molecule. One of these positively charged clusters probably interacts with cytochrome oxidase.

SUMMARY

Oxidative phosphorylation is the process in which ATP is formed as electrons are transferred from NADH or $FADH_2$ to O_2 by a series of electron carriers. The respiratory assembly carrying out this process is an integral part of the inner mitochondrial membrane. Two electrons from NADH are transferred to O_2 by a series of electron carriers, starting with the FMN prosthetic group of NADH dehydrogenase. These electrons are transferred to CoQ, a quinone, and then to the cytochromes, which are a group of heme proteins. There are five kinds of cytochromes between CoQ and O_2 in the electron-transport chain: cytochromes b, c_1, c, a, and a_3. The complex of cytochromes a and a_3 is sometimes called cytochrome oxidase. ATP is generated at three sites as electrons flow from NADH to O_2. Site 1 is between NADH and CoQ, site 2 is between cytochromes b and c, and site 3 is between cytochrome c and O_2. Three ATP are generated per NADH oxidized, whereas only two ATP are formed per $FADH_2$ oxidized because the electrons from this carrier enter the chain at CoQ, after the first phosphorylation site. Also, only two ATP are generated in the oxidation of NADH formed in the cytosol because one ATP is lost when the glycerol phosphate shuttle carries these electrons into mitochondria. The entry of ADP into mitochondria is coupled to the exit of ATP by a process called facilitated exchange diffusion. Thirty-six ATP are generated when a molecule of glucose is completely oxidized to CO_2 and H_2O.

Electron transport is normally tightly coupled to phosphorylation. NADH and $FADH_2$ are oxidized only if ADP is simultaneously phosphorylated to ATP. This coupling, called respiratory control, can be disrupted by uncouplers such as DNP. The mechanism of coupling is not yet known. Three coupling mechanisms have been proposed: chemical, conformational, and chemiosmotic. The distinctive feature of the chemiosmotic hypothesis is its proposal that the movement of protons across the inner membrane of mitochondria is the primary energy-conserving event in oxidative phosphorylation. Submitochondrial particles and reconstituted assemblies are providing insight into the molecular basis of oxidative phosphorylation.

SELECTED READINGS

Where to start:

Racker, E., 1974. Inner mitochondrial membranes: basic and applied aspects. *Hosp. Pract.* 9:87–93. [A lively introduction to oxidative phosphorylation.]

Racker, E., 1968. The membrane of the mitochondrion. *Sci. Amer.* 218(2):32–39. [Available as *Sci. Amer.* Offprint 1101.]

Dickerson, R. E., 1972. The structure and history of an ancient protein. *Sci. Amer.* 226(4):58–72. [Offprint 1245. A beautifully illustrated account of the conformation and evolution of cytochrome *c*.]

Books:

Racker, E., (ed.), 1971. *Membranes of Mitochondria and Chloroplasts*. Reinhold.

Ernster, L., and Drahota, Z., (eds.), 1969. *Mitochondrial Structure and Function*. Academic Press.

Sato, S., (ed.), 1972. *Mitochondria*. University Park Press. [Reprints of many important papers on mitochondria and oxidative phosphorylation, including contributions by G. E. Palade, A. L. Lehninger, B. Chance, P. Mitchell, and E. Racker.]

Okunuki, K., Kamen, M. D., and Sekuzu, I., (eds.), 1968. *Structure and Function of Cytochromes*. University Park Press.

Lehninger, A. L., 1966. *The Mitochondrion: Molecular Basis of Structure and Function*. Benjamin.

Reviews:

Slater, E. C., 1971. The coupling between energy-yielding and energy-utilizing reactions in mitochondria. *Quart. Rev. Biophys.* 4:35–71.

Wilson, D. F., Erecinska, M., and Dutton, P. L., 1974. Thermodynamic relationships in mitochondrial oxidative phosphorylation. *Annu. Rev. Biophys. Bioeng.* 3:203–230.

van Dam, K., and Meyer, A. J., 1971. Oxidation and energy conservation by mitochondria. *Annu. Rev. Biochem.* 40:115–160.

Greville, G. D., 1969. A scrutiny of Mitchell's chemiosmotic hypothesis of respiratory chain and photosynthetic phosphorylation. *Curr. Top. Bioenergetics* 3:1–78.

Defective respiratory control:

Luft, R., Ikkos, D., Palmieri, G., Ernster, L., and Afzelius, B., 1962. A case of severe hypermetabolism of nonthyroid origin with a defect in the maintenance of mitochondrial respiratory control: a correlated clinical, biochemical, and morphological study. *J. Clin. Invest.* 41:1776–1804.

Historical aspects:

Keilin, D., 1966. *The History of Cell Respiration and Cytochromes*. Cambridge University Press.

Kalckar, H. M., (ed.), 1969. *Biological Phosphorylations: Development of Concepts*. Prentice-Hall. [A collection of classical papers on oxidative phosphorylation and other aspects of bioenergetics.]

Fruton, J. S., 1972. *Molecules and Life: Historical Essays on the Interplay of Chemistry and Biology*. Wiley-Interscience. [Includes an excellent account of cellular respiration, starting on page 262.]

PROBLEMS

1. What is the yield of ATP when each of the following substrates is completely oxidized by a cell homogenate? Assume that glycolysis, the citric acid cycle, and oxidative phosphorylation are fully active.
   (a) Pyruvate.
   (b) NADH.
   (c) Fructose 1,6-diphosphate.
   (d) Phosphoenolpyruvate.
   (e) Glucose.
   (f) Dihydroxyacetone phosphate.

2. (a) Write an equation for the oxidation of reduced glutathione (see p. 366) by O_2. Calculate $\Delta E_0'$ and $\Delta G°'$ for this reaction using the data in Table 14-1.
   (b) What is $\Delta E_0'$ and $\Delta G°'$ for the reduction of oxidized glutathione by NADPH?

3. What is the effect of each of the following inhibitors on electron transport and ATP formation by the respiratory chain?
   (a) Azide.
   (b) Atractyloside.
   (c) Rotenone.
   (d) DNP.
   (e) Carbon monoxide.
   (f) Antimycin A.

4. The addition of oligomycin to mitochondria markedly decreases both the rate of electron transfer from NADH to O_2 and the rate of ATP formation. The subsequent addition of DNP leads to an increase in the rate of electron transfer without changing the rate of ATP formation. What does oligomycin inhibit?

5. Compare the $\Delta G°'$ for the oxidation of succinate by NAD^+ and by FAD. Use the data given in Table 14-1, and assume that E_0' for the $FAD/FADH_2$ redox couple is nearly 0 volt. Why is FAD rather than NAD^+ the electron acceptor in the reaction catalyzed by succinate dehydrogenase?

6. The immediate administration of nitrite is a highly effective treatment for cyanide poisoning. What is the basis for the action of this antidote? (Hint: Nitrite oxidizes ferrohemoglobin to ferrihemoglobin.)

For additional problems, see W. B. Wood, J. H. Wilson, R. M. Benbow, and L. E. Hood. *Biochemistry: A Problems Approach* (Benjamin, 1974), Chapter 11.

CHAPTER 15

PENTOSE PHOSPHATE PATHWAY AND GLUCONEOGENESIS

Reduced nicotinamide adenine dinucleotide phosphate (NADPH)

The preceding chapters on glycolysis, the citric acid cycle, and oxidative phosphorylation were primarily concerned with the generation of ATP, starting with glucose as the fuel. We now turn to the generation of a different type of metabolic energy—reducing power. Some of the electrons and hydrogen atoms of fuel molecules must be conserved for biosynthetic purposes rather than transferred to O_2 to generate ATP. *The currency of readily available reducing power in cells is NADPH.* The phosphoryl group on C-2 of one of the ribose units of NADPH distinguishes it from NADH. As mentioned previously (p. 263), there is a *fundamental distinction between NADPH and NADH in most biochemical reactions. NADH is oxidized by the respiratory chain to generate ATP, whereas NADPH serves as a hydrogen and electron donor in reductive biosyntheses.*

THE PENTOSE PHOSPHATE PATHWAY GENERATES NADPH AND SYNTHESIZES FIVE-CARBON SUGARS

In the pentose phosphate pathway, NADPH is generated as glucose 6-phosphate is oxidized to ribose 5-phosphate. This five-carbon sugar and its derivatives are components of such important bio-molecules as ATP, CoA, NAD^+, FAD, RNA, and DNA.

$$\text{Glucose 6-phosphate} + 2\,NADP^+ + H_2O \longrightarrow$$
$$\text{ribose 5-phosphate} + 2\,NADPH + 2\,H^+ + CO_2$$

The pentose phosphate pathway also catalyzes the interconversion of three-, four-, five-, six-, and seven-carbon sugars in a series of nonoxidative reactions. All of these reactions occur in the cytosol. In plants, part of the pentose phosphate pathway also participates in the formation of hexoses from CO_2 in photosynthesis (Chapter 19).

The pentose phosphate pathway is sometimes called *the pentose shunt, the hexose monophosphate pathway,* or *the phosphogluconate oxidative pathway.* The discovery of glucose 6-phosphate dehydrogenase, the first enzyme in the pathway, by Otto Warburg in 1931 led to its complete elucidation by Fritz Lipmann, Frank Dickens, Bernard Horecker, and Efraim Racker.

TWO NADPH ARE GENERATED IN THE CONVERSION OF GLUCOSE 6-PHOSPHATE TO RIBULOSE 5-PHOSPHATE

The pentose phosphate pathway starts with the dehydrogenation of glucose 6-phosphate at C-1, a reaction catalyzed by *glucose 6-phosphate dehydrogenase* (Figure 15-1). This enzyme is highly specific for $NADP^+$; the K_M for NAD^+ is about 1000 times greater than for $NADP^+$. The product is *6-phosphoglucono-δ-lactone,* which is an intramolecular ester between the C-1 carboxyl group and the C-5 hydroxyl group. The next step is the hydrolysis of 6-phospho-glucono-δ-lactone by a specific *lactonase* to give *6-phosphogluconate.* This six-carbon sugar is then oxidatively decarboxylated by *6-phos-phogluconate dehydrogenase,* yielding *ribulose 5-phosphate.* $NADP^+$ is again the electron acceptor.

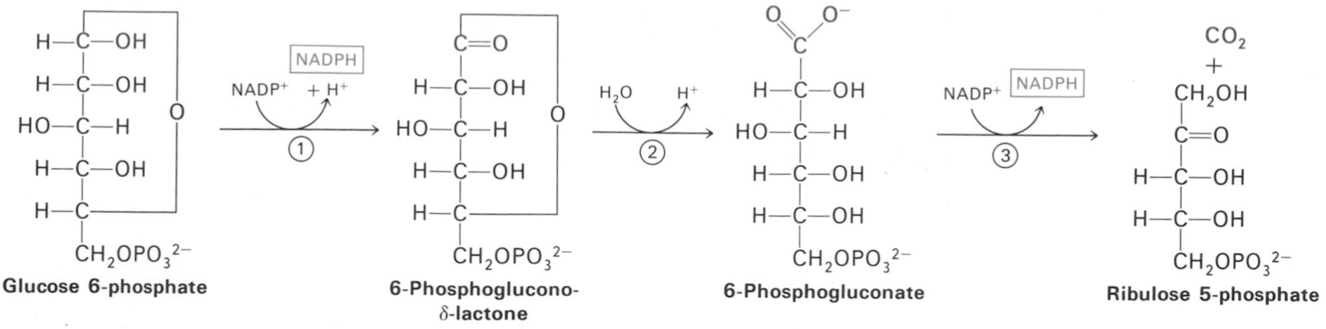

Glucose 6-phosphate

6-Phosphoglucono-δ-lactone

6-Phosphogluconate

Ribulose 5-phosphate

Figure 15-1
Oxidative branch of the pentose phosphate pathway. These three reactions are catalyzed by glucose 6-phosphate dehydrogenase, lactonase, and 6-phosphogluconate dehydrogenase.

RIBULOSE 5-PHOSPHATE IS ISOMERIZED TO RIBOSE 5-PHOSPHATE VIA AN ENEDIOL INTERMEDIATE

The final step in the synthesis of ribose 5-phosphate is the isomerization of ribulose 5-phosphate by *phosphopentose isomerase*.

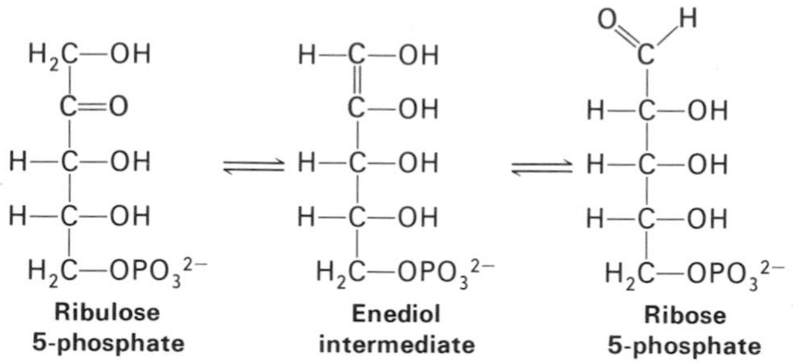

Ribulose 5-phosphate

Enediol intermediate

Ribose 5-phosphate

This reaction is similar to the glucose 6-phosphate $\rightleftharpoons$ fructose 6-phosphate and to the dihydroxyacetone phosphate $\rightleftharpoons$ glyceraldehyde 3-phosphate reactions in glycolysis. *These three ketose-aldose isomerizations proceed through an enediol intermediate.*

THE PENTOSE PHOSPHATE PATHWAY AND GLYCOLYSIS ARE LINKED BY TRANSKETOLASE AND TRANSALDOLASE

The preceding reactions yield two NADPH and one ribose 5-phosphate for each glucose 6-phosphate oxidized. However, many cells need much more NADPH for reductive biosyntheses than ribose

5-phosphate for incorporation into nucleotides and nucleic acids. Under these conditions, ribose 5-phosphate is converted into glyceraldehyde 3-phosphate and fructose 6-phosphate by *transketolase* and *transaldolase*. *These enzymes create a reversible link between the pentose phosphate pathway and glycolysis* by catalyzing these three reactions:

$$C_5 + C_5 \xrightleftharpoons{\text{transketolase}} C_3 + C_7$$

$$C_7 + C_3 \xrightleftharpoons{\text{transaldolase}} C_4 + C_6$$

$$C_5 + C_4 \xrightleftharpoons{\text{transketolase}} C_3 + C_6$$

The sum of these reactions is the *formation of two hexoses and one triose from three pentoses.*

The essence of these reactions is that *transketolase transfers a two-carbon unit whereas transaldolase transfers a three-carbon unit.* The sugar that donates the two- or three-carbon unit is always a ketose, whereas the acceptor is always an aldose.

The first of the three reactions linking the pentose phosphate pathway and glycolysis is the formation of *glyceraldehyde 3-phosphate* and *sedoheptulose 7-phosphate* from two pentoses.

The donor of the two-carbon unit in this reaction is xylulose 5-phosphate, which is an epimer of ribulose 5-phosphate. A ketose is a substrate of transketolase only if its hydroxyl group at C-3 has the configuration of xylulose rather than of ribulose. Ribulose

5-phosphate is converted into the appropriate epimer for the trans-ketolase reaction by *phosphopentose epimerase.*

$$
\begin{array}{ccc}
\text{CH}_2\text{OH} & & \text{CH}_2\text{OH} \\
| & & | \\
\text{C}=\text{O} & \xrightarrow{\text{Phosphopentose epimerase}} & \text{C}=\text{O} \\
| & & | \\
\text{H}-\text{C}-\text{OH} & & \text{HO}-\text{C}-\text{H} \\
| & & | \\
\text{H}-\text{C}-\text{OH} & & \text{H}-\text{C}-\text{OH} \\
| & & | \\
\text{CH}_2\text{OPO}_3^{2-} & & \text{CH}_2\text{OPO}_3^{2-} \\
\textbf{Ribulose 5-phosphate} & & \textbf{Xylulose 5-phosphate}
\end{array}
$$

Glyceraldehyde 3-phosphate and sedoheptulose 7-phosphate then react to form *fructose 6-phosphate* and *erythrose 4-phosphate.* This synthesis of a four-carbon sugar and a six-carbon sugar is catalyzed by *transaldolase.*

$$
\begin{array}{ccccccc}
\text{CH}_2\text{OH} & & & & & & \text{CH}_2\text{OH} \\
| & & \text{O}\!\!\diagdown\!\!\diagup\text{H} & & \text{O}\!\!\diagdown\!\!\diagup\text{H} & & | \\
\text{C}=\text{O} & & \text{C} & & \text{C} & & \text{C}=\text{O} \\
| & & | & \xrightarrow{\text{Transaldolase}} & | & & | \\
\text{HO}-\text{C}-\text{H} & + & \text{H}-\text{C}-\text{OH} & & \text{H}-\text{C}-\text{OH} & + & \text{HO}-\text{C}-\text{H} \\
| & & | & & | & & | \\
\text{H}-\text{C}-\text{OH} & & \text{CH}_2\text{OPO}_3^{2-} & & \text{H}-\text{C}-\text{OH} & & \text{H}-\text{C}-\text{OH} \\
| & & & & | & & | \\
\text{H}-\text{C}-\text{OH} & & & & \text{CH}_2\text{OPO}_3^{2-} & & \text{H}-\text{C}-\text{OH} \\
| & & & & & & | \\
\text{H}-\text{C}-\text{OH} & & & & & & \text{CH}_2\text{OPO}_3^{2-} \\
| & & & & & & \\
\text{CH}_2\text{OPO}_3^{2-} & & & & & & \\
\textbf{Sedoheptulose} & & \textbf{Glyceraldehyde} & & \textbf{Erythrose} & & \textbf{Fructose} \\
\textbf{7-phosphate} & & \textbf{3-phosphate} & & \textbf{4-phosphate} & & \textbf{6-phosphate}
\end{array}
$$

In the third reaction, transketolase catalyzes the synthesis of *fructose 6-phosphate* and *glyceraldehyde 3-phosphate* from erythrose 4-phosphate and xylulose 5-phosphate.

$$
\begin{array}{ccccccc}
& & & & & & \text{CH}_2\text{OH} \\
\text{CH}_2\text{OH} & & \text{O}\!\!\diagdown\!\!\diagup\text{H} & & \text{O}\!\!\diagdown\!\!\diagup\text{H} & & | \\
| & & \text{C} & & \text{C} & & \text{C}=\text{O} \\
\text{C}=\text{O} & & | & \xrightarrow{\text{Transketolase}} & | & & \text{HO}-\text{C}-\text{H} \\
| & & \text{H}-\text{C}-\text{OH} & & \text{H}-\text{C}-\text{OH} & + & | \\
\text{HO}-\text{C}-\text{H} & + & | & & | & & \text{H}-\text{C}-\text{OH} \\
| & & \text{H}-\text{C}-\text{OH} & & \text{CH}_2\text{OPO}_3^{2-} & & | \\
\text{H}-\text{C}-\text{OH} & & | & & & & \text{H}-\text{C}-\text{OH} \\
| & & \text{CH}_2\text{OPO}_3^{2-} & & & & | \\
\text{CH}_2\text{OPO}_3^{2-} & & & & & & \text{CH}_2\text{OPO}_3^{2-} \\
\textbf{Xylulose} & & \textbf{Erythrose} & & \textbf{Glyceraldehyde} & & \textbf{Fructose} \\
\textbf{5-phosphate} & & \textbf{4-phosphate} & & \textbf{3-phosphate} & & \textbf{6-phosphate}
\end{array}
$$

The sum of these reactions is:

2 Xylulose 5-phosphate + ribose 5-phosphate $\rightleftharpoons$

2 fructose 6-phosphate + glyceraldehyde 3-phosphate

Since xylulose 5-phosphate can be formed from ribose 5-phosphate by the sequential action of phosphopentose isomerase and phosphopentose epimerase, the net reaction starting from ribose 5-phosphate is:

3 Ribose 5-phosphate $\rightleftharpoons$

2 fructose 6-phosphate + glyceraldehyde 3-phosphate

Thus, excess ribose 5-phosphate formed by the pentose phosphate pathway can be quantitatively converted to glycolytic intermediates.

THE FLOW OF GLUCOSE 6-PHOSPHATE IS DETERMINED BY THE NEED FOR NADPH AND RIBOSE 5-PHOSPHATE

Let us follow the fate of glucose 6-phosphate in three different situations:

1. *Much more ribose 5-phosphate than NADPH is required.* Most of the glucose 6-phosphate is converted to fructose 6-phosphate and glyceraldehyde 3-phosphate by the glycolytic pathway. Transaldolase and transketolase then convert two molecules of fructose 6-phosphate and one molecule of glyceraldehyde 3-phosphate into three molecules of ribose 5-phosphate by a reversal of the reactions described above.

2. *The need for NADPH and ribose 5-phosphate is balanced.* The predominant reaction under these conditions is the formation of two NADPH and one ribose 5-phosphate from glucose 6-phosphate by the oxidative branch of the pentose phosphate pathway.

3. *Much more NADPH than ribose 5-phosphate is required.* Three groups of reactions are active in this situation. First, two NADPH and one ribose 5-phosphate are formed by the oxidative branch of the pentose phosphate pathway. Then, ribose 5-phosphate is converted into fructose 6-phosphate and glyceraldehyde 3-phosphate by transketolase and transaldolase. Finally, glucose 6-phosphate is

resynthesized from fructose 6-phosphate and glyceraldehyde 3-phosphate by the gluconeogenic pathway (discussed later in this chapter). The stoichiometries of these three sets of reactions are:

$$6 \text{ Glucose 6-phosphate} + 12 \text{ NADP}^+ + 6 \text{ H}_2\text{O} \longrightarrow$$
$$6 \text{ ribose 5-phosphate} + 12 \text{ NADPH} + 12 \text{ H}^+ + 6 \text{ CO}_2$$

$$6 \text{ Ribose 5-phosphate} \longrightarrow$$
$$4 \text{ fructose 6-phosphate} + 2 \text{ glyceraldehyde 3-phosphate}$$

$$4 \text{ Fructose 6-phosphate} + 2 \text{ glyceraldehyde 3-phosphate} +$$
$$\text{H}_2\text{O} \longrightarrow 5 \text{ glucose 6-phosphate} + \text{P}_i$$

The sum of these reactions is:

$$\text{Glucose 6-phosphate} + 12 \text{ NADP}^+ + 7 \text{ H}_2\text{O} \longrightarrow$$
$$6 \text{ CO}_2 + 12 \text{ NADPH} + 12 \text{ H}^+ + \text{P}_i$$

Thus, glucose 6-phosphate can be completely oxidized to CO_2 with the concomitant generation of NADPH. The essence of these reactions is that *the ribose 5-phosphate produced by the pentose phosphate pathway is recycled into glucose 6-phosphate* by transketolase, transaldolase, and some of the enzymes of the gluconeogenic pathway.

THE PENTOSE PHOSPHATE PATHWAY IS MUCH MORE ACTIVE IN ADIPOSE TISSUE THAN IN MUSCLE

Radioactive-labeling experiments can provide an estimate of how much glucose 6-phosphate is metabolized by the pentose phosphate pathway and how much by the combined action of glycolysis and the citric acid cycle. One aliquot of a tissue homogenate is incubated with glucose labeled with ^{14}C at C-1, the other with glucose labeled with ^{14}C at C-6. The radioactivity of the CO_2 produced by the two samples is then compared. The rationale of this experiment is that only C-1 is decarboxylated by the pentose phosphate pathway, whereas C-1 and C-6 are decarboxylated to an equal extent when glucose is metabolized by the glycolytic pathway, pyruvate dehydrogenase complex, and citric acid cycle. The reason for the equivalence of C-1 and C-6 in the latter set of reactions is that glyceraldehyde 3-phosphate and dihydroxyacetone phosphate are rapidly interconverted by triose phosphate isomerase.

Table 15-1
Pentose phosphate pathway

Reaction	Enzyme
OXIDATIVE BRANCH	
Glucose 6-phosphate + NADP$^+$ $\rightleftharpoons$ 6-phosphoglucono-δ-lactone + NADPH + H$^+$	Glucose 6-phosphate dehydrogenase
6-Phosphoglucono-δ-lactone + H$_2$O $\longrightarrow$ 6-phosphogluconate + H$^+$	Lactonase
6-Phosphogluconate + NADP$^+$ $\longrightarrow$ ribulose 5-phosphate + CO$_2$ + NADPH	6-Phosphogluconate dehydrogenase
NONOXIDATIVE BRANCH	
Ribulose 5-phosphate $\rightleftharpoons$ ribose 5-phosphate	Phosphopentose isomerase
Ribulose 5-phosphate $\rightleftharpoons$ xylulose 5-phosphate	Phosphopentose epimerase
Xylulose 5-phosphate + ribose 5-phosphate $\rightleftharpoons$ sedoheptulose 7-phosphate + glyceraldehyde 3-phosphate	Transketolase
Sedoheptulose 7-phosphate + glyceraldehyde 3-phosphate $\rightleftharpoons$ fructose 6-phosphate + erythrose 4-phosphate	Transaldolase
Xylulose 5-phosphate + erythrose 4-phosphate $\rightleftharpoons$ fructose 6-phosphate + glyceraldehyde 3-phosphate	Transketolase

This experimental approach has shown that *the activity of the pentose phosphate pathway is very low in skeletal muscle, whereas it is very high in adipose tissue*. These findings support the idea that a major role of the pentose phosphate pathway is to generate NADPH for reductive biosyntheses. Large amounts of NADPH are consumed by adipose tissue in the reductive synthesis of fatty acids from acetyl CoA (see Chapter 17).

TPP, THE PROSTHETIC GROUP OF TRANSKETOLASE, TRANSFERS ACTIVATED ALDEHYDES

Transketolase contains a tightly bound thiamine pyrophosphate (TPP) as its prosthetic group. We have previously encountered this prosthetic group in the decarboxylation of pyruvate by the pyruvate dehydrogenase complex. The mechanism of catalysis of transketolase is similar in that an *activated aldehyde unit is transferred to an acceptor*. The acceptor in the transketolase reaction is an aldose, whereas in the pyruvate dehydrogenase reaction it is lipoamide.

In both reactions, the site of addition of the keto substrate is the *thiazole ring* of the prosthetic group. The C-2 carbon atom is highly acidic and readily ionizes to give a *carbanion*.

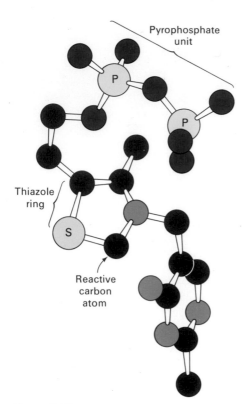

Figure 15-2
Molecular model of thiamine pyrophosphate (TPP).

This carbanion adds to the carbonyl group of the ketose substrate (e.g., xylulose 5-phosphate, fructose 6-phosphate, and sedoheptulose 7-phosphate).

This addition compound loses its R—CHOH moiety to yield a negatively charged, *activated glycoaldehyde* unit. The positively charged nitrogen in the thiazole ring acts as an electron sink to promote the development of a negative charge on the activated intermediate.

The carbonyl group of a suitable aldehyde acceptor then condenses with the activated glycoaldehyde unit to form a new ketose, which is released from the enzyme.

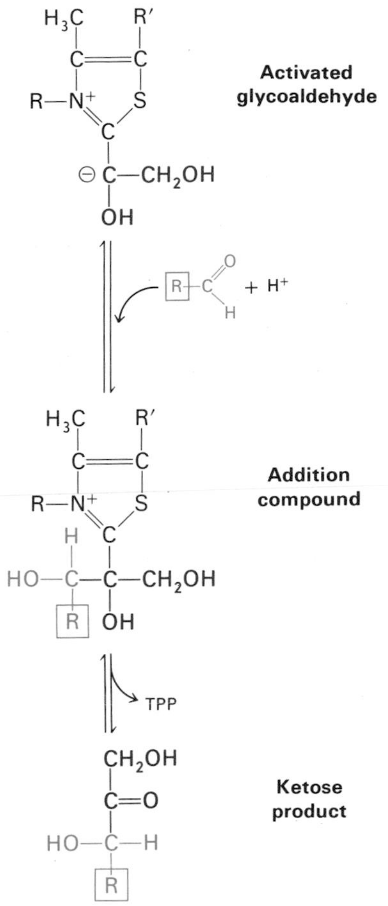

ACTIVATED DIHYDROXYACETONE IS CARRIED BY TRANSALDOLASE AS A SCHIFF BASE

Transaldolase transfers a three-carbon *dihydroxyacetone* unit from a ketose donor to an aldose acceptor. Transaldolase, in contrast to transketolase, does not contain a prosthetic group. Rather, *a Schiff base is formed between the carbonyl group of the ketose substrate and the ε-amino group of a lysine residue at the active site of the enzyme.* We have previously encountered this kind of covalent enzyme-substrate (ES) intermediate in fructose diphosphate aldolase in the glycolytic pathway.

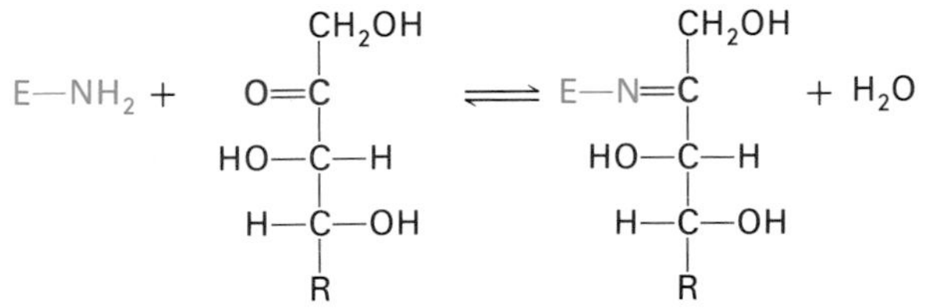

Ketose substrate   Schiff base

This Schiff base becomes protonated, the bond between C-3 and C-4 is split, and an aldose is released.

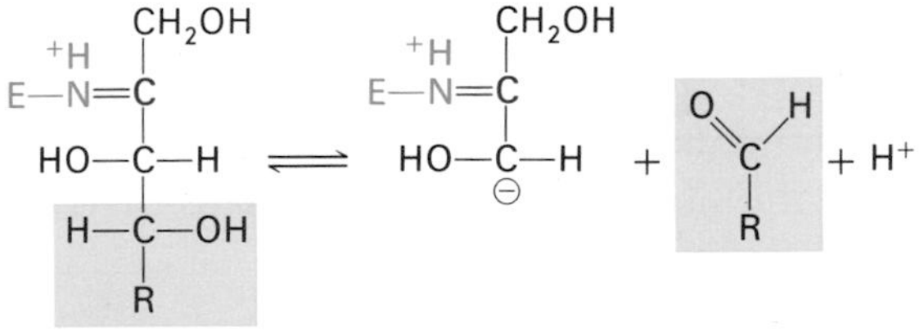

Protonated Schiff base   Carbanion   Aldose

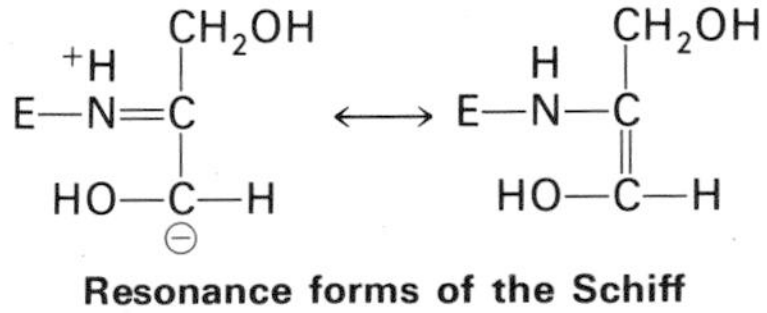

The negative charge on the dihydroxyacetone moiety is stabilized by resonance. The positively charged nitrogen atom of the Schiff base acts as an electron sink. This nitrogen atom plays the same role in transaldolase as does the thiazole ring nitrogen in transketolase.

Activated glycoaldehyde

Addition compound

Ketose product

Resonance forms of the Schiff base carbanion

The Schiff base between dihydroxyacetone and transaldolase is stable until a suitable aldose becomes bound. The carbanion of the dihydroxyacetone moiety then reacts with the carbonyl group of the aldose. The ketose product is released by hydrolysis of the Schiff base.

Carbanion   **Aldose substrate**   **Schiff base**   **Ketose product**

GLUCOSE 6-PHOSPHATE DEHYDROGENASE DEFICIENCY CAUSES A DRUG-INDUCED HEMOLYTIC ANEMIA

An antimalarial drug, pamaquine, was introduced in 1926. Most patients tolerated the drug well, but a few developed severe symptoms within a few days after therapy was started. The urine turned black, jaundice developed, and the hemoglobin content of the blood dropped sharply. In some cases, massive destruction of red blood cells caused death.

The basis of this *drug-induced hemolytic anemia* was elucidated in 1956. The primary defect is a *deficiency in glucose 6-phosphate dehydrogenase in red cells.* This block in the pentose phosphate pathway in red cells leads to a diminished production of NADPH. In turn, there is a *lowered concentration of reduced glutathione,* a tripeptide that contains a sulfhydryl group. Reduced glutathione serves as a *sulfhydryl buffer,* which maintains protein cysteine residues in a reduced form. Reduced glutathione is regenerated from oxidized glutathione by reaction with NADPH.

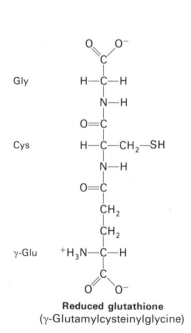

Reduced glutathione
(γ-Glutamylcysteinylglycine)

$$\begin{array}{c} \gamma\text{-Glu--Cys--Gly} \\ | \\ S \\ | \\ S \\ | \\ \gamma\text{-Glu--Cys--Gly} \end{array} + NADPH + H^+ \rightleftharpoons \begin{array}{c} 2\ \gamma\text{-Glu--Cys--Gly} \\ | \\ SH \end{array} + NADP^+$$

Oxidized glutathione   **Reduced glutathione**

Reduced glutathione appears to be essential for the maintenance of normal red cell structure. Cells with a lowered level of reduced glutathione are more susceptible to hemolysis for reasons that are not yet understood. Drugs such as pamaquine may distort the surface of red cells in the absence of reduced glutathione, making them more liable to destruction and removal by the spleen.

Glucose 6-phosphate dehydrogenase deficiency is not a rare disease. It is inherited as a sex-linked trait. Female heterozygotes have two populations of red cells: one has normal enzymatic activity, whereas the other is deficient in glucose 6-phosphate dehydrogenase. The glucose 6-phosphate dehydrogenase in most other organs is specified by a different gene. The incidence of type A deficiency of glucose 6-phosphate dehydrogenase, characterized by a tenfold reduction in enzymatic activity in red cells, is 11% among black Americans. This high gene frequency suggests that the deficiency may be advantageous under certain environmental conditions. Indeed, *glucose 6-phosphate dehydrogenase deficiency in red cells appears to protect an individual from falciparum malaria,* because the parasites that cause this disease require the pentose phosphate pathway and reduced glutathione for optimal growth. Thus, glucose 6-phosphate dehydrogenase deficiency and sickle-cell trait are parallel mechanisms of protection against malaria, which accounts for their high gene frequencies in malaria-infested regions of the world.

The occurrence of glucose 6-phosphate dehydrogenase deficiency clearly demonstrates that *atypical reactions to drugs may have a genetic basis.* This inherited enzymatic deficiency is relatively benign until certain drugs are administered. We see here the interplay of heredity and environment in the production of disease. Galactosemia, hereditary fructose intolerance, phenylketonuria, and succinylcholine sensitivity also illustrate this interaction in a striking way.

GLUCOSE CAN BE SYNTHESIZED FROM NONCARBOHYDRATE PRECURSORS

We now turn to the *synthesis of glucose from noncarbohydrate precursors,* a process called *gluconeogenesis.*

The starting point of the gluconeogenic pathway is pyruvate, whereas the terminus is glucose. The principal points of entry into the pathway are pyruvate and oxaloacetate. The most important

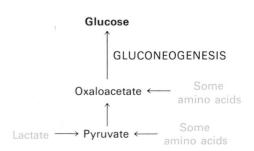

noncarbohydrate precursors of glucose are *lactate* and most *amino acids*. In animals, fatty acids cannot be converted into glucose.

The major site of gluconeogenesis is the *liver.* Gluconeogenesis also occurs in the cortex of the *kidney,* but the total amount of glucose formed there is about one-tenth of that formed in the liver because of its smaller mass. Very little gluconeogenesis takes place in the brain, skeletal muscle, or heart muscle. On the contrary, brain and skeletal muscle have a high demand for glucose. In fact, *gluconeogenesis by liver and kidney serves to maintain the glucose level in the blood so that brain and muscle can extract sufficient glucose from it to meet their metabolic demands.*

GLUCONEOGENESIS IS NOT A REVERSAL OF GLYCOLYSIS

In glycolysis, glucose is converted to pyruvate; in gluconeogenesis, pyruvate is converted to glucose. However, gluconeogenesis is not a reversal of glycolysis. The stoichiometry is different and there are some new reactions. A different pathway is required because glycolysis is a highly exergonic process, with a $\Delta G^{\circ\prime}$ of about -20 kcal/mol for the reaction:

$$\text{Glucose} + 2\,\text{ADP} + 2\,\text{P}_i + 2\,\text{NAD}^+ \longrightarrow$$
$$2\,\text{pyruvate} + 2\,\text{ATP} + 2\,\text{NADH} + 2\,\text{H}_2\text{O}$$

In fact, *there are three essentially irreversible steps in glycolysis:*

$$\text{Phosphoenolpyruvate} + \text{ADP} \xrightarrow{\text{pyruvate kinase}} \text{pyruvate} + \text{ATP}$$

$$\text{Fructose 6-phosphate} + \text{ATP} \xrightarrow{\text{phosphofructokinase}}$$
$$\text{fructose 1,6-diphosphate} + \text{ADP}$$

$$\text{Glucose} + \text{ATP} \xrightarrow{\text{hexokinase}} \text{glucose 6-phosphate} + \text{ADP}$$

DISTINCTIVE REACTIONS OF GLUCONEOGENESIS

In gluconeogenesis, these virtually irreversible reactions are bypassed by the following new ones:

1. *Phosphoenolpyruvate is formed from pyruvate via oxaloacetate.* First, pyruvate is carboxylated to oxaloacetate at the expense of an ATP. Then, oxaloacetate is decarboxylated and phosphorylated to yield phosphoenolpyruvate, at the expense of a second high-energy phosphate bond.

$$Pyruvate + CO_2 + ATP + H_2O \rightleftharpoons$$
$$oxaloacetate + ADP + P_i + 2\,H^+$$

$$Oxaloacetate + GTP \rightleftharpoons phosphoenolpyruvate + GDP + CO_2$$

The first reaction is catalyzed by *pyruvate carboxylase,* and the second by *phosphoenolpyruvate carboxykinase.* The sum of these reactions is:

$$Pyruvate + ATP + GTP + H_2O \rightleftharpoons$$
$$phosphoenolpyruvate + ADP + GDP + P_i + 2\,H^+$$

This pathway for the formation of phosphoenolpyruvate from pyruvate is thermodynamically feasible, since $\Delta G°'$ is $+0.2$ kcal/mol in contrast with $+7.5$ kcal/mol for the reaction catalyzed by pyruvate kinase. This much more favorable $\Delta G°'$ results from the input of an additional high-energy phosphate bond.

2. *Fructose 6-phosphate is formed from fructose 1,6-diphosphate by hydrolysis of the phosphate ester at C-1.* Fructose 1,6-diphosphatase catalyzes this exergonic hydrolysis.

$$Fructose\ 1,6\text{-}diphosphate + H_2O \longrightarrow fructose\ 6\text{-}phosphate + P_i$$

3. *Glucose is formed by hydrolysis of glucose 6-phosphate,* a reaction catalyzed by glucose 6-phosphatase.

$$Glucose\ 6\text{-}phosphate + H_2O \longrightarrow glucose + P_i$$

Table 15-2
Enzymatic differences between glycolysis and gluconeogenesis

Glycolysis	*Gluconeogenesis*
Hexokinase	Glucose 6-phosphatase
Phosphofructokinase	Fructose 1,6-diphosphatase
Pyruvate kinase	Pyruvate carboxylase
	Phosphoenolpyruvate carboxykinase

SIX HIGH-ENERGY PHOSPHATE BONDS ARE SPENT IN SYNTHESIZING GLUCOSE FROM PYRUVATE

The stoichiometry of gluconeogenesis is:

$$2 \text{ Pyruvate} + 4 \text{ ATP} + 2 \text{ GTP} + 2 \text{ NADH} + 2 \text{ H}_2\text{O} \longrightarrow$$
$$\text{glucose} + 4 \text{ ADP} + 2 \text{ GDP} + 6 \text{ P}_i + 2 \text{ NAD}^+$$
$$\Delta G^{\circ\prime} = -9 \text{ kcal/mol}$$

Note that *six* high-energy phosphate bonds are used to synthesize glucose from pyruvate in gluconeogenesis, whereas only *two* ATP are generated in glycolysis in the conversion of glucose to pyruvate. Thus, the extra price of gluconeogenesis is four high-energy phosphate bonds per glucose synthesized from pyruvate. The four extra high-energy phosphate bonds are needed to turn an energetically unfavorable process (the reversal of glycolysis, $\Delta G^{\circ\prime} = +20$ kcal/mol) into a favorable one (gluconeogenesis, $\Delta G^{\circ\prime} = -9$ kcal/mol).

BIOTIN IS A MOBILE CARRIER OF ACTIVATED CO_2

The finding that mitochondria can form phosphoenolpyruvate from pyruvate led to the discovery of pyruvate carboxylase by Merton Utter in 1960. This enzyme is of especial interest because of its catalytic and allosteric properties. Pyruvate carboxylase contains a covalently attached prosthetic group, *biotin*, which serves as a *carrier of activated CO_2*. The carboxyl terminus of biotin is linked to the ϵ-amino group of a specific lysine residue by an amide bond.

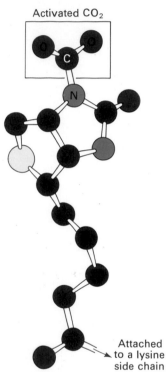

Activated CO$_2$

Figure 15-3
Molecular model of carboxybiotin.

Note that biotin is attached to pyruvate carboxylase by a *long, flexible chain* like that of lipoamide in the pyruvate dehydrogenase complex.

The carboxylation of pyruvate occurs in two stages:

$$\text{Biotin-enzyme} + \text{ATP} + \text{HCO}_3^- \xrightleftharpoons{\overset{\text{Mg}^{2+}}{\text{acetyl CoA}}}$$

$$\text{CO}_2 \sim \text{biotin-enzyme} + \text{ADP} + \text{P}_i$$

$$\text{CO}_2 \sim \text{biotin-enzyme} + \text{pyruvate} \xrightleftharpoons{\text{Mn}^{2+}}$$

$$\text{biotin-enzyme} + \text{oxaloacetate}$$

The carboxyl group in the carboxybiotin-enzyme intermediate is bonded to the N-1 nitrogen atom of the biotin ring, according to some investigators; others regard the adjacent carbonyl group as a more plausible site for the bound CO_2. The carboxyl group in this carboxybiotin intermediate is *activated*. The $\Delta G°'$ for its cleavage

$$\text{CO}_2 \sim \text{biotin-enzyme} + \text{H}^+ \longrightarrow \text{CO}_2 + \text{biotin-enzyme}$$

is -4.7 kcal/mol, which enables carboxybiotin to transfer CO_2 to acceptors without the input of additional free energy.

The activated carboxyl group is then transferred from carboxybiotin to pyruvate to form oxaloacetate. The long, flexible link between biotin and the enzyme enables this prosthetic group to rotate from one active site of the enzyme (the ATP-bicarbonate site) to the other (the pyruvate site).

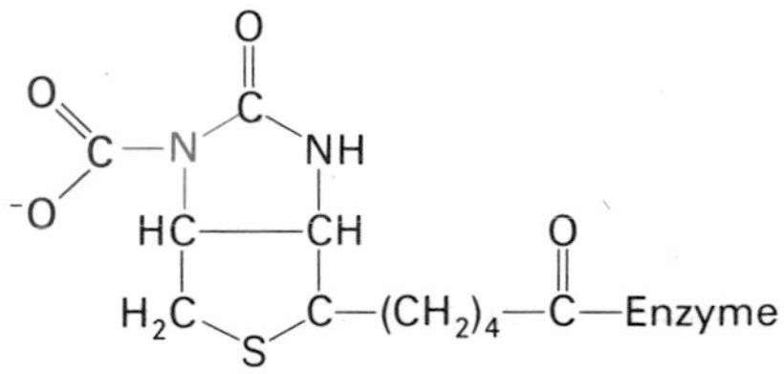

Carboxybiotin-enzyme
intermediate

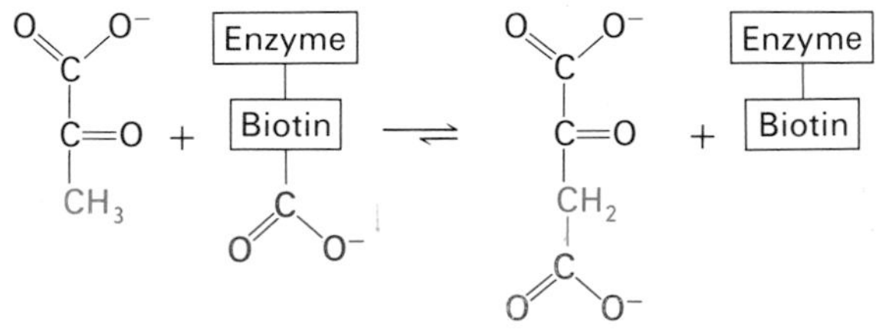

Pyruvate

Oxaloacetate

PYRUVATE CARBOXYLASE IS ACTIVATED BY ACETYL CoA

The activity of pyruvate carboxylase depends on the presence of acetyl CoA. *Biotin is not carboxylated unless acetyl CoA (or a closely related acyl CoA) is bound to the enzyme.* The second partial reaction is not affected by acetyl CoA. The allosteric activation of pyruvate carboxylase by acetyl CoA is an important physiological control mechanism. Oxaloacetate, the product of the pyruvate carboxylase reaction, is both a stoichiometric intermediate in gluconeogenesis and a catalytic intermediate in the citric acid cycle. *A high level of acetyl CoA signals the need for more oxaloacetate.* If there is a surplus of ATP, oxaloacetate will be consumed in gluconeogenesis. If there is a deficiency of ATP, oxaloacetate will enter the citric acid cycle upon condensing with acetyl CoA.

Thus, pyruvate carboxylase is important not only in gluconeogenesis. It also plays a *critical role in maintaining the level of citric acid cycle intermediates.* These intermediates need to be replenished because they are consumed in some biosynthetic reactions, such as heme synthesis. This role of pyruvate carboxylase is termed *anaplerotic,* meaning to fill up.

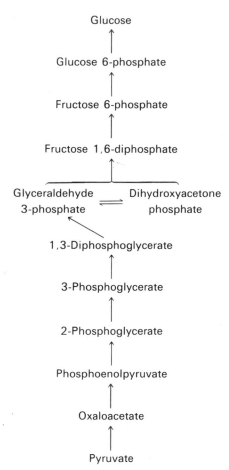

Figure 15-4
Pathway of gluconeogenesis. The distinctive reactions of gluconeogenesis are denoted by red arrows. The other reactions are common to glycolysis.

OXALOACETATE IS SHUTTLED INTO THE CYTOSOL AND CONVERTED INTO PHOSPHOENOLPYRUVATE

Pyruvate carboxylase is a mitochondrial enzyme, whereas the other enzymes of gluconeogenesis are cytoplasmic. Oxaloacetate, the product of the pyruvate carboxylase reaction, is transported across the mitochondrial membrane in the form of *malate*. Oxaloacetate is reduced to malate inside the mitochondrion by an NADH-linked malate dehydrogenase. Malate diffuses across the mitochondrial membrane and is reoxidized to oxaloacetate by an NAD^+-linked malate dehydrogenase in the cytoplasm.

Oxaloacetate is simultaneously *decarboxylated* and *phosphorylated* by phosphoenolpyruvate carboxykinase.

$$\text{Oxaloacetate} + \text{GTP} \rightleftharpoons \text{Phosphoenolpyruvate} + CO_2 + GDP$$

Oxaloacetate   **Phosphoenolpyruvate**

The CO_2 that was added to pyruvate by pyruvate carboxylase comes off in this step. In fact, the phosphorylation reaction is made energetically feasible by the concomitant decarboxylation. *Decarboxylations often drive reactions that would otherwise be highly endergonic.* We will encounter this device again in fatty acid synthesis.

The pathway from phosphoenolpyruvate to glucose is shown in Figure 15-4. All of the enzymes in this sequence except for fructose 1,6-diphosphatase and glucose 6-phosphatase are shared with the glycolytic pathway.

LACTATE FORMED BY CONTRACTING MUSCLE IS CONVERTED TO GLUCOSE BY THE LIVER

The major raw material for gluconeogenesis is lactate produced by active skeletal muscle. The rate of production of pyruvate by glycolysis exceeds the rate of oxidation of pyruvate by the citric acid cycle in contracting skeletal muscle. Moreover, the rate of formation of NADH in glycolysis in active muscle is greater than the rate of its oxidation by the respiratory chain. Continued glycolysis depends on the availability of NAD^+ for the oxidation of glyceraldehyde 3-phosphate. This is achieved by lactate dehydrogenase, which oxidizes NADH to NAD^+ as it reduces pyruvate to lactate.

Lactate is a dead end in metabolism. It must go back to pyruvate before it can be metabolized. The only purpose of the reduction of pyruvate to lactate is to regenerate NADH so that glycolysis can proceed in active skeletal muscle. *The formation of lactate buys time and shifts part of the metabolic burden from muscle to liver.*

The plasma membrane of most cells is highly permeable to lactate and pyruvate. Both substances diffuse out of active skeletal muscle into the blood and are carried to the liver. Much more lactate than pyruvate is carried because of the high $NADH/NAD^+$ ratio in contracting skeletal muscle. The lactate that enters the liver is oxidized to pyruvate, a reaction favored by the low $NADH/NAD^+$ in the cytosol of liver. Pyruvate is then converted to glucose by the gluconeogenic pathway in liver. Glucose then enters the blood and is taken up by skeletal muscle. Thus, *liver furnishes glucose to contracting skeletal muscle, which derives ATP from the glycolytic conversion of glucose to lactate. Glucose is then synthesized from lactate by the liver. These conversions constitute the Cori cycle* (Figure 15-5).

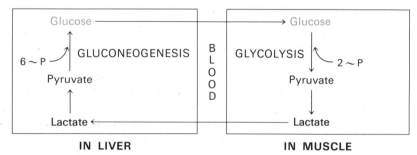

Figure 15-5
The Cori cycle. Lactate formed by active muscle is converted to glucose by the liver. This cycle shifts part of the metabolic burden of active muscle to the liver.

These conversions are facilitated by differences in the catalytic properties of lactate dehydrogenase enzymes in skeletal muscle and liver. Lactate dehydrogenase is a tetramer of 35,000 dalton subunits. There are two kinds of polypeptide chains called M and H, which can form five types of tetramers: M_4, M_3H, M_2H_2, M_1H_3, and H_4. These species are called *isoenzymes* (or isozymes). The M_4 isoenzyme has a much higher affinity for pyruvate than does the H_4 isoenzyme. The other isoenzymes have intermediate affinities. The principal isoenzyme in skeletal muscle is M_4, whereas the main one in liver and heart muscle is H_4. Thus, skeletal muscle is rich in the isoenzyme that has the highest affinity for pyruvate and converts it most rapidly to lactate.

SUMMARY

The pentose phosphate pathway generates NADPH and ribose 5-phosphate in the cytosol. NADPH is used in reductive biosyntheses, whereas ribose 5-phosphate is used in the synthesis of RNA, DNA, and nucleotide coenzymes. The pentose phosphate pathway starts with the dehydrogenation of glucose 6-phosphate to form a lactone, which is hydrolyzed to give 6-phosphogluconate and then oxidatively decarboxylated to yield ribulose 5-phosphate. $NADP^+$ is the electron acceptor in both of these oxidations. The last step is the isomerization of ribulose 5-phosphate (a ketose) to ribose 5-phosphate (an aldose). A second mode of the pathway is active when cells need much more NADPH than ribose 5-phosphate. Under these conditions, ribose 5-phosphate is converted to glyceraldehyde 3-phosphate and fructose 6-phosphate by transketolase and transaldolase. Transketolase contains TPP as its prosthetic group. These enzymes create a reversible link between the pentose phosphate pathway and glycolysis. Xylulose 5-phosphate, sedoheptulose 7-phosphate, and erythrose 4-phosphate are intermediates in these interconversions. In this way, twelve NADPH can be generated for each glucose 6-phosphate that is completely oxidized to CO_2. Alternatively, only the nonoxidative branch of the pathway is active when much more ribose 5-phosphate than NADPH needs to be synthesized. Under these conditions, fructose 6-phosphate and glyceraldehyde 3-phosphate (formed by the glycolytic pathway) are converted to ribose 5-phosphate without the formation of NADPH.

Gluconeogenesis is the synthesis of glucose from noncarbohydrate precursors. The major point of entry into this pathway is pyruvate. Several of the reactions that convert pyruvate to glucose are common to glycolysis. However, gluconeogenesis requires four new reactions to bypass the essential irreversibility of the corresponding reactions in glycolysis. Pyruvate is carboxylated to oxaloacetate, which in turn is decarboxylated and phosphorylated to phosphoenolpyruvate. Two high-energy phosphate bonds are consumed in these reactions, which are catalyzed by pyruvate carboxylase and phosphoenolpyruvate carboxykinase. Pyruvate carboxylase contains a biotin prosthetic group. The other distinctive reactions of gluconeogenesis are the hydrolysis of fructose 1,6-diphosphate and glucose 6-phosphate, which are catalyzed by specific phosphatases.

SELECTED READINGS

Reviews of the pentose phosphate pathway and gluconeogenesis:

Pontremoli, S., and Grazi, E., 1969. Hexose monophosphate oxidation. *Compr. Biochem.* 17:163–189.

Pontremoli, S., and Grazi, E., 1968. Gluconeogenesis. *In* Dickens, F., Randle, P. J., and Whelan, W. J., (eds.), *Carbohydrate Metabolism and Its Disorders,* vol. 1, pp. 259–295. Academic Press.

Enzymes and reaction mechanisms:

Horecker, B. L., 1964. Transketolase and transaldolase. *Compr. Biochem.* 15:48–70.

Scrutton, M. C., and Young, M. R., 1972. Pyruvate carboxylase. *In* Boyer, P. D., (ed.), *The Enzymes* (3rd ed.), vol. 6, pp. 1–35, Academic Press.

Moss, J., and Lane, M. D., 1971. The biotin-dependent enzymes. *Advan. Enzymol.* 35: 321–422.

Regulation:

Scrutton, M. C., and Utter, M. F., 1968. The regulation of glycolysis and gluconeogenesis in animal tissue. *Annu. Rev. Biochem.* 37:249–302.

Newsholme, E. A., and Start, C., 1973. *Regulation in Metabolism.* Wiley. [Chapter 6 deals with the control of glycolysis and gluconeogenesis in liver and kidney cortex.]

Soling, H. D., and Willms, B., (eds.), 1971. *Regulation of Gluconeogenesis.* Academic Press.

Glucose 6-phosphate dehydrogenase deficiency:

Beutler, E., 1972. Glucose 6-phosphate dehydrogenase deficiency. *In* Stanbury, J. B., Wyngaarden, J. B., and Fredrickson, D. S., (eds.), *The Metabolic Basis of Inherited Disease* (3rd ed.), pp. 1358–1388. McGraw-Hill.

Luzzatto, L., Usanga, E. A., and Reddy, S., 1969. Glucose 6-phosphate dehydrogenase deficient red cells: resistance to infection by malarial parasites. *Science* 164:839–842.

PROBLEMS

1. What is the stoichiometry of the synthesis of ribose 5-phosphate from glucose 6-phosphate without the concomitant generation of NADPH? What is the stoichiometry of the synthesis of NADPH from glucose 6-phosphate without the concomitant formation of pentose sugars?

2. Glucose labeled with ^{14}C at C-6 is added to a solution containing the enzymes and cofactors of the oxidative branch of the pentose phosphate pathway. What is the fate of the radioactive label?

3. Which reaction in the citric acid cycle is most analogous to the oxidative decarboxylation of 6-phosphogluconate to ribulose 5-phosphate? What kind of enzyme-bound intermediate is formed in both reactions?

4. Ribose 5-phosphate labeled with ^{14}C at C-1 is added to a solution containing transketolase, transaldolase, phosphopentose epimerase, phos-phopentose isomerase, and glyceraldehyde 3-phosphate. What is the distribution of the radioactive label in the erythrose 4-phosphate and fructose 6-phosphate that are formed in this reaction mixture?

5. Avidin, a 70,000-dalton protein in egg white, has a very high affinity for biotin. In fact, it is a highly specific inhibitor of biotin enzymes. Which of the following conversions would be blocked by the addition of avidin to a cell homogenate?
   (a) Glucose $\longrightarrow$ pyruvate.
   (b) Pyruvate $\longrightarrow$ glucose.
   (c) Oxaloacetate $\longrightarrow$ glucose.
   (d) Glucose $\longrightarrow$ ribose 5-phosphate.
   (e) Pyruvate $\longrightarrow$ oxaloacetate.
   (f) Ribose 5-phosphate $\longrightarrow$ glucose.

6. Design a chemical experiment to identify the lysine residue that forms a Schiff base at the active site of transaldolase.

For additional problems, see W. B. Wood, J. H. Wilson, R. M. Benbow, and L. E. Hood, *Biochemistry: A Problems Approach* (Benjamin, 1974), Chapter 9.

CHAPTER 16

GLYCOGEN AND DISACCHARIDE METABOLISM

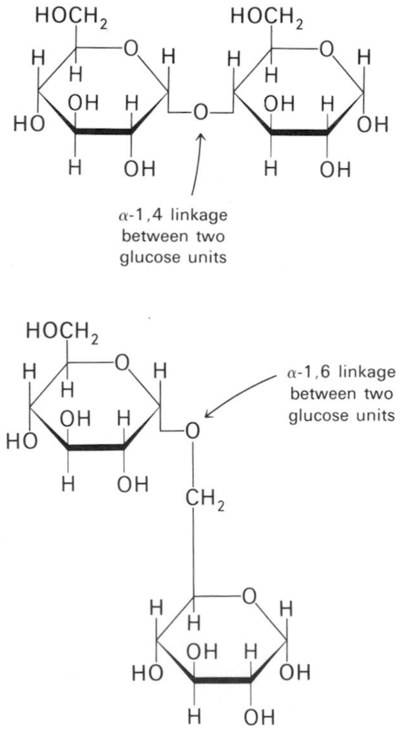

α-1,4 linkage
between two
glucose units

α-1,6 linkage
between two
glucose units

Glycogen is a *readily mobilized storage form of glucose*. It is a branched polymer of glucose residues (Figure 16-1) and has a high molecular weight. Most of the glucose residues in glycogen are linked by α-1,4-glycosidic bonds. The branches are created by α-1,6-glycosidic bonds, which occur at an average of once in ten residues.

The presence of glycogen greatly increases the amount of glucose that is immediately available between meals and during muscular activity. The amount of glucose in the body fluids of an average 70-kg man has an energy content of only 40 kcal, whereas the total body glycogen has an energy content of more than 600 kcal, even after an overnight fast. The two major sites of glycogen storage are the liver and skeletal muscle. The concentration of glycogen is higher in the liver than in muscle, but more glycogen is stored in skeletal muscle because of its much greater mass. Glycogen is present in the cytosol in the form of granules with a diameter ranging from about 100 to 400 Å. This range in size reflects the fact that glycogen molecules do not have a unique size; a typical distribution is centered on a molecular weight of several million. Glycogen granules have a dense appearance in electron micrographs (Figure 16-3). They contain the enzymes that carry out the synthesis and degradation of glycogen and some of the enzymes that regulate

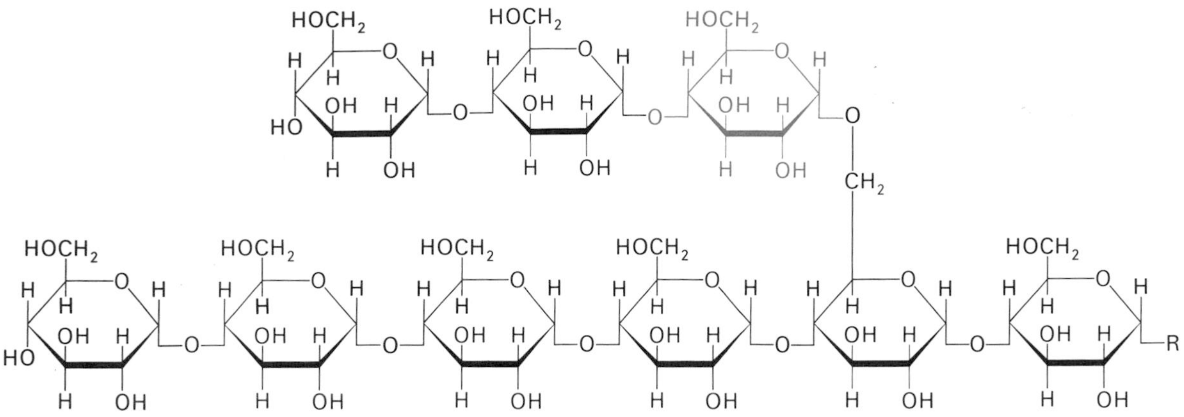

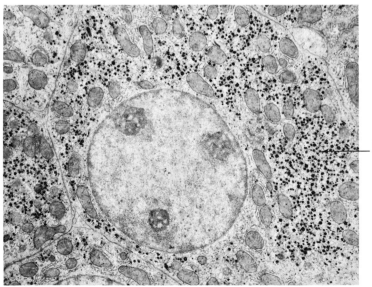

Figure 16-1
Structure of two outer branches of a glycogen particle.
The residues at the nonreducing ends are shown in red.
The residue that starts a branch is shown in green.
The rest of the glycogen molecule is represented by R.

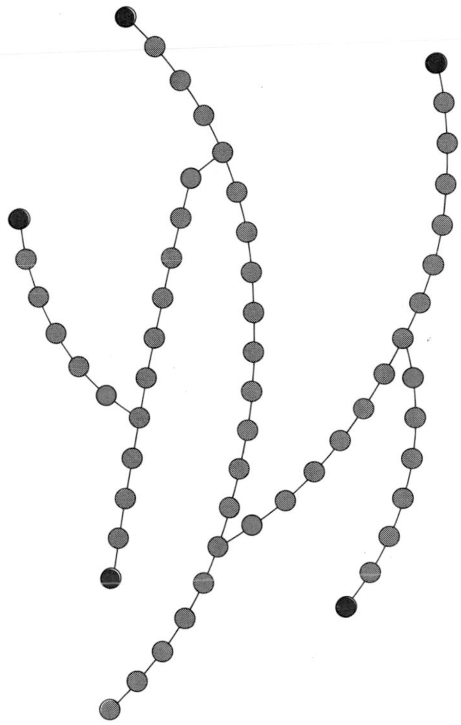

Figure 16-2
Diagram of a cross section of a
glycogen molecule. (Residues are
differentiated by the same colors as
in Figure 16-1.)

Glycogen
granules

Figure 16-3
Electron micrograph of a liver cell. The dense
particles in the cytoplasm are glycogen granules.
[Courtesy of Dr. George Palade.]

these processes. However, a glycogen granule differs from a multi-enzyme complex (such as the pyruvate dehydrogenase complex) in that the bound enzymes are not present in defined stoichiometric ratios. Also, the degree of structural organization in a glycogen granule is less than in a multienzyme complex.

The synthesis and degradation of glycogen is considered here in some detail for several reasons. First, these processes are important because they *regulate the blood glucose level* and provide a *reservoir of glucose* for strenuous muscular activity. Second, the synthesis and degradation of glycogen occur by *different reaction pathways,* which illustrates an important principle of biochemistry. Third, the hormonal regulation of glycogen metabolism is mediated by mechanisms that are of general significance. The *role of cyclic AMP* in the coordinated control of glycogen synthesis and breakdown is well understood and provides insight into the action of hormones in a variety of other systems. Fourth, a number of inherited enzyme defects resulting in impaired glycogen metabolism have been characterized. Some of these *glycogen storage diseases* are lethal in infancy, whereas others have a relatively mild clinical course. The final part of this chapter deals with the metabolism of the common disaccharides: *lactose, maltose, and sucrose.*

PHOSPHORYLASE CATALYZES THE PHOSPHOROLYTIC CLEAVAGE OF GLYCOGEN TO GLUCOSE 1-PHOSPHATE

The pathway of glycogen breakdown was elucidated by the incisive studies of Carl Cori and Gerty Cori. They showed that glycogen is *cleaved by orthophosphate* to yield a new kind of phosphorylated sugar, which they identified as *glucose 1-phosphate.* The Coris also isolated and crystallized *glycogen phosphorylase,* the enzyme that catalyzes this reaction.

$$\text{Glycogen} + \text{P}_i \rightleftharpoons \text{glucose 1-phosphate} + \text{glycogen}$$
$$(n \text{ residues}) \qquad\qquad\qquad\qquad (n-1 \text{ residues})$$

Phosphorylase catalyzes the sequential removal of glycosyl residues from the nonreducing end of the glycogen molecule. The glycosidic linkage between C-1 of the terminal residue and C-4 of the adjacent one is split by orthophosphate. Specifically, the bond between the C-1 carbon atom and the glycosidic oxygen atom is cleaved by orthophosphate, and the α configuration at C-1 is retained.

Glycogen (n residues)   **Glucose 1-phosphate**   **Glycogen** ($n - 1$ residues)

This reaction probably proceeds through a carbonium ion inter-mediate. The breaking of the C_1—O bond, the retention of con-figuration at C-1, and the likely participation of a carbonium ion intermediate are reminiscent of the lysozyme-catalyzed cleavage of chitin.

The reaction catalyzed by phosphorylase is readily reversible in vitro. At pH 6.8, the equilibrium ratio of orthophosphate to glucose 1-phosphate is 3.6. The $\Delta G^{\circ\prime}$ for this reaction is small because a glycosidic bond is replaced by a phosphate ester bond that has a nearly equal transfer potential. However, this reaction proceeds far in the direction of glycogen breakdown in vivo because the $[P_i]/[\text{glucose 1-phosphate}]$ ratio is usually greater than 100.

The phosphorolytic cleavage of glycogen is energetically advan-tageous because the released sugar is phosphorylated. In contrast, a hydrolytic cleavage would yield glucose, which would have to be phosphorylated at the expense of an ATP to enter the glycolytic pathway. An additional advantage of phosphorolytic cleavage for muscle cells is that glucose 1-phosphate cannot diffuse out of the cell, whereas glucose can. The significance of the retention of phosphorylated sugars by muscle will be discussed shortly.

A DEBRANCHING ENZYME IS ALSO NEEDED FOR THE BREAKDOWN OF GLYCOGEN

Glycogen is degraded to a limited extent by phosphorylase alone. The α-1,6-glycosidic bonds at the branch points are not susceptible to cleavage by phosphorylase. Indeed, phosphorylase stops cleaving

α-1,4 linkages when it reaches a terminal residue four away from a branch point. The action of phosphorylase on two outer branches of a glycogen particle is shown in Figure 16-4. Five α-1,4-glycosidic bonds on one branch and three on the other are cleaved by phosphorylase. Cleavage by phosphorylase stops at this stage because terminal residues *a* and *d* are four away from the branch point *h*. A new enzymatic activity is required at this stage. *A transferase shifts a block of three glycosyl residues from one outer branch to the other.* The

GLYCOGEN DEGRADATION

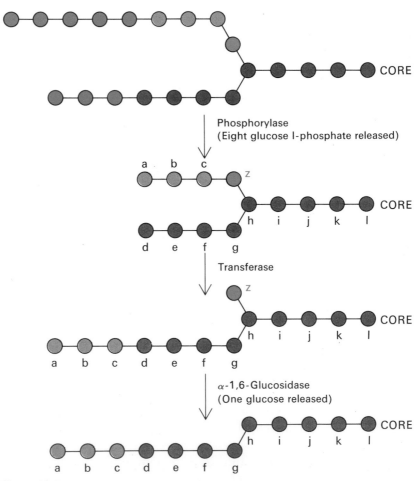

Figure 16-4
Steps in the degradation of glycogen.

α-1,4-glycosidic link between c and z is broken, and a new α-1,4 link between c and d is formed. This transfer exposes residue z to the action of a third degradative enzyme, an *α-1,6-glucosidase*, which is also known as the *debranching enzyme*. This enzyme hydrolyzes the α-1,6-glycosidic bond between residues z and h.

Thus, the transferase and the debranching enzyme (α-1,6-glucosidase) convert the branched structure into a linear one, which paves the way for further cleavage by phosphorylase. The hydrolysis of z renders all of the residues a through l susceptible to phosphorylase. It is interesting to note that the transferase and the α-1,6-glucosidase activities have not been separated, which suggests that they arise from a single enzyme or from two tightly associated enzymes.

PHOSPHOGLUCOMUTASE CONVERTS GLUCOSE 1-PHOSPHATE TO GLUCOSE 6-PHOSPHATE

The glucose 1-phosphate formed in the phosphorylytic cleavage of glycogen is converted to glucose 6-phosphate by *phosphoglucomutase*. The equilibrium mixture contains 95% glucose 6-phosphate. The catalytic site of an active enzyme molecule contains a phosphorylated serine residue. In catalysis, this phosphoryl group is probably transferred to the hydroxyl group at C-6 of glucose 1-phosphate to form glucose 1,6-diphosphate. This intermediate then transfers its C-1 phosphoryl group to the serine residue at the active site of the enzyme, yielding glucose 6-phosphate and regenerating the phosphoenzyme.

Glucose 1-phosphate   Glucose 1,6-diphosphate   Glucose 6-phosphate

The phosphoryl group on the mutase is slowly lost by hydrolysis. It is restored by phosphoryl transfer from glucose 1,6-diphosphate, which is formed from glucose 1-phosphate and adenosine triphosphate (ATP) in a reaction catalyzed by phosphoglucokinase. These reactions are similar to those of *phosphoglyceromutase,* a glycolytic enzyme (p. 289). 2,3-Diphosphoglycerate (2,3-DPG) plays a similar role in the interconversion of 2-phosphoglycerate and 3-phosphoglycerate as does glucose 1,6-diphosphate in the interconversion of the phosphoglucoses. Furthermore, a phosphoenzyme intermediate participates in both reactions.

LIVER CONTAINS GLUCOSE 6-PHOSPHATASE, A HYDROLYTIC ENZYME ABSENT FROM MUSCLE

One of the functions of the liver is to maintain a relatively constant level of glucose in the blood. The liver releases glucose into the blood during muscular activity and in the interval between meals. The released glucose is taken up primarily by the brain and by skeletal muscle. Phosphorylated glucose, in contrast to glucose, cannot readily diffuse out of cells. The liver contains a hydrolytic enzyme, *glucose 6-phosphatase,* that enables glucose to leave that organ. This enzyme is essential for gluconeogenesis (p. 369).

$$\text{Glucose 6-phosphate} + H_2O \longrightarrow \text{glucose} + P_i$$

Glucose 6-phosphatase is also present in the kidneys and intestine, but *it is absent from muscle and brain.* The consequence is that glucose 6-phosphate is retained by muscle and brain, which need large amounts of fuel for the generation of ATP.

GLYCOGEN IS SYNTHESIZED AND DEGRADED BY DIFFERENT PATHWAYS

The reaction catalyzed by glycogen phosphorylase is readily reversed in vitro, since the $\Delta G^{\circ\prime}$ for the elongation of glycogen by glucose 1-phosphate is -0.5 kcal/mol. In fact, the Coris were able to synthesize glycogen from glucose 1-phosphate using phosphorylase and a branching enzyme. However, a number of subsequent experimental observations indicated that glycogen is synthesized in vivo by a different pathway. First, the reaction catalyzed by phosphorylase is at equilibrium when the $[P_i]/[\text{glucose 1-phosphate}]$ ratio is 3.6 at neutral pH, whereas this ratio in cells is usually greater than 100. Hence, the phosphorylase reaction in vivo must proceed in the direction of glycogen degradation. Second, hormones that lead to an increase in phosphorylase activity always elicit glycogen breakdown. Third, patients who lack muscle phosphorylase entirely (in McArdle's disease, see p. 395) are able to synthesize muscle glycogen.

In 1957, Luis Leloir and his coworkers showed that glycogen is synthesized by a different pathway. The glycosyl donor is uridine diphosphate glucose (UDP-glucose) rather than glucose 1-phosphate. *The synthetic reaction is not a reversal of the degradative reaction:*

Synthesis: $\text{Glycogen}_n + \text{UDP-glucose} \longrightarrow \text{glycogen}_{n+1} + \text{UDP}$

Degradation: $\text{Glycogen}_{n+1} + P_i \longrightarrow \text{glycogen}_n + \text{glucose 1-phosphate}$

We now know that *biosynthetic and degradative pathways in biological systems* are almost always distinct. Glycogen metabolism provided the first example of this important principle. *Separate pathways afford much greater flexibility, in terms of both energetics and control.* The cell is no longer at the mercy of mass action; glycogen can be synthesized despite a high ratio of orthophosphate to glucose 1-phosphate.

**Uridine diphosphate glucose
(UDP-glucose)**

UDP-GLUCOSE IS AN ACTIVATED FORM OF GLUCOSE

UDP-glucose, the glucose donor in the biosynthesis of glycogen, is an *activated form of glucose,* just as ATP and acetyl CoA are activated forms of orthophosphate and acetate, respectively. The C-1 carbon atom of the glucosyl unit of UDP-glucose is activated because its hydroxyl group is esterified to the diphosphate moiety of UDP.

UDP-glucose is synthesized from glucose 1-phosphate and uridine triphosphate (UTP) in a reaction catalyzed by *UDP-glucose pyrophosphorylase.* The pyrophosphate liberated in this reaction comes from the outer two phosphoryl residues of UTP.

Glucose 1-phosphate   UDP-glucose

This reaction is readily reversible. However, pyrophosphate is rapidly hydrolyzed in vivo to orthophosphate by an inorganic pyrophosphatase. The essentially irreversible hydrolysis of pyrophosphate drives the synthesis of UDP-glucose.

$$\text{Glucose 1-phosphate} + \text{UTP} \rightleftharpoons \text{UDP-glucose} + \text{PP}_i$$
$$\text{PP}_i + \text{H}_2\text{O} \longrightarrow 2\,\text{P}_i$$
$$\overline{\text{Glucose 1-phosphate} + \text{UTP} + \text{H}_2\text{O} \longrightarrow \text{UDP-glucose} + 2\,\text{P}_i}$$

The synthesis of UDP-glucose exemplifies a recurring theme in biochemistry: *biosynthetic reactions are often driven by the hydrolysis of pyrophosphate*. Another aspect of this reaction has broad significance. Nucleoside diphosphate sugars serve as glycosyl donors in the biosynthesis of many disaccharides and polysaccharides.

GLYCOGEN SYNTHETASE CATALYZES THE TRANSFER OF GLUCOSE FROM UDP-GLUCOSE TO A GROWING CHAIN

New glucosyl units are added to the nonreducing terminal residues of glycogen. The activated glucosyl unit of UDP-glucose is transferred to the hydroxyl group at a C-4 terminus of glycogen to form an α-1,4-glycosidic linkage. In this elongation reaction, UDP is displaced by this terminal hydroxyl group of the growing glycogen molecule. This reaction is catalyzed by *glycogen synthetase,* which can add glucosyl residues only if the polysaccharide chain already contains more than four residues. Thus, glycogen synthesis requires a *primer.*

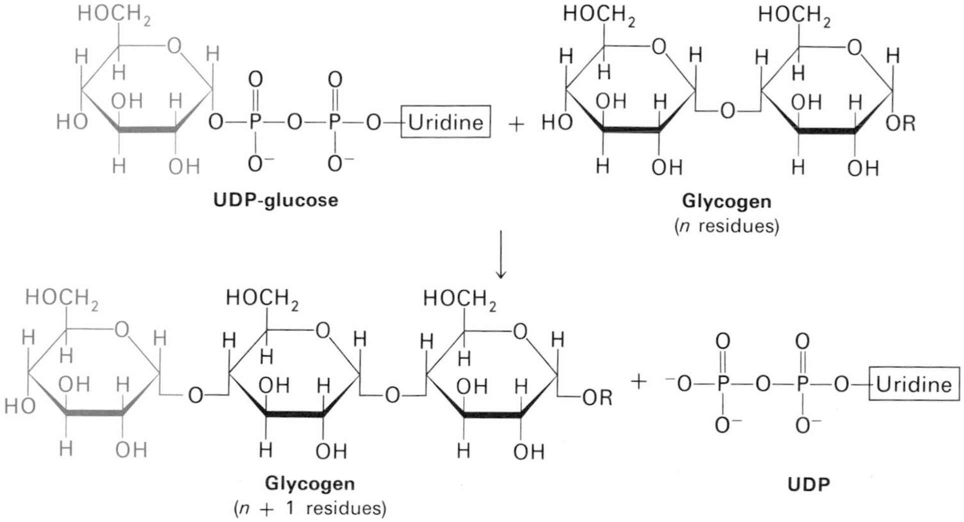

UDP-glucose

Glycogen
(*n* residues)

Glycogen
(*n* + 1 residues)

UDP

A BRANCHING ENZYME FORMS ALPHA-1,6 LINKAGES

Glycogen synthetase catalyzes only the synthesis of α-1,4 linkages. Another enzyme is needed to form α-1,6 linkages that make glycogen a branched polymer. *Branching is important because it increases the solubility of glycogen.* Furthermore, branching creates a large number of nonreducing terminal residues, which are the sites of action of glycogen phosphorylase and synthetase. Thus, *branching increases the rate of glycogen synthesis and degradation.*

Branching occurs after a number of glucosyl residues are joined in α-1,4 linkage by glycogen synthetase. A branch is created by the breaking of an α-1,4 link and the formation of an α-1,6 link, which is a different reaction from debranching. A block of residues, typically seven in number, is transferred to a more interior site. The *branching enzyme* that catalyzes this reaction is quite exacting. The block of seven or so residues must include the nonreducing terminus and come from a chain at least eleven residues long. In addition, the new branch point must be at least four residues away from a preexisting one.

GLYCOGEN IS A VERY EFFICIENT STORAGE FORM OF GLUCOSE

What is the cost of converting glucose 6-phosphate into glycogen and back to glucose 6-phosphate? The pertinent reactions have already been discussed, except for reaction (5) below, which is the regeneration of UTP. UDP is phosphorylated by ATP in a reaction catalyzed by *nucleoside diphosphokinase.*

(1)   Glucose 6-phosphate $\longrightarrow$ glucose 1-phosphate

(2)   Glucose 1-phosphate + UTP $\longrightarrow$ UDP-glucose + PP_i

(3) $PP_i + H_2O \longrightarrow 2 P_i$

(4)   UDP-glucose + glycogen$_n$ $\longrightarrow$ glycogen$_{n+1}$ + UDP

(5)   UDP + ATP $\longrightarrow$ UTP + ADP

Sum: Glucose 6-phosphate + ATP + glycogen$_n$ + $H_2O \longrightarrow$
$$\text{glycogen}_{n+1} + \text{ADP} + 2 P_i$$

Thus, one high-energy phosphate bond is spent in incorporating glucose 6-phosphate into glycogen. The energy yield from the breakdown of glycogen is highly efficient. About 90% of the residues are phosphorolytically cleaved to glucose 1-phosphate, which is converted at no cost to glucose 6-phosphate. The other 10% are branch residues, which are hydrolytically cleaved. One ATP is then used to phosphorylate each of these glucose molecules to glucose 6-phosphate. Since the complete oxidation of glucose 6-phosphate yields 37 ATP and storage consumes slightly more than 1 ATP per glucose 6-phosphate, *the overall efficiency of storage is nearly 97%.*

CYCLIC AMP IS CENTRAL TO THE COORDINATED CONTROL OF GLYCOGEN SYNTHESIS AND BREAKDOWN

The occurrence of separate pathways for the synthesis and degradation of glycogen means that they must be rigorously controlled. ATP would be wastefully hydrolyzed if both sets of reactions were simultaneously active. In fact, glycogen synthesis and degradation are *coordinately controlled so that glycogen synthetase is switched off when phosphorylase is fully active, and vice versa.*

Glycogen metabolism is profoundly affected by specific hormones. Muscular activity or its anticipation leads to the release of epinephrine by the adrenal medulla. *Epinephrine* markedly stimulates glycogen breakdown in muscle and, to a lesser extent, in liver. The liver is more responsive to *glucagon,* a polypeptide hormone that is secreted by the α cells of the pancreas when the blood sugar level is low. This hormone increases the blood sugar level by stimulating the breakdown of glycogen in the liver.

Epinephrine

$$^+H_3N\text{-}His\text{-}Ser\text{-}Glu\text{-}Gly\text{-}Thr\text{-}Phe\text{-}Thr\text{-}Ser\text{-}Asp\text{-}Tyr\text{-} \quad 10$$

$$\text{-}Ser\text{-}Lys\text{-}Tyr\text{-}Leu\text{-}Asp\text{-}Ser\text{-}Arg\text{-}Arg\text{-}Ala\text{-}Gln\text{-} \quad 20$$

$$\text{-}Asp\text{-}Phe\text{-}Val\text{-}Gln\text{-}Trp\text{-}Leu\text{-}Met\text{-}Asn\text{-}Thr\text{-}COO^- \quad 29$$

Glucagon

Earl Sutherland discovered that the action of epinephrine and glucagon metabolism is mediated by *cyclic adenosine monophosphate* (cyclic AMP). This discovery led to the recognition that cyclic AMP is ubiquitous in all forms of life and plays a key role in controlling biological processes (see Chapter 34). The synthesis of this regulatory molecule from ATP is catalyzed by *adenyl cyclase*, an enzyme associated with the plasma membrane. The synthesis of cyclic AMP is accelerated by the subsequent hydrolysis of pyrophosphate.

ATP   **Cyclic AMP**

Epinephrine and glucagon do not enter their target cells. Rather, they bind to the plasma membrane and stimulate adenyl cyclase. The increased intracellular level of cyclic AMP triggers a series of reactions that activate phosphorylase and inhibit glycogen synthetase. We will now consider the structural basis of the control of these enzymes, and then turn to the reaction cascade that links cyclic AMP to these crucial enzymes of glycogen metabolism.

PHOSPHORYLASE IS ACTIVATED BY PHOSPHORYLATION OF A SPECIFIC SERINE RESIDUE

Skeletal muscle phosphorylase exists in two forms: an *active* phosphorylase *a* and a usually *inactive* phosphorylase *b*. Phosphorylase *b* is a dimer, whereas phosphorylase *a* is primarily a tetramer. The monomer has a molecular weight of 92,000. Phosphorylase *b* is converted to phosphorylase *a* by the phosphorylation of a specific serine residue in each subunit. This covalent modification is catalyzed by a specific enzyme, *phosphorylase kinase,* which was discovered by Edmond Fischer and Edwin Krebs.

Muscle phosphorylase *b* is active only in the presence of high concentrations of AMP, which acts allosterically. AMP binds to

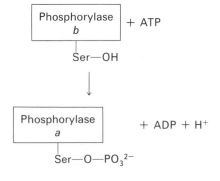

the nucleotide binding site and alters the conformation of phosphorylase *b*. ATP acts as a negative allosteric effector by competing with AMP. Glucose 6-phosphate also inhibits phosphorylase *b*, presumably by binding to another site. Under most physiological conditions, *phosphorylase* b *is inactive because of the inhibitory effects of ATP and glucose 6-phosphate.* In contrast, *phosphorylase* a *is fully active,* irrespective of the levels of AMP, ATP, and glucose 6-phosphate.

PHOSPHORYLASE KINASE IS ALSO ACTIVATED BY PHOSPHORYLATION

The activity of phosphorylase kinase is also regulated by covalent modification. Phosphorylase kinase, like phosphorylase, is converted from *a low-activity to a high-activity form by phosphorylation.* The enzyme that catalyzes this activation reaction is a component of the hormone-cyclic-AMP system, which will be discussed shortly. Phosphorylase kinase can be partially activated in a different way by Ca^{2+} levels of the order of 10^{-7} M. This mode of activation of the kinase is biologically significant because muscle contraction is triggered by the release of Ca^{2+} (Chapter 35). Thus, *glycogen breakdown and muscle contraction are linked by a transient increase in the* Ca^{2+} *level in the cytoplasm.*

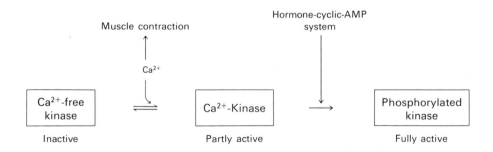

GLYCOGEN SYNTHETASE ACTIVITY IS INACTIVATED BY PHOSPHORYLATION OF A SPECIFIC SERINE RESIDUE

Phosphorylation converts glycogen synthetase into a less active *dependent form* (D form). The phosphorylated D form requires a high level of glucose 6-phosphate for activity, whereas the dephosphory-lated enzyme does not depend on glucose 6-phosphate for activity. Hence, it is called the *independent form* (I form). Thus, *phosphorylation has opposite effects on the enzymatic activities of glycogen synthetase and phosphorylase.*

A REACTION CASCADE CONTROLS THE PHOSPHORYLATION OF GLYCOGEN SYNTHETASE AND PHOSPHORYLASE

We turn now to the link between the hormones that affect glycogen metabolism and the phosphorylation reactions that determine the activities of glycogen synthetase and phosphorylase. The reaction sequence (Figure 16-5) is:

1. Epinephrine binds to the plasma membrane of the muscle cell and stimulates adenyl cyclase.

2. Adenyl cyclase in the plasma membrane catalyzes the formation of cyclic AMP from ATP.

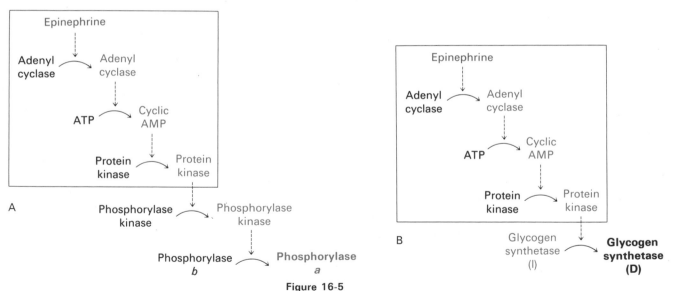

Figure 16-5
Reaction cascades for the control of glycogen metabolism: (A) glycogen degradation; (B) glycogen synthesis. Inactive forms are shown in red, and active ones in green. The sequence of reactions leading to the activation of the protein kinase is the same for the regulation of glycogen degradation and synthesis.

3. The increased intracellular level of cyclic AMP activates a *protein kinase*. This kinase is inactive in the absence of cyclic AMP. The binding of cyclic AMP allosterically stimulates the protein kinase (see Chapter 34).

4. The cyclic-AMP-dependent protein kinase phosphorylates both phosphorylase kinase and glycogen synthetase. *The phosphorylation of both enzymes is the basis of the coordinated regulation of glycogen synthesis and breakdown.* Phosphorylation by this cyclic-AMP-dependent kinase *switches on phosphorylase* (via phosphorylase kinase) and simultaneously *switches off glycogen synthetase* (directly).

PHOSPHATASES REVERSE THE REGULATORY EFFECTS OF KINASES

The changes in enzymatic activity produced by phosphorylation can be reversed by the hydrolytic removal of the phosphoryl group. For example, the conversion of phosphorylase *a* to *b* is catalyzed by *phosphorylase phosphatase*.

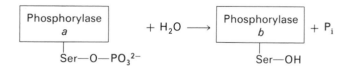

In fact, *for every kinase, there is a corresponding phosphatase.* The other two phosphatases in the control of glycogen metabolism are *phosphorylase kinase phosphatase* and *glycogen synthetase phosphatase*.

The enzymes controlling glycogen metabolism are not simultaneously phosphorylated and dephosphorylated, since this would only lead to a wasteful hydrolysis of ATP. The activities of the phosphatases also appear to be regulated. For example, the combination of Ca^{2+} and Mg-ATP inhibits phosphorylase phosphatase, whereas it activates phosphorylase kinase. Muscle glycogen synthetase phosphatase is inhibited by glycogen. Thus, at high glycogen levels, muscle glycogen synthetase will remain in the phosphorylated dependent form.

The signal arising from cyclic AMP can also be switched off. The phosphodiester bond in cyclic AMP is hydrolyzed by a specific phosphodiesterase, yielding AMP, which does not activate the protein kinase. This highly exergonic reaction has a $\Delta G°'$ of -11.9 kcal/mol.

Cyclic AMP   AMP

THE REACTION CASCADE AMPLIFIES THE HORMONAL SIGNAL

The enzymatic cascade in the control of glycogen metabolism is analogous to the proteolytic cascade in blood clotting (see p. 193). In both processes, the enzymatic cascade provides a *high degree of amplification*. In glycogen breakdown, there are three enzymatically catalyzed control stages, whereas in glycogen synthesis, there are two control stages. If glycogen phosphorylase and synthetase were directly regulated by the binding of epinephrine, more than a thousand times as much epinephrine would be required to elicit glycogen breakdown than in the presence of the amplifying cascade.

It is significant that two types of control mechanisms—covalent modification and noncovalent allosteric control—are used to regulate glycogen metabolism. This combination gives greater flexibility than either kind of mechanism alone. For example, epinephrine is not the only signal for stimulating glycogen breakdown in muscle. A low energy charge also has this effect since phosphorylase *b* is activated by the reversible binding of AMP under these conditions. Glycogen breakdown is also stimulated by an elevated concentration of Ca^{2+}, which occurs during muscle contraction.

A VARIETY OF GENETICALLY DETERMINED
GLYCOGEN STORAGE DISEASES ARE KNOWN

The first glycogen storage disease was described by Edgar von Gierke in 1929. Patients with this disease have a huge abdomen caused by a *massive enlargement of the liver*. There is a pronounced *hypoglycemia* between meals. Furthermore, the blood glucose level does not rise on administration of epinephrine and glucagon. Infants with this glycogen storage disease may have convulsions because their blood glucose level is low.

The enzymatic defect in von Gierke's disease was elucidated by the Coris in 1952. They found that *glucose 6-phosphatase was missing from the liver of a patient with this disease*. This was the first demonstration of an inherited deficiency of a liver enzyme. The liver glycogen is normal in structure but present in abnormally large amounts. The absence of glucose 6-phosphatase in the liver causes hypoglycemia because glucose cannot be formed from glucose 6-phosphate. There is a compensatory increase in glycolysis in the liver, leading to a high level of lactate and pyruvate in the blood. These patients also have an increased dependence on fat metabolism.

A number of glycogen storage disease have been characterized (Table 16-1). The Coris elucidated the biochemical defect in another glycogen storage disease (type III), which cannot be distinguished from von Gierke's disease (type I) by physical examination alone. In type III disease, the structure of liver and muscle glycogen is abnormal and the amount is markedly increased. Most striking, the outer branches of the glycogen are very short. *Patients having this type lack the debranching enzyme (α-1,6-glucosidase), and so only the outermost branches of glycogen can be effectively utilized.* Thus, only a small fraction of this abnormal glycogen is functionally active as an accessible store of glucose.

A defect in glycogen metabolism confined to muscle is found in McArdle's disease (type V). *Muscle phosphorylase activity is absent,* and the patient exhibits a limited capacity to perform strenuous exercise because of painful muscle cramps. The patient is otherwise normal and well developed. Thus, effective utilization of muscle glycogen is not essential for life.

Table 16-1
Glycogen storage diseases

Type	Defective enzyme	Organ affected	Glycogen in the affected organ	Clinical features
I VON GIERKE'S DISEASE	Glucose 6-phosphatase	Liver and kidney	Increased amount; normal structure.	Massive enlargement of the liver. Failure to thrive. Severe hypoglycemia, ketosis, hyperuricemia, hyperlipemia.
II POMPE'S DISEASE	α-1,4-Glucosidase (lysosomal)	All organs	Massive increase in amount; normal structure.	Cardiorespiratory failure causes death, usually before age 2.
III CORI'S DISEASE	Amylo-1,6-glucosidase (debranching enzyme)	Muscle and liver	Increased amount; short outer branches.	Like Type I, but milder course.
IV ANDERSEN'S DISEASE	Branching enzyme (α-1,4 $\longrightarrow$ α-1,6)	Liver and spleen	Normal amount; very long outer branches.	Progressive cirrhosis of the liver. Liver failure causes death usually before age 2.
V McARDLE'S DISEASE	Phosphorylase	Muscle	Moderately increased amount; normal structure.	Limited ability to perform strenuous exercise because of painful muscle cramps. Otherwise patient is normal and well developed.
VI HERS' DISEASE	Phosphorylase	Liver	Increased amount.	Like Type I, but milder course.
VII	Phosphofructokinase	Muscle	Increased amount; normal structure.	Like Type V.
VIII	Phosphorylase kinase	Liver	Increased amount; normal structure.	Mild liver enlargement. Mild hypoglycemia.

Note: Types I through VII are inherited as autosomal recessives. Type VIII is sex-linked.

STARCH IS THE STORAGE POLYSACCHARIDE IN PLANTS

We turn now to the other common polysaccharides. The nutritional reservoir in plants is *starch*, of which there are two forms. *Amylose*, the unbranched type of starch, consists of glucose residues in α-1,4 linkage. *Amylopectin*, the branched form, has about one α-1,6 linkage per thirty α-1,4 linkages, and so it is like glycogen except for its lower degree of branching.

More than half of the carbohydrate ingested by humans is starch. Both amylopectin and amylose are rapidly hydrolyzed by *α-amylase,* which is secreted by the salivary glands and the pancreas. Alpha-amylase hydrolyzes internal α-1,4 linkages to yield *maltose, malto-triose,* and *α-dextrin.* Maltose consists of two glucose residues in α-1,4 linkage (Figures 16-6 and 16-7) and maltotriose consists of three such residues. Alpha-dextrin is made up of several glucose units joined by an α-1,6 linkage in addition to α-1,4 linkages. Maltose and maltotriose are hydrolyzed to glucose by *maltase,* whereas α-dextrin is hydrolyzed to glucose by *α-dextrinase.* There is a different kind of amylase in malt, called *β-amylase,* which hydrolyzes starch into maltose. Beta-amylase acts only on residues at the nonreducing terminus.

The other major polysaccharide of plants is *cellulose,* which serves a structural rather than a nutritional role. In fact, *cellulose is the most abundant organic compound in the biosphere,* containing more than half of all the organic carbon. Cellulose is an unbranched polymer of glucose residues joined by β-1,4 linkages. Mammals do not have cellulases and therefore cannot digest wood and vegetable fibers. However, some ruminants harbor cellulase-producing bacteria in their digestive tracts and thus can digest cellulose.

Dextran is another polysaccharide made up of glucose residues only, mainly in α-1,6 linkage. Occasional branches are formed by α-1,2, α-1,3, or α-1,4 linkages, depending on the species. Dextran is a storage polysaccharide in yeasts and bacteria. The exoskeletons of insects and crustacea contain *chitin,* which consists of *N*-acetyl-glucosamine in β-1,4 linkage. Thus, chitin is like cellulose except that the substituent at C-2 is an acetylated amino group rather than a hydroxyl group.

MALTOSE, SUCROSE, AND LACTOSE ARE THE COMMON DISACCHARIDES

The structures of the three common disaccharides are shown in Figure 16-7. As mentioned previously, *maltose* arises from the hydrolysis of starch and is then hydrolyzed to glucose by maltase. *Sucrose,* the common table sugar, is obtained commercially from cane or beet. The anomeric carbon atoms of a glucose and a fructose residue are in α-glycosidic linkage in sucrose. Consequently, sucrose

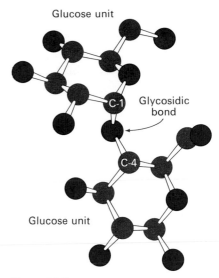

Glucose unit

Glycosidic bond

C-1

C-4

Glucose unit

Figure 16-6
Model of maltose.

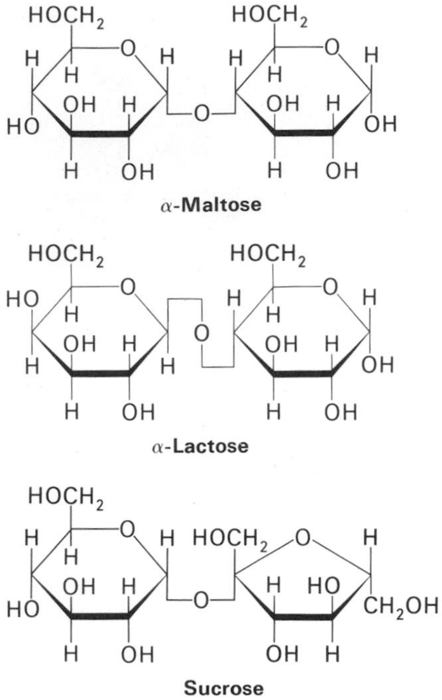

Figure 16-7
Structure of the common disaccharides: maltose, lactose, and sucrose.

has no reducing end group, in contrast to most other sugars. The hydrolysis of sucrose to glucose and fructose is catalyzed by *sucrase*. *Lactose* is the disaccharide of milk. It is not found elsewhere in appreciable amounts. Lactose is hydrolyzed to galactose and glucose by *lactase*. Lactase, sucrase, maltase, and α-dextrinase are bound to mucosal cells lining the small intestine.

UDP-sugars are the activated intermediates in the synthesis of sucrose and lactose, just as UDP-glucose is the glucosyl donor in the synthesis of glycogen. In fact, *nucleoside diphosphate sugars are the activated intermediates in nearly all syntheses of glycosidic linkages.* For example, cellulose is synthesized from either adenosine diphosphoglucose (ADP-glucose), cytidine diphosphoglucose (CDP-glucose), or guanosine diphosphoglucose (GDP-glucose), depending on the kind of plant. *Sucrose* is synthesized by the transfer of glucose from UDP-glucose to fructose 6-phosphate to form sucrose 6-phosphate, which is then hydrolyzed to sucrose.

LACTOSE SYNTHESIS IS CONTROLLED BY A MODIFIER SUBUNIT

The control of lactose synthesis is especially interesting. Lactose synthetase consists of a catalytic subunit and a modifier subunit. The isolated catalytic subunit, called *galactosyl transferase,* is catalytically active in transferring galactose from UDP-galactose to *N*-acetylglucosamine to form *N*-acetyllactosamine.

$$\text{UDP-galactose} + N\text{-acetylglucosamine}$$
$$\downarrow \text{catalytic subunit alone}$$
$$\text{UDP} + N\text{-acetyllactosamine}$$

The specificity of the catalytic subunit changes on binding α-*lactalbumin,* the modifier subunit. The resulting complex, called *lactose synthetase,* transfers galactose to glucose rather than to *N*-acetylglucosamine, yielding lactose.

$$\text{UDP-galactose} + \text{glucose}$$
$$\downarrow \text{lactose synthetase (catalytic and modifier subunits)}$$
$$\text{UDP} + \text{lactose}$$

Galactosyl transferase is present in most tissues, where it partici-pates in the synthesis of the carbohydrate moiety of glycoproteins. Lactose synthetase, in contrast, is found only in the mammary gland. During pregnancy, galactosyl transferase is synthesized and stored in the mammary gland, but there is little modifier subunit. At parturition, abrupt changes in the levels of some hormones elicit the synthesis of large amounts of the modifier subunit. The resulting lactose synthetase complex then forms large amounts of lactose. *The control of lactose synthesis illustrates that hormones may exert their physiological effects by switching enzymatic specificity.*

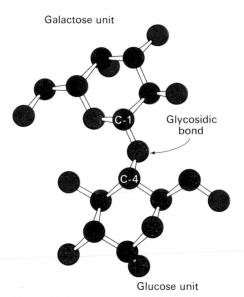

Figure 16-8
Model of lactose.

ENTRY OF FRUCTOSE AND GALACTOSE INTO GLYCOLYSIS

Fructose comprises a significant portion of the carbohydrate in the diet. A typical daily intake is 100 grams, both as the free sugar and as a moiety of sucrose. A large part of the ingested fructose is metabolized by the liver, using the *fructose 1-phosphate pathway.* The first step in this pathway is the phosphorylation of fructose to fructose 1-phosphate by fructokinase. Fructose 1-phosphate is then split into glyceraldehyde and dihydroxyacetone phosphate. This aldol cleavage is catalyzed by a specific fructose 1-phosphate aldo-lase. Glyceraldehyde is then phosphorylated to glyceraldehyde 3-phosphate by triose kinase so that it too can enter glycolysis. *Alternatively, fructose can be phosphorylated to fructose 6-phosphate by hexokinase.* However, the affinity of hexokinase for glucose is twenty times as high as it is for fructose. In the liver, the glucose level is high and so there is little phosphorylation of fructose to fructose 6-phosphate. However, in adipose tissue, the fructose level is much higher than that of glucose, and so the formation of fructose 6-phosphate is not competitively inhibited to an appreciable extent. Hence, most of the fructose in adipose tissue is metabolized via fructose 6-phosphate.

Galactose arises from the hydrolysis of lactose, the major carbohy-drate of milk. Galactose is then converted to glucose 1-phosphate in four steps. The first reaction in the *galactose-glucose interconversion pathway* is the phosphorylation of galactose to galactose 1-phosphate by galactokinase.

$$\text{Galactose} + \text{ATP} \longrightarrow \text{galactose 1-phosphate} + \text{ADP} + \text{H}^+$$

UDP-galactose is then formed in an interchange reaction that is catalyzed by *galactose 1-phosphate uridyl transferase.*

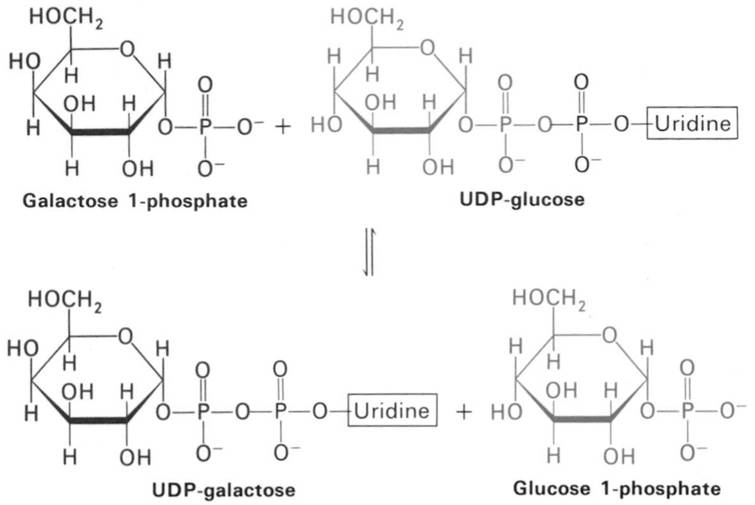

HOCH₂

Galactose 1-phosphate   **UDP-glucose**

HOCH₂

UDP-galactose   **Glucose 1-phosphate**

Galactose is then epimerized to glucose while attached to UDP. The configuration of the hydroxyl group at C-4 is inverted by *UDP-galactose-4-epimerase,* which contains a tightly bound NAD⁺. Fluorescence studies have shown that this NAD⁺ is transiently reduced to NADH during catalysis. NAD⁺ probably accepts the hydrogen atom attached to C-4 of the sugar, and a 4-keto sugar intermediate is formed. NADH then transfers its hydrogen to the other side of C-4 to form the other epimer.

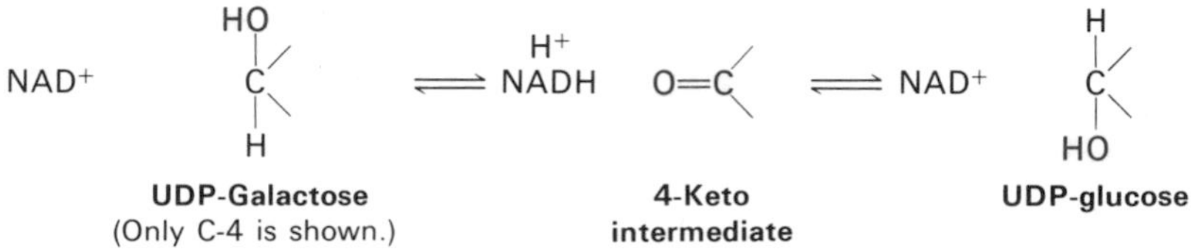

NAD⁺   H⁺   NADH   ⇌   NAD⁺

UDP-Galactose   **4-Keto**   **UDP-glucose**
(Only C-4 is shown.)   **intermediate**

The sum of the reactions catalyzed by galactokinase, the transferase, and the epimerase is:

$$\text{Galactose} + \text{ATP} \longrightarrow \text{glucose 1-phosphate} + \text{ADP} + \text{H}^+$$

The reaction catalyzed by the epimerase is important not only for the utilization of galactose as a fuel molecule. *The conversion of UDP-glucose to UDP-galactose is essential for the synthesis of galactosyl residues in complex polysaccharides and glycoproteins if the amount of galactose in the diet is inadequate to meet these needs.*

GALACTOSE IS HIGHLY TOXIC IF THE TRANSFERASE IS MISSING

The absence of galactose 1-phosphate uridyl transferase causes *galactosemia,* a severe disease that is inherited as an autosomal recessive. The metabolism of galactose in individuals who have this disease is blocked at galactose 1-phosphate. Afflicted infants fail to thrive. Vomiting or diarrhea occurs following the onset of milk consumption, and enlargement of the liver and jaundice are common. Furthermore, many galactosemics are mentally retarded. The blood galactose level is markedly elevated and galactose is found in the urine. In addition, the concentration of galactose 1-phosphate in red cells is elevated. The absence of the transferase in red blood cells is a definitive diagnostic criterion.

Galactosemia is treated by the exclusion of galactose from the diet. A galactose-free diet leads to a striking regression of virtually all of the clinical symptoms, except for mental retardation, which may not be reversible. Continued galactose intake may lead to death in some patients. *The damage in galactosemia is caused by an accumulation of toxic substances, rather than by the absence of an essential compound.* Patients are able to synthesize UDP-galactose from UDP-glucose since their epimerase activity is normal. One of the toxic substances is *galactitol,* which is formed by reduction of galactose. A high level of galactitol leads to the formation of cataracts in the lenses. Galactose 1-phosphate may be a precursor of some of the other toxic agents, or it may itself be toxic at high levels.

$$
\begin{array}{c}
CH_2OH \\
| \\
H-C-OH \\
| \\
HO-C-H \\
| \\
HO-C-H \\
| \\
H-C-OH \\
| \\
CH_2OH
\end{array}
$$

Galactitol

SUMMARY

Glycogen, a readily mobilized fuel store, is a branched polymer of glucose residues. Most of the glucose units in glycogen are linked by α-1,4-glycosidic bonds. At about every tenth residue, a branch is created by an α-1,6-glycosidic bond. Glycogen is present in large amounts in muscle and in the liver, where it is stored in the cytoplasm in the form of hydrated granules. Most of the glycogen molecule is degraded to glucose 1-phosphate by the action of phosphorylase. The glycosidic linkage between C-1 of a terminal residue and C-4 of the adjacent one is split by orthophosphate to give glucose 1-phosphate, which can be reversibly converted to glucose 6-phosphate. Branch points are degraded by the concerted action of two other enzymes, a transferase and an α-1,6-glucosidase. The latter enzyme (also known as the debranching enzyme) catalyzes the hydrolysis of α-1,6 linkages, yielding free glucose. Glycogen is synthesized by a different pathway. UDP-glucose, the activated intermediate in glycogen synthesis, is formed from glucose 1-phosphate and UTP. Glycogen synthetase catalyzes the transfer of glucose from UDP-glucose to the C-4 hydroxyl group of a terminal residue in the growing glycogen molecule. A branching enzyme converts some of the α-1,4 linkages into α-1,6 linkages.

Glycogen synthesis and degradation are coordinately controlled by an amplifying cascade of reactions. Glycogen synthetase is inactive when phosphorylase is active, and vice versa. Epinephrine and glucagon stimulate glycogen breakdown and inhibit its synthesis by increasing the intracellular level of cyclic AMP, which activates a protein kinase. Glycogen synthetase is then phosphorylated, which inactivates it. Phosphorylase is also phosphorylated, but in this case the enzyme is converted into a more active form. The phosphoryl groups on these enzymes can be removed by specific phosphatases. Glycogen synthetase and phosphorylase are also regulated by noncovalent allosteric interactions.

The nutritional reservoir in plants is starch, which is like glycogen except that it is less branched. Starch is hydrolyzed to maltose, which consists of two glucose residues joined by an α-1,4 linkage. In turn, maltose is hydrolyzed to glucose. The other common disaccharides are sucrose, which is hydrolyzed to glucose and fructose, and lactose, which is hydrolyzed to galactose and glucose. Galactose and glucose are interconvertible when linked to UDP. UDP-galactose is the activated intermediate in the synthesis of lactose.

SELECTED READINGS

Where to start:
    Cori, C. F., and Cori, G. T., 1947. Polysaccharide phosphorylase. In *Nobel Lectures: Physiology or Medicine (1942–1962)*, pp. 186–206.

American Elsevier (1964).
Leloir, L. F., 1971. Two decades of research on the biosynthesis of saccharides. *Science* 172:1299–1302.

Books and reviews on glycogen and disaccharide metabolism:

Dickens, F., Randle, P. J., and Whelan, W. J., (eds.), 1968. *Carbohydrate Metabolism and Its Disorders,* vols. 1 and 2. Academic Press. [A valuable collection of review articles. Volume 1 contains reviews on glycogen metabolism and hormonal control of carbohydrate metabolism in liver and muscle. Volume 2 contains reviews on glycogen storage diseases and other disorders of glycogen metabolism.]

Newsholme, E. A., and Start, C., 1973. *Regulation in Metabolism.* Academic Press. [Chapter 4 deals with the regulation of glycogen metabolism.]

Fischer, E. H., Heilmeyer, L. M. G., Jr., and Haschke, R., 1971. Phosphorylase and the control of glycogen degradation. *Curr. Top. Cell Regul.* 4:211–251.

Larner, J., and Villar-Palasi, C., 1971. Glycogen synthase and its control. *Curr. Top. Cell Regul.* 3:196–236.

Stalmans, W., and Hers, H. G., 1973. Glycogen synthesis from UDPG. *In* Boyer, P. D., (ed.), *The Enzymes* (3rd ed.), vol. 9, pp. 310–361. Academic Press.

Inherited disease of glycogen and galactose metabolism:

Howell, R. R., 1972. The glycogen storage diseases. *In* Stanbury, J. B., Wyngaarden, J. B., and Fredrickson, D. S., (eds.), *The Metabolic Basis of Inherited Disease* (3rd ed.), pp. 149–173. McGraw-Hill.

Segal, S., 1972. Disorders of galactose metabolism. *In* Stanbury, J. B., Wyngaarden, J. B., and Fredrickson, D. S., (eds.), *The Metabolic Basis of Inherited Disease* (3rd ed.), pp. 131–148.

PROBLEMS

1. Write a balanced equation for the formation of glycogen from galactose.

2. Write a balanced equation for the formation of glucose from fructose in the liver.

3. A sample of glycogen from a patient with liver disease is incubated with orthophosphate, phosphorylase, the transferase, and the debranching enzyme. The ratio of glucose 1-phosphate to glucose formed in this mixture is 100. What is the most likely enzymatic deficiency in this patient?

4. Suggest an explanation for the fact that the amount of glycogen in type I glycogen storage disease (von Gierke's disease) is increased.

5. Patients deficient in galactokinase have a much less severe disease than those deficient in galactose 1-phosphate uridyl transferase.
   (a) What do you infer about the relative toxicity of high intracellular levels of galactose and galactose 1-phosphate?
   (b) By analogy, predict the clinical severity of a deficiency of fructose 1-phosphate aldolase relative to a deficiency of fructokinase.

CHAPTER 17

FATTY ACID METABOLISM .

We turn now from the metabolism of carbohydrates to that of fatty acids, a class of compounds containing a long hydrocarbon chain and a terminal carboxylate group. Fatty acids have two major physiological roles. First, *they are building blocks of phospholipids and glycolipids*. These amphipathic molecules are important components of biological membranes, as was discussed in Chapter 10. Second, *fatty acids are fuel molecules*. They are stored as *triacylglycerols*, which are uncharged esters of glycerol. Triacylglycerols are also called neutral fats or triglycerides.

$$H_3C-(CH_2)_7-\overset{\displaystyle H}{C}=\overset{\displaystyle H}{C}-(CH_2)_7-\overset{\displaystyle O}{\underset{}{C}}-O-\overset{\displaystyle CH_2-O-\overset{O}{\underset{}{C}}-(CH_2)_{14}-CH_3}{\underset{\displaystyle CH_2-O-\underset{O}{\overset{}{C}}-(CH_2)_{16}-CH_3}{CH}}$$

A triacylglycerol

NOMENCLATURE OF FATTY ACIDS

A brief comment on the nomenclature of fatty acids is appropriate before discussing fatty acid metabolism. The systematic name for a fatty acid is derived from the name of its parent hydrocarbon by the substitution of *oic* for the final *e*. The C_{18} saturated fatty acid is called octadecanoic acid since the parent hydrocarbon is octadecane. A C_{18} fatty acid with one double bond is called octadec*enoic* acid; with two double bonds, octadeca*dienoic* acid; and with three double bonds, octadeca*trienoic* acid. The symbol 18:0 denotes a C_{18} fatty acid with no double bonds, whereas 18:2 signifies that there are two double bonds. Fatty acid carbon atoms are numbered starting at the carboxyl terminus:

$$H_3\underset{\omega}{C}-(CH_2)_n-\underset{\beta}{\overset{3}{C}}-\underset{\alpha}{\overset{2}{C}}-\overset{1}{C}\diagup\!\!\!\!\overset{O}{\underset{OH}{}}$$

Carbon atoms 2 and 3 are often referred to as α and β, respectively. The methyl carbon atom at the distal end of the chain is called the ω carbon. The position of a double bond is represented by the symbol Δ followed by a superscript number. For example, *cis*-Δ^9 means that there is a *cis* double bond between carbon atoms 9 and 10; *trans*-Δ^2 means that there is a *trans* double bond between carbon atoms 2 and 3. Since fatty acids are ionized at physiological pH, it is appropriate to refer to them according to their carboxylate form: for example, palmitate or hexadecanoate.

FATTY ACIDS VARY IN CHAIN LENGTH AND DEGREE OF UNSATURATION

Fatty acids in biological systems (Table 17-1) usually contain an even number of carbon atoms, typically between 14 and 24. The 16- and 18-carbon fatty acids are most common. The hydrocarbon chain is almost invariably unbranched in animal fatty acids. The alkyl chain may be saturated or it may contain one or more double bonds. The configuration of the double bonds in most unsaturated fatty acids is *cis*. The double bonds in polyunsaturated fatty acids are separated by at least one methylene group.

The properties of fatty acids and of lipids derived from them are markedly dependent on their chain length and on the degree to which they are unsaturated. Unsaturated fatty acids have a lower melting point than saturated fatty acids of the same length. For example, the melting point of stearic acid is 69.6°C, whereas that of oleic acid (which contains one *cis* double bond) is 13.4°C. The

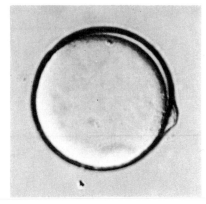

Figure 17-1
Photomicrograph of a fat cell. A large globule of fat is surrounded by a thin rim of cytoplasm and a bulging nucleus. [Courtesy of Dr. Pedro Cuatrecasas.]

Table 17-1
Some naturally occurring fatty acids in animals

Number of carbons	Number of double bonds	Common name	Systematic name	Formula
12	0	Laurate	*n*-Dodecanoate	$CH_3(CH_2)_{10}COO^-$
14	0	Myristate	*n*-Tetradecanoate	$CH_3(CH_2)_{12}COO^-$
16	0	Palmitate	*n*-Hexadecanoate	$CH_3(CH_2)_{14}COO^-$
18	0	Stearate	*n*-Octadecanoate	$CH_3(CH_2)_{16}COO^-$
20	0	Arachidate	*n*-Eicosanoate	$CH_3(CH_2)_{18}COO^-$
22	0	Behenate	*n*-Docosanoate	$CH_3(CH_2)_{20}COO^-$
24	0	Lignocerate	*n*-Tetracosanoate	$CH_3(CH_2)_{22}COO^-$
16	1	Palmitoleate	*cis*-Δ^9-Hexadecenoate	$CH_3(CH_2)_5CH{=}CH(CH_2)_7COO^-$
18	1	Oleate	*cis*-Δ^9-Octadecenoate	$CH_3(CH_2)_7CH{=}CH(CH_2)_7COO^-$
18	2	Linoleate	*cis, cis*-Δ^9,Δ^{12}-Octadecadienoate	$CH_3(CH_2)_4(CH{=}CHCH_2)_2(CH_2)_6COO^-$
18	3	Linolenate	all *cis*-$\Delta^9,\Delta^{12},\Delta^{15}$-Octadecatrienoate	$CH_3CH_2(CH{=}CHCH_2)_3(CH_2)_6COO^-$
20	4	Arachidonate	all *cis*-$\Delta^5,\Delta^8,\Delta^{11},\Delta^{14}$-Eicosatetraenoate	$CH_3(CH_2)_4(CH{=}CHCH_2)_4(CH_2)_2COO^-$

melting points of polyunsaturated fatty acids of the C_{18} series are even lower. Chain length also affects the melting point, as illustrated by the fact that palmitic acid (C_{16}) melts at a temperature 6.5° lower than the melting point of stearic acid (C_{18}). *Thus, short chain length and unsaturation enhance the fluidity of fatty acids and of their derivatives.* The significance of membrane fluidity has already been discussed (Chapter 10).

TRIACYLGLYCEROLS ARE HIGHLY CONCENTRATED ENERGY STORES

Triacylglycerols are highly concentrated stores of metabolic energy because they are *reduced* and *anhydrous*. The yield from the complete oxidation of fatty acids is about 9 kcal/g, in contrast with about 4 kcal/g for carbohydrate and protein. The basis of this large difference in caloric yield is that fatty acids are much more highly reduced. Furthermore, triacylglycerols are very nonpolar and so they are stored in a nearly anhydrous form, whereas proteins and carbohydrates are much more polar and hence more highly hydrated. In fact, a gram of dry glycogen binds about two grams of water. *Consequently, a gram of nearly anhydrous fat stores more than six times as much energy as a gram of hydrated glycogen,* which is the reason that triacylglycerols rather than glycogen were selected in evolution as the major energy reservoir. Consider a typical 70-kg man who has fuel reserves of 100,000 kcal in triacylglycerols, 25,000 kcal in protein (mostly in muscle), 600 kcal in glycogen, and 40 kcal in glucose. Triacylglycerols constitute about 11 kg of his total body weight. If this amount of energy were stored in glycogen, his total body weight would be 55 kg greater.

In mammals, the major site of accumulation of triacylglycerols is the cytoplasm of *adipose cells (fat cells)*. Droplets of triacylglycerol coalesce to form a large globule, which may occupy most of the cell volume. Adipose cells are specialized for the synthesis and storage of triacylglycerols and for their mobilization into fuel molecules that are transported to other tissues by the blood.

TRIACYLGLYCEROLS ARE HYDROLYZED BY CYCLIC-AMP-REGULATED LIPASES

The initial event in the utilization of fat as an energy source is the hydrolysis of triacylglycerol by lipases.

Triacylglycerol   Glycerol   Fatty acids

The activity of adipose cell lipase is regulated by hormones. Epinephrine, norepinephrine, glucagon, and adrenocorticotrophic hormone stimulate the adenyl cyclase of adipose cells. The increased level of cyclic adenosine monophosphate (cyclic AMP) then stimulates a protein kinase, which activates the lipase by phosphorylating it. Thus, epinephrine, norepinephrine, glucagon, and adrenocorticotrophic hormone cause lipolysis. In contrast, insulin inhibits adenyl cyclase and thereby diminishes lipolysis. Thus, *cyclic AMP is a second messenger in the regulation of lipolysis in adipose cells*, which is analogous to its role in the regulation of glycogen breakdown (Chapter 16).

The glycerol formed by lipolysis can enter the glycolytic pathway by being converted to dihydroxyacetone phosphate.

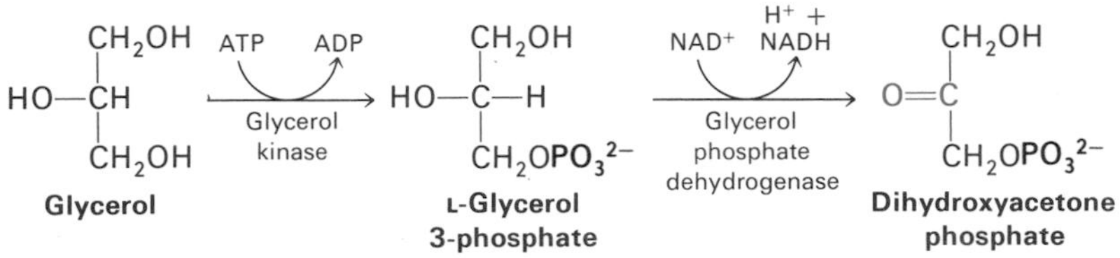

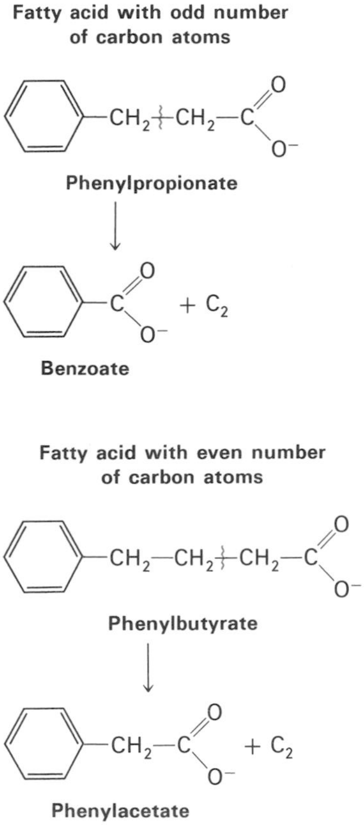

Fatty acid with odd number of carbon atoms

Phenylpropionate

Benzoate

Fatty acid with even number of carbon atoms

Phenylbutyrate

Phenylacetate

Figure 17-2
Knoop's experiment showing that fatty acids are degraded by the removal of two-carbon units.

FATTY ACIDS ARE DEGRADED BY THE SEQUENTIAL REMOVAL OF TWO-CARBON UNITS

In 1904, Franz Knoop made a critical contribution to the elucidation of the mechanism of fatty acid oxidation. He fed dogs straight-chain fatty acids in which the ω-carbon atom was joined to a phenyl group. Knoop found that the urine of these dogs contained a derivative of phenylacetic acid when they were fed phenylbutyrate. In contrast, a derivative of benzoic acid was formed when they were fed phenylpropionate. In fact, phenylacetic acid was produced whenever a fatty acid containing an even number of carbon atoms was fed, whereas benzoic acid was formed whenever a fatty acid containing an odd number was fed (Figure 17-2). Knoop deduced from these findings that *fatty acids are degraded by oxidation at the β-carbon*. These experiments are a landmark in biochemistry because they were the first to use a synthetic label to elucidate reaction mechanisms. Deuterium and radioisotopes came into biochemistry several decades later.

FATTY ACIDS ARE LINKED TO COENZYME A BEFORE THEY ARE OXIDIZED

Eugene Kennedy and Albert Lehninger showed in 1949 that fatty acids are oxidized in mitochondria. Subsequent work demonstrated that they are activated prior to their entry into the mitochondrial matrix. Adenosine triphosphate (ATP) drives the formation of a thioester linkage between the carboxyl group of a fatty acid and the sulfhydryl group of CoA. This activation reaction occurs on the outer mitochondrial membrane, where it is catalyzed by *acyl CoA synthetase* (also called fatty acid thiokinase).

$$R-\overset{\overset{O}{\|}}{C}-O^- + ATP + HS-CoA \rightleftharpoons R-\overset{\overset{O}{\|}}{C}-S-CoA + AMP + PP_i$$

Paul Berg showed that the activation of a fatty acid occurs in two steps. First, the fatty acid reacts with ATP to form an *acyl adenylate*. In this mixed anhydride, the carboxyl group of a fatty acid is bonded to the phosphoryl group of AMP. The other two phosphoryl groups of the ATP substrate are released as pyrophosphate. The sulfhydryl group of CoA then attacks the acyl adenylate, which is tightly bound to the enzyme, to form acyl CoA and AMP.

$$R-\overset{\overset{O}{\|}}{C}-O-\overset{\overset{O}{\underset{O^-}{\|}}}{P}-O-Ribose-Adenine$$

Acyl adenylate
(Acyl-AMP)

$$\underset{\textbf{Fatty acid}}{R-\overset{\overset{O}{\|}}{C}-O^-} + ATP \rightleftharpoons \underset{\textbf{Acyl adenylate}}{R-\overset{\overset{O}{\|}}{C}-AMP} + PP_i$$

$$R-\overset{\overset{O}{\|}}{C}-AMP + HS-CoA \rightleftharpoons \underset{\textbf{Acyl CoA}}{R-\overset{\overset{O}{\|}}{C}-S-CoA} + AMP$$

These partial reactions are freely reversible. In fact, the equilibrium constant for the sum of these reactions is close to 1.

$$R-COO^- + CoA + ATP \rightleftharpoons acyl\ CoA + AMP + PP_i$$

One high-energy bond is broken (between PP_i and AMP) and one high-energy bond is formed (the thioester linkage in acyl CoA).

How is this reaction driven forward? The answer is that pyrophosphate is rapidly hydrolyzed by a pyrophosphatase.

$$R\text{—}COO^- + CoA + ATP + H_2O \longrightarrow$$
$$\text{acyl CoA} + AMP + 2\,P_i$$

This makes the overall reaction irreversible because two high energy bonds are consumed, whereas only one is formed. We see here another example of a recurring theme in biochemistry: *many biosynthetic reactions are made irreversible by the hydrolysis of inorganic pyrophosphate.*

There is another recurring motif in this activation reaction. The occurrence of an enzyme-bound acyl adenylate intermediate is not unique to the synthesis of acyl CoA. *Acyl adenylates are frequently formed when carboxyl groups are activated in biochemical reactions.* For example, the activation of amino acids in protein synthesis occurs by a similar mechanism.

CARNITINE CARRIES LONG-CHAIN ACTIVATED FATTY ACIDS INTO THE MITOCHONDRIAL MATRIX

Fatty acids are activated on the outer mitochondrial membrane, whereas they are oxidized in the mitochondrial matrix. Since long-chain acyl CoA molecules do not readily traverse the inner mitochondrial membrane, a special transport mechanism is needed. Activated long-chain fatty acids are carried across the inner mitochondrial membrane by *carnitine.* The acyl group is transferred from the sulfur atom of CoA to the hydroxyl group of carnitine to form *acyl carnitine,* which diffuses across the inner mitochondrial membrane. On the matrix side of this membrane, the acyl group is transferred back to CoA, which is thermodynamically feasible

Acyl CoA   Carnitine   Acyl carnitine

because the *O*-acyl link in carnitine has a high group-transfer potential. These transacylation reactions are catalyzed by *fatty acyl CoA: carnitine fatty acid transferase.*

A defect in the transferase or a deficiency of carnitine might be expected to impair the oxidation of long-chain fatty acids. Such a disorder has in fact been found in identical twins who had had aching muscle cramps since early childhood. The aches were precipitated by fasting, exercise, or a high-fat diet; fatty acid oxidation is the major energy-yielding process in these three states. The enzymes of glycolysis and glycogenolysis were found to be normal. Lipolysis of triacylglycerols was normal, as evidenced by a rise in the concentration of unesterified fatty acids in the plasma after fasting. Assay of a muscle biopsy showed that the long-chain acyl CoA synthetase was fully active. Furthermore, medium-chain (C_8 and C_{10}) fatty acids were normally metabolized. It is known that carnitine is not required for the permeation of medium-chain acyl CoAs into the mitochondrial matrix. This case report demonstrates in a striking way that *the impaired flow of a metabolite from one compartment of a cell to another can cause disease.*

ACETYL CoA, NADH, AND FADH$_2$ ARE GENERATED IN EACH ROUND OF FATTY ACID OXIDATION

A saturated acyl CoA is degraded by a recurring sequence of four reactions: oxidation linked to flavin adenine dinucleotide (FAD), hydration, oxidation linked to NAD^+, and thiolysis by CoA (Figure 17-3). The fatty acyl chain is shortened by two carbon atoms as a result of these reactions, and FADH$_2$, NADH, and acetyl CoA are generated. David Green, Severo Ochoa, and Feodor Lynen made important contributions to the elucidation of this series of reactions, which is called the *β-oxidation pathway.*

The first reaction in each round of degradation is the *oxidation* of acyl CoA by an acyl CoA dehydrogenase to give an enoyl CoA with a *trans* double bond between C-2 and C-3.

$$\text{Acyl CoA} + \text{E-FAD} \longrightarrow trans\text{-}\Delta^2\text{-enoyl CoA} + \text{E-FADH}_2$$

It is interesting to note than the dehydrogenation of acyl CoA is very similar to the dehydrogenation of succinate in the citric acid

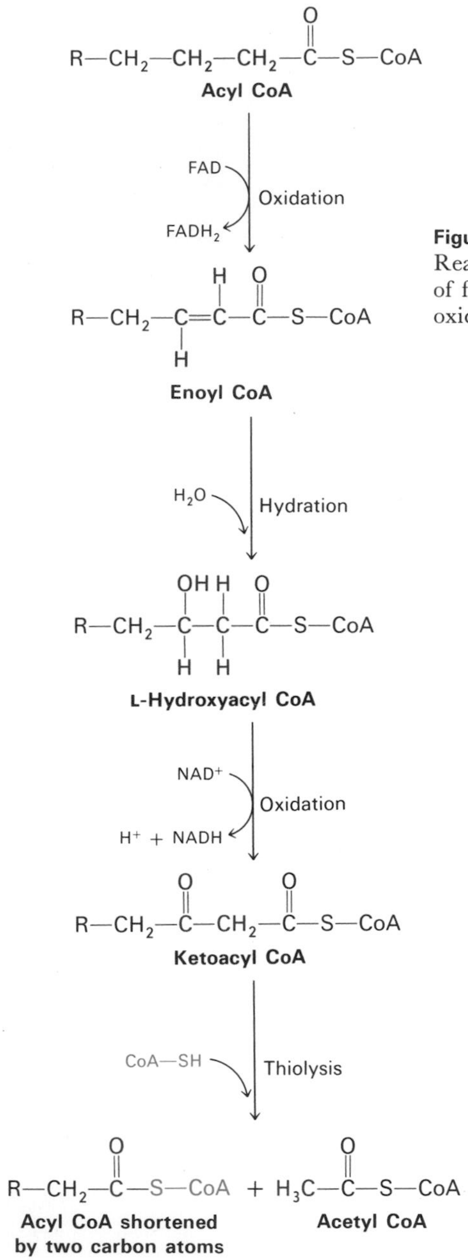

412

Acyl CoA

$$R-CH_2-CH_2-CH_2-\overset{\overset{\displaystyle O}{\|}}{C}-S-CoA$$

FAD ⟶
FADH₂ ↙ | Oxidation

Enoyl CoA

$$R-CH_2-\overset{\overset{\displaystyle H}{|}}{C}=\overset{\overset{\displaystyle H}{|}}{C}-\overset{\overset{\displaystyle O}{\|}}{C}-S-CoA$$

H_2O ↘ | Hydration

L-Hydroxyacyl CoA

$$R-CH_2-\overset{\overset{\displaystyle OH}{|}}{\underset{\underset{\displaystyle H}{|}}{C}}-\overset{\overset{\displaystyle H}{|}}{\underset{\underset{\displaystyle H}{|}}{C}}-\overset{\overset{\displaystyle O}{\|}}{C}-S-CoA$$

NAD^+ ↘ | Oxidation
$H^+ + NADH$ ↙

Ketoacyl CoA

$$R-CH_2-\overset{\overset{\displaystyle O}{\|}}{C}-CH_2-\overset{\overset{\displaystyle O}{\|}}{C}-S-CoA$$

CoA—SH ↘ | Thiolysis

$$R-CH_2-\overset{\overset{\displaystyle O}{\|}}{C}-S-CoA + H_3C-\overset{\overset{\displaystyle O}{\|}}{C}-S-CoA$$

Acyl CoA shortened by two carbon atoms   **Acetyl CoA**

Figure 17-3
Reaction sequence in the degradation of fatty acids: oxidation, hydration, oxidation, and thiolysis.

Figure 17-4
First three rounds in the degradation of palmitate. Two-carbon units are sequentially removed from the carboxyl end of the fatty acid.

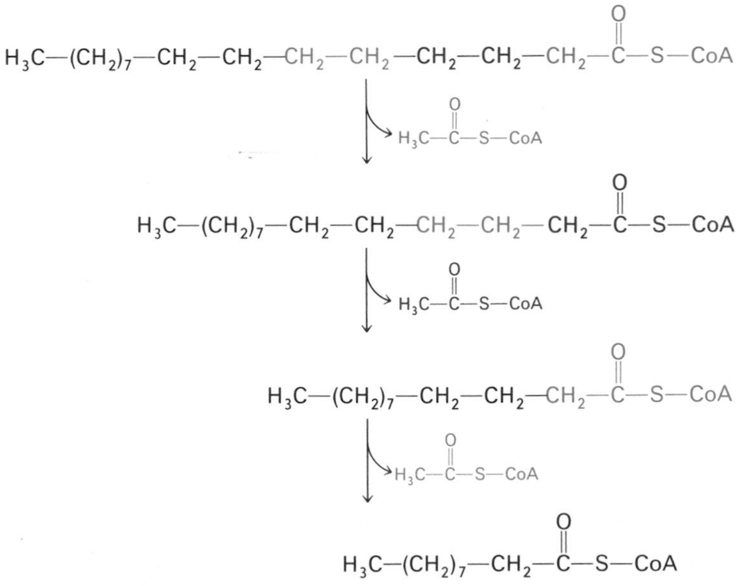

$$H_3C-(CH_2)_7-CH_2-CH_2-CH_2-CH_2-CH_2-CH_2-CH_2-\overset{\overset{\displaystyle O}{\|}}{C}-S-CoA$$

↓ $H_3C-\overset{\overset{\displaystyle O}{\|}}{C}-S-CoA$

$$H_3C-(CH_2)_7-CH_2-CH_2-CH_2-CH_2-CH_2-\overset{\overset{\displaystyle O}{\|}}{C}-S-CoA$$

↓ $H_3C-\overset{\overset{\displaystyle O}{\|}}{C}-S-CoA$

$$H_3C-(CH_2)_7-CH_2-CH_2-CH_2-\overset{\overset{\displaystyle O}{\|}}{C}-S-CoA$$

↓ $H_3C-\overset{\overset{\displaystyle O}{\|}}{C}-S-CoA$

$$H_3C-(CH_2)_7-CH_2-\overset{\overset{\displaystyle O}{\|}}{C}-S-CoA$$

cycle. In fact, the first three reactions in each round of fatty acid degradation closely resemble the last steps in the citric acid cycle:

Acyl CoA $\longrightarrow$ enoyl CoA $\longrightarrow$ hydroxyacyl CoA $\longrightarrow$ ketoacyl CoA

Succinate $\longrightarrow$ fumarate $\longrightarrow$ malate $\longrightarrow$ oxaloacetate

The next step is the *hydration* of the double bond between C-2 and C-3 by enoyl CoA hydratase.

$$trans\text{-}\Delta^2\text{-Enoyl CoA} + H_2O \rightleftharpoons \text{L-3-hydroxyacyl CoA}$$

The hydration of the enoyl CoA is stereospecific, as are the hydrations of fumarate and aconitate. Only the L-isomer of 3-hydroxyacyl CoA is formed when the *trans*-Δ^2 double bond is hydrated. The enzyme also hydrates a *cis*-Δ^2 double bond, but the product then is the D-isomer. We will return to this point shortly in a discussion of the oxidation of unsaturated fatty acids.

The hydration of enoyl CoA is the prelude to the second *oxidation* reaction, which converts the hydroxyl group at C-3 to a keto group and generates NADH. This oxidation is catalyzed by L-3-hydroxyacyl CoA dehydrogenase, which is absolutely specific for the L-isomer of the hydroxyacyl substrate.

$$\text{L-3-Hydroxyacyl CoA} + NAD^+ \rightleftharpoons$$
$$\text{3-ketoacyl CoA} + NADH + H^+$$

The preceding reactions have oxidized the methylene group at C-3 to a keto group. The final step is the *cleavage* of 3-ketoacyl CoA by the thiol group of a second molecule of CoA, which yields acetyl CoA and an acyl CoA shortened by two carbon atoms. This thiolytic cleavage is catalyzed by β-ketothiolase.

$$\underset{(n \text{ carbons})}{\text{3-Ketoacyl CoA}} + \text{HS-CoA} \rightleftharpoons \text{acetyl CoA} + \underset{(n - 2 \text{ carbons})}{\text{acyl CoA}}$$

The shortened acyl CoA then undergoes another cycle of oxidation, starting with the reaction catalyzed by acyl CoA dehydrogenase (Figure 17-4). Beta-ketothiolase, hydroxyacyl dehydrogenase, and enoyl CoA hydratase have broad specificity with respect to the length of the acyl group.

Table 17-2
Principal reactions in fatty acid oxidation

Step	Reaction	Enzyme
1	Fatty acid + CoA + ATP $\rightleftharpoons$ acyl CoA + AMP + PP_i	Acyl CoA synthetase (also called fatty acid thiokinase; fatty acid: CoA ligase [AMP])
2	Carnitine + acyl CoA $\rightleftharpoons$ acyl carnitine + CoA	Fatty acyl CoA: carnitine fatty acid transferase
3	Acyl CoA + E-FAD $\longrightarrow$ $trans$-Δ^2-enoyl CoA + E-FADH$_2$	Acyl CoA dehydrogenases (several enzymes having different chain-length specificity)
4	$trans$-Δ^2-Enoyl CoA + H$_2$O $\rightleftharpoons$ L-3-hydroxyacyl CoA	Enoyl CoA hydratase (also called crotonase or 3-hydroxyacyl CoA hydrolyase)
5	L-3-Hydroxyacyl CoA + NAD$^+$ $\rightleftharpoons$ 3-ketoacyl CoA + NADH + H$^+$	L-3-Hydroxyacyl CoA dehydrogenase
6	3-Ketoacyl CoA + CoA $\rightleftharpoons$ acetyl CoA + acyl CoA (shortened by C$_2$)	β-Ketothiolase (also called thiolase)

THE COMPLETE OXIDATION OF PALMITATE YIELDS 129 ATP

We can calculate the energy yield derived from the oxidation of a fatty acid. In each reaction cycle, an acyl CoA is shortened by two carbons, and one FADH$_2$, NADH, and acetyl CoA are formed.

$$C_n\text{-acyl CoA} + \text{FAD} + \text{NAD}^+ + \text{H}_2\text{O} + \text{CoA} \longrightarrow$$
$$C_{n-2}\text{-acyl CoA} + \text{FADH}_2 + \text{NADH} + \text{acetyl CoA} + \text{H}^+$$

The degradation of palmitoyl CoA (C$_{16}$-acyl CoA) requires seven reaction cycles. In the seventh cycle, the C$_4$-ketoacyl CoA is thiolyzed to two molecules of acetyl CoA. Hence, the stoichiometry of oxidation of palmitoyl CoA is:

$$\text{Palmitoyl CoA} + 7\text{ FAD} + 7\text{ NAD}^+ + 7\text{ CoA} + 7\text{ H}_2\text{O} \longrightarrow$$
$$8\text{ acetyl CoA} + 7\text{ FADH}_2 + 7\text{ NADH} + 7\text{ H}^+$$

Three ATP are generated when each of these NADH are oxidized by the respiratory chain, whereas two ATP are formed for each FADH$_2$, since their electrons enter the chain at the level of CoQ. Recall that the oxidation of acetyl CoA by the citric acid cycle yields 12 ATP. Hence, the number of ATP formed in the oxidation of palmitoyl CoA is 14 from the 7 $FADH_2$, 21 from the 7 NADH, and 96 from the 8 molecules of acetyl CoA, giving a total of 131. Two high-energy phosphate bonds are consumed in the activation of palmitate, in which ATP is split to AMP and 2 P_i. Thus, *the net yield from the complete oxidation of palmitate is 129 ATP.*

The efficiency of energy conservation in fatty acid oxidation can be estimated from the number of ATP formed and from the free energy of oxidation of palmitic acid to CO_2 and H_2O, as determined by calorimetry. The standard free energy of hydrolysis of 129 ATP is −940 kcal (129 × −7.3 kcal). The standard free energy of oxidation of palmitic acid is −2,340 kcal. Hence, *the efficiency of energy conservation in fatty acid oxidation under standard conditions is about 40%,* a value similar to that of glycolysis, the citric acid cycle, and oxidative phosphorylation.

AN ISOMERASE AND AN EPIMERASE ARE REQUIRED FOR THE OXIDATION OF UNSATURATED FATTY ACIDS

We turn now to the oxidation of unsaturated fatty acids. Many of the reactions are the same as those for saturated fatty acids. In fact, only two additional enzymes—an isomerase and an epimerase—are needed to degrade a wide range of unsaturated fatty acids.

Consider the oxidation of palmitoleate. This C_{16} unsaturated fatty acid, which has one double bond between C-9 and C-10, is activated and transported across the inner mitochondrial membrane in the same way as palmitate. Palmitoleoyl CoA then undergoes three cycles of degradation, which are carried out by the same enzymes as in the oxidation of saturated fatty acids. However, the *cis*-Δ^3-enoyl CoA formed in the third round is not a substrate for acyl CoA dehydrogenase. The presence of a double bond between C-3 and C-4 prevents the formation of another double bond between C-2 and C-3. This impasse is resolved by a new reaction that shifts the position and configuration of the *cis*-Δ^3 double bond. *An isomerase converts this double bond into a* trans-Δ^2 *double bond.* The subsequent reactions are those of the saturated fatty acid oxidation pathway, in which the *trans*-Δ^2-enoyl CoA is a regular substrate.

A second accessory enzyme is needed for the oxidation of polyunsaturated fatty acids. The activated C_{18} unsaturated fatty acid

cis-Δ^3-**Enoyl CoA**

Isomerase

trans-Δ^2-**Enoyl CoA**

with $cis\text{-}\Delta^6$ and $cis\text{-}\Delta^9$ double bonds undergoes two rounds of degradation by the saturated fatty acid oxidation pathway. Then $cis\text{-}\Delta^2\Delta^5$-enoyl CoA is hydrated by enoyl CoA hydratase, the same enzyme that hydrates $trans\text{-}\Delta^2$ double bonds in the saturated pathway. However, hydration of a $cis\text{-}\Delta^2$ double bond yields the D-isomer of 3-hydroxyacyl CoA, which is not a substrate for L-3-hydroxyacyl CoA dehydrogenase. This hurdle is overcome by an *epimerase that inverts the configuration of the hydroxyl group at C-3.*

$$H_3C-(CH_2)_7-\overset{H}{C}=\overset{H}{C}-CH_2-\underset{\underset{OH}{|}}{\overset{\overset{H}{|}}{C}}-CH_2-\overset{\overset{O}{\|}}{C}-S-CoA$$

D-3-Hydroxy-
cis-Δ^5-enoyl CoA

$\Big\updownarrow$ Epimerase

$$H_3C-(CH_2)_7-\overset{H}{C}=\overset{H}{C}-CH_2-\underset{\underset{H}{|}}{\overset{\overset{OH}{|}}{C}}-CH_2-\overset{\overset{O}{\|}}{C}-S-CoA$$

L-3-Hydroxy-
cis-Δ^5-enoyl CoA

ODD-CHAIN FATTY ACIDS YIELD PROPIONYL COENZYME A IN THE FINAL THIOLYSIS STEP

Fatty acids having an odd number of carbon atoms are minor species. They are oxidized in the same way as fatty acids having an even number, except that propionyl CoA and acetyl CoA, rather than two molecules of acetyl CoA, are produced in the final round of degradation. The activated three-carbon unit in propionyl CoA enters the citric acid cycle after it is converted to succinyl CoA. The pathway from propionyl CoA to succinyl CoA will be discussed in the next chapter (p. 444) since propionyl CoA is also formed in the oxidation of several amino acids.

$$H_3C-CH_2-\overset{\overset{O}{\|}}{C}-S-CoA$$

Propionyl CoA

KETONE BODIES ARE FORMED FROM ACETYL COENZYME A IF FAT BREAKDOWN PREDOMINATES

The acetyl CoA formed in fatty acid oxidation enters the citric acid cycle if fat and carbohydrate degradation are appropriately balanced. The entry of acetyl CoA depends on the availability of oxaloacetate for the formation of citrate. However, if fat breakdown predominates, acetyl CoA undergoes a different fate. The reason is that the concentration of oxaloacetate is lowered if carbohydrate is not available or is not properly utilized. In fasting or in diabetes, oxaloacetate is used to form glucose and is thus unavailable for condensation with acetyl CoA. Under these conditions, acetyl CoA is diverted to the formation of acetoacetate and D-3-hydroxy-butyrate. Acetoacetate, D-3-hydroxybutyrate, and acetone are sometimes referred to as *ketone bodies*.

Acetoacetate is formed from acetyl CoA in three steps (Figure 17-5). Two molecules of acetyl CoA condense to form acetoacetyl CoA. This reaction, which is catalyzed by thiolase, is a reversal of the thiolysis step in the oxidation of fatty acids. Acetoacetyl CoA then reacts with acetyl CoA and water to give 3-hydroxy-3-methyl-glutaryl CoA and CoA. The unfavorable equilibrium in the formation of acetoacetyl CoA is compensated for by the favorable equilibrium of this reaction, which is due to the hydrolysis of a thioester linkage. 3-Hydroxy-3-methylglutaryl CoA is then cleaved to acetyl CoA and acetoacetate. The sum of these reactions is:

$$2 \text{ Acetyl CoA} + H_2O \longrightarrow \text{acetoacetate} + 2 \text{ CoA} + H^+$$

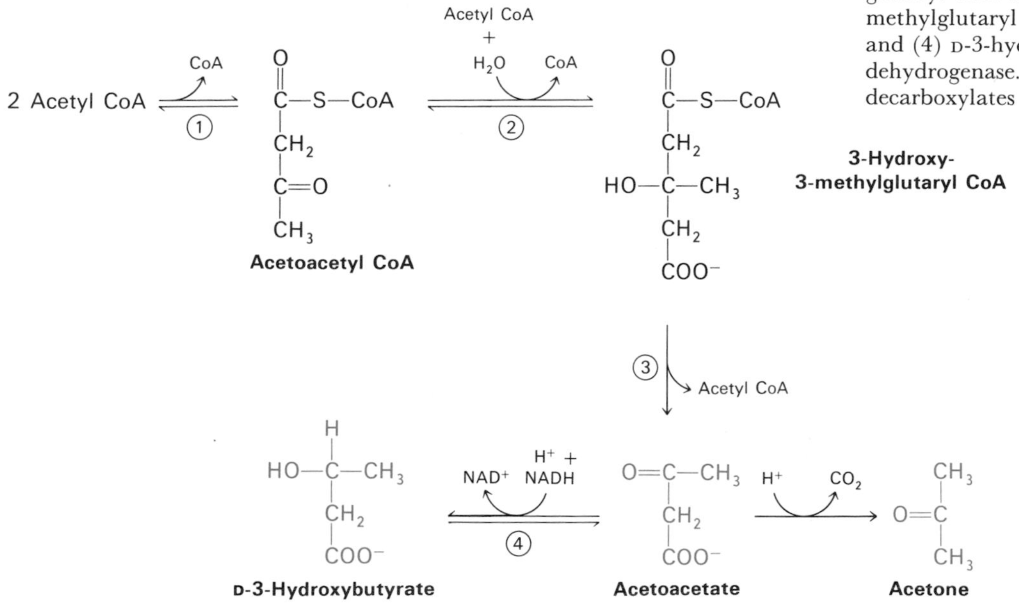

Figure 17-5
Formation of acetoacetate, D-3-hydroxy-butyrate, and acetone from acetyl CoA. Enzymes catalyzing these reactions are: (1) 3-ketothiolase, (2) hydroxymethyl-glutaryl CoA synthetase, (3) hydroxy-methylglutaryl CoA cleavage enzyme, and (4) D-3-hydroxybutyrate dehydrogenase. Acetoacetate spontaneously decarboxylates to form acetone.

3-Hydroxybutyrate is formed by reduction of acetoacetate in the mitochondrial matrix. The ratio of hydroxybutyrate to acetoacetate depends on the $NADH/NAD^+$ ratio inside mitochondria. Acetoacetate also undergoes a slow, spontaneous decarboxylation to acetone. The odor of acetone may be detected in the breath of a person who has a high level of acetoacetate in the blood.

The major site of production of acetoacetate and 3-hydroxybutyrate is the liver. These substances diffuse from the liver mitochondria into the blood and are transported to peripheral tissues. *Acetoacetate and 3-hydroxybutyrate are normal fuels of respiration and are quantitatively important as sources of energy.* Indeed, heart muscle and renal cortex use acetoacetate in preference to glucose. In contrast, glucose is the major fuel for the brain in well-nourished individuals on a balanced diet. However, the brain adapts to the utilization of acetoacetate during starvation and diabetes.

Acetoacetate can be activated by transfer of CoA from succinyl CoA in a reaction catalyzed by thiophorase. Acetoacetyl CoA is then cleaved by thiolase to yield two molecules of acetyl CoA. The liver can supply acetoacetate to other organs because it lacks thiophorase.

It is important to note that *animals are unable to convert fatty acids into glucose.* Specifically, acetyl CoA cannot be converted to pyruvate or oxaloacetate in animals. The two carbon atoms of the acetyl group of acetyl CoA enter the citric acid cycle, but two carbon atoms leave the cycle in the decarboxylations catalyzed by isocitrate dehydrogenase and α-ketoglutarate dehydrogenase. Consequently, oxaloacetate is regenerated but it is not formed de novo when the acetyl unit of acetyl CoA is oxidized by the citric acid cycle.

FATTY ACIDS ARE SYNTHESIZED AND DEGRADED BY DIFFERENT PATHWAYS

Fatty acid synthesis does not occur by a reversal of the degradative pathway. Rather, a new set of reactions are involved, exemplifying the principle that *synthetic and degradative pathways in biological systems are usually distinct.* Some salient features of the pathway for the biosynthesis of fatty acids are:

1. Synthesis occurs in the *cytosol,* in contrast to degradation, which occurs in the mitochondrial matrix.

2. The intermediates in fatty acid synthesis are covalently linked to the sulfhydryl groups of an *acyl carrier protein* (ACP), whereas the intermediates in fatty acid breakdown are bonded to coenzyme A.

3. Many of the enzymes of fatty acid synthesis in higher organisms are organized into a *multienzyme complex* called the *fatty acid synthetase*. In contrast, the degradative enzymes do not appear to be associated.

4. The growing fatty acid chain is elongated by the *sequential addition of two-carbon units* derived from acetyl CoA. The activated donor of two-carbon units in the elongation step is *malonyl-ACP*. The elongation reaction is driven by the release of CO_2.

5. The reductant in fatty acid synthesis is *NADPH*.

6. Elongation by the fatty acid synthetase complex stops upon formation of *palmitate* (C_{16}). Further elongation and the insertion of double bonds are carried out by other enzyme systems.

THE FORMATION OF MALONYL COENZYME A IS THE COMMITTED STEP IN FATTY ACID SYNTHESIS

Salih Wakil's finding that bicarbonate is required for fatty acid biosynthesis was an important clue in the elucidation of this process. In fact, fatty acid synthesis starts with the carboxylation of acetyl CoA to *malonyl CoA*. This irreversible reaction is the committed step in fatty acid synthesis.

$$H_3C-\overset{O}{\underset{\|}{C}}-S-CoA + ATP + HCO_3^- \longrightarrow \;^-O-\overset{O}{\underset{\|}{C}}-CH_2-\overset{O}{\underset{\|}{C}}-S-CoA + ADP + P_i + H^+$$

Acetyl CoA　　　　　　　　　　　**Malonyl CoA**

The synthesis of malonyl CoA is catalyzed by *acetyl CoA carboxylase*, which contains a biotin prosthetic group. The carboxyl group of biotin is covalently attached to the ε-amino group of a lysine residue, as in pyruvate carboxylase (p. 370). Another similarity between acetyl CoA carboxylase and pyruvate carboxylase is that the carboxylation of acetyl CoA occurs in two stages. First, a carboxybiotin intermediate is formed at the expense of an ATP.

The activated CO_2 group in this intermediate is then transferred to acetyl CoA to form malonyl CoA.

$$\text{Biotin-enzyme} + \text{ATP} + \text{HCO}_3^- \rightleftharpoons$$
$$CO_2 \sim \text{biotin-enzyme} + \text{ADP} + P_i$$

$$CO_2 \sim \text{biotin-enzyme} + \text{acetyl CoA} \longrightarrow$$
$$\text{malonyl CoA} + \text{biotin-enzyme}$$

Acetyl CoA carboxylase has been separated into subunits that catalyze partial reactions. Biotin is covalently attached to a small protein (9,000 daltons) called the *biotin carboxyl carrier protein.* The carboxylation of the biotin unit in this carrier protein is catalyzed by *biotin carboxylase,* a second subunit. The third component of the system is a *transcarboxylase,* which catalyzes the transfer of the activated CO_2 unit from carboxybiotin to acetyl CoA. The length and flexibility of the link between biotin and its carrier protein is probably important in enabling the activated carboxyl group to move from one active site to another in this enzyme complex.

The activity of acetyl CoA carboxylase is allosterically regulated, as might be expected since it catalyzes the first committed step in fatty acid synthesis. *Citrate and isocitrate increase the* V_{max} *of acetyl CoA carboxylase approximately fifteenfold,* probably by altering the conformation in the vicinity of the biotin prosthetic group. Acetyl CoA carboxylase exists as an enzymatically inactive protomer or as an active, filamentous polymer comprising from ten to twenty protomers. Citrate and isocitrate bind preferentially to the polymeric form of the enzyme, and so they enhance its activity. *Acetyl CoA carboxylase is inhibited by palmitoyl CoA* and other long-chain acyl CoAs, with a K_i of the order of 10^{-6} M. However, the physiological significance of this inhibition is uncertain, whereas the activating effects of citrate and isocitrate are probably important in vivo. In the presence of these activators, malonyl CoA is formed at a rate equal to that of its utilization in fatty acid synthesis.

INTERMEDIATES IN FATTY ACID SYNTHESIS ARE ATTACHED TO AN ACP

P. Roy Vagelos discovered that the intermediates in fatty acid synthesis are linked to an acyl carrier protein. Specifically, they are linked to the sulfhydryl terminus of a phosphopantetheine group

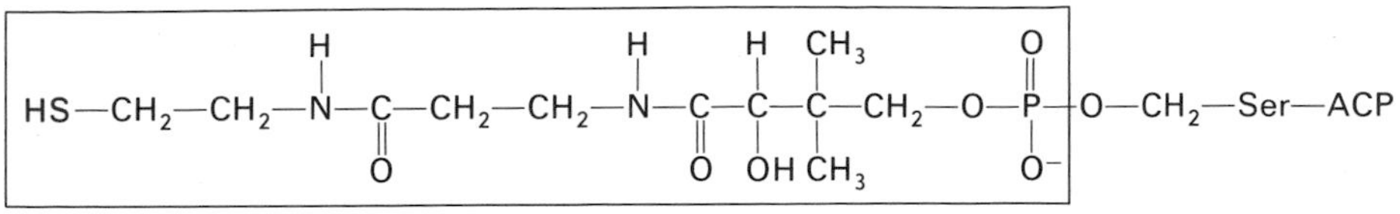

Phosphopantetheine prosthetic group of ACP

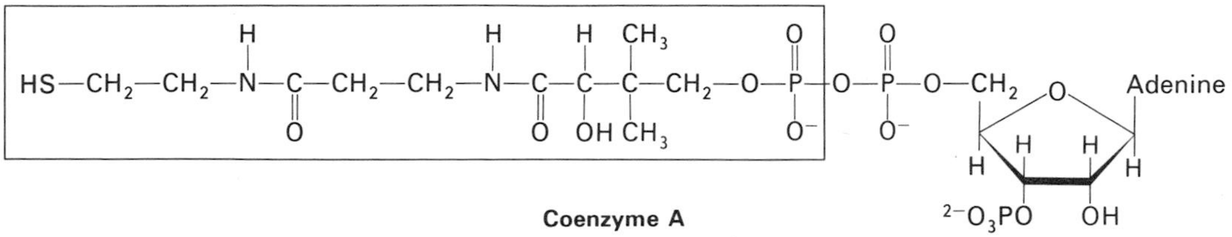

Coenzyme A

Figure 17-6
Phosphopantetheine is the reactive unit of ACP and of CoA.

(Figure 17-6). In the degradation of fatty acids, this unit is part of CoA, whereas, in synthesis, it is attached to a serine residue of the ACP. The ACP, which is a single polypeptide chain of 77 residues, can be regarded as a giant prosthetic group, a "macro CoA."

THE ELONGATION CYCLE IN FATTY ACID SYNTHESIS

The enzyme system that catalyzes the synthesis of saturated long-chain fatty acids from acetyl CoA, malonyl CoA, and NADPH is called the *fatty acid synthetase*. It is present in higher organisms as a multienzyme complex. In contrast, the constituent enzymes of bacterial fatty acid synthetases are dissociated when the cells are disrupted. The availability of these isolated enzymes has facilitated the elucidation of the steps in fatty acid synthesis (Table 17-3). In fact, the reactions leading to fatty acid synthesis in higher organisms are very much like those of bacteria.

The elongation phase of fatty acid synthesis starts with the formation of acetyl-ACP and malonyl-ACP. *Acetyl transacylase* and *malonyl transacylase* catalyze these reactions.

$$\text{Acetyl CoA} + \text{ACP} \rightleftharpoons \text{acetyl-ACP} + \text{CoA}$$
$$\text{Malonyl CoA} + \text{ACP} \rightleftharpoons \text{malonyl-ACP} + \text{CoA}$$

Table 17-3
Principal reactions in fatty acid synthesis

Step	Reaction	Enzyme
1	Acetyl CoA + HCO_3 + ATP $\longrightarrow$ malonyl CoA + ADP + P_i + H^+	Acetyl CoA carboxylase
2	Acetyl CoA + ACP $\rightleftharpoons$ acetyl-ACP + CoA	Acetyl transacylase
3	Malonyl CoA + ACP $\rightleftharpoons$ malonyl-ACP + CoA	Malonyl transacylase
4	Acetyl-ACP + malonyl-ACP $\longrightarrow$ acetoacetyl-ACP + ACP + CO_2	Acyl-malonyl-ACP condensing enzyme
5	Acetoacetyl-ACP + NADPH + H^+ $\rightleftharpoons$ D-3-hydroxybutyryl-ACP + $NADP^+$	β-Ketoacyl-ACP-reductase
6	D-3-Hydroxybutyryl-ACP $\rightleftharpoons$ crotonyl-ACP + H_2O	3-Hydroxyacyl-ACP-dehydratase
7	Crotonyl-ACP + NADPH + H^+ $\longrightarrow$ butyryl-ACP + $NADP^+$	Enoyl-ACP reductase

Malonyl transacylase is highly specific, whereas acetyl transacylase can transfer acyl groups other than the acetyl unit, though at a much slower rate. Fatty acids with an odd number of carbon atoms are synthesized starting with propionyl-ACP, which is formed from propionyl CoA by acetyl transacetylase.

Acetyl-ACP and malonyl-ACP react to form acetoacetyl-ACP. This condensation reaction is catalyzed by the *acyl-malonyl-ACP condensing enzyme.*

$$\text{Acetyl-ACP} + \text{malonyl-ACP} \longrightarrow \text{acetoacetyl-ACP} + \text{ACP} + CO_2$$

In the condensation reaction, a four-carbon unit is formed from a two-carbon unit and a three-carbon unit, and CO_2 is released. Why isn't the four-carbon unit formed from two two-carbon units? In other words, why are the reactants acetyl-ACP and malonyl-ACP rather than two molecules of acetyl-ACP? The answer is that the equilibrium is highly unfavorable for the synthesis of acetoacetyl-ACP from two molecules of acetyl-ACP. In contrast, *the equilibrium is favorable if malonyl-ACP is a reactant because its decarboxylation contributes a substantial decrease in free energy.* In effect, the condensation reaction is driven by ATP, though ATP does not directly participate in the condensation reaction. Rather, ATP is used to form an energy-rich substrate in the carboxylation of acetyl CoA to malonyl CoA. The free energy stored in malonyl CoA in the carboxylation reaction is released in the decarboxylation accompanying the formation of acetoacetyl-ACP. Although HCO_3^- is required for fatty acid synthesis, its carbon atom does not appear in the product. *Rather, all of the carbon atoms of fatty acids containing an even number of them are derived from acetyl CoA.*

The next three steps in fatty acid synthesis reduce the keto group at C-3 to a methylene group (Figure 17-7). First, acetoacetyl-ACP is reduced to D-3-hydroxybutyryl-ACP. This reaction differs from the corresponding one in fatty acid degradation in two respects. The D- rather than the L-epimer is formed. NADPH is the reducing agent, whereas NAD⁺ is the oxidizing agent in β-oxidation. This difference exemplifies the general principle that *NADPH is consumed in biosynthetic reactions, whereas NADH is generated in energy-yielding reactions.* Then D-3-hydroxybutyryl-ACP is *dehydrated* to form

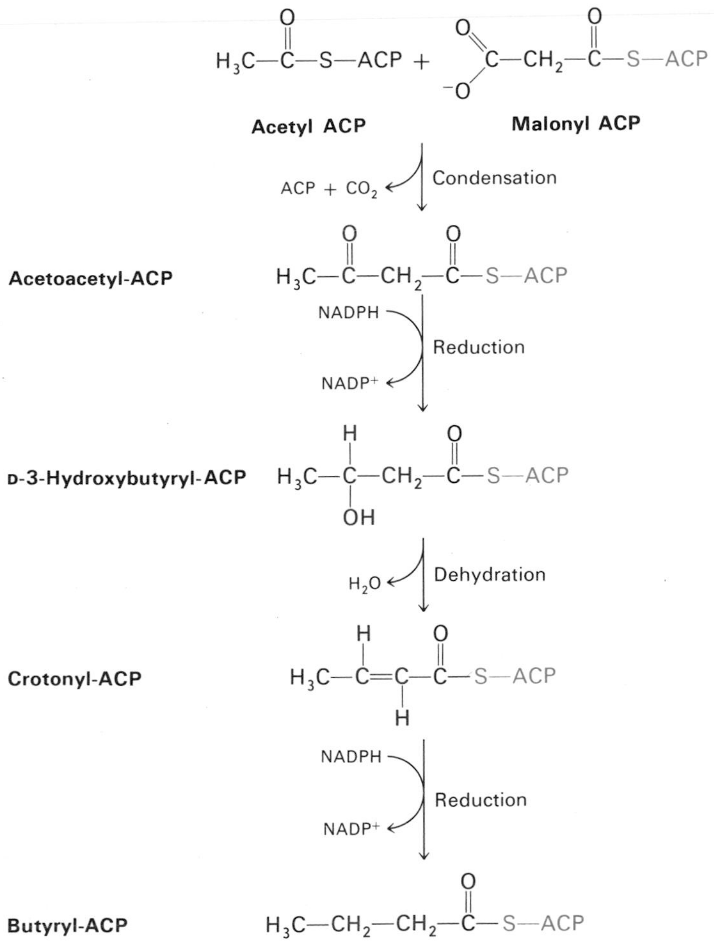

Figure 17-7
Reaction sequence in the synthesis of fatty acids: condensation, reduction, dehydration, and reduction. The intermediates shown here occur in the first round of synthesis.

crotonyl-ACP, which is a *trans*-Δ^2-enoyl-ACP. The final step in the cycle *reduces* crotonyl-ACP to butyryl-ACP. NADPH is again the reductant, whereas FAD^+ is the oxidant in the corresponding reaction in β-oxidation. These last three reactions—a reduction, a dehydration, and a second reduction—convert acetoacetyl-ACP into butyryl-ACP, which completes the first elongation cycle.

In the second round of fatty acid synthesis, butyryl-ACP condenses with malonyl-ACP to form a C_6-β-ketoacyl-ACP. This reaction is like the one in the first round, in which acetyl-ACP condensed with malonyl-ACP to form a C_4-β-ketoacyl-ACP. Reduction, dehydration, and a second reduction convert the C_6-β-ketoacyl-ACP to a C_6-acyl-ACP, which is ready for a third round of elongation. The elongation cycles continue until C_{16}-acyl-ACP is formed. This intermediate is not a substrate for the condensing enzyme. Rather, it is hydrolyzed to yield palmitate and ACP.

STOICHIOMETRY OF FATTY ACID SYNTHESIS

The stoichiometry of the synthesis of palmitate is:

$$\text{Acetyl CoA} + 7\,\text{malonyl CoA} + 14\,\text{NADPH} + 7\,H^+ \longrightarrow$$
$$\text{palmitate} + 7\,CO_2 + 14\,\text{NADP}^+ + 8\,\text{CoA} + 6\,H_2O$$

The equation for the synthesis of the malonyl CoA used in the above reaction is:

$$7\,\text{Acetyl CoA} + 7\,CO_2 + 7\,\text{ATP} \longrightarrow$$
$$7\,\text{malonyl CoA} + 7\,\text{ADP} + 7\,P_i + 7\,H^+$$

Hence, the overall stoichiometry for the synthesis of palmitate is:

$$8\,\text{Acetyl CoA} + 7\,\text{ATP} + 14\,\text{NADPH} \longrightarrow$$
$$\text{palmitate} + 14\,\text{NADP}^+ + 8\,\text{CoA} + 6\,H_2O + 7\,\text{ADP} + 7\,P_i$$

FATTY ACID SYNTHETASE IS A MULTIENZYME COMPLEX

The fatty acid synthetase from yeast has been isolated as a multienzyme complex by Lynen and his associates. Their studies of this organized enzyme assembly are highly pertinent to an understanding of similar complexes in higher organisms. The yeast complex has a molecular weight of 2,300,000. In electron micrographs, the complex appears ellipsoidal with a length of 250 Å and a cross-sectional diameter of 210 Å. Attempts to dissociate the complex into its enzymatically active subunits have not been successful thus far. However, an ACP has been separated from the complex. The complex displays the enzymatic activities of the fatty acid synthesizing system in *E. coli:* acetyl transacylase, malonyl transacylase, condensing enzyme, β-ketoacyl reductase, β-hydroxyacyl dehydratase, and enoyl reductase.

Two different types of sulfhydryl residues participate in the synthetic process. One of them, called the "peripheral" thiol, reacts rapidly with iodoacetamide and can be protected by prior incubation with acetyl CoA. The other, called the "central" thiol, reacts much more slowly with iodoacetamide. The names "peripheral" and "central" should not be taken to indicate their position in the complex. The "peripheral" thiol is the sulfhydryl group of a cysteine residue, whereas the "central" thiol is the sulfhydryl group of a phosphopantetheinyl moiety. It seems likely that the "peripheral" thiol corresponds to the sulfhydryl group of the condensing enzyme, whereas the "central" thiol is the sulfhydryl group of the ACP.

A significant feature of the mechanism proposed by Lynen is that the growing fatty acid chain is transferred from the ACP to the condensing enzyme and back again in each round of elongation. The first *translocation* makes room for the incoming malonyl unit, whereas the second occurs in the condensation step. It is interesting to note that analogous translocations occur in protein synthesis.

The flexibility and 20-Å maximal length of the phosphopantetheinyl moiety appear to be critical for the function of the multienzyme complex, since it enables the growing fatty acid chain to come into close contact with the active site of each enzyme in the complex. The enzyme subunits need not undergo large structural rearrangements to interact with the substrate. Instead, the substrate is on a long, flexible arm that can reach each of the active sites. Recall that biotin and lipoamide are also on long, flexible arms in their multienzyme complexes. The organized structure of the fatty acid synthetases of yeast and higher organisms enhances the efficiency of the overall process because intermediates are directly transferred from one active site to the next. The reactants are not diluted in the cytosol. Moreover, they do not have to find each other by random diffusion. Another advantage of such a multienzyme complex is that the covalently bound intermediates are sequestered and protected from competing reactions.

CITRATE CARRIES ACETYL GROUPS FROM MITOCHONDRIA TO THE CYTOSOL FOR FATTY ACID SYNTHESIS

The synthesis of palmitate requires the input of 8 molecules of acetyl CoA, 14 NADPH, and 7 ATP. Fatty acids are synthesized in the cytosol, whereas acetyl CoA is formed from pyruvate in mitochondria. Hence, acetyl CoA must be transferred from mitochondria to the cytosol for fatty acid synthesis. However, mitochondria are not readily permeable to acetyl CoA. Recall that carnitine carries only long-chain fatty acids. *The barrier to acetyl CoA is bypassed by citrate, which carries acetyl groups across the inner mitochondrial membrane.* Citrate is formed in the mitochondrial matrix by the condensation of acetyl CoA and oxaloacetate. It then diffuses to the cytosol, where it is cleaved by *citrate lyase:*

$$\text{Citrate} + \text{ATP} + \text{CoA} \longrightarrow$$
$$\text{acetyl CoA} + \text{ADP} + \text{P}_i + \text{oxaloacetate}$$

Thus, acetyl CoA and oxaloacetate are transferred from mitochondria to the cytosol at the expense of an ATP.

SOURCES OF NADPH FOR FATTY ACID SYNTHESIS

Oxaloacetate formed in the transfer of acetyl groups to the cytosol must now be returned to the mitochondria. The inner mitochondrial membrane is impermeable to oxaloacetate. Hence, a series of bypass reactions are needed. Most important, these reactions generate much of the NADPH needed for fatty acid synthesis. First, oxaloacetate is reduced to malate by NADH. This reaction is catalyzed by a *malate dehydrogenase* in the cytosol.

$$\text{Oxaloacetate} + \text{NADH} + \text{H}^+ \rightleftharpoons \text{malate} + \text{NAD}^+$$

Second, malate is oxidatively decarboxylated by an *NADP+-linked*

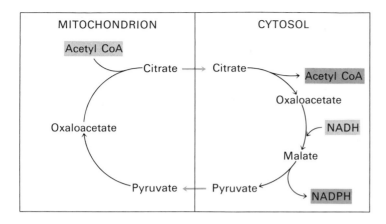

Figure 17-8
Acetyl CoA is transferred from mitochondria to the cytosol, and NADH is concomitantly converted to NADPH by this series of reactions.

malate enzyme. We have not previously mentioned this reaction.

$$\text{Malate} + \text{NADP}^+ \longrightarrow \text{pyruvate} + \text{CO}_2 + \text{NADPH}$$

The pyruvate formed in this reaction readily diffuses into mitochondria, where it is carboxylated to oxaloacetate by pyruvate carboxylase.

$$\text{Pyruvate} + \text{CO}_2 + \text{ATP} + \text{H}_2\text{O} \longrightarrow$$
$$\text{oxaloacetate} + \text{ADP} + \text{P}_i + 2\,\text{H}^+$$

The sum of these three reactions is:

$$\text{NADP}^+ + \text{NADH} + \text{ATP} + \text{H}_2\text{O} \longrightarrow$$
$$\text{NADPH} + \text{NAD}^+ + \text{ADP} + \text{P}_i + \text{H}^+$$

Thus, *one NADPH is generated for each acetyl CoA that is transferred from the mitochondria to the cytosol.* Hence, 8 NADPH are formed when 8 molecules of acetyl CoA are transferred to the cytosol for the synthesis of palmitate. *The additional 6 NADPH required for this process come from the pentose phosphate pathway.*

ELONGATION AND UNSATURATION OF FATTY ACIDS ARE CARRIED OUT BY ACCESSORY ENZYME SYSTEMS

The major product of the fatty acid synthetase is palmitate. Longer fatty acids are formed by elongation reactions catalyzed by two different systems, one in mitochondria, the other in microsomes.

In both systems, two-carbon units are added. In higher organisms, double bonds can be introduced into long-chain acyl CoAs by *microsomal enzymes*. For example, in the conversion of stearoyl CoA to oleoyl CoA, a *cis*-Δ^9 double bond is inserted by an oxidase that employs *molecular oxygen* and *NADH* (or *NADPH*):

$$\text{Stearoyl CoA} + \text{NADH} + \text{H}^+ + \text{O}_2 \longrightarrow$$
$$\text{oleoyl CoA} + \text{NAD}^+ + 2\,\text{H}_2\text{O}$$

A variety of unsaturated fatty acids can be formed from oleate by a combination of elongation and unsaturation reactions. For example, oleate can be elongated to a 20:1 *cis*-Δ^{11} fatty acid. Alternatively, a second double bond can be inserted to yield an 18:2 *cis*-Δ^6, Δ^9 fatty acid. Similarly, palmitate (16:0) can be oxidized to palmitoleate (16:1 *cis*-Δ^9), which can then be elongated to *cis*-vaccenate (18:1 *cis*-Δ^{11}).

Mammals lack the enzymes to introduce double bonds at carbon atoms beyond C-9 in the fatty acid chain. Hence, mammals cannot synthesize linoleate (18:2 *cis*-Δ^9, Δ^{12}) and linolenate (18:3 *cis*-Δ^9, Δ^{12}, Δ^{15}). *Linoleate and linolenate are the two essential fatty acids.* The term "essential" means that they must be supplied in the diet because they are required by the organism and cannot be endogeneously synthesized. Linoleate and linolenate furnished by the diet are the starting points for the synthesis of a variety of the other unsaturated fatty acids. Unsaturated fatty acids in mammals are derived from either palmitoleate (16:1), oleate (18:1), linoleate (18:2), or linolenate (18:3). The number of methylene carbons between the ω-CH_3 group of an unsaturated fatty acid and the nearest double bond identifies its precursor:

Precursor	Formula
Linolenate	$CH_3-(CH_2)_1-CH{=}CH-R$
Linoleate	$CH_3-(CH_2)_4-CH{=}CH-R$
Palmitoleate	$CH_3-(CH_2)_5-CH{=}CH-R$
Oleate	$CH_3-(CH_2)_7-CH{=}CH-R$

SUMMARY

Fatty acids are physiologically important both as components of phospholipids and glycolipids and as fuel molecules. They are stored in adipose tissue as triacylglycerols (neutral fat), which can be mobilized by the hydrolytic action of lipases that are under hormonal control. Fatty acids are activated to acyl CoAs, transported across the inner mitochondrial membrane by carnitine, and degraded in the mitochondrial matrix by a recurring sequence of four reactions: oxidation linked to FAD, hydration, oxidation linked to NAD^+, and thiolysis by CoA. The $FADH_2$ and NADH formed in the oxidation steps transfer their electrons to O_2 via the respiratory chain, whereas the acetyl CoA formed in the thiolysis step normally enters the citric acid cycle by condensing with oxaloacetate. When the concentration of oxaloacetate is insufficient, acetyl CoA gives rise to acetoacetate and 3-hydroxybutyrate, which are normal fuel molecules. In starvation and in diabetes, large amounts of acetoacetate, 3-hydroxybutyrate, and acetone (collectively known as ketone bodies) accumulate in the blood. Mammals are unable to convert fatty acids into glucose because there is no pathway for the net production of oxaloacetate, pyruvate, or other gluconeogenic intermediate from acetyl CoA. In higher organisms, the enzymes of fatty acid synthesis are organized in a multienzyme complex.

Fatty acids are synthesized in the cytosol by a different pathway from that of β-oxidation. Synthesis starts with the carboxylation of acetyl CoA to malonyl CoA. This ATP-driven reaction is catalyzed by acetyl CoA carboxylase, a biotin-enzyme. The intermediates in fatty acid synthesis are linked to an acyl carrier protein (ACP), specifically to the sulfhydryl terminus of its phosphopantetheine prosthetic group. Acetyl-ACP is formed from acetyl CoA, and malonyl-ACP is formed from malonyl CoA. Acetyl-ACP and malonyl-ACP condense to form acetoacetyl-ACP, a reaction driven by the release of CO_2 from the activated malonyl unit. This is followed by a reduction, a dehydration, and a second reduction. NADPH is the reductant in these steps. The butyryl-ACP formed in this way is ready for a second round of elongation starting with the addition of a two-carbon unit from malonyl-ACP. Seven rounds of elongation yield palmitoyl-ACP, which is hydrolyzed to palmitate. The synthesis of palmitate requires eight molecules of acetyl

CoA, fourteen NADPH, and seven ATP. A reaction cycle based on the cleavage of citrate carries acetyl groups from mitochondria to the cytosol and generates some of the required NADPH. The rest is formed by the pentose phosphate pathway. Fatty acids are elongated and unsaturated by enzyme systems in mitochondria and in the microsomes. Mammals lack the enzymes to introduce double bonds distal to C-9, and so they require linoleate and linolenate in their diet.

SELECTED READINGS

Where to start:

Lynen, F., 1972. The pathway from "activated acetic acid" to the terpenes and fatty acids. In *Nobel Lectures: Physiology or Medicine (1963–1970)*, pp. 103–138. American Elsevier (1973). [An account of the author's pioneering work on fatty acid degradation and synthesis, and of the roles of acetyl CoA and biotin.]

General reviews:

Wakil, S. J., and Barnes, E. M., Jr., 1971. Fatty acid metabolism. *Compr. Biochem.* 18S:57–104.

Newsholme, E. A., and Start, C., 1973. *Regulation in Metabolism.* Wiley. [Chapters 4 and 7 deal with the regulation of fat metabolism.]

Fatty acid oxidation:

Bressler, R., 1970. Fatty acid oxidation. *Compr. Biochem.* 18:331–359.

Garland, P. B., Shepherd, D. Nicholls, D. G., Yates, D. W., and Light, P. A., 1969. Interactions between fatty acid oxidation and the tricarboxylic acid cycle. *In* Lowenstein, J. M., (ed.), *Citric Acid Cycle: Control and Compartmentation,* pp. 163–212. Dekker.

Carnitine:

Fritz, I. B., 1968. The metabolic consequences of the effects of carnitine on long-chain fatty acid oxidation. *In* Gran, F. C., (ed.), *Symposium on Cellular Compartmentalization and Control of Fatty Acid Metabolism,* pp. 39–63. Academic Press.

Engel, W. K., Vick, N. A., Glueck, C. J., and Levy, R. I., 1970. A skeletal-muscle disorder associated with intermittent symptoms and a possible defect of lipid metabolism. *N. Engl. J. Med.* 282:697–704.

Fatty acid synthesis:

Volpe, J. J., and Vagelos, P. R., 1973. Saturated fatty acid biosynthesis and its regulation. *Annu. Rev. Biochem.* 42:21–60.

Lynen, F., Osterholt, D., Schweiger, E., and Willecke, K., 1968. The biosynthesis of fatty acids. *In* Gran, F. C., (ed.), *Symposium on Cellular Compartmentalization and Control of Fatty Acid Metabolism,* pp. 1–24.

Utter, M. F., 1969. Metabolic roles of oxaloacetate. *In* Lowenstein, J. M., (ed.), *Citric Acid Cycle: Control and Compartmentation,* pp.

249–296. [Includes a discussion of the role of oxaloacetate in the transport of acetyl groups to the cytosol for fatty acid synthesis.]

Enzyme mechanisms:

Moss, J., and Lane, M. D., 1971. The biotin-dependent, enzymes. *Advan. Enzymol.* 35:321–442.

Alberts, A. W., and Vagelos, P. R., 1972. Acyl-CoA carboxylases. *In* Boyer, P. D., (ed.), *The Enzymes,* (3rd ed.), vol. 6, pp. 37–82. Academic Press.

Vagelos, P. R., 1973. Acyl group transfer (acyl carrier protein). *In* Boyer, P. D., (ed.), *The Enzymes,* (3rd ed.), vol. 8, pp. 155–199.

PROBLEMS

1. Write a balanced equation for the conversion of glycerol to pyruvate. Which enzymes are required in addition to those of the glycolytic pathway?

2. Write a balanced equation for the conversion of stearate to acetoacetate.

3. Compare the following aspects of fatty acid oxidation and synthesis:
    (a) Site of the process.
    (b) Acyl carrier.
    (c) Reductants and oxidants.
    (d) Stereochemistry of the intermediates.
    (e) Direction of synthesis or degradation.
    (f) Organization of the enzyme system.

4. For each of the following unsaturated fatty acids, indicate whether the biosynthetic precursor in animals is palmitoleate, oleate, linoleate, or linolenate.
    (a) $18:1$ *cis* Δ^{11}
    (b) $18:3$ *cis* Δ^6, Δ^9, Δ^{12}
    (c) $20:2$ *cis* Δ^{11}, Δ^{14}
    (d) $20:3$ *cis* Δ^5, Δ^8, Δ^{11}
    (e) $22:1$ *cis* Δ^{13}
    (f) $22:6$ *cis* Δ^4, Δ^7, Δ^{10}, Δ^{13}, Δ^{16}, Δ^{19}

5. Consider a cell extract that actively synthesizes palmitate. Suppose that a fatty acid synthetase in this preparation forms one palmitate in about five minutes. A large amount of malonyl CoA labeled with ^{14}C in each carbon of its malonyl unit is suddenly added to this system, and fatty acid synthesis is stopped a minute later by altering the pH. The fatty acids in the supernatant are analyzed for radioactivity. Is C-1 or C-16 of the palmitate formed by this system more radioactive?

6. The citric acid cycle is modified in some plants and microorganisms. They contain isocitrase and malate synthetase, two enzymes absent from animals. Isocitrase catalyzes the aldol cleavage of isocitrate to succinate and glyoxylate ($OHC—COO^-$), whereas malate synthetase catalyzes the formation of malate from acetyl CoA, glyoxylate, and water.
    (a) If isocitrate dehydrogenase is absent and malate dehydrogenase is present, can oxaloacetate be formed from acetyl CoA in these organisms? If so, write a balanced equation for this conversion.
    (b) Can these organisms synthesize glucose from acetyl CoA?

CHAPTER 18

AMINO ACID DEGRADATION
AND THE UREA CYCLE

Amino acids in excess of those needed for the synthesis of proteins and other biomolecules cannot be stored, in contrast to fatty acids and glucose, nor are they excreted. Rather, surplus amino acids are used as metabolic fuel. Most of the amino groups of surplus amino acids are converted to urea, whereas their carbon skeletons are transformed into acetyl CoA, acetoacetyl CoA, pyruvate, or one of the intermediates of the citric acid cycle. Hence *fatty acids, ketone bodies, and glucose can be formed from amino acids.*

ALPHA-AMINO GROUPS ARE CONVERTED TO AMMONIUM ION BY OXIDATIVE DEAMINATION OF GLUTAMATE

The major site of amino acid degradation in mammals is the liver. The fate of the α-amino group will be considered first, followed by that of the carbon skeleton. The α-amino group of many amino acids is transferred to α-ketoglutarate to form *glutamate,* which is then oxidatively deaminated to yield NH_4^+.

$$^+H_3N-\overset{\overset{\displaystyle H}{|}}{\underset{\underset{\displaystyle COO^-}{|}}{C}}-R \longrightarrow {}^+H_3N-\overset{\overset{\displaystyle H}{|}}{\underset{\underset{\displaystyle COO^-}{|}}{C}}-CH_2-CH_2-COO^- \longrightarrow NH_4^+$$

Amino acid   **Glutamate**

Transaminases catalyze the transfer of an α-amino group from an α-amino acid to an α-keto acid.

$$
\underset{\boxed{R_1}}{\overset{NH_3^+}{\underset{|}{H-C-COO^-}}} + \underset{\textcircled{R_2}}{\overset{O}{\overset{\|}{C-COO^-}}} \rightleftharpoons \underset{\boxed{R_1}}{\overset{O}{\overset{\|}{C-COO^-}}} + \underset{\textcircled{R_2}}{\overset{NH_3^+}{\underset{|}{H-C-COO^-}}}
$$

Glutamate transaminase, the most important of these enzymes, catalyzes the transfer of an amino group to α-ketoglutarate.

α-Amino acid $+$ α-ketoglutarate $\rightleftharpoons$ α-keto acid $+$ glutamate

Alanine transaminase, which is also prevalent in mammalian tissue, catalyzes the transfer of an amino group to pyruvate.

α-Amino acid $+$ pyruvate $\rightleftharpoons$ α-keto acid $+$ alanine

The alanine formed in this step can transfer its amino group to α-ketoglutarate to form glutamate. *These two transaminases funnel α-amino groups from a variety of amino acids into glutamate for conversion to NH_4^+.*

Ammonium ion is formed from glutamate by oxidative deamination. This reaction is catalyzed by *glutamate dehydrogenase,* which is unusual in being able to utilize either NAD^+ or $NADP^+$.

$$
\begin{array}{c}
\overset{NH_3^+}{\underset{|}{H-C-COO^-}} \\
\underset{|}{CH_2} \\
\underset{|}{CH_2} \\
COO^-
\end{array}
+ \underset{(or\ NADP^+)}{NAD^+} + H_2O \rightleftharpoons NH_4^+ +
\begin{array}{c}
\overset{O}{\overset{\|}{C-COO^-}} \\
\underset{|}{CH_2} \\
\underset{|}{CH_2} \\
COO^-
\end{array}
+ \underset{(or\ NADPH)}{NADH} + H^+
$$

Glutamate α-**Ketoglutarate**

The activity of glutamate dehydrogenase is allosterically regulated. The enzyme consists of six identical subunits, which can polymerize further. Guanosine triphosphate (GTP) and adenosine triphosphate (ATP) are allosteric inhibitors, whereas guanosine diphosphate (GDP) and adenosine diphosphate (ADP) are allosteric activators. Hence, *a lowering of the energy charge accelerates the oxidation of amino acids.*

The sum of the reactions catalyzed by the transaminases and by glutamate dehydrogenase is:

$$\alpha\text{-Amino acid} + \underset{(\text{or NADP}^+)}{\text{NAD}^+} + H_2O \rightleftharpoons$$

$$\alpha\text{-keto acid} + NH_4^+ + \underset{(\text{or NADPH})}{\text{NADH}} + H^+$$

In terrestrial vertebrates, NH_4^+ is converted into urea, which is excreted. The synthesis of urea will be discussed shortly.

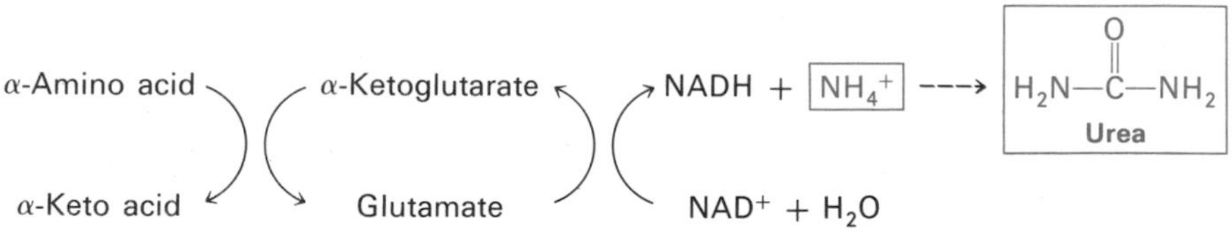

PYRIDOXAL PHOSPHATE, THE PROSTHETIC GROUP IN TRANSAMINASES, FORMS SCHIFF-BASE INTERMEDIATES

The prosthetic group of all transaminases is *pyridoxal phosphate* (PLP), which is derived from *pyridoxine* (Vitamin B$_6$). During transamination, pyridoxal phosphate is transiently converted to *pyridoxamine phosphate* (PMP).

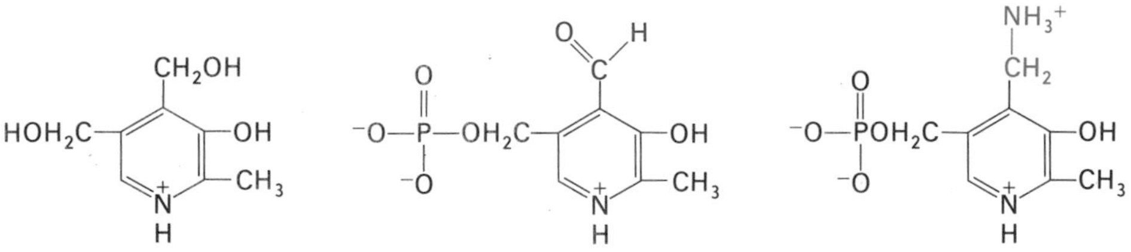

Pyridoxine
(Vitamin B$_6$)

Pyridoxal phosphate
(PLP)

Pyridoxamine phosphate
(PMP)

PLP enzymes form covalent Schiff-base intermediates with their substrates. In the absence of substrate, the aldehyde group of PLP is in Schiff-base linkage with the ε-*amino group of a specific lysine residue at the active site.* A new Schiff-base linkage is formed on addition of an amino acid substrate. *The α-amino group of the amino acid substrate displaces the ε-NH₂ group of the active-site lysine.* The amino acid-PLP Schiff base that is formed remains tightly bound to the enzyme by noncovalent forces.

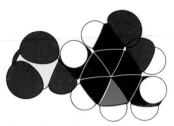

Figure 18-1
Space-filling model of pyridoxal phosphate (PLP).

This Schiff base and the one between PLP and the active-site lysine are *aldimines*. During catalysis, the double bond in the amino acid–PLP Schiff-base linkage shifts position to form a *ketimine,* which is then hydrolyzed to *PMP* and an *α-keto acid*. A reaction mechanism proposed by Esmond Snell and Alexander Braunstein is shown in Figure 18-2. The active-site lysine or another suitably positioned basic group probably facilitates the aldimine-ketimine conversion by serving as an electron sink.

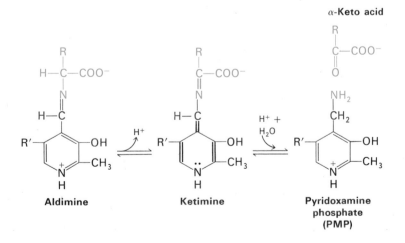

Figure 18-2
Proposed mechanism for transamination reactions.

These steps constitute half of the overall reaction:

$$\text{Amino acid}_1 + \text{E-PLP} \rightleftharpoons \alpha\text{-keto acid} + \text{E-PMP}$$

The second half of the overall reaction occurs by a reversal of the above pathway. A second α-keto acid reacts with the enzyme–pyridoxamine phosphate complex (E-PMP) to yield a second amino acid and regenerate the enzyme–pyridoxal phosphate complex (E-PLP).

$$\alpha\text{-Keto acid}_2 + \text{E-PMP} \rightleftharpoons \text{amino acid}_2 + \text{E-PLP}$$

The sum of these partial reactions is:

$$\text{Amino acid}_1 + \alpha\text{-keto acid}_2 \rightleftharpoons \text{amino acid}_2 + \alpha\text{-keto acid}_1$$

The catalytic versatility of PLP enzymes is remarkable. Transamination is just one of a wide range of amino acid transformations that are catalyzed by PLP enzymes. These reactions include decarboxylations, deaminations, racemizations, and aldol cleavages of amino acids. PLP enzymes labilize one of three bonds in an amino acid substrate (Figure 18-3).

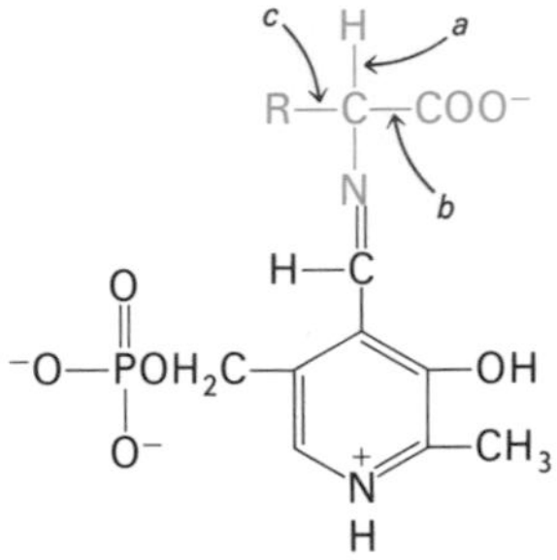

Figure 18-3
A pyridoxal phosphate enzyme labilizes one of three bonds of an amino acid substrate. For example, bond *a* is labilized by transaminases, bond *b* by decarboxylases, and bond *c* by aldolases (such as threonine aldolase).

SERINE AND THREONINE CAN BE DIRECTLY DEAMINATED

The α-amino groups of serine and threonine can be directly converted to NH_4^+ because these amino acids contain a hydroxyl group in their side chains. These direct deaminations are catalyzed by *serine dehydratase* and *threonine dehydratase,* which also contain PLP as their prosthetic group.

$$\text{Serine} \longrightarrow \text{pyruvate} + NH_4^+$$

$$\text{Threonine} \longrightarrow \alpha\text{-ketobutyrate} + NH_4^+$$

These enzymes are called dehydratases because dehydration precedes deamination. Serine loses a hydrogen atom from its α-carbon and a hydroxyl group from its β-carbon, to yield aminoacrylate. This unstable compound reacts with H_2O to give pyruvate and NH_4^+.

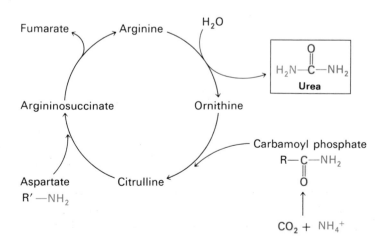

NH_4^+ IS CONVERTED TO UREA IN MOST TERRESTRIAL VERTEBRATES AND THEN EXCRETED

Some of the NH_4^+ formed in the breakdown of amino acids is consumed in the biosynthesis of nitrogen compounds. In most terrestrial vertebrates, the excess NH_4^+ is converted to urea and then excreted. In birds and terrestrial reptiles, NH_4^+ is converted to uric acid for excretion, whereas in many aquatic animals, NH_4^+ itself is excreted. These three classes of organisms are called *ureotelic, uricotelic,* and *ammonotelic.*

In terrestrial vertebrates, urea is synthesized by the *urea cycle.* This series of reactions was proposed by Hans Krebs and Kurt Henseleit in 1932, five years before the elucidation of the citric acid cycle.

Figure 18-4
The urea cycle.

In fact, the urea cycle was the first cyclic metabolic pathway to be discovered. One of the nitrogen atoms of the urea synthesized by this pathway comes from ammonia, whereas the other nitrogen atom comes from aspartate. The carbon atom of urea is derived from CO_2. *Ornithine* is the carrier of these carbon and nitrogen atoms in the urea cycle.

The immediate precursor of urea is *arginine,* which is hydrolyzed to urea and ornithine by *arginase.*

Arginine   **Urea**   **Ornithine**

The other reactions of the urea cycle are devoted to synthesizing arginine from ornithine. First, a carbamoyl group is transferred to ornithine to form *citrulline,* a reaction catalyzed by ornithine transcarbamoylase. The carbamoyl donor in this reaction is carbamoyl phosphate, which has a high transfer potential because of its anhydride bond.

Ornithine   **Carbamoyl phosphate**   **Citrulline**

Argininosuccinate synthetase then catalyzes the condensation of citrulline and aspartate. This synthesis of *argininosuccinate* is driven by the cleavage of ATP into AMP and pyrophosphate, and the subsequent hydrolysis of pyrophosphate.

Citrulline   **Aspartate**   **Argininosuccinate**

Finally, argininosuccinase cleaves argininosuccinate into *arginine* and fumarate. Note that the carbon skeleton of aspartate is preserved in these reactions, which transfer the amino group of aspartate to form arginine.

Argininosuccinate   **Arginine**   **Fumarate**

Carbamoyl phosphate is synthesized from NH_4^+, CO_2, ATP, and H_2O in a complex reaction that is catalyzed by *carbamoyl phosphate synthetase*. An unusual feature of this enzyme is that it requires *N*-acetyglutamate for activity.

Carbamoyl phosphate

The consumption of two molecules of ATP makes this synthesis of carbamoyl phosphate essentially irreversible.

THE UREA CYCLE IS LINKED TO THE CITRIC ACID CYCLE

The stoichiometry of urea synthesis is:

$$CO_2 + NH_4^+ + 3\,ATP + aspartate + 2\,H_2O \longrightarrow$$
$$urea + 2\,ADP + 2\,P_i + AMP + PP_i + fumarate$$

Since pyrophosphate is rapidly hydrolyzed, four high-energy phosphate bonds are used to synthesize one molecule of urea. *The synthesis of fumarate by the urea cycle is important since it links the urea cycle and the citric acid cycle* (Figure 18-5). Fumarate is hydrated to malate, which is in turn oxidized to oxaloacetate. This key intermediate has several possible fates: (1) oxaloacetate can be transaminated to aspartate; (2) it can be converted to glucose by the gluconeogenic pathway; or (3) it can condense with acetyl CoA to form citrate.

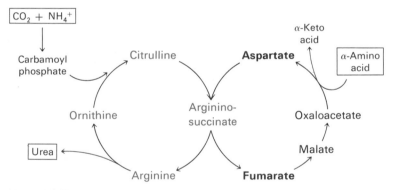

Figure 18-5
The urea cycle, the citric acid cycle, and the transamination of oxaloacetate are linked by fumarate and aspartate.

INHERITED ENZYMATIC DEFECTS OF THE UREA CYCLE CAUSE HYPERAMMONEMIA

High levels of NH_4^+ are toxic to man. The synthesis of urea in the liver is the major route of removal of NH_4^+. A complete block of any of the steps of the urea cycle is probably incompatible with life, since there is no known alternative pathway for the synthesis of urea. Inherited disorders caused by a partial block of each of the urea-cycle reactions have been diagnosed. The common finding is an *elevated level of NH_4^+ in the blood* (hyperammonemia), which can lead to *mental retardation.* Other clinical symptoms include vomiting in infancy and an aversion to protein-rich foods. A low-protein diet leads to a lowering of the blood ammonium level and to clinical improvement. The molecular basis of ammonia toxicity is not yet known.

FATE OF THE CARBON ATOMS OF DEGRADED AMINO ACIDS

We turn now to the fate of the carbon skeleton in the breakdown of amino acids. Some or all of the carbon atoms of a degraded amino acid emerge in acetyl CoA, acetoacetyl CoA, pyruvate, or a citric acid cycle intermediate. Amino acids that are degraded to acetyl CoA or acetoacetyl CoA are termed *ketogenic* because they give rise to ketone bodies. In contrast, amino acids that are degraded to pyruvate, α-ketoglutarate, succinyl CoA, fumarate, or oxaloacetate are termed *glucogenic*. Net synthesis of glucose from these amino acids is feasible because these citric acid cycle intermediates and pyruvate can be converted to phosphoenolpyruvate and then to glucose (p. 367). Recall that mammals lack a pathway for the net synthesis of glucose from acetyl CoA or acetoacetyl CoA.

Of the basic set of twenty amino acids, only leucine is purely ketogenic. Isoleucine, lysine, phenylalanine, tryptophan, and tyrosine are both ketogenic and glucogenic. Some of their carbon atoms emerge in acetyl CoA or acetoacetyl CoA, whereas others appear in potential precursors of glucose. The other fourteen amino acids are purely glucogenic.

PYRUVATE IS A POINT OF ENTRY FOR SEVERAL GLUCOGENIC AMINO ACIDS

Alanine, serine, cysteine, and glycine can give rise to pyruvate. *Alanine* is converted to pyruvate by transamination:

Alanine + α-ketoglutarate $\rightleftharpoons$ pyruvate + glutamate

As mentioned previously, L-glutamate is then oxidatively deaminated, yielding NH_4^+ and regenerating α-ketoglutarate. The sum of these reactions is:

$$\text{Alanine} + NAD^+ \longrightarrow \text{pyruvate} + NH_4^+ + NADH + H^+$$

Serine is directly deaminated and converted to pyruvate by serine dehydratase.

$$\text{Serine} \longrightarrow \text{pyruvate} + NH_4^+$$

Serine can be formed from *glycine* by enzymatic addition of a hydroxymethyl group (p. 506). *Cysteine* yields pyruvate and H_2S by a variety of pathways.

ALPHA-KETOGLUTARATE IS THE POINT OF ENTRY FOR GLUTAMATE

The oxidative deamination of glutamate by glutamate dehydrogenase yields α-ketoglutarate, as has already been discussed. Alpha-ketoglutarate is also the point of entry for glutamine, proline, arginine, and histidine since these amino acids are readily converted to glutamate (Figures 18-6 and 18-7). *Glutamine* is hydrolyzed to glutamate and NH_4^+ by glutaminase. *Proline* and *arginine* are converted to glutamate-γ-semialdehyde, which is then oxidized to glutamate. *Histidine* is converted to 4-imidazolone-5-propionate. The amide bond in the ring of this intermediate is hydrolyzed to the *N*-formimino derivative of glutamate, which is then converted to glutamate by transfer of its formimino group to tetrahydrofolate, a carrier of actived one-carbon units (see p. 507).

SUCCINYL COENZYME A IS A POINT OF ENTRY FOR SOME AMINO ACIDS

Succinyl CoA is the point of entry for some of the carbon atoms of methionine, isoleucine, threonine, and valine. Methylmalonyl CoA is an intermediate in the breakdown of these four amino acids (Figure 18-8).

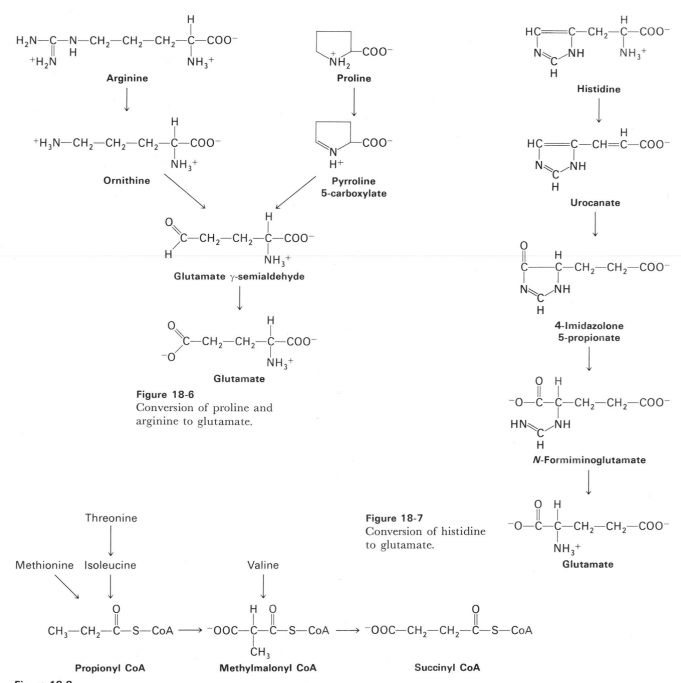

Figure 18-6
Conversion of proline and
arginine to glutamate.

Figure 18-7
Conversion of histidine
to glutamate.

Figure 18-8
Conversion of methionine, isoleucine, threonine, and valine to succinyl CoA.

The pathway from propionyl CoA to succinyl CoA is especially interesting. Propionyl CoA is carboxylated at the expense of an ATP to yield the D-isomer of methylmalonyl CoA. This carboxylation reaction is catalyzed by *propionyl CoA carboxylase,* a biotin-enzyme that has a catalytic mechanism like that of acetyl CoA carboxylase and pyruvate carboxylase. The D-isomer of methylmalonyl CoA is racemized to the L-isomer, which is the substrate for the mutase enzyme that converts it to succinyl CoA.

$$
\begin{array}{ccc}
\text{Propionyl CoA} & \xrightarrow[\text{ATP}]{\text{HCO}_3^- +} \xrightarrow{\text{AMP} + \text{PP}_i} & \text{D-Methylmalonyl CoA} \rightleftharpoons \text{L-Methylmalonyl CoA}
\end{array}
$$

Propionyl CoA

H—C—CH$_3$ / H (top) ; C—S—CoA ; O

D-Methylmalonyl CoA COO$^-$; H—C—CH$_3$; C—S—CoA ; O

L-Methylmalonyl CoA COO$^-$; H$_3$C—C—H ; C—S—CoA ; O

Succinyl CoA is formed from L-methylmalonyl CoA by an *intramolecular rearrangement.* The —CO—S—CoA group migrates from C-2 to C-3 in exchange for a hydrogen atom. *This very unusual isomerization is catalyzed by methylmalonyl CoA mutase, one of the two mammalian enzymes known to contain a derivative of Vitamin B_{12} as its coenzyme.*

L-Methylmalonyl CoA ⇌ **Succinyl CoA**

This pathway from propionyl CoA to succinyl CoA also participates in the oxidation of *fatty acids that have an odd number of carbon atoms.* The final thiolytic cleavage of an odd-numbered acyl CoA yields acetyl CoA and propionyl CoA (p. 416).

DEOXYADENOSYLCOBALAMIN IS THE COENZYME OF METHYLMALONYL MUTASE

Cobalamin (Vitamin B_{12}) has been a challenging problem in biochemistry and medicine since the discovery by George Minot and William Murphy in 1926 that pernicious anemia can be treated by

feeding the patient large amounts of liver. Cobalamin was purified
and crystallized in 1948 and its complex three-dimensional structure was elucidated in 1956 by Dorothy Hodgkin. The core of
cobalamin consists of a *corrin ring with a central cobalt atom* (Figure
18-9). The corrin ring, like a porphyrin, has *four pyrrole units*. Two
of them (rings A and D) are directly bonded to each other, whereas
the others are joined by methene bridges, as in porphyrins. The
substituents on the pyrrole rings are methyl, propionamide, and
acetamide groups.

A cobalt atom is bonded to the four pyrrole nitrogens. *The fifth
substituent* (below the corrin plane in Figure 18-10) is a derivative
of *dimethylbenzimidazole* that contains ribose 3-phosphate and aminoisopropanol. One of the nitrogen atoms of dimethylbenzimidazole
is linked to cobalt. The amino group of aminoisopropanol is in
amide linkage with a side chain of the D ring. The *sixth substituent*
on the cobalt atom (located above the corrin plane in Figure 18-10)
can be CN^-, $-CH_3$, OH^-, or deoxyadenosyl. The presence of the
cyanide ion at the sixth coordination position in cyanocobalamin,

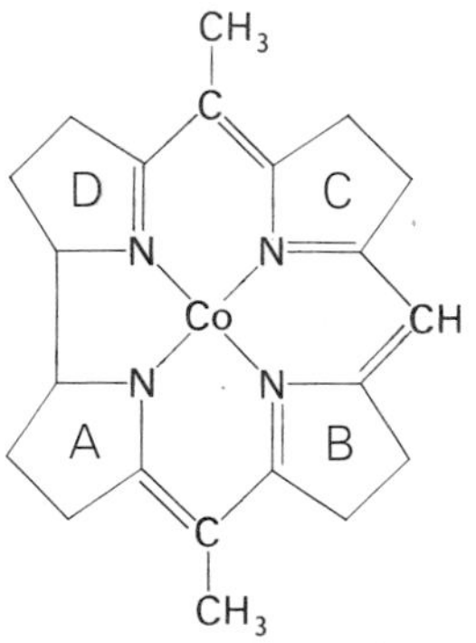

Figure 18-9
Corrin core of cobalamin. Substituents
on the pyrroles and the other two cobalt
ligands are not shown in this diagram.

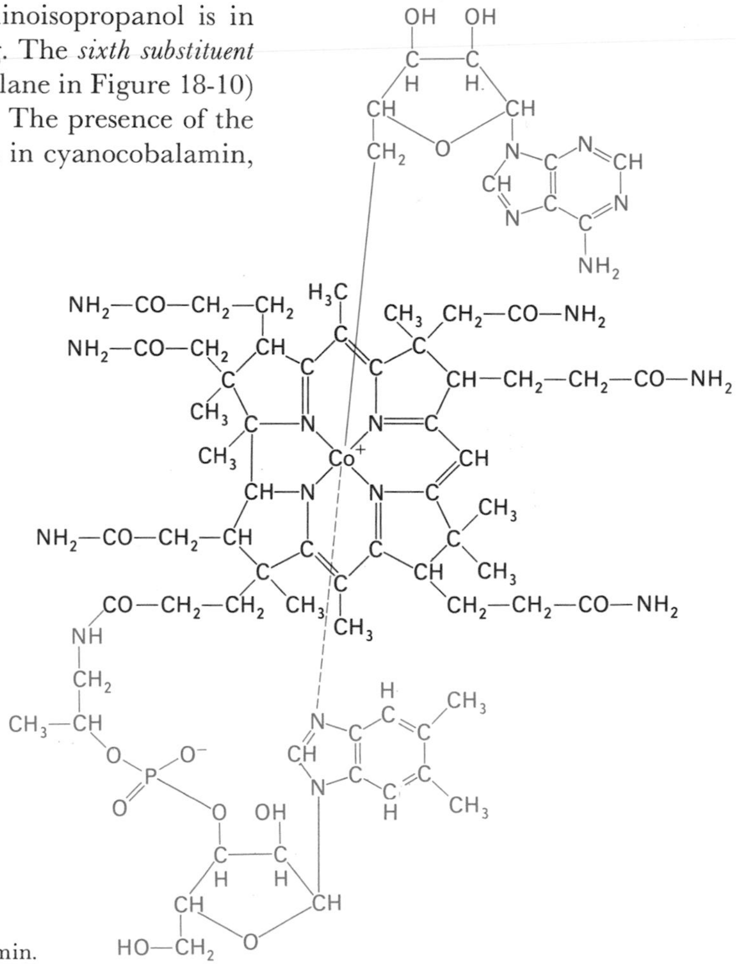

Figure 18-10
Structure of
5'-deoxyadenosylcobalamin.

the most common commercial form of the vitamin, results from the isolation procedure. It is not combined with cobalamin in vivo.

The cobalt atom in cobalamin can have a +1, +2, or +3 oxidation state. The cobalt atom is in the +3 state in hydroxocobalamin (where OH^- occupies the sixth coordination site). This form, called B_{12a} (Co^{3+}), is reduced to a divalent state called B_{12r} (Co^{2+}) by a flavoprotein reductase. The B_{12r} (Co^{2+}) form is reduced by a second flavoprotein reductase to B_{12s} (Co^+). NADH is the reductant in both reactions. The B_{12s} form is the substrate for the final enzymatic reaction that yields the active coenzyme. The 5'-deoxy-adenosyl group is transferred from ATP to Vitamin B_{12s} (Co^+) to give *5'-deoxyadenosylcobalamin, which is the coenzyme of methylmalonyl CoA mutase* (Figure 18-10).

$$B_{12a}(Co^{3+}) \longrightarrow B_{12r}(Co^{2+}) \longrightarrow B_{12s}(Co^+) \longrightarrow$$
$$5'\text{-deoxyadenosylcobalamin}$$

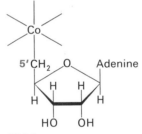

Figure 18-11
The 5' carbon of 5'-deoxyadenosine is coordinated to the cobalt atom in 5'-deoxyadenosylcobalamin. This is the only known example of a carbon-metal bond in a biomolecule.

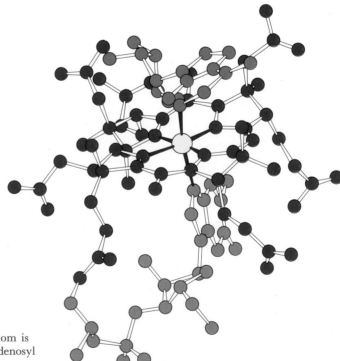

Figure 18-12
Model of 5'-deoxyadenosylcobalamin. The cobalt atom is shown in yellow, the corrin unit in red, the deoxyadenosyl unit in blue, and the benzimidazole unit in green.

ABSORPTION OF COBALAMIN IS IMPAIRED IN PERNICIOUS ANEMIA

There is a specialized transport system for the absorption of cobalamin. The stomach secretes a glycoprotein called *intrinsic factor,* which binds cobalamin in the intestinal lumen. This complex is subsequently bound by a *specific receptor in the lining of the ileum.* The complex of cobalamin and intrinsic factor is then dissociated by a *releasing factor* and actively transported across the ileal membrane into the bloodstream. *Pernicious anemia is caused by a deficiency of intrinsic factor, which leads to impaired absorption of cobalamin.* This disease was originally treated by feeding patients large amounts of liver, a rich source of cobalamin, so that enough of the vitamin was absorbed even in the absence of intrinsic factor. The most reliable therapy is intramuscular injection of cobalamin at monthly intervals.

Animals and plants are unable to synthesize cobalamin. *This vitamin is unique in that it appears to be synthesized only by microorganisms,* particularly anaerobic bacteria. Normal individuals require less than 10 micrograms of cobalamin per day. Nutritional deficiency of cobalamin is rare because this vitamin is found in virtually all animal tissues.

SEVERAL INHERITED DEFECTS OF METHYLMALONYL COENZYME A METABOLISM ARE KNOWN

Several *inherited disorders of methylmalonyl CoA metabolism* have recently been characterized. These disorders usually become evident during the first year of life, when the striking symptom is *acidosis.* The pH of arterial blood is about 7.0, rather than the normal value of 7.4. *Large amounts of methylmalonate appear in the urine of patients who have these disorders.* Normal individuals excrete less than 5 mg of methylmalonate per day, whereas patients with defects in methylmalonyl CoA metabolism may excrete 1 g or more. About half of the patients with methylmalonic aciduria show marked improvement when large doses of cobalamin are administered parenterally. The arterial blood pH returns to normal and there is a marked decrease in the excretion of methylmalonate. These patients appear to have a defect in the transferase that catalyzes the synthesis of deoxyadenosylcobalamin:

$$B_{12s}(Co^+) \longrightarrow\!\!\!| \longmapsto \text{deoxyadenosylcobalamin}$$

In contrast, other patients with impaired methylmalonyl CoA metabolism do not respond to large doses of cobalamin. Some of them may have a defective methylmalonyl CoA mutase apoenzyme. This form of methylmalonic aciduria is frequently lethal.

FUMARATE AND OXALOACETATE ARE POINTS OF ENTRY FOR THE OTHER GLUCOGENIC AMINO ACIDS

The two other points of entry in the citric acid cycle are fumarate and oxaloacetate. *Aspartate* is converted to fumarate in the urea cycle. Aspartate also enters the citric acid cycle following transamination to oxaloacetate:

$$\text{Aspartate} + \alpha\text{-ketoglutarate} \rightleftharpoons \text{oxaloacetate} + \text{glutamate}$$

Asparagine is hydrolyzed by asparaginase to aspartate and NH_4^+.

Tyrosine and *phenylalanine* are both glucogenic and ketogenic. Half of their carbon atoms emerge in fumarate, the other half in acetoacetate. The degradation of these aromatic amino acids will be discussed shortly.

LEUCINE IS DEGRADED TO ACETYL COENZYME A AND ACETOACETYL COENZYME A

Leucine is the only one of the basic set of twenty amino acids that is purely ketogenic. It is degraded by reactions that we have already encountered in fatty acid degradation and the citric acid cycle. First, leucine is transaminated to the corresponding α-keto acid, *α-ketoisocaproate*. This α-keto acid is then *oxidatively decarboxylated* to *isovaleryl CoA*. This reaction is analogous to the oxidative decarboxylation of pyruvate to acetyl CoA and of α-ketoglutarate to succinyl CoA.

Leucine   α-Ketoisocaproate   Isovaleryl CoA

Isovaleryl CoA is *dehydrogenated* to yield *β-methylcrotonyl CoA*. This oxidation is catalyzed by isovaleryl CoA dehydrogenase, in which the hydrogen acceptor is FAD, as in the analogous reaction in fatty acid oxidation that is catalyzed by acyl CoA dehydrogenase. *Beta-methylglutaconyl CoA* is formed by *carboxylation* of β-methylcrotonyl CoA at the expense of an ATP. As might be expected, the mechanism of carboxylation by β-methylcrotonyl CoA carboxylase is very similar to that of pyruvate carboxylase and acetyl CoA carboxylase. In fact, much of our present knowledge of the mechanism of biotin-dependent carboxylations comes from Feodor Lynen's pioneering work on this enzyme.

$$
\begin{array}{ccccc}
O{=}C{-}S{-}CoA & & O{=}C{-}S{-}CoA & & O{=}C{-}S{-}CoA \\
| & & | & & | \\
CH_2 & & C{-}H & & C{-}H \\
| & \longrightarrow & \parallel & \longrightarrow & \parallel \\
H_3C{-}C{-}H & & H_3C{-}C & & H_3C{-}C \\
| & & | & & | \\
CH_3 & & CH_3 & & CH_2{-}COO^- \\
\textbf{Isovaleryl CoA} & & \textbf{β-Methylcrotonyl CoA} & & \textbf{β-Methylglutaconyl CoA}
\end{array}
$$

Beta-methylglutaconyl CoA is then *hydrated* to form *β-hydroxy-β-methylglutaryl CoA*, which is cleaved to *acetyl CoA* and *acetoacetate*. This reaction has already been discussed in regard to the formation of ketone bodies from fatty acids (p. 417).

$$
\begin{array}{ccccc}
 & & & & \textbf{Acetyl CoA} \\
O{=}C{-}S{-}CoA & & O{=}C{-}S{-}CoA & & O{=}C{-}S{-}CoA \\
| & & | & & | \\
C{-}H & & H{-}C{-}H & & CH_3 \\
\parallel & \longrightarrow & | & \longrightarrow & \\
H_3C{-}C & & H_3C{-}C{-}OH & & + \\
| & & | & & \\
CH_2{-}COO^- & & CH_2{-}COO^- & & H_3C{-}C{=}O \\
 & & & & | \\
 & & & & CH_2{-}COO^- \\
\textbf{β-Methylglutaconyl CoA} & & \textbf{β-Hydroxy-β-methylglutaryl CoA} & & \textbf{Acetoacetate}
\end{array}
$$

It is interesting to note that many coenzymes participate in the degradation of leucine to acetyl CoA and acetoacetate: *PLP* in transamination; *TPP, lipoate, FAD, and NAD$^+$* in oxidative decarboxylation; FAD again in dehydrogenation; and *biotin* in carboxylation. *Coenzyme A* is the acyl carrier in these reactions.

The degradative pathways of valine and isoleucine resemble that of leucine. All three amino acids are initially transaminated to the corresponding α-keto acid, which is then oxidatively decarboxylated to yield a derivative of CoA. The subsequent reactions are like those of fatty acid oxidation. Isoleucine yields acetyl CoA and propionyl CoA, whereas valine yields methylmalonyl CoA. There is an inborn error of metabolism that affects the degradation of valine, isoleucine, and leucine. In *maple syrup urine disease,* the oxidative decarboxylation of these three amino acids is blocked. The amounts of leucine, isoleucine, and valine in blood and urine are markedly elevated, which results in a corresponding increase in the α-keto acids derived from thse amino acids. The urine of patients having this disease has the odor of maple syrup, hence the name of the disease.

PHENYLALANINE AND TYROSINE ARE DEGRADED TO ACETOACETATE AND FUMARATE

Phenylalanine and tyrosine have a common degradative pathway. This pathway starts with the hydroxylation of phenylalanine to tyrosine. This reaction is catalyzed by *phenylalanine hydroxylase,* which is a mixed function oxidase.

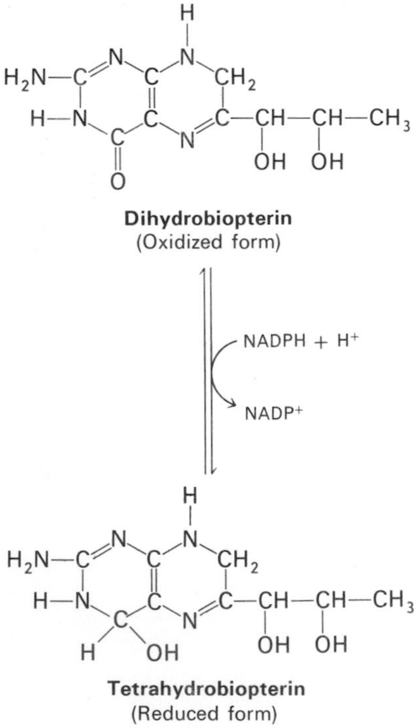

Dihydrobiopterin
(Oxidized form)

NADPH + H$^+$

NADP$^+$

Tetrahydrobiopterin
(Reduced form)

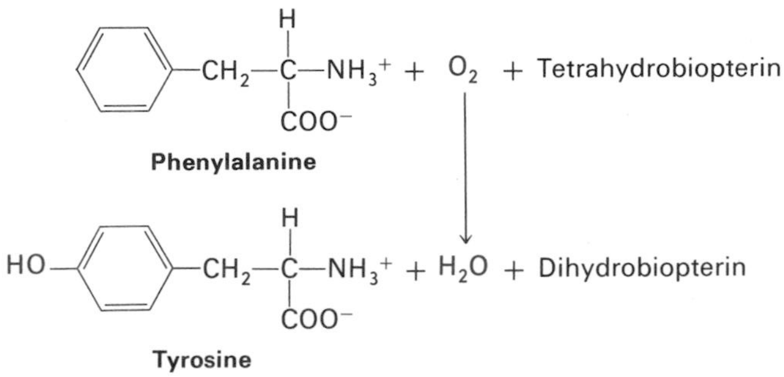

Phenylalanine

Tyrosine

The reductant here is *tetrahydrobiopterin,* an electron carrier that has not been previously discussed. The oxidized form of this electron carrier is *dihydrobiopterin.* NADPH reduces dihydrobiopterin to regenerate tetrahydrobiopterin, a reaction catalyzed by dihydrobiopterin reductase. The sum of the reactions catalyzed by phenylalanine hydroxylase and dihydrobiopterin reductase is:

Phenylalanine + O$_2$ + NADPH + H$^+$ $\longrightarrow$

tyrosine + NADP$^+$ + H$_2$O

The next step is the transamination of tyrosine to p-*hydroxyphenylpyruvate* (Figure 18-13). This α-keto acid is oxidized to homogentisate. Ascorbate is essential for this complex reaction, which is catalyzed by hydroxyphenylpyruvate oxidase, a copper-containing protein. The aromatic ring of homogentisate is then cleaved by O_2, yielding *4-maleylacetoacetate,* in a reaction catalyzed by homogentisate oxidase. Both atoms of molecular oxygen appear in the product. 4-Maleylacetoacetate is isomerized to *4-fumarylacetoacetate,* which is finally hydrolyzed to *fumarate* and *acetoacetate.*

GARROD'S DISCOVERY OF INBORN ERRORS OF METABOLISM

Alcaptonuria is an inherited metabolic disorder caused by the absence of homogentisate oxidase. Homogentisate accumulates and is excreted in the urine, which turns dark on standing as homogentisate is oxidized and polymerized to a melaninlike substance. Alcaptonuria is a relatively benign condition, as described by Zacutus Lusitanus in 1649:

> The patient was a boy who passed black urine and who, at the age of fourteen years, was submitted to a drastic course of treatment which had for its aim the subduing of the fiery heat of his viscera, which was supposed to bring about the condition in question by charring and blackening his bile. Among the measures prescribed were bleedings, purgation, baths, a cold and watery diet and drugs galore. None of these had any obvious effect, and eventually the patient, who tired of the futile and superfluous therapy, resolved to let things take their natural course. None of the predicted evils ensued, he married, begat a large family, and lived a long and healthy life, always passing urine black as ink.

In 1902, Archibald Garrod showed that alcaptonuria is transmitted as a single recessive Mendelian trait. Furthermore, he recognized that homogentisate is a normal intermediate in the degradation of phenylalanine and tyrosine and that it accumulates in alcaptonuria because its degradation is blocked. He concluded that "the splitting of the benzene ring in normal metabolism is the work of a special enzyme, that in congenital alcaptonuria this enzyme is wanting." Garrod perceived the direct relationship between genes

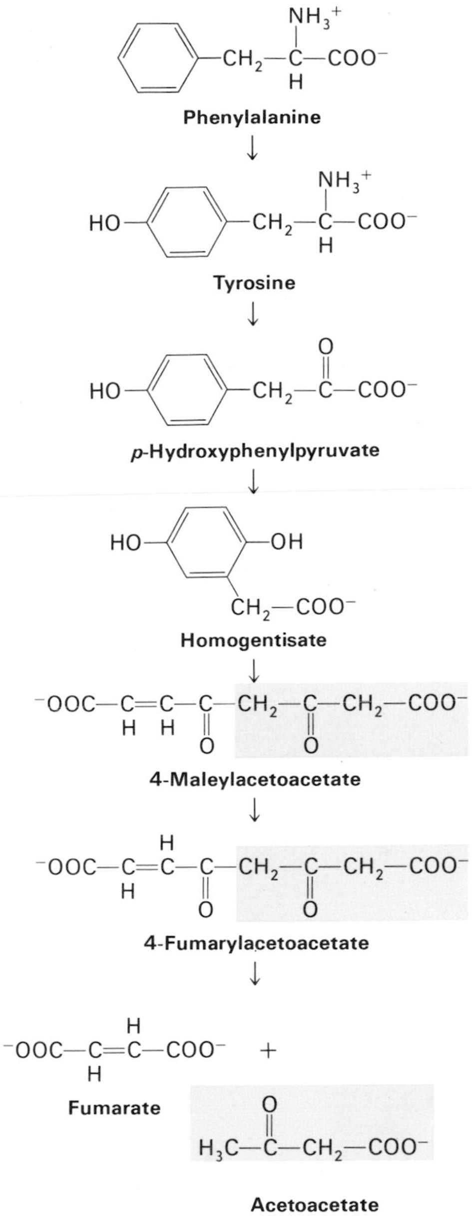

Figure 18-13
Pathway for the degradation of phenylalanine and tyrosine.

and enzymes, and recognized the importance of chemical individuality. His book *Inborn Errors of Metabolism* was a most imaginative and important contribution to biology and medicine.

A BLOCK IN THE HYDROXYLATION OF PHENYLALANINE CAN LEAD TO SEVERE MENTAL RETARDATION

Phenylketonuria, an inborn error of phenylalanine metabolism, can have devastating effects, in contrast to alcaptonuria. Untreated individuals with phenylketonuria are nearly always *severely mentally retarded.* In fact, about 1% of patients in mental institutions have phenylketonuria. The weight of the brain of these individuals is below normal, myelination of their nerves is defective, and their reflexes are hyperactive. The life expectancy of untreated phenylketonurics is drastically shortened. Half are dead by age twenty, and three-quarters by age thirty.

Phenylalanine hydroxylase is missing or deficient in phenylketonurics. Hence, phenylalanine cannot be converted to tyrosine. One consequence of this block is that there is an *increased concentration of phenylalanine in all body fluids.* Some fates of phenylalanine that are quantitatively insignificant in normal individuals become prominent in phenylketonurics. The most evident of these is the transamination of phenylalanine to form *phenylpyruvate.* The disease acquired its name from the high levels of this phenylketone in urine. Phenyllactate, phenylacetate, and *o*-hydroxylphenylacetate are derived from phenylpyruvate. The α-amino group of glutamine forms an amide bond with the carboxyl group of phenylacetate, yielding phenylacetylglutamine.

There are many other derangements of amino acid metabolism associated with phenylketonuria, particularly among the aromatic compounds. Phenylketonurics have a lighter skin and hair color than their siblings. The hydroxylation of tyrosine is the first step in the formation of the pigment melanin. In phenylketonurics this reaction is competitively inhibited by the high levels of phenylalanine, and so less melanin is formed. *The biochemical basis of mental retardation in untreated phenylketonuria is an enigma.*

Phenylketonurics appear normal at birth but are severely defective by age one if untreated. The therapy for phenylketonuria is a *low phenylalanine diet.* The aim is to provide just enough phenylalanine to meet the needs for growth and replacement. Proteins that have an initially low content of phenylalanine, such as casein from milk, are hydrolyzed and phenylalanine is removed by adsorption. A low phenylalanine diet must be started very soon after birth to prevent irreversible brain damage. The average I.Q. of phenylketonurics treated within a few weeks after birth was 93; a control group of siblings treated starting at age one had an average I.Q. of 53.

Thus, *early diagnosis of the disease is essential,* and can be accomplished by mass screening programs. The urine of newborns is assayed for phenylpyruvate or phenylalanine. Addition of $FeCl_3$ to the urine of an untreated phenylketonuric gives an olive green color because of the presence of phenylpyruvate. The incidence of phenylketonuria is about 1 in 20,000 newborns. The disease is inherited as an *autosomal recessive.* Heterozygotes, which comprise about 1.5% of a typical population, appear normal. Carriers of the phenylketonuric gene have a reduced level of phenylalanine hydroxylase, as reflected in an increased level of phenylalanine in the blood. However, these criteria are not absolute, since the blood levels of phenylalanine of carriers and normal individuals overlap to some extent. The measurement of the kinetics of disappearance of intravenouly administered phenylalanine is a more definitive test for the carrier state.

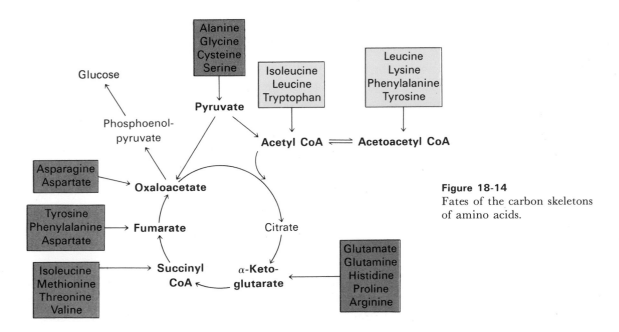

Figure 18-14
Fates of the carbon skeletons of amino acids.

SUMMARY

Surplus amino acids are used as metabolic fuel. The degradation of most surplus amino acids starts with the removal of their α-amino groups by transamination to an α-keto acid. Pyridoxal phosphate is the coenzyme in all transaminases. These α-amino groups funnel into α-ketoglutarate to form glutamate, which is then oxidatively deaminated by glutamate dehydrogenase to give NH_4^+ and α-ketoglutarate. NAD^+ or $NADP^+$ is the electron acceptor in this reaction. In terrestrial vertebrates, NH_4^+ is converted to urea by the urea cycle. Urea is formed by the hydrolysis of arginine. The subsequent reactions of the urea cycle synthesize arginine from ornithine, the other product of the hydrolysis reaction. First, ornithine is carbamoylated to citrulline by carbamoyl phosphate. Citrulline then condenses with aspartate to form argininosuccinate, which is cleaved to arginine and fumarate. The carbon atom and one nitrogen atom of urea come from carbamoyl phosphate, which is synthesized from CO_2, NH_4^+, and ATP. The other nitrogen atom of urea comes from aspartate. Four high-energy phosphate bonds are consumed in the synthesis of one molecule of urea.

The carbon atoms of degraded amino acids are converted into pyruvate, acetyl CoA, acetoacetate, or a citric acid cycle intermediate. Most amino acids are purely glucogenic, one is purely ketogenic, and a few are both ketogenic and glucogenic. Alanine, glycine, cysteine, and serine are degraded to pyruvate. Leucine is degraded to acetoacetate and acetyl CoA. Alpha-ketoglutarate is the point of entry for glutamate and four amino acids (glutamine, histidine, proline, and arginine) that can be converted into glutamate. Succinyl CoA is the point of entry for some of the carbon atoms of the four amino acids (methionine, isoleucine, threonine, and valine) that are degraded via methylmalonyl CoA. Deoxyadenosylcobalamin, a Vitamin B_{12} derivative, is required for the isomerization of methylmalonyl CoA to succinyl CoA. Some of the carbon atoms of tyrosine, phenylalanine, and aspartate enter at fumarate, whereas the other carbon atoms of tyrosine and phenylalanine emerge in acetoacetate. Finally, the carbon atoms of asparagine and aspartate are converted to oxaloacetate.

SELECTED READINGS

Books on amino acid metabolism:

Meister, A., 1965. *Biochemistry of the Amino Acids* (2nd ed.), vols. 1 and 2. Academic Press.

Nyhan, W. L., (ed.), 1967. *Amino Acid Metabolism and Genetic Variations*. McGraw-Hill.

Reaction mechanisms:

Snell, E. E., and DiMari, S. J., 1970. Schiff base intermediates in enzyme catalyis. *In* Boyer, P. D., (ed.), *The Enzymes* (3rd ed.), vol. 2, pp. 335–370. Academic Press.

Barker, H. A., 1972. Coenzymes B_{12}-dependent mutases causing carbon chain rearrangements. *In* Boyer, P. D., (ed.), *The Enzymes* (3rd ed.), vol. 6, pp. 509–537.

Dunathan, H. C., 1971. Stereochemical aspects of pyridoxal phosphate catalysis. *Advan. Enzymol.* 35:79–134.

Inherited disorders:

Shih, V. E., and Efron, M. L., 1972. Urea cycle disorders. *In* Stanbury, J. B., Wyngaarden, J. B., and Frederickson, D. S., (eds.), *The Metabolic Basis of Inherited Disease* (3rd ed.), pp. 370–392. McGraw-Hill.

Rosenberg, L. E., 1972. Disorders of propionate, methylmalonate, and vitamin B_{12} metabolism. *In* Stanbury, J. B., Wyngaarden, J. B., and Fredrickson, D. S., (eds.), *The Metabolic Basis of Inherited Disease* (3rd ed.), pp. 440–458.

Knox, W. E., 1972. Phenylketonuria. *In* Stanbury, J. B., Wyngaarden, J. B., and Fredrickson, D. S., (eds.), *The Metabolic Basis of Inherited Disease* (3rd ed.), pp. 266–295.

Garrod, A. E., 1909. *Inborn Errors in Metabolism.* Oxford University Press (reprinted in 1963 with a supplement by H. Harris).

Childs, B., 1970. Sir Archibald Garrod's conception of chemical individuality: a modern appreciation. *N. Engl. J. Med.* 282:71–78.

PROBLEMS

1. Name the α-keto acid that is formed by transamination of each of these amino acids:
   - (a) Alanine.
   - (b) Aspartate.
   - (c) Glutamate.
   - (d) Leucine.
   - (e) Phenylalanine.
   - (f) Tyrosine.
2. Write a balanced equation for the conversion of aspartate to glucose via oxaloacetate. Cite the coenzymes that participate in these steps.
3. Write a balanced equation for the conversion of aspartate to oxaloacetate via fumarate.
4. Consider the mechanism of the conversion of L-methylmalonyl CoA to succinyl CoA by L-methylmalonyl CoA mutase.

   - (a) Design an experiment to distinguish between the migration of the COO^- group and the —CO—S—CoA group in this reaction.
   - (b) In fact, it is the —CO—S—CoA group that migrates. Now design an experiment to distinguish between intermolecular and intramolecular transfer of the —CO—S—CoA group.
   - (c) What is the significance of the finding that no tritium is incorporated into succinyl CoA when the mutase reaction is carried out in tritiated water?

CHAPTER 19

PHOTOSYNTHESIS

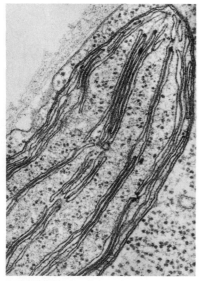

Figure 19-1
Electron micrograph of part of a chloroplast in a green alga. The thylakoid membranes pile on top of each other to form grana. [Courtesy of Dr. George Palade.]

All free energy consumed by biological systems arises from solar energy that is trapped by the process of photosynthesis. The basic equation of photosynthesis is simple—indeed deceptively so:

$$H_2O + CO_2 \xrightarrow{\text{light}} (CH_2O) + O_2$$

In this equation, (CH_2O) represents carbohydrate. The mechanism of photosynthesis is complex and involves the interplay of many macromolecules and small molecules. Photosynthesis in green plants occurs in chloroplasts, which are specialized organelles. The energy conversion apparatus is an integral part of the *thylakoid* membrane system in chloroplasts (Figure 19-1). The first step in photosynthesis is the absorption of light by a *chlorophyll molecule*. The energy is transferred from one chlorophyll to another until it reaches a chlorophyll with special properties at a site called the *reaction center*. Light is converted into chemically useful energy at two kinds of reaction centers. In fact, *the cooperation of two light reactions* is required for photosynthesis. One of them, called *photosystem I*, generates reducing power in the form of NADPH, whereas the other, called *photosystem II*, splits water to produce O_2 and generates a reductant. *ATP* is produced as electrons flow through an electron-transfer chain linking the two photosystems. Alternatively, ATP can be generated without the concomitant formation of NADPH.

The NADPH and ATP generated by light are then used to reduce CO_2 to carbohydrate by a series of dark reactions called the *Calvin cycle*. These reactions occur in the soluble component of chloroplasts. The first step is the reaction of CO_2 with ribulose diphosphate to form two molecules of *3-phosphoglycerate*. Hexose is formed from 3-phosphoglycerate by the gluconeogenic pathway, and ribulose diphosphate is regenerated by the action of transketolase, aldolase, and several other enzymes. In one round of the cycle, 3 ATP and 2 NADPH take CO_2 to the level of a hexose phosphate.

DISCOVERY OF THE BASIC EQUATION OF PHOTOSYNTHESIS

Most of the basic equation of photosynthesis could have been written at the end of the eighteenth century. The production of oxygen in photosynthesis was discovered by Joseph Priestley in 1780. He found that plants could "restore air which has been injured by the burning of candles." He placed a sprig of mint in an inverted glass jar in a vessel of water and found several days later "that the air would neither extinguish a candle, nor was it all inconvenient to a mouse which I put into it." This great chemist of the eighteenth century was also a nonconformist English minister. Indeed, theology, philosophy, and politics were his primary interests. In 1791, Priestley was forced to leave England because of his sympathies with the French Revolution. He went to France and then to the United States, where he died in 1804, after a few quiet years high on the banks of the Susquehanna. A quite different fate was suffered by a contemporary of Priestley, who set the groundwork for the elucidation of the basic process of photosynthesis. Antoine Lavoisier devised methods for the handling of gases and discovered the concept of oxidation and the law of conservation of mass in chemical reactions. He was associated with the monarchy and was executed in 1794 in the course of the French Revolution. The judge who pronounced sentence commented that "the Republic has no need for scientists."

Figure 19-2
Priestley's classic experiment on photosynthesis. [Based on E. I. Rabinowitch. Photosynthesis. Copyright © 1948 by Scientific American, Inc. All rights reserved.]

The next major contribution to the elucidation of photosynthesis was made by Jan Ingenhousz, a Dutchman, who was court physician to the Austrian empress. Ingenhousz was a worldly man who liked to visit London. He once heard a discussion of Priestley's experiments on the restoration of air by plants and was so taken by it that he decided that he must do some experiments at the "earliest opportunity." This came six years later, when Ingenhousz rented a villa near London and spent a summer feverishly performing more than five hundred experiments. He discovered the role of light in photosynthesis:

> I observed that plants not only have the faculty to correct bad air in six or ten days, by growing in it, as the experiments of Dr. Priestley indicate, but that they perform this important office in a complete manner in a few hours; that *this wonderful operation is by no means owing to the vegetation of the plant, but to the influence of light of the sun upon the plant.*

Ingenhousz was quick to publish his findings because he was concerned that someone else would beat him to it. By the end of the summer, he had published a book entitled *Experiments Upon Vegetables, Discovering Their Great Power of Purifying the Common Air in Sunshine and Injuring it in the Shade and at Night.*

Ingenhousz's fear of being scooped was justified. Similar experiments were being carried out in Geneva by Jean Senebier, a Swiss pastor. His distinctive contribution was to show that "fixed air"—namely, CO_2—is taken up in photosynthesis. The role of water in photosynthesis was demonstrated by Theodore de Saussure, also a Genevan. He showed that the sum of the weights of organic matter produced by plants and of oxygen evolved is much more than the weight of CO_2 consumed. From Lavoisier's law of the conservation of mass, de Saussure concluded that another substance was utilized. The only inputs in his system were CO_2, water, and light. Hence, de Saussure concluded that the other reactant must be water.

The final contribution to the basic equation of photosynthesis came nearly a half-century later. Julius Robert Mayer, a German surgeon, discovered the law of conservation of energy in 1842. Mayer recognized that plants convert solar energy into chemical free energy:

> The plants take in one form of power, light; and produce another power, chemical difference.

The amount of energy stored by photosynthesis is enormous. More than 10^{12} kcal of free energy is stored annually by photosynthesis on earth, corresponding to the assimilation of more than 10^{10} tons of carbon into carbohydrate and other forms of organic matter.

CHLOROPHYLLS ARE THE PHOTORECEPTOR MOLECULES

Mayer stated, "Nature has put itself the problem of how to catch in flight light streaming to the earth and to store the most elusive of all powers in rigid form." What is the mechanism of trapping this most elusive of all powers? The first step is the absorption of light by a photoreceptor molecule. The principal photoreceptor in the chloroplasts of green plants is *chlorophyll* a, a substituted tetrapyrrole. The four nitrogen atoms of the pyrroles are coordinated to a magnesium atom. Thus, chlorophyll is a *magnesium porphyrin,* whereas heme is an iron porphyrin. The porphyrin ring structure

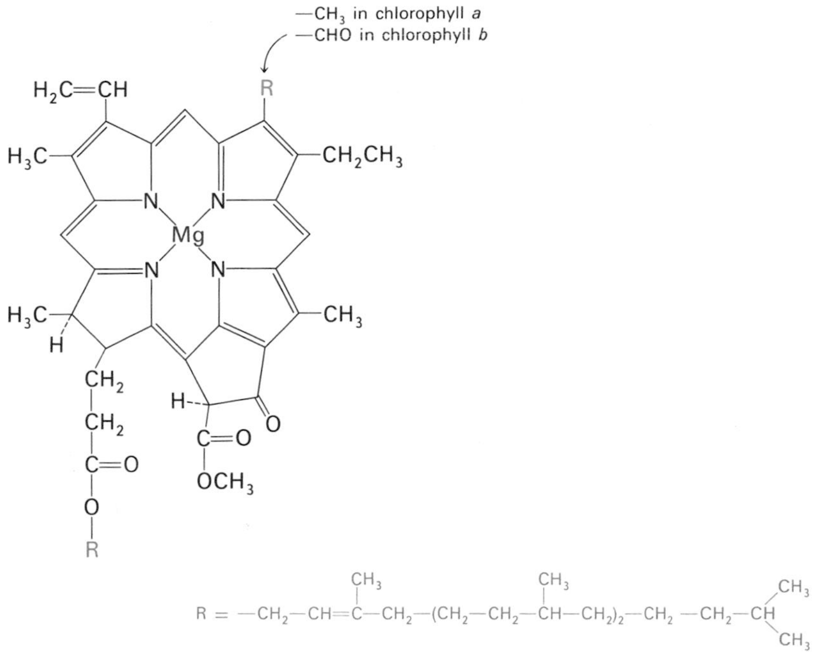

Figure 19-3
Formulas of chlorophyll *a* and *b*.

of chlorophyll differs from that of heme in several ways: (1) one of the pyrroles is partially reduced; (2) a cyclopentanone ring is fused to one of the pyrroles; (3) both acid side chains in chlorophyll are esterified, whereas in heme they are free. One of the acid chains in chlorophyll is a methyl ester, whereas the other is an ester of phytol ($C_{20}H_{39}OH$). This long-chain alcohol consists of four isoprene units, which makes it highly hydrophobic. *Chlorophyll* b differs from chlorophyll *a* in a substituent on one of the pyrroles. In chlorophyll *b,* there is a formyl group in place of a methyl group.

These chlorophylls are very effective photoreceptors because they contain a network of alternating single and double bonds. In other words, they are *polyenes.* They have a very strong absorption bands in the visible region of the spectrum, where the solar output reaching the earth is also maximal. The peak extinction coefficients of chlorophyll *a* and *b* are higher than 10^5 cm^{-1} M^{-1}, among the highest observed for organic compounds.

The absorption spectra of chlorophyll *a* and *b* are different (Figure 19-4). Light that is not appreciably absorbed by chlorophyll *a*, at 460 nm for example, is captured by chlorophyll *b,* which has intense absorption at that wavelength. Thus, *these two kinds of chlorophyll complement each other in absorbing the incident sunlight.* There is still a large spectral region, from 500 to 600 nm, where the absorption of light is relatively weak. However, most plants need not capture light in that spectral region, since enough is absorbed in the blue and red parts of the spectrum.

THE PRIMARY EVENTS OF PHOTOSYNTHESIS OCCUR IN A HIGHLY ORGANIZED MEMBRANE SYSTEM

Chloroplasts, the organelles of photosynthesis, are typically 5 μm long. Like mitochondria, they have an outer membrane and a complex inner-membrane system (Figure 19-5). The inner membrane is folded to form *thylakoids,* which are flattened sacs. A pile of these sacs constitutes a *granum.* Different grana are connected by membrane regions called stroma lamellae. The thylakoid membranes contain the chlorophyll molecules and the other components of the energy-transducing machinery. They contain nearly equal amounts of protein and lipid. Rather little is known about their detailed structure. Electron micrographs of freeze-etched prepara-

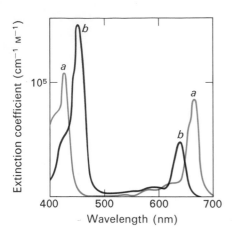

Figure 19-4
Absorption spectra of chlorphyll *a* and *b*.

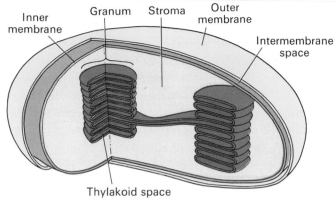

Figure 19-5
Diagram of a chloroplast. [Based on S. L. Wolfe, *Biology of the Cell,* p. 130. © 1972 by Wadsworth Publishing Company, Inc. Adapted by permission of the publisher.]

tions of chloroplasts show two kinds of particles, with diameters of 175 Å and 110 Å.

The soluble component of chloroplasts contains the enzymes that convert CO_2 to sugar. Chloroplasts have ribosomes and actively participate in protein synthesis. In addition, they contain their own DNA, which is distinct from nuclear DNA. Thus, *the chloroplast is an organelle with considerable autonomy.*

THE PHOTOSYNTHETIC UNIT: PHOTONS FUNNEL INTO A REACTION CENTER

When the rate of photosynthesis is measured as a function of the intensity of illumination, it increases linearly at low intensities and reaches a saturating value at high intensities (Figure 19-6). A saturating value is observed in strong light because chemical reactions that utilize the absorbed photons become rate-limiting. Thus, *photosynthesis can be separated into light reactions and dark reactions.* As will be discussed shortly, the light reactions generate NADPH and ATP, whereas the dark reactions use these energy-rich molecules to reduce CO_2.

In 1932, Robert Emerson and William Arnold measured the oxygen yield of photosynthesis when *Chlorella* cells were exposed

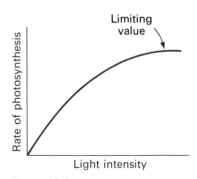

Figure 19-6
The rate of photosynthesis reaches a limiting value when the light intensity suffices to excite only a small fraction of the chlorophyll molecules.

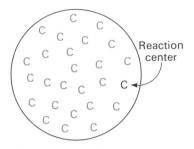

Photosynthetic unit

Figure 19-7
Diagram of a photosynthetic
unit. The antenna chlorophyll
molecules (denoted by green
C's) transfer their excitation
energy to a specialized
chlorophyll at the reaction
center (denoted by a red C).

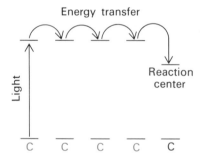

Figure 19-8
Diagram of the energy level
of the excited state of the
antenna chlorophylls and of
the reaction center.

to light flashes lasting a few microseconds. They expected to find
that the yield per flash would increase with the flash intensity until
each chlorophyll molecule absorbed a photon, which would then
be used in dark reactions. Their experimental observation was
entirely unexpected: the yield per flash reached a maximum when
just 1 out of 2500 chlorophylls absorbed a quantum during a flash.

This experiment led to the concept of the *photosynthetic unit*. Hans
Gaffron proposed that light is absorbed by hundreds of chlorophyll
molecules, which then transfer their excitation energy to a site at
which chemical reactions occur (Figure 19-7). This site is called
a *reaction center*. Thus, the function of most chlorophyll molecules
in the photosynthetic unit is to absorb light. Only a small proportion of chlorophylls, those at reaction centers, mediate the transformation of light into chemical energy. The chlorophylls at the
reaction center are chemically identical to the other chlorophylls
in the photosynthetic unit, but they have special properties because
they are located in a unique environment. One difference is that
the energy level of their excited state is lower than that of the other
chlorophylls, which makes them energy traps. The energy absorbed
by chlorophyll molecules hops around the photosynthetic unit until
it reaches a reaction-center chlorophyll. The transfer of energy to
the reaction center is a very rapid process that occurs in less than
10^{-10} sec.

THE OXYGEN EVOLVED IN PHOTOSYNTHESIS COMES FROM WATER

Let us turn now to the chemical changes in photosynthesis. The
source of the evolved oxygen in green plants has important implications for the mechanism of photosynthesis. Comparative studies
of photosynthesis in many organisms provided the answer as early
as 1931. Some photosynthetic bacteria convert hydrogen sulfide to
sulfur in the presence of light. Cornelis Van Niel perceived that
the reactions for photosynthesis in green plants and green sulfur
bacteria are very similar:

$$CO_2 + 2\ H_2O \xrightarrow{\text{light}} (CH_2O) + O_2 + H_2O$$

$$CO_2 + 2\ H_2S \xrightarrow{\text{light}} (CH_2O) + 2\ S + H_2O$$

The sulfur that is formed by the photosynthetic bacteria is analogous to the oxygen that is evolved in plants. Van Niel proposed a general formula for photosynthesis:

$$CO_2 \;+\; 2\,H_2A \xrightarrow{\text{light}} (CH_2O) \;+\; 2\,A \;+\; H_2O$$

| Hydrogen acceptor | Hydrogen donor | Reduced acceptor | Dehydrogenated donor | |

The hydrogen donor H_2A is H_2O in green plants and H_2S in the photosynthetic sulfur bacteria. Thus, photosynthesis in plants could be formulated as a reaction in which CO_2 is reduced by hydrogen derived from water. Oxygen evolution would then be the necessary consequence of this dehydrogenation process. The essence of this view of photosynthesis is that *water is split by light*.

The availability in 1941 of a heavy isotope of oxygen, namely ^{18}O, made it feasible to test this concept directly. In fact, ^{18}O appeared in the evolved oxygen when photosynthesis was carried out in water enriched in this isotope. This result confirmed the proposal that the O_2 formed in photosynthesis comes from water.

$$H_2{}^{18}O + CO_2 \xrightarrow{\text{light}} (CH_2O) + {}^{18}O_2$$

HILL REACTION: ILLUMINATED CHLOROPLASTS EVOLVE OXYGEN AND REDUCE AN ARTIFICIAL ELECTRON ACCEPTOR

In 1939, Robert Hill discovered that isolated chloroplasts evolve oxygen when they are illuminated in the presence of a suitable electron acceptor, such as ferricyanide. There is a concomitant reduction of ferricyanide to ferrocyanide.

The Hill reaction is a landmark in the elucidation of the mechanism of photosynthesis for several reasons:

1. It dissected photosynthesis by showing that oxygen evolution can occur without the reduction of CO_2. Artificial electron acceptors such as ferricyanide can substitute for CO_2.

2. It confirmed that the evolved oxygen comes from water rather than from CO_2, since no CO_2 was present.

$$2\,H_2O + 4\,Fe^{3+}$$
$$\downarrow \text{illuminated chloroplasts}$$
$$O_2 + 4\,H^+ + 4\,Fe^{2+}$$

3. It showed that isolated chloroplasts can perform a significant partial reaction of photosynthesis.

4. It revealed that a primary event in photosynthesis is the *light-activated transfer of an electron from one substance to another against a chemical-potential gradient.* The reduction of ferric to ferrous ion by light is a conversion of light into chemical energy.

PHOTOSYNTHESIS REQUIRES THE INTERACTIONS OF TWO KINDS OF PHOTOSYSTEMS

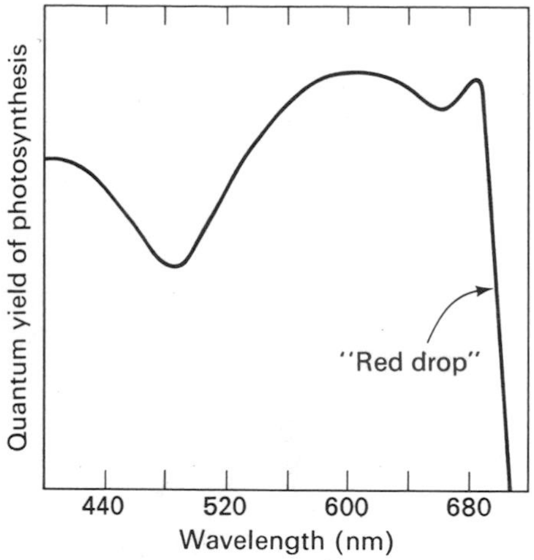

A number of experimental observations led to the finding that there are two different photosystems in chloroplasts. The rate of photosynthesis was investigated as a function of the wavelength of the incident light. The photosynthetic rate divided by the number of quanta observed at each wavelength gives the relative quantum efficiency of the process. For a single kind of photoreceptor, the quantum efficiency is expected to be independent of wavelength over its entire absorption band. This is not the case in photosynthesis: the quantum efficiency of photosynthesis drops sharply at wavelengths longer than 680 nm, although chlorophyll still absorbs light in the range from 680 to 700 nm (Figure 19-9). However, the rate of photosynthesis using long-wavelength light can be enhanced by adding light of a shorter wavelength, such as 600 nm. The photosynthetic rate in the presence of both 600-nm and 700-nm light is greater than the sum of the rates obtained when the two wavelengths are given separately. These observations, called the red drop and the enhancement phenomenon, led Emerson to propose that *photosynthesis requires the interaction of two light reactions, both of which can be driven by light of less than 680 nm but only one by light of longer wavelength.*

Figure 19-9
The quantum yield of photosynthesis drops abruptly when the excitation wavelength is greater than 680 nm.

ROLES OF THE TWO PHOTOSYSTEMS

Photosystem I, which can be excited by light of wavelength shorter than 700 nm, generates a strong reductant that leads to the formation of NADPH. In contrast, photosystem II, which requires light of shorter wavelength than 680 nm, produces a strong oxidant that leads to the formation of O_2. In addition, photosystem I produces a weak oxidant, whereas photosystem II produces a weak reductant. The interaction of these species results in the formation of ATP. This part of photosynthesis, which was discovered by Daniel Arnon, is called photosynthetic phosphorylation or *photophosphorylation.*

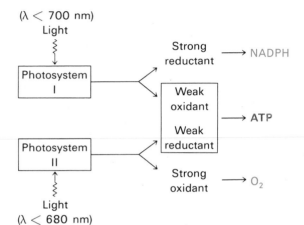

Figure 19-10
Interaction of photosystems I and II in photosynthesis.

Photosystems I and II are structurally distinct. When thylakoid membranes are treated with detergents, photosystem I particles are preferentially released. By density-gradient centrifugation, it is feasible to separate particles that have only photosystem-I activity from others that are enriched in photosystem-II activity. Both kinds of particles contain chlorophyll *a* and *b,* but in different ratios. The compositions of these particles differ in other respects too.

PHOTOSYSTEM I GENERATES NADPH VIA REDUCED FERREDOXIN

The reaction center of photosystem I (Figure 19-11) is a chlorophyll *a* molecule in a unique environment. The absorption maximum is shifted from 678 to 700 nm. Consequently, this reaction center is called *P700* (P stands for pigment).

Many chlorophyll molecules absorb light and transfer the excitation energy to P700. When P700 is excited, it transfers an electron

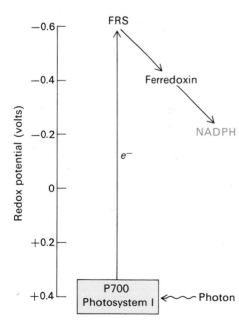

Figure 19-11
Photosystem I.

to an acceptor called ferredoxin-reducing substance (FRS). In the dark, P700 has an oxidation-reduction potential of $+0.4$ volt. When P700 is excited by light, it has a different distribution of electrons that changes its oxidation-reduction potential to about -0.6 volt. Thus, *light pumps an electron in photosystem I from a potential of about $+0.4$ to -0.6 V.* The energy of a red photon, 1.8 V, is sufficient to elevate an electron by 1.0 V. The transfer of an excited electron from P700 to FRS makes P700 electron deficient. This oxidized form of P700 must regain an electron before P700 can again function as a reaction center. The source of the restoring electrons will be discussed shortly.

The reduced form of FRS then transfers its electron to *ferredoxin,* a nonheme iron protein of molecular weight 11,600. The iron atom at the active site of ferredoxin is alternately oxidized and reduced. Reduced ferredoxin transfers its electron to NADP to form NADPH. This reaction is catalyzed by *ferredoxin-NADP reductase,* which contains FAD as its prosthetic group. Note that the reduction of NADP$^+$ to NADPH is a two-electron process, whereas ferredoxin is a one-electron carrier, Hence, the electrons from two reduced ferredoxins must converge to form one NADPH.

$$2\ \text{Ferredoxin}_{\text{reduced}} + \text{H}^+ + \text{NADP}^+$$

$$\Big\downarrow \begin{array}{l} \text{ferredoxin-NADP} \\ \text{reductase} \end{array}$$

$$2\ \text{Ferredoxin}_{\text{oxidized}} + \text{NADPH}$$

PHOTOSYSTEM II GENERATES A STRONG OXIDANT THAT SPLITS WATER

Relatively little is known about the reaction center and the primary electron acceptor of photosystem II (Figure 19-12). The oxidation-reduction potential of the reaction center is about $+0.8$ volt. Light creates a very strong oxidant (the oxidized form of the reaction center) and a weak reductant (called Q) at this center. The strong oxidant extracts electrons from water, thereby forming O_2. Manganese plays an important role in this step.

$$2\ \text{H}_2\text{O} \longrightarrow 4\ \text{H}^+ + \text{O}_2 + 4\ e^-$$

ATP IS FORMED AS ELECTRONS FLOW FROM
PHOTOSYSTEM II TO I

The two photosystems are linked by a series of electron carriers
that have two roles:

1. Electrons flow through this link from photosystem II to I.
These electrons are needed to *regenerate the reduced form of P700,* the
reaction center of photosystem I.

2. ATP is formed from ADP and P_i as electrons flow through
this link. In other words, *phosphorylation is coupled to electron transport*
(Figures 19-12 and 19-13).

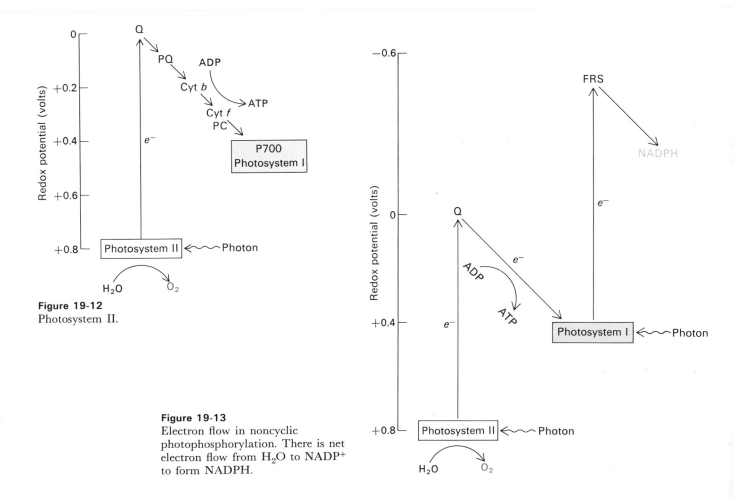

Figure 19-12
Photosystem II.

Figure 19-13
Electron flow in noncyclic
photophosphorylation. There is net
electron flow from H_2O to $NADP^+$
to form NADPH.

The electron-transport chain between the two photosystems contains a number of proteins and lipids. Two cytochromes, a *b* and an *f* type, are involved, as is *plastocyanin,* a copper-containing protein. In addition, a quinone with a long lipid tail, called *plasto-quinone,* is a component of this electron-transport system. It is interesting to note that plastoquinone resembles ubiquinone, a constituent of the electron-transport chain in mitochondria.

$$H_3C \quad \begin{array}{c} O \\ \| \end{array} \quad \left(\begin{array}{c} CH_3 \\ | \end{array} \right)$$

$$H_3C \quad\quad\quad \left(CH_2 - CH = C - CH_2 \right)_n H$$

$$O$$

$$n = 6-10$$

Plastoquinone

How is ATP formed as electrons flow through these carriers from a potential of 0 V to $+0.4$ V? The answer is not yet known, but it seems likely that the mechanisms of photophosphorylation and of oxidative phosphorylation are similar.

ATP CAN ALSO BE FORMED BY CYCLIC ELECTRON FLOW THROUGH PHOTOSYSTEM I

There is an alternate pathway for electrons arising from P700, the reaction center of photosystem I. The high-potential electron in the reduced form of FRS or ferredoxin can be transferred to cytochrome b_6 rather than to $NADP^+$. This electron then flows back to the oxidized form of P700 through cytochrome f and plastocyanin. In other words, there is a cyclic flow of electrons. ATP is generated as electrons return to the reaction center via cytochrome b_6. Hence, this process is called *cyclic photophosphorylation* (Figure 19-14). In this mode, *ATP is generated without the concomitant formation of NADPH.* Photosystem II does not participate in cyclic photophosphorylation, and so O_2 is not formed from H_2O during this process. Cyclic photophosphorylation is operative when there is insufficient $NADP^+$ to accept electrons from reduced FRS or reduced ferredoxin. This condition exists when the ratio of NADPH to $NADP^+$ is high.

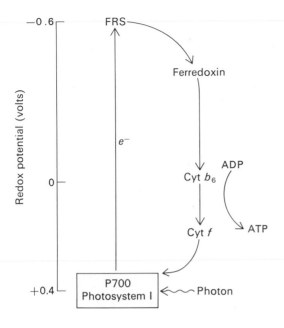

Figure 19-14
Electron flow in cyclic photophosphorylation. Electrons are transferred from P700, the reaction center of photosystem I, to FRS upon absorption of a photon. The electrons return to P700.

ELUCIDATION OF THE PATH OF CARBON BY RADIOACTIVE PULSE LABELING

In 1945, Melvin Calvin and his colleagues started a series of investigations that resulted in the elucidation of the dark reactions of photosynthesis. They used the unicellular green alga *Chlorella* in their work because it was easy to culture these organisms in a highly reproducible way. Their findings later proved to be pertinent to a wide variety of photosynthetic organisms ranging from photosynthetic bacteria to higher plants.

The aim of their work was to determine the pathway by which CO_2 becomes fixed into carbohydrate. The experimental strategy was to use radioactive ^{14}C to trace the fate of CO_2. Radioactive $^{14}CO_2$ was injected into an illuminated suspension of algae that had been carrying out photosynthesis with normal CO_2. The algae were killed after a preselected time by dropping the suspension into alcohol, which also stopped the enzymatic reactions.

The radioactive compounds in the algae were separated and identified by two-dimensional paper chromatography. The paper

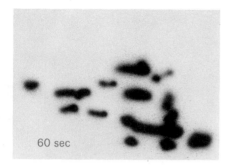

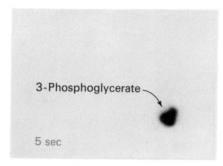

Figure 19-15
Radiochromatograms of illuminated suspensions of algae 5 sec and 60 sec after the injection of $^{14}CO_2$. [Based on J. A. Bassham. The path of carbon in photosynthesis. Copyright © 1962 by Scientific American, Inc. All rights reserved.]

chromatogram was then pressed against photographic film, which became black where the paper contained a radioactive spot. In his Nobel Lecture, Calvin noted that their primary data resided "in the number, position, and intensity—that is, radioactivity—of the blackened areas. The paper ordinarily does not print out the names of these compounds, unfortunately, and our principal chore for the succeeding ten years was to properly label those blacked areas on the film."

CO$_2$ REACTS WITH RIBULOSE DIPHOSPHATE TO FORM TWO PHOSPHOGLYCERATES

The radiochromatogram after sixty seconds of illumination was so complex (Figure 19-15) that it was not feasible to detect the earliest intermediate in the fixation of CO_2. However, the pattern after only five seconds of illumination was much simpler. In fact, there was just one prominent radioactive spot, which proved to be *3-phosphoglycerate*.

The formation of 3-phosphoglycerate as the first detectable radioactive intermediate suggested that a two-carbon compound is the acceptor for the CO_2. This proved not to be so. The actual reaction sequence is more complex:

$$C_5 \xrightarrow{CO_2} C_6 \xrightarrow{H_2O} C_3 + C_3$$

The CO_2 molecule condenses with ribulose 1,5-diphosphate to form a transient six-carbon compound, which is rapidly hydrolyzed to two molecules of 3-phosphoglycerate. The overall reaction is highly exergonic ($\Delta G^{\circ\prime} = -12.4$ kcal/mol).

These reactions are catalyzed by ribulose 1,5-diphosphate carboxylate (also called ribulose diphosphate carboxydismutase). This enzyme is very abundant in chloroplasts, comprising more than 15% of the total protein of chloroplasts. The enzyme is located on the surface of the thylakoid membranes.

FORMATION OF FRUCTOSE 6-PHOSPHATE AND REGENERATION OF RIBULOSE 1,5-DIPHOSPHATE

The steps in the conversion of 3-phosphoglycerate to fructose 6-phosphate (Figure 19-16) are like those of the gluconeogenic pathway (p. 369), except that glyceraldehyde 3-phosphate dehydrogenase in chloroplasts is specific for NADPH, rather than NADH. These reactions bring CO_2 to the level of a hexose. The remaining task is to regenerate ribulose diphosphate, the acceptor of CO_2 in the first dark step. The problem is to construct a five-carbon sugar from six-carbon and three-carbon sugars. This is accomplished by reactions that are catalyzed by *transketolase* and *aldolase*.

$$C_6 + C_3 \xrightarrow{\text{transketolase}} C_4 + C_5$$

$$C_4 + C_3 \xrightarrow{\text{aldolase}} C_7$$

$$C_7 + C_3 \xrightarrow{\text{transketolase}} C_5 + C_5$$

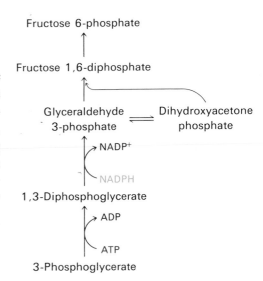

Figure for the conversion of 3-phosphoglycerate to fructose 6-phosphate in chloroplasts.

Figure 19-16
Pathway for the conversion of 3-phosphoglycerate to fructose 6-phosphate in chloroplasts.

We have already encountered transketolase in the pentose phosphate pathway (p. 359). Recall that transketolase, a thiamine pyrophosphate (TPP) enzyme, transfers a two-carbon unit (CH_2—OH—CO—) from a ketose to an aldose. Aldolase carries out an aldol condensation between dihydroxyacetone and an aldehyde. This enzyme is highly specific for dihydroxyacetone phosphate but it accepts a wide variety of aldehydes. The specific reactions catalyzed by transketolase and aldolase in the Calvin cycle are:

Fructose 6-phosphate + glyceraldehyde 3-phosphate $\xrightarrow{\text{transketolase}}$

xylulose 5-phosphate + erythrose 4-phosphate

Erythrose 4-phosphate + dihydroxyacetone phosphate $\xrightarrow{\text{aldolase}}$ sedoheptulose 1,7-diphosphate

Sedoheptulose 7-phosphate + glyceraldehyde 3-phosphate $\xrightarrow{\text{transketolase}}$

ribose 5-phosphate + xylulose 5-phosphate

Four additional enzymes are needed for the dark reactions of photosynthesis. One, a *phosphatase,* hydrolyzes sedoheptulose 1,7-diphosphate to sedoheptulose 7-phosphate. A second, *phosphopentose epimerase,* converts xylulose 5-phosphate to ribulose 5-phosphate. A third, *phosphopentose isomerase,* converts ribose 5-phosphate to ribulose 5-phosphate. Recall that the second and third enzymes also participate in the pentose phosphate pathway (p. 358). The sum of the preceding reactions is:

Fructose 6-phosphate + 2 glyceraldehyde 3-phosphate +

dihydroxyacetone phosphate $\longrightarrow$ 3 ribulose 5-phosphate

Finally, the fourth, *phosphoribulose kinase,* catalyzes the phosphorylation of ribulose 1-phosphate to regenerate ribulose 1,5-diphosphate, the acceptor of CO_2.

Ribulose 5-phosphate + ATP $\longrightarrow$

ribulose 1,5-diphosphate + ADP + H^+

THREE ATP AND TWO NADPH BRING CO_2 TO THE LEVEL OF A HEXOSE

We now write a balanced equation for the net reaction of the Calvin cycle:

$6 CO_2$ + 18 ATP + 12 NADPH + 12 H_2O $\longrightarrow$

$C_6H_{12}O_6$ + 18 ADP + 18 P_i + 12 $NADP^+$ + 6 H^+

Thus, three molecules of ATP and two of NADPH are consumed in converting CO_2 to the level of a hexose such as glucose or fructose. The efficiency of photosynthesis can be estimated in the following way:

1. The $\Delta G^{\circ\prime}$ for the reduction of CO_2 to the level of hexose is 114 kcal/mol.

2. The reduction of $NADP^+$ is a two-electron process. Hence, the formation of two NADPH requires the pumping of four photons by photosystem I. The electrons given up by photosystem I are replenished by photosystem II, which needs to absorb an equal number of photons. Hence eight photons are needed to generate

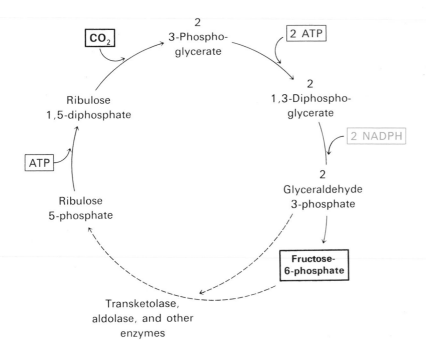

Figure 19-17
The Calvin cycle. The formation of ribulose 5-phosphate from three-carbon and six-carbon sugars is not explicitly shown in the diagram.

the required NADPH. More than the three molecules of ATP needed to convert CO_2 to a hexose are formed concurrently.

3. A mole of photons of 600-nm wavelength has an energy content of 47.6 kcal. Thus, the energy input of eight moles of photons is 381 kcal.

4. Thus, the overall efficiency of photosynthesis under standard conditions is at least 114/381 or 30%. In fact, it is slightly higher, since more ATP is formed in the light reactions than consumed in the fixation of CO_2.

SUMMARY

The first step in photosynthesis is the absorption of light by chlorophyll molecules, which are organized into photosynthetic units in the thylakoid membranes of chloroplasts. The excitation energy is transferred from one chlorophyll molecule to another until it is trapped by a reaction center. The critical event at a reaction center is the light-activated transfer of an electron to an acceptor against a chemical-potential gradient. Photosynthesis in green plants requires the interaction of two light reactions. Photosystem I generates a strong reductant (ferredoxin-reducing substance) that leads to the formation of NADPH. Photosystem II produces a strong oxidant that forms O_2 from H_2O. ATP is generated as electrons flow through an electron-transport chain from photosystem II to photosystem I. Thus, light causes the flow of electrons from H_2O to NADPH with the concomitant generation of ATP. Alternatively, ATP can be generated without the formation of NADPH by a process called cyclic photophosphorylation following the absorption of light by photosystem I.

The ATP and NADPH formed in the light reactions of photosynthesis are used to convert CO_2 into hexoses and other organic compounds. The dark phase of photosynthesis, called the Calvin cycle, starts with the reaction of CO_2 and ribulose 1,5-diphosphate to form two molecules of 3-*phosphoglycerate*. The steps in the conversion of 3-phosphoglycerate to fructose 6-phosphate and glucose 6-phosphate, glyceraldehyde 3-phosphate, and dihydroxyacetone aldehyde 3-phosphate in chloroplasts is specific for NADPH rather than NADH. Ribulose diphosphate is regenerated from fructose 6-phosphate, glyceraldehyde 3-phosphate, and dihydroxyacetone phosphate in a complex series of reactions. Several of the steps in the regeneration of ribulose 1,5-diphosphate are like those of the pentose phosphate pathway. Three ATP and two NADPH are consumed in converting CO_2 to the level of a hexose. Four photons are absorbed by photosystem I and another four by photosystem II to generate two NADPH. The calculated efficiency of photosynthesis under standard conditions is about 30%.

SELECTED READINGS

Where to start:

Levine, R. P., 1969. The mechanism of photosynthesis. *Sci. Amer.* 221(6):58–70. [Available as *Sci. Amer.* Offprint 1163.]

Bassham, J. A., 1962. The path of carbon in photosynthesis. *Sci. Amer.* 206(6):88–100. [Offprint 122.]

Books and reviews:

Gregory, R. P. F., 1971. *Biochemistry of Photosynthesis.* Wiley.

Rabinowitch, E., and Govindjee, 1969. *Photosynthesis.* Wiley.

Dienes, M., (ed.), 1967. *Energy Conversion by the Photosynthetic Apparatus.* Brookhaven Symposia in Biology, vol. 19.

Racker, E., (ed.), 1970. *Membranes of Mitochondria and Chloroplasts.* Van Nostrand Reinhold.

Hill, R., 1965. The biochemists' green mansions: the photosynthetic electron-transport chain in plants. *Essays Biochem.* 1:121–151.

Boardman, N. K., 1968. The photochemical system of photosynthesis. *Advan. Enzymol.* 30:1–80.

Arnon, D. I., 1971. The light reactions of photosynthesis. *Proc. Nat. Acad. Sci.* 68:2883–2892. [A very readable review in which Arnon presents his view that photosystems I and II are not in series and that there are three light reactions.]

Clayton, R. K., 1973. Primary processes in bacterial photosynthesis. *Annu. Rev. Biophys. Bioeng.* 2:131–156.

Walker, D. A., and Crofts, A. R., 1970. Photosynthesis. *Annu. Rev. Biochem.* 39:389–428.

PROBLEMS

1. Calculate the $\Delta E_0'$ and $\Delta G°'$ for the reduction of $NADP^+$ by FRS.

2. Sedoheptulose 1;7-diphosphate is an intermediate in the Calvin cycle but not in the pentose phosphate pathway. What is the enzymatic basis of this difference?

3. Suppose an illuminated suspension of *Chlorella* was actively carrying out photosynthesis when the light was suddenly switched off. How would the levels of 3-phosphoglycerate and ribulose 1,5-diphosphate change during the next minute?

4. Suppose an illuminated suspension of *Chlorella* was actively carrying out photosynthesis in the presence of 1% CO_2 when the concentration of CO_2 was abruptly reduced to .003%. What effect would this have on the levels of 3-phosphoglycerate and ribulose 1,5-diphosphate during the next minute?

5. The photosynthetic apparatus of blue-green algae contains large amounts of phycoerythrin and phycocyanin in addition to chlorophyll *a*. Phycoerythrin absorbs maximally between 480 and 600 nm, whereas phycocyanin absorbs maximally near 620 nm. Suggest a role for these accessory photosynthetic pigments.

6. The Van Niel equation for photosynthesis in higher plants is:

$$6\,CO_2 + 12\,H_2O \longrightarrow$$
$$C_6H_{12}O_6 + 6\,H_2O + 6\,O_2$$

   (a) H_2O appears on both sides of this equation. Is this merely a formalism or does it express an important aspect of the mechanism of photosynthesis?

   (b) The conventional equation for respiration is:

$$C_6H_{12}O_6 + 6\,O_2 \longrightarrow 6\,CO_2 + 6\,H_2O$$

   Is the Van Niel equation (in reverse) more revealing regarding the mechanism of respiration? In other words, does the combustion of glucose require the input of six molecules of H_2O?

For additional problems, see W. B. Wood, J. H. Wilson, R. J. Benbow, and L. E. Hood, *Biochemistry: A Problems Approach* (Benjamin, 1974), Chapter 12.

On the facing page: Model of *S*-adenosylmethionine, a methyl donor in many biosyntheses.

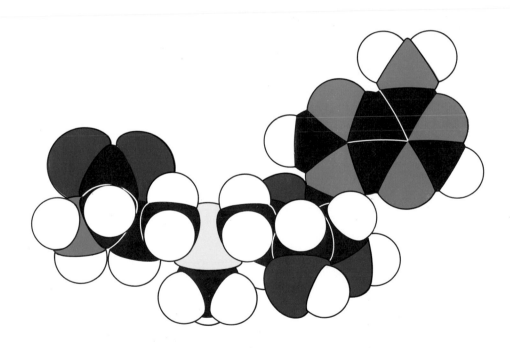

PART III

BIOSYNTHESIS OF
MACROMOLECULAR PRECURSORS

CHAPTER 20

BIOSYNTHESIS OF MEMBRANE LIPIDS AND STEROID HORMONES

We turn now from the generation of metabolic energy to the biosynthesis of macromolecular precursors and related biomolecules. This chapter deals with the biosynthesis of phosphoglycerides, sphingolipids, and cholesterol, three important components of biological membranes (Chapter 10). The synthesis of triacylglycerols is also considered here because this pathway overlaps that of the phosphoglycerides; and the formation of steroid hormones is presented since they are derived from cholesterol. The next two chapters are centered on the biosynthesis of amino acids and nucleotides.

> *Anabolism—*
>   Biosynthetic processes.
> *Catabolism—*
>   Degradative processes.
>   Derived from the Greek
>   words *ana,* up; *cata,* down;
>   *ballein,* to throw.

PHOSPHATIDATE IS AN INTERMEDIATE IN THE SYNTHESIS OF PHOSPHOGLYCERIDES AND TRIACYLGLYCEROLS

Phosphatidate (diacylglycerol 3-phosphate) is a common intermediate in the synthesis of phosphoglycerides and triacylglycerols. The starting point is *glycerol 3-phosphate,* which is formed mainly by reduction of dihydroxyacetone phosphate and to a lesser extent by phosphorylation of glycerol. Glycerol 3-phosphate is acylated by acyl CoA to form *lysophosphatidate,* which is again acylated by acyl CoA to yield *phosphatidate.* These acylations are catalyzed by glycerolphosphate acyl transferase.

Glycerol 3-phosphate   Lysophosphatidate   Phosphatidate

The pathways diverge at phosphatidate. In the synthesis of triacylglycerols, phosphatidate is hydrolyzed by a specific phosphatase to give a *diacylglycerol.* This intermediate is acylated to a *triacylglycerol* in a reaction that is catalyzed by diglyceride acyl transferase. Recent studies suggest that these enzymes are associated in a membrane-bound *triacylglycerol synthetase complex.*

Phosphatidate   Diacylglycerol   Triacylglycerol

CDP-DIACYLGLYCEROL IS THE ACTIVATED INTERMEDIATE IN THE DE NOVO SYNTHESIS OF PHOSPHOGLYCERIDES

There are several routes for the synthesis of phosphoglycerides. The de novo pathway starts with the formation of *cytidine diphosphodiacylglycerol* (CDP-diacylglycerol) from phosphatidate and cytidine triphosphate (CTP). This reaction is driven forward by the hydrolysis of pyrophosphate.

Phosphatidate ⇌ (CTP, PP$_i$) **CDP-diacylglycerol**

The activated phosphatidyl unit then reacts with the hydroxyl group of a polar alcohol. The products are *phosphatidyl serine* and cytidine monophosphate (CMP) when the polar alcohol is serine.

CDP-diacylglycerol + Serine ⇌ Phosphatidyl serine + CMP

Thus, a cytidine nucleotide plays the same role in the synthesis of phosphoglycerides as does a uridine nucleotide in the formation of glycogen. In both biosyntheses, an activated intermediate (UDP-glucose or CDP-diacylglycerol) is formed from a phosphorylated substrate (glucose 1-phosphate or phosphatidate) and a

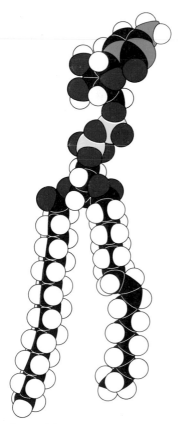

Figure 20-1
Space-filling model of CDP-diacylglycerol.

nucleoside triphosphate (UTP or CTP). The activated intermediate then reacts with a hydroxyl group (the 4-OH terminus of glycogen or the hydroxyl side chain of serine).

PHOSPHATIDYL ETHANOLAMINE AND PHOSPHATIDYL CHOLINE CAN BE FORMED FROM PHOSPHATIDYL SERINE

Decarboxylation of phosphatidyl serine by a pyridoxal phosphate enzyme yields *phosphatidyl ethanolamine*. The amino group of this phosphoglyceride is then methylated three times to form *phosphatidyl choline*. S-*adenosylmethionine*, the methyl donor in this and many other methylations, will be discussed later (p. 508).

PHOSPHATIDYL ETHANOLAMINE AND PHOSPHATIDYL CHOLINE CAN BE FORMED FROM PHOSPHATIDYL SERINE figure: Phosphatidyl serine → (H+, CO2) → Phosphatidyl ethanolamine → (3 ~ CH3, 3H+) → Phosphatidyl choline

PHOSPHOGLYCERIDES CAN ALSO BE SYNTHESIZED BY A SALVAGE PATHWAY

Phosphatidyl choline can also be synthesized by a pathway that makes use of choline. Hence, these reactions are called the *salvage pathway* (Figure 20-2). Choline is phosphorylated by ATP to *phosphorylcholine*, which then reacts with CTP to form *CDP-choline*. The phosphorylcholine unit of CDP-choline is then transferred to a diacylglycerol to form *phosphatidyl choline*. Note that the activated species in this salvage pathway is the cytidine derivative of phosphorylcholine rather than of phosphatidate.

Phosphatidyl ethanolamine can be synthesized from ethanolamine by analogous reactions. *Plasmalogens* are also formed by the salvage pathway. A plasmalogen differs from a phosphoglyceride in that the substituent at C-1 of the glycerol moiety is an α, β unsaturated ether. The final step in the synthesis of phosphatidal choline (the plasmalogen corresponding to phosphatidyl choline) is the transfer of phosphorylcholine from CDP-choline to an alkenyl glycerol ether.

An alkenyl glycerol ether

Phosphatidal choline

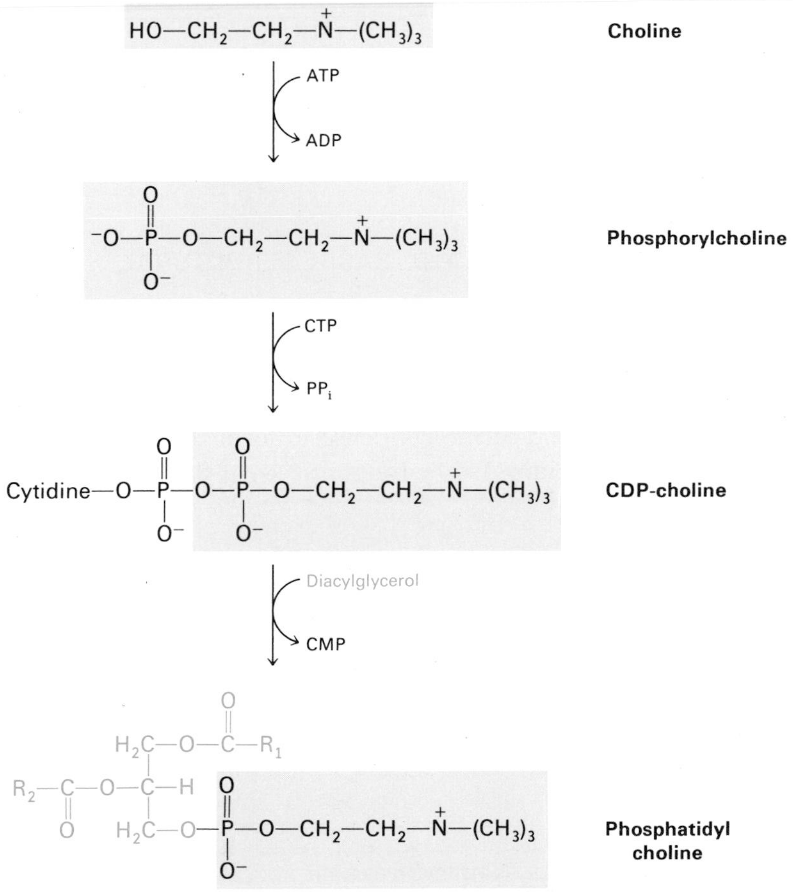

Choline

Phosphorylcholine

CDP-choline

Phosphatidyl choline

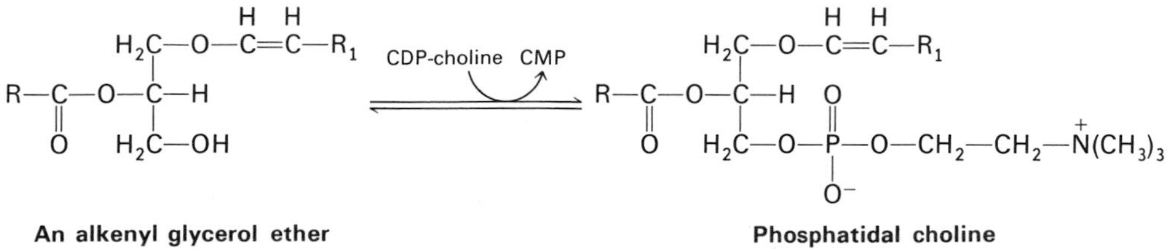

Figure 20-2
Synthesis of phosphatidyl choline by the salvage pathway.

SEVERAL SPECIFIC PHOSPHOLIPASES HAVE BEEN ISOLATED

Enzymes can be used as highly specific biochemical reagents. For example, we have already seen that trypsin is an invaluable tool in studies of protein structure. Specific phospholipases are assuming comparable significance in studies of biological membranes and of their phospholipid constituents. Phospholipases are grouped according to their specificity. The bonds hydrolyzed by phospholipases A_1, A_2, C, and D are shown in Figure 20-3.

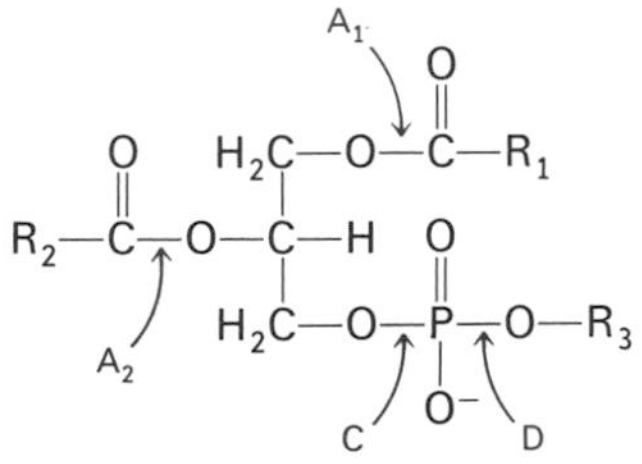

Figure 20-3
Specificity of phospholipases.

SYNTHESIS OF CERAMIDE, THE BASIC STRUCTURAL UNIT OF SPHINGOLIPIDS

We turn now from phosphoglycerides to sphingolipids. The backbone is *sphingosine,* a long-chain aliphatic amine, rather than glycerol. Palmitoyl CoA and serine condense to form dihydrosphingosine, which is then oxidized to sphingosine by a flavoprotein.

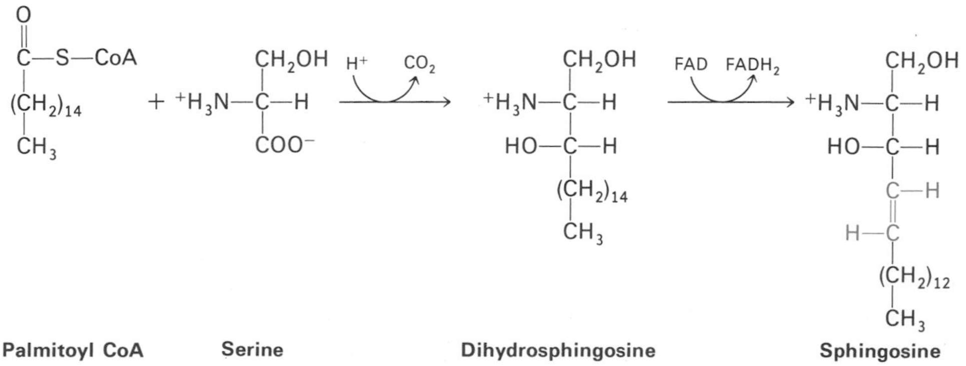

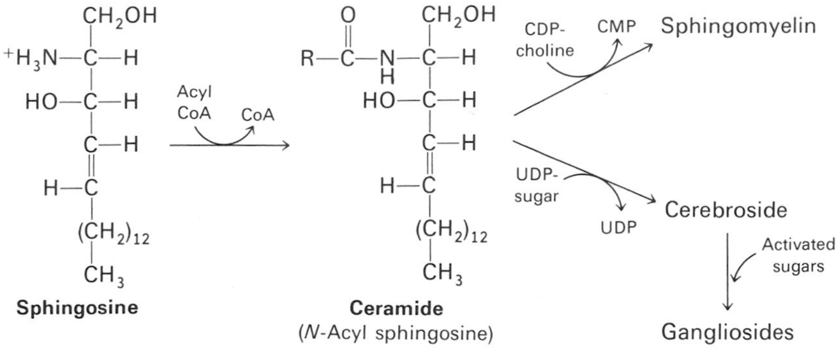

Figure 20-4
Ceramide is the precursor of sphingomyelin and of gangliosides.

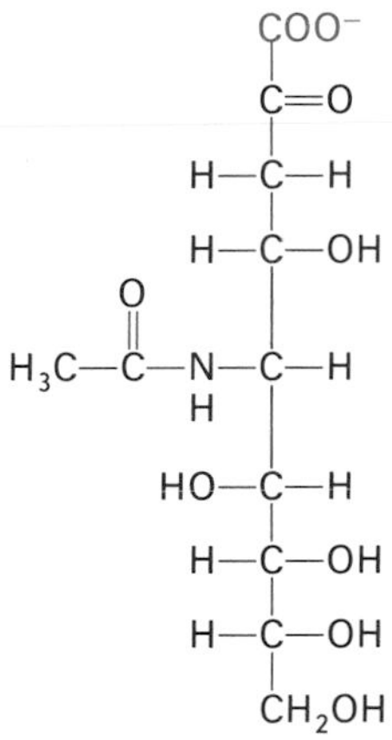

The amino group of sphingosine is acylated in all sphingolipids (Figure 20-4). A long-chain acyl CoA reacts with sphingosine to form *ceramide* (*N*-acyl sphingosine). The terminal hydroxyl group is also substituted in sphingolipids. In *sphingomyelin*, the substituent is phosphorylcholine, which comes from CDP-choline. In *cerebrosides*, glucose or galactose is linked to the terminal hydroxyl group of ceramide. UDP-glucose or UDP-galactose is the sugar donor in the synthesis of cerebrosides. In *gangliosides*, an oligosaccharide is linked to ceramide by a glucose residue (Figure 20-5).

GANGLIOSIDES ARE CARBOHYDRATE-RICH SPHINGOLIPIDS THAT CONTAIN ACIDIC SUGARS

Gangliosides are the most complex sphingolipids. An *oligosaccharide chain* containing at least one acidic sugar is attached to ceramide. The acidic sugar is N-*acetylneuraminate* or *N*-glycolylneuraminate. These acidic sugars are called *sialic acids*. Their 9-carbon backbone is synthesized from phosphoenolpyruvate (a C_3 unit) and *N*-acetyl-mannosamine 6-phosphate (a C_6 unit).

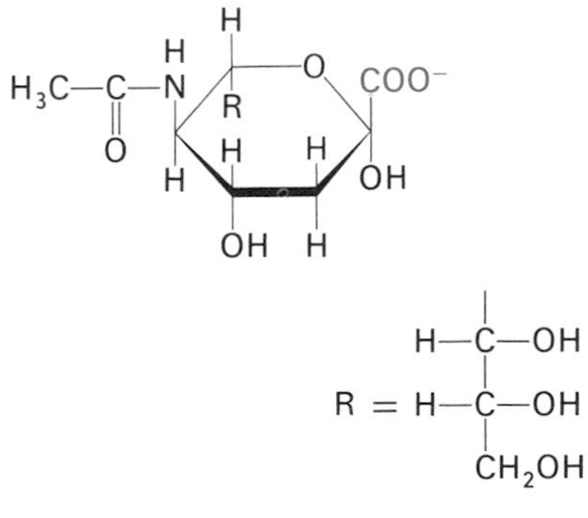

Gangliosides are synthesized by the ordered, step-by-step addition of sugar residues to ceramide. UDP-glucose, UDP-galactose, and UDP-*N*-acetylgalactosamine are the activated donors of these sugar residues. The CMP derivative of *N*-acetylneuraminate is the activated donor of this acidic sugar. CMP-*N*-acetylneuraminate is

N-Acetylneuraminate
(Open-chain form)

N-Acetylneuraminate
(Pyranose form)

synthesized from CTP and *N*-acetylneuraminate; C-1 is the activated carbon, as in the UDP-sugars. The structure of the resulting ganglioside is determined by the specificity of the glycosyl transferases in the cell. More than 15 different gangliosides have been characterized. The structure of ganglioside G_{M1} is shown in Figure 20-5.

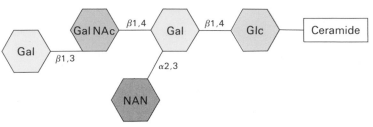

Figure 20-5
Structure of ganglioside G_{M1}. (Abbreviations: Gal, galactose; GalNAc, *N*-acetylgalactosamine; Glc, glucose; NAN, *N*-acetylneuraminate. The types of linkages between the sugars—such as $\beta1,4$—are noted next to the bonds joining them.)

TAY-SACHS DISEASE: AN INHERITED DISORDER OF GANGLIOSIDE BREAKDOWN

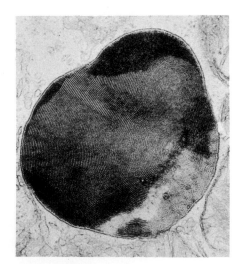

Figure 20-6
Electron micrograph of a lysosome.
[Courtesy of Dr. George Palade.]

Gangliosides are found in highest concentration in the nervous system, particularly in gray matter, where they constitute 6% of the lipids. Gangliosides are continually synthesized and degraded by the sequential removal of their terminal sugar. The glycosyl hydrolases that catalyze these reactions are highly specific. Ganglioside breakdown occurs inside *lysosomes*. These organelles contain many types of degradative enzymes and are specialized for the orderly destruction of cellular components.

Disorders of ganglioside breakdown can have serious clinical consequences. In Tay-Sachs disease, the symptoms are usually evident before an affected infant is a year old. Weakness, retardation in development, and difficulty in feeding are typical presenting symptoms. Blindness usually follows a few months later. Tay-Sachs disease is usually fatal before age three. The striking pathological changes in this disease occur in the nervous system, where the ganglion cells of the cerebral cortex and some other parts of the brain become enormously swollen. Also, there is a distinctive cherry-red spot in the retina.

The ganglioside content of the brain of an infant with Tay-Sachs disease is greatly elevated. Specifically, *the concentration of ganglioside G_{M2} is many times higher than normal.* The abnormally high level of this ganglioside is caused by a deficiency of the enzyme that removes its terminal *N*-acetylgalactosamine residue. The missing or deficient enzyme is a specific *β*-N-*acetylhexosaminidase.*

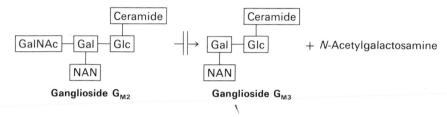

Ganglioside G_{M2} Ganglioside G_{M3}

Tay-Sachs disease is inherited as an *autosomal recessive.* The carrier rate is 1/30 in Jewish Americans and 1/300 in non-Jewish Americans. Consequently, the incidence of the disease is about 100 times higher in Jewish Americans. Tay-Sachs disease can be diagnosed during fetal development. Amniotic fluid is obtained by *amniocentesis* and assayed for *β*-N-acetylhexosaminidase activity.

CHOLESTEROL IS SYNTHESIZED FROM ACETYL COENZYME A

We now turn to the synthesis of cholesterol, a steroid that is present in the membranes of eucaryotes but not procaryotes. Cholesterol is also the precursor of steroid hormones such as progesterone, testosterone, estradiol, and cortisol. An early and important clue concerning the synthesis of cholesterol came from the work of Konrad Bloch in the 1940s. Acetate isotopically labeled in its carbon atoms was prepared and fed to rats. The cholesterol that was synthesized by these rats contained the isotopic label, showing that acetate is a precursor of cholesterol. In fact, *all twenty-seven carbon atoms of cholesterol are derived from acetyl CoA.* Further insight into the synthesis of cholesterol came from studies that utilized acetate labeled in either its methyl or carboxyl carbon as the precursor. Degradation of cholesterol synthesized from acetate labeled in one of its carbons showed the origin of each atom of

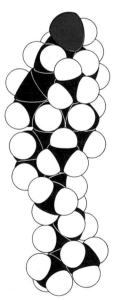

Figure 20-7
Space-filling model of cholesterol.

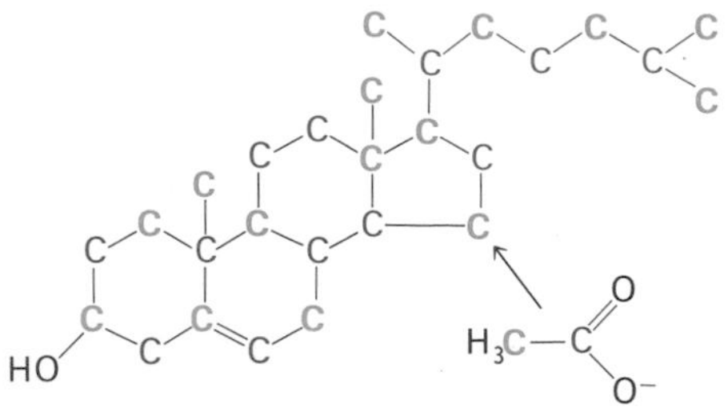

Figure 20-8
Isotope-labeling pattern of cholesterol synthesized from acetate labeled in its methyl (green) or carboxyl (red) carbon atom.

the cholesterol molecule (Figure 20-8). This pattern played a crucial role in the generation and testing of hypotheses concerning the pathway of cholesterol synthesis.

MEVALONATE AND SQUALENE ARE INTERMEDIATES IN THE SYNTHESIS OF CHOLESTEROL

The next major clue came from the discovery that *squalene*, a C_{30} hydrocarbon, is an intermediate in the synthesis of cholesterol. Squalene consists of six *isoprene* units.

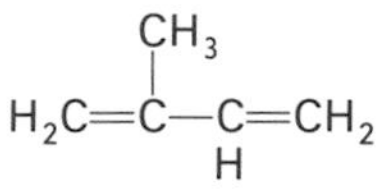

This finding raised the question of how isoprene units are formed from acetate.

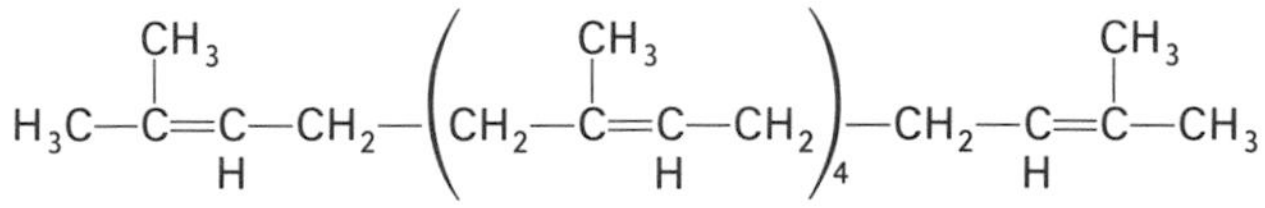

$$\text{Acetate} \longrightarrow [\text{isoprene}] \longrightarrow \text{squalene} \longrightarrow \text{cholesterol}$$
$$\quad C_2 \qquad\qquad C_5 \qquad\qquad C_{30} \qquad\qquad C_{27}$$

The answer came unexpectedly from unrelated studies of bacterial mutants. It was found that mevalonate could substitute for acetate in meeting the nutritional needs of acetate-requiring mutants.

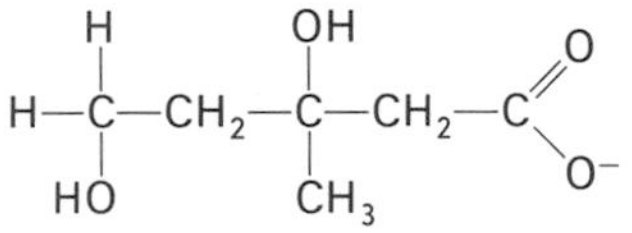

The discovery of mevalonate was a key step in the elucidation of the pathway of cholesterol biosynthesis, since it was soon recognized that this C_6 acid could decarboxylate to yield the postulated C_5 isoprene intermediate. Isotope-labeling studies then showed that mevalonate is indeed a precursor of squalene and that it could be formed from acetate. The activated isoprene intermediate proved to be *isopentenyl pyrophosphate*, which is formed by decarboxylation of a derivative of mevalonate. A pathway for the synthesis of cholesterol from acetate could then be outlined:

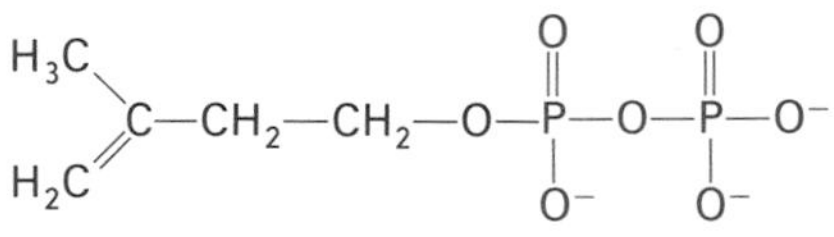

Mevalonate

Isopentenyl pyrophosphate

$$\text{Acetate} \longrightarrow \text{mevalonate} \longrightarrow \text{isopentenyl pyrophosphate} \longrightarrow$$
$$\quad C_2 \qquad\qquad C_6 \qquad\qquad\qquad C_5$$

$$\text{squalene} \longrightarrow \text{cholesterol}$$
$$\quad C_{30} \qquad\qquad C_{27}$$

SYNTHESIS OF ISOPENTENYL PYROPHOSPHATE, AN ACTIVATED INTERMEDIATE IN CHOLESTEROL FORMATION

The first stage in the synthesis of cholesterol is the formation of isopentenyl pyrophosphate from acetyl CoA. This set of reactions starts with the formation of 3-hydroxy-3-methylglutaryl CoA from acetyl CoA and acetoacetyl CoA. One of the fates of 3-hydroxy-3-methylglutaryl CoA, its cleavage to acetyl CoA and acetoacetate, was discussed previously in regard to the formation of ketone bodies (p. 417). Alternatively, 3-hydroxyl-3-methylglutaryl CoA can be reduced to mevalonate (Figure 20-9).

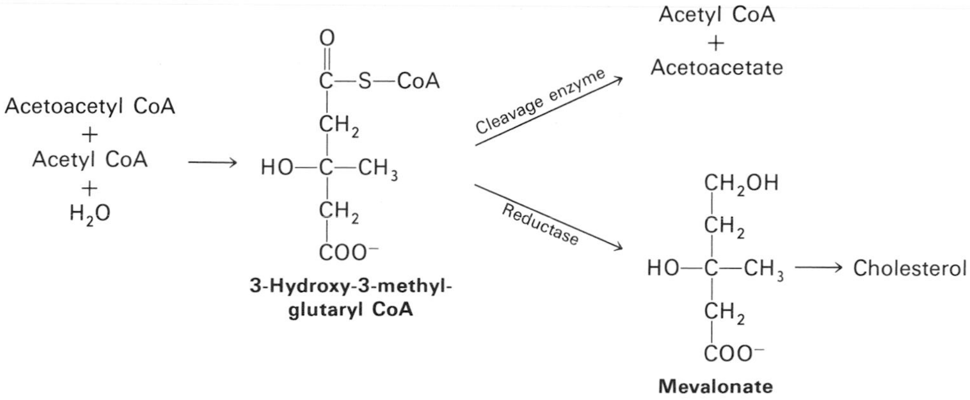

Figure 20-9
Synthesis and fates of 3-hydroxy-3-methylglutaryl CoA.

The synthesis of mevalonate is the committed step in cholesterol formation. The enzyme catalyzing this irreversible step, 3-hydroxy-3-methylglutaryl CoA reductase, is an important control site in cholesterol biosynthesis, as will be discussed shortly.

$$\text{3-Hydroxy-3-methylglutaryl CoA} + 2\,\text{NADPH} + 2\,\text{H}^+ \longrightarrow$$
$$\text{mevalonate} + 2\,\text{NADP}^+ + \text{CoA}$$

Mevalonate is converted to 3-phospho-5-pyrophosphomevalonate by three consecutive phosphorylations. This labile intermediate loses CO_2 and P_i, yielding *3-isopentenyl pyrophosphate* (Figure 20-10).

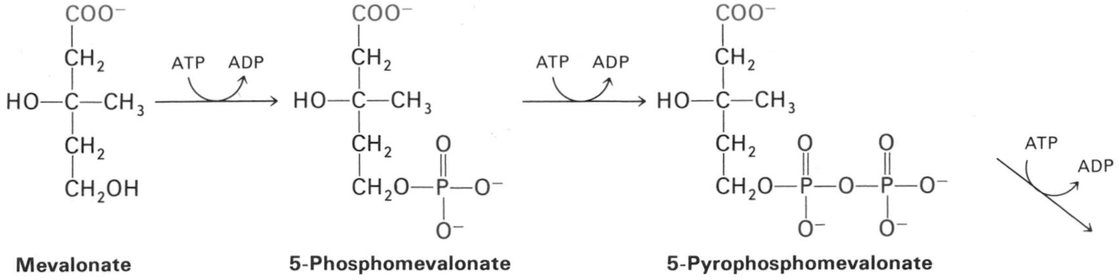

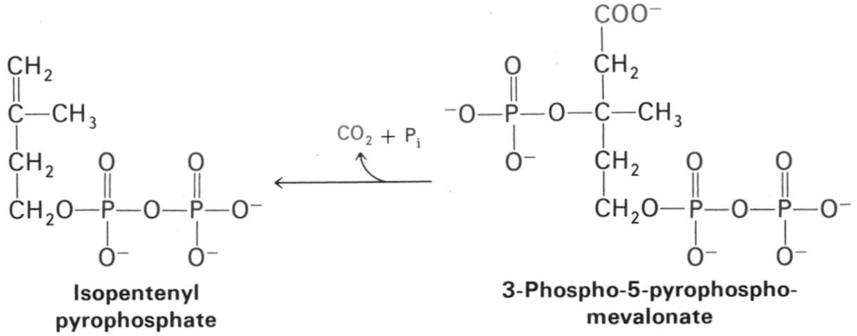

490

Figure 20-10
Synthesis of isopentenyl pyrophosphate
from mevalonate.

SYNTHESIS OF SQUALENE FROM ISOPENTENYL PYROPHOSPHATE

Squalene is synthesized from isopentenyl pyrophosphate by the reaction sequence

$$C_5 \longrightarrow C_{10} \longrightarrow C_{15} \longrightarrow C_{30}$$

This stage in the synthesis of cholesterol (Figure 20-11) starts with the isomerization of *isopentenyl pyrophosphate* to *dimethylallyl pyrophosphate*.

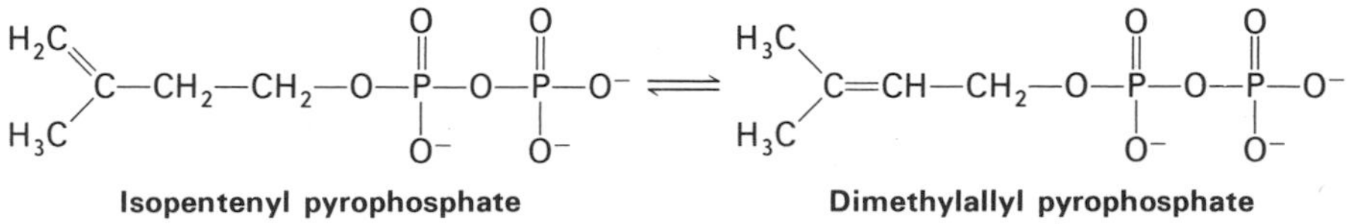

These two C_5 units condense to form *geranyl pyrophosphate* (C_{10}), which condenses with another molecule of isopentenyl pyrophosphate to form *farnesyl pyrophosphate* (C_{15}). The last step in the synthesis of *squalene* is a reductive condensation of two molecules of farnesyl pyrophosphate:

$$2 \text{ Farnesyl pyrophosphate} + \text{NADPH} \longrightarrow \text{squalene} + 2 \text{ PP}_i + \text{NADP}^+ + \text{H}^+$$

$$\underset{C_{15}}{\phantom{2 \text{ Farnesyl pyrophosphate}}} \qquad \underset{C_{30}}{\phantom{\text{squalene}}}$$

Dimethylallyl pyrophosphate

Isopentenyl pyrophosphate

$\rightarrow$ PP$_i$

Geranyl pyrophosphate

Isopentenyl pyrophosphate

$\rightarrow$ PP$_i$

Farnesyl pyrophosphate

Farnesyl pyrophosphate + NADPH

$\rightarrow$ NADP$^+$ + 2 PP$_i$ + H$^+$

Squalene

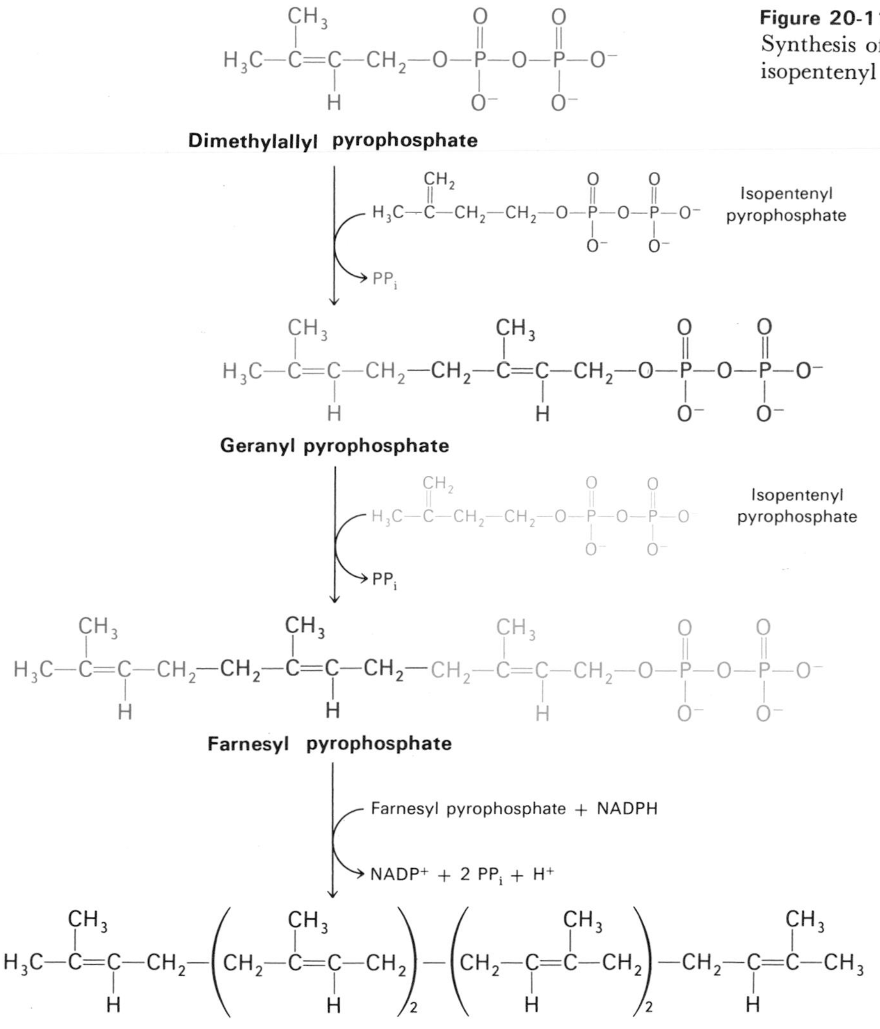

Figure 20-11
Synthesis of squalene from isopentenyl pyrophosphate.

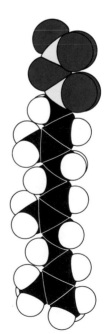

Figure 20-12
Space-filling model of farnesyl pyrophosphate.

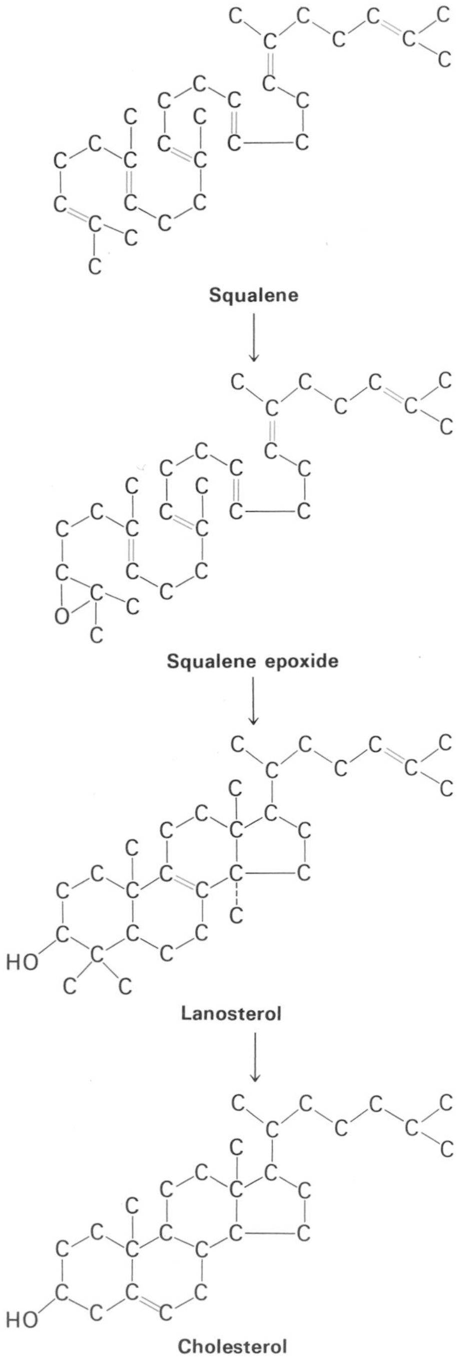

Figure 20-13
Synthesis of cholesterol from squalene.

SQUALENE EPOXIDE CYCLIZES TO LANOSTEROL, WHICH IS CONVERTED TO CHOLESTEROL

The final stage of cholesterol biosynthesis starts with the cyclization of squalene (Figure 20-13). *Squalene epoxide,* the reactive intermediate, is formed in a reaction that uses O_2 and NADPH. Squalene epoxide is then cyclized to *lanosterol* by a cyclase. There is a concerted movement of electrons through four double bonds and a migration of two methyl groups in this remarkable closure. Finally, lanosterol is converted to *cholesterol* by the removal of three methyl groups, the reduction of one double bond by NADPH, and the migration of the other double bond.

CHOLESTEROL INHIBITS ITS OWN SYNTHESIS BY BLOCKING THE FORMATION OF MEVALONATE

The major site of cholesterol synthesis is the liver. In vitro studies reveal that the amount of cholesterol formed from acetate by liver slices depends on the amount of cholesterol in the diet of the experimental animal. Cholesterol synthesis is half-inhibited if the diet contains about 0.15% cholesterol. Cholesterol blocks the conversion of 3-hydroxy-3-methylglutaryl CoA to mevalonate. This committed step in cholesterol biosynthesis is catalyzed by 3-hydroxy-3-methylglutaryl CoA reductase. *Cholesterol acts by inhibiting the synthesis of the reductase rather than by diminishing its activity.* This effect is mediated by serum lipoproteins that bind to the plasma membrane.

BILE SALTS DERIVED FROM CHOLESTEROL FACILITATE THE DIGESTION OF LIPIDS

Bile salts are polar derivatives of cholesterol. These compounds are highly effective *detergents* because they contain polar and nonpolar regions. Bile salts are synthesized in the liver, stored and concentrated in the gallbladder, and then released into the small intestine. Bile salts, the major constituent of bile, *solubilize dietary lipids.* The resulting increase in the surface area of the lipids has two consequences: it promotes their hydrolysis by lipases and facilitates their absorption. Bile salts are also the major breakdown products of cholesterol.

Cholesterol is converted to trihydroxycoprostanoate and then to *cholyl CoA,* the activated intermediate in the synthesis of most bile salts (Figure 20-14). The activated carboxyl carbon of cholyl CoA then reacts with the amino group of glycine to form *glycocholate,* or with the amino group of taurine (H_2N—CH_2—CH_2—SO_3^-) to form *taurocholate. Glycocholate is the major bile salt.*

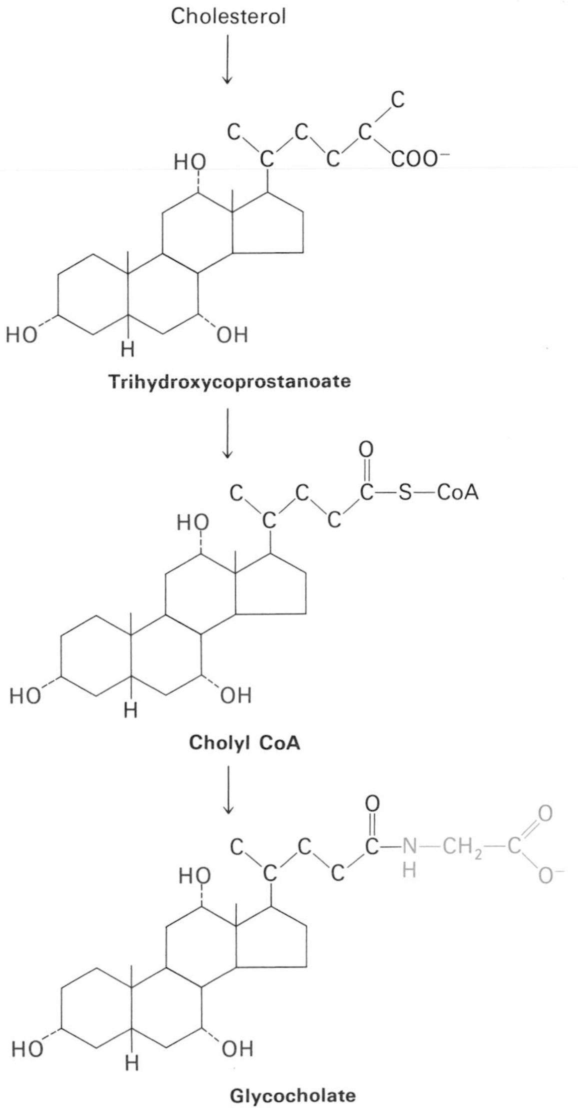

Figure 20-14
Synthesis of glycocholate, the major bile salt.

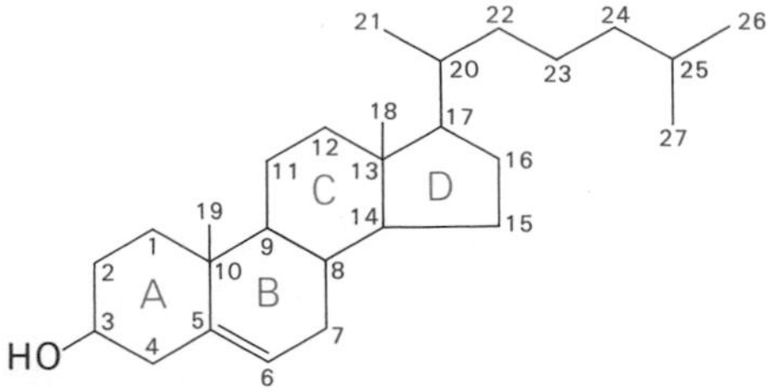

Figure 20-15
Numbering of the carbon atoms of cholesterol.

NOMENCLATURE OF STEROIDS

A few comments concerning steroid nomenclature are appropriate before turning to the synthesis of steroid hormones. Carbon atoms in steroids are numbered as shown in Figure 20-15 for cholesterol. The rings in steroids are named A, B, C, and D. Cholesterol contains two angular methyl groups: the C-19 methyl group is attached to C-10, and the C-18 methyl group is attached to C-13. A line above C-10 or C-13 denotes a methyl group. By definition, the C-18 and C-19 methyl groups of cholesterol are *above* the plane containing the four rings. A substituent that is above the plane is termed *β-oriented,* and is shown by a *solid* bond. In contrast, a substituent that is below the plane is *α-oriented* and denoted by a *dashed* or *dotted* line.

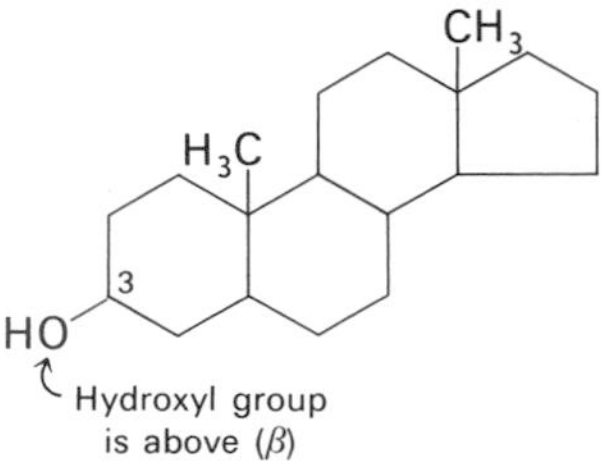

3β-Hydroxy

Hydroxyl group
is above (β)

3α-Hydroxy

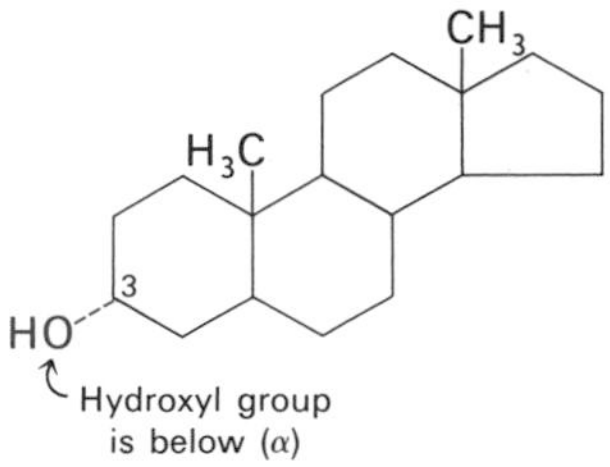

Hydroxyl group
is below (α)

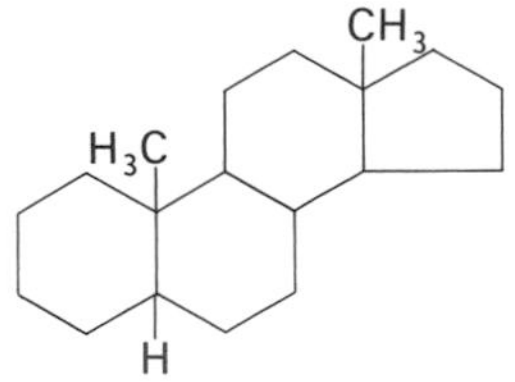

5β-Hydrogen
(A *cis* fusion)

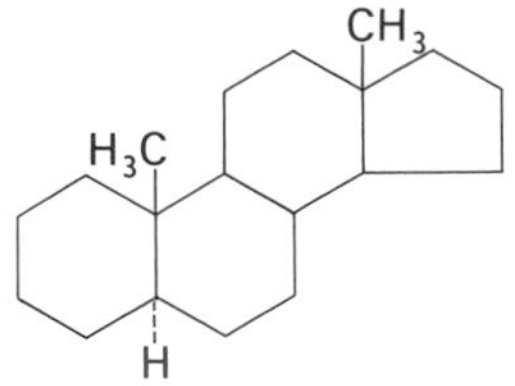

5α-Hydrogen
(A *trans* fusion)

A hydrogen atom attached to C-5 can be α- or β-oriented. When this hydrogen atom is *α-oriented,* the A and B rings are fused in a *trans* conformation, whereas a *β-orientation* yields a *cis* fusion. The absence of a symbol for the C-5 hydrogen atom implies a *trans* fusion. The C-5 hydrogen atom is α-oriented in all steroid hormones that contain a hydrogen atom in that position. In contrast, bile salts have a *β*-oriented hydrogen atom at C-5. Thus, *a cis fusion is characteristic of the bile salts, whereas a* trans *fusion is characteristic of all steroid hormones that possess a hydrogen atom at C-5.* A *trans* fusion yields a nearly planar structure, whereas a *cis* fusion gives a buckled structure.

STEROID HORMONES ARE DERIVED FROM CHOLESTEROL

Cholesterol is the precursor of the five major classes of steroid hormones: progestagens, glucocorticoids, mineralocorticoids, androgens, and estrogens (Figure 20-16). *Progesterone,* a *progestagen,* prepares the lining of the uterus for implantation of an ovum. Progesterone is also essential for the maintenance of pregnancy. *Androgens* (such as *testosterone*) are responsible for the development of male secondary sex characteristics, whereas *estrogens* (such as *estrone*) are required for the development of female secondary sex characteristics. Estrogens also participate in the ovarian cycle. *Glucocorticoids* (such as *cortisol*) promote gluconeogenesis and the formation of glycogen, and also enhance the degradation of fat and protein. *Mineralocorticoids* (such as *aldosterone*) cause increased reabsorption of Na^+, Cl^-, and HCO_3^- by the kidney, which leads to an increase in blood volume and blood pressure. The major sites of synthesis of these classes of hormones are: progestagens, corpus luteum; estrogens, ovary; androgens, testis; glucocorticoids and mineralocorticoids, adrenal cortex.

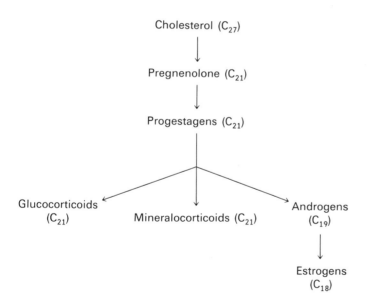

Figure 20-16
Biosynthetic relationships of steroid hormones.

STEROIDS ARE HYDROXYLATED BY MIXED-FUNCTION OXIDASES THAT UTILIZE NADPH AND O_2

Hydroxylation reactions play a very important role in the conversion of cholesterol to steroid hormones and bile salts. All of these hydroxylations require *NADPH* and O_2. The oxygen atom of the incorporated hydroxyl group comes from O_2 rather than from H_2O, as shown by the use of ^{18}O-labeled O_2 and H_2O. One oxygen atom of the O_2 molecule goes into the substrate, whereas the other is reduced to water. The enzymes catalyzing these reactions are therefore called *mixed-function oxidases.*

$$RH + O_2 + NADPH + H^+ \longrightarrow ROH + H_2O + NADP^+$$

Hydroxylation requires the activation of oxygen, which is accomplished by a specialized cytochrome called P_{450} because its complex with CO has an absorption maximum at 450 nm. Cytochrome P_{450} is the terminal component of an *electron-transport chain* that has been found in adrenal mitochondria and liver microsomes. The role of this assembly is hydroxylation rather than oxidative phosphorylation. NADPH transfers its high-potential electrons to a flavoprotein in this chain, which are then conveyed to *adrenodoxin,* a nonheme iron protein. Adrenodoxin transfers an electron to the oxidized form of cytochrome P_{450}. The reduced form of P_{450} then activates O_2.

PREGNENOLONE IS FORMED FROM CHOLESTEROL BY CLEAVAGE OF ITS SIDE CHAIN

Steroid hormones contain 21 or fewer carbon atoms, whereas cholesterol contains 27. The first stage in the synthesis of steroid hormones is the removal of a C_6 unit from the side chain of cholesterol to form *pregnenolone.* The side chain of cholesterol is hydroxylated at C-20 and then at C-22, followed by the cleavage of the bond between C-20 and C-22. The latter reaction is catalyzed by desmolase. All three reactions utilize NADPH and O_2.

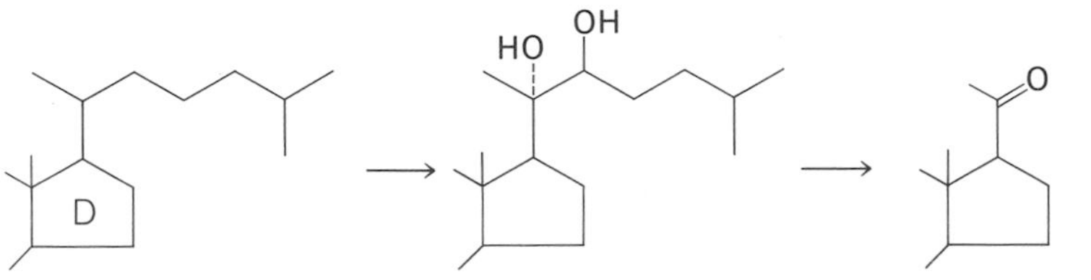

Cholesterol
(Only ring D and
the side chain are shown.)

20α,22-Dihydroxy-
cholesterol

Pregnenolone

Adrenocorticotrophic hormone (ACTH), a polypeptide synthesized by the anterior pituitary gland, stimulates the conversion of cholesterol to pregnenolone, which is the precursor of all steroid hormones.

SYNTHESIS OF PROGESTERONE AND CORTICOIDS

Progesterone is synthesized from pregnenolone in two steps. The 3-hydroxyl group of pregnenolone is oxidized to a 3-keto group, and the Δ^5 double bond is isomerized to a Δ^4 double bond. *Cortisol,* the major glucocorticoid, is synthesized from progesterone by hydroxylations at C-17, C-21, and C-11; C-17 must be hydroxylated before C-21, whereas hydroxylation at C-11 may occur at any stage. The enzymes catalyzing these hydroxylations are highly specific, as shown by some inherited disorders of steroid metabolism. The initial step in the synthesis of *aldosterone,* the major mineralocorticoid, is the hydroxylation of progesterone at C-21. The resulting deoxycorticosterone is hydroxylated at C-11. The C-18 angular methyl group is then oxidized to an aldehyde, yielding aldosterone.

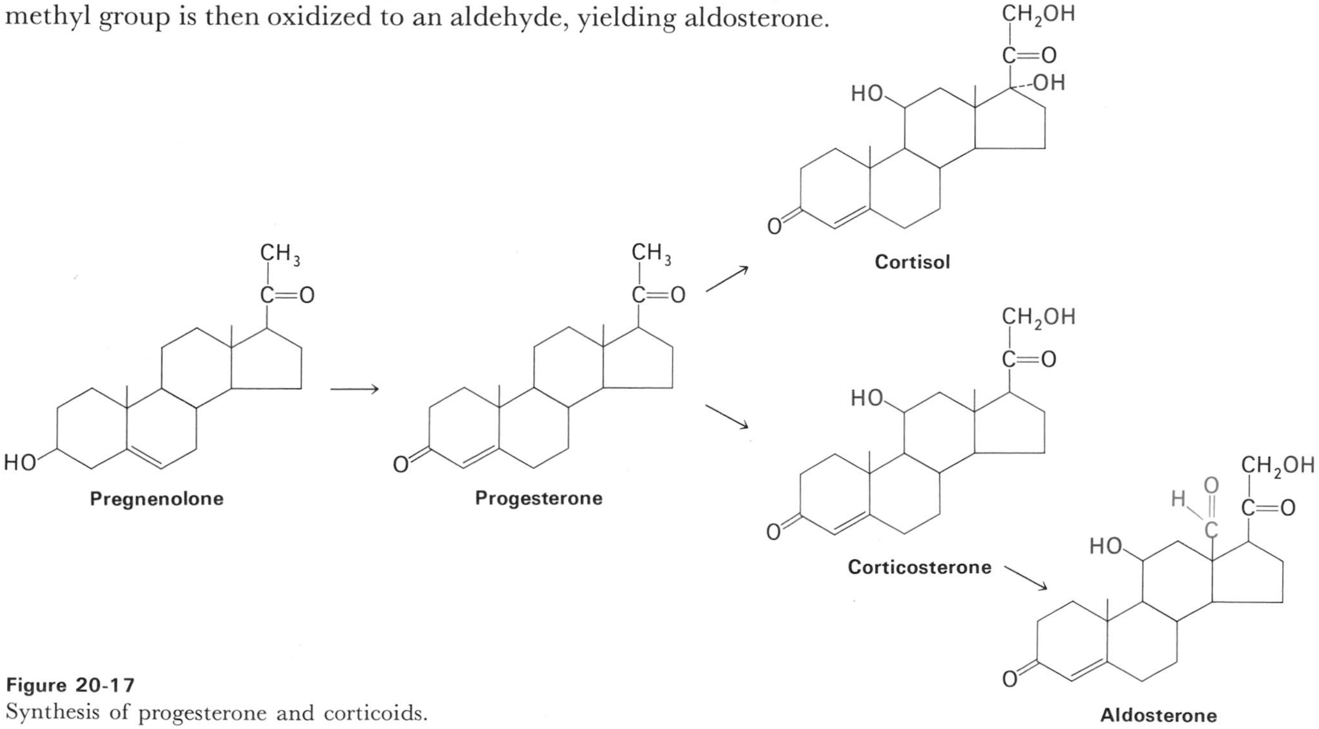

Figure 20-17
Synthesis of progesterone and corticoids.

SYNTHESIS OF ANDROGENS AND ESTROGENS

The synthesis of androgens (Figure 20-18) starts with the hydroxylation of progesterone at C-17. The side chain consisting of C-20 and C-21 is then cleaved to yield *androstenedione,* an androgen. *Testosterone,* another androgen, is formed by reduction of the 17-keto group of androstenedione. Androgens contain nineteen carbon atoms. Estrogens are synthesized from androgens by the loss of the C-19 angular methyl group and the formation of an aromatic A ring. These reactions require NADPH and O_2. *Estrone,* an estrogen, is derived from androstenedione, whereas *estradiol,* another estrogen, is formed from testosterone.

Figure 20-18
Synthesis of androgens and estrogens.

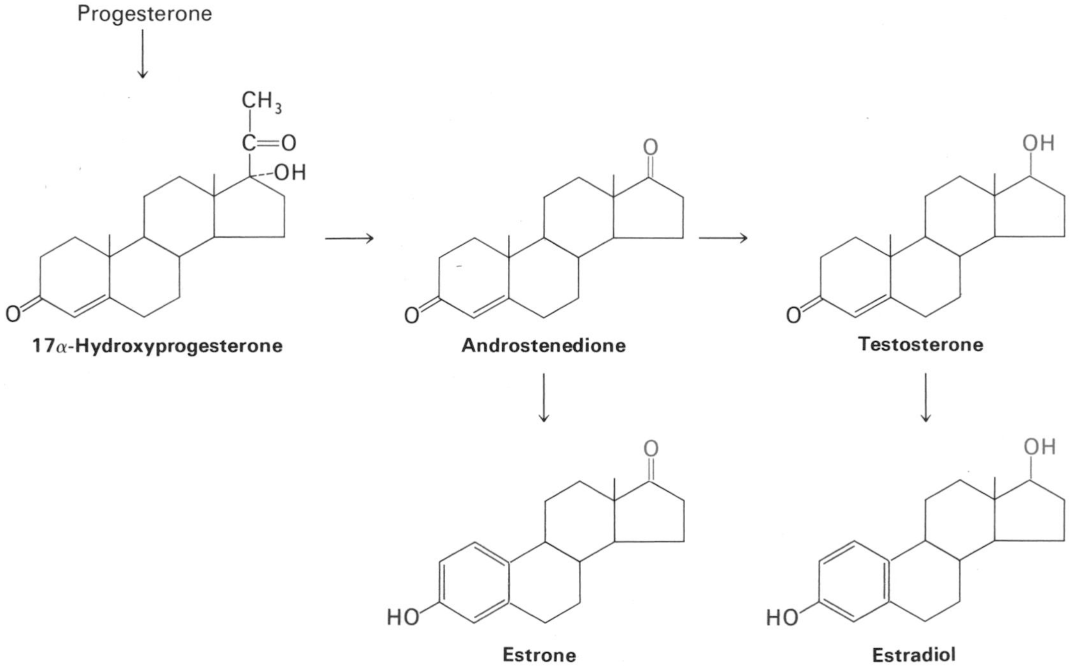

DEFICIENCY OF THE 21-HYDROXYLASE CAUSES VIRILIZATION AND ENLARGEMENT OF THE ADRENALS

The most common inherited disorder of steroid hormone synthesis is a deficiency of the *21-hydroxylase,* an enzyme needed for the synthesis of glucocorticoids and mineralocorticoids. The diminished production of glucocorticoids leads to an increased secretion of ACTH by the anterior pituitary gland. This response is an expression of a normal feedback mechanism that controls adrenal cortical activity. *The adrenal glands enlarge because of the high level of ACTH in the blood, and more pregnenolone is synthesized.* Consequently, the concentrations of progesterone and 17α-hydroxyprogesterone increase. In turn, there is a *marked increase in the amount of androgens,* since they are derived from 17α-hydroxyprogesterone.

The striking clinical finding in 21-hydroxylase deficiency is *virilization* caused by high levels of androgens. In affected females, virilization is usually evident at birth. Androgens secreted during the development of the female fetus produce a masculinization of the external genitalia. In males, the sexual organs appear normal at birth. Sexual precocity becomes apparent several months later. There is accelerated growth and very early bone maturations so that short stature is the typical final result. About half of the patients with 21-hydroxylase deficiency *persistently lose Na+ in the urine.* These individuals have very low levels of aldosterone, the principal mineralocorticoid. Loss of salt leads to hydration and hypotension, which may lead to shock and sudden death.

Effective therapy is available for 21-hydroxylase deficiency. The administration of a glucocorticoid provides this needed hormone and concomitantly eliminates the excessive secretion of ACTH. Excessive formation of androgens is thereby stopped. A mineralocorticoid may also be given to patients who lose salt. Some of the symptoms of 21-hydroxylase deficiency are reversed if therapy is started in the first two years of life.

Several other inherited defects of steroid hormone synthesis are known. The affected enzymes include 11-hydroxylase, 17-hydroxylase, 3β-dehydrogenase, and desmolase. All of these enzymatic lesions lead to a compensatory enlargement of the adrenal gland. Hence, the clinical term for this group of disorders is *congenital adrenal hyperplasia.* Like 21-hydroxylase deficiency, 11-hydroxylase deficiency is accompanied by virilization.

SUMMARY

Phosphatidate, an intermediate in the synthesis of phosphoglycerides and triacylglycerols, is formed by the acylation of glycerol 3-phosphate by acyl CoA. Hydrolysis of its phosphoryl group followed by acylation yields a triacylglycerol. CDP-diacylglycerol, the activated intermediate in the de novo synthesis of phosphoglycerides, is formed from phosphatidate and CTP. The activated phosphatidyl unit is then transferred to the hydroxyl group of a polar alcohol such as serine, forming phosphatidyl serine. Decarboxylation of this phosphoglyceride yields phosphatidyl ethanolamine, which is methylated by S-adenosylmethionine to form phosphatidyl choline. This phosphoglyceride can also be synthesized by a salvage pathway that utilizes preformed choline. CDP-choline is the activated intermediate in this route. Sphingolipids are synthesized from ceramide, which is formed by the acylation of sphingosine. Gangliosides are sphingolipids that contain an oligosaccharide unit having at least one residue of N-acetylneuraminate or a related sialic acid. Gangliosides are synthesized by the step-by-step addition of activated sugars such as UDP-glucose to ceramide.

Cholesterol, a steroid component of eucaryotic membranes and a precursor of steroid hormones, is formed from acetyl CoA. The synthesis of mevalonate from 3-hydroxy-3-methylglutaryl CoA (derived from acetyl CoA and acetoacetyl CoA) is the committed step in the formation of cholesterol. Mevalonate is converted to isopentenyl pyrophosphate (C_5), which condenses with its isomer, dimethylallyl pyrophosphate (C_5), to form geranyl pyrophosphate (C_{10}). Addition of a second molecule of isopentenyl pyrophosphate yields farnesyl pyrophosphate (C_{15}), which condenses with itself to form squalene (C_{30}). This intermediate cyclizes to lanosterol (C_{30}), which is modified to yield cholesterol (C_{27}). Five major classes of steroid hormones are derived from cholesterol: progestagens, glucocorticoids, mineralocorticoids, androgens, and estrogens. Hydroxylations by mixed-function oxidases that use NADPH and O_2 play an important role in the synthesis of steroid hormones and bile salts from cholesterol. Pregnenolone (C_{21}), a key intermediate in the synthesis of steroid hormones, is formed by scission of the side chain of cholesterol. Progesterone (C_{21}), synthesized from pregnenolone, is the precursor of cortisol and aldosterone. Cleavage of the side chain of progesterone yields androstenedione, an androgen (C_{19}). Estrogens (C_{18}) are synthesized from androgens by the loss of an angular methyl group and the formation of an aromatic A ring.

SELECTED READINGS

Where to start:

Bloch, K., 1965. The biological synthesis of cholesterol. *Science* 150:19–28.

Books and general reviews:

Wakil, S., (ed.), 1970. *Lipid Metabolism*. Academic Press.

Lennarz, W. J., 1970. Lipid metabolism. *Annu. Rev. Biochem.* 39:359–388.

Stoffel, W., 1971. Sphingolipids. *Annu. Rev. Biochem.* 40:57–82.

McMurray, W. C., and Magee, W. L., 1972. Phospholipid metabolism. *Annu. Rev. Biochem.* 41:129–160.

Gatt, S., and Barenholz, Y., 1973. Enzymes of complex lipid metabolism. *Annu. Rev. Biochem.* 42:61–85.

Regulation and reaction mechanisms:

Goldstein, J. L., and Brown, M. S., 1973. Familial hypercholesterolemia: identification of a defect in the regulation of 3-hydroxy-3-methylglutaryl coenzyme A reductase activity associated with overproduction of cholesterol. *Proc. Nat. Acad. Sci.* 70:2804–2808.

Brown, M. S., and Goldstein, J. L., 1974. Familial hypercholesterolemia: defective binding of lipoproteins to cultured fibroblasts associated with impaired regulation of 3-hydroxy-3-methylglutaryl coenzyme A reductase activity. *Proc. Nat. Acad. Sci.* 71:788–792.

Popják, G., and Cornforth, J. W., 1960. The biosynthesis of cholesterol. *Advan. Enzymol.* 22:281–335.

van Tamelen, E. E., 1968. Bioorganic chemistry: sterols and acyclic terpene terminal epoxides. *Accounts Chem. Res.* 1:111–120.

Hayaishi, O., (ed.), 1974. *Molecular Mechanisms of Oxygen Activation.* Academic Press. [Includes several articles on mixed-function oxygenases.]

Inherited diseases:

Brady, R. O., 1973. Inborn errors of lipid metabolism. *Advan. Enzymol.* 38:293–316.

Sloan, H. R., and Fredrickson, D. S., 1972. G_{M2} gangliosidoses: Tay-Sachs disease. *In* Stanbury, J. B., Wyngaarden, J. B., and Fredrickson, D. S., (eds.), *The Metabolic Basis of Inherited Disease,* (3rd ed.), pp. 615–638. McGraw-Hill.

Bongiovanni, A. M., 1972. Disorders of adrenocortical steroid biogenesis (the adrenogenital syndrome associated with congenital adrenal hyperplasia). *In* Stanbury, J. B., Wyngaarden, J. B., and Fredrickson, D. S., (eds.), *The Metabolic Basis of Inherited Disease,* (3rd ed.), pp. 857–885.

PROBLEMS

1. Write a balanced equation for the synthesis of a triacylglycerol starting from glycerol and fatty acids.

2. Write a balanced equation for the synthesis of phosphatidyl serine by the de novo pathway starting from serine, glycerol, and fatty acids.

3. What is the activated reactant in each of these biosyntheses?
    (a) Phosphatidyl serine from serine.
    (b) Phosphatidyl ethanolamine from ethanolamine.
    (c) Ceramide from sphingosine.

   (d) Sphingomyelin from ceramide.
   (e) Cerebroside from ceramide.
   (f) Ganglioside G_{M1} from ganglioside G_{M2}.
   (g) Farnesyl pyrophosphate from geranyl pyrophosphate.

4. What is the distribution of isotopic labeling in cholesterol synthesized from each of these precursors?
    (a) Mevalonate labeled with ^{14}C in its carboxyl carbon atom.
    (b) Malonyl CoA labeled with ^{14}C in its carboxyl carbon atom.

CHAPTER 21

BIOSYNTHESIS OF AMINO ACIDS AND HEME

This chapter deals with the biosynthesis of amino acids and some biomolecules that are derived from amino acids. Humans can synthesize only half of the basic set of twenty amino acids (Table 21-1). These amino acids are called *nonessential,* whereas the ones that must be supplied in the diet are called *essential.* These terms refer to the needs of an organism under a particular set of conditions. For example, enough arginine is synthesized by the urea cycle to meet the needs of an adult but not those of a growing child. A deficiency of even one amino acid results in a *negative nitrogen balance.* In this state, more protein is degraded than is synthesized, and so more nitrogen is excreted than is ingested.

We will start with a brief discussion of nitrogen fixation, followed by a presentation of the biosynthesis of the nonessential amino acids. These amino acids are synthesized from citric acid cycle intermediates and other common metabolic intermediates by quite simple reactions. Then, we will turn to the essential amino acids, which are synthesized by more complex routes. The biosynthesis of the aromatic amino acids is examined in detail. The final section of this chapter is concerned with the synthesis and degradation of the heme group.

Table 21-1
Basic set of twenty amino acids

Nonessential	Essential
Alanine	Arginine
Asparagine	Histidine
Aspartate	Isoleucine
Cysteine	Leucine
Glutamate	Lysine
Glutamine	Methionine
Glycine	Phenylalanine
Proline	Threonine
Serine	Tryptophan
Tyrosine	Valine

MICROORGANISMS USE ATP AND A POWERFUL REDUCTANT TO CONVERT N_2 TO NH_4^+

The nitrogen atoms of amino acids, purines, pyrimidines, and other biomolecules come from NH_4^+. Higher organisms are unable to convert N_2 to organic form. This conversion, called *nitrogen fixation*, is carried out by bacteria and blue-green algae. Some of these microorganisms—namely, the *Rhizobium* bacteria—invade the roots of leguminous plants and form root nodules, in which nitrogen fixation takes place. The conversion of N_2 to NH_4^+ in all the organisms studied thus far *requires ATP and a powerful reductant*. In some species, the reductant is *ferredoxin,* an electron carrier that also participates in photosynthesis (p. 466). In turn, the reduction of ferredoxin is coupled to photosynthesis or oxidative processes, depending on the particular species.

$$N_2 + 6\,e^- + 12\,ATP + 12\,H_2O \longrightarrow$$
$$2\,NH_4^+ + 12\,ADP + 12\,P_i + 4\,H^+$$

This process is carried out by a *nitrogenase* enzyme complex, which consists of at least two kinds of subunits. One of them, the Mo-Fe-protein, contains one molybdenum, fourteen irons, and sixteen inorganic sulfides per molecule. The other component, termed the Fe-protein, contains four irons and four inorganic sulfides. The mechanism of nitrogen fixation is a challenging area of inquiry.

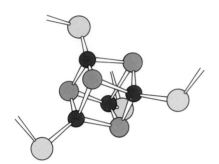

Figure 21-1
Structure of the iron-sulfide cluster in the high-potential iron protein from *Chromatium*. The sulfur atoms of cysteine residues are shown in yellow, whereas the inorganic sulfides are shown in green. Iron atoms are shown in red. This cluster probably occurs in numerous nonheme iron sulfide proteins, including nitrogenase. [Based on C. W. Carter, Jr., S. T. Freer, N. H. Xuong, R. A. Alden, and J. Kraut. *Cold Spring Harbor Symp. Quant. Biol.* 36(1971):383.]

GLUTAMATE IS THE PRECURSOR OF SEVERAL NONESSENTIAL AMINO ACIDS

Glutamate plays a pivotal role in the biosynthesis of several amino acids. *Most of the nonessential amino acids derive their α-amino groups from glutamate.* In turn, glutamate is synthesized from α-ketoglutarate and NH_4^+:

$$NH_4^+ + \text{α-ketoglutarate} + NADPH + H^+ \rightleftharpoons$$
$$\text{L-glutamate} + NADP^+ + H_2O$$

This reductive amination is catalyzed by glutamate dehydrogenase, which can also use NADH as the reductant.

Alanine and *aspartate* are synthesized from the corresponding α-keto acids by transamination reactions in which glutamate is the donor of the α-amino group. As discussed previously (p. 434), pyridoxal phosphate (PLP) is the coenzyme in transamination reactions.

$$\text{Pyruvate} + \text{glutamate} \rightleftharpoons \text{alanine} + \alpha\text{-ketoglutarate}$$
$$\text{Oxaloacetate} + \text{glutamate} \rightleftharpoons \text{aspartate} + \alpha\text{-ketoglutarate}$$

Glutamine is synthesized from glutamate and NH_4^+ in a reaction catalyzed by *glutamine synthetase*. A high-energy phosphate bond is consumed in the formation of this amide linkage. *Asparagine* is synthesized in a similar way.

$$\text{Glutamate} + NH_4^+ + \text{ATP} \longrightarrow \text{glutamine} + \text{ADP} + P_i + H^+$$
$$\text{Aspartate} + NH_4^+ + \text{ATP} \longrightarrow \text{asparagine} + \text{ADP} + P_i + H^+$$

The regulation of glutamine synthetase is discussed later (p. 518).
Proline is synthesized from glutamate by reduction of the carboxylate group to an aldehyde. Cyclization with loss of water yields Δ^1-pyrroline-5-carboxylate, which is then reduced to proline.

Glutamate   Glutamic-γ-semialdehyde   Δ'-Pyrroline-5-carboxylate   Proline

SERINE IS SYNTHESIZED FROM 3-PHOSPHOGLYCERATE

Serine is synthesized from 3-phosphoglycerate, an intermediate in glycolysis. The first step is an oxidation to 3-phosphohydroxypyruvate. This α-keto acid is transaminated to 3-phosphoserine, which is then hydrolyzed to yield serine.

3-Phosphoglycerate   3-Phosphohydroxy-pyruvate   3-Phosphoserine   Serine

Alternatively, hydrolysis of the phosphate group may precede oxidation and transamination:

3-Phosphoglycerate $\longrightarrow$ glycerate $\longrightarrow$

hydroxypyruvate $\longrightarrow$ serine

Glycine is derived from serine in a complex reaction that is catalyzed by serine transhydroxymethylase. Two coenzymes participate: *PLP* and *tetrahydrofolate*. The bond between the α and β carbon atoms of serine is labilized following formation of a Schiff base between serine and pyridoxal phosphate. The β carbon atom of serine is then transferred to tetrahydrofolate, a carrier of one-carbon units.

Serine + tetrahydrofolate $\rightleftharpoons$

glycine + methylenetetrahydrofolate

TETRAHYDROFOLATE CARRIES ACTIVATED ONE-CARBON UNITS AT SEVERAL OXIDATION LEVELS

We turn now to *tetrahydrofolate* (also called tetrahydropteroylglutamate), a carrier of activated one-carbon units. Tetrahydrofolate consists of three groups: a substituted pteridine, *p*-aminobenzoate, and glutamate. Mammals are unable to synthesize a pteridine ring. They obtain tetrahydrofolate from microorganisms in the intestinal tract.

Tetrahydrofolate

Table 21-2
One-carbon groups carried by tetrahydrofolate

Oxidation state	Group	
Most reduced	$-CH_3$	Methyl
Intermediate	$-CH_2-$	Methylene
Most oxidized	$-CHO$	Formyl
	$-CHNH$	Formimino
	$-CH=$	Methenyl

The one-carbon group carried by tetrahydrofolate is bonded to its N-5 or N-10 nitrogen atom (denoted as N^5 and N^{10}) or to both. This unit can exist in three oxidation states (Table 21-2). The most reduced form carries a *methyl* group, whereas the intermediate form carries a *methylene* group. The most oxidized forms carry a *methenyl,* *formyl,* or *formimino* group. The most oxidized one-carbon unit, CO_2, is carried by biotin (p. 371) rather than by tetrahydrofolate.

These one-carbon units are interconvertible (Figure 21-2). N^5, N^{10}-*Methylene*tetrahydrofolate can be reduced to N^5-*methyl*tetra-hydrofolate or oxidized to N^5-*methenyl*tetrahydrofolate. N^5, N^{10}-*Methenyl*tetrahydrofolate can be converted into N^5-*formimino*tetra-hydrofolate and N^{10}-*formyl*tetrahydrofolate, which are at the same

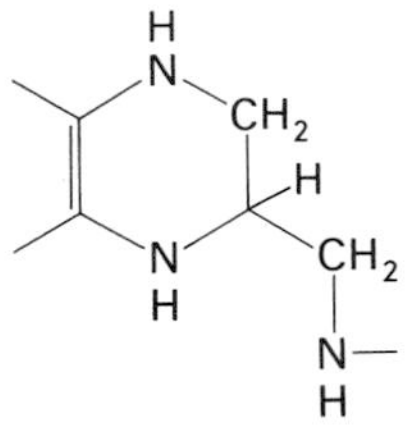

Reactive part of tetrahydrofolate

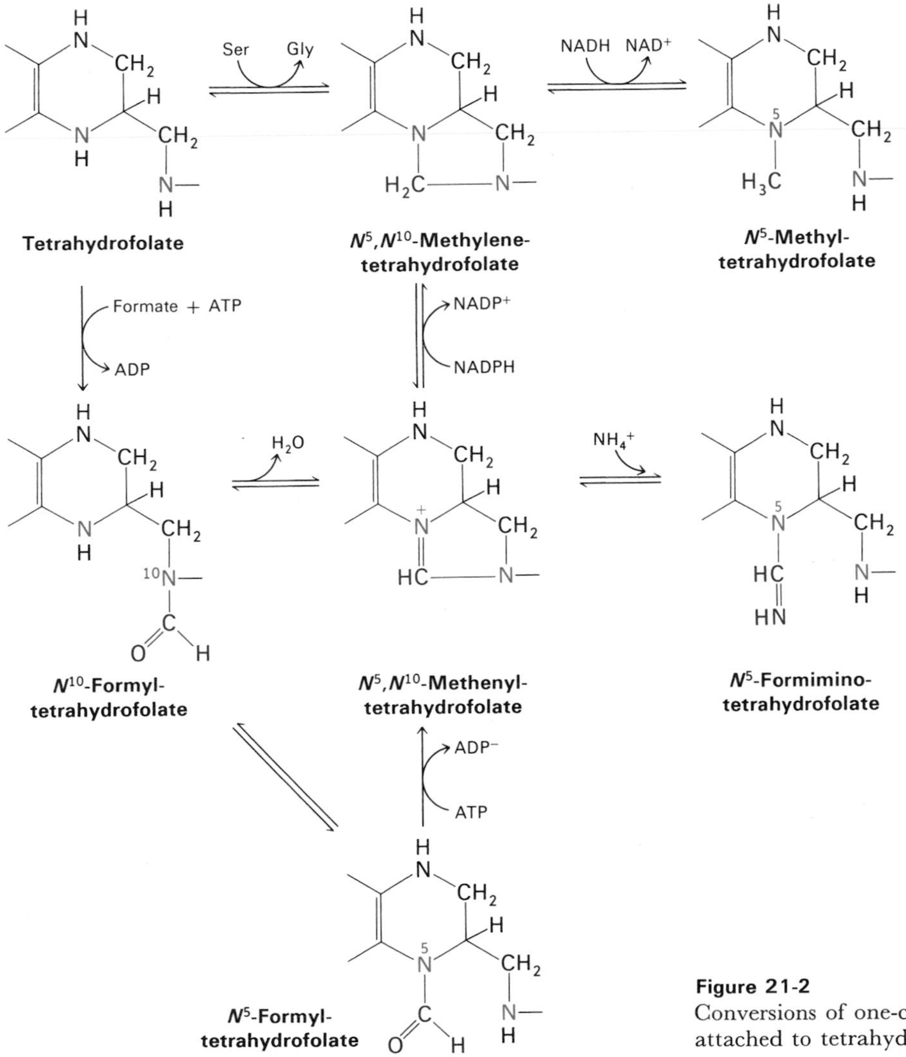

Tetrahydrofolate

N^5,N^{10}-**Methylene-tetrahydrofolate**

N^5-**Methyl-tetrahydrofolate**

N^{10}-**Formyl-tetrahydrofolate**

N^5,N^{10}-**Methenyl-tetrahydrofolate**

N^5-**Formimino-tetrahydrofolate**

N^5-**Formyl-tetrahydrofolate**

Figure 21-2
Conversions of one-carbon units attached to tetrahydrofolate.

oxidation level. N^{10}-Formyltetrahydrofolate can also be synthesized from formate and ATP:

$$\text{Formate} + \text{ATP} + \text{tetrahydrofolate} \rightleftharpoons$$
$$N^{10}\text{-formyltetrahydrofolate} + \text{ADP} + \text{P}_i$$

These tetrahydrofolate derivatives serve as donors of one-carbon units in a variety of biosyntheses. Methionine is synthesized from homocysteine by transfer of the methyl group of N^5-methyltetrahydrofolate, as will be discussed shortly. Some of the carbon atoms of *purines* are derived from the N^5, N^{10}-methenyl and the N^{10}-formyl derivatives of tetrahydrofolate. The methyl group of *thymine*, a pyrimidine, comes from N^5, N^{10}-methylenetetrahydrofolate. Thus, one-carbon units at each of the three oxidation levels are utilized in biosyntheses. In turn, *tetrahydrofolate serves as an acceptor of one-carbon units in degradative reactions.* N^5, N^{10}-Methylenetetrahydrofolate is formed in the conversion of *serine* to glycine, as previously mentioned. The breakdown of *histidine* yields *N*-formiminoglutamate, which transfers its formimino group to tetrahydrofolate to form the N^5-derivative.

S-ADENOSYLMETHIONINE IS THE MAJOR DONOR OF METHYL GROUPS

Tetrahydrofolate is not the only carrier of activated methyl groups. In fact, the major activated methyl donor is S-*adenosylmethionine*, which we have already encountered in the conversion of phosphatidyl ethanolamine to phosphatidyl choline (p. 482). *S*-Adenosylmethionine is synthesized by the transfer of an adenosyl group from ATP to the sulfur atom of methionine. The methyl group of the methionine unit in this sulfonium compound is activated.

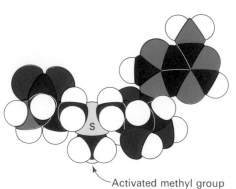

Figure 21-3
Space-filling model of
S-adenosylmethionine.

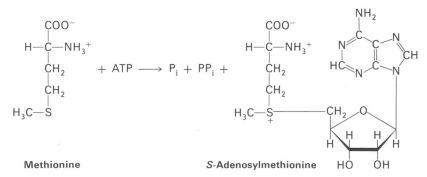

Methionine   *S*-Adenosylmethionine

This reaction is unusual in that the triphosphate group of ATP is split into pyrophosphate and orthophosphate. Pyrophosphate is then hydrolyzed. Thus, three high-energy phosphate bonds are consumed in the synthesis of S-adenosylmethionine.

S-*Adenosylhomocysteine* is formed when the methyl group of S-adenosylmethionine is transferred to an acceptor such as phosphatidyl ethanolamine. S-Adenosylhomocysteine is then hydrolyzed to *homocysteine* and adenosine.

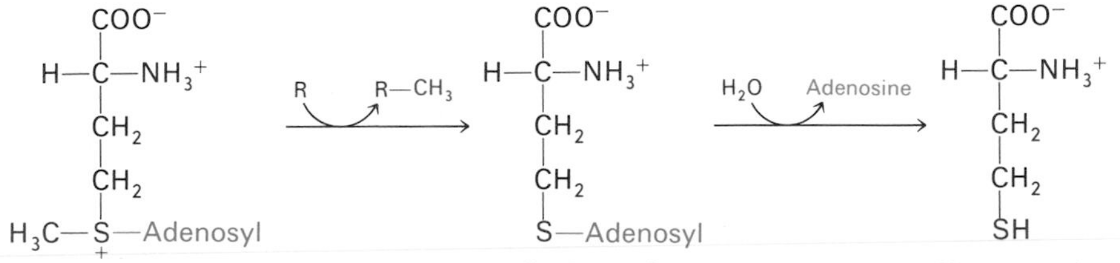

Homocysteine can be methylated to form methionine by a variety of methyl donors. One of these methyl donors is *betaine*, which is formed by the oxidation of choline.

$$(CH_3)_3 - \overset{+}{N} - CH_2 - COO^-$$
Betaine

Methionine can also be regenerated by the transfer of a methyl group from N^5-*methyltetrahydrofolate*. Thus, the two carriers of activated methyl groups are linked. *Methylcobalamin* is the coenzyme of homocysteine transmethylase, the enzyme that catalyzes this transfer of a methyl group.

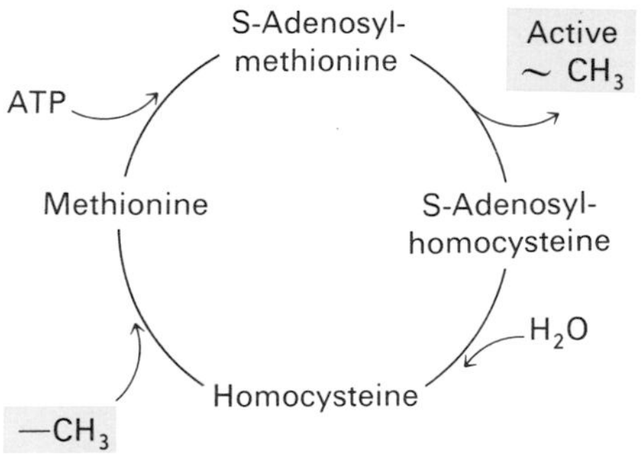

Figure 21-4
Activated methyl cycle.

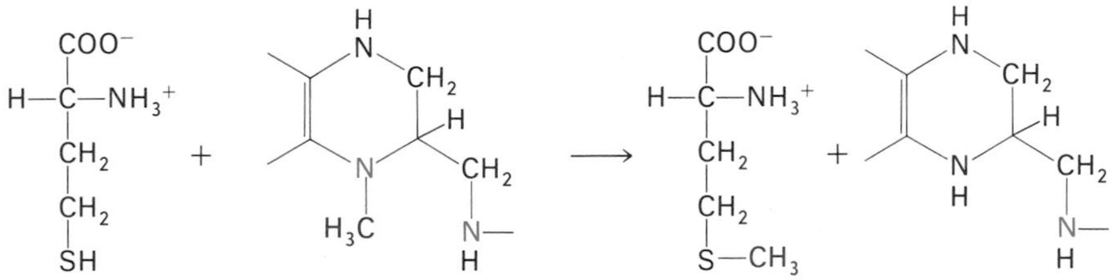

Finally, S-adenosylmethionine is formed from methionine and ATP, as discussed previously. The *activated methyl cycle* is summarized in Figure 21-4.

CYSTEINE IS SYNTHESIZED FROM SERINE AND HOMOCYSTEINE

Homocysteine is an intermediate in the synthesis of cysteine, in addition to being a precursor of methionine in the activated methyl cycle. Serine and homocysteine condense to form *cystathionine* (Figure 21-5). This reaction is catalyzed by cystathionine synthetase, a PLP enzyme. Cystathionine is then deaminated and cleaved to cysteine and α-ketobutyrate by *cystathioninase,* another PLP enzyme. The net reaction is:

$$\text{Homocysteine} + \text{serine} \longrightarrow \text{cysteine} + \text{α-ketobutyrate}$$

Note that the sulfur atom of cysteine is derived from homocysteine, whereas the carbon skeleton comes from serine.

This completes our consideration of the biosynthesis of the nonessential amino acids. The formation of tyrosine by the hydroxylation of phenylalanine was discussed earlier (p. 450).

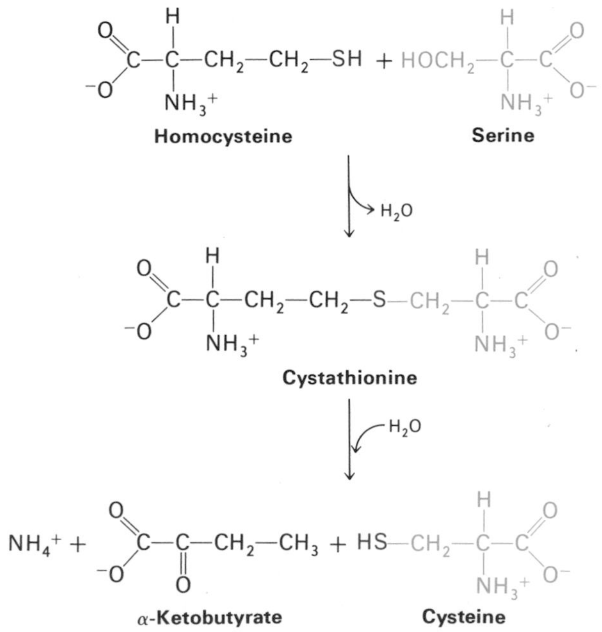

Figure 21-5
Synthesis of cysteine.

SHIKIMATE AND CHORISMATE ARE INTERMEDIATES IN THE BIOSYNTHESIS OF AROMATIC AMINO ACIDS

We turn now to the biosynthesis of essential amino acids, which are formed by much more complex routes than are the nonessential amino acids. Two pathways have been selected for discussion here—those of the aromatic amino acids and of histidine.

Phenylalanine, tyrosine, and tryptophan are synthesized by a common pathway in *E. coli* (Figure 21-6). The initial step is the condensation of phosphoenolpyruvate (a glycolytic intermediate) and erythrose 4-phosphate (a pentose phosphate pathway intermediate). The resulting C_7 open-chain sugar loses its phosphoryl group and cyclizes to 5-dehydroquinate. Dehydration then yields 5-dehydroshikimate, which is reduced by NADPH to *shikimate* (Figure 21-7). A second molecule of phosphoenolpyruvate then condenses with 5-phosphoshikimate to give an intermediate that loses its phosphoryl group, yielding *chorismate*.

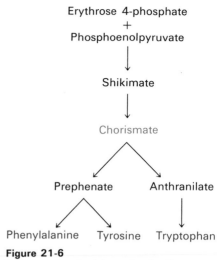

Figure 21-6
Pathway for the biosynthesis of aromatic amino acids in *E. coli.*

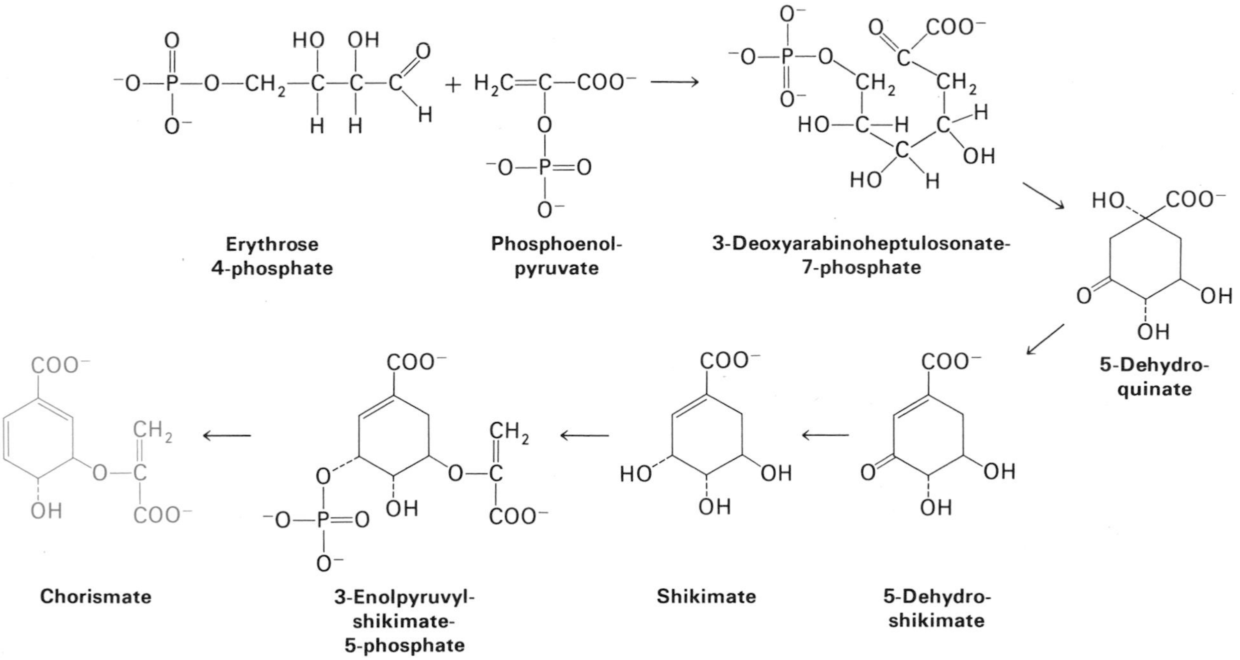

Figure 21-7
Synthesis of chorismate, an intermediate in the biosynthesis of phenylalanine, tyrosine, and tryptophan in *E. coli.*

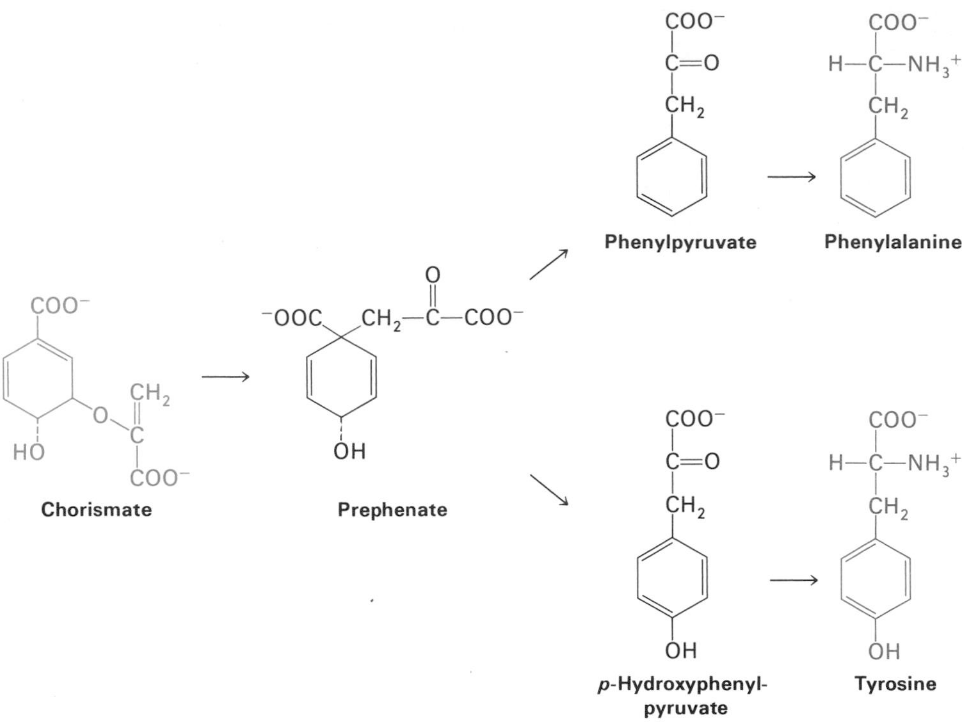

Figure 21-8
Synthesis of tyrosine and tryptophan from chorismate.

The pathway bifurcates at chorismate. Let us first follow the *prephenate branch* (Figure 21-8). A mutase converts chorismate to prephenate, the immediate precursor of the aromatic ring of phenylalanine and tyrosine. Dehydration and decarboxylation yield *phenylpyruvate*. Alternatively, prephenate can be oxidatively decarboxylated to yield *p-hydroxyphenylpyruvate*. These α-keto acids are then transaminated, yielding *phenylalanine* and *tyrosine*, respectively.

The branch starting with *anthranilate* leads to the synthesis of *tryptophan*. Chorismate acquires an amino group from the side chain of glutamine to form anthranilate. In fact, *glutamine serves as an amino donor in many biosynthetic reactions*. Anthranilate then condenses with *phosphoribosylpyrophosphate* (PRPP), *an activated form of ribose phosphate.* PRPP is also a key intermediate in the synthesis of histidine, purine nucleotides, and pyrimidine nucleotides (see Chapter 22). The C-1 atom of ribose 5-phosphate becomes bonded to the nitrogen atom of anthranilate in a reaction that is driven by the hydrolysis of pyrophosphate.

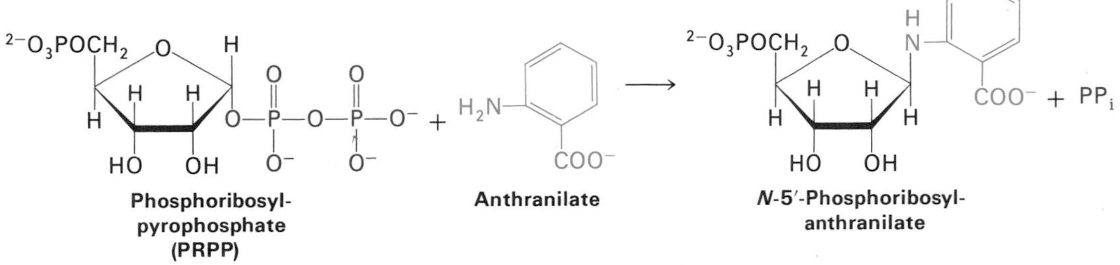

The ribose moiety of phosphoribosylanthranilate undergoes rearrangement (Figure 21-9) yielding 1-(*o*-carboxyphenylamino)-1-deoxyribulose 5-phosphate. This intermediate is dehydrated and decarboxylated to form *indole-3-glycerol phosphate*. Finally, indole-3-glycerol phosphate reacts with serine to form *tryptophan*. The

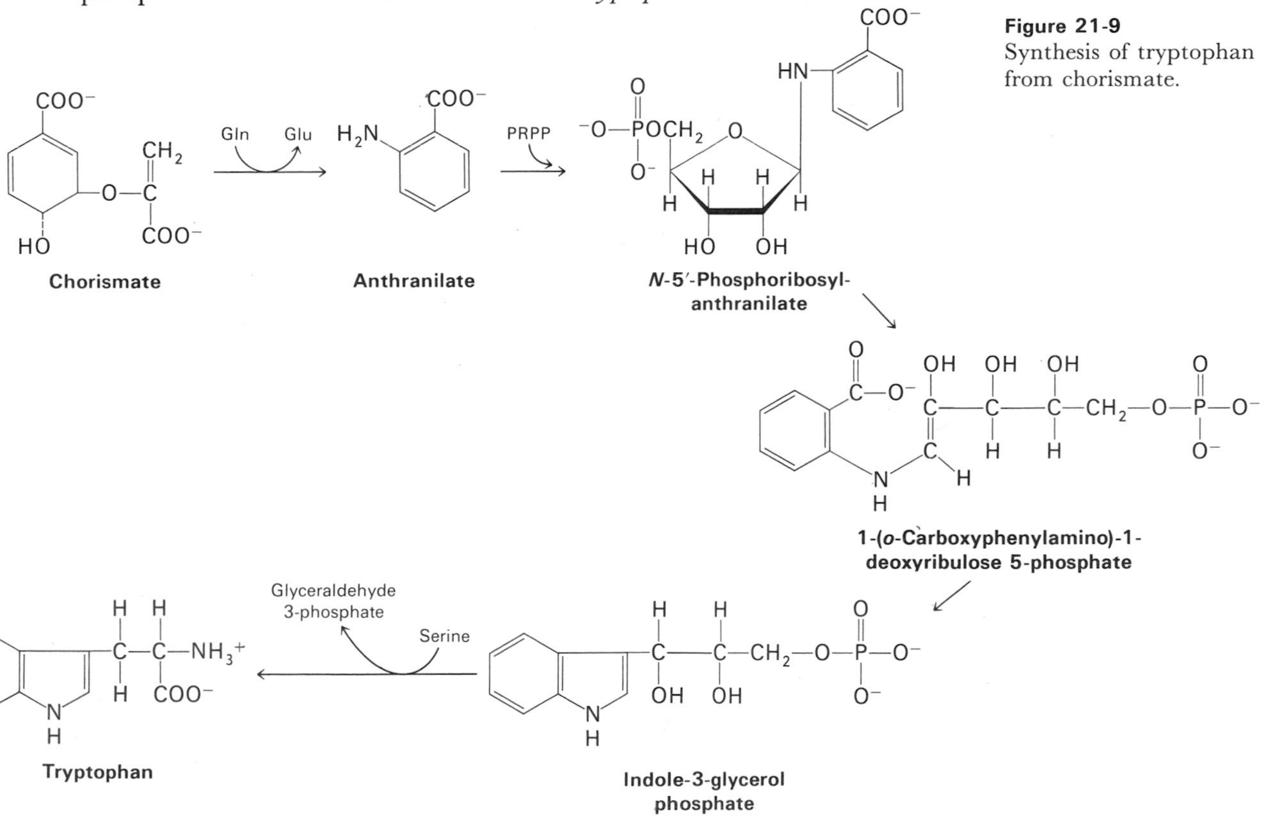

Figure 21-9
Synthesis of tryptophan from chorismate.

glycerol phosphate side chain of indole-3-glycerol phosphate is replaced by the carbon skeleton and amino group of serine. This reaction is catalyzed by tryptophan synthetase.

Tryptophan synthetase of *E. coli* has the subunit structure $\alpha_2\beta_2$. The enzyme can be dissociated into two α subunits and a β_2 subunit. The isolated subunits catalyze partial reactions that lead to the synthesis of tryptophan:

$$\text{Indole-3-glycerol phosphate} \xrightarrow{\alpha \text{ subunit}} \text{indole} + \text{glyceraldehyde 3-phosphate}$$

$$\text{Indole} + \text{serine} \xrightarrow{\beta_2 \text{ subunit}} \text{tryptophan} + H_2O$$

Each active site on the β_2 subunit contains a PLP prosthetic group. *The catalytic properties of the α and β_2 subunits are markedly altered on the formation of the $\alpha_2\beta_2$ complex.* The rates of the partial reactions are more than ten times greater for the $\alpha_2\beta_2$ complex than for the isolated subunits. Furthermore, the $\alpha_2\beta_2$ complex synthesizes tryptophan by a concerted mechanism. Indole formed by the first partial reaction reacts immediately with serine, so that indole is not released from the $\alpha_2\beta_2$ complex. Thus, the catalytic properties of a multisubunit enzyme can be altered by interactions between its subunits.

HISTIDINE IS SYNTHESIZED FROM ATP, PRPP, AND GLUTAMINE

The pathway for histidine biosynthesis in *E. coli* and *Salmonella* contains many complex and novel features (Figure 21-10). The reaction sequence starts with the condensation of ATP and PRPP, in which N-1 of the purine ring becomes bonded to C-1 of the ribose unit of PRPP. In fact, five carbon atoms of histidine come from PRPP. The adenine unit of ATP provides a nitrogen and a carbon atom of the imidazole ring of histidine. The other nitrogen atom of the imidazole ring comes from the side chain of glutamine. A noteworthy aspect of this pathway is that 5-aminoimidazole-4-carboxamide ribonucleotide, which is produced in the cleavage reaction that forms the imidazole ring, is an intermediate in purine biosynthesis (p. 534). Thus, histidine biosynthesis and purine biosynthesis are linked.

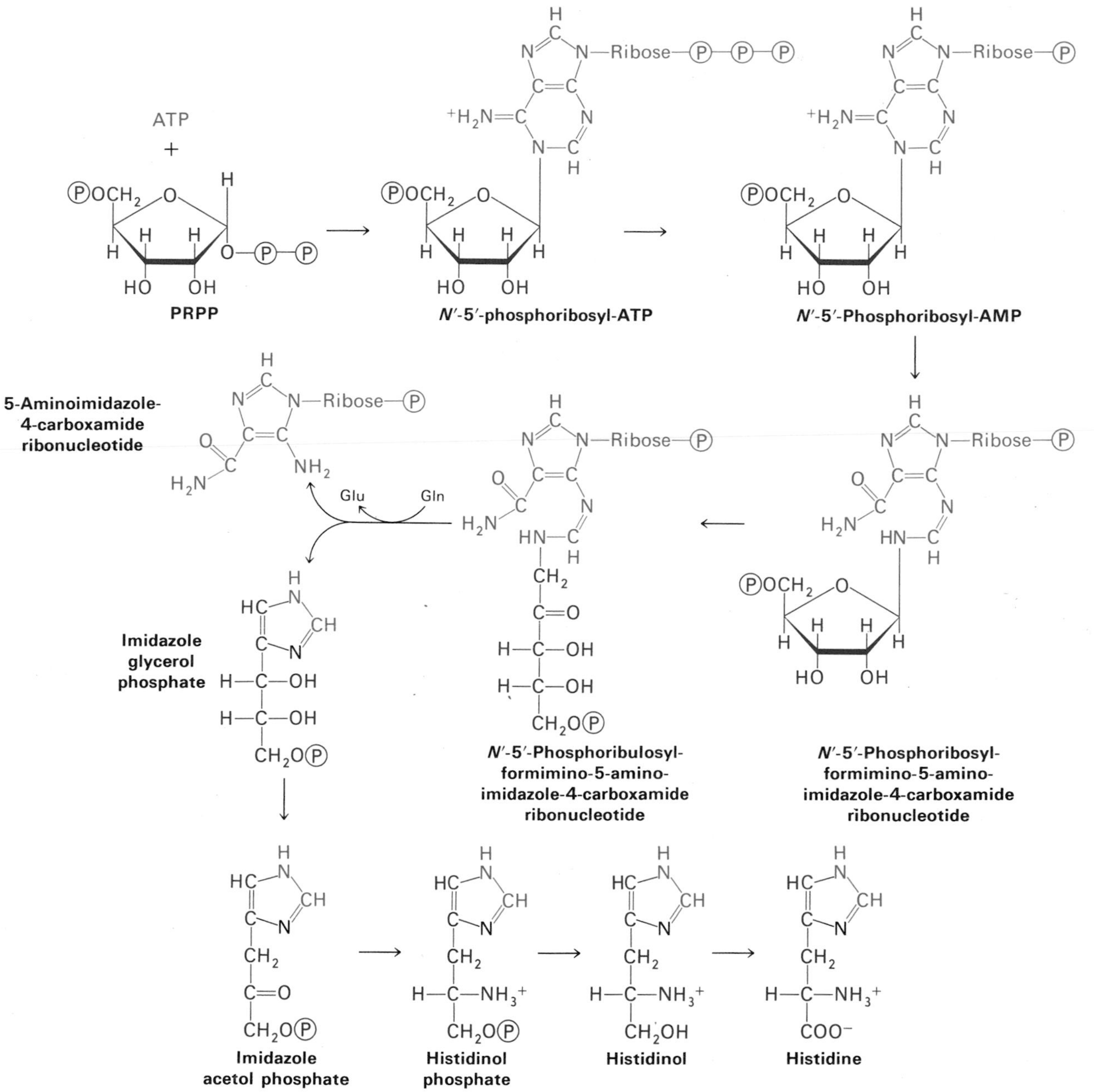

Figure 21-10

Pathway for the biosynthesis of histidine in *E. coli* and *Salmonella* (Ⓟ denotes a phosphoryl group).

AMINO ACID BIOSYNTHESIS IS REGULATED BY FEEDBACK INHIBITION

The rate of synthesis of amino acids depends mainly on the *amounts* of the biosynthetic enzymes and on their enzymatic *activities*. We will now consider the control of enzymatic activity. The regulation of enzyme synthesis will be discussed in Chapter 27.

The first irreversible reaction in a biosynthetic pathway, called the committed step, is usually an important regulatory site. *The final product of the pathway (Z) often inhibits the enzyme that catalyzes the committed step (A → B).* This kind of control is essential for the conservation of building blocks and metabolic energy. The first example of this important principle of metabolic control came from studies of the biosynthesis of isoleucine in *E. coli*. The dehydration and deamination of threonine to α-ketobutyrate is the committed step in the synthesis of isoleucine. *Threonine deaminase,* the PLP enzyme that catalyzes this reaction, is allosterically inhibited by isoleucine.

Consider a branched biosynthetic pathway in which Y and Z are the final products. Suppose that high levels of Y *or* Z completely inhibit the first common step (A ⟶ B). Then, high levels of Y would prevent the synthesis of Z even if there were a deficiency of Z. Such a regulatory scheme is obviously not optimal. In fact, several intricate control mechanisms have been found in branched biosynthetic pathways:

1. *Sequential feedback control.* The first common step (A ⟶ B) is not inhibited directly by Y and Z. Rather, these final products inhibit the reactions leading away from the point of branching: Y inhibits the C ⟶ D step, and Z inhibits the C ⟶ F step.

In turn, high levels of C inhibit the A $\longrightarrow$ B step. Thus, the first common reaction is blocked only if both final products are present in excess.

Sequential feedback control is used to regulate the synthesis of aromatic amino acids in *Bacillus subtilis*. The first divergent steps in the synthesis of phenylalanine, tyrosine, and tryptophan are inhibited by the respective final product. If all three are present in excess, chorismate and prephenate accumulate. These branch-point intermediates in turn inhibit the first common step in the overall pathway, which is the condensation of phosphoenolpyruvate and erythrose 4-phosphate.

2. *Enzyme multiplicity.* The distinguishing feature of this mechanism is that the first common step (A $\longrightarrow$ B) is catalyzed by two different enzymes. One of them is inhibited by Y, and the other by Z. Thus, both Y and Z must be present at high levels to prevent the conversion of A to B completely. The other aspect of this control scheme is like that in sequential feedback control: Y inhibits the C $\longrightarrow$ D step and Z inhibits the C $\longrightarrow$ F step.

Differential inhibition of multiple enzymes is used to control a variety of biosynthetic pathways in microorganisms. In *E. coli,* the condensation of phosphoenolpyruvate and erythrose 4-phosphate is catalyzed by three different enzymes. One is inhibited by phenylalanine, another by tyrosine, and the third by tryptophan. Furthermore, there are two different mutases that convert chorismate to prephenate. One of them is inhibited by phenylalanine, the other by tyrosine.

3. *Concerted feedback control.* The first common step (A $\longrightarrow$ B) is inhibited only if high levels of Y and Z are simultaneously present. A high level of either product alone does not inhibit the A $\longrightarrow$ B step. As in the control schemes discussed above, Y inhibits the C $\longrightarrow$ D step and Z inhibits the C $\longrightarrow$ F step.

An example of concerted feedback control is the inhibition of aspartyl kinase by threonine and lysine, the final products.

4. *Cumulative feedback control.* The first common step (A $\longrightarrow$ B) is partially inhibited by each of the final products. Each final product acts independently of the others. Suppose that a high level of Y decreases the rate of the A $\longrightarrow$ B step from 100 to 60 sec^{-1} and that Z alone decreases the rate from 100 to 40 sec^{-1}. Then, the rate of the A $\longrightarrow$ B step in the presence of high levels of A and B is 24 sec^{-1} (0.6 $\times$ 0.4 $\times$ 100 sec^{-1}).

The regulation of the activity of *glutamine synthetase* in *E. coli* is a striking example of cumulative feedback inhibition. Glutamine is synthesized from glutamate, NH_4^+, and ATP (p. 505). The amide group of glutamine is a source of nitrogen in the biosyntheses of a variety of compounds such as tryptophan, histidine, carbamoyl phosphate, glucosamine 6-phosphate, CTP, and AMP. Glutamine

Figure 21-11
Electron micrograph of glutamine synthetase molecules. The enzyme consists of 12 identical subunits, which are arranged in a double hexagonal ring. The diameter of the hexagon is about 140 Å. [Courtesy of Dr. David Eisenberg and Dr. Terrence Frey.]

synthetase is cumulatively inhibited by each of these final products of glutamine metabolism, and also by alanine and glycine. There seem to be specific binding sites for each of these inhibitors. The enzymatic activity of glutamine synthetase is almost completely switched off when all eight final products are bound to the enzyme.

THE ACTIVITY OF GLUTAMINE SYNTHETASE IS MODULATED BY ADENYLYLATION

Another interesting feature of glutamine synthetase from *E. coli* is that its activity is altered by *reversible covalent modification* (Figure 21-12). We previously encountered this type of control in the synthesis and degradation of glycogen. Phosphorylation activates glycogen phosphorylase and inactivates glycogen synthetase. The activity of glutamine synthetase is regulated in part by the covalent attachment of an *AMP unit* to the hydroxyl group of a specific tyrosine residue in each subunit. *This adenylylated enzyme is more susceptible to cumulative feedback inhibition than the deadenylylated form.*

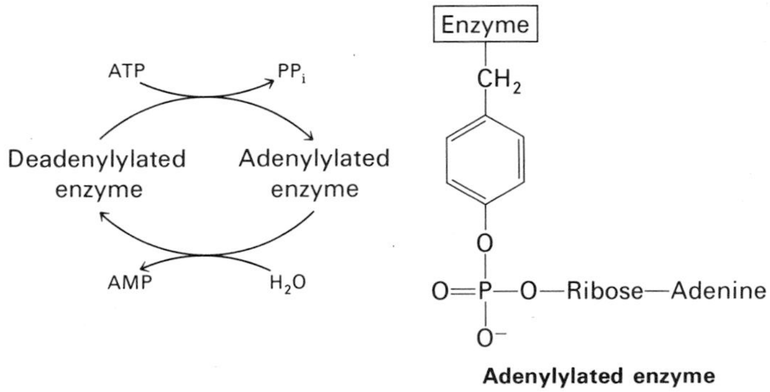

Figure 21-12
Control of the activity of glutamine synthetase by adenylylation and deadenylylation.

The attachment of AMP to the enzyme is catalyzed by *adenylyl transferase*. The AMP unit can be hydrolyzed off the adenylyated enzyme by a *deadenylylating* enzyme. Glutamine activates adenylyl transferase and inhibits the deadenylylating enzyme, whereas α-ketoglutarate has the opposite effect. If the supply of activated nitrogen is low, adenylylation is inhibited and deadenylyation is stimulated. Glutamine synthetase then becomes less susceptible to cumulative feedback inhibition, and the supply of glutamine consequently increases.

AMINO ACIDS ARE PRECURSORS OF A VARIETY OF BIOMOLECULES

Amino acids are the building blocks of proteins and peptides. They also serve as precursors of many kinds of small molecules that have important biological roles. Let us briefly survey some of the biomolecules that are derived from amino acids (Figure 21-13). *Purines* and *pyrimidines* are derived in part from amino acids. The biosynthesis of these precursors of DNA, RNA, and numerous coenzymes is discussed in detail in the next chapter. Six of the nine atoms of the purine ring and four of the six atoms of the pyrimidine ring are derived from amino acids. The reactive terminus of *sphingosine*, an intermediate in the synthesis of sphingolipids, comes from serine.

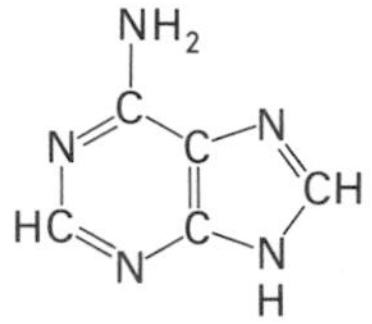

Adenine
(A purine)

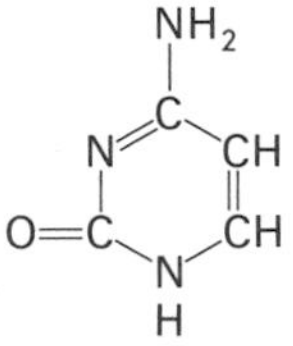

Cytosine
(A pyrimidine)

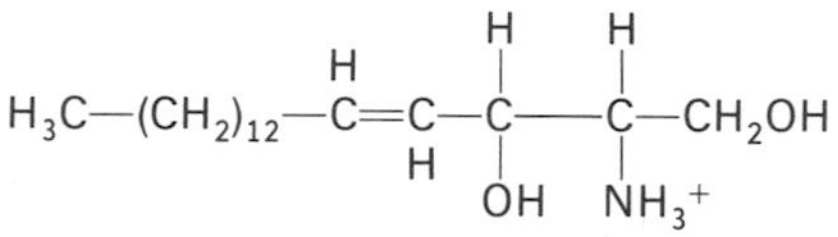

Sphingosine

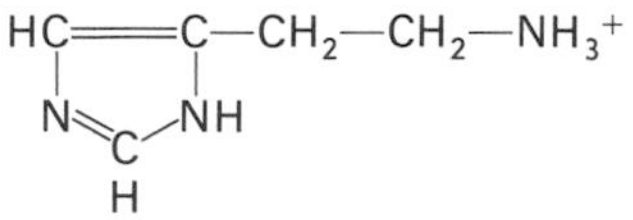

Histamine

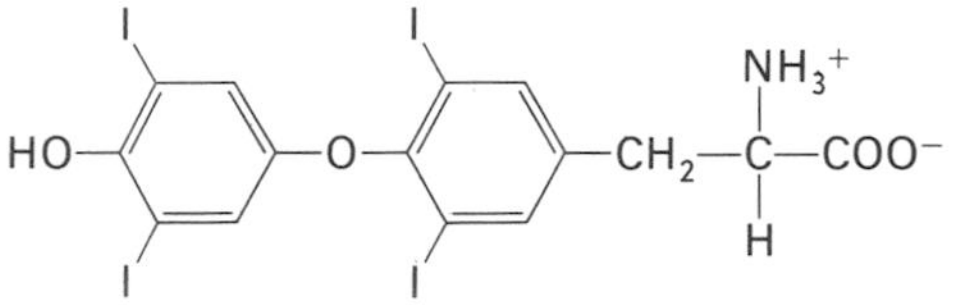

Thyroxine
(Tetraiodothyronine)

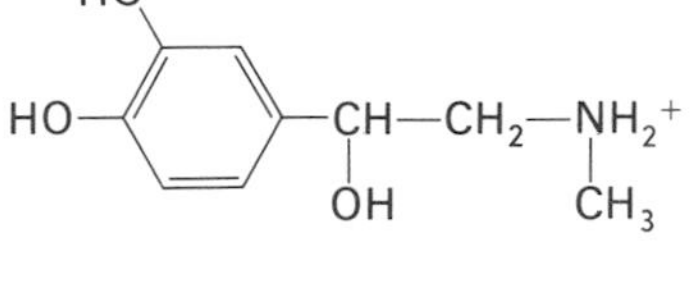

Epinephrine

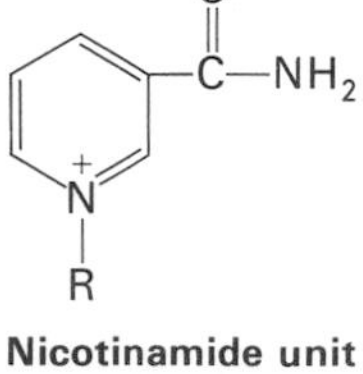

Nicotinamide unit of NAD⁺

Figure 21-13
Biomolecules derived from amino acids.

Histamine, a potent vasodilator, is derived from histidine by decarboxylation. Tyrosine is a precursor of the hormones *thyroxine* (tetraiodothyronine) and *epinephrine* and of *melanin,* a polymeric pigment. The neurotransmitter 5-*hydroxytryptamine* (*serotonin*) and the *nicotinamide ring* of NAD^+ are synthesized from tryptophan. Glutamine contributes the amide group of the nicotinamide moiety.

PORPHYRINS ARE SYNTHESIZED FROM GLYCINE AND SUCCINYL COENZYME A

The porphyrin ring in hemes and chlorophylls is derived from glycine and succinyl CoA. In fact, the first step in the biosynthesis of porphyrins is the condensation of these precursors to form δ-*aminolevulinate*.

Succinyl CoA   **Glycine**   **δ-Aminolevulinate**

These reactions are catalyzed by δ-aminolevulinate synthetase, a PLP enzyme in mitochondria. As expected, this committed step in the biosynthesis of porphyrins is an important control site. Two molecules of δ-aminolevulinate then condense to form *porphobilinogen.* This dehydration reaction is catalyzed by δ-aminolevulinate dehydrase.

δ-Aminolevulinate   **Porphobilinogen**

522

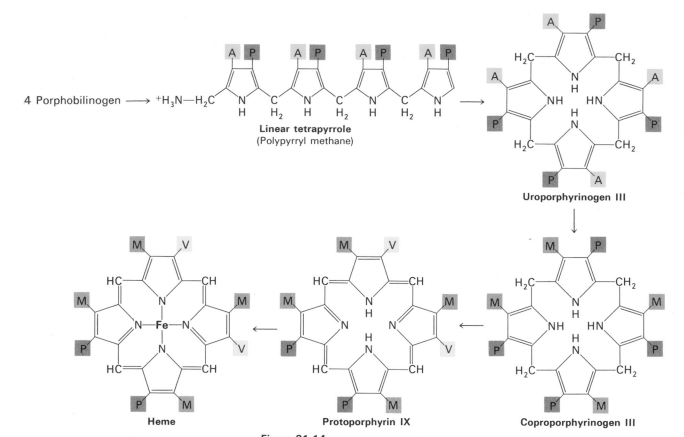

4 Porphobilinogen ⟶ Linear tetrapyrrole (Polypyrryl methane) ⟶ Uroporphyrinogen III ⟶ Coproporphyrinogen III ⟶ Protoporphyrin IX ⟶ Heme

Figure 21-14
Pathway for the synthesis of heme from porphobilinogen. (Abbreviations: A, acetate; M, methyl; P, propionate; V, vinyl.)

Figure 21-15
Space-filling model of protoporphyrin IX, the immediate precursor of heme.

Four porphobilinogens condense head-to-tail to form a *linear tetrapyrrole*, which remains bound to the enzyme (Figure 21-14). An ammonium ion is released for each methylene bridge formed. This linear tetrapyrrole cyclizes by losing NH_4^+. The cyclic product is uroporphyrinogen III, which has an asymmetric arrangement of side chains. These reactions require a *synthetase* and a *cosynthetase*. In the presence of synthetase alone, uroporphyrinogen I, the symmetric isomer, is produced. The cosynthetase is essential for isomerizing one of the pyrrole rings to yield the asymmetric uroporphyrinogen III.

The porphyrin skeleton is now formed. The subsequent reactions alter the side chains and the degree of saturation of the porphyrin ring (Figure 21-14). *Coproporphyrinogen III* is formed by decarboxylation of the acetate side chains. Unsaturation of the porphyrin ring and conversion of two of the propionate side chains to vinyl groups yield *protoporphyrin IX*. Chelation of iron finally gives *heme*, the prosthetic group of proteins such as myoglobin, hemoglobin, catalase, peroxidase, and cytochrome *c*. The insertion of iron is catalyzed by ferrochelatase.

Several factors that regulate heme biosynthesis in animals have been elucidated. *Delta-aminolevulinate synthetase, the enzyme that catalyzes the committed step in this pathway, is feedback inhibited by heme,* as is δ-aminolevulinate dehydrase and ferrochelatase. Regulation also occurs at the level of enzyme synthesis. *Heme represses the synthesis of δ-aminolevulinate synthetase.*

PORPHYRINS ACCUMULATE IN SOME INHERITED DISORDERS OF PORPHYRIN METABOLISM

Several inherited disorders of porphyrin metabolism are known. In *congenital erythropoietic porphyria,* there is a deficiency of uroporphyrinogen III cosynthetase, the isomerase that yields the asymmetric isomer on cyclization of the linear tetrapyrrole. The synthesis of the required amount of uroporphyrinogen III is accompanied by the formation of very large quantities of uroporphyrinogen I, the symmetric isomer that has no physiologic role. Uroporphyrin I, coproporphyrin I, and other symmetric derivatives also accumulate. The disease is transmitted as an autosomal recessive. The urine of patients having this disease is red because of the excretion of large amounts of uroporphyrin I. Their teeth exhibit a strong red fluorescence under ultraviolet light because of the deposition of porphyrins. Photosensitivity and enlargement of the spleen are other characteristic signs.

Acute intermittent porphyria is a quite different disease. The liver, rather than the red cells, is affected. *The activity of δ-aminolevulinate synthetase is markedly elevated,* and the concentrations of δ-aminolevulinate and porphobilinogen in the liver are increased. Large amounts of these two compounds are excreted in the urine. The disease is inherited as an *autosomal dominant.* The striking clinical

Uroporphyrinogen I

Uroporphyrinogen III

symptoms are intermittent abdominal pain and neurologic disturbances. As its name implies, the disease is episodic in its clinical expression. Acute attacks are sometimes precipitated by drugs such as barbiturates and estrogens.

Recent historical research suggests that the mania of King George III, who reigned from 1760 to 1811, resulted from variegate porphyria, which is similar to acute intermittent porphyria. This inference is based on the following findings. First, George had only three short episodes of mental derangement, each time preceded by bouts of acute abdominal pain. Second, his physicians repeatedly noted that his urine was red or dark in color. Third, many of his ancestors and descendants displayed symptoms consistent with porphyria. These signs can be traced back to Mary, Queen of Scots. Her son James described his urine as having the color of his favorite Alicante wine. Fourth, laboratory tests on some of George III's present day descendants are consistent with a diagnosis of variegate porphyria.

BILIVERDIN AND BILIRUBIN ARE INTERMEDIATES IN THE BREAKDOWN OF HEME

The normal human erythrocyte has a life span of about 120 days. Old cells are removed from the circulation and degraded by the spleen. The apoprotein of hemoglobin is hydrolyzed to its constituent amino acids. The first step in the degradation of the heme group to bilirubin (Figure 21-16) is the cleavage of its α-methene bridge to form *biliverdin*, a linear tetrapyrrole. This reaction is catalyzed by *heme oxygenase*. Two aspects of this enzyme are noteworthy. First, it is a mixed-function oxidase: O_2 and NADPH are required for the cleavage reaction. Second, it is a microsomal enzyme that is coupled to the cytochrome P_{450} electron-transport chain. We have already considered this microsomal electron-transport chain in regard to its participation in the hydroxylation of steroids (p. 496). The central methene bridge of biliverdin is then reduced by biliverdin reductase to form *bilirubin*. Again, the reductant is NADPH.

Bilirubin complexed to serum albumin is transported to the liver, where it is rendered more soluble by the attachment of sugar residues to its propionate side chains. The solubilizing sugar is

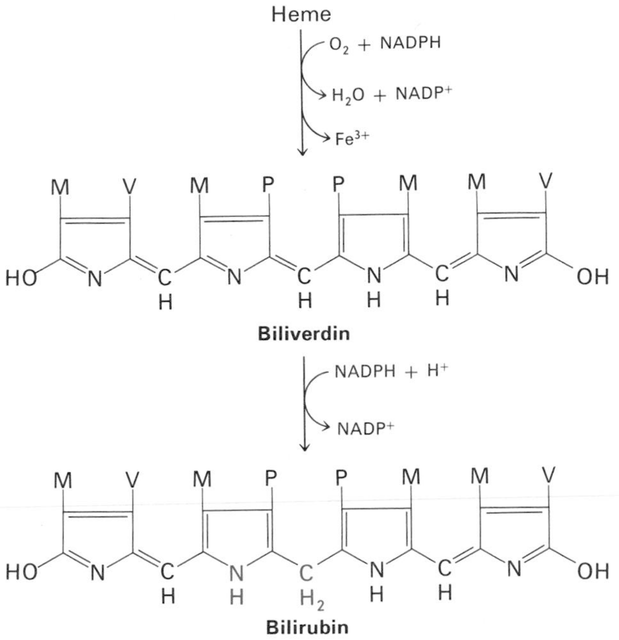

Figure 21-16
Degradation of heme to bilirubin.

glucuronate, which differs from glucose in having a COO^- group at C-6 rather than a CH_2OH group. The conjugate of bilirubin and two glucuronates, called *bilirubin diglucuronide,* is secreted into bile. *UDP-glucuronate, derived from the oxidation of UDP-glucose, is the activated intermediate in the synthesis of bilirubin diglucuronide.*

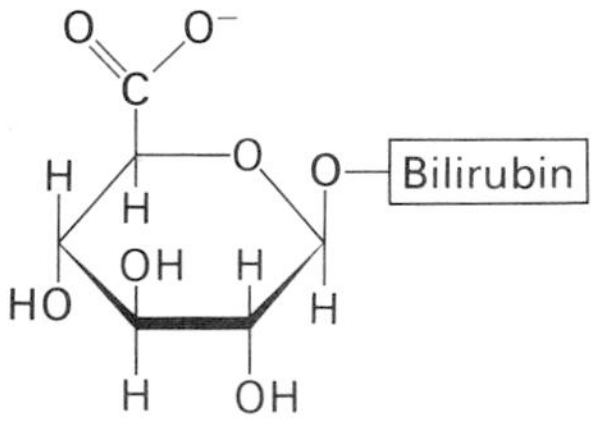

**A glucuronate unit
in bilirubin diglucuronide**

SUMMARY

Microorganisms use ATP and a powerful reductant to convert N_2 to NH_4^+, which is consumed by higher organisms in the synthesis of amino acids, nucleotides, and other biomolecules. The major points of entry of NH_4^+ into intermediary metabolism are glutamine, glutamate, and carbamoyl phosphate. Humans can synthesize only half of the basic set of twenty amino acids. These amino acids are called nonessential, in contrast to the essential ones, which must be supplied in the diet. The pathways for the synthesis of nonessential amino acids are quite simple. Glutamate dehydrogenase catalyzes the reductive amination of α-ketoglutarate to glutamate. Alanine and aspartate are synthesized by transamination of pyruvate and oxaloacetate, respectively. Glutamine is synthesized from NH_4^+ and glutamate, and asparagine is synthesized similarly. Proline is derived from glutamate. Serine, formed from 3-phosphoglycerate, is the precursor of glycine and cysteine. Tyrosine is synthesized by the hydroxylation of phenylalanine, an essential amino acid. The pathways for the biosynthesis of essential amino acids are much more complex than for the nonessential ones. Most of these pathways are regulated by feedback inhibition, in which the committed step is allosterically inhibited by the final product.

Tetrahydrofolate, a carrier of activated one-carbon units, plays an important role in amino acid and nucleotide metabolism. This coenzyme carries one-carbon units at three oxidation states, which are interconvertible: most reduced—methyl; intermediate—methylene; most oxidized—formyl, formimino, and methenyl. The other donor of activated methyl groups is S-adenosylmethionine, which is synthesized by the transfer of an adenosyl group from ATP to the sulfur atom of methionine. S-Adenosylhomocysteine is formed when the activated methyl group is transferred to an acceptor. It is hydrolyzed to adenosine and homocysteine, which is then methylated to methionine to complete the activated methyl cycle.

Amino acids are precursors of a variety of biomolecules. Porphyrins are synthesized from glycine and succinyl CoA, which condense to give δ-aminolevulinate. This intermediate condenses with itself to form porphobilinogen. Four porphobilinogens combine to form a linear tetrapyrrole, which cyclizes to form uroporphyrinogen III. Oxidation and side-chain modifications lead to the synthesis of protoporphyrin IX, which acquires an iron atom to form heme. Delta-aminolevulinate synthetase, the enzyme that catalyzes the committed step in this pathway, is feedback inhibited by heme.

SELECTED READINGS

Books:

Meister, A., 1965. *Biochemistry of the Amino Acids*, (2nd ed.), vols. 1 and 2. Academic Press. [A comprehensive and authoritative treatise on amino acid metabolism.]

Cohen, G. N., 1967. *Biosynthesis of Small Molecules*. Harper and Row.

Prusiner, S., and Stadtman, E. R., (eds.), 1973. *The Enzymes of Glutamine Metabolism*. Academic Press. [A valuable collection of articles on the role of glutamine in nitrogen metabolism.]

Biosynthetic pathways and control:

Umbarger, H. E., 1969. Regulation of amino acid metabolism. *Annu. Rev. Biochem.* 38:327–370.

Truffa-Bachi, P., and Cohen, G. N., 1973. Amino acid metabolism. *Annu. Rev. Biochem.* 42:113–134. [Emphasizes the synthesis of histidine, glutamine, and the aromatic amino acids.]

Pittard, J., and Gibson, F., 1970. The regulation of biosynthesis of aromatic amino acids and vitamins. *Curr. Top. Cell Regul.* 2:29–63.

Yanofsky, C., and Crawford, I. P., 1972. Tryptophan synthetase. *In* Boyer, P. D., (ed.), *The Enzymes* (3rd ed.), vol. 7, pp. 1–31.

Inherited diseases of heme metabolism:

Macalpine, I., and Hunter, R., 1969. Porphyria and King George III. *Sci. Amer.* 221 (1):38–46. [Available as *Sci. Amer.* Offprint 1149.]

Tschudy, D. P., and Schmid, R., 1972. The porphyrias. *In* Stanbury J. B., Wyngaarden, J. B., and Fredrickson, D. S., (eds.), *The Metabolic Basis of Inherited Disease*, (3rd ed.), pp. 1087–1140. McGraw-Hill.

Nitrogen fixation:

Streicher, S. L., and Valentine, R. C., 1973. Comparative biochemistry of nitrogen fixation. *Annu. Rev. Biochem.* 42:279–302.

Postgate, J. R., (ed.), 1972. *The Chemistry and Biochemistry of Nitrogen Fixation*. Plenum.

Dalton, H., and Mortenson, L. E., 1972. Dinitrogen (N_2) fixation (with a biochemical emphasis). *Bacteriol. Rev.* 36:231–260.

PROBLEMS

1. Write a balanced equation for the synthesis of alanine from glucose.

2. What are the intermediates in the flow of nitrogen from N_2 to heme?

3. What derivative of tetrahydrofolate is a reactant in each of the following conversions:
   (a) Glycine ⟶ serine.
   (b) Histidine ⟶ glutamate.
   (c) Homocysteine ⟶ methionine.

4. In the reaction catalyzed by glutamine synthetase, an oxygen atom is transferred from the side chain of glutamate to orthophosphate, as shown by ^{18}O-labeling studies. Propose an interpretation for this finding.

5. The isolated β_2 subunit of tryptophan synthetase catalyzes the deamination of serine to pyruvate. Propose a catalytic intermediate for this reaction that might also participate in the synthesis of tryptophan.

CHAPTER 22

BIOSYNTHESIS OF NUCLEOTIDES

This chapter deals with the biosynthesis of nucleotides. These compounds participate in nearly all biochemical processes:

1. They are the *activated precursors of DNA and RNA*.

2. Nucleotide derivatives are *activated intermediates in many biosyntheses*. For example, UDP-glucose and CDP-diacylglycerol are precursors of glycogen and phosphoglycerides, respectively.

3. ATP, an adenine nucleotide, is the *universal currency of energy in biological systems*.

4. Adenine nucleotides are *components of three major coenzymes:* NAD$^+$, FAD, and CoA.

5. Nucleotides are *metabolic regulators*. Cyclic AMP, the most ubiquitous one, mediates the action of many hormones. Covalent modifications introduced by ATP alter the activities of some enzymes, as exemplified by the phosphorylation of glycogen synthetase and the adenylylation of glutamine synthetase.

NOMENCLATURE OF BASES, NUCLEOSIDES, AND NUCLEOTIDES

A nucleotide consists of a nitrogeneous base, a sugar, and one or more phosphate groups. The nitrogeneous base is a *purine* or *pyrimidine* derivative.

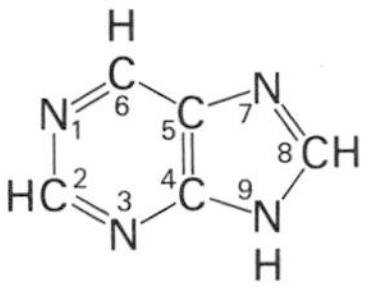

Purine 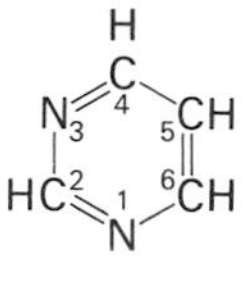   **Pyrimidine**

The two major purines are *adenine* and *guanine,* and the three major pyrimidines are *cytosine, uracil,* and *thymine* (Figure 22-1).

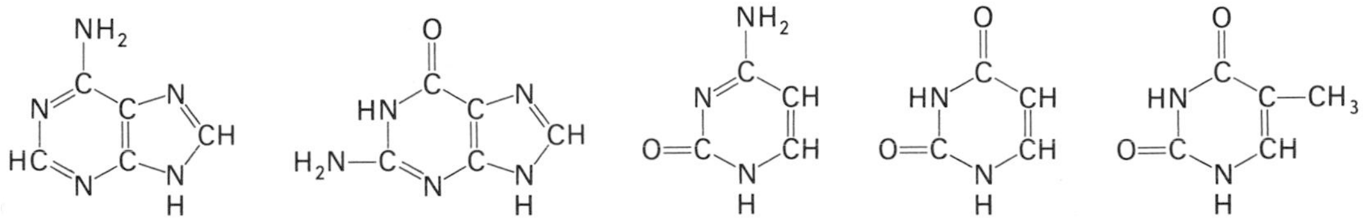

Adenine   **Guanine**   **Cytosine**   **Uracil**   **Thymine**

Figure 22-1
Structures of the major purines and pyrimidines.

A *nucleoside* consists of a purine or pyrimidine base linked to a pentose. The pentose is D-*ribose* or *2-deoxy-D-ribose.*

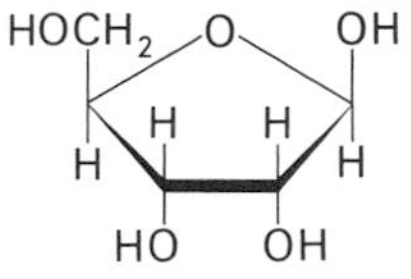

β-D-**Ribose** 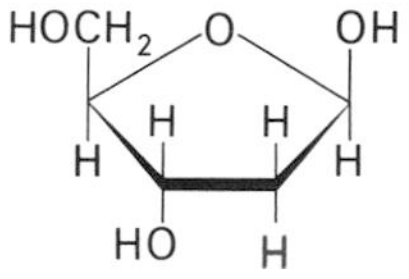   *β*-D-**2-Deoxyribose**

In a nucleoside, the glycosidic C-1 carbon atom of the pentose is bonded to N-1 of the pyrimidine or N-9 of the purine base. The configuration of this *N*-glycosidic linkage is *β* in all naturally occurring nucleosides.

In a ribonucleoside, the pentose is ribose, whereas in a deoxyribonucleoside it is deoxyribose. The major ribonucleosides are *adenosine, guanosine, uridine,* and *cytidine.*

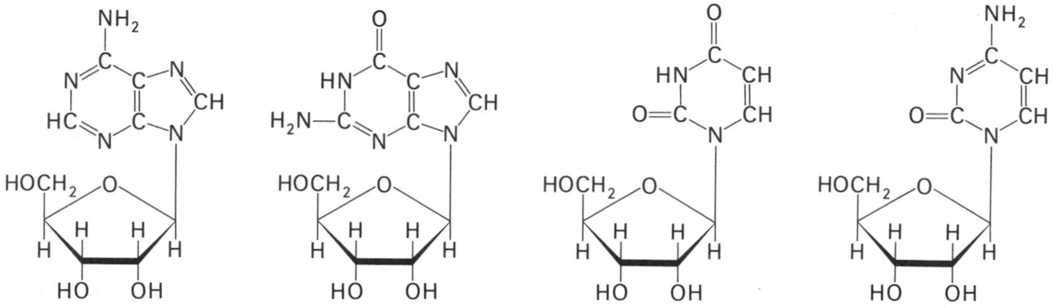

Adenosine **Guanosine** **Uridine** **Cytidine**

Figure 22-2
Structures of the major ribonucleosides.

The major deoxyribonucleosides are *deoxyadenosine, deoxyguanosine, deoxythymidine,* and *deoxycytidine.* Note that uracil is replaced by thymine, its methylated analog, in the deoxy series.

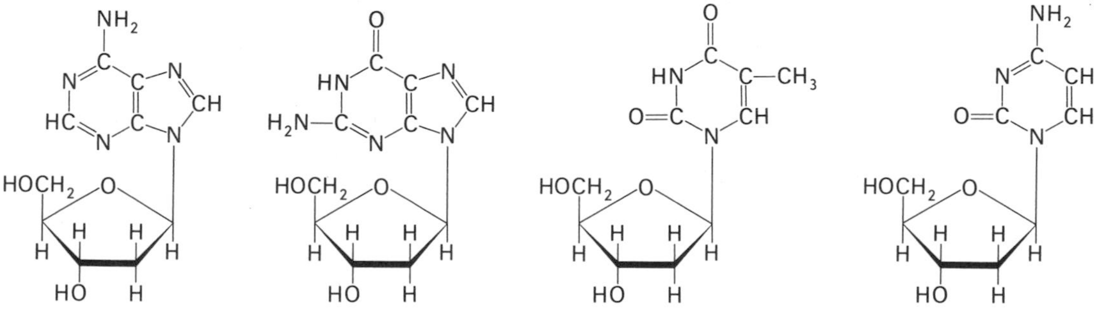

Deoxyadenosine **Deoxyguanosine** **Deoxythymidine** **Deoxycytidine**

Figure 22-3
Structures of the major deoxyribonucleosides.

A *nucleotide* is a phosphate ester of a nucleoside. At least one of the hydroxyl groups of the pentose moiety of a nucleotide is esterified. The most common site of esterification is the hydroxyl group attached to C-5 of the pentose. Such a compound is called a *nucleoside 5′-phosphate* or a *5′-nucleotide.* A primed number designates an atom of the pentose, whereas an unprimed number designates an atom of the purine or pyrimidine ring. The type of pentose is denoted by the prefix in the terms *5'-ribonucleotide* and *5'-deoxyribonucleotide*.

The nucleotide derived by esterification of the 5'-hydroxyl group of adenosine is *adenosine 5'-phosphate*, which is often called *adenylate*. The name adenylic acid is sometimes used, but adenylate is preferred because the phosphate group is ionized at physiological pH. The standard abbreviation for this compound is AMP (for adenosine monophosphate). The common names of the other major 5'-ribonucleotides are *guanylate* (GMP), *uridylate* (UMP), and *cytidylate* (CMP). The major 5'-dexoyribonucleotides are called *deoxyadenylate* (dAMP), *deoxyguanylate* (dGMP), *deoxythymidylate* (dTMP), and *deoxycytidylate* (dCMP). The first letter in their abbreviations is a *d* to denote that they are 2'-deoxyribonucleotides.

In a *nucleoside 5'-diphosphate,* a diphosphate group is esterified to the 5'-hydroxyl of the pentose, whereas in a *nucleoside 5'-triphosphate,* a triphosphate group is linked to this hydroxyl. Thus, the adenine ribonucleotide series is called adenosine 5'-monophosphate (AMP), adenosine 5'-diphosphate (ADP), and adenosine 5'-triphosphate (ATP). The corresponding deoxyribonucleotides are deoxyadenosine 5'-monophosphate (dAMP), deoxyadenosine 5'-diphosphate (dADP), and deoxyadenosine 5'-triphosphate (dATP). The nomenclature of bases, nucleosides, and nucleotides is summarized in Table 22-1.

Adenosine 5'-phosphate (AMP)

Thymidine 5'-phosphate (TMP)

Table 22-1
Nomenclature of bases, nucleosides, and nucleotides

Base	Ribonucleoside	Ribonucleotide (5'-monophosphate)
Adenine (A)	Adenosine	Adenylate (AMP)
Guanine (G)	Guanosine	Guanylate (GMP)
Uracil (U)	Uridine	Uridylate (UMP)
Cytosine (C)	Cytidine	Cytidylate (CMP)

Base	Deoxyribonucleoside	Deoxyribonucleotide (5'-monophosphate)
Adenine (A)	Deoxyadenosine	Deoxyadenylate (dAMP)
Guanine (G)	Deoxyguanosine	Deoxyguanylate (dGMP)
Thymine (T)	Deoxythymidine	Deoxythymidylate (dTMP)
Cytosine (C)	Deoxycytidine	Deoxycytidylate (dCMP)

THE PURINE RING IS SYNTHESIZED FROM AMINO ACIDS, TETRAHYDROFOLATE DERIVATIVES, AND CO_2

The purine ring in purine nucleotides is assembled from a variety of precursors (Figure 22-4). *Glycine* provides C-4, C-5, and N-7. The N-1 atom comes from *aspartate*. The other two nitrogen atoms, N-3 and N-9, come from the amide group of the side chain of *glutamine*. Activated derivatives of *tetrahydrofolate* furnish C-2 and C-8, whereas CO_2 is the source of C-6.

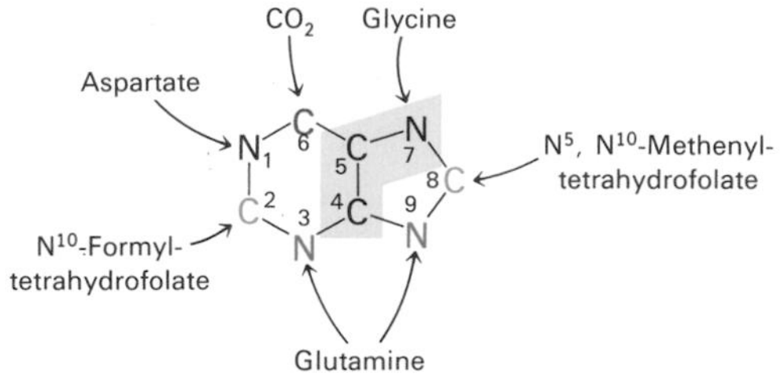

Figure 22-4
Origin of the atoms in the purine ring.

PRPP IS THE DONOR OF THE RIBOSE PHOSPHATE MOIETY OF NUCLEOTIDES

The pathway of purine biosynthesis was elucidated in the 1950s by John Buchanan, G. Robert Greenberg, and others. The ribose phosphate portion of purine and pyrimidine nucleotides comes from *5-phosphoribosyl-1-pyrophosphate* (PRPP). PRPP is also a key intermediate in the biosynthesis of histidine and tryptophan. PRPP is synthesized from ATP and ribose 5-phosphate, which is primarily formed by the pentose phosphate pathway (p. 358). The pyrophosphate group is transferred from ATP to C-1 of ribose 5-phosphate. PRPP has an α configuration.

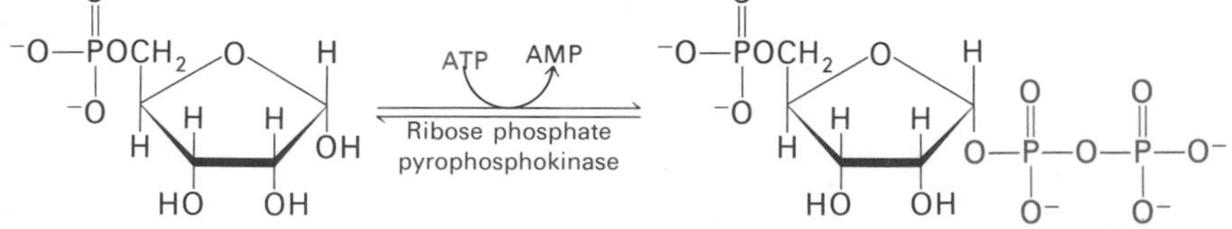

Ribose 5-phosphate **5-Phosphoribosyl-1-pyrophosphate (PRPP)**

THE PURINE RING IS ATTACHED TO RIBOSE PHOSPHATE DURING ITS ASSEMBLY

The committed step in the de novo synthesis of purine nucleotides is the formation of *5-phosphoribosylamine* from PRPP and glutamine. The amino group from the side chain of glutamine displaces the pyrophosphate group attached to C-1 of PRPP with inversion of configuration. The resulting C–N glycosidic bond has the β configuration that is characteristic of naturally occurring nucleotides. This reaction is driven forward by the hydrolysis of pyrophosphate.

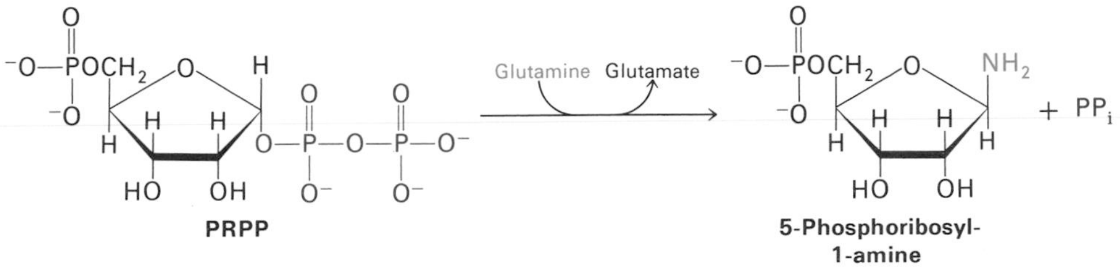

Glycine joins phosphoribosylamine to yield *glycinamide ribonucleotide* (Figure 22-5). An ATP is consumed in the formation of an amide bond between the carboxyl group of glycine and the amino group of phosphoribosylamine. The α-amino terminus of the glycine residue is then formylated by methenyl tetrahydrofolate to give α-N-*formylglycinamide ribonucleotide*. The amide group in this compound is converted into an amidine group. The nitrogen atom is donated by the side chain of glutamine in a reaction that consumes an ATP. *Formylglycinamidine ribonucleotide then undergoes ring closure to form 5-aminoimidazole ribonucleotide.* This intermediate contains the complete five-membered ring of the purine skeleton.

Figure 22-5
First stage of purine biosynthesis: formation of 5-aminoimidazole ribonucleotide from PRPP. The essence of these reactions is (1) displacement of PP_i by the side-chain amino group of glutamine, (2) addition of glycine, (3) formylation by methenyl tetrahydrofolate, (4) transfer of a nitrogen atom from glutamine, and (5) dehydration and ring closure.

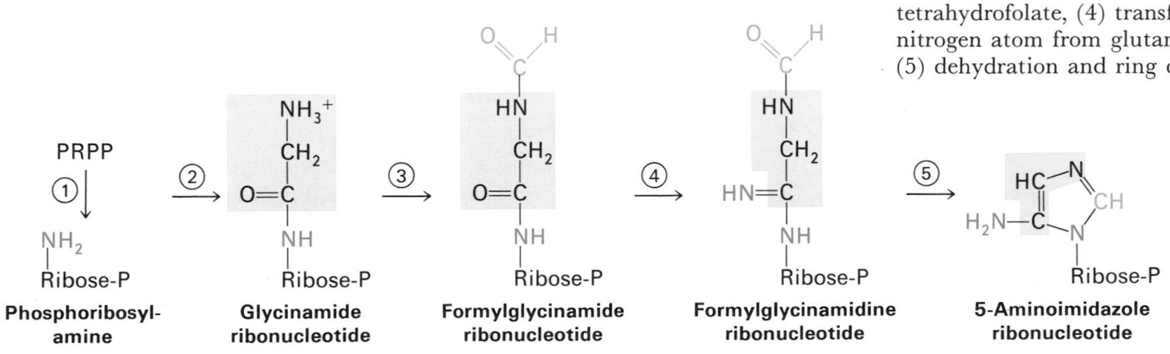

The next phase in the synthesis of the purine skeleton, the formation of a six-membered ring, starts at this point (Figure 22-6). Three of the six atoms of this ring are already present in amino-imidazole ribonucleotide. The other three come from CO_2, aspartate, and formyl tetrahydrofolate. The next carbon atom in the six-membered ring is introduced by the carboxylation of amino-imidazole ribonucleotide, yielding *5-aminoimidazole-4-carboxylate ribonucleotide.*

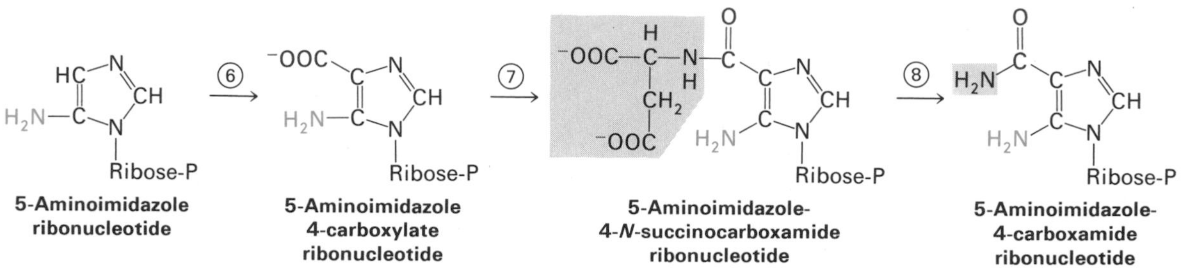

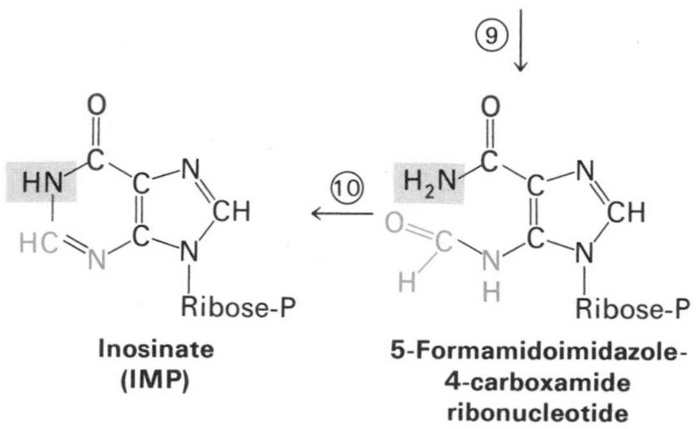

Figure 22-6
Second stage of purine biosynthesis: formation of inosinate from 5-aminoimidazole ribonucleotide. The essence of these reactions is (6) carboxylation, (7) addition of aspartate, (8) elimination of fumarate (leaving the amino group of aspartate), (9) formylation by N^{10}-formyltetrahydrofolate, and (10) dehydration and ring closure.

The amino group of aspartate then reacts with the carboxyl group of this intermediate to form *5-aminoimidazole-4-*N-*succinocarboxamide ribonucleotide.* An ATP is consumed in the formation of this amide bond. The carbon skeleton of the aspartate moiety comes off as fumarate in the next reaction, yielding *5-aminoimidazole-4-carboxamide ribonucleotide.* Note that the result of these two reactions is the conversion of a carboxylate to an amide. Thus, *aspartate contributes only its nitrogen atom to the purine ring.* The final atom of the purine ring is contributed by N^{10}-formyltetrahydrofolate. The resulting

5-formamidoimidazole-4-carboxamide ribonucleotide undergoes dehydra-tion and ring closure to form *inosinate* (IMP), which contains a complete purine ring. The purine-base part of inosinate is called *hypoxanthine*.

AMP AND GMP ARE FORMED FROM IMP

Inosinate is the precursor of AMP and GMP (Figure 22-7). *Adenylate* is synthesized from inosinate by the insertion of an amino group at C-6 in place of the carbonyl oxygen. Aspartate again contributes its amino group by addition of this amino acid followed by elimi-nation of fumarate. GTP is the donor of a high-energy phosphate bond in the synthesis of *adenylosuccinate* from inosinate and aspartate. The removal of fumarate from adenylosuccinate and from 5-amino-imidazole-4-*N*-succinocarboxamide ribonucleotide are catalyzed by the same enzyme.

Guanylate (GMP) is synthesized by the oxidation of inosinate, followed by the insertion of an amino group at C-2. NAD^+ is the hydrogen acceptor in the oxidation of inosinate to xanthylate (XMP). The amino group in the side chain of glutamine is then

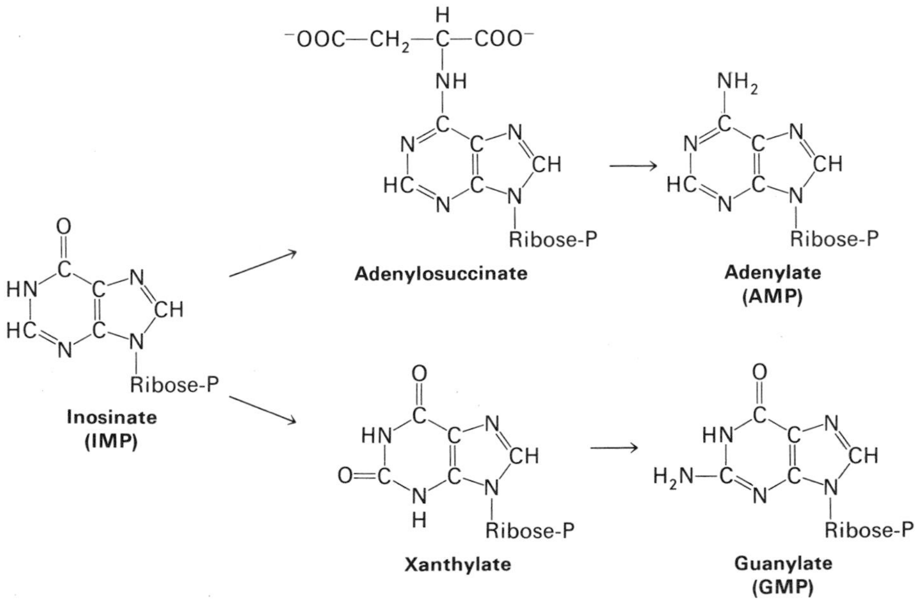

Figure 22-7
AMP and GMP are synthesized from IMP.

transferred to xanthylate. Two high-energy phosphate bonds are consumed in this reaction, since ATP is cleaved to AMP and PP_i, which is subsequently hydrolyzed.

PURINE BASES CAN BE RECYCLED BY SALVAGE REACTIONS THAT UTILIZE PRPP

Free purine bases are formed by the hydrolytic degradation of nucleic acids and nucleotides. Purine nucleotides can be synthesized from these preformed bases by a *salvage reaction*, which is simpler and much less costly than the reactions of the *de novo pathway* discussed above. In the salvage reaction, the ribose phosphate moiety of PRPP is transferred to the purine to form the corresponding nucleotide:

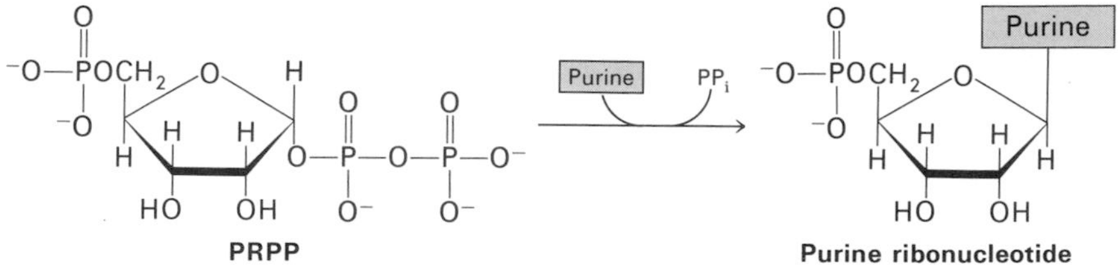

There are two salvage enzymes with different specificities. *Adenine phosphoribosyl transferase* catalyzes the formation of adenylate:

$$\text{Adenine} + \text{PRPP} \longrightarrow \text{adenylate} + PP_i$$

whereas *hypoxanthine-guanine phosphoribosyl transferase* catalyzes the formation of inosinate and guanylate.

$$\text{Hypoxanthine} + \text{PRPP} \longrightarrow \text{inosinate} + PP_i$$

$$\text{Guanine} + \text{PRPP} \longrightarrow \text{guanylate} + PP_i$$

The versatile and efficient use of the purine ring is also evident in the biosynthesis of histidine. The six-membered portion of the purine ring of ATP contributes part of the imidazole ring of histidine (p. 515). The rest of the purine skeleton is not discarded. Rather, it is conserved in 5-aminoimidazole-4-carboxamide ribonucleotide, an intermediate in the de novo pathway of purine biosynthesis.

AMP AND GMP ARE FEEDBACK INHIBITORS OF PURINE NUCLEOTIDE BIOSYNTHESIS

The synthesis of purine nucleotides is controlled by feedback inhibition at several sites (Figures 22-8).

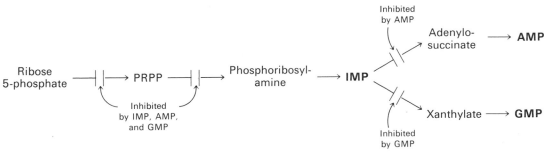

Figure 22-8
Control of purine biosynthesis.

1. Feedback inhibition of *5-phosphoribosyl-1-pyrophosphate synthetase* by purine nucleotides regulates the level of PRPP. This synthetase is inhibited by AMP, GMP, and IMP.

2. The committed step in purine nucleotide biosynthesis is the conversion of PRPP to phosphoribosylamine by transfer of the side-chain amino group of glutamine. *Glutamine PRPP amidotransferase is feedback inhibited by many purine ribonucleotides.* It is noteworthy that AMP and GMP, the final products of the pathway, are synergistic in inhibiting this enzyme.

3. Inosinate is the branch point in the synthesis of AMP and GMP. *The reactions leading away from inosinate are sites of feedback inhibition.* AMP inhibits the conversion of inosinate to adenylosuccinate, its immediate precursor. Similarly, GMP inhibits the conversion of inosinate to xanthylate, its immediate precursor.

4. GTP is a substrate in the synthesis of AMP, whereas ATP is a substrate in the synthesis of GMP. This *reciprocal substrate relationship* tends to balance the synthesis of adenine and guanine ribonucleotides.

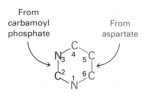

Figure 22-9
Origin of the atoms in the pyrimidine ring. C-2 and N-3 come from carbamoyl phosphate, whereas the other atoms of the ring come from aspartate.

THE PYRIMIDINE RING IS SYNTHESIZED FROM CARBAMOYL PHOSPHATE AND ASPARTATE

The pyrimidine ring is assembled first and then linked to ribose phosphate to form a pyrimidine nucleotide, in contrast to the reaction sequence in the de novo synthesis of purine nucleotides. PRPP is the donor of ribose phosphate in the synthesis of pyrimidine nucleotides and purine nucleotides. The precursors of the pyrimidine ring are carbamoyl phosphate and aspartate (Figure 22-9).

The synthesis of carbamoyl phosphate is compartmentalized. Carbamoyl phosphate consumed in the synthesis of pyrimidines is formed in the cytosol, whereas carbamoyl phosphate used in the synthesis of urea is formed in mitochondria. Furthermore, glutamine instead of NH_4^+ is the nitrogen donor in the cytosol synthesis of carbamoyl phosphate.

$$\text{Glutamine} + 2\,\text{ATP} + \text{HCO}_3^- \longrightarrow$$
$$\text{carbamoyl phosphate} + 2\,\text{ADP} + \text{P}_i + \text{glutamate}$$

The committed step in the biosynthesis of pyrimidines is the formation of N-*carbamoylaspartate from aspartate and carbamoyl phosphate.* This carbamoylation is catalyzed by *aspartate transcarbamoylase,* an especially interesting regulatory enzyme (see p. 540).

Carbamoyl phosphate   **Aspartate**   **N-Carbamoylaspartate**

The pyrimidine ring is formed in the next reaction, in which carbamoylaspartate cyclizes with loss of water to yield *dihydroorotate. Orotate* is then formed by dehydrogenation of dihydroorotate.

N-Carbamoylaspartate   **Dihydroorotate**   **Orotate**

OROTATE ACQUIRES A RIBOSE PHOSPHATE MOIETY FROM PRPP

The next step in the synthesis of pyrimidine nucleotides is the *acquisition of a ribose phosphate group*. Orotate (a free pyrimidine) reacts with PRPP to form *orotidylate* (a pyrimidine nucleotide). This reaction, which is catalyzed by orotidylate pyrophosphorylase, is driven forward by the hydrolysis of pyrophosphate. Orotidylate is then decarboxylated to yield *uridylate* (UMP), a major pyrimidine nucleotide.

Orotate   Orotidylate   Uridylate
                                                           (UMP)

NUCLEOSIDE MONO-, DI-, AND TRIPHOSPHATES ARE INTERCONVERTIBLE

The active forms of nucleotides in biosyntheses and energy conversions are the diphosphates and triphosphates. Nucleoside monophosphates are converted by specific *nucleoside monophosphate kinases* that utilize ATP as the phosphoryl donor. For example, UMP is phosphorylated by *UMP kinase.*

$$UMP + ATP \rightleftharpoons UDP + ADP$$

AMP, ADP, and ATP are interconverted by *adenylate kinase* (also called myokinase). The equilibrium constants of these reactions are close to 1.

$$AMP + ATP \rightleftharpoons ADP + ADP$$

Nucleoside diphosphates and triphosphates are interconverted by *nucleoside diphosphate kinase,* an enzyme that has broad specificity, in contrast to the monophosphate kinases. In the equation below, X and Y can be any of several ribonucleosides or deoxyribonucleosides.

$$XDP + YTP \rightleftharpoons XTP + YDP$$

For example,

$$UDP + ATP \rightleftharpoons UTP + ADP$$

CTP IS FORMED BY AMINATION OF UTP

Cytidine triphosphate (CTP) is derived from uridine triphosphate (UTP), the other major pyrimidine ribonucleotide. The carbonyl oxygen at C-4 is replaced by an amino group. In mammals, the side chain of glutamine is the amino donor, whereas NH_4^+ is used in this reaction in *E. coli*. An ATP is consumed in both amination reactions.

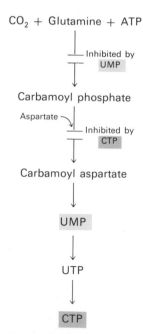

Figure 22-10
Control of pyrimidine biosynthesis.

PYRIMIDINE NUCLEOTIDE BIOSYNTHESIS IS REGULATED BY FEEDBACK INHIBITION

The committed step in pyrimidine nucleotide biosynthesis in *E. coli* is the formation of *N*-carbamoylaspartate from aspartate and carbamoyl phosphate. *Aspartate transcarbamoylase, the enzyme that catalyzes this reaction, is feedback inhibited by CTP, the final product in the pathway.* A second control site is *carbamoyl phosphate synthetase,* which is feedback inhibited by UMP (Figure 22-10).

The allosteric properties of aspartate transcarbamoylase (ATCase) have been intensively investigated by John Gerhart and

Howard Schachman. *The binding of carbamoyl phosphate and aspartate are cooperative,* as reflected in the sigmoidal dependence of the reaction velocity on substrate concentration (Figure 22-11). *CTP inhibits the enzyme by decreasing its affinity for substrates* without affecting its V_{max}. The extent of inhibition exerted by CTP, which may reach 90%, depends on the concentrations of the substrates. In contrast, *ATP is an activator of ATCase.* The affinity of the enzyme for its substrates is enhanced by ATP, whereas V_{max} is unaffected. Furthermore, the binding of ATP and CTP to the regulatory site of ATCase is competitive. High levels of ATP displace CTP from the enzyme so that it cannot exert its inhibitory effect.

The biological significance of the activation of ATCase by ATP is twofold. First, *it tends to equalize the rates of formation of purine and pyrimidine nucleotides.* Comparable quantities of these two types of nucleotides are needed for the synthesis of nucleic acids. Second, *activation by ATP signals its availability as a substrate* for some of the reactions of pyrimidine nucleotide biosynthesis, such as the synthesis of carbamoyl phosphate and the phosphorylations of UMP to UTP.

ASPARTATE TRANSCARBAMOYLASE CONSISTS OF SEPARABLE CATALYTIC AND REGULATORY SUBUNITS

The regulatory properties of ATCase vanish when the enzyme is treated with mercurials such as *p*-hydroxymercuribenzoate. ATP and CTP no longer have any effect on catalytic activity. Furthermore, the binding of substrates becomes noncooperative. However, the modified enzyme has full catalytic activity. This loss of regulatory properties with retention of enzymatic activity is called *desensitization.*

The desensitization of ATCase by mercurials is accompanied by its *dissociation into two kinds of subunits,* as shown by ultracentrifuge studies (Figure 22-12). The sedimentation coefficient of the native enzyme is 11.6 S, whereas that of the dissociated subunits is 2.8 S and 5.8 S. These subunits can readily be separated by ion-exchange chromatography because they differ markedly in charge, or by centrifugation in a sucrose density gradient because they differ in size. *p*-Hydroxymercuribenzoate, the dissociating agent, can be removed after the subunits are separated. The larger of the subunits,

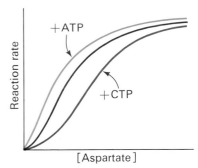

Figure 22-11
Allosteric effects in aspartate transcarbamoylase. ATP is an activator, whereas CTP is an inhibitor. [Based on J. C. Gerhart. *Curr. Top. Cell Regul.* 2(1970):275.]

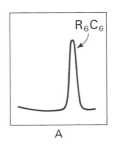

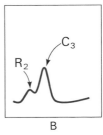

Figure 22-12
Sedimental velocity patterns of (A) native ATCase and (B) the enzyme dissociated by a mercurial into regulatory and catalytic subunits. [Based on J. C. Gerhart and H. K. Schachman. *Biochemistry* 4(1965):1054.]

called the *catalytic subunit,* is catalytically active. However, the activity of the isolated catalytic subunit is not affected by ATP and CTP. The smaller of the subunits, called the *regulatory subunit,* is devoid of catalytic activity but contains specific binding sites for CTP and ATP. The catalytic subunit consists of 3 polypeptide chains (each 35,000 daltons), whereas the regulatory subunit is made up of two polypeptide chains (each 17,000 daltons).

The catalytic and regulatory subunits combine rapidly when they are mixed. The resulting complex has the same structure, R_6C_6, as the native enzyme.

$$3\,R_2 + 2\,C_3 \longrightarrow R_6C_6$$

Furthermore, *the reconstituted enzyme has the same allosteric properties as the native enzyme*.

X-ray crystallographic studies of ATCase are in progress in William Lipscomb's laboratory. An electron-density map at 5.5-Å resolution shows that the two catalytic trimers (C_3) are above and below an equatorial belt of three regulatory dimers (R_2) (Figure 22-13). A distinctive feature of the molecule is that it contains a large central cavity, which is accessible through several channels.

Figure 22-13
A schematic view of ATCase expanded along the vertical axis, based on an electron-density map at 5.5-Å resolution. The catalytic subunits (shown in blue) are above and below the regulatory subunits (shown in red). There is a large central aqueous cavity (shown in yellow) about 50 by 50 by 25 Å in size. The molecule is about 100 Å long. This diagram does not indicate the mode of assembly of the enzyme complex. [Based on S. G. Warren, B. F. P. Edwards, D. R. Evans, D. C. Wiley, and W. N. Lipscomb. *Proc. Nat. Acad. Sci.* 70(1973):1118.]

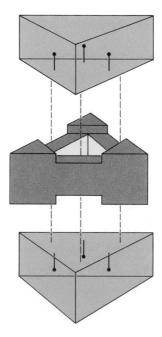

DEOXYRIBONUCLEOTIDES ARE SYNTHESIZED BY REDUCTION OF RIBONUCLEOSIDE DIPHOSPHATES

The substrates in the synthesis of deoxyribonucleotides are the ribonucleoside diphosphates. The 2′-hydroxyl group on the sugar moiety is replaced by a hydrogen atom. The net stoichiometry is:

Ribonucleoside diphosphate + NADPH + H⁺ ⟶
deoxyribonucleoside diphosphate + NADP⁺ + H₂O

Peter Reichard has shown that this reaction is catalyzed by an enzyme system composed of *four proteins: thioredoxin, thioredoxin reductase, and enzymes B₁ and B₂* (Figure 22-14). The immediate hydrogen donor in the reduction of the ribose moiety is *reduced thioredoxin*. This small protein contains two sulfhydryl residues in close proximity. These sulfhydryls are oxidized to a disulfide as the ribose unit is reduced to deoxyribose. This reaction is catalyzed by *enzymes B₁ and B₂*. NADPH then reacts with oxidized thioredoxin to regenerate reduced thioredoxin. This reaction is catalyzed by *thioredoxin reductase,* a flavoprotein. Thus, *thioredoxin is a protein cofactor in ribonucleotide reduction.* It is alternately reduced by thioredoxin reductase and oxidized by enzymes B₁ and B₂.

   The reduction of ribonucleoside diphosphates is allosterically controlled. Enzymes B₁ and B₂ bind a number of nucleotide effectors. *dATP and dTTP are feedback inhibitors of the reduction of all four ribonucleoside diphosphates.* In contrast, ATP enhances the reduction of ADP and CDP, and dGTP accelerates the reduction of ADP and GDP. This complex pattern of regulation tends to balance the supply of the four deoxyribonucleotides needed for the synthesis of DNA.

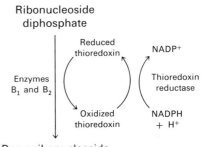

Figure 22-14
Enzyme system for the reduction of ribonucleoside diphosphates to deoxyribonucleoside diphosphates.

DEOXYTHYMIDYLATE IS FORMED BY METHYLATION OF DEOXYURIDYLATE

Uracil is not a component of DNA. Rather, DNA contains *thymine,* the methylated analog of uracil. Deoxyuridylate (dUMP) is methylated to deoxythymidylate (dTMP) by N^5, N^{10}-methylenetetrahydrofolate. This coenzyme serves both as an *electron donor* and as a *one-carbon donor* in the methylation reaction. Dihydrofolate, rather than tetrahydrofolate, is a product.

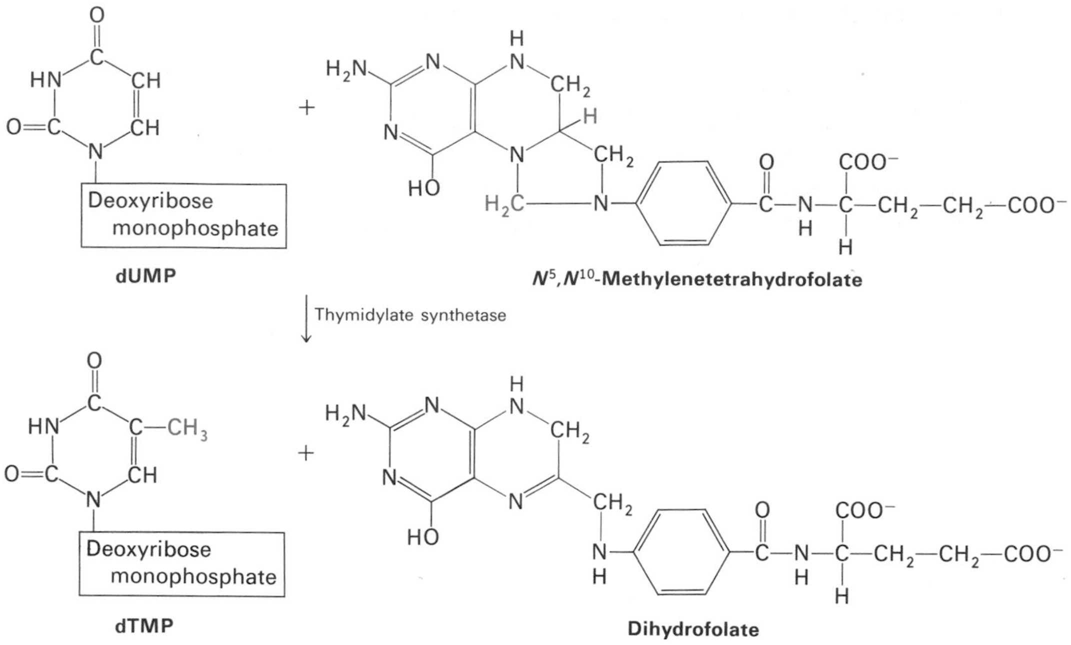

dUMP + **N^5,N^{10}-Methylenetetrahydrofolate**

Thymidylate synthetase

dTMP + **Dihydrofolate**

Tetrahydrofolate is regenerated by the reduction of dihydrofolate.

$$\text{Dihydrofolate} + \text{NADPH} + \text{H}^+ \longrightarrow \text{tetrahydrofolate} + \text{NADP}^+$$

This reaction is catalyzed by *dihydrofolate reductase*. Analogs of dihydrofolate, such as aminopterin and amethopterin (methotrexate), are potent inhibitors of this enzyme. *Amethopterin is a valuable drug in the treatment of acute leukemia and choriocarcinoma.* The rationale for using antifolate drugs in the chemotherapy of cancers is that rapidly growing cells require large amounts of thymidylate for incorporation into DNA. The conversion of dUMP to dTMP is blocked if tetrahydrofolate cannot be regenerated.

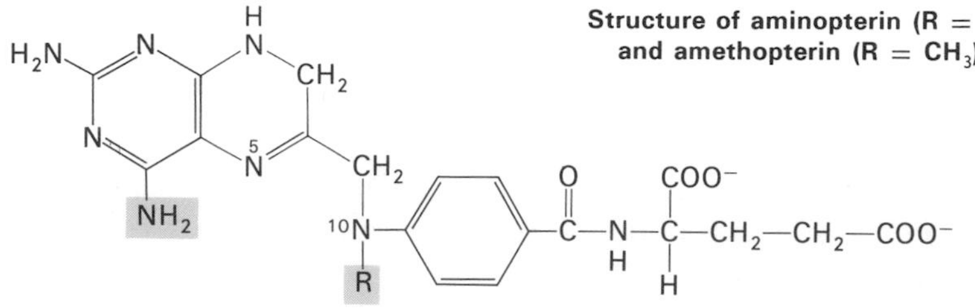

Structure of aminopterin (R = H) and amethopterin (R = CH$_3$)

ATP IS A PRECURSOR OF NAD⁺, FAD, AND COENZYME A

The *biosynthesis of NAD⁺* starts with the formation of *nicotinate ribonucleotide* from nicotinate and PRPP. *Nicotinate* (also called *niacin*) is derived from tryptophan. Humans can synthesize the required amount of nicotinate if the supply of tryptophan in the diet is adequate. However, an exogeneous supply of nicotinate is required if the dietary intake of tryptophan is low. *Pellagra* is a deficiency disease caused by a dietary insufficiency of tryptophan and nicotinate.

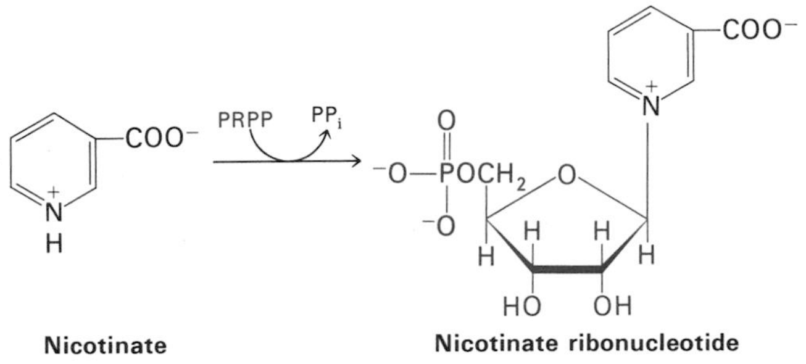

Nicotinate   **Nicotinate ribonucleotide**

An AMP moiety is transferred from ATP to nicotinate ribonucleotide to form *desamido-NAD⁺*. The final step is the transfer of the amide group of glutamine to the nicotinate carboxyl group to form NAD⁺ (Figure 22-15). NADP⁺ is derived from NAD⁺ by phosphorylation of the 2'-hydroxyl group of the adenine ribose moiety. This transfer of a phosphoryl group is catalyzed by NAD⁺ kinase.

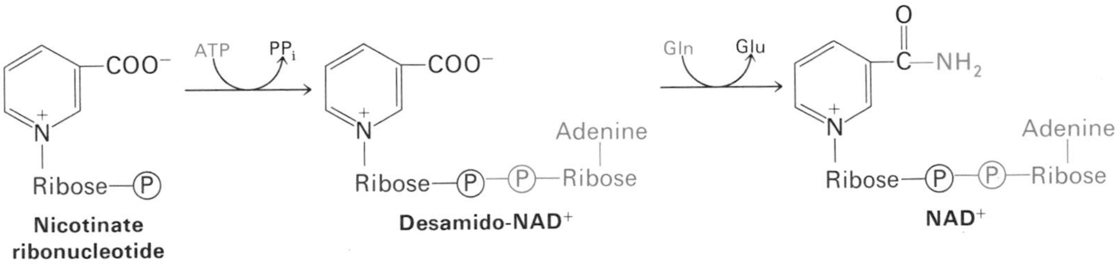

Nicotinate ribonucleotide   **Desamido-NAD⁺**   **NAD⁺**

Figure 22-15
Synthesis of NAD⁺ from nicotinate ribonucleotide.

Flavin adenine dinucleotide (FAD) is synthesized from riboflavin and two molecules of ATP. Riboflavin is phosphorylated by ATP to give *riboflavin 5'-phosphate* (also called *flavin mononucleotide*). FAD is then

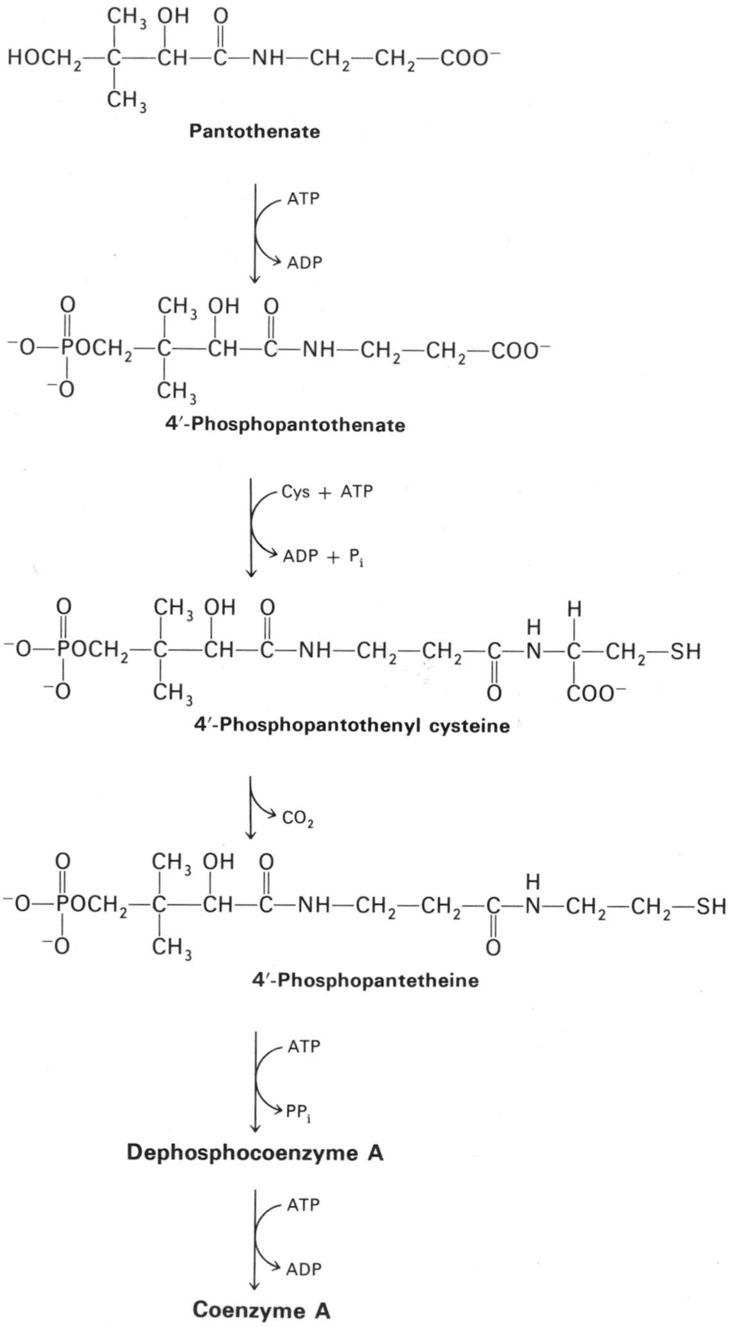

Figure 22-16
Synthesis of coenzyme A from pantothenate.

formed by the transfer of an AMP moiety from a second molecule of ATP to riboflavin 5′-phosphate.

$$\text{Riboflavin} + \text{ATP} \longrightarrow \text{Riboflavin 5′-phosphate} + \text{ADP}$$

$$\text{Riboflavin 5′-phosphate} + \text{ATP} \rightleftharpoons$$
$$\text{Flavin adenine dinucleotide} + \text{PP}_i$$

The *synthesis of coenzyme A* in animals starts with the phosphorylation of *pantothenate* (Figure 22-16). Pantothenate is required in the diet of animals. It is synthesized by plants and microorganisms. A peptide bond is formed between the carboxyl group of 4′-phosphopantothenate and the amino group of cysteine. The carboxyl group of the cysteine moiety is lost, yielding *4′-phosphopantotheine*. The AMP moiety of ATP is then transferred to this intermediate, forming *dephosphocoenzyme A*. Finally, phosphorylation of its 3′-hydroxyl group yields coenzyme A.

A common feature of the biosyntheses of NAD^+, FAD, and CoA is the *transfer of the AMP moiety of ATP to the phosphate group of a phosphorylated intermediate*. The pyrophosphate formed in this reaction is hydrolyzed to orthophosphate. This is a recurring motif in biochemistry: *biosynthetic reactions are frequently driven by the hydrolysis of the released pyrophosphate*.

PURINES ARE DEGRADED TO URATE IN HUMANS

The nucleotides of a cell undergo continuous turnover. Nucleotides are hydrolytically degraded to nucleosides by *nucleotidases*. Phosphorolytic cleavage of nucleosides to free bases and ribose 1-phosphate (or deoxyribose 1-phosphate) is catalyzed by *nucleoside phosphorylases*. Ribose 1-phosphate is isomerized by *phosphoribomutase* to ribose 5-phosphate, a substrate in the synthesis of PRPP. Some of the bases are reused to form nucleotides by salvage pathways.

The pathway for the degradation of AMP (Figure 22-17) includes an additional step. AMP is deaminated to IMP by *adenylate deaminase*. The subsequent reactions leading to the free base hypoxanthine follow the general pattern. *Xanthine oxidase*, a molybdenum and iron-containing flavoprotein, then oxidizes hypoxanthine to *xanthine* and then to *urate*. Molecular oxygen, the oxidant in both reactions, is reduced to H_2O_2, which is decomposed to H_2O and

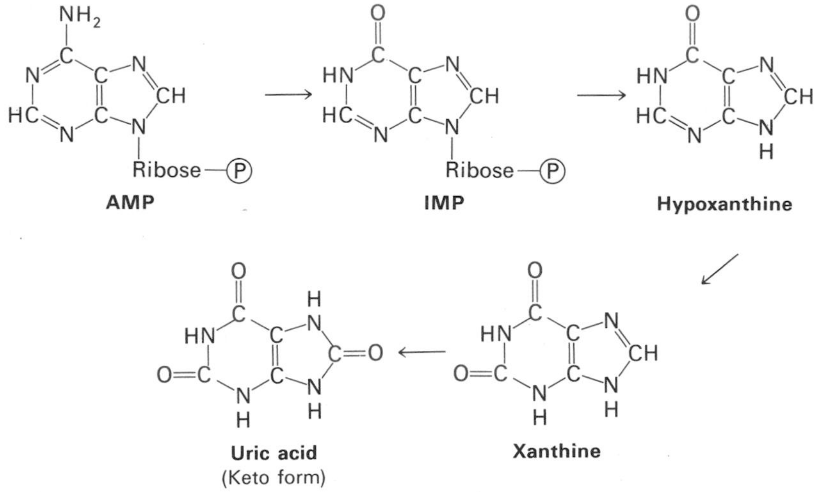

Figure 22-17
Degradation of AMP to uric acid.

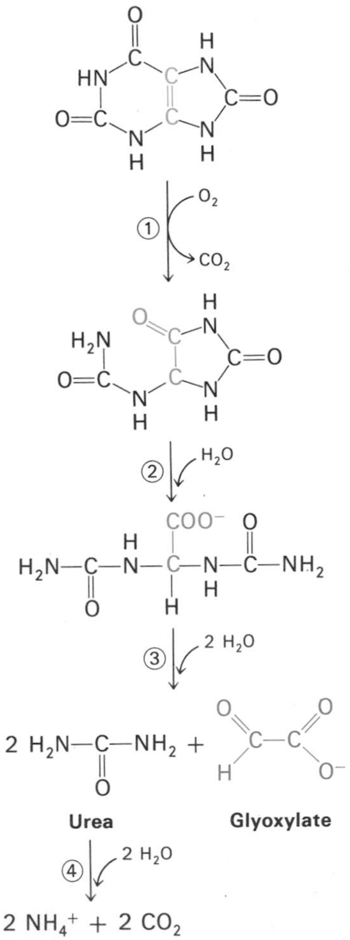

O$_2$ by catalase. Xanthine is also an intermediate in the formation of urate from guanine. *In humans, urate is the final product of purine degradation and is excreted in the urine.*

URATE IS FURTHER DEGRADED IN SOME ORGANISMS

The breakdown of purines proceeds further in some species (Figure 22-18). Mammals other than primates excrete *allantoin,* which is formed by oxidation of urate. Teleost fish excrete *allantoate,* which is formed by hydration of allantoin. The degradation proceeds a step further in amphibians and most fish. Allantoate is hydrolyzed to two molecules of *urea* and one of *glyoxylate.* Finally, some marine invertebrates hydrolyze urea to NH_4^+ and CO_2. It seems likely that the enzymes catalyzing these reactions were progressively lost in the evolution of primates.

BIRDS AND TERRESTRIAL REPTILES EXCRETE URATE INSTEAD OF UREA TO CONSERVE WATER

In terrestrial reptiles and birds, urea is not the final product of amino nitrogen metabolism. These animals synthesize purines from their excess amino nitrogen and then degrade these purines to urate.

Figure 22-18
Degradation of urate to NH_4^+ and CO_2. These reactions are catalyzed by (1) uricase, (2) allantoinase, (3) allantoicase, and (4) urease.

This circuitous path for the elimination of amino nitrogen serves a vital function: the conservation of water. Urate is the vehicle for the excretion of amino nitrogen because of its very low solubility at acid pH. The pK$_a$ of the most acidic group in uric acid is 5.4.

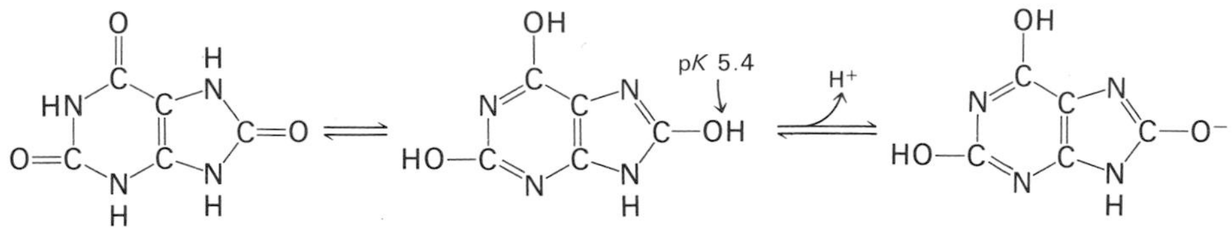

Uric acid (Keto form)   Uric acid (Enol form)   Urate

The acidic urine of birds and terrestrial reptiles consists of a paste of crystals of uric acid. Little water accompanies the excretion of these crystals. In contrast, the excretion of a comparable amount of highly soluble urea would be accompanied by a large efflux of water.

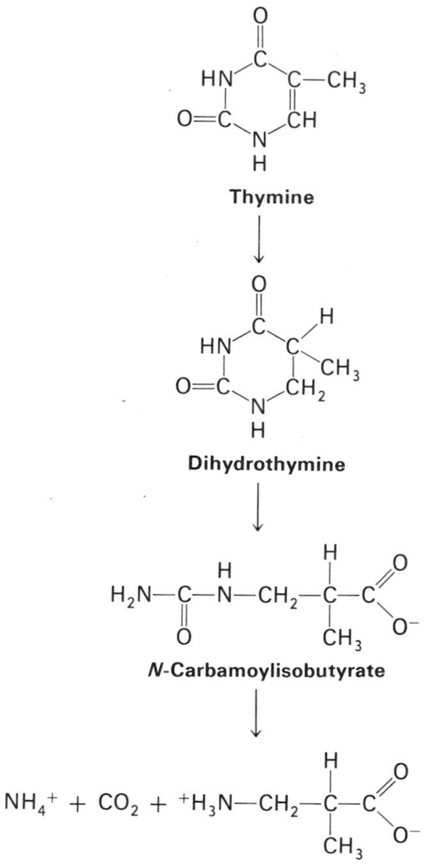

DEGRADATION OF PYRIMIDINES

The degradation of thymine (Figure 22-19) is illustrative of the breakdown of pyrimidines. Thymine is degraded to β-aminoiso-butyrate, which is metabolized as though it were an amino acid. The amino group is removed by transamination to yield methyl-malonate semialdehyde, which is converted to methylmalonyl CoA. The conversion of methylmalonyl CoA to succinyl CoA, the point of entry into the citric acid cycle, has already been discussed (p. 444).

EXCESSIVE PRODUCTION OF URATE IS A CAUSE OF GOUT

Gout is a disease that affects the joints and leads to arthritis. The major biochemical feature of gout is an *elevated level of urate in the serum.* Inflammation of the joints is triggered by the *precipitation of sodium urate crystals.* Kidney disease may also occur because of the deposition of urate crystals in that organ. Gout primarily affects adult males. The biochemical lesion in most cases of gout has not

Thymine

Dihydrothymine

N-Carbamoylisobutyrate

β-Aminoisobutyrate

Figure 22-19
Degradation of thymine.

been elucidated. It seems likely that gout is an expression of a variety of inborn errors of metabolism in which *excessive production of urate* is a common finding. Some patients with this abnormality have a partial deficiency of *hypoxanthine-guanine phosphoribosyl transferase,* the enzyme that catalyzes the salvage synthesis of IMP and GMP.

$$\text{Hypoxanthine} + \text{PRPP} \longrightarrow\!\!\!|\,|\!\longrightarrow \quad \text{IMP} \quad + \text{PP}_i$$
$$\text{(or guanine)} \qquad\qquad\qquad\qquad \text{(or GMP)}$$

Deficiency of this enzyme leads to reduced synthesis of GMP and IMP by the salvage pathway and an increase in the level of PRPP. There is a *marked acceleration of purine biosynthesis by the de novo pathway.* *Allopurinol,* an analog of hypoxanthine, is used to treat gout.

Allopurinol   **Hypoxanthine**

Allopurinol is an *inhibitor of xanthine oxidase.* The synthesis of urate from xanthine decreases soon after the administration of this drug.

The serum concentrations of hypoxanthine and xanthine increase after administration of allopurinol, whereas that of urate drops. The formation of uric acid stones is virtually abolished by allopurinol, and there is some improvement in the arthritis. Also, *there is a decrease in the total rate of purine biosynthesis.*

LESCH-NYHAN SYNDROME: SELF-MUTILATION, MENTAL RETARDATION, AND EXCESSIVE PRODUCTION OF URATE

A nearly total absence of hypoxanthine-guanine phosphoribosyl transferase has devastating consequences. The most striking expression of this inborn error of metabolism, called the *Lesch-Nyhan syndrome,* is *compulsive self-destructive behavior.* At age two or three, children with this disease begin to bite their fingers and lips. The tendency to self-mutilate is so extreme that it is necessary to protect these patients by such measures as wrapping their hands in gauze. These individuals also tend to be aggressive toward others. *Mental deficiency and spasticity* are other characteristics of the Lesch-Nyhan syndrome. Elevated levels of urate in the serum lead to the formation of stones early in life, followed by the symptoms of *gout* years later. The disease is inherited as a sex-linked recessive.

The biochemical consequences of the virtual absence of hypoxanthine-guanine phosphoribosyl transferase are an *overproduction of urate* and an *elevated concentration of PRPP.* Also, there is a marked increase in the rate of purine biosynthesis by the de novo pathway. The relationships between the absence of the transferase and the bizarre neurologic signs are an enigma. The brain may be very dependent on the salvage pathway for the synthesis of IMP and GMP. The normal level of hypoxanthine-guanine phosphoribosyl transferase is higher in the brain than in any other tissue. In contrast, the activity of the amidotransferase that catalyzes the committed step in the de novo pathway is rather low in the brain. Allopurinol is effective in diminishing urate synthesis in the Lesch-Nyhan syndrome. However, it has no effect on the rate of de novo synthesis of purines and it fails to alleviate the neurologic expressions of the disease.

The Lesch-Nyhan syndrome demonstrates that the salvage pathway for the synthesis of IMP and GMP is not gratuitous. The salvage pathway evidently serves a critical role that is not yet fully understood. Furthermore, the interplay between the de novo and salvage pathways of purine synthesis remains to be elucidated. Moreover, the Lesch-Nyhan syndrome reveals that abnormal behavior such as self-mutilation and extreme hostility can be caused by the absence of a single enzyme. This finding has important implications for the future development of psychiatry.

SUMMARY

The purine ring is assembled from a variety of precursors: glutamine, glycine, aspartate, methenyltetrahydrofolate, N^{10}-formyltetrahydrofolate, and CO_2. The committed step in the de novo synthesis of purine nucleotides is the formation of 5-phosphoribosylamine from PRPP and glutamine. The purine ring is attached to ribose phosphate during its assembly. The addition of glycine, followed by formylation, amination, and ring closure, yields 5-aminoimidazole ribonucleotide. This intermediate contains the completed five-membered ring of the purine skeleton. The addition of CO_2, the nitrogen atom of aspartate, and a formyl group, followed by ring closure, yields inosinate (IMP), a purine ribonucleotide. AMP and GMP are formed from IMP. Purine ribonucleotides can also be synthesized by a salvage pathway in which a preformed base reacts directly with PRPP. Feedback inhibition of 5-phosphoribosyl-1-pyrophosphate synthetase and of glutamine-PRPP amidotransferase by purine nucleotides is important in regulating their biosynthesis.

The pyrimidine ring is assembled first and then linked to ribose phosphate to form a pyrimidine nucleotide, in contrast to the sequence in the de novo synthesis of purine nucleotides. PRPP is again the donor of the ribose phosphate moiety. The synthesis of the pyrimidine ring starts with the formation of carbamoylaspartate from carbamoyl phosphate and aspartate, a reaction catalyzed by aspartate transcarbamoylase. Dehydration, cyclization, and oxidation yields orotate, which reacts with PRPP to give orotidylate. Decarboxylation of this pyrimidine nucleotide yields UMP. CTP is then formed by amination of UTP. Pyrimidine biosynthesis in *E. coli* is regulated by feedback inhibition of aspartate transcarbamoylase, the enzyme that catalyzes the committed step. CTP inhibits and ATP stimulates this enzyme. Aspartate transcarbamoylase consists of separable regulatory and catalytic subunits.

Deoxyribonucleotides, the precursors of DNA, are formed by the reduction of ribonucleoside diphosphates. These conversions are catalyzed by an enzyme system consisting of thioredoxin, thioredoxin reductase, and enzymes B_1 and B_2. dTMP is formed by the methylation of dUMP. The one-carbon and electron donor in this reaction is N^5, N^{10}-methylenetetrahydrofolate, which is converted to dihydrofolate. In turn, tetrahydrofolate is regenerated by the reduction of dihydrofolate. Dihydrofolate reductase, which catalyzes this reaction, is inhibited by folate analogs such as aminopterin and amethopterin (methotrexate). These compounds are used as antitumor drugs.

SELECTED READINGS

Books and reviews:

Henderson, J. F., and Paterson, A. R. P., 1973. *Nucleotide Metabolism: An Introduction.* Academic Press.

Hartman, S. C., 1970. Purines and pyrimidines. *In* Greenberg, D. M., (ed.), *Metabolic Pathways,* (3rd ed.), vol. 4, pp. 1–68. Academic Press

Blakley, R. L., 1969. *The Biochemistry of Folic Acid and Related Pterdines.* North-Holland.

Blakley, R. L., and Vitols, E., 1968. Control of nucleotide biosynthesis. *Annu. Rev. Biochem.* 37:201–224.

Raivio, K. O., and Seegmiller, J. E., 1970. The role of phosphoribosyl transferases in purine metabolism. *Curr. Top. Cell Regul.* 2:201–225.

Murray, A. W., 1971. The biological significance of purine salvage. *Annu. Rev. Biochem.* 40:811–826.

Larsson, A., and Reichard, P., 1967. Enzymatic reduction of ribonucleotides. *Progr. Nucl. Acid Res. Mol. Biol.* 7:303–347.

Jones, M. E., 1971. Regulation of pyrimidine and arginine biosynthesis in mammals. *Advan. Enzyme Regul.* 9:19–49.

Aspartate transcarbamoylase:

Gerhart, J. C., 1970. A discussion of the regulatory properties of aspartate transcarbamylase from *Escherichia coli. Curr. Top. Cell Regul.* 2:275–325.

Jacobson, G. R., and Stark, G. R., 1973. Aspartate transcarbamylases. *In* Boyer, P. D., (ed.), *The Enzymes,* (3rd ed.), vol. 9, pp. 225–308. Academic Press.

Warren, S. G., Edwards, B. F. P., Evans, D. R., Wiley, D. C., and Lipscomb, W. N., 1973. Aspartate transcarbamoylase from *Escherichia coli*: electron density at 5.5 Å resolution. *Proc. Nat. Acad. Sci.* 70:1117–1121.

Inborn errors of purine metabolism:

Wyngaarden, J. B., and Kelley, W. N., 1972. Gout. *In* Stanbury, J. B., Wyngaarden, J. B., and Fredrickson, D. S., (eds.), *The Metabolic Basis of Inherited Disease,* (3rd ed.), pp. 889–968. McGraw-Hill.

Kelley, W. N., and Wyngaarden, J. B., 1972. The Lesch-Nyhan syndrome. *In* Stanbury, J. B., Wyngaarden, J. B., and Fredrickson, D. S., (eds.), *The Metabolic Basis of Inherited Disease,* (3rd ed.), pp. 969–991.

PROBLEMS

1. Write a balanced equation for the synthesis of PRPP from glucose via the oxidative branch of the pentose phosphate pathway.

2. Write a balanced equation for the synthesis of orotate from glutamine, CO_2, and aspartate.

3. What is the activated reactant in the biosynthesis of each of these compounds?
   (a) Phosphoribosylamine.
   (b) Carbamoylaspartate.
   (c) Orotidylate (from orotate).
   (d) Nicotinate ribonucleotide.
   (e) Phosphoribosyl anthranilate.

4. Amidotransferases are inhibited by the antibiotic azaserine (O-diazoacetyl-L-serine), which is an analog of glutamine.

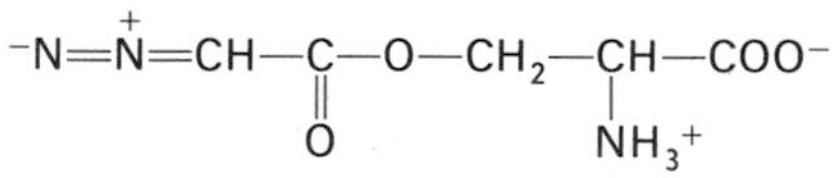

$$^{-}N{=}\overset{+}{N}{=}CH{-}\underset{\underset{O}{\|}}{C}{-}O{-}CH_2{-}\underset{\underset{NH_3^+}{|}}{CH}{-}COO^{-}$$

Azaserine

Which intermediates in purine biosynthesis would accumulate in cells treated with azaserine?

5. Write a balanced equation for the synthesis of dTMP from dUMP that is coupled to the conversion of serine to glycine.

6. Bacterial growth is inhibited by sulfanilamide and related sulfa drugs, and there is a concomitant accumulation of 5-aminoimidazole-4-carboxamide ribonucleotide. This inhibition is reversed by the addition of p-aminobenzoate.

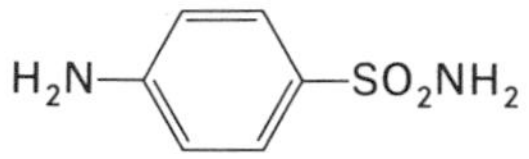

$$H_2N{-}\langle\rangle{-}SO_2NH_2$$

Sulfanilamide

Propose a mechanism for the inhibitory effect of sulfanilamide.

For additional problems see W. B. Wood, J. H. Wilson, R. M. Benbow, and L. E. Hood, *Biochemistry: A Problems Approach* (Benjamin, 1974), Chapter 15.

On the facing page: Model of a pair of deoxyribonucleotide units of a DNA double helix. Adenine is hydrogen-bonded to thymine in this base pair.

PART IV

INFORMATION

storage, transmission, and expression of genetic information

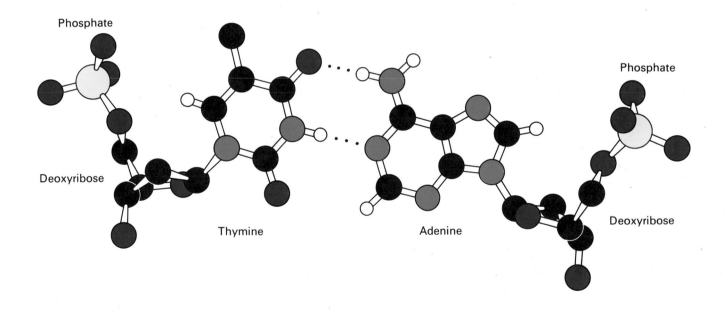

CHAPTER 23

DNA: GENETIC ROLE, STRUCTURE, AND REPLICATION

Deoxyribonucleic acid (DNA) is the molecule of heredity. DNA is a very long, threadlike macromolecule made up of a large number of deoxyribonucleotides. *The purine and pyrimidine bases of DNA carry genetic information, whereas the sugar and phosphate groups perform a structural role.*

COVALENT STRUCTURE AND NOMENCLATURE OF DNA

The *backbone* of DNA, which is constant throughout the molecule, consists of deoxyriboses linked by phosphodiester bridges. The 3′-hydroxyl of the sugar moiety of one deoxyribonucleotide is joined to the 5′-hydroxyl of the adjacent sugar by a phosphodiester bond. The *variable* part of DNA is its *sequence of bases*. DNA contains four kinds of bases. The two purines are *adenine* (A) and *guanine* (G).

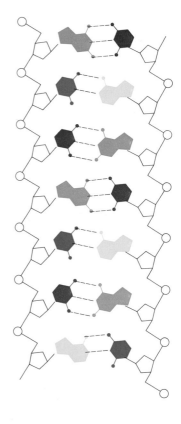

Figure 23-1
A schematic diagram of the structure of DNA. The sugar-phosphate backbone is shown in black, whereas the purine and pyrimidine bases are shown in color. [Based on A. Kornberg. The synthesis of DNA. Copyright © 1968 by Scientific American, Inc. All rights reserved.]

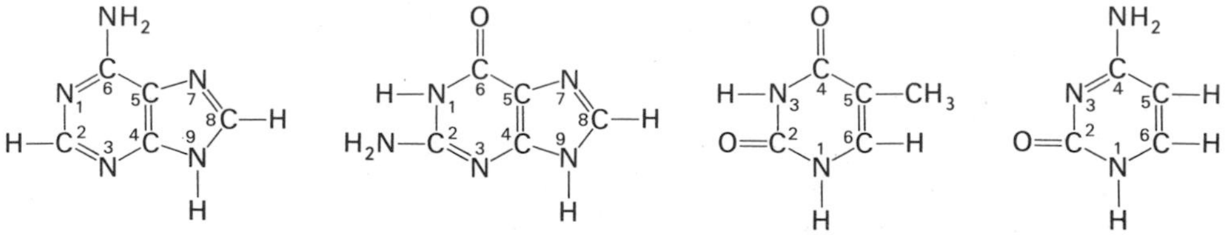

Adenine
(A)

Guanine
(G)

Thymine
(T)

Cytosine
(C)

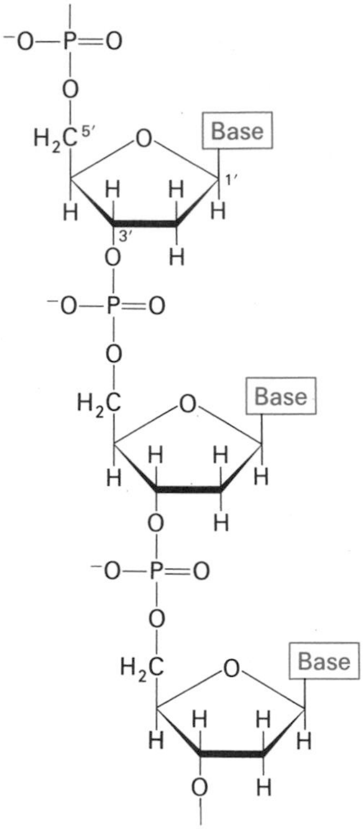

Figure 23-2
Structure of part of
a DNA chain.

The two pyrimidines are *thymine* (T) and *cytosine* (C). The structure of a DNA chain is shown in Figure 23-2.

The structure of a DNA chain can be concisely represented in the following way. The symbols for the four principal deoxyribonucleosides are:

A   G   C   T

The bold line refers to the sugar, whereas A, G, C, and T represent the bases. The Ⓟ within the diagonal line in the diagram below denotes a phosphodiester bond. This diagonal line joins the end of one bold line and the middle of another. These junctions refer to the 5'-OH and 3'-OH, respectively. In this example, the symbol Ⓟ indicates that deoxyadenylate is linked to deoxycytidine by a phosphodiester bridge. The 3'-OH of deoxyadenylate is joined to the 5'-OH of deoxycytidine by a phosphoryl group.

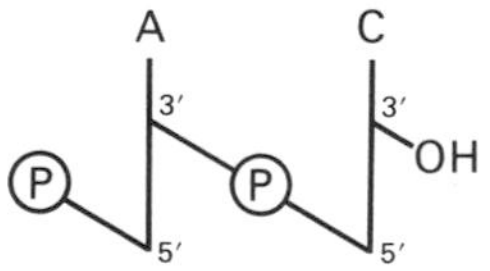

Now suppose that deoxyguanylate becomes linked to the deoxycytidine unit in this dinucleotide. The resulting trinucleotide can be represented by

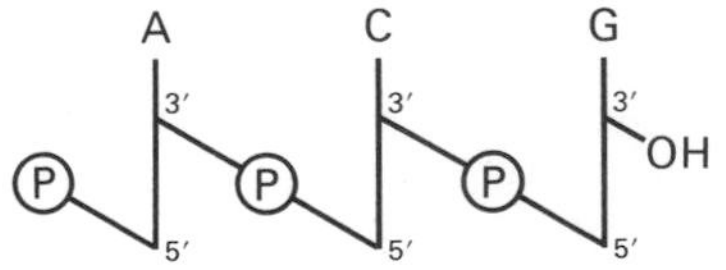

An abbreviated notation for this trinucleotide is pApCpG or ACG.

A DNA chain has polarity. One end of the chain has a 5'-OH group and the other a 3'-OH group that is not linked to another nucleotide. By convention, the symbol ACG means that the unlinked 5'-OH group is on deoxyadenosine, whereas the unlinked 3'-OH group is on deoxyguanosine. Thus, *the base sequence is written in the 5' $\longrightarrow$ 3' direction.* Recall that the amino acid sequence of a protein is written in the amino $\longrightarrow$ carboxyl direction. Note that ACG and GCA refer to different compounds, just as Glu-Phe-Ala differs from Ala-Phe-Glu.

TRANSFORMATION OF PNEUMOCOCCI BY DNA REVEALED THAT GENES ARE MADE OF DNA

The pneumococcus bacterium played an important part in the discovery of the genetic role of DNA. Pneumococci are normally surrounded by a slimy, glistening polysaccharide capsule. This outer layer is essential for the pathogenicity of the bacterium, which causes pneumonia in humans and other susceptible mammals. Mutants devoid of a polysaccharide coat are not pathogenic. The normal bacterium is referred to as the S form (because it forms smooth colonies), whereas mutants without capsules are called R forms (because they form rough colonies). One group of R mutants lacks the dehydrogenase enzyme that converts UDP-glucose to UDP-glucuronate. This enzyme is required for the synthesis of the capsular polysaccharide, which in one type of pneumococcus consists of an alternating sequence of glucose and glucuronate residues.

In 1928, Fred Griffith discovered that a nonpathogenic R mutant could be *transformed* into the pathogenic S form in the following way. He injected mice with a mixture of live R and heat-killed

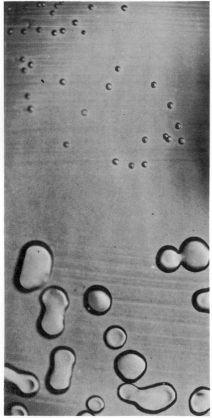

Figure 23-3
Transformation of nonpathogenic R
pneumococci (small colonies) to
pathogenic S pneumococci (large
glistening colonies) by DNA from
heat-killed S pneumococci. [From O. T.
Avery, C. M. MacLeod, and M. McCarty.
J. Exp. Med. 79(1944):137.]

S pneumococci. The striking finding was that this mixture was
lethal to the mice, whereas either live R or heat-killed S pneumococci alone was not. The blood of the dead mice contained live
S pneumococci. Thus, the heat-killed S pneumococci had somehow
transformed live R pneumococci into live S pneumococci. This
change was permanent: the transformed pneumococci yielded
pathogenic progeny of the S form. It was then found that this
R ⟶ S transformation can occur in vitro. Some of the cells in
a growing culture of the R form were transformed into the S form
by the addition of a *cell-free extract* of heat-killed S pneumococci.
This finding set the stage for the elucidation of the chemical nature
of the "transforming principle."

The cell-free extract of heat-killed S pneumococci was fractionated and the transforming activity of its components was assayed
(Figure 23-3). In 1944, Oswald Avery, Colin MacLeod, and Maclyn
McCarty published their discovery that "*a nucleic acid of the deoxyribose type is the fundamental unit of the transforming principle of Pneumococcus Type III.*" The experimental basis for their conclusion was:
(1) the purified, highly active transforming principle gave an ele-
mental chemical analysis that agreed closely with that calculated
for DNA; (2) the optical, ultracentrifugal, diffusive, and electrophoretic properties of the purified material were like those of DNA;
(3) there was no loss of transforming activity upon extraction of
protein or lipid; (4) trypsin and chymotrypsin did not affect transforming activity; (5) ribonuclease (known to digest ribonucleic acid)
had no effect on the transforming principle; and (6) in contrast,
transforming activity was lost following the addition of deoxyribonuclease.

This work is a landmark in the development of biochemistry.
Until 1944, it was generally assumed that chromosomal proteins
carry genetic information and that DNA plays a secondary role.
This prevailing view was decisively shattered by the rigorously
documented finding that *purified DNA has genetic specificity.* Avery
gave a vivid description of this research and of its implications in
a letter he wrote in 1943 to his brother, a medical microbiologist
at another university (Figure 23-4).

Further support for the genetic role of DNA came from the
studies of a virus that infects *E. coli.* The T2 bacteriophage consists
of a core of DNA surrounded by a protein coat. In 1951, Roger
Herriott suggested that "the virus may act like a little hypodermic

For the past two years, first with MacLeod and now with Dr. McCarty, I have been trying to find out what is the chemical nature of the substance in the bacterial extract which induces this specific change. The crude extract of Type III is full of capsular polysaccharide, C (somatic) carbohydrate, nucleoproteins, free nucleic acids of both the yeast and thymus type, lipids, and other cell constituents. Try to find in the complex mixtures the active principle! Try to isolate and chemically identify the particular substance that will by itself, when brought into contact with the R cell derived from Type II, cause it to elaborate Type III capsular polysaccharide and to acquire all the aristocratic distinctions of the same specific type of cells as that from which the extract was prepared! Some job, full of headaches and heartbreaks. But at last perhaps we have it.

. . . if we prove to be right—and of course that is a big if—then it means that both the chemical nature of the inducing stimulus is known and the chemical structure of the substance produced is also known, the former being thymus nucleic acid, the latter Type III polysaccharide, and both are thereafter reduplicated in the daughter cells and after innumerable transfers without further addition of the inducing agent and the same active and specific transforming substance can be recovered far in excess of the amount originally used to induce the reaction. Sounds like a virus—may be a gene. But with mechanisms I am not now concerned. One step at a time and the first step is what is the chemical nature of the transforming principle? Some one else can work out the rest. Of course the problem bristles with implications. It touches the biochemistry of the thymus type of nucleic acids which are known to constitute the major part of chromosomes but have been thought to be alike regardless of origin and species. It touches genetics, enzyme chemistry, cell metabolism and carbohydrate synthesis. But today it takes a lot of well documented evidence to convince anyone that the sodium salt of deoxyribose nucleic acid, protein free, could possibly be endowed with such biologically active and specific properties and that is the evidence we are now trying to get. It is lots of fun to blow bubbles but it is wiser to prick them yourself before someone else tries to.

Figure 23-4
Part of a letter from Oswald Avery to his brother Roy, written in May 1943. [From R. D. Hotchkiss. In *Phage and the Origins of Molecular Biology*, J. Cairns, G. S. Stent, and J. D. Watson, eds. (Cold Spring Harbor Laboratory, 1966), pp. 185–186.]

needle full of transforming principles; the virus as such never enters the cell; only the tail contacts the host and perhaps enzymatically cuts a small hole through the outer membrane and then the nucleic acid of the virus head flows into the cell." This idea was tested

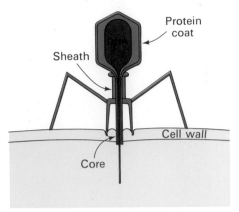

Figure 23-5
Diagram of a T2 bacteriophage injecting its DNA into a bacterial cell. [Based on W. B. Wood and R. S. Edgar. Building a bacterial virus. Copyright © 1967 by Scientific American, Inc. All rights reserved.]

by Alfred Hershey and Martha Chase in the following way. Phage DNA was labeled with the radioisotope ^{32}P, whereas the protein coat was labeled with ^{35}S. These labels are highly specific since DNA does not contain sulfur and the protein coat is devoid of phosphorus. A sample of an *E. coli* culture was infected with labeled phage, which became attached to the bacteria during a short incubation period. The suspension was spun for a few minutes in a Waring Blendor at 10,000 rpm. This treatment subjected the phage-infected cells to very strong shearing forces, which severed the connections between the viruses and bacteria. The resulting suspension was centrifuged at a speed sufficient to throw the bacteria to the bottom of the tube. Thus, the pellet contained the infected bacteria, whereas the supernatant contained smaller particles. These fractions were analyzed for ^{32}P and ^{35}S to determine the location of the phage DNA and the protein coat. The results of these experiments were:

1. Most of the phage DNA was found in the bacteria.

2. Most of the phage protein was found in the supernatant.

3. The blender treatment had almost no effect on the competence of the infected bacteria to produce progeny virus.

Additional experiments showed that less than 1% of the ^{35}S was transferred from the parental phage to the progeny phage. In contrast, 30% of the parental ^{32}P appeared in the progeny. These simple, incisive experiments led to the conclusion that "*a physical separation of the phage T2 into genetic and non-genetic parts is possible. . . .* The sulfur-containing protein of resting phage particles is confined to a protective coat that is responsible for the adsorption of bacteria, and functions as an instrument for the injection of the phage DNA into the cell. This protein probably has no function in the growth of intracellular phage. The DNA has some function. Further chemical inferences should not be drawn from the experiments presented."

The cautious tone of this conclusion did not detract from its impact. The genetic role of DNA soon became a generally accepted fact. The experiments of Hershey and Chase strongly reinforced what Avery, MacLeod, and McCarty had found eight years earlier in a different system. Additional support came from studies of the

DNA content of single cells, which showed that in a given species *the DNA content is the same for all cells that have a diploid set of chromosomes. Haploid cells were found to have half as much DNA.*

THE GENES OF SOME VIRUSES ARE MADE OF RNA

Genes in all procaryotic and eucaryotic organisms are made of DNA. In viruses, genes are made either of DNA or of RNA. Tobacco mosaic virus, which infects the leaves of tobacco plants, is one of the best-characterized RNA viruses. It consists of a single molecule of RNA surrounded by a protein coat of 2,130 identical subunits (see Chapter 29 for a discussion of its structure and assembly). The protein can be separated from the RNA by treatment of the virus with phenol. *The isolated viral RNA is infective, whereas the viral protein is not.* Synthetic hybrid virus particles provide additional evidence that the genetic specificity of the virus resides exclusively in its RNA. There are a variety of strains of tobacco mosaic virus. A synthetic hybrid virus was prepared from the RNA of strain 1 and the protein of strain 2. Another was prepared from the RNA of strain 2 and the protein of strain 1. After infection, *the progeny virus always consisted of RNA and protein corresponding to the specificity of the RNA in the infecting hybrid virus.*

THE WATSON-CRICK DNA DOUBLE HELIX

In 1953, James Watson and Francis Crick deduced the three-dimensional structure of DNA and immediately inferred its mechanism of replication. This brilliant accomplishment ranks as one of the most significant in the history of biology because it led the way to an understanding of gene function in molecular terms. Watson and Crick analyzed x-ray diffraction photographs of DNA fibers taken by Rosalind Franklin and Maurice Wilkins and derived a structural model that has proven to be essentially correct. The salient features of their model DNA are:

1. Two helical polynucleotide chains are coiled around a common axis. The chains run in opposite directions (Figure 23-7).

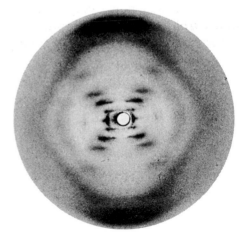

Figure 23-6
X-ray diffraction photograph of a hydrated DNA fiber (form B). The central cross is diagnostic of a helical structure. The strong arcs on the meridian arise from the stack of base pairs, which are 3.4 Å apart. [Courtesy of Dr. Maurice Wilkins.]

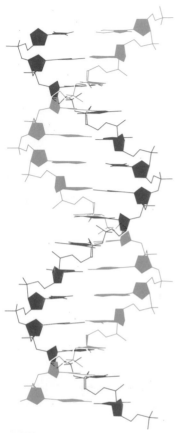

Figure 23-7
A skeletal model of double-helical DNA. The structure repeats at intervals of 34 Å, which corresponds to ten residues on each chain.

2. The purine and pyrimidine bases are on the inside of the helix, whereas the phosphate and deoxyribose units are on the outside (Figure 23-8). The planes of the bases are perpendicular to the helix axis. The planes of the sugars are nearly at right angles to those of the bases.

3. The diameter of the helix is 20 Å. Adjacent bases are separated by 3.4 Å along the helix axis and related by a rotation of 36 degrees. Hence, the helical structure repeats after ten residues on each chain; that is, at intervals of 34 Å.

4. The two chains are held together by hydrogen bonds between pairs of bases. Adenine is always paired with thymine. Guanine is always paired with cytosine (Figures 23-9 and 23-10).

5. The sequence of bases along a polynucleotide chain is not restricted in any way. *The precise sequence of bases carries the genetic information.*

The most important aspect of the DNA double helix is the specificity of the pairing of bases. Watson and Crick deduced that adenine must pair with thymine, and guanine with cytosine, because of steric and hydrogen-bonding factors. The steric restriction is imposed by the regular helical nature of the sugar-phosphate backbone of each polynucleotide chain. The glycosidic bonds that are attached to a bonded pair of bases are

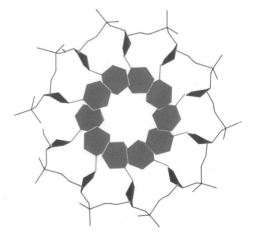

Figure 23-8
Diagram of one of the strands of a DNA double helix, viewed down the helix axis. The bases (all pyrimidines here) are inside, whereas the sugar-phosphate backbone is outside. The tenfold symmetry is evident. The bases are shown in blue and the sugars in red.

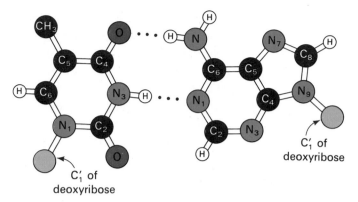

Figure 23-9
Model of an
adenine-thymine
base pair.

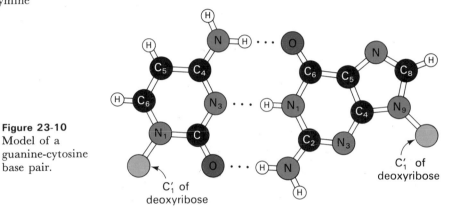

Figure 23-10
Model of a
guanine-cytosine
base pair.

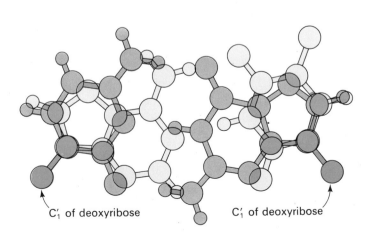

C'$_1$ of deoxyribose

C'$_1$ of deoxyribose

Figure 23-11
Superposition of an A-T base pair (shown in
yellow) on a G-C base pair (shown in blue).
Note that the positions of the glycosidic
bonds and of the C-1′ atom of deoxyribose
(shown in green) are almost identical for the
two base pairs.

always 10.85 Å apart (Figure 23-11). A purine-pyrimidine base pair fits perfectly in this space. In contrast, there is insufficient room for two purines. There is more than enough space for two pyrimidines, but they would be too far apart to form hydrogen bonds. Hence, one member of a base pair in a DNA helix must always be a purine, the other a pyrimidine, because of steric factors. The base pairing is further restricted by hydrogen-bonding requirements. The hydrogen atoms in the purine and pyrimidine bases have well-defined positions. Adenine cannot pair with cytosine because there would be two hydrogens near one of the bonding positions and none at the other. Likewise, guanine cannot pair with thymine. In contrast, adenine forms two hydrogen bonds with thymine, whereas guanine forms three with cytosine (Figures 23-9 and 23-10). The orientations and distances of these hydrogen bonds are optimal for achieving strong interaction between the bases.

This base-pairing scheme was strongly supported by the results of earlier studies of the base compositions of DNAs from different species. In 1950, Erwin Chargaff found that the *ratios of adenine to thymine and of guanine to cytosine were nearly 1.0 in all species studied.* The meaning of these equivalences was not evident until the Watson-Crick model was proposed. Only then could it be seen that they reflect an essential facet of DNA structure and function—the specificity of base pairing.

THE COMPLEMENTARY CHAINS ACT AS TEMPLATES FOR EACH OTHER IN DNA REPLICATION

The double helical model of DNA immediately suggested a mechanism for the replication of DNA. Watson and Crick (1953*b*) published their hypothesis a month after they had presented their structural model in a beautifully simple and lucid paper:

> . . . If the actual order of the bases on one of the pairs of chains were given, one could write down the exact order of the bases on the other one, because of the specific pairing. Thus one chain is, as it were, the complement of the other, and it is this feature which suggests how the deoxyribonucleic acid molecule might duplicate itself.

Previous discussions of self-duplication have usually involved the concept of a template, or mould. Either the template was supposed to copy itself directly or it was to produce a "negative," which in its turn was to act as a template and produce the original "positive" once again. In no case has it been explained in detail how it would do this in terms of atoms and molecules.

Now our model for deoxyribonucleic acid is, in effect, a *pair* of templates, each of which is complementary to the other. We imagine that prior to duplication the hydrogen bonds are broken, and the two chains unwind and separate. Each chain then acts as a template for the formation on to itself of a new companion chain, so that eventually we shall have *two* pairs of chains, where we only had one before. Moreover, the sequence of the pairs of bases will have been duplicated exactly.

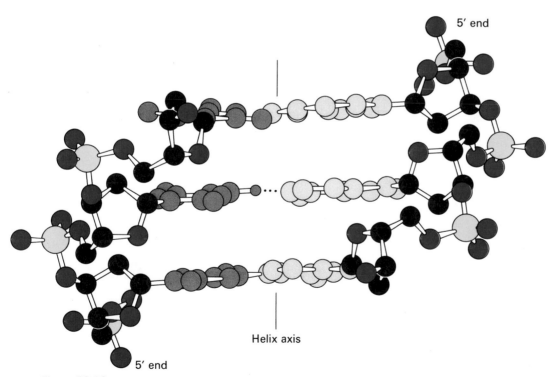

Figure 23-12
Model of a double-helical DNA molecule showing three base pairs. Note that the two strands run in opposite directions.

DNA REPLICATION IS SEMICONSERVATIVE

Watson and Crick proposed that one of the strands of each daughter DNA molecule is newly synthesized, whereas the other is derived from the parent DNA molecule. This distribution of parental atoms is called *semiconservative*. A critical test of this hypothesis was carried out by Matthew Meselson and Franklin Stahl. The parent DNA was labeled with ^{15}N, the heavy isotope of nitrogen, to make it denser than ordinary DNA. This was accomplished by growing *E. coli* for many generations in a medium that contained $^{15}NH_4Cl$ as the sole nitrogen source. The bacteria were abruptly transferred to a medium that contained ^{14}N, the ordinary isotope of nitrogen. The question asked was: What is the distribution of ^{14}N and ^{15}N in the DNA molecules after successive rounds of replication?

The distribution of ^{14}N and ^{15}N was revealed by the newly developed technique of *density-gradient equilibrium sedimentation*. A small amount of DNA was dissolved in a concentrated solution of cesium chloride having a density close to that of the DNA (~1.7 g/cm^3). This solution was centrifuged until it was nearly at equilibrium. The opposing processes of sedimentation and diffusion created a gradient in the concentration of cesium chloride across the centrifuge cell. The result was a stable density gradient, ranging from 1.66 to 1.76 g/cm^3. The DNA molecules in this density gradient were driven by centrifugal force into the region where the solution density was equal to their own buoyant density. High-molecular-weight DNA yielded a sharp band that was detected by its absorption of ultraviolet light. The ^{14}N DNA and ^{15}N DNA molecules were clearly resolved in the density gradient since they differ in density by about 1% (Figure 23-14).

DNA was extracted from the bacteria at various times after they were transferred from a ^{15}N to a ^{14}N medium. Analysis of these samples by the density-gradient technique showed that there was a single band of DNA after one generation (Figure 23-15). The density of this band was precisely halfway between those of ^{14}N DNA and ^{15}N DNA. *The absence of ^{15}N DNA indicated that parental DNA was not preserved as an intact unit on replication.* The absence of ^{14}N DNA indicated that all of the daughter DNA molecules derived some of their atoms from the parent DNA. This proportion had to be one-half, since the density of the hybrid DNA band was halfway between those of ^{14}N DNA and ^{15}N DNA.

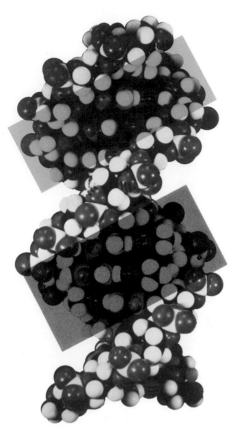

Figure 23-13
Space-filling model of a DNA double helix. Two kinds of grooves are evident: a major groove (red) and a minor groove (blue). Note that part of each base is accessible for interaction with other molecules. [Courtesy of Dr. Sung-Hou Kim.]

A

B

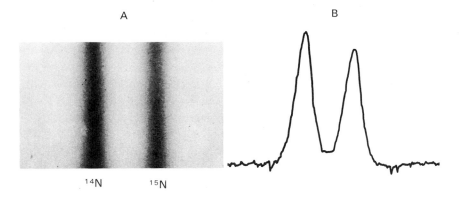

Figure 23-14
Resolution of ^{14}N DNA and ^{15}N DNA
by density-gradient centrifugation: (A)
ultraviolet absorption photograph of a
centrifuge cell; (B) densitometric
tracing of the absorption photograph.
[From M. Meselson and F. W. Stahl.
Proc. Nat. Acad. Sci. 44(1958):671.]

^{14}N ^{15}N

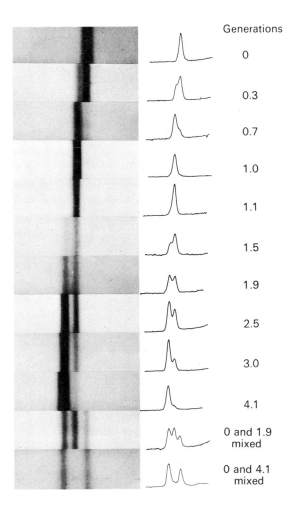

Generations

0

0.3

0.7

1.0

1.1

1.5

1.9

2.5

3.0

4.1

0 and 1.9
mixed

0 and 4.1
mixed

Figure 23-15
Demonstration of semiconservative
replication in *E. coli* by density-
gradient centrifugation. The position
of a band of DNA depends on its
content of ^{14}N and ^{15}N. After 1.0
generation, all of the DNA molecules
are hybrids containing equal amounts
of ^{14}N and ^{15}N. No parental DNA
(^{15}N) is left after 1.0 generation. [From
M. Meselson and F. W. Stahl. *Proc.
Nat. Acad. Sci.* 44(1958):671.]

570

Part IV / INFORMATION

After two generations, there were equal amounts of two bands of DNA. One was hybrid DNA, the other was ¹⁴N DNA. Meselson and Stahl concluded from these incisive experiments "*that the nitrogen of a DNA molecule is divided equally between two physically continuous subunits; that, following duplication, each daughter molecule receives one of these; and that the subunits are conserved through many duplications.*" Their results agreed perfectly with the Watson-Crick model for DNA replication (Figure 23-16).

DNA MOLECULES ARE VERY LONG

Let us consider some properties of DNA before turning to its mechanism of replication, which is an intricate enzymatic process. A striking characteristic of naturally occurring DNA molecules is their length. *The* E. coli *chromosome is a single molecule of double helical DNA consisting of 3.4 million base pairs.* The molecular weight of this DNA molecule is 2.3×10^9. It has a highly *asymmetric shape:* its contour length is 12×10^6 Å, whereas its diameter is 20 Å. The 1.2-mm contour length of this DNA molecule corresponds to a macroscopic dimension, whereas its width of 20 Å is on the atomic scale. Recently, Bruno Zimm has found that the largest chromosome of *Drosophila melanogaster* contains a single DNA molecule of 6.2×10^7 base pairs, which have a contour length of 2.1 cm. Such highly asymmetric DNA molecules are very susceptible to cleavage

Original parent molecule

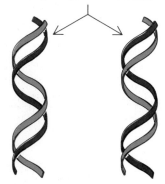

First generation daughter molecules

Second generation daughter molecules

Figure 23-16
A schematic diagram of semiconservative replication. Parental DNA is shown in green and newly synthesized DNA in red. [Based on M. Meselson and F. W. Stahl. *Proc. Nat. Acad. Sci.* 44(1958):671.]

Table 23-1
Sizes of DNA molecules from several viruses and bacteria

Source	Number of base pairs	Length (μm)	Molecular weight
Viruses:			
Polyoma or SV40	4,600	1.6	3.1×10^6
λ phage	53,000	16	31×10^6
T2 phage	185,000	63	122×10^6
Vaccinia	240,000	81	157×10^6
Bacteria:			
Mycoplasma	760,000	260	504×10^6
E. coli	3,400,000	1,200	$2,320 \times 10^6$

Note: 1 base pair corresponds to 0.66×10^3 daltons and 1 μm of double helix corresponds to 2.94×10^3 base pairs or 1.94×10^6 daltons.

by shearing forces. They are easily degraded into segments whose molecular weight is one thousand times less than that of the original molecule unless special precautions are taken in their handling.

The DNA molecules of many bacteria and viruses have been directly visualized by electron microscopy (Figure 23-17). The dimensions of some of these DNA molecules are given in Table 23-1. It should be noted that even the smallest DNA molecules are highly elongate. The DNA from polyoma virus, for example, consists of 4,600 base pairs and has a contour length of 1.6 μm (16,000 Å). Recall that hemoglobin has a diameter of 65 Å and that collagen, one of the longest proteins, has a length of 3,000 Å. This comparison emphasizes the remarkable length and asymmetry of DNA molecules.

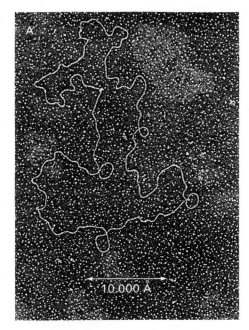

SOME DNA MOLECULES ARE CIRCULAR

Electron microscopy has shown that intact DNA molecules from many sources are circular (Figure 23-17). The finding that *E. coli* has a circular chromosome was anticipated by genetic studies that revealed that the *gene-linkage map of this bacterium is circular*. The term "circular" refers to the continuity of the DNA chain, not to its geometrical form. DNA molecules in vivo necessarily have a very compact shape. Note that the contour length of the *E. coli* chromosome is about 1000 times as long as the greatest diameter of the bacterium.

Not all DNA molecules are circular. DNA from the T7 bacteriophage, for example, is *linear*. The DNA molecules of some viruses, such as the λ bacteriophage, *interconvert between linear and circular forms*. The linear form is present inside the virus particle, whereas the circular form is present in the host cell (see p. 720).

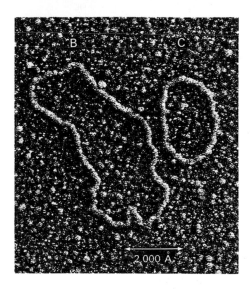

THE DNA IN A FEW VIRUSES IS SINGLE-STRANDED FOR A PART OF THEIR LIFE CYCLE

DNA is not always double-stranded. Robert Sinsheimer discovered that the DNA in *φX174, a small virus that infects* E. coli, *is single-stranded*. Several experimental results led to this unexpected conclusion. First, the base ratios of φX174 DNA do not conform to the rule that [A] = [T] and [G] = [C]. Second, a solution of φX174 DNA is much less viscous than a solution of the same concentration

Figure 23-17
Electron micrographs of some DNA molecules: (A) DNA from λ bacteriophage (RF II form); (B) DNA from φX174 bacteriophage (RF II form); and (C) plasmid τ from *E. coli.* [Courtesy of Dr. Thomas Broker.]

of *E. coli* DNA. The hydrodynamic properties of φX174 DNA are like those of a randomly coiled polymer. In contrast, the DNA double helix behaves hydrodynamically as a quite rigid rod. Third, the amino groups of the bases of φX174 DNA react readily with formaldehyde, whereas the bases in double helical DNA are virtually inaccessible to this reagent.

The finding of this single-stranded DNA raised doubts concerning the universality of the semiconservative replicative scheme proposed by Watson and Crick. However, it was soon shown that φX174 DNA is single-stranded for only a part of its life cycle. Sinsheimer found that infected *E. coli* cells contain a *double-stranded form of φX174 DNA*. This double-helical DNA is called the *replicative form* because it serves as the template for the synthesis of the DNA of the progeny virus. Viruses that contain single-stranded RNA also replicate via a double-stranded replicative form. The detailed mechanisms of these processes will be discussed in Chapter 29. The important point now is that these studies reinforced the generality of the Watson-Crick scheme for replication. *Double-helical DNA (or RNA) is the replicative form of all known genes.*

DISCOVERY OF A DNA POLYMERASE

We turn now to the molecular events in the replication of DNA. The search for an enzyme that synthesizes DNA was initiated by Arthur Kornberg and his associates in 1955. This search soon proved fruitful, largely because three appropriate choices were made in the design of their experiments:

1. What are the *activated precursors* of DNA? They correctly deduced that *deoxyribonucleoside 5′-triphosphates* are the activated intermediates in DNA synthesis. This inference was based on two clues. First, pathways of purine and pyrimidine biosynthesis lead to nucleoside 5′-phosphates rather than to nucleoside 3′-phosphates. Second, ATP is the activated intermediate in the synthesis of a pyrophosphate bond in coenzymes such as NAD^+, FAD, and CoA.

2. What is the criterion of *DNA synthesis?* It was expected that the net amount of DNA synthesis might be very small in the initial experiments, especially where nucleases are prevalent. Hence, a sensitive assay was essential. This was accomplished by using *radio-*

active precursor nucleotides. The incorporation of these precursors into DNA was detected by measuring the radioactivity of an *acid-precipitate of the incubation mixture.* The basis of this technique is that DNA is precipitated by acids, such as trichloroacetic acid, whereas precursor nucleotides stay in solution.

3. Which *kinds of cells* should be analyzed? *E. coli* bacteria were used after initial experiments with animal-cell extracts were negative. *E. coli* was chosen because it has a generation time of only twenty minutes and can be harvested in large quantities. As expected, this bacterium is a choice source of enzymes that synthesize DNA.

An extract of *E. coli* was incubated with radioactive deoxythymidine 5′-triphosphate. The level of radioactivity in this ^{14}C-labeled precursor was one million counts per minute. The acid-precipitate of this incubation mixture contained just 50 counts. Only a few picomoles of DNA were synthesized, but it was a start. Kornberg wrote: "Although the amount of nucleotide incorporated into nucleic acid was miniscule, it was nonetheless significantly above the level of background noise. Through this tiny crack we tried to drive a wedge. The hammer was enzyme purification, a technique that had matured during the elucidation of alcoholic fermentation."

This new enzyme was named DNA polymerase. It is now called *DNA polymerase I* because other DNA polymerases have since been isolated. After a decade of effort in Kornberg's laboratory, DNA polymerase I has been purified to homogeneity and characterized in detail. An appreciation of the magnitude of the task can be gained by noting that five hundred milligrams of pure enzyme were obtained from one hundred kilograms of *E. coli* cells.

DNA polymerase I is a single polypeptide chain whose molecular weight is 109,000. It catalyzes the *step-by-step addition of deoxyribonucleotide units to a DNA chain:*

$$(\text{DNA})_{n\text{ residues}} + \text{dNTP} \rightleftharpoons (\text{DNA})_{n+1} + \text{PP}_i$$

DNA polymerase I requires the following components to synthesize a chain of DNA:

1. All four *deoxyribonucleoside 5′-triphosphates*—dATP, dGTP, dTTP, and dCTP—must be present. We will use the abbreviation

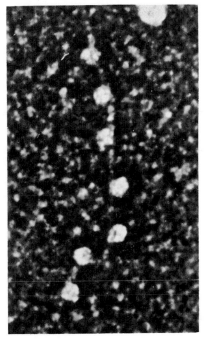

Figure 23-18
Electron micrograph of DNA polymerase molecules (spheres) bound to a DNA molecule (thin thread). [Courtesy of Dr. Jack Griffith.]

dNTP to refer to these deoxyribonucleoside triphosphates. Mg^{2+} is also required.

2. DNA polymerase I adds deoxyribonucleotides to the 3'-OH terminus of a preexisting DNA (or RNA) strand. In other words, a *primer* chain with a free 3'-OH group is required.

3. A DNA *template* is essential. The template can be single- or double-stranded DNA. Double-stranded DNA is an effective template only if its sugar-phosphate backbone is broken at one or more sites.

The chain-elongation reaction catalyzed by DNA polymerase occurs by means of *a nucleophilic attack of the 3'-OH terminus of the primer on the innermost phosphorus atom of the incoming deoxyribonucleoside triphosphate*. A phosphodiester bridge is formed and pyrophosphate is concomitantly released (Figure 23-19). The subsequent hydrolysis of pyrophosphate drives the polymerization forward. This displacement of the overall equilibrium could not occur if the activated intermediates were nucleoside diphosphates. We see here a compelling reason for the prevalence of nucleoside triphosphates rather than diphosphates as activated precursors in biosynthetic reactions. *The elongation of the DNA chain proceeds in the 5' ⟶ 3' direction*. About 1000 nucleotides are added per minute per molecule of DNA polymerase I.

Figure 23-19
Chain-elongation reaction catalyzed by DNA polymerases.

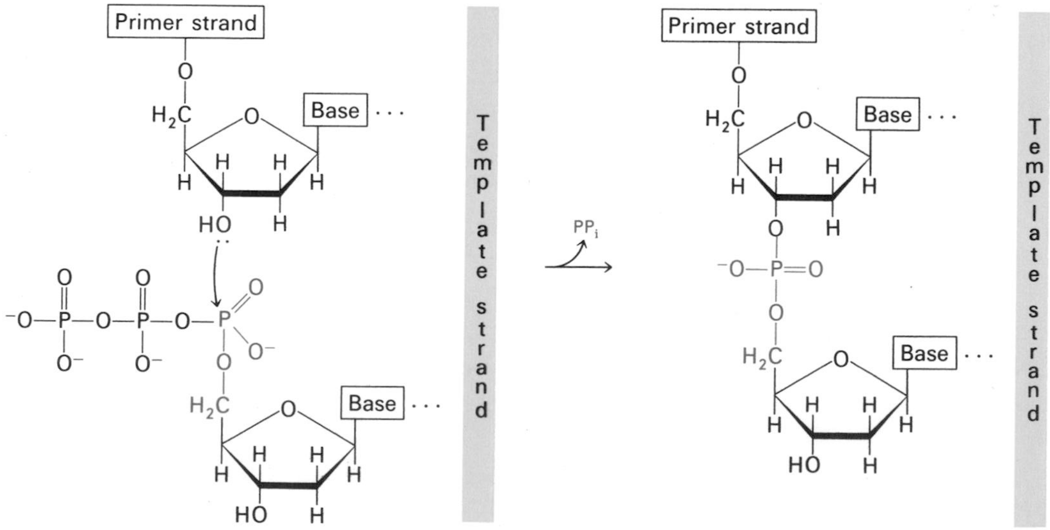

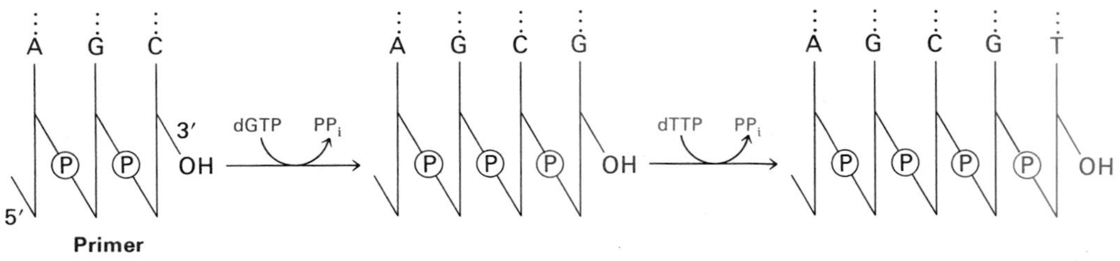

Figure 23-20
DNA polymerases catalyze elongation
of DNA chains in the 5′ ⟶ 3′
direction.

DNA POLYMERASE TAKES INSTRUCTIONS FROM A TEMPLATE

DNA polymerase catalyzes the formation of a phosphodiester bond only if the base on the incoming nucleotide is complementary to the base on the template strand. The probability of making a covalent link is very low unless the incoming base forms a Watson-Crick type of base pair with the base on the template strand. Thus, DNA polymerase is a *template-directed enzyme*. In fact, it was the first enzyme of this type to be discovered.

Several kinds of experiments have shown that DNA polymerase takes instructions from a template:

1. The earliest evidence was the finding that *appreciable synthesis of DNA occurs only if all four deoxyribonucleoside triphosphates and a DNA template are present.*

2. DNA polymerase can incorporate certain base analogs. For example, uracil or 5-bromouracil can substitute for thymine. Hypoxanthine can substitute for guanine. The pattern of substitution is highly specific. *The rule is that a base analog can be incorporated only if it can form a Watson-Crick type of base pair with the complement of the base that it replaces.* Thus, hypoxanthine can substitute for guanine (but not for A, T, or C) because it can form an appropriate base pair with cytosine.

3. The *base composition* of newly synthesized DNA depends on the nature of the template, not on the relative proportions of the four precursor nucleotides. The product DNA has the same base composition as the double-helical template DNA. This result indicates that both strands of the template DNA are replicated by DNA polymerase.

4. It is still difficult to determine the base sequence of a long DNA chain. However, the proportions of the sixteen possible dinucleotide sequences in DNA can readily be measured by a technique discussed below. *The dinucleotide frequencies (nearest-neighbor frequencies) of newly synthesized DNA and template DNA are the same.* This implies that DNA polymerase faithfully replicates the base sequence of the template DNA.

5. The most compelling evidence for the fidelity of replication by DNA polymerase I comes from the finding that this enzyme can replicate the DNA from φX174 phage in vitro. *The synthetic DNA formed by this system is fully infective, indicating that the error rate of the enzyme must be very low.*

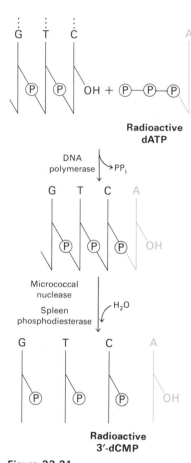

Figure 23-21
Method for determining nearest-neighbor frequencies in DNA.

NEAREST-NEIGHBOR ANALYSIS BY SPECIFIC HYDROLYSIS OF LABELED DNA

In DNA, there can be sixteen dinucleotide sequences, since each of the four bases can have four kinds of neighbors (AA, AT, AG, AC, and so on). The frequencies of these dinucleotides in several kinds of DNA have been determined in the following way (Figure 23-21):

1. DNA was synthesized using radioactive dATP. The other three precursor nucleotides were not radioactive. The innermost (α) phosphorus atom attached to the 5′-OH group of dATP was labeled with ^{32}P. Consequently, the phosphorus atoms in the phosphodiester bridges on the 5′ side of dA residues in the new DNA were radioactive.

2. The DNA product was hydrolyzed by enzymes (micrococcal nuclease and spleen phosphodiesterase) that yield deoxyribonucleoside 3′-monophosphates.

3. The result is that the ^{32}P is no longer linked to the nucleotide that brought it into the DNA. Rather, the ^{32}P has been transferred to the neighboring nucleotide on the 5′ side. In the example considered here, ^{32}P enters the DNA via dATP and emerges in cytidine 3′-monophosphate. Hence, the amounts of radioactive 3′-GMP, 3′-AMP, and 3′-TMP reflect the frequencies of GA, AA, and TA, respectively.

Nearest-neighbor frequencies can reveal whether the two strands in a DNA double helix are parallel or antiparallel. Consider the sequence $\overrightarrow{AG}$, where the arrow denotes the $5' \longrightarrow 3'$ direction. The complementary sequence is $\overrightarrow{TC}$ if the strands are parallel or $\overrightarrow{CT}$ if the strands are antiparallel. Hence, [AG] = [CT] if the strands are antiparallel. There are several other diagnostic equivalences (Table 23-2). *The observed nearest-neighbor frequencies showed that the strands in the DNA double helix are antiparallel,* as depicted in the Watson-Crick model.

Table 23-2
Diagnostic equivalences of nearest-neighbor frequencies

Antiparallel strands	Parallel strands
[CA] = [TG]	[CA] = [GT]
[AC] = [GT]	[AC] = [TG]
[GA] = [TC]	[GA] = [CT]
[AG] = [CT]	[AG] = [TC]

DNA LIGASE JOINS DNA FRAGMENTS

DNA polymerase I can add deoxyribonucleotides to a primer chain, but it cannot catalyze the joining of two DNA chains or the closure of a single DNA chain. The finding of circular DNA indicated that such an enzyme must exist. In 1967, researchers in several laboratories simultaneously discovered *DNA ligase,* an enzyme that catalyzes the formation of a phosphodiester bond between two DNA chains (Figure 23-22). This enzyme requires a *free OH group at the 3' end* of one DNA chain and a phosphate group at the *5'* end of the other. The formation of a phosphodiester bond between these groups is an endergonic reaction. Hence, *an energy source is required for the joining reaction.* In *E. coli* and other bacteria, *NAD$^+$* serves this role, whereas in some animal cells and bacteriophage, the reaction is driven by *ATP.*

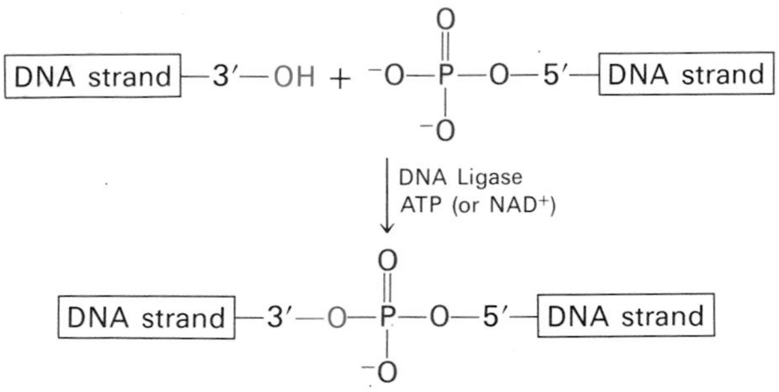

Figure 23-22
DNA ligase catalyzes the joining of two DNA chains that are part of a double-helical molecule.

Let us look at the mechanism of this reaction (Figure 23-23). ATP or NAD$^+$ reacts with DNA ligase to form a *covalent enzyme-AMP complex* in which the AMP is linked to the ϵ-amino group of a lysine residue of the enzyme through a phosphoamide bond. The AMP moiety activates the phosphate group at the 5' terminus of the DNA. The final step is a *nucleophilic attack of the 3'-OH group on this activated phosphorus atom.* A phosphodiester bond is formed and AMP is released. This sequence of reactions is driven by the hydrolysis of the pyrophosphate that was released in the formation of the enzyme-adenylate complex. Thus, *two high-energy phosphate bonds are spent in forming a phosphodiester bridge in the DNA backbone if ATP is the energy source.*

$$E + ATP \text{ (or NAD}^+) \rightleftharpoons E\text{-}AMP + PP_i \text{ (or NMN)}$$

$$E\text{-}AMP + \textcircled{P}\text{—}5'\text{-}DNA \rightleftharpoons E + AMP\text{—}\textcircled{P}\text{—}5'\text{-}DNA$$

$$DNA\text{-}3'\text{-}OH + AMP\text{—}\textcircled{P}\text{—}5'\text{-}DNA \rightleftharpoons DNA\text{-}3'\text{-}O\text{—}\textcircled{P}\text{—}5'\text{-}DNA + AMP$$

$$DNA\text{-}3'\text{-}OH + \textcircled{P}\text{—}5'\text{-}DNA + ATP \text{ (or NAD}^+) \rightleftharpoons DNA\text{-}3'\text{-}O\text{—}\textcircled{P}\text{—}5'\text{-}DNA + AMP + PP_i \text{ (or NMN)}$$

Figure 23-23
Mechanism for the reaction catalyzed by DNA ligase.

One other aspect of DNA ligase merits attention. DNA ligase cannot link two molecules of single-stranded DNA. Rather, *the DNA chains joined by DNA ligase must be part of a double-helical DNA molecule.* Studies of model systems suggest that DNA ligase forms a phosphodiester bridge only if there are at least several base pairs in the region of this link. In fact, DNA ligase closes nicks in the backbone of double-helical DNA. This joining process is essential for the *normal synthesis of DNA,* for *repair of damaged DNA,* and for the *splicing of DNA chains in genetic recombination.*

DNA POLYMERASE I CORRECTS MISTAKES IN DNA

An additional enzymatic property of DNA polymerase is its capacity to hydrolyze DNA chains under certain conditions. DNA polymerase can hydrolyze DNA progressively from the 3'-hydroxyl terminus of a DNA chain. The products are mononucleotides. Thus, *DNA polymerase I is a 3′ ⟶ 5′ exonuclease* (Figure 23-24). The nucleotide that is removed must have a free 3′-OH terminus, and it must not be part of a double helix. Is this exonuclease activity an undesirable side effect of the enzyme or does it contribute to the biological action of DNA polymerase? Experiments using chemically synthesized polynucleotides with a mismatched residue at the primer terminus have indicated that the *3′ ⟶ 5′ nuclease activity has an editing function in polymerization.* Consider the polymer shown in Figure 23-24 in which a sequence of dT residues forms a double helix with a longer polymer of dA. At the 3′ terminus of this poly dT sequence there is a single dC residue that is not hydrogen-bonded because it is not complementary to dA. Upon addition of DNA polymerase and dTTP, this mismatched dC residue was hydrolyzed before dT residues were added. Experiments with a variety of synthetic polymers have shown that *DNA polymerase I always removes mismatched residues at the primer terminus before proceeding with polymerization.* There is little or no hydrolysis if the terminus is properly matched and activated precursors are on hand. Polymerization prevents hydrolysis from the 3′ end.

It seems likely that DNA replication is very accurate because *base pairs are checked twice.* Polymerization usually does not occur unless the base pair fits into a double helix. However, if an error is made at this stage, it can be corrected before the next nucleotide is added. In effect, DNA polymerase I examines the result of each polymerization it catalyzes before going on to the next.

DNA polymerase I can also hydrolyze DNA starting from the 5′ end of a chain (Figure 23-24). This *5′ ⟶ 3′ nuclease* activity

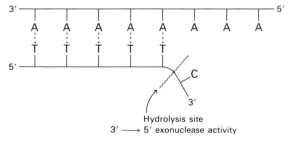

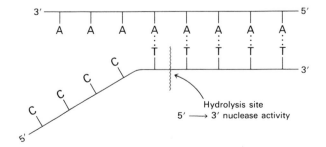

Figure 23-24
Nuclease activities of DNA polymerase I.

is very different from the 3′ ⟶ 5′ exonuclease activity discussed above. First, the cleaved bond must be in a double-helical region. Second, cleavage can occur at the terminal phosphodiester bond or at a bond several residues away from the 5′ terminus (which can be a free hydroxyl group or phosphorylated). Third, 5′ ⟶ 3′ nuclease activity is enhanced by concomitant DNA synthesis. Fourth, the active site for the 5′ ⟶ 3′ nuclease activity is entirely different from the active sites for polymerization and 3′ ⟶ 5′ hydrolysis. DNA polymerase I can be split into 75,000- and 36,000-dalton fragments by controlled proteolytic cleavage. The small fragment contains all of the original 5′ ⟶ 3′ nuclease activity, whereas the larger fragment retains the polymerase and 3′ ⟶ 5′ nuclease activities. Thus, *DNA polymerase I contains at least two distinct enzymes in one polypeptide chain.*

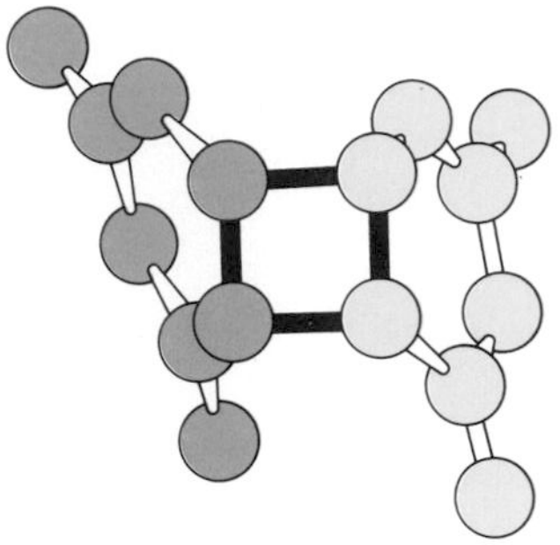

What is the physiological role of the 5′ ⟶ 3′ nuclease activity? It probably complements the 3′ ⟶ 5′ nuclease activity by correcting errors of a different type. *The 5′ ⟶ 3′ nuclease may play an important role in the excision of pyrimidine dimers that are formed following exposure of DNA to ultraviolet light.* Adjacent pyrimidine residues on a DNA strand can become covalently linked under these conditions. Such a pyrimidine dimer cannot fit into a double helix, and so replication is blocked unless the lesion is removed. Four enzymatic activities are essential for this repair process. First, an *endonuclease* detects the distorted region and makes a nick at or near the pyrimidine dimer. The segment containing the dimer peels away, thereby allowing *DNA polymerase* to carry out repair synthesis in the 5′ ⟶ 3′ direction. The intact complementary strand serves as the template. Then, the pyrimidine dimer region is excised by the *5′ ⟶ 3′ nuclease* activity of DNA polymerase. Finally, the newly synthesized stretch of DNA and the original DNA chain are joined by *DNA ligase.*

DISCOVERY OF OTHER DNA POLYMERASES

We have seen that DNA polymerase I can carry out two kinds of reactions in the test tube: (1) the synthesis of biologically active DNA and (2) the repair of DNA that contains mismatched bases or pyrimidine dimers. Does DNA polymerase I perform these functions in vivo? This question is pertinent because an enzyme may or may not carry out the same reaction *in vivo* as it does *in*

Figure 23-25
Model of a dimer of uracil produced by ultraviolet irradiation. A thymine dimer has nearly the same structure.

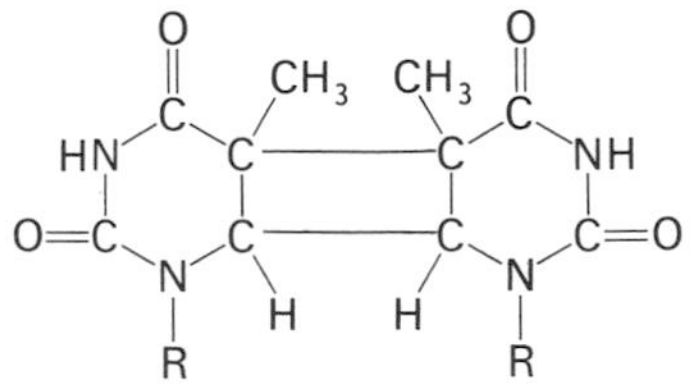

Thymine dimer

vitro. The reaction conditions inside a cell may be different and other enzymes may be present. In fact, *E. coli* contains at least two other DNA polymerases, named II and III. These polymerases were found approximately fifteen years after the discovery of DNA polymerase I. Why this lag? The reason is that the occurrence of polymerases II and III was masked by the high levels of activity due to polymerase I.

The breakthrough came in 1969, when Paula DeLucia and John Cairns isolated a mutant of *E. coli*. Its extract had 0.5 to 1% of the normal polymerizing activity of DNA polymerase I. This mutant (named *polA 1*) multiplied at the same rate as the parent strain. Furthermore, it was as susceptible to infection by a variety of phages as the parent strain. This mutant differed from the parent strain in two ways: it was much more readily killed by ultraviolet light, and it was more susceptible to the lethal effects of methylmethane-sulfonate, a chemical mutagen. DeLucia and Cairns inferred that DNA replication was normal in their *polA 1* mutant, whereas DNA repair was markedly impaired. They suggested that a polymerase different from DNA polymerase I might be essential for DNA synthesis.

The low background level of polymerase I in the *polA1* mutant facilitated the discovery of new DNA polymerases. Two such enzymes were soon isolated and characterized in several laboratories. *DNA polymerases II and III are like polymerase I in these respects:*

1. They catalyze a template-directed synthesis of DNA from deoxyribonucleoside triphosphate precursors. A primer with a free 3'-hydroxyl group is required. Synthesis is in the 5' $\longrightarrow$ 3' direction.

2. They possess 3' $\longrightarrow$ 5' exonuclease activity. Unpaired nucleotides at the 3' terminus of a DNA chain are hydrolyzed. The significance of this editing function has already been discussed in connection with DNA polymerase I.

DNA polymerase II and III *differ* from polymerase I in two important ways:

1. They lack a 5' $\longrightarrow$ 3' nuclease activity. Recall that this enzymatic activity plays an essential role in the repair of DNA containing pyrimidine dimers.

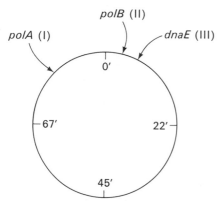

Figure 23-26
Locations of the structural genes for the three DNA polymerases in *E. coli*. DNA polymerase I maps at the *polA* locus, II at *polB*, and III at *dnaE*.

2. These enzymes have a strong preference for a template of double-stranded DNA with short gaps. In contrast, extensive single-stranded regions near double-helical regions are optimal DNA templates for polymerase I.

The physiological role of DNA polymerase II is uncertain. A mutant of *E. coli* whose extracts contain only about 0.1% of the normal activity of this enzyme has been isolated. This mutant (called *polB1*) grows normally and has the same sensitivity to ultraviolet radiation and chemical mutagens as does the wild type. In contrast, *DNA polymerase III is clearly essential for DNA replication in* E. coli. The evidence comes from a group of mutants that replicate DNA normally at 30°C but not at 45°C. Some of these temperature-sensitive mutants for DNA synthesis map at a locus called *dnaE*. It was found that the DNA polymerase III activities of these *dnaE* mutants are temperature-sensitive: essentially normal at 30°C but very low or absent at 45°C. Polymerase I and II activities are normal at both temperatures in these *dnaE* mutants. The close correspondence between the effect of temperature on DNA replication and on polymerase III activity in these mutants is strong evidence for the critical participation of DNA polymerase III in the duplication of DNA in *E. coli*.

We will return later in this chapter to the role of DNA polymerase I. For the moment, note that *DNA polymerase I is probably also essential for DNA replication*. A viable mutant totally devoid of polymerase I has not been isolated.

THE SYNTHESIS OF NEW DNA IS CLOSELY COUPLED TO THE UNWINDING OF PARENTAL DNA

DNA in the midst of replication has been visualized by autoradiography and by electron microscopy. In *autoradiography*, an image is formed by the decay of a suitable radioisotope, such as tritium. The emitted electrons affect the silver grains of a photographic emulsion. Black spots appear when the film is developed. The resolution of autoradiography is relatively low, of the order of several hundred angstroms. However, the technique is attractive because only what is labeled can be seen. DNA is made visible in autoradiographs by the incorporation of tritiated thymine or thymidine.

Autoradiographs and electron micrographs show the form of replicating DNA from *E. coli* to be a closed circle with an inner loop (Figure 23-27). Such forms are called theta structures because of their resemblance to the Greek letter θ. The theta structures show that the *DNA molecule maintains its circular form while it is being replicated.* The resolution of the technique does not suffice to display free ends. However, it is clear that long stretches of single-stranded DNA are absent. Thus, these pictures rule out a replication mechanism in which the parental DNA strands unwind completely before serving as templates for the synthesis of new DNA. Rather, *the synthesis of new DNA is closely coupled to the unwinding of parental DNA.* The site of simultaneous unwinding and synthesis is called the *replicating fork.*

DNA REPLICATION STARTS AT A UNIQUE ORIGIN AND PROCEEDS SEQUENTIALLY IN OPPOSITE DIRECTIONS

Does DNA replication in *E. coli* start anywhere in the chromosome or is there a specific initiation site? Since DNA replication is a rigorously controlled process, a specific initiation site seems much more likely a priori. In fact, DNA replication in *E. coli* starts at a unique origin as shown by studies of the relative amounts of different genes under conditions of rapid DNA synthesis. Consider two genes, *a* and *b*. Suppose that *a* is located near the starting point of replication, whereas *b* is near the finish. Then, gene *a* will be replicated well ahead of *b*. In a rapidly growing culture, there will be nearly two *a* genes for each *b* gene. In contrast, if DNA replication starts randomly, there will be equal amounts of *a* and *b*. The relative frequencies of a number of genes were determined by hybridization, a technique that will be discussed later (p. 600). The results of these experiments clearly showed that the relative gene frequency does depend on the map position as shown in Figure 23-28. These data revealed that:

1. *Replication starts at a unique site,* near the *ilv* gene, which is located at 74' on the standard genetic map of *E. coli.*

2. Replication proceeds simultaneously in the two opposite directions, at about the same velocity. In other words, *there are two replicating forks: one moves clockwise, the other counterclockwise.*

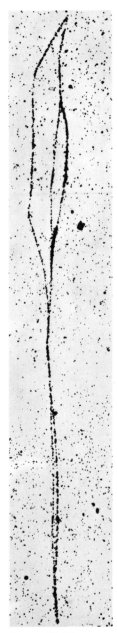

Figure 23-27
Autoradiograph of replicating DNA from *E. coli.* [Courtesy of Dr. John Cairns.]

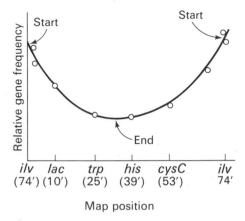

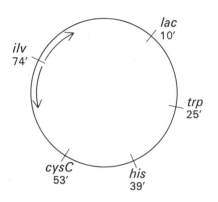

Figure 23-28
Relative amounts of different genes under
conditions of rapid DNA synthesis in *E.
coli* as a function of position on the
genetic map.

3. The two replicating forks meet near *trp* (25' on the map), a point diametrically opposite the origin of replication.

Further evidence for the bidirectionality of replication of *E. coli* DNA comes from autoradiography. Replication was initiated in a medium containing *moderately* radioactive tritiated thymine. After a few minutes incubation, the bacteria were transferred to a medium containing *highly* radioactive tritiated thymidine. Two levels of radioactivity were used to create two kinds of grain tracks in the autoradiographs: a low-density track corresponding to the DNA initially synthesized and a high-density track corresponding to the DNA synthesized later. If replication were unidirectional, tracks with a low grain density at one end and a high grain density at the other end would be seen. On the other hand, if replication were bidirectional, the middle of a track would have a low density (Figure 23-29). The autoradiographic patterns (Figure 23-30) provided a vivid answer. All of the grain tracks were denser on both ends than in the middle, indicating that *replication of the* E. coli *chromosome is bidirectional.*

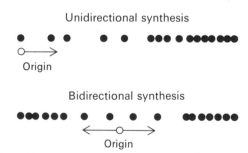

Figure 23-29
Autoradiographic pattern expected for unidirectional
replication and bidirectional replication, when
bacteria are transferred from a medium containing
moderately radioactive thymine to one containing
highly radioactive thymine.

Figure 23-30
Autoradiograph of *E. coli* DNA during replication (under
conditions described in Figure 23-29). The observed grain
pattern indicates that replication is bidirectional. [From D.
M. Prescott and P. L. Kuempel. *Proc. Nat. Acad. Sci.*
69(1972):2842.]

DNA IS SYNTHESIZED AS FRAGMENTS, WHICH THEN BECOME JOINED

Let us turn now to the enzymatic events at the replicating forks. At a replicating fork, both strands of parental DNA serve as templates for the synthesis of new DNA. Recall that the parental strands are antiparallel. Hence, the overall direction of DNA synthesis must be 5′ ⟶ 3′ for one daughter strand and 3′ ⟶ 5′ for the other (Figure 23-31). However, the three known DNA polymerases synthesize DNA in the 5′ ⟶ 3′ direction but not in the 3′ ⟶ 5′ direction. How then does one of the daughter DNA strands *appear at low resolution* to grow in the 3′ ⟶ 5′ direction?

This dilemma was resolved by Reiji Okazaki, who found that a *large proportion of newly synthesized DNA exists as small fragments.* These units of about 1000 nucleotides (sometimes called *Okazaki fragments*) occur momentarily in the vicinity of the replicating fork. As replication proceeds, these fragments become covalently joined into long daughter chains by a DNA ligase. Thus, *DNA replication is a discontinuous process.*

A plausible model for the synthesis of DNA is shown in Figure 23-32. In this scheme, fragments of both daughter strands are synthesized in the 5′ ⟶ 3′ direction by the same kind of DNA polymerase. Nascent fragments that are base-paired to the same template strand are then joined by DNA ligase. The overall result is that one daughter strand grows in the 5′ ⟶ 3′ direction whereas the other *appears* to grow in the 3′ ⟶ 5′ direction.

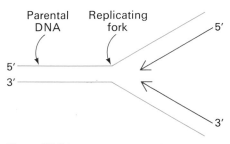

Figure 23-31
At *low resolution,* the *apparent* direction of DNA replication is 5′ ⟶ 3′ for one daughter strand and 3′ ⟶ 5′ for the other. Actually, both strands are synthesized in the 5′ ⟶ 3′ direction, as shown in Figure 23-32.

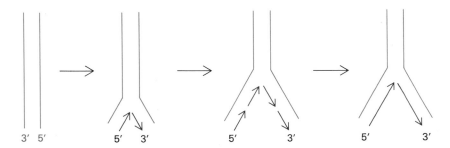

Figure 23-32
A model for DNA replication. Both daughter strands are synthesized in the 5′ ⟶ 3′ direction. The unwinding of parental DNA (blue) is coupled to the synthesis of new DNA (red). Nascent fragments that are base-paired to the same template strand become linked by DNA ligase.

DNA SYNTHESIS IS PRIMED BY RNA

How does DNA synthesis begin? Recall that all three DNA polymerases require a primer with a free 3'-hydroxyl group for the initiation of DNA synthesis. What is the primer in vivo? Recent studies have shown that RNA synthesis is essential for the initiation of DNA synthesis. Furthermore, a short stretch of RNA is covalently linked to newly synthesized DNA fragments. Thus, *RNA primes the synthesis of DNA*.

It has been proposed that DNA replication in *E. coli* is initiated in the following way:

1. An RNA polymerase synthesizes a short stretch of RNA (~ 100 nucleotides) that is complementary to one of the DNA template strands. RNA polymerase, in contrast to DNA polymerase, does not require a primer for the initiation of polynucleotide synthesis.

2. The 3'-hydroxyl group of the terminal ribonucleotide of this RNA chain serves as the primer for the synthesis of new DNA (a fragment of $\sim 1,000$ nucleotides), probably by DNA polymerase III.

3. The RNA portion of the RNA-DNA chain is then hydrolyzed (probably by DNA polymerase I), leaving only the newly synthesized DNA.

Figure 23-33
Proposed scheme for the initiation of DNA synthesis: (A) RNA polymerase synthesizes a short complementary stretch of RNA; (B) this RNA serves as the primer for the synthesis of new DNA; and (C) the RNA part of the new chain is hydrolyzed, leaving a gap that is then filled.

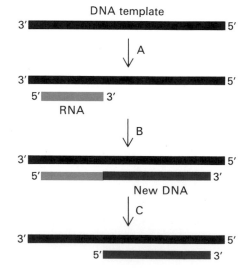

The removal of RNA from the nascent chains leaves substantial gaps between the DNA fragments. These gaps might be filled by means of the action of DNA polymerase I, which is well suited for synthesizing DNA in response to instructions from a single-stranded template. The resulting DNA fragments are then joined by DNA ligase. It is significant in this regard that the *polA1* mutant described previously (which has 0.5%–1% of the polymerizing activity of DNA polymerase I, a normal percentage of the $5' \longrightarrow 3'$ exonuclease activity, and a normal amount of DNA ligase) is defective in converting newly replicated DNA fragments into high-molecular-weight DNA.

MANY PROTEINS ARE REQUIRED FOR DNA SYNTHESIS

The replication of DNA is a complex process. Thus far the roles of the DNA polymerases, RNA polymerase, and DNA ligase have been discussed. Many other proteins participate also. In fact, in *E. coli*, more than twenty genes code for proteins that are essential for DNA replication. One of these proteins is probably an *endonuclease*. Closed circular DNA cannot serve as a template because its strands cannot unwind and separate. Cleavage of the backbone of at least one template strand seems essential for the initiation of DNA replication. The opening of local regions of double-helical DNA is facilitated by specific *unwinding proteins*. These proteins bind preferentially to single-stranded DNA. Furthermore, the process is cooperative: the binding of one unwinding protein to DNA enhances the binding of a second, resulting in more unwinding of the double helix. Unwinding proteins are likely to be active participants in replicating forks.

It will be interesting to see whether the enzymes and other proteins that participate in DNA replication are associated in vivo in an organized assembly, like the fatty acid synthetase complex of *E. coli*. A related area of inquiry is the role of membranes in DNA replication. Bacterial chromosomes are associated with an infolding of the cell membrane. In fact, the region of attachment is near the site on DNA for the initiation of replication. Future insights into the control of DNA replication will probably depend to a large extent on a deeper understanding of biological membranes.

Figure 23-34
An electron micrograph of the chromosome of *E. coli* attached to two fragments of the cell membrane. This single molecule of double-helical DNA is intact and supercoiled. [From H. Delius and A. Worcel. *J. Mol. Biol.* 82(1974):108.]

The focus of this chapter has been on the structure and replication of DNA in bacterial cells. Much less is known about the DNA of higher organisms. However, it is evident that the enzymatic processes involving DNA in *E. coli* are very pertinent to what occurs in mammalian cells. Mammalian DNA polymerases and DNA ligases, for example, are like their bacterial counterparts. On the other hand, the control of DNA replication and gene expression may be quite different in eucaryotes and procaryotes. The structure and function of eucaryotic chromosomes will be discussed in Chapter 28.

A SKIN CANCER SEEMS TO BE CAUSED BY DEFECTIVE REPAIR OF DNA

Xeroderma pigmentosum is a rare skin disease in humans. It is genetically transmitted as an autosomal recessive trait. The skin in an affected homozygote is extremely sensitive to sunlight or ultraviolet light. In infancy, severe changes in the skin become evident and worsen with time. The skin becomes dry and there is a marked atrophy of the dermis. Keratoses appear, the eyelids become scarred, and the cornea ulcerates. Skin cancer usually develops at several sites. Many patients die before age thirty from metastases of these malignant skin tumors.

Ultraviolet light produces pyrimidine dimers in human DNA as it does in *E. coli* DNA. Furthermore, the repair mechanisms seem to be similar. Studies of skin fibroblasts from patients with xeroderma pigmentosum have revealed the biochemical defect in this disease. In normal fibroblasts, half of the pyrimidine dimers produced by ultraviolet radiation are excised in less than 24 hours. In contrast, there is almost no excision of dimers in this amount of time in fibroblasts derived from patients with xeroderma pigmentosum. Which step in repair is blocked? The answer came from studies of the molecular weights of DNA strands derived from UV-irradiated fibroblasts. In normal cells, there is a marked reduction of the molecular weight of single-stranded DNA within a few hours after radiation. This reduction in molecular weight is caused by the first step in the repair process, namely cleavage of a DNA chain near the pyrimidine dimer. In contrast, a reduction of molecular weight is not evident after UV-irradiation of xeroderma pigmentosum cells. Thus, *this skin disease seems to be caused by a defect in the endonuclease that hydrolyzes the DNA backbone near a pyrimidine dimer. The drastic clinical consequences of this enzymatic defect emphasize the critical importance of the DNA repair process.*

SUMMARY

DNA is the molecule of heredity in all procaryotic and eucaryotic organisms. In viruses, the genetic material is either DNA or RNA. All cellular DNA consists of two very long, helical polynucleotide chains coiled around a common axis. The two strands of the double helix run in opposite directions. Most DNA molecules are circular. The sugar-phosphate backbone of each strand is on the outside of the double helix, whereas the purine and pyrimidine bases are inside. The two chains are held together by hydrogen bonds between pairs of bases. Adenine (A) is always paired with thymine (T), and guanine (G) is always paired with cytosine (C). Hence, one strand of a double helix is the complement of the other. Genetic information is encoded in the precise sequence of bases along a strand.

In the replication of DNA, the two strands of a double helix unwind and separate as new chains are synthesized. Each parent strand acts as a template for the formation of a new complementary strand. Thus, the replication of DNA is semiconservative—each daughter molecule receives one strand from the parent DNA molecule. The replication of DNA is a complex process carried out by many proteins, including three kinds of DNA polymerases and a DNA ligase. The activated precursors in the synthesis of DNA are the four deoxyribonucleoside 5'-triphosphates. The new strand is synthesized in the 5' $\longrightarrow$ 3' direction by a nucleophilic attack of the 3'-hydroxyl terminus of the primer strand on the innermost phosphorus atom of the incoming deoxyribonucleoside triphosphate. Most important, DNA polymerases catalyze the formation of a phosphodiester bond only if the base on the incoming nucleotide is complementary to the base on the template strand. In other words, DNA polymerases are template-directed enzymes. DNA polymerases I, II, and III also have a 3' $\longrightarrow$ 5' exonuclease activity that enhances the fidelity of replication by removing mismatched residues. In addition, DNA polymerase I has a 5' $\longrightarrow$ 3' nuclease activity that may be essential for the repair of DNA and for DNA replication.

DNA replication in *E. coli* starts at a unique origin and proceeds sequentially in opposite directions. At the replicating fork (where two strands of DNA become four), both strands of parental DNA serve as templates for the synthesis of new DNA. The replication of DNA is a discontinuous process. Fragments of both daughter strands are synthesized in the 5' $\longrightarrow$ 3' direction. The synthesis of these DNA fragments is preceded by the synthesis of RNA, which serves as the primer. The RNA part of the nascent RNA-DNA chain is then hydrolyzed and replaced by DNA. Nascent DNA fragments that are base-paired to the same template strand then become linked by DNA ligase in a reaction driven by NAD^+.

SELECTED READINGS

Where to start:

Kornberg, A., 1968. The synthesis of DNA. *Sci. Amer.* 219(4):64–70. [Available as *Sci. Amer. Offprint* 1124.]

Kornberg, A., 1974. *DNA Synthesis.* W. H. Freeman and Company. [An outstanding and highly readable book.]

Books on DNA:

Davidson, J. N., 1972. *The Biochemistry of the Nucleic Acids.* Academic Press.

Bloomfield, V. A., Crothers, D. M., and Tinoco, I., Jr., 1974. *Physical Chemistry of Nucleic Acids.* Harper and Row.

Cantoni, G. L., and Davies, D. R., (eds.), 1971. *Procedures in Nucleic Acid Research,* vol. 2. Harper and Row.

Discovery of the major concepts:

Avery, O. T., MacLeod, C. M., and McCarty, M., 1944. Studies on the chemical nature of the substance inducing transformation of pneumococcal types. Induction of transformation by a deoxyribonucleic acid fraction isolated from Pneumococcus Type III. *J. Exp. Med.* 79:137–158.

Hershey, A. D., and Chase, M., 1952. Independent functions of viral protein and nucleic acid in growth of bacteriophage. *J. Gen. Physiol.* 36:39–56.

Watson, J. D., and Crick, F. H. C., 1953a. Molecular structure of nucleic acid. A structure for deoxyribose nucleic acid. *Nature* 171:737–738.

Watson, J. D., and Crick, F. H. C., 1953b. Genetic implications of the structure of deoxyribonucleic acid. *Nature* 171:964–967.

Kornberg, A., 1960. Biologic synthesis of deoxyribonucleic acid. *Science* 131:1503–1508.

Meselson, M., and Stahl, F. W., 1958. The replication of DNA in *Escherichia coli. Proc. Nat. Acad. Sci.* 44:671–682.

Taylor, J. H., (ed.), 1965. *Selected Papers on Molecular Genetics.* Academic Press. [Contains the classic papers listed above.]

DNA polymerases:

Smith, D. W., 1973. DNA synthesis in prokaryotes. *Progr. Biophys. Mol. Biol.* 26:321–408.

DeLucia, P., and Cairns, J., 1969. Isolation of an *E. coli* strain with a mutation affecting DNA polymerase. *Nature* 224:1164–1169. [Discovery of the *polA1* mutant.]

Lehman, I. R., and Chien, J. R., 1973. Persistence of DNA polymerase I and its $5' \longrightarrow 3'$ exonuclease activity in *polA* mutants of *E. coli* K12. *J. Biol. Chem.* 248:7717–7723.

Gefter, M. L., Hirota, Y., Kornberg, T., Wechsler, J. A., and Barnoux, C., 1971.

Analysis of DNA polymerases II and III in mutants of *E. coli* thermosensitive for DNA synthesis. *Proc. Nat. Acad. Sci.* 68:3150–3153.

Hurwitz, J., and Wickner, S., 1974. Involvement of two protein factors and ATP in *in vitro* DNA synthesis catalyzed by DNA polymerase III of *E. coli. Proc. Nat. Acad. Sci.* 71:6–10.

Tait, R. C., and Smith, D. W., 1974. Roles for *E. coli* DNA polymerases I, II, and III in DNA replication. *Nature* 249:116–119.

Sugino, A., and Okazaki, R., 1973. RNA-linked DNA fragments *in vitro. Proc. Nat. Acad. Sci.* 70:88–92.

Origin and direction of DNA replication:

Prescott, D. M., and Kuempel, P. L., 1972. Bidirectional replication of the chromosome in *E. coli. Proc. Nat. Acad. Sci.* 69:2842–2845.

Bird, R. E., Louarn, J., Martuscelli, J., and Caro, L., 1972. Origin and sequence of chromosome replication in *E. coli. J. Mol. Biol.* 70:549–566.

DNA repair and xeroderma pigmentosum:

Kelley, R. B., Atkinson, M. R., Huberman, J. A., and Kornberg, A., 1969. Excision of thymine dimers and other mismatched sequences by DNA polymerase of *Escherichia coli. Nature* 224:495–501.

Setlow, R. B., and Setlow, J. K., 1972. Effects of radiation on polynucleotides. *Annu. Rev. Biophys. Bioeng.* 1:293–346.

Cleaver, J. E., 1968. Defective repair replication of DNA in xeroderma pigmentosum. *Nature* 218:652–656.

Setlow, R. B., Regan, J. D., German, J., and Carrier, W. L., 1969. Evidence that xeroderma pigmentosum cells do not perform the first step in the repair of ultraviolet damage to DNA. *Proc. Nat. Acad. Sci.* 64:1035–1041.

Reminiscences:

Cairns, J., Stent, G. S., and Watson, J. D., (eds.), 1966. *Phage and the Origins of Molecular Biology.* Cold Spring Harbor Laboratory. [A fascinating collection of reminiscences by some of the architects of molecular biology.]

Watson, J. D., 1968. *The Double Helix.* Atheneum. [A lively, personal account of the discovery of the structure of DNA and its biological implications.]

PROBLEMS

1. Write the complementary sequence (in the standard 5′ ⟶ 3′ notation) for:
   (a) GATCAA.
   (b) TCGAAC.
   (c) ACGCGT.
   (d) TACCAT.

2. The composition (in mole fraction units) of one of the strands of a double-helical DNA is [A] = 0.30 and [G] = 0.24.
   (a) What can you say about [T] and [C] for the same strand?
   (b) What can you say about [A], [G], [T], and [C] of the complementary strand?

3. The DNA of a deletion mutant of λ bacteriophage has a contour length of 15 μm instead of 17 μm. How many base pairs are missing from this mutant?

4. For a double-helical DNA, list the nearest-neighbor frequencies that must be equal to [AG], [GC], [TA], [GT], [CA], and [GA].

5. The double-helical structure of DNA is disrupted by heating. The transition from the helical to the disordered form takes place within a narrow temperature range, indicating that the process is cooperative. The temperature at the midpoint of the transition is called the *melting temperature* (T_m). For many DNA molecules, there is a linear relationship between T_m and the mole fraction of G-C base pairs (f_{GC}) In 0.2 M Na$^+$, this relationship is

$$T_m(°C) = 69.3 + 41 f_{GC}$$

Propose a structural basis for the dependence of T_m on f_{GC}.

6. What result would Meselson and Stahl have obtained if the replication of DNA were conservative? Give the expected distribution of DNA molecules after 1.0 and 2.0 generations for conservative replication (i.e., the parental double helix stays together).

7. How many high-energy phosphate bonds are consumed in the synthesis of DNA starting from ATP and the four deoxyribonucleoside 5′-monophosphates?

8. The joining of two DNA chains by known DNA ligases is driven by NAD$^+$ or ATP, depending on the species. Suppose that a new DNA ligase requiring a different energy donor is found. Propose a plausible substitute for NAD$^+$ or ATP in this reaction.

For additional problems, see W. B. Wood, J. H. Wilson, R. M. Benbow, and L. E. Hood, *Biochemistry: A Problems Approach* (Benjamin, 1974), Chapters 16 and 17.

CHAPTER 24

MESSENGER RNA
AND TRANSCRIPTION

We turn now to the expression of the genetic information contained in DNA. Genes act by determining the kinds of proteins that are made by cells. However, DNA is not the direct template for protein synthesis. Rather, the templates for protein synthesis are RNA (ribonucleic acid) molecules. The flow of genetic information in normal cells is:

$$\text{DNA} \xrightarrow{\text{transcription}} \text{RNA} \xrightarrow{\text{translation}} \text{protein}$$

In this chapter, evidence is presented that *certain RNA molecules are the information-carrying intermediates in protein synthesis, whereas other RNA molecules are part of the machinery of protein synthesis*. The synthesis of RNA according to instructions given by a DNA template is then considered. This process of *transcription* is followed by *translation,* in which RNA templates specify the synthesis of proteins, as will be discussed in Chapter 26.

The participation of RNA in protein synthesis seemed likely by 1940, several years before it was shown that DNA is the genetic material. Torbjörn Caspersson ascertained the distribution of RNA and DNA in single eucaryotic cells by light microscopy, making use of the fact that both RNA and DNA strongly absorb 260-nm ultraviolet light, whereas only DNA is intensely stained by the Feulgen reagent. He found that almost all of the DNA is in the nucleus, whereas most of the RNA is in the cytoplasm. Jean Brachet came to the same conclusion by separating cells into nuclear and cytoplasmic fractions and chemically analyzing them for DNA and RNA. Furthermore, he showed that the RNA in the cytoplasmic fraction is located in small particles, in association with protein. These particles consisting of RNA and protein were shown to be the sites of protein synthesis. They are now called *ribosomes*.

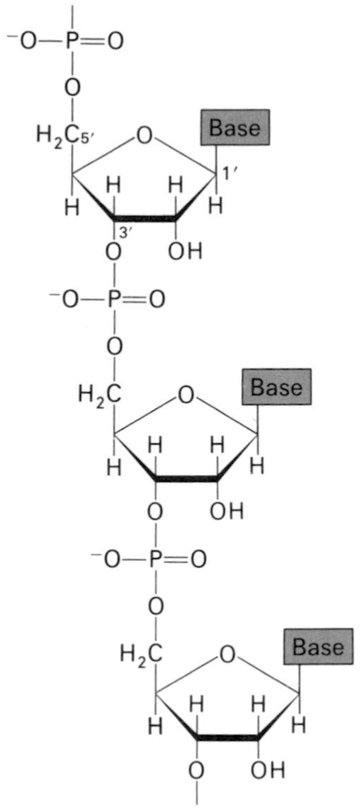

STRUCTURE OF RNA

RNA, like DNA, is a long, unbranched macromolecule consisting of nucleotides joined by $3' \longrightarrow 5'$ phosphodiester bonds. The number of nucleotides in RNA ranges from as few as 75 to many thousands. The covalent structure of RNA differs from that of DNA in two ways. As indicated by its name, the sugar unit in RNA is *ribose* rather than deoxyribose. Ribose contains a 2'-hydroxyl group not present in deoxyribose. The other difference is that one of the four major bases in RNA is *uracil* (U) instead of thymine. Uracil, like thymine, can form a base pair with adenine. However, it lacks the methyl group present in thymine.

RNA molecules are single-stranded, except in some viruses. Consequently, an RNA molecule need not have complementary base ratios. In fact, the proportion of adenine differs from that of uracil, and the proportion of guanine differs from that of cytosine, in most RNA molecules. However, RNA molecules do contain regions of double-helical structure that are produced by the formation of hairpin loops (Figure 24-2). In these regions, A pairs with U and

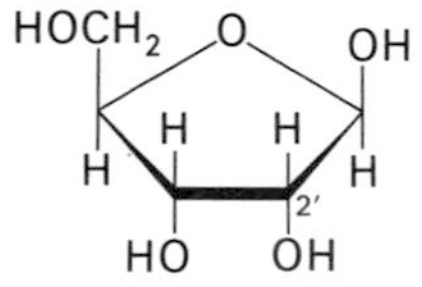

Figure 24-1
Structure of part of an RNA chain.

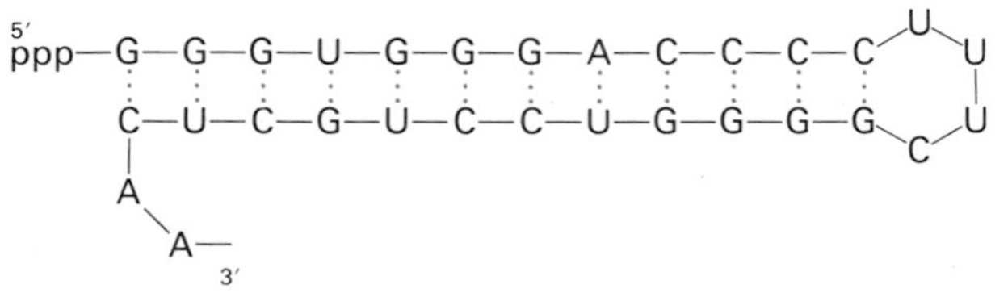

Figure 24-2.
RNA can fold back on itself to form double-helical regions.

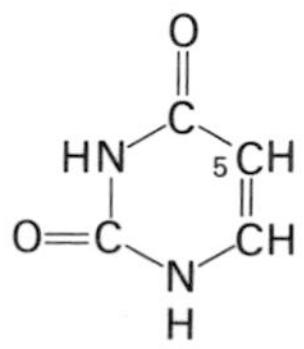

Ribose

Uracil

G pairs with C. G can also form a base pair with U, but it is less strong than the GC base pair (p. 656). The base pairing in RNA hairpins is frequently imperfect. Some of the apposing bases may not be complementary, and one or more bases along a single strand may be looped out to facilitate the pairing of the others. The proportion of helical regions in different kinds of RNA varies over a wide range; a value of 50% is typical.

CELLS CONTAIN THREE TYPES OF RNA: RIBOSOMAL, TRANSFER, AND MESSENGER

Cells contain three kinds of RNA (Table 24-1). Messenger RNA (mRNA) is the template for protein synthesis. There is an mRNA molecule corresponding to each gene or group of genes that is being expressed. Consequently, mRNA is a very heterogeneous class of molecules. In *E. coli*, the average length of an mRNA molecule is about 1200 nucleotides (400,000 daltons). *Transfer RNA* (tRNA) carries amino acids in an activated form to the ribosome for peptide-bond formation, in a sequence determined by the mRNA template. There is at least one kind of tRNA for each of the twenty amino acids. Transfer RNA consists of about 75 nucleotides (25,000 daltons), which makes it the smallest of the RNA molecules. *Ribosomal RNA* (rRNA) is the major component of ribosomes, but its precise role in protein synthesis is not yet known. In *E. coli*, there are three kinds of rRNA, called 23S, 16S, and 5S RNA because of their sedimentation behavior. One molecule of each of these species

Table 24-1
RNA molecules in *E. coli*

Type	Relative amount (%)	Sedimentation coefficient(S)	Molecular weight	Number of nucleotides
Ribosomal RNA (rRNA)	80	23	1.2×10^6	3,700
		16	0.55×10^6	1,700
		5	3.6×10^4	120
Transfer RNA (tRNA)	15	4	2.5×10^4	75
Messenger RNA (mRNA)	5	Heterogeneous		

of rRNA is present in each ribosome. Ribosomal RNA is the most abundant of the three types of RNA. Transfer RNA comes next, followed by messenger RNA, which comprises only 5% of the total RNA.

FORMULATION OF THE CONCEPT OF MESSENGER RNA

The concept of mRNA was formulated by Francois Jacob and Jacques Monod in a classical paper published in 1961. Since proteins are synthesized in the cytoplasm rather than in the nucleus of eucaryotic cells, it was evident that there must be a chemical intermediate specified by the genes, which they called the structural messenger. What is the nature of this intermediate? An important clue came from their studies of the control of protein synthesis in *E. coli* (Chapter 27). Certain enzymes in *E. coli,* such as those that participate in the uptake and utilization of lactose, are inducible— that is, the amount of these enzymes increases more than a thousandfold if an inducer (such as isopropylthiogalactoside) is present. The kinetics of induction were very revealing. The addition of an inducer elicited maximal synthesis of the lactose enzymes within a few minutes. Furthermore, the removal of the inducer resulted in the cessation of the synthesis of these enzymes in an equally short time. These experimental findings were incompatible with the presence of stable templates for the formation of these enzymes. Hence, Jacob and Monod surmised that *the messenger must be a very short-lived intermediate.*

They then proposed that the messenger should have the following properties:

1. The messenger should be a polynucleotide.

2. The base composition of the messenger should reflect the base composition of the DNA that specifies it.

3. The messenger should be very heterogeneous in size because genes (or groups of genes) vary in length. They correctly assumed that three nucleotides code for one amino acid and calculated that the molecular weight of a messenger should be at least a half million.

4. The messenger should be transiently associated with ribosomes, the sites of protein synthesis.

5. The messenger should be synthesized and degraded very rapidly. A corollary is that base analogs such as 5-fluorouracil should be incorporated into the messenger within a few minutes.

It was apparent to Jacob and Monod that none of the known RNA fractions at that time met these criteria. Ribosomal RNA, then generally assumed to be the template for protein synthesis, was too homogeneous in size. Also, its base composition was similar in species that had very different DNA base ratios. Furthermore, rRNA was stable and it incorporated 5-fluorouracil at a slow rate. Transfer RNA also seemed an unlikely candidate for the same reasons. In addition, it was too small. However, there were suggestions in the literature of a third class of RNA that appeared to meet the above criteria for the messenger. In *E. coli* infected with T2 bacteriophage, there was a new RNA fraction of appropriate size that had a very short half-life. Most interesting, the base composition of this new RNA fraction was like that of the viral DNA rather than like that of *E. coli* DNA.

EXPERIMENTAL EVIDENCE FOR MESSENGER RNA, THE INFORMATIONAL INTERMEDIATE IN PROTEIN SYNTHESIS

The hypothesis of a short-lived messenger RNA as the information-carrying intermediate in protein synthesis was tested shortly after the concept was formulated. Sydney Brenner, Francois Jacob, and Matthew Meselson carried out experiments on *E. coli* infected with T2 bacteriophage. Nearly all of the proteins made after infection are genetically determined by the phage. The synthesis of these proteins is not accompanied by net synthesis of RNA. However, a minor RNA fraction with a short half-life appears soon after infection. In fact, as mentioned above, this RNA fraction has a nucleotide composition like that of the phage DNA. This system appeared optimal for a test of the messenger hypothesis because there was a sudden switch in the kinds of proteins synthesized.

Another attractive feature was that rRNA and tRNA are not synthesized after infection.

How did this switch in the kinds of proteins made after infection occur? One possibility was that the phage DNA specifies a new set of ribosomes. In this model, genes control the synthesis of specialized ribosomes, and each ribosome can make only one kind of protein. An alternative model, the one proposed by Jacob and Monod, was that ribosomes are nonspecialized structures that receive genetic information from the gene in the form of an unstable messenger RNA. The experiments of Brenner, Jacob, and Meselson were designed to determine whether new ribosomes are synthesized after infection or whether new RNA joins preexisting ribosomes.

Bacteria were grown in a medium containing heavy isotopes (^{15}N and ^{13}C), infected with phage, and then immediately transferred to a medium containing light isotopes (^{14}N and ^{12}C). Constituents synthesized before and after infection could be separated by density-gradient centrifugation since their densities differed ("heavy" and "light," respectively). New RNA was labeled by the radioisotope ^{32}P or ^{14}C uracil and new protein by ^{35}S. These experiments showed that:

1. *Ribosomes were not synthesized after infection,* as evidenced by the absence of "light" ribosomes.

2. RNA was synthesized after infection. Most of the radioactively labeled RNA emerged in the "heavy" ribosome peak. Thus, *most of the new RNA was associated with preexisting ribosomes.* Additional experiments showed that this new RNA turns over rapidly during the growth of phage.

3. The radioisotope ^{35}S appeared transiently in the "heavy" ribosome peak, showing that *new proteins were synthesized in preexisting ribosomes.*

These experiments led to the conclusion that "*ribosomes are nonspecialized structures which synthesize, at a given time, the protein dictated by the messenger they happen to contain.*" Studies of uninfected bacterial cells also showed that messenger RNA is the information-carrying link between gene and protein. In a very short time, the concept of messenger RNA became a central facet of molecular biology.

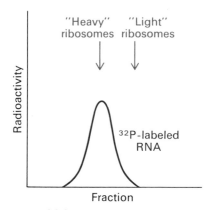

Figure 24-3
Density-gradient centrifugation of ribosomes and RNA of normal and T4 phage-infected bacteria. ^{32}P-labeled RNA synthesized *after* infection banded with ribosomes formed prior to infection ("heavy" ribosomes). Ribosomes were not synthesized after infection.

HYBRIDIZATION STUDIES SHOWED THAT MESSENGER RNA IS COMPLEMENTARY TO ITS DNA TEMPLATE

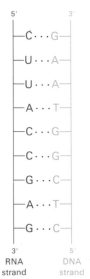

Figure 24-4
An RNA-DNA hybrid can be formed if they have complementary sequences.

In 1961, Sol Spiegelman developed a new technique called *hybridization* to answer this question: Is the base sequence of the RNA synthesized after infection with T2 phage complementary to the base sequence of T2 DNA? It was known from the work of Julius Marmur and Paul Doty that heating double-helical DNA above its melting temperature results in the formation of single strands. These strands reassociate to form a double-helical structure with biological activity if the mixture is cooled slowly. They also found that double-helical molecules are formed only from strands derived from the same or from closely related organisms. This observation suggested to Spiegelman that a double-stranded DNA-RNA hybrid might be formed from a mixture of single-stranded DNA and RNA if their base sequences were complementary. The experimental design was:

1. The RNA synthesized after infection of *E. coli* with T2 phage (T2 mRNA) was labeled with ^{32}P. T2 DNA labeled with ^{3}H was prepared in a separate experiment.

2. A mixture of T2 mRNA and T2 DNA was heated to 100°C, which melted the double-helical DNA into single strands. This solution of single-stranded RNA and DNA was slowly cooled to room temperature.

3. The cooled mixture was analyzed by density-gradient centrifugation. The samples were centrifuged for several days in swinging bucket rotors. The plastic sample tubes were then punctured at the bottom and drops were collected for analysis.

Three bands were found (Figure 24-5). The densest of these was single-stranded RNA. A second band contained double-helical DNA. A third band consisting of double-stranded DNA-RNA hybrid molecules was present near the DNA band. Thus, T2 mRNA formed a hybrid with T2 DNA. In contrast, T2 mRNA did not hybridize with DNA derived from a variety of bacteria and unrelated viruses, even if their base ratios were like those of T2 DNA. Subsequent experiments showed that the mRNA fraction of uninfected cells hybridized with the DNA derived from that

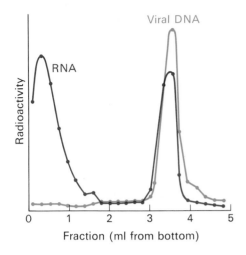

Figure 24-5
This hybridization experiment shows that the RNA produced after *E. coli* is infected with T2 phage is complementary to the viral DNA. RNA was labeled with ^{32}P, whereas T2 DNA was labeled with ^{3}H. The distribution of radioactivity in this cesium chloride density gradient shows that much of the RNA synthesized after infection bands with the T2 DNA. [Based on S. Spiegelman. Hybrid nucleic acids. Copyright © 1964 by Scientific American, Inc. All rights reserved.]

particular organism but not from unrelated ones. These incisive experiments revealed that *the base sequence of mRNA is complementary to that of its DNA template. Furthermore, a powerful tool for tracing the flow of genetic information in cells and for determining whether two nucleic acid molecules are similar was developed.*

RIBOSOMAL RNA AND TRANSFER RNA ARE ALSO SYNTHESIZED ON DNA TEMPLATES

The hybridization technique was then used to determine whether rRNA and tRNA are also synthesized on DNA templates. The formation of RNA-DNA hybrids was detected by a filter assay rather than by density-gradient centrifugation because it is simpler, more sensitive, and much more rapid. Single-stranded RNA passes through a nitrocellulose filter, whereas double-helical DNA and RNA-DNA hybrids are retained by this filter. RNA from *E. coli* labeled with ^{32}P was added to unlabeled *E. coli* DNA. This mixture was heated, slowly cooled, and then filtered through nitrocellulose. The radioactivity retained on the filter was counted. The results were unequivocal: *RNA-DNA hybrids were formed with both rRNA (5S, 16S, and 23S) and tRNA, showing that complementary sequences for these RNA molecules are present in the* E. coli *genome.*

ALL CELLULAR RNA IS SYNTHESIZED BY RNA POLYMERASE

The concept of mRNA stimulated the search for an enzyme that synthesizes RNA according to instructions given by a DNA template. The experimental strategy was like the one used in the search for DNA polymerase I. In 1960, researchers in several laboratories reported the discovery of such an enzyme, which was called RNA polymerase. RNA polymerase from *E. coli* requires the following components for the synthesis of RNA:

1. *Template.* The preferred template is *double-stranded DNA*. Single-stranded DNA can also serve as a template. RNA, whether single or double-stranded, is not an effective template, nor are RNA-DNA hybrids.

2. *Activated precursors.* All four *ribonucleoside triphosphates*—ATP, GTP, UTP, and CTP—are required.

3. *Divalent metal ion.* Mg^{2+} or Mn^{2+} are effective. Mg^{2+} meets this requirement in vivo.

RNA polymerase catalyzes the step-by-step elongation of an RNA chain. The reaction catalyzed by this enzyme is:

$$(RNA)_{n \text{ residues}} + \text{ribonucleoside triphosphate} \rightleftharpoons$$
$$(RNA)_{n+1 \text{ residues}} + PP_i$$

The synthesis of RNA is like that of DNA in several respects (Figure 24-6). First, the direction of synthesis is $5' \longrightarrow 3'$, as will be discussed shortly. Second, the mechanism of elongation appears to be similar. There is a nucleophilic attack of the 3'-OH group at the terminus of the growing chain on the innermost nucleotidyl phosphate of the incoming nucleoside triphosphate. Third, the synthesis is driven forward by the hydrolysis of pyrophosphate.

RNA synthesis differs from that of DNA in several important ways. First, RNA polymerase does not require a primer. Second, the DNA template is fully conserved in RNA synthesis, whereas it is semiconserved in DNA synthesis. Third, RNA polymerase has no known nuclease activities.

All three types of cellular RNA—mRNA, tRNA, and rRNA—are synthesized in *E. coli* by the same RNA polymerase according to

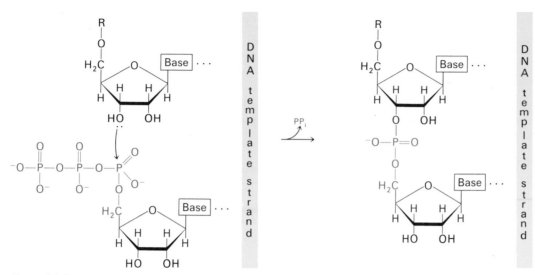

Figure 24-6
Mechanism of the chain-elongation reaction catalyzed by RNA polymerase.

instructions given by a DNA template. In mammalian cells, there may be some division of labor among several different kinds of RNA polymerases. It is also significant that some viruses code for RNA synthesizing enzymes that are very different from those of their host cells, as exemplified by the RNA polymerase specified by T7, a DNA phage, and by the RNA replicase that is encoded by Qβ, an RNA phage. The Qβ replicase is an RNA-dependent RNA polymerase since it takes instructions from an RNA template, rather than from a DNA template (Chapter 29). In contrast, the cellular enzymes that synthesize RNA are *DNA-dependent RNA polymerases.*

RNA POLYMERASE TAKES INSTRUCTIONS FROM A DNA TEMPLATE

Several lines of evidence have shown that RNA polymerase takes instructions from a DNA template. First the *base composition* of the newly synthesized RNA is the complement of that of the DNA template strand. If poly (dT), a synthetic polydeoxyribonucleotide containing only thymidylate residues, is used as the template, then only one ribonucleoside triphosphate (ATP) is incorporated into

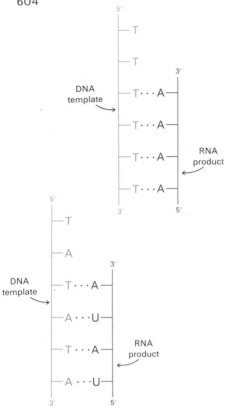

Figure 24-7
RNA polymerase takes instructions from a DNA template. The template DNA strand is shown in green and the new RNA strand in red.

Table 24-3
Nearest-neighbor frequencies of RNA synthesized from a viral DNA template

DNA template (plus strand of φX174)		RNA product	
CG	22	23	CG
CT	34	35	AG
CA	24	23	UG
CC	20	19	GG

the new polyribonucleotide chain. The product is polyriboadenylate (abbreviated as poly rA). If the alternating copolymer poly d(A-T) is used as the template for RNA polymerase, then UTP and ATP are incorporated and the product is poly r(A-U). The base composition of RNA synthesized using single-stranded φX174 DNA as template further demonstrates the complementary relationship between RNA product and DNA template (Table 24-2).

Table 24-2
Base composition of RNA synthesized from a viral DNA template

DNA template (plus strand of φX174)		RNA product	
Adenine	0.25	0.25	Uracil
Thymine	0.33	0.32	Adenine
Guanine	0.24	0.23	Cytosine
Cytosine	0.18	0.20	Guanine

Further information is provided by the *nearest-neighbor frequencies* of the newly synthesized RNA. The experimental technique used to determine the nearest-neighbor frequencies of an RNA molecule is similar to the one employed for DNA (p. 576). One of the four ribonucleoside triphosphates in the incubation mixture is labeled with ^{32}P in its innermost phosphate. The resulting RNA is split into a mixture of 2′- and 3′-ribonucleoside monophosphates by alkaline hydrolysis. Recall that a specific enzyme was used to obtain 3′-nucleoside monophosphates in DNA. RNA, in contrast to DNA, can be readily hydrolyzed by alkali because it contains a 2′-hydroxyl group that acts as an intramolecular catalyst in this reaction. The nearest-neighbor frequencies obtained in this way for RNA synthesized using the plus strand of φX174 DNA as template were revealing. The results given in Table 24-3 demonstrated that the sequence of the newly synthesized RNA is *complementary* and *antiparallel* to that of the DNA template strand, the same result as that obtained for DNA synthesis.

The third line of evidence that RNA polymerase takes instructions from a DNA template came from hybridization experiments. Such an analysis of the RNA synthesized in bacteria infected with T4 phage was particularly informative. First, ^{32}P-labeled RNA was extracted from T4-infected *E. coli* that was grown in a medium to which radioactive phosphate was added shortly after infection. Double-helical DNA from T4 phage was then heated to melt the helical structure and separate the complementary strands. Next, ^{32}P-labeled RNA was added to this hot solution of dissociated DNA strands and the mixture was then gradually cooled. Hybrids consisting of a phage DNA strand and a ^{32}P-labeled RNA strand then formed, as shown by their binding to a nitrocellulose filter, which does not bind RNA alone. In contrast, when *E. coli* DNA was used, there was little hybrid formed with ^{32}P-RNA from infected cells. This experiment, together with appropriate controls, showed that the *base sequences of most of the RNA formed after infection with T4 phage is complementary to the DNA of the infecting phage rather than to the host cell DNA.*

USUALLY ONLY ONE STRAND OF DNA IS TRANSCRIBED IN A PARTICULAR REGION OF THE GENOME

Are both strands of a double-helical DNA template transcribed or is transcription restricted to one of them? A priori, it seems unlikely that both strands of a particular region of DNA could code for functional proteins. Hybridization studies of *E. coli* infected with the φX174 phage provided one of the earliest answers to this question. The φX174 phage particles contain a single strand of DNA called the plus strand. After the plus strand enters the bacterium, a complementary minus strand of DNA is synthesized, yielding a circular, double-stranded DNA molecule termed the replicative form (RF). This RF DNA directs the synthesis of mRNA that in turn specifies the synthesis of phage proteins. Radioactive phage mRNA was prepared by adding ^{32}P-phosphate to the medium shortly after *E. coli* was infected with φX174. This labeled mRNA was isolated. Also, RF DNA was separated into its constituent plus and minus strands, which is feasible because they differ in base composition and hence in buoyant density. A hybridization assay was then carried out to determine whether the newly synthesized mRNA was complementary to the plus strand or to the minus strand or to both strands. A decisive result was obtained: only the minus strand of the RF DNA formed a hybrid with the labeled mRNA. Furthermore, the nearest-neighbor base frequencies of this mRNA were the complement of the minus strand of the RF DNA. Thus, *only one strand of the RF DNA of φX174 serves as a template for transcription in vivo.*

Only one strand of the DNA, also of viruses such as T7, SP8, and α, is transcribed in vivo. The situation for viruses such as the T4 and λ phages is more complex. In part of their genome, one strand serves as the template, whereas, in a different region, the other strand may be the template. The shift in the template in going from one group of genes to another also occurs in *E. coli.*

Figure 24-8
Electron micrograph showing the process of transcription. [From O. L. Miller, Jr., and Barbara A. Hamkalo. Visualization of RNA synthesis on chromosomes. *Int. Rev. Cytol.* 33(1972):1.]

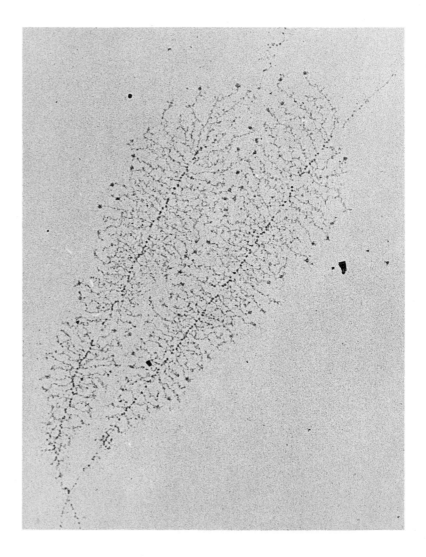

RNA POLYMERASE FROM E. COLI CONSISTS OF SUBUNITS

RNA polymerase from *E. coli* is a very large and complex enzyme. The molecular weight of the whole enzyme is nearly a half million. RNA polymerase is made up of subunits (Table 24-4), as can be shown by dissociating the enzyme in a concentrated urea solution. The subunit composition of the entire enzyme, called the *holoenzyme*, is $\alpha_2\beta\beta'\sigma$. As will be discussed shortly, the σ subunit dissociates from the enzyme following the initiation of RNA synthesis. RNA polymerase without the sigma subunit is called the *core enzyme* ($\alpha_2\beta\beta'$). The catalytic site of RNA polymerase is contained in the $\alpha_2\beta\beta'$ core. The β' and β subunits have been implicated in the binding of the DNA template. The σ subunit of the holoenzyme participates in the selection of initiation sites for transcription. It is interesting to note that the RNA polymerase encoded by the T7 bacteriophage is a much simpler enzyme than the one in *E. coli*.

The synthesis of RNA in vitro by *E. coli* RNA polymerase occurs in a series of steps, which are discussed below: (1) binding of the DNA template, (2) chain initiation, (3) chain elongation, and (4) chain termination.

Table 24-4

Subunits of RNA polymerase from *E. coli*

Subunit	Number	Molecular weight
α	2	40,000
β	1	155,000
β'	1	165,000
σ	1	95,000

THE SIGMA SUBUNIT INITIATES TRANSCRIPTION AT PROMOTER SITES

How does RNA polymerase know where to start transcribing the DNA template? The discovery of the sigma subunit provided insight into this initiation process. The sigma factor was discovered in the following way. When RNA polymerase was purified by chromatography on a phosphocellulose column, it was found that the resulting enzyme was nearly inactive when assayed with T4 DNA as template but still quite active with calf-thymus DNA as template. In contrast, RNA polymerase purified by centrifugation on a glycerol gradient was highly active with both templates. This observation suggested that some factor might be missing from the phosphocellulose-purified RNA polymerase. This was indeed the case (Figure 24-9). The activity of the phosphocellulose purified enzyme could be markedly enhanced by the addition of another fraction of the effluent of this column. This stimulatory fraction, called the σ factor, had no catalytic activity by itself. Further

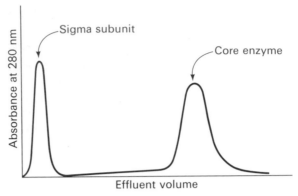

Figure 24-9
Resolution of RNA polymerase into the σ subunit and
the core enzyme $(\alpha_2\beta\beta')$ on a phosphocellulose column.

experiments showed that the phosphocellulose-purified enzyme
lacked the σ subunit, whereas the glycerol gradient-purified enzyme
retained this subunit. Thus, the enzyme that was fully active with
the T4 DNA template was the $\alpha_2\beta\beta'\sigma$ holoenzyme, whereas the
$\alpha_2\beta\beta'$ core enzyme was ineffective in transcribing this DNA. The
addition of σ to the core enzyme led to the reconstitution of the
fully active holoenzyme.

The sigma subunit stimulates the rate of initiation of new RNA
chains and enhances the specificity of transcription. The holoenzyme asymmetrically transcribes phage DNA templates, whereas
the core enzyme transcribes both strands. Thus, transcription by
the holoenzyme more closely resembles what occurs in vivo. *The
sigma subunit enhances the rate and specificity of transcription by enabling
RNA polymerase to recognize start signals on the DNA template.* These
start signals on DNA are called *promoter regions.*

The interaction of RNA polymerase with the DNA template is
thought to occur in several stages. First, RNA polymerase binds
to random sites on the DNA that it encounters by diffusion. The
binding of the enzyme to these sites is rapidly reversible. Then,
within a few seconds, RNA polymerase finds a promoter region on
the DNA. The affinity of the enzyme for the relatively few promoter
regions is much higher than for nonspecific sites. *The binding of RNA
polymerase at the promoter site then leads to a local unwinding of the DNA
helix.* The stage is set for the formation of the first phosphodiester
bond of the new RNA chain.

The σ subunit dissociates from the rest of the enzyme soon after the initiation of transcription in vitro. The core polymerase continues to transcribe the DNA template. The dissociated σ subunit can then be used by another polymerase molecule to initiate transcription. Thus, σ *catalyzes the initiation of transcription in vitro.* It is not yet known whether σ also acts this way *in vivo.*

RNA CHAINS START WITH pppG OR pppA

Most newly synthesized RNA chains carry a tag that reveals an aspect of how the RNA was made. The 5′ end of a new RNA chain is highly distinctive: *the molecule starts with either pppG or pppA.* In contrast to DNA synthesis, primer is not needed. *RNA chains can be formed de novo.*

This telltale tag at the 5′ terminus was discovered in two ways. It was found that RNA chains incorporated ^{32}P when the incubation mixture contained ^{32}P γ-labeled ATP. Clearly, the incorporated label had to be in a terminal position since only the α phosphorus atom of a nucleoside triphosphate can become part of internal phosphodiester bridges in the RNA. Furthermore, alkaline hydrolysis of newly synthesized RNA yielded three kinds of products: nucleosides, nucleoside 2′- (or 3′-) monophosphates, and nucleoside tetraphosphates. When ^{32}P γ-labeled ATP was used as a substrate, adenosine 3′-phosphate-5′-triphosphate was formed. The γ-phosphorus atom of this nucleoside tetraphosphate was labeled. The corresponding result was obtained when ^{32}P γ-labeled GTP was used. However, radioactivity was not incorporated when γ-labeled UTP or CTP were used. Hence, the structure of a new RNA chain is:

which yields the following species on alkaline hydrolysis.

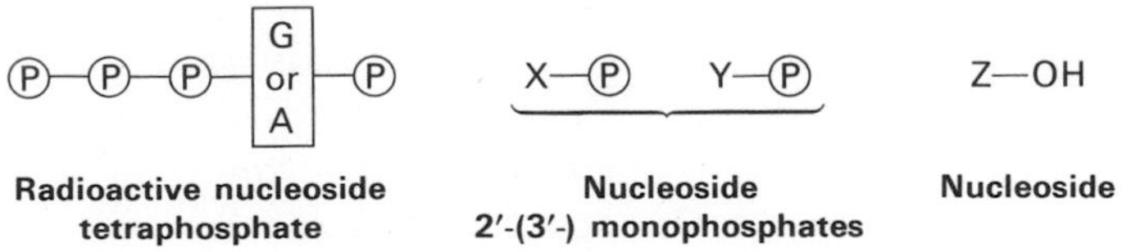

Radioactive nucleoside   **Nucleoside**   **Nucleoside**
tetraphosphate   **2'-(3'-) monophosphates**

Thus, a new RNA chain has a triphosphate group at its 5' terminus and a free hydroxyl group at its 3' terminus.

RNA CHAINS ARE SYNTHESIZED IN THE 5' ⟶ 3' DIRECTION

Is RNA synthesized in the $5' \longrightarrow 3'$ direction or in the $3' \longrightarrow 5'$ direction? Two contrasting mechanisms of chain growth are depicted in Figure 24-10. For $5' \longrightarrow 3'$ growth, the triphosphate terminus is inserted at the start of the synthesis of a chain, whereas in $3' \longrightarrow 5'$ growth, the triphosphate terminus comes from the last residue that is incorporated. The kinetics of incorporation of radioactivity into RNA synthesized from ^{32}P γ-labeled GTP (or ATP) distinguished between these alternatives. When ^{32}P γ-labeled GTP was used as a substrate, the ratio of ^{32}P incorporation to total nucleotide incorporation was highest shortly after the components

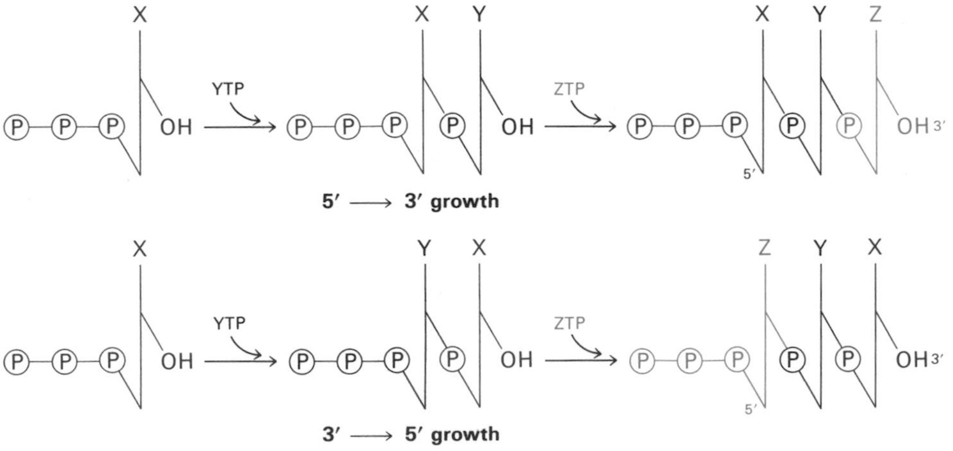

Figure 24-10
Location of ^{32}P expected for $5' \longrightarrow 3'$ growth and for $3' \longrightarrow 5'$ growth. The observed location of label shows that RNA chains are synthesized in the $5' \longrightarrow 3'$ direction.

were mixed and then decreased progressively with time. Furthermore, the total radioactivity of the RNA already labeled was not reduced by the subsequent addition of a large excess of nonradioactive GTP to the incubation mixture. Thus, ^{32}P entered the RNA molecule at the start of its synthesis rather than at the last step. *Hence, the growth of an RNA chain is in the 5'* $\longrightarrow$ *3' direction, as in DNA synthesis.* RNA polymerase moves along the DNA template strand in the 3' $\longrightarrow$ 5' direction, since the DNA template strand is antiparallel to the newly synthesized RNA strand. The region of DNA that has been transcribed regains its double-helical conformation, as the next section of DNA unwinds.

RHO, A PROTEIN FACTOR, PARTICIPATES IN CHAIN TERMINATION

There are specific signals on DNA to terminate transcription. Some of these signals can be recognized by RNA polymerase itself, whereas others are sensed by a specific protein called rho (ρ). Rho, a tetramer of about 200,000 daltons, binds to RNA polymerase. RNA molecules synthesized in vitro in the presence of ρ are shorter than those made in its absence. For example, RNA synthesized from fd phage DNA in the absence of ρ has a sedimentation coefficient of 26S. In contrast, 10S RNA is made in the presence of ρ. Further information about the action of ρ was obtained by adding this termination factor to the incubation mixture at various times after

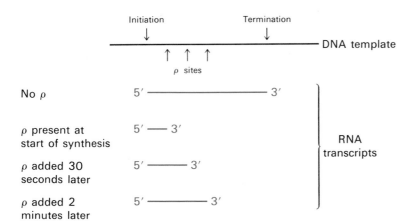

Figure 24-11
Effect of rho factor on the size of RNA transcripts.

the initiation of RNA synthesis. Species with sedimentation coefficients of 13S, 17S, and 23S were obtained when ρ was added a few seconds, two minutes, and ten minutes, respectively, after initiation. This result suggests that the template contains at least three termination sites that respond to ρ (yielding 10S, 13S, and 17S RNA), and one termination site that does not require ρ (yielding 23S RNA). Thus, specific termination can occur in the absence of ρ. However, ρ detects additional termination signals that are not read by RNA polymerase alone.

SOME RNA MOLECULES ARE CHEMICALLY MODIFIED AND CLEAVED AFTER TRANSCRIPTION

RNA molecules synthesized by RNA polymerase are frequently not the final products. Some nascent RNA chains are chemically modified or split in highly selective ways. These posttranscriptional events are sometimes called *maturation* or *processing*. For example, the 28S and 18S rRNA molecules in mammalian cells are not the direct products of transcription. Rather, the product of transcription is a 45S precursor (Figure 24-12), which is then methylated and cleaved. The 2'-hydroxyl group of approximately one of one

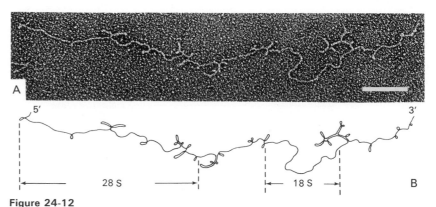

Figure 24-12
The 45S precursor of ribosomal RNA in HeLa cells has a highly distinctive pattern of hairpin loops when spread and examined by electron microscopy. This pattern makes it feasible to map the linear arrangement of 28S and 18S RNA molecules derived from this precursor. A. Electron micrograph of the 45S precursor (bar = 2000 Å). B. A tracing of the molecule shown in part A. [From P. K. Wellauer and I. B. Dawid. *Proc. Nat. Acad. Sci.* 70(1973):2828.]

hundred ribose units of this precursor is *methylated by S-adenosyl-methionine*. Specific enzymes then cleave this methylated RNA molecule in several steps to yield the 28S and 18S RNA molecules that are constituents of mammalian ribosomes. In bacteria, rRNA precursors are also cleaved and methylated, but *the predominant sites of methylation are the bases* rather than the ribose units. The functional significance of the methylation of RNA is a puzzle.

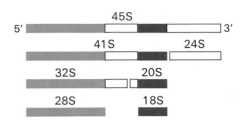

Figure 24-13
Formation of 28S and 18S mammalian ribosomal RNA from a 45S precursor.

HOCH$_2$ OH

2'

HO

CH$_3$

2'-O-Methylribose

H$_3$C   CH$_3$
N
C
N   C   N
6
CH
HC   C   N
N   H

6-Dimethyladenine

Some striking posttranscriptional modifications occur in the biosynthesis of tRNA. Transfer RNA molecules contain a monophosphate rather than a triphosphate group at the 5′ terminus, which suggests that they are formed by cleavage of larger precursors. Indeed they are. For example, in *E. coli* a precursor with 129 residues is trimmed by specific nucleases to give a tRNA molecule with 85 nucleotides. Furthermore, all tRNAs contain a number of modified bases in addition to the standard four ribonucleotides, as will be discussed in detail in the next chapter. These bases are formed by enzymatic modification of the tRNA molecule, perhaps after it is trimmed to its final size. Pseudouridylate and ribothymidylate, for example, are formed by modification of uridylate residues after transcription.

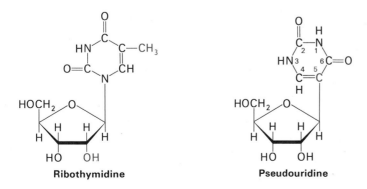

Ribothymidine

Pseudouridine

ANTIBIOTIC INHIBITORS OF TRANSCRIPTION: RIFAMYCIN AND ACTINOMYCIN

Rifamycin B

Antibiotics are interesting molecules because many of them are highly specific inhibitors of biological processes. Actinomycin and rifamycin are two antibiotics that inhibit transcription in quite different ways. Rifamycin, which is derived from *Streptomyces,* and rifampicin, a semisynthetic derivative, *specifically inhibit the initiation of RNA synthesis.* The binding of RNA polymerase to the DNA template is not blocked by this antibiotic. Rather, rifampicin interferes with the formation of the first phosphodiester bond in the RNA chain. In contrast, chain elongation is not appreciably affected by this antibiotic. This high degree of selectivity in its inhibitory action makes rifampicin a very useful experimental tool. For example, it can be used to block the initiation of new RNA chains without interfering with the transcription of chains that are already being synthesized. The site of action of rifampicin appears to be the *β subunit of RNA polymerase.* Mutants of *E. coli* that are resistant to rifampicin (called *rif-r mutants*) have been isolated. In some of these mutants, the *β* subunit has an altered electrophoretic mobility.

Actinomycin D, a polypeptide-containing antibiotic from *Streptomyces,* inhibits transcription by an entirely different mechanism from that of rifampicin. *Actinomycin D binds tightly to double-helical DNA and thereby prevents it from being an effective template for RNA synthesis.* It consists of two identical cyclic peptides joined to a phenoxazone ring system (Figure 24-14). The composition of these cyclic peptides is unusual in that sarcosine, methylvaline, and D-valine are present. Also, there is an ester bond between the hydroxyl group of threonine and the carboxyl group of methylvaline.

Actinomycin D binds tightly to double-helical DNA but not to single-stranded DNA or RNA, double-stranded RNA, or RNA-DNA hybrids. Furthermore, the binding of actinomycin to DNA is markedly enhanced by the *presence of guanine residues.* Spectroscopic and hydrodynamic studies of complexes of actinomycin D and DNA have suggested that the phenoxazone ring of actinomycin slips in between neighboring base pairs in DNA. This mode of binding is called *intercalation* (Figure 24-16). At low concentrations, *actinomycin D inhibits transcription without appreciably affecting*

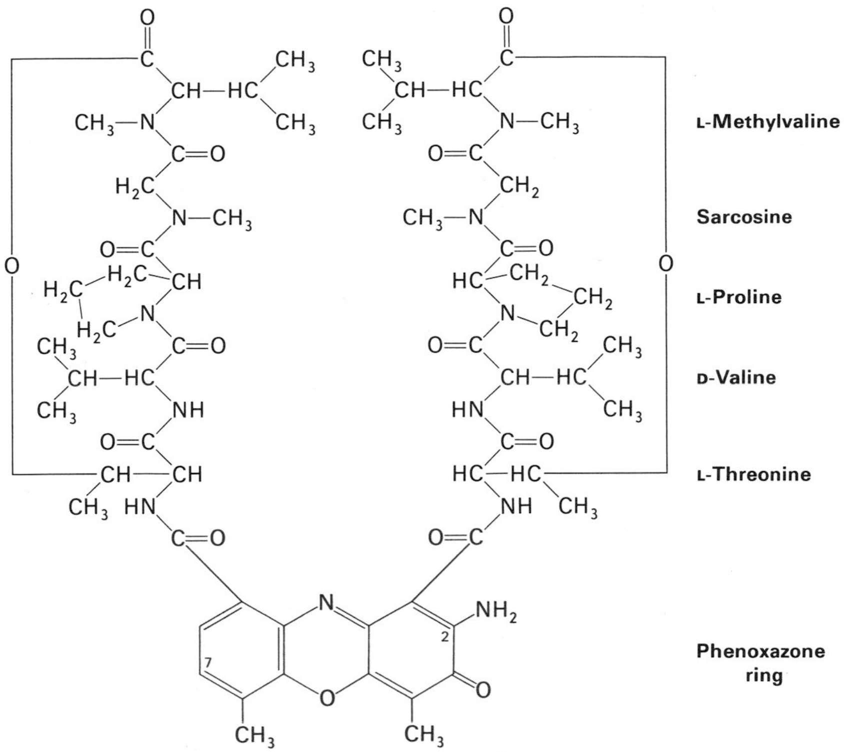

L-Methylvaline

Sarcosine

L-Proline

D-Valine

L-Threonine

Phenoxazone ring

Actinomycin D

Figure 24-14
Structure of actinomycin D.

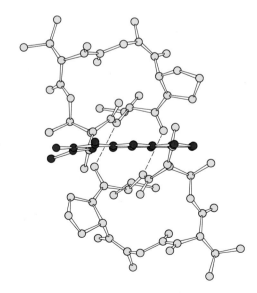

Figure 24-15
Model of actinomycin D. The phenoxazone ring is shown in red and the cyclic peptides in yellow. [Courtesy of Dr. Henry Sobell.]

DNA replication. Also, protein synthesis is not directly inhibited by actino-mycin. For these reasons, actinomycin D has been extensively used as a highly specific inhibitor of the formation of new RNA in both procaryotic and eucaryotic cells.

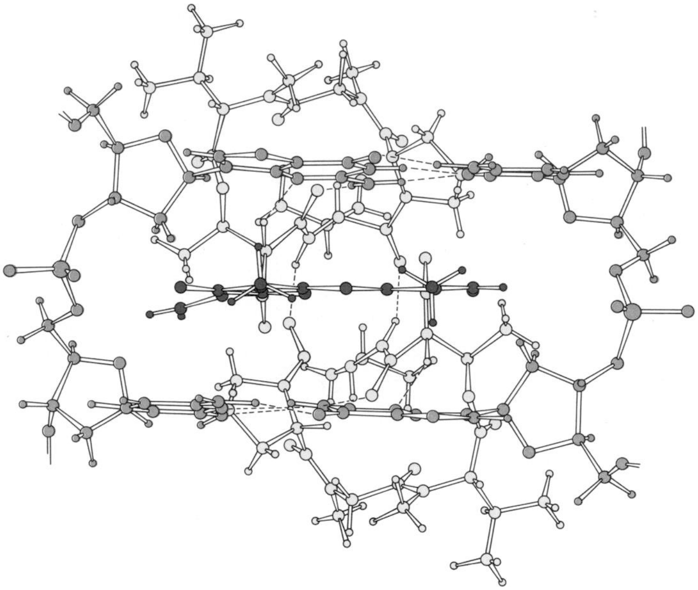

Figure 24-16
Proposed structure of the complex of actinomycin D and DNA. The phenoxazone ring of actinomycin D (shown in red) is intercalated between two GC base pairs of the DNA (shown in blue). The cyclic peptides of actinomycin D (shown in yellow) bind to the narrow groove of the DNA helix. There are several hydrogen bonds between actinomycin D and the adjacent guanines. The axis of symmetry relating the subunits on actinomycin coincides with the axis of symmetry relating the sugar-phosphate backbone and base sequence on the DNA helix. [Courtesy of Dr. Henry Sobell.]

The structure of a crystalline complex of one actinomycin molecule and two deoxyguanosine molecules has recently been solved at atomic resolution (Figure 24-16). The phenoxazone ring of actinomycin is sandwiched between the two guanine rings in this complex. One of the cyclic peptides is located above the phenoxazone ring, the other below. Each of these peptides forms a strong hydrogen bond with the 2-amino group of a guanine residue. There are numerous favorable van der Waals interactions between the antibiotic and the nucleosides. A significant feature of this complex is that it is nearly symmetric. There is a twofold axis of symmetry along the line joining the central O and N atoms of the phenoxazone ring. The conformation of this complex suggested that *actinomycin recognizes the base sequence GpC in DNA.* Note that if the sequence along one strand of DNA is 5′ GpC 3′ then it must be 3′

CpG 5′ along the complementary strand. It seems likely that actinomycin intercalates between two GC base pairs in DNA and interacts with the G residues in much the same way as it does in its complex with deoxyguanosine. In this model, the cyclic peptide units are located in the minor groove of the DNA helix. *A key aspect of this model is that the symmetry of actinomycin matches the symmetry of DNA.* We will return to the role of symmetry in protein-DNA interactions in a discussion of the control of gene expression (p. 685).

POWERFUL METHODS FOR THE SEQUENCING OF RNA CHAINS HAVE BEEN DEVELOPED

We have already seen that the determination of the amino acid sequences of proteins has greatly contributed to our understanding of the function and evolution of proteins. In recent years, important advances have also been made in the sequencing of RNA. The presence of four common bases rather than twenty makes nucleic acid sequencing more difficult because there is a much higher probability of recurrence of a particular base sequence in an RNA molecule than of an amino acid sequence in a protein. For example, the sequence AAUG occurs many times in a long RNA chain, whereas the sequence Ala-Ala-Glu-Lys has a very low probability of occuring more than once in a long polypeptide chain. This difficulty is less severe for tRNA, where there are a number of distinctive minor bases, such as pseudouridine. Indeed, the first RNA molecule to be sequenced was a tRNA molecule. This landmark accomplishment by Robert Holley in 1965 will be discussed in Chapter 26.

The overall strategy of RNA sequencing is like that of protein sequencing:

1. The molecule is specifically split into small fragments.

2. The fragments are separated.

3. The composition and then the sequence of the small fragments are determined.

4. The order of these small fragments is established by obtaining a second set of fragments that overlaps the first.

Table 24-5

Enzymes used in RNA sequencing

Enzyme	Type	Substrate	Products
Ribonuclease T_1	Endonuclease (and exonuclease)	—XpGpYpZ—	—XpGp + YpZ—
Pancreatic ribonuclease	Endonuclease (and exonuclease)	—XpUpYpZ—	—XpUp + YpZ—
		—XpCpYpZ—	—XpCp + YpZ—
Ribonuclease U_2	Endonuclease (and exonuclease)	—XpGpYpZ—	—XpGp + YpZ—
		—XpApYpZ—	—XpAp + YpZ—
Alkaline phosphatase (from *E. coli*)	Phosphatase	—YpZp	—YpZ + P_i
		pXpY—	XpY— + P_i
Bovine spleen phosphodiesterase	Exonuclease	XpYpZ—	Xp + YpZ—
Snake venom phosphodiesterase	Exonuclease	—XpYpZ	—XpY + pZ
Polynucleotide kinase	Kinase	ATP + XpYp—	pXpYp— + ADP

Specific fragments of RNA are obtained by enzymatic hydrolysis (Table 24-5). For example, ribonuclease T_1 (isolated from a mold) cleaves RNA chains on the 3′ side of G residues, yielding fragments that terminate with -Gp.

Pancreatic ribonuclease, on the other hand, hydrolyzes RNA on the 3′ side of pyrimidine residues to give fragments that terminate with either -Up or -Cp. Enzymes that remove or add a terminal phosphoryl group (Table 24-5) are also very helpful in sequence determination. RNA chains can be fragmented in a nonspecific way by alkaline hydrolysis.

A powerful fingerprinting method for the separation and identification of oligonucleotide fragments has been developed by Frederick Sanger. Fragments obtained from an enzymatic digest of ^{32}P-labeled RNA are first separated by electrophoresis on cellulose acetate and then subjected to ion-exchange chromatography and electrophoresis on paper in a perpendicular direction. This sheet of paper is then placed on photographic film. After development, black spots appear on the film where it was in contact with a radioactive fragment (Figure 24-17). The base composition of each fragment can be deduced from its position in the fingerprint. Furthermore, the fingerprinting procedure does not destroy the oligonucleotide, which can be eluted. The sequence of each of these small fragments is then readily determined.

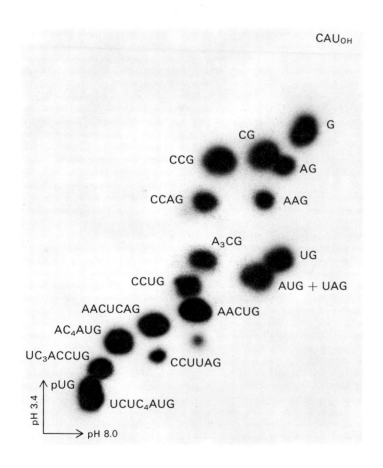

Figure 24-17
Fingerprint of a sample of 5S RNA that was digested by ribonuclease T_1. [Courtesy of Dr. Beverly Griffin.]

The complete sequence of a 5S rRNA, which contains 120 nucleotides, has been elucidated. Also, the complete sequences of many tRNA molecules and the partial sequences of a variety of other RNA molecules are known. This information is making a great impact on our understanding of the molecular basis of gene expression. The direct sequencing of DNA, on the other hand, is not as far along, in part because of the paucity of enzymes for cleavage at specific bases. However, DNA can be sequenced by determining the sequence of its RNA transcript. This experimental approach has been used, for example, to determine the base sequence of the operator region of the lactose operon in *E. coli* (p. 684).

SUMMARY

The flow of genetic information in normal cells is from DNA to RNA to protein. The synthesis of an RNA from a DNA template is called transcription, whereas the synthesis of a protein from an RNA template is termed translation. There are three kinds of RNA molecules: messenger RNA (mRNA), transfer RNA (tRNA), and ribosomal RNA (rRNA). These RNA molecules are single-stranded. Transfer RNA and ribosomal RNA contain double-helical regions that arise from the folding of the chain into hairpins. The smallest RNA molecules are the tRNAs, which contain about 75 nucleotides, whereas the largest ones are among the mRNAs, which may have more than 5000 nucleotides.

All cellular RNA is synthesized by RNA polymerase according to instructions given by a DNA template. The activated intermediates are ribonucleoside triphosphates. The direction of RNA synthesis is $5' \longrightarrow 3'$, like that of DNA synthesis. RNA polymerase differs from DNA polymerase in not requiring a primer and in not having nuclease activities. Another difference is that the DNA template is fully conserved in RNA synthesis, whereas it is semiconserved in DNA synthesis. RNA polymerase from *E. coli* has the subunit structure $\alpha_2\beta\beta'\sigma$ and a molecular weight of nearly a half million. The core enzyme consists of $\alpha_2\beta\beta'$. The σ subunit enhances the rate and specificity of transcription by enabling RNA polymerase to recognize start signals on the DNA template. The binding of RNA polymerase to these promoter regions leads to a local unwinding of the DNA template, setting the stage for the formation of the first phosphodiester bond. RNA chains start with pppG or pppA. There also are specific signals on DNA to terminate transcription. Some of them are sensed by a specific protein called rho, which binds to RNA polymerase. Some RNA molecules are

cleaved and chemically modified (e.g., methylated) after transcription. Transcription can be blocked by specific inhibitors. For
example, the antibiotic rifampicin inhibits the initiation of RNA
synthesis, whereas actinomycin D blocks the elongation of RNA.
The sequences of long RNA chains can be elucidated by the use
of specific hydrolytic enzymes and fingerprinting methods.

SELECTED READINGS

Where to start:

Miller, O. L., Jr., 1973. The visualization of
genes in action. *Sci. Amer.* 228(3):34–42.
[Available as *Sci. Amer.* Offprint 1267.]

Spiegelman, S., 1964. Hybrid nucleic acids.
Sci. Amer. 210(5):48–56. [Offprint 183.]

Sobell, H. M., 1974. How actinomycin binds
to DNA. *Sci. Amer.* 231(2):82–91. [Offprint
1303.]

Books and reviews:

Cold Spring Harbor Laboratory, 1970. *Transcription of Genetic Material* (Cold Spring Harbor Symposia of Quantitative Biology, vol.
35). [A most valuable collection of 97 articles on the transcription of bacterial, animal, and viral DNA.]

Chamberlin, M. J., 1974. The selectivity of
transcription. *Annu. Rev. Biochem.* 43:721–
776.

Losick, R., 1972. In vitro transcription. *Annu.
Rev. Biochem.* 41:409–446.

Chamberlin, M., 1970. Transcription 1970: a
summary. *Cold Spring Harbor Symp. Quant.
Biol.* 35:851–873.

Weinberg, T. A., 1973. Nuclear RNA metabolism. *Annu. Rev. Biochem.* 42:329–354.

Brownlee, G., 1972. *Determination of Sequences in
RNA*. American Elsevier.

Discovery of messenger RNA:

Jacob, F., and Monod, J., 1961. Genetic regulatory mechanisms in the synthesis of proteins. *J. Mol. Biol.* 3:318–356.

Brenner, S., Jacob, F., and Meselson, M., 1961.
An unstable intermediate carrying information from genes to ribosomes for protein
synthesis. *Nature* 190:576–581.

Hall, B. D., and Spiegelman, S., 1961. Sequence complementarity of T2-DNA and
T2-specific RNA. *Proc. Nat. Acad. Sci.*
47:137–146.

RNA polymerases:

Berg, P., and Chamberlin, M., 1962. Deoxyribonucleic acid-directed synthesis of ribonucleic acid by an enzyme from *Escherichia
coli*. *Proc. Nat. Acad. Sci.* 48:81–94.

Chamberlin, M., and McGrath, J., 1970.
Characterization of a T7-specific RNA
polymerase isolated from *E. coli* infected
with T7 phage. *Cold Spring Harbor Symp.
Quant. Biol.* 35:259–262.

Initiation and termination of RNA synthesis:

Hayashi, M., Hayashi, M. N., and Spiegelman, S., 1963. Restriction of in vivo genetic
transcription to one of the complementary
strands of DNA. *Proc. Nat. Acad. Sci.*
50:664–672.

Taylor, K., Hradecna, Z., and Szybalski, W., 1967. Asymmetric distribution of the transcribing regions on the complementary strands of coliphage λ DNA. *Proc. Nat. Acad. Sci.* 57:1618–1625.

Burgess, R. R., Travers, A. A., Dunn, J. J., and Bautz, E. K. F., 1969. Factor stimulating transcription by RNA polymerase. *Nature* 221:43–46.

DeCrombrugghe, B., Adhya, S., Gottesman, M., and Pastan, I., 1973. Effect of rho on transcription of bacterial operons. *Nature New Biol.* 241:260–264.

Processing of RNA:

Altman, S., and Robertson, H. D., 1973. RNA precursor molecules and ribonucleases in *E. coli. Mol. Cell. Biochem.* 1:83–93.

Antibiotic inhibitors of RNA synthesis:

Goldberg, I. H., and Friedman, P. A., 1971. Antibiotics and nucleic acids. *Annu. Rev. Biochem.* 40:775–810.

Sobell, H. M., 1973. The stereochemistry of actinomycin binding to DNA and its implications in molecular biology. *Progr. Nucl. Acid Res. Mol. Biol.* 13:153–190.

PROBLEMS

1. Compare DNA polymerase I and RNA polymerase from *E. coli* in regard to each of the following features:
   (a) Subunit structure.
   (b) Activated precursors.
   (c) Direction of chain elongation.
   (d) Nuclease activities.
   (e) Conservation of the template.
   (f) Need for a primer.
   (g) Energetics of the elongation reaction.

2. Write the sequence of the mRNA molecule synthesized from a DNA template strand having the sequence

   5' ATCGTACCGTTA 3'

3. Which nearest-neighbor frequencies of mRNA are equal to each of the following dinucleotide frequencies of the template DNA strand?
   (a) AT   (b) GC   (c) CA   (d) TG

4. RNA is readily hydrolyzed by alkali, whereas DNA is not. Why?

5. How does cordycepin (3'-deoxyadenosine) block the synthesis of RNA?

6. What are the expected products of the digestion of the oligoribonucleotide

   5' pGCAGUACUGUC 3'

   by each of the following enzymes?

   (a) Pancreatic ribonuclease.
   (b) Ribonuclease T_1.
   (c) Ribonuclease U_2.

7. Hydrolysis of an oligoribonucleotide by pancreatic ribonuclease yields Cp, 2 Up, AGCp, and GAAUp. Hydrolysis of the same oligonucleotide by ribonuclease T_1 gives AAUp, UAGp, and CCUGp? What is its sequence?

For additional problems, see W. B. Wood, J. H. Wilson, R. M. Benbow, and L. E. Hood, *Biochemistry: A Problems Approach* (Benjamin, 1974), Chapters 16 and 17.

CHAPTER 25

THE GENETIC CODE AND
GENE-PROTEIN RELATIONSHIPS

In the preceding chapter, we saw that messenger RNA is the information-carrying intermediate between the gene and its polypeptide product. In this chapter, we continue our consideration of the flow of information from gene to protein. The central theme here is the genetic code, which is the relationship between the sequence of bases in DNA (or its mRNA transcript) and the sequence of amino acids in a protein. The code, which is the same in all organisms, is beautiful in its simplicity. Three bases, called a codon, specify an amino acid. Codons are read sequentially by transfer RNA (tRNA) molecules, which serve as adaptors in protein synthesis. The complete elucidation of the genetic code in the 1960s is one of the outstanding accomplishments of modern biology.

TRANSFER RNA IS THE ADAPTOR MOLECULE
IN PROTEIN SYNTHESIS

We have seen that mRNA is the template for protein synthesis. How does it direct amino acids to become joined in the correct sequence? In 1958, Francis Crick wrote:

> One's first naive idea is that the RNA will take up a configuration capable of forming twenty different "cavities," one for the side chain of each of the twenty amino acids. If this were so, one might expect

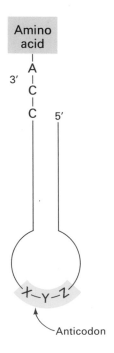

Transfer RNA

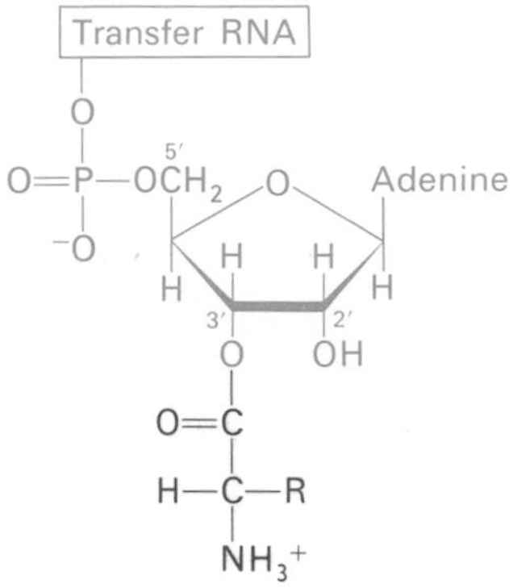

Figure 25-1
Mode of attachment of an amino acid
(shown in red) to a tRNA molecule. The
amino acid is esterified to the 3'-hydroxyl
group of the terminal adenosine of tRNA.
A tRNA having an attached amino acid
is called an aminoacyl-tRNA or a
"charged" tRNA, whereas a tRNA
without an attached amino acid is called
"uncharged."

to be able to play the problem backwards—that is, to find the configuration of RNA by trying to form such cavities. All attempts to
do this have failed, and on physical-chemical grounds the idea does
not seem in the least plausible.

He observed that RNA does not have the knobby hydrophobic
surfaces to distinguish valine from leucine and isoleucine, nor does
it have properly positioned charged groups to discriminate between
positively and negatively charged amino acid side chains. Crick
then proposed an entirely different mechanism for recognition of
the mRNA template:

> . . . RNA presents mainly a sequence of sites where hydrogen
> bonding could occur. One would expect, therefore, that whatever went
> on to the template in a *specific* way did so by forming hydrogen bonds.
> It is therefore a natural hypothesis that the amino acid is carried to
> the template by an adaptor molecule, and that the adaptor is the part
> which actually fits onto the RNA. In its simplest form one would
> require twenty adaptors, one for each amino acid.

This highly innovative hypothesis soon became an established
fact. *The adaptor in protein synthesis is tRNA.* The structure and
reactions of these remarkable adaptor molecules will be considered
in detail in the next chapter. For the moment it suffices to note
that tRNA contains an *amino acid attachment site* and a *template
recognition site* (Figures 25-1 and 25-2). A tRNA molecule carries a
specific amino acid in an activated form to the site of protein
synthesis. The carboxyl group of this amino acid is esterified to the
3'- (or 2'-) hydroxyl group of the ribose unit at the 3' end of the
tRNA chain. The esterified amino acid may migrate between the
2'- and 3'-hydroxyl groups during protein synthesis. The joining
of an amino acid to a tRNA to form an aminoacyl-tRNA is catalyzed by a specific enzyme called an aminoacyl-tRNA synthetase
(or activating enzyme). This esterification reaction is driven by
ATP. There is at least one specific synthetase for each of the twenty
amino acids. The template recognition site on tRNA is a sequence

Figure 25-2
A symbolic diagram of an aminoacyl-tRNA,
showing the amino acid attachment site and the
anticodon, which is the template recognition site.

of three bases called the *anticodon* (Figure 25-2). The anticodon on tRNA recognizes a complementary sequence of three bases on mRNA, called the *codon*.

AMINO ACIDS ARE CODED BY GROUPS OF THREE BASES STARTING FROM A FIXED POINT

The *genetic code* is the relationship between the sequence of bases in DNA (or its RNA transcripts) and the sequence of amino acids in proteins. Experiments by Francis Crick, Sydney Brenner, and others established the following features of the genetic code by 1961:

1. *What is the coding ratio?* A single base can specify only four kinds of amino acids since there are four kinds of bases in DNA. Sixteen kinds of amino acids can be specified by two bases ($4 \times 4 = 16$), whereas sixty-four kinds of amino acids can be determined by three bases ($4 \times 4 \times 4 = 64$). Since proteins are built from a basic set of twenty amino acids, it was evident from this simple calculation that three or more bases are probably needed to specify one amino acid. Genetic experiments then showed that *a group of three bases in fact codes for one amino acid*. This group of bases is called a *codon*.

2. *Is the code nonoverlapping or overlapping?* In a nonoverlapping triplet code, each group of three bases specifies only one amino acid, whereas in a completely overlapping triplet code, ABC specifies the first amino acid, BCD the second, CDE the third, and so forth. These alternatives were distinguished by studies of the sequence of amino acids in mutants. Suppose that the base C is mutated to C'. In a nonoverlapping code, only amino acid 1 will be changed. In a completely overlapping code, amino acids 1, 2, and 3 will be altered by a mutation of C to C'. Studies of the sequence of amino acids in the coat protein of mutants of tobacco mosaic virus showed that only a single amino acid was usually altered. Also, recall from our discussion of abnormal hemoglobins (Chapter 5) that only one amino acid is changed in most mutants. Hence, it was concluded that the *genetic code is nonoverlapping*.

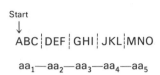

3. *How is the correct group of three bases read?* One possibility a priori is that one of the four bases (denoted as Q) serves as a "comma" between groups of three bases:

$$\ldots \text{QABCQDEFQGHIQJKLQ} \ldots$$

This turned out not to be the case. Rather, *the sequence of bases is read sequentially from a fixed starting point.* There are no commas. Suppose that a mutation deletes base G:

Start
↓   G (deleted)
ABC ┊ DEF ┊ HIJ ┊ KLM ┊ NOP

aa₁—aa₂—aa₃—aa₄—aa₅
   Normal   Altered

The first two amino acids in this polypeptide chain will be normal, but the rest of this base sequence will be read incorrectly because the *reading frame has been shifted* by the deletion of G. Now suppose that a base Z has been added between F and G:

Start
↓   Insertion
ABC ┊ DEF ┊ ZGH ┊ IJK ┊ LMN

aa₁—aa₂—aa₃—aa₄—aa₅
   Normal   Altered

This addition mutation also disrupts the reading frame starting at the codon for amino acid 3. In fact, genetic studies of recombinants of addition and deletion mutants revealed many of the features of the genetic code.

4. As mentioned above, there are sixty-four possible base triplets and twenty amino acids. Is there just one triplet for each of the twenty amino acids or are some amino acids coded by more than one triplet? Genetic studies showed that most of the sixty-four triplets do code for amino acids. Subsequent biochemical studies demonstrated that sixty-one of the sixty-four triplets specify a particular amino acid. Thus, *for most amino acids, there is more than one code word.* In other words, the genetic code is *degenerate.*

DECIPHERING THE GENETIC CODE: SYNTHETIC RNA
CAN SERVE AS MESSENGER

What is the relationship between sixty-four kinds of code words and the twenty kinds of amino acids? In principle, this question can be directly answered by comparing the amino acid sequence of a protein with the corresponding base sequence of its gene or mRNA. However, this approach was not experimentally feasible in 1961 because the base sequences of genes and mRNA molecules were entirely unknown. The breaking of the genetic code then did not seem imminent, but the situation suddenly changed. Marshall Nirenberg discovered that *the addition of polyuridylate (poly U) to a cell-free protein-synthesizing system led to the synthesis of polyphenylalanine.* Poly U evidently served as a messenger RNA. The first code word was deciphered: UUU codes for phenylalanine. This remarkable experiment pointed the way to the complete elucidation of the genetic code.

Let us consider this landmark experiment in more detail. The two essential components were a cell-free system that actively synthesizes protein and a synthetic polyribonucleotide that serves as the messenger RNA. A cell-free protein-synthesizing system was obtained from *E. coli* in the following way. Bacterial cells were gently broken open by grinding them with finely powdered alumina to yield a cell sap. Cell-wall and cell-membrane fragments were then removed by centrifugation. The resulting extract contained DNA, mRNA, tRNA, ribosomes, enzymes, and other cell constituents. Protein was synthesized by this cell-free system when it was supplemented with ATP, GTP, and amino acids. At least one of the added amino acids was radioactive so that its incorporation into protein could be detected. This mixture was incubated at 37°C for about an hour. Trichloroacetic acid was then added to terminate the reaction and precipitate the proteins, leaving the free amino acids in the supernatant. The precipitate was washed and its radioactivity was then counted to determine how much labeled amino acid was incorporated into newly synthesized protein. A crucial feature of this system is that protein synthesis can be halted by the addition of deoxyribonuclease, which destroys the template for the synthesis of new mRNA. The mRNA present at the time of addition of deoxyribonuclease is labile, and so protein synthesis stops within a few minutes. Nirenberg then found that protein synthesis resumed on addition of a crude fraction of mRNA. Thus, here was a *cell-free protein-synthesizing system that was responsive to the addition of mRNA*.

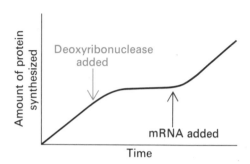

Figure 25-3
Protein synthesis in a cell-free system stops a few minutes after the addition of deoxyribonuclease and resumes following the addition of mRNA.

The other critical component in this experiment was a synthetic polyribonucleotide—namely, poly U. Poly U was synthesized by using *polynucleotide phosphorylase,* an enzyme discovered in 1955 by Marianne Grunberg-Manago and Severo Ochoa. This enzyme catalyzes the synthesis of polyribonucleotides from ribonucleoside diphosphates:

$$(RNA)_n + \text{ribonucleoside diphosphate} \rightleftharpoons (RNA)_{n+1} + P_i$$

The reactions catalyzed by this enzyme and by RNA polymerase are very different. In this reaction, the activated substrates are ribonucleoside diphosphates rather than triphosphates. Orthophosphate is a product instead of pyrophosphate. Hence, this reaction cannot be driven to the right by the hydrolysis of pyrophosphate. Indeed, the equilibrium in vivo is in the direction of RNA degradation, not synthesis. A critical difference is that *polynucleotide phosphorylase does not utilize a template*. The RNA synthesized by this enzyme has a composition dictated by the ratios of the ribonucleotides present in the incubation mixture and a sequence that is essentially random. For this reason, polynucleotide phosphorylase proved to be a most valuable experimental tool in deciphering the genetic code. Thus, poly U was synthesized by incubating high concentrations of UDP in the presence of this enzyme. Copolymers of two ribonucleotides, say U and A, with a random sequence were prepared by incubating UDP and ADP with this enzyme.

A variety of synthetic ribonucleotides were added to this cell-free protein-synthesizing system and the incorporation of ^{14}C-labeled L-phenylalanine was measured. The results were striking:

Polyribonucleotide added	[14]C counts per minute
None	44
Poly A	50
Poly C	38
Poly U	39,800

The same experiment was carried out with a different ^{14}C-labeled amino acid in each incubation mixture. It was found that *poly A led to the synthesis of polylysine* and that *poly C led to the synthesis of polyproline*. Thus, three code words were deciphered:

Code word	Amino acid
UUU	Phenylalanine
AAA	Lysine
CCC	Proline

The code word GGG could not be deciphered in the same way because poly G does not act as a template, probably because it forms a triple-stranded helical structure. Polyribonucleotides with extensive regions of ordered structure are ineffective templates for protein synthesis.

CODON COMPOSITIONS FOR MANY AMINO ACIDS WERE DETERMINED BY USING COPOLYMERS AS TEMPLATES

Polyribonucleotides consisting of two kinds of bases were used as templates to elucidate the genetic code further. For example, a random copolymer of U and G contains eight different triplets: UUU, UUG, UGU, GUU, UGG, GUG, GGU, and GGG. The expected incidence of these triplets can be readily calculated from the mole proportions of U and G in the copolymer, which were 0.76 and 0.24, respectively (Table 25-1). The extent of incorporation of various amino acids elicited by this template is given in Table 25-2. Phenylalanine was incorporated to the greatest extent, as expected, since the triplet UUU was the most prevalent one. Valine, leucine, and cysteine came next. The extent to which they were incorporated was sightly more than a third of that of phenylalanine, which corresponds with the calculated incidence of triplets that contain one U and two G. The incorporation of other amino acids was very low. Thus, *it was concluded that valine, leucine, and cysteine are coded by codons that contain 2U and 1G, whereas tryptophan and glycine are coded by codons that contain 1U and 2G.*

The same type of experiment was carried out with other random copolymers—namely, UA, UC, AC, AG, and CG, and also UGC, AGC, UAC, and UAG. The *compositions* of codons corresponding to each of the twenty amino acids were deduced in this way in the laboratories of Nirenberg and of Ochoa.

Table 25-1
Expected incidence of triplets in a random copolymer of U (0.76) and G (0.24)

Triplet	Probability	Relative incidence
UUU	$.76 \times .76 \times .76 = .439$	100
UUG	$.76 \times .76 \times .24 = .139$	31.6
UGU	$.76 \times .24 \times .76 = .139$	31.6
GUU	$.24 \times .76 \times .24 = .139$	31.6
UGG	$.76 \times .24 \times .24 = .0438$	10.0
GUG	$.24 \times .76 \times .24 = .0438$	10.0
GGU	$.24 \times .24 \times .76 = .0438$	10.0
GGG	$.24 \times .24 \times .24 = .0138$	3.1

Table 25-2
Amino acid incorporation stimulated by a random copolymer of U (0.76) and G (0.24)

Amino acid	Relative amount incorporated	Inferred codon composition
Phenylalanine	100	UUU
Valine	37	2U, 1G
Leucine	36	2U, 1G
Cysteine	35	2U, 1G
Tryptophan	14	1U, 2G
Glycine	12	1U, 2G

TRINUCLEOTIDES PROMOTE THE BINDING OF SPECIFIC TRANSFER RNA MOLECULES TO RIBOSOMES

The use of mixed copolymers as templates revealed the *composition but not the sequence of codons* corresponding to particular amino acids (except for UUU, AAA, and CCC). As described above, valine is coded by a triplet with 2U and 1G. Is it UUG, UGU, or GUU? The answer to this question was provided by two entirely different experimental approaches: (1) the use of synthetic polyribonucleotides with an ordered sequence; and (2) the codon-dependent binding of specific tRNA molecules to ribosomes.

In 1964, Nirenberg discovered that *trinucleotides promote the binding of specific tRNA molecules to ribosomes in the absence of protein synthesis.* For example, the addition of pUpUpU led to the binding of phenylalanine tRNA, whereas pApApA markedly enhanced the binding of lysine tRNA, as did pCpCpC for proline tRNA. Dinucleotides were ineffective in stimulating the binding of tRNA to ribosomes. These studies showed that a *trinucleotide (like a triplet in mRNA) specifically binds the particular tRNA molecule for which it is the code word.* A simple and rapid binding assay was devised: tRNA molecules bound to ribosomes are retained by cellulose nitrate filters, whereas unbound tRNA passes through the filter. The kind of tRNA retained by the filter was identified by using tRNA molecules charged with a particular ^{14}C-labeled amino acid.

All sixty-four trinucleotides were synthesized by organic chemical or enzymatic techniques. For each trinucleotide, the binding of tRNA molecules corresponding to all twenty amino acids was assayed. For example, it was found that pUpUpG stimulated the binding of leucine tRNA only, pUpGpU stimulated the binding of cysteine tRNA only, and pGpUpU stimulated the binding of valine tRNA only. Hence, it was concluded that the codons UUG, UGU, and GUU correspond to leucine, cysteine, and valine, respectively. For a few codons, no tRNA was strongly bound, whereas, for a few others, more than one kind of tRNA was bound. For most codons, clean-cut binding results were obtained. In fact, about fifty codons were deciphered by this simple and elegant experimental approach.

COPOLYMERS WITH A DEFINED SEQUENCE WERE ALSO INSTRUMENTAL IN BREAKING THE CODE

At about the same time, H. Gobind Khorana succeeded in synthesizing polyribonucleotides with a defined repeating sequence. A variety of copolymers with two, three, and four kinds of bases were synthesized by a combination of organic chemical and enzymatic techniques. Let us consider the strategy for the synthesis of poly (GUA), for example. This ordered copolymer has the sequence:

GUAGUAGUAGUAGUAGUAGUAGUA...

First, Khorana synthesized by organic chemical methods two complementary deoxyribonucleotides, each with nine residues: $d(TAC)_3$ and $d(GTA)_3$. These two oligonucleotides then served as templates for the synthesis of long DNA chains from the four deoxyribonucleoside triphosphates by DNA polymerase I. Either oligonucleotide alone was not an effective template. If both were present, $d(TAC)_3$ was the template for the enzymatic synthesis of poly $d(GTA)$, whereas $d(GTA)_3$ was the template for the synthesis of poly $d(TAC)$. These long complementary DNA chains formed a double-helical structure. The next step was to obtain long polyribonucleotide chains with a sequence corresponding to poly $d(TAC)$ and poly $d(GTA)$. This was accomplished by using the poly $d(TAC)$:poly $d(GTA)$ duplex as the template for synthesis by RNA polymerase. The DNA strand to be transcribed could be selected by adding three appropriate ribonucleoside triphosphates. When GTP, UTP, and ATP were added to the incubation mixture, the polyribonucleotide product synthesized from the poly $d(TAC)$ template strand was poly (GUA). The other strand was not transcribed because CTP, one of the required substrates, was missing. When CTP, UTP, and ATP were added, poly (UAC) was synthesized from the other template strand. Thus, *organic synthesis, followed by template-directed syntheses carried out by DNA polymerase and then RNA polymerase, yielded two long polyribonucleotides having defined repeating sequences* (Figure 25-4).

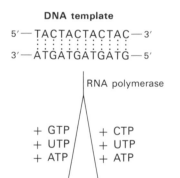

DNA template

5′—TACTACTACTAC—3′

3′—ATGATGATGATG—5′

RNA polymerase

+ GTP   + CTP
+ UTP   + UTP
+ ATP   + ATP

5′-GUAGUAGUAGUA-3′   5′-UACUACUACUAC-3′

Figure 25-4
The strand of this double-helical DNA template to be transcribed is determined by the choice of ribonucleoside triphosphates in the incubation mixture.

These regular copolymers were used as templates in the cell-free protein-synthesizing system. Let us examine some of the results. A copolymer consisting of an alternating sequence of two bases P and Q

PQP ┊ QPQ ┊ PQP ┊ QPQ ┊ PQP ┊ . . .

contains two codons, PQP and QPQ. The polypeptide product

should therefore contain an alternating sequence of two kinds of amino acids (abbreviated aa$_1$ and aa$_2$):

$$aa_1—aa_2—aa_1—aa_2—aa_1—$$

Whether aa$_1$ or aa$_2$ is amino-terminal in the polypeptide product depends on whether the reading frame starts at P or Q. When poly (UG) was the template, a polypeptide with an alternating sequence of cysteine and valine was synthesized. This result unequivocally confirmed the triplet nature of the code and showed that either UGU or GUG codes for cysteine and that the other of these two triplets codes for valine. When considered together with tRNA binding data, it was evident that UGU codes for cysteine and GUG codes for valine. The polypeptides synthesized in response to several other alternating copolymers of two bases were:

UGU﹒GUG﹒UGU﹒GUG﹒UGU﹒GUG

Cys—Val—Cys—Val—Cys—Val

Template	Product	Codon Assignments			
Poly (UC)	Poly (Ser-Leu)	UCU ⟶ Ser	CUC ⟶ Leu		
Poly (AG)	Poly (Arg-Gln)	AGA ⟶ Arg	GAG ⟶ Gln		
Poly (AC)	⁻Poly (Thr-His)	ACA ⟶ Thr	CAC ⟶ His		

Now let us consider a template consisting of three bases in a repeating sequence, poly (PQR). If the reading frame starts at P, the resulting polypeptide should contain only one kind of amino acid, the one coded by the triplet PQR:

Start
↓
PQR﹒PQR﹒PQR﹒PQR﹒PQR﹒PQR. . .

$$aa_1—aa_1—aa_1—aa_1—aa_1—aa_1$$

However, if the reading frame starts at Q, the polypeptide synthesized should contain a different amino acid, the one coded by the triplet QRP:

Start
↓
P﹒QRP﹒QRP﹒QRP﹒QRP﹒QRP﹒QRP. . .

$$aa_2—aa_2—aa_2—aa_2—aa_2—aa_2$$

If the reading frame starts at R, a third kind of polypeptide should be formed, containing only the amino acid coded by the triplet RPQ:

Start
↓
PQ ┊ RPQ ┊ RPQ ┊ RPQ ┊ RPQ ┊ RPQ ┊ RPQ . . .

$aa_3 - aa_3 - aa_3 - aa_3 - aa_3 - aa_3$

Table 25-3

Homopolypeptides synthesized using messengers containing repeating trinucleotide sequences

Messenger	Homopolypeptides synthesized
Poly (UUC)	Phe; Ser; Leu
Poly (AAG)	Lys; Glu; Arg
Poly (UUG)	Cys; Leu; Val
Poly (CCA)	Gln; Thr; Asn
Poly (GUA)	Val; Ser
Poly (UAC)	Tyr; Thr; Leu
Poly (AUC)	Ile; Ser; His
Poly (GAU)	Met; Asp

Thus, the expected products are three different homopolypeptides. In fact, this was observed for most templates consisting of repeating sequences of three nucleotides. For example, poly (UUC) led to the synthesis of polyphenylalanine, polyserine, and polyleucine. This result, taken together with the outcome of other experiments, showed that UUC codes for phenylalanine, UCU codes for serine, and CUU codes for leucine. The polypeptides synthesized in response to other templates of this type are given in Table 25-3. Note that poly (GUA) and poly (GAU) each elicited the synthesis of two rather than three homopolypeptides. The reason will be evident shortly.

Khorana also synthesized several copolymers with repeating tetranucleotides, such as poly (UAUC). This template led to the synthesis of a polypeptide with the repeating sequence Tyr-Leu-Ser-Ile, irrespective of the reading frame:

UAU ┊ CUA ┊ UCU ┊ AUC ┊ UAU ┊ CUA ┊ UCU ┊ AUC

Tyr — Leu — Ser — Ile — Tyr — Leu — Ser — Ile

Four codon assignments were deduced from this result.

A very different result was obtained when poly (GUAA) was the template. The only products were dipeptides and tripeptides. Why not longer chains? The reason is that one of the triplets present in this polymer—namely, UAA—codes not for an amino acid but rather for the termination of protein synthesis:

GUA ┊ AGU ┊ AAG ┊ UAA ┊ GUA ┊ A . . .

Val — Ser — Lys — Stop

Only di- and tripeptides were formed when poly (AUAG) was the template because UAG is a second signal for chain termination:

AUA⎮GAU⎮ AGA ⎮UAG⎮AUA⎮G...

Ile—Asp—Arg—Stop

Now let us look again at Table 25-3. Two rather than three homopolypeptides were synthesized with poly (GUA) as a template because the third reading frame corresponds to the sequence

G⎮UAG ⎮ UAG ⎮ UAG...

Stop—Stop—Stop

which is a repeating sequence of termination signals. What about poly (GAU)? Again, only two homopolypeptides were synthesized with this template, because the third reading frame corresponds to

GA⎮UGA ⎮ UGA ⎮ UGA⎮U...

Stop—Stop—Stop

UGA is yet another signal for chain termination. In fact, *UAG, UAA, and UGA are the only three codons that do not specify an amino acid.*

Khorana's synthesis of polynucleotides with a defined sequence was an outstanding accomplishment. The use of these polymers as templates for protein synthesis, together with Nirenberg's studies of the trinucleotide-stimulated binding of tRNA to ribosomes, resulted in the complete elucidation of the genetic code by 1966, an outcome that would have been deemed a quixotic dream only six years earlier. The powerful synthetic methods designed by Khorana also made possible another feat: the total synthesis of a DNA molecule corresponding in sequence to a transfer RNA molecule.

MAJOR FEATURES OF THE GENETIC CODE

All sixty-four codons have been deciphered (Table 25-4). Sixty-one triplets correspond to particular amino acids, whereas three code for chain termination. Since there are twenty amino acids and sixty-one triplets that code for them, it is evident that the code is highly *degenerate*. In other words, *many amino acids are designated by more than one triplet*. Only tryptophan and methionine are coded by just one triplet. The other eighteen amino acids are coded by two or more. Indeed, leucine, arginine, and serine are specified by six codons each. Under normal physiological conditions, *the code is not ambiguous:* a codon designates only one amino acid.

Codons that specify the same amino acid are called *synonyms*. For example, CAU and CAC are synonyms for histidine. Note that synonyms are not distributed haphazardly throughout the table of the genetic code (Table 25-4). An amino acid specified by two or more synonyms occupies a single box (unless there are more than four synonyms). The amino acids in a box are specified by codons that have the same first two bases but differ in the third base, as exemplified by GUU, GUC, GUA, and GUG. Thus, *most synonyms differ only in the last base of the triplet*. Inspection of the code shows

Table 25-4
The genetic code

First position (5′ end)	Second position				Third position (3′ end)
	U	C	A	G	
U	Phe	Ser	Tyr	Cys	U
	Phe	Ser	Tyr	Cys	C
	Leu	Ser	Stop	Stop	A
	Leu	Ser	Stop	Trp	G
C	Leu	Pro	His	Arg	U
	Leu	Pro	His	Arg	C
	Leu	Pro	Gln	Arg	A
	Leu	Pro	Gln	Arg	G
A	Ile	Thr	Asn	Ser	U
	Ile	Thr	Asn	Ser	C
	Ile	Thr	Lys	Arg	A
	Met	Thr	Lys	Arg	G
G	Val	Ala	Asp	Gly	U
	Val	Ala	Asp	Gly	C
	Val	Ala	Glu	Gly	A
	Val	Ala	Glu	Gly	G

Note: Given the position of the bases in a codon, it is possible to find the corresponding amino acid. For example, the codon 5′ AUG 3′ on mRNA specifies methionine, whereas CAU specifies histidine. UAA, UAG, and UGA are termination signals. AUG is part of the initiation signal, in addition to coding for internal methionines.

that XYC and XYU always code for the same amino acid, whereas XYG and XYA usually code for the same amino acid. The structural basis for these equivalences of codons will become evident when we consider the nature of the anticodons of tRNA molecules (p. 655).

What is the biological significance of the extensive degeneracy of the genetic code? One possibility is that *degeneracy minimizes the deleterious effects of mutations*. If the code were not degenerate, then twenty codons would designate amino acids and forty-four would lead to chain termination. The probability of mutating to chain termination would therefore be much higher with a nondegenerate code than with the actual code. It is important to recognize that chain-termination mutations usually lead to inactive proteins, whereas substitutions of one amino acid for another are usually relatively harmless. *Degeneracy of the code may also be significant in permitting DNA base composition to vary over a wide range without altering the amino acid sequence of the proteins encoded by the DNA.* The [G] + [C] content of bacterial DNA ranges from less than 30% to more than 70%. DNA molecules with quite different [G] + [C] contents could code for the same proteins if different synonyms were consistently used.

START AND STOP SIGNALS FOR PROTEIN SYNTHESIS

It has already been mentioned that *UAA, UAG, and UGA designate chain termination*. These codons are read not by tRNA molecules but rather by specific proteins called release factors. The start signal for protein synthesis is more complex. Polypeptide chains in bacteria start with a modified amino acid—namely, formylmethionine (fMet). There is a specific tRNA that carries fMet. This fMet-tRNA recognizes the codon AUG (or less frequently, GUG). However, AUG is also the codon for an internal methionine, and GUG is the codon for an internal valine. This means that the signal for the first amino acid in the polypeptide chain must be more complex than for all subsequent ones. AUG (or GUG) is *part of the initiation signal*. An as yet unknown feature of the conformation of the mRNA around AUG (or GUG) determines whether it is to be read as a chain initiation signal or as a codon for an internal methionine (or valine) residue. The mechanisms of initiation and termination of protein synthesis will be discussed in the next chapter.

Formylmethionine (fMet)

THE GENETIC CODE IS UNIVERSAL

The genetic code was deciphered by studies of trinucleotides and synthetic mRNA templates in cell-free systems derived from bacteria. Two questions arise: Is the genetic code the same in vivo as in vitro? Is the genetic code the same in all organisms? A definitive answer to these questions has come from analyses of mutations in viruses, bacteria, and higher organisms. A large number of amino acid substitutions arising from mutations of the genes for hemoglobin in man, the coat protein in tobacco mosaic virus (TMV), and the α chain of tryptophan synthetase in *E. coli* are known. *Nearly all of the amino acid substitutions can be accounted for by a change of a single base* (Table 25-5). *This is a stringent test of the correctness of the entire genetic code and of its universality.*

Table 25-5
Mutations in human hemoglobin, *E. coli* tryptophan synthetase, and tobacco mosaic virus (TMV) coat protein

Protein	Amino acid substitution	Inferred codon change
Hemoglobin	Glu ⟶ Val*	GAA ⟶ GUA
Hemoglobin	Glu ⟶ Lys*	GAA ⟶ AAA
Hemoglobin	Glu ⟶ Gly	GAA ⟶ GGA
Tryptophan synthetase	Gly ⟶ Arg	GGA ⟶ AGA
Tryptophan synthetase	Gly ⟶ Glu	GGA ⟶ GAA
Tryptophan synthetase	Glu ⟶ Ala	GAA ⟶ GCA
TMV coat protein	Leu ⟶ Phe	CUU ⟶ UUU
TMV coat protein	Glu ⟶ Gly	GAA ⟶ GGA
TMV coat protein	Pro ⟶ Ser	CCC ⟶ UCC

*The substitutions Glu ⟶ Val and Glu ⟶ Lys occur at position 6 in the β chain of human hemoglobin S and C, respectively.

Why has the code remained invariant over billions of years of evolution? Consider the effect of a mutation that would alter the reading of mRNA. Such a mutation would change the amino acid sequence of most, if not all, of the proteins synthesized by that particular organism. Many of these changes would undoubtedly be deleterious, and so there would be strong selection against such a mutation.

THE BASE SEQUENCE OF A GENE AND THE AMINO ACID SEQUENCE OF ITS POLYPEPTIDE PRODUCT ARE COLINEAR

We turn now to relationships between genes and proteins. High-resolution genetic-mapping studies carried out by Seymour Benzer revealed that genes are unbranched structures. This important result was in harmony with the established fact that DNA consists of a linear sequence of base pairs. Polypeptide chains also have unbranched structures. The following question was therefore asked in the early 1960s: *Is there a linear correspondence between a gene and its polypeptide product?*

Charles Yanofsky's approach to this problem was to use mutants of *E. coli* that produced an altered enzyme molecule. Numerous mutants of the α chain of tryptophan synthetase were isolated and their positions on the genetic map for the α chain were determined by carrying out recombination studies with a transducing phage. Some of these mutants were very close to each other on the genetic map, whereas others were far apart within the same gene. The next major task was to determine the location of the amino acid substitution for each of these ten mutants. First, the amino acid sequence of the 168 residues of the wild-type α chain was determined. Then, the site and kind of amino acid substitution in each mutant was established by fingerprinting. *The order of the mutants on the genetic map was the same as the order of the corresponding changes in amino acid sequence of the polypeptide product* (Figure 25-5). In other words, *the gene for the α chain and its polypeptide product are colinear.*

Figure 25-5
Colinearity of the gene and amino acid sequence of the α chain of tryptophan synthetase. The locations of mutations in DNA (shown in yellow) were determined by genetic-mapping techniques. The positions of the altered amino acids in the amino acid sequence (shown in blue) are in the same order as the corresponding mutations. [Based on C. Yanofsky. Gene structure and protein structure. Copyright © 1967 by Scientific American, Inc. All rights reserved.]

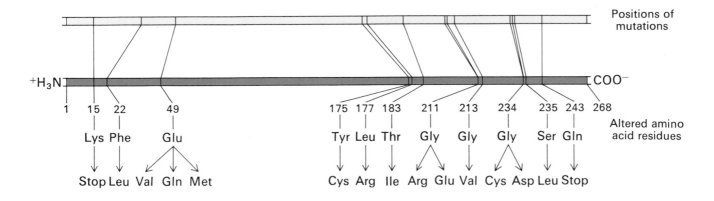

MUTATIONS ARE PRODUCED BY CHANGES
IN THE BASE SEQUENCE OF DNA

There have been no problems in St. Louis created by pre-trial publicity because the newspapers have exorcised sound judgment and restraint in what they have printed.
—*St. Louis Globe-Democrat.*
Substitution

In analyzing the pictures relayed by the Mariner 10 spacecraft en route to Mercury, the scientists said the planet becomes much more dramatic looking when viewed through an ultra-violent lens.
—*Boston Globe.*
Insertion

"I can speak just as good nglish as you," Gorbulove corrected in a merry voice.
—*Seattle Times.*
Deletion

Tomorrow: "Give Baby Time to Learn to Swallow Solid Food." etaoin-oshrdlucmfwypvbgkq
—*Youngstown (Ohio) Vindicator.*
Nonsense

Figure 25-6
Types of mutations, as illustrated by typographical errors. [Based on S. Benzer. *Harvey Lectures,* Ser. 56 (Academic Press, 1961), p. 3.]

There are several types of mutations: (1) the *substitution* of one base pair for another or of several pairs; (2) the *deletion* of one or more base pairs; and (3) the *insertion* of one or more base pairs (Figure 25-6). *The substitution of one base pair for another is the most common type of mutation.* For T4 bacteriophage, the spontaneous mutation rate has been estimated to be 1.7×10^{-8} per base per replication. The corresponding values for *E. coli* and *Drosophila melanogaster* are 4×10^{-10} and 7×10^{-11}, respectively.

There are two kinds of single base-pair substitutions. A *transition* is the replacement of one purine by another purine, or of one pyrimidine by another pyrimidine. In contrast, a *transversion* is the replacement of a purine by a pyrimidine, or of a pyrimidine by a purine.

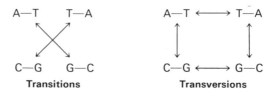

A mechanism for the spontaneous occurrence of transitions was suggested by Watson and Crick in their classical paper on the DNA double helix. They noted that some of the hydrogen atoms on each of the four bases can change their locations. *These tautomeric forms, which have a low probability of occurrence, can form base pairs other than A—T and G—C.* For example, the imino tautomer of adenine can pair with cytosine (Figure 25-7). In the next round of replication, this adenine will probably assume its normal tautomeric state, and so it will pair with thymine, whereas cytosine will pair with guanine. Thus, one of the daughter DNA molecules will contain a G-C base pair in place of A-T base pair (Figure 25-8).

Mutations can also be caused by *defective DNA polymerases.* There is a mutant of *E. coli* that has a mutation rate that is a thousand times as high as normal, probably owing to an altered polymerase. A distinctive feature of the mutations in this organism is that nearly all of them are transversions of A—T to C—G.

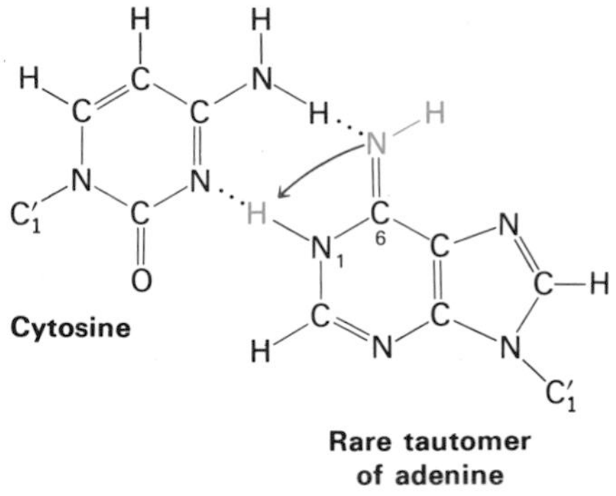

Cytosine

**Rare tautomer
of adenine**

Figure 25-7
The rare tautomer of adenine pairs
with cytosine instead of thymine. This
tautomer is formed by the shift of a
proton from the 6-amino group to N-1.

$$A—T \rightarrow \begin{cases} A—T \rightarrow A—T \\ A—T \end{cases}$$
$$A^*—C \rightarrow \begin{cases} A—T \\ G—C \end{cases}$$

Figure 25-8
The pairing of the rare
tautomer of adenine (A*)
with cytosine leads to a
G—C base pair in the
next generation.

SOME CHEMICAL MUTAGENS ARE QUITE SPECIFIC

Base analogs such as 5-bromouracil and 2-aminopurine can be
incorporated into DNA. They lead to transitions as a consequence
of altered base pairing in the course of their incorporation into DNA
or in a subsequent round of replication (Figure 25-9). 5-*Bromouracil,*
an analog of thymine, normally pairs with adenine. However, the
proportion of the enol tautomer of 5-bromouracil is higher than
it is for thymine, presumably because of the highly electronegative
character of the bromine atom compared with a methyl group at
C-5. The enol form of 5-bromouracil pairs with guanine, thereby
causing transitions of A—T $\longleftrightarrow$ G—C. 2-*Aminopurine* normally
pairs with thymine. In contrast to adenine, the normal tautomer
of 2-aminopurine can form a single hydrogen bond with cytosine.
Hence, 2-aminopurine can produce transitions of A—T $\longleftrightarrow$
G—C.

Other mutagens act by chemically modifying the bases of DNA. For
example, nitrous acid reacts with bases that contain amino groups.
Adenine is oxidatively deaminated to hypoxanthine, cytosine to
uracil, and guanine to xanthine. Hypoxanthine pairs with cytosine
rather than with thymine. Cytosine pairs with adenine rather than

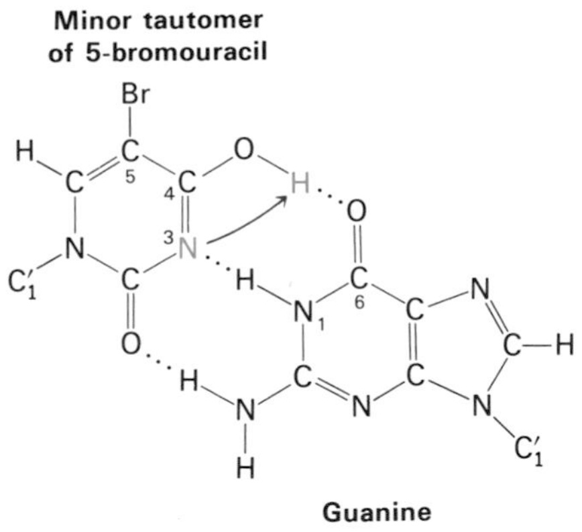

**Minor tautomer
of 5-bromouracil**

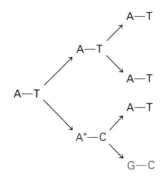

Guanine

Figure 25-9
5-Bromouracil, an analog of thymine,
occasionally pairs with guanine instead
of adenine. The presence of bromine at
C-5 increases the proportion of the
rare tautomer formed by the shift of a
proton from N-3 to the C-4 oxygen
atom.

with guanine. Xanthine, like guanine, pairs with cytosine. Consequently, nitrous acid causes transitions of A—T $\longleftrightarrow$ G—C. *Hydroxylamine* (NH_2OH) is a highly specific mutagen. It reacts almost exclusively with cytosine, giving a derivative that pairs with adenine rather than with guanine. Hydroxylamine produces a unidirectional transition of C—G to A—T.

A different kind of mutation is produced in bacteriophage by *acridines* such as proflavine. These dyes intercalate in DNA—that is, they slip in between adjacent base pairs in the DNA double helix (p. 616). *Acridine dyes lead to the insertion or deletion of one or more base pairs.* The effect of such mutations is to alter the reading frame, unless an integral multiple of three base pairs is inserted or deleted. In fact, it was the analysis of such mutants that revealed the triplet nature of the genetic code.

SUMMARY

The base sequence of a gene is colinear with the amino acid sequence of its polypeptide product. The genetic code is the relationship between the sequence of bases in DNA (or its RNA transcript) and the sequence of amino acids in proteins. Amino acids are coded by groups of three bases (called codons) starting from a fixed point. Sixty-one of the sixty-four codons specify a particular amino acid, whereas the other three codons (UAA, UAG, and UGA) are signals for chain termination. Thus, for most amino acids there is more than one code word. In other words, the code is degenerate. Codons specifying the same amino acid are called synonyms. Most synonyms differ only in the last base of the triplet.

The genetic code, which is the same in all organisms, was deciphered by three kinds of experiments. First, synthetic polyribonucleotides (formed by polynucleotide phosphorylase) were used as mRNAs in cell-free protein-synthesizing systems, following the discovery that poly U codes for polyphenylalanine. Second, a trinucleotide (like a codon in mRNA) specifically binds the particular tRNA for which it is a code word. All sixty-four trinucleotides were synthesized and tested for their capacity to stimulate the binding of specific tRNAs to ribosomes. Third, synthetic polyribonucleotides with a defined repeating sequence were used as mRNAs. For example, poly (UAUC) led to the synthesis of poly (Tyr-Leu-Ser-Ile).

Mutations are caused by changes in the base sequence of DNA. The major types are substitutions, deletions, and insertions. The substitution of one base pair for another is the most common type of mutation. A transition is the replacement of one purine by another purine, or of one pyrimidine by another pyrimidine. A transversion is the replacement of a purine by a pyrimidine, or vice versa. Mutations may be produced by the spontaneous tautomerization of a base, or by the action of base analogs (e.g., 5-bromouracil) or other chemical mutagens (e.g., nitrous acid.)

SELECTED READINGS

Where to start:

Yanofsky, C., 1967. Gene structure and protein structure. *Sci. Amer.* 216(5):80–94. [Available as *Sci. Amer.* Offprint 1074. Presents the evidence for colinearity.]

Crick, F. H. C., 1962. The genetic code. *Sci. Amer.* 207(4):66–74. [Offprint 123. Describes the experiments that revealed the triplet nature of the code.]

Nirenberg, M. W., 1963. The genetic code II. *Sci. Amer.* 208(3):80–94. [Offprint 153. Describes using synthetic polyribonucleotides to decipher the code.]

Crick, F. H. C., 1966. The genetic code III. *Sci. Amer.* 215(4):55–62. [Offprint 1052. A view of the code when it was nearly completely elucidated.]

Colinearity of gene and protein:

Yanofsky, C., Carlton, B. C., Guest, J. R., Helinski, D. R., and Henning, U., 1964. On the colinearity of gene structure and protein structure. *Proc. Nat. Acad. Sci.* 51:266–272.

Sarabhai, A. S., Stretton, O. W., Brenner, S., and Bolle, A., 1964. Colinearity of gene with polypeptide chain. *Nature* 201:13–17.

The adaptor (tRNA) hypothesis:

Crick, F. H. C., 1958. On protein synthesis. *Symp. Soc. Exp. Biol.* 12:138–163. [A brilliant anticipatory view of the problem of protein synthesis. The adaptor hypothesis is presented in this article.]

The genetic code (original articles and reviews):

Crick, F. H. C., Barnett, L., Brenner, S., and Watts-Tobin, R. J., 1961. General nature of the genetic code for proteins. *Nature* 192:1227–1232.

Khorana, H. G., 1968. Nucleic acid synthesis in the study of the genetic code. In *Nobel Lectures: Physiology or Medicine* (1963–1970), pp. 341–369. American Elsevier (1973).

Nirenberg, M., 1968. The genetic code. In *Nobel Lectures: Physiology or Medicine* (1963–1970), pp. 372–395. American Elsevier (1973).

Garen, A., 1968. Sense and nonsense in the genetic code. *Science* 160:149–159.

The genetic code (books):

Ycas, M., 1969. *The Biological Code.* Wiley.

Woese, C. R., 1967. *The Genetic Code.* Harper and Row.

Cold Spring Harbor Laboratory, 1966. *The Genetic Code* (Cold Spring Harbor Symposia on Quantitative Biology, vol. 31). [An outstanding collection of papers establishing the major features of the code.]

Mutagenesis:

Hayes, W., 1968. *The Genetics of Bacteria and Their Viruses* (2nd ed.). Wiley. [Chapter 13 on the nature of mutations is lucid and concise.]

Drake, J. W., 1970. *The Molecular Basis of Mutation.* Holden-Day.

PROBLEMS

1. What amino acid sequence is encoded by the following base sequence of an mRNA molecule? Assume that the reading frame starts at the 5′ end.

   5′ -UUGCCUAGUGAUUGGAUG- 3′

2. What is the sequence of the polypeptide formed on addition of poly (UUAC) to a cell-free protein-synthesizing system?

3. Suppose that the single-stranded RNA from tobacco mosaic virus was treated with a chemical mutagen, that mutants were obtained having serine or leucine instead of proline at a specific position, and that further treatment of these mutants with the same mutagen yielded phenylalanine at this position.

   (a) What are the plausible codon assignments for these four amino acids?

   (b) Was the mutagen 5-bromouracil, nitrous acid, or an acridine dye?

4. A protein chemist told a molecular geneticist that he found a new mutant hemoglobin in which aspartate replaced lysine. The molecular geneticist expressed surprise and sent his friend scurrying back to the laboratory.

   (a) Why was the molecular geneticist dubious about the reported amino acid substitution?

   (b) Which amino acid substitutions would have been more palatable to the molecular geneticist?

5. The amino acid sequences of part of lysozyme from wild-type T4 bacteriophage and a mutant are:

   Wild-type
   -Thr-Lys-Ser-Pro-Ser-Leu-Asn-Ala-Ala-Lys-
   -Thr-Lys-Val-His-His-Leu-Met-Ala-Ala-Lys-
   Mutant

   (a) Could this mutant have arisen by the change of a single base pair in T4 DNA? If not, how might this mutant have been produced?

   (b) What is the base sequence of the mRNA that codes for the five amino acids in the wild-type that are different in the mutant?

For additional problems, see W. B. Wood, J. H. Wilson, R. M. Benbow, and L. E. Hood, *Biochemistry: A Problems Approach* (Benjamin, 1974), Chapter 19.

CHAPTER 26

PROTEIN SYNTHESIS

In the preceding chapter, we saw that the sequence of amino acids in a protein is determined by the sequence of codons in mRNA as read by tRNA molecules. We turn now to the mechanism of protein synthesis, a process called *translation* because the language consisting of the four base-pairing letters of nucleic acids is turned into that comprising the twenty letters of proteins. As might be expected, translation is a more complex process than either the replication or the transcription of DNA, which take place within the framework of a common base-pairing language. In fact, translation involves the coordinated interplay of more than 100 kinds of macromolecules. Transfer RNA molecules, activating enzymes, soluble factors, and mRNA are required in addition to ribosomes.

Let us consider the overall process of protein synthesis before it is described in more detail. Proteins are synthesized in the amino-to-carboxyl direction by the sequential addition of amino acids to the carboxyl end of the growing peptide chain. The activated precursors are aminoacyl-tRNAs, in which the carboxyl group of an amino acid is joined to the 3' terminus of a tRNA. The linking of an amino acid to its corresponding tRNA is catalyzed by an aminoacyl-tRNA synthetase. This activation reaction, which is analogous to the activation of fatty acids, is driven by ATP. For each amino acid, there is at least one kind of tRNA and activating enzyme. There are three stages in protein synthesis: initiation, elongation, and termination. *Initiation* results in the binding of the initiator tRNA to the start signal of mRNA. The initiator tRNA occupies the P site on the ribosome, which is one of two binding sites for tRNA. *Elongation* starts with the binding of an aminoacyl-tRNA to the other tRNA site on the ribosome, which is called the A site. A peptide bond is then formed between the amino group of the incoming aminoacyl-tRNA and the carboxyl group of the fMet carried by the initiator tRNA. The resulting dipeptidyl-tRNA is then translocated from the A site to the P site, whereas the other tRNA molecule leaves the ribosome. The binding of aminoacyl-tRNA, the movement of peptidyl-tRNA from the A site to the P site, and the associated movement of the mRNA to the next codon require the hydrolysis of GTP. An aminoacyl-tRNA then binds to the vacant A site to start another round of elongation, which proceeds as described above. *Termination* occurs when a stop signal on the mRNA is read by a protein release factor, which leads to the release of the completed polypeptide chain from the ribosome.

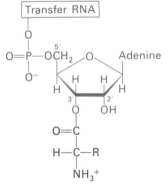

Figure 26-1
An amino acid is esterified to the 2'- or the 3'-hydroxyl group of the terminal adenosine in an aminoacyl-tRNA.

AMINO ACIDS ARE ACTIVATED AND LINKED TO TRANSFER RNA BY SPECIFIC SYNTHETASES

The formation of a peptide bond between the amino group of one amino acid and the carboxyl group of another is thermodynamically unfavorable. This thermodynamic barrier is overcome by activating the carboxyl group of the precursor amino acids. *The activated intermediates in protein synthesis are amino acid esters,* in which the carboxyl group of an amino acid is linked to either the 2'- or the 3'-hydroxyl group of the ribose unit at the 3' end of tRNA. The aminoacyl group can migrate rapidly between the 2'- and 3'-hydroxyl groups. This activated intermediate is called an *aminoacyl-tRNA* (Figure 26-1).

The attachment of an amino acid to a tRNA is important not only for activating its carboxyl group so that it can form a peptide. Amino acids by themselves cannot recognize the codons on mRNA. Rather, amino acids are carried to the ribosomes by specific tRNAs, which do recognize codons on mRNA and thereby act as adaptor molecules.

The activation of amino acids and their subsequent linkage to tRNAs is catalyzed by specific *aminoacyl-tRNA synthetases,* which are also called *activating enzymes.* For some synthetases, the first step is the formation of an *aminoacyl-adenylate* from an amino acid and ATP. This activated species is a mixed anhydride in which the carboxyl group of the amino acid is linked to the phosphoryl group of AMP. For other synthetases, the reaction of ATP, amino acid, and tRNA occurs without a detectable aminoacyl-adenylate intermediate.

$$
\overset{\text{H}}{\underset{\text{R}}{\text{+H}_3\text{N}-\text{C}}}-\text{C}\overset{\text{O}}{\underset{\text{O}^-}{}} + \text{ATP} \rightleftharpoons \overset{\text{H}}{\underset{\text{R}}{\text{+H}_3\text{N}-\text{C}}}-\overset{\text{O}}{\text{C}}-\text{O}-\overset{\text{O}}{\underset{\text{O}^-}{\text{P}}}-\text{O}-\text{Ribose}-\text{Adenine} + \text{PP}_i
$$

Aminoacyl-adenylate
(Aminoacyl-AMP)

The next step is the transfer of the aminoacyl group of aminoacyl-AMP to a tRNA molecule to form *aminoacyl-tRNA,* the activated intermediate in protein synthesis. Recent studies indicate that the aminoacyl group is transferred first to the 2′-hydroxyl and then to the 3′-hydroxyl of the ribose unit at the 3′ end of tRNA. We will refer to aminoacyl-tRNA as a 3′ derivative. However, it must be borne in mind that acyl migration between the 2′- and the 3′-hydroxyl can occur rapidly.

$$\text{Aminoacyl-AMP} + \text{tRNA} \rightleftharpoons \text{aminoacyl-tRNA} + \text{AMP}$$

The sum of these activation and transfer steps is:

$$\text{Amino acid} + \text{ATP} + \text{tRNA} \rightleftharpoons$$
$$\text{aminoacyl-tRNA} + \text{AMP} + \text{PP}_i$$

The $\Delta G°'$ of this reaction is close to 0, since the free energy of hydrolysis of the ester bond of aminoacyl-tRNA is similar to that of the terminal phosphoryl group of ATP. What then drives the synthesis of aminoacyl-tRNA? As expected, the reaction is driven by the hydrolysis of pyrophosphate. The sum of these three reactions is highly exergonic:

$$\text{Amino acid} + \text{ATP} + \text{tRNA} + \text{H}_2\text{O} \rightleftharpoons$$
$$\text{aminoacyl-tRNA} + \text{AMP} + 2\,\text{P}_i$$

Thus, *two high-energy phosphate bonds are consumed in the synthesis of an aminoacyl-tRNA*. One of them is consumed in forming the ester linkage of aminoacyl-tRNA, whereas the other is consumed in driving the reaction forward.

The activation and transfer steps for a particular amino acid are catalyzed by the same aminoacyl-tRNA synthetase. In fact, *the aminoacyl-AMP intermediate does not dissociate from the synthetase.* Rather, it is tightly bound to the active site of the enzyme by noncovalent interactions. The aminoacyl-AMP is normally a transient intermediate in the synthesis of aminoacyl-tRNA, but it is quite stable and readily isolated if tRNA is absent from the reaction mixture.

We have previously encountered an acyl adenylate intermediate in fatty acid activation (p. 409). It is interesting to note that Paul Berg first discovered this intermediate in fatty acid activation, and then recognized that it also occurs in amino acid activation. The major difference between these reactions, of course, is that the acceptor of the acyl group is CoA in the former and tRNA in the latter. The energetics of these biosyntheses is very similar: both are made irreversible by the hydrolysis of pyrophosphate.

SYNTHETASES ARE SPECIFIC IN BOTH THE ACTIVATION AND THE TRANSFER REACTIONS

There is at least one aminoacyl-tRNA synthetase for each amino acid. These enzymes differ in size, subunit structure, and amino acid composition (Table 26-1). Synthetases consisting of a single polypeptide chain typically have a molecular weight of about 100,000.

The correct translation of the genetic message depends on the high degree of specificity of aminoacyl-tRNA synthetases. These enzymes are highly selective in their recognition of the amino acid to be activated and of the prospective tRNA acceptor. As will be discussed shortly, tRNA molecules that accept different amino acids have different base sequences. However, it is not yet known in detail how synthetases recognize specific tRNAs.

An interesting error-correction mechanism is present in the synthesis of an aminoacyl-tRNA. The synthetase specific for isoleucine can also activate valine to form an enzyme-bound valine-AMP complex. However, the activated valine is not transferred to the

Table 26-1
Properties of some aminoacyl-tRNA synthetases

Source	Amino acid specificity	Molecular weight	Subunit structure
E. coli	Histidine	85,000	α_2
E. coli	Isoleucine	114,000	One chain
E. coli	Lysine	104,000	α_2
E. coli	Glycine	227,000	$\alpha_2\beta_2$
Yeast	Lysine	138,000	α_2
Yeast	Phenylalanine	270,000	$\alpha_2\beta_2$
Beef pancreas	Tryptophan	108,000	α_2

tRNA specific for isoleucine. Instead, this particular tRNA promotes the hydrolysis of valine-AMP. This hydrolysis reaction prevents the erroneous incorporation of valine into proteins and also frees the synthetase for activation of the correct amino acid. In general, the transfer step in the synthesis of aminoacyl-tRNAs is more specific than the activation step.

TRANSFER RNA MOLECULES HAVE A COMMON DESIGN

The base sequence of a transfer RNA molecule was first determined by Robert Holley in 1965, the culmination of seven years of effort. Indeed, this work was the first determination of the complete sequence of a nucleic acid. Holley's work was important both in providing insight into the biological activity of tRNA and in creating techniques for the sequencing of nucleic acids in general. The first step was to obtain a pure species of tRNA. This was accomplished by employing the technique of *countercurrent distribution,* which takes advantage of the fact that different tRNA molecules partition differently between two immiscible solvents, such as formamide and isopropanol. A mixture of tRNAs from yeast was sequentially partitioned and repartitioned from one solvent phase to another in a series of 200 tubes, using an automated apparatus. The contents of these tubes were then analyzed for total RNA (by the absorbance at 260 nm) and for their capacity to accept particular amino acids when incubated with ATP and the appropriate

650

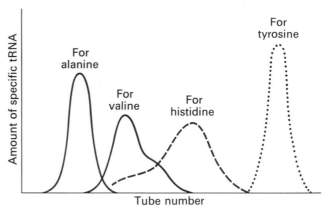

Figure 26-2
Partial resolution of different tRNAs by countercurrent distribution. [Based on R. W. Holley. The nucleotide sequence of a nucleic acid. Copyright © 1966 by Scientific American, Inc. All rights reserved.]

Anticodon

$$—\overset{3'}{C}—\overset{}{G}—\overset{5'}{I}—$$
$$—\underset{5'}{G}—\underset{}{C}—\underset{3'}{C}—$$

Codon

activating enzymes. As shown in Figure 26-2, the tRNA molecules specific for alanine, valine, histidine, and tyrosine were fractionated by this procedure, but they were not yet pure. Additional countercurrent distribution of the fractions containing alanine tRNA resulted in a pure preparation, which was the starting point for the elucidation of its sequence. Some procedures for the sequencing of RNA have already been discussed (p. 617).

The sequence of yeast alanine tRNA is shown in Figure 26-3. The molecule is a single chain of seventy-seven ribonucleotides. The 5′ terminus is phosphorylated (pG), whereas the 3′ terminus has a free hydroxyl group. A distinctive feature of the molecule is its high content of bases other than A, U, G, and C. There are nine unusual nucleosides in this tRNA: inosine, pseudouridine, dihydrouridine, ribothymidine, and methylated derivatives of guanosine and inosine. The amino acid attachment site is the 3′-hydroxyl group of the adenosine residue at the 3′ terminus of the molecule. The sequence IGC in the middle of the molecule is the anticodon. It is complementary to GCC, one of the codons for alanine.

The sequences of several other tRNA molecules were determined a short time later. More than forty sequences are now known. The striking finding is that all of these sequences can be written in a cloverleaf pattern in which about half the residues are base-paired. Furthermore, different kinds of tRNA molecules can crystallize together. The unit-cell dimensions of these mixed crystals are like those of the crystals of the pure species. Hence, *tRNA molecules have many common structural features*. This finding is not unexpected, since all tRNA molecules must be able to interact in nearly the same way with ribosomes and mRNAs. Specifically, all tRNA molecules must fit into the A and P sites on the ribosomes, and must interact with the enzyme that catalyzes peptide-bond formation.

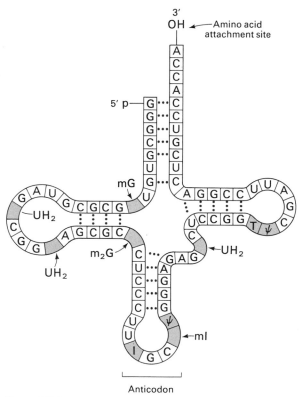

Figure 26-3
Base sequence of yeast alanine tRNA. Modified nucleosides (shown in green) are abbreviated as follows: inosine (I), methylinosine (mI), dihydrouridine (UH₂), ribothymidine (T), pseudouridine (ψ), methylguanosine (mG), and dimethylguanosine (m₂G).

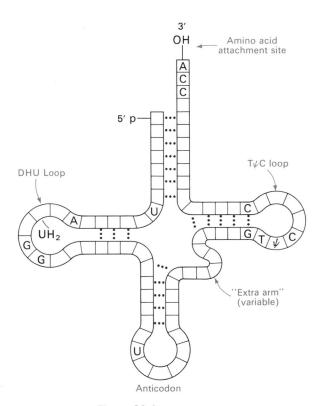

Figure 26-4
Common features of tRNA molecules.

All transfer RNA molecules have these features in common:

1. They are single chains containing between *73 and 93 ribonucleotides,* which correspond to a molecular weight of about 25,000.

2. They contain *many unusual bases,* typically between 7 and 15 per molecule. Many of these unusual bases are methylated or dimethylated derivates of A, U, C, and G that are formed by enzymatic modification of a precursor tRNA (p. 613). The role of these unusual bases is uncertain. One possibility is that methylation prevents the formation of certain base pairs, thereby rendering some of the bases accessible for other interactions. Also, methylation imparts a hydrophobic character to some regions of tRNAs, which may be important for their interaction with the synthetases and with ribosomal proteins.

3. The 5′ end of tRNAs is phosphorylated. The 5′ terminal residue is usually pG.

4. The base sequence at the 3′ end of tRNAs is CCA. The activated amino acid is attached to the 3′-hydroxyl of the terminal adenosine.

5. About half of the nucleotides in tRNAs are base-paired to form double helices. Five groups of bases are not base-paired: the *3′ CCA terminal region;* the *TψC loop,* which acquired its name from the sequence ribothymine-pseudouracil-cytosine; the *"extra arm,"* which contains a variable number of residues; the *DHU loop,* which contains several dihydrouracil residues; and the *anticodon loop.*

6. The *anticodon loop* consists of seven bases, with the following sequence:

$$
\begin{array}{l}
5' \\
\text{—Pyrimidine—Pyrimidine—}\underbrace{\text{X—Y—Z—}}_{\text{Anticodon}}\text{Modified purine} \text{—} \overset{3'}{\text{Variable base}} \text{—}
\end{array}
$$

TRANSFER RNA IS L-SHAPED

A most significant recent development has been the elucidation of the three-dimensional structure of a tRNA molecule in the laboratories of Alexander Rich and Aaron Klug. Their independent x-ray crystallographic studies of yeast phenylalanine tRNA were carried out at a resolution of 3 Å. Studies at a higher resolution are in progress. The entire polynucleotide chain could be readily traced in the 3 Å electron-density map. The major findings are:

1. The molecule is *L-shaped* (Figure 26-5).

2. There are *two segments of double helix*. Each of these helices contains about ten base pairs, which correspond to one turn of helix. These helical segments are perpendicular to each other, thus forming an L. The base pairing given in the cloverleaf model, which was based on sequence studies, is correct.

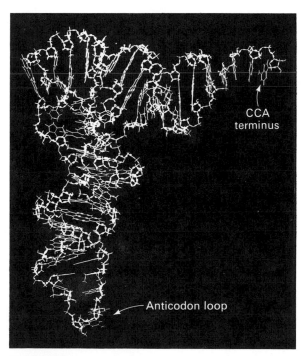

Figure 26-5
Photograph of a skeletal model of yeast phenylalanine tRNA based on an electron-density map at 3-Å resolution. [Courtesy of Dr. Sung-Hou Kim.]

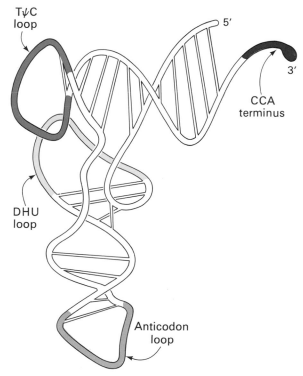

Figure 26-6
Schematic diagram of the three-dimensional structure of yeast phenylalanine tRNA. [Based on a drawing kindly provided by Dr. Sung-Hou Kim.]

3. The CCA terminus containing the *amino acid attachment site* is at one end of the L. The other end of the L is the *anticodon loop*. Thus, *the amino acid in aminoacyl-tRNA is far from the anticodon* (about 80 Å). The DHU and TψC loops form the corner of the L.

4. The CCA terminus and the adjacent helical region do not interact strongly with the rest of the molecule. This part of the molecule may change conformation during amino acid activation and also during protein synthesis on the ribosome.

The determination of the three-dimensional structure of tRNA is an important step toward elucidating the process of translation in precise molecular terms. This achievement seemed only a remote possibility a decade ago. The thoughts and efforts of investigators in this field are now turning to even more demanding goals, such as solving the three-dimensional structure of tRNA-synthetase complexes and even of tRNA bound to mRNA and ribosomal proteins.

THE CODON IS RECOGNIZED BY THE ANTICODON RATHER THAN BY THE ACTIVATED AMINO ACID

It has already been mentioned that the anticodon on tRNA is the recognition site for the codon on mRNA and that recognition occurs by base pairing. Does the amino acid attached to the tRNA play a role in this recognition process? This question was answered in the following way. First, cysteine was attached to its cognate tRNA (denoted by tRNA^{Cys}). The attached cysteine unit was then converted to alanine by reacting Cys-tRNA^{Cys} with Raney nickel, which removed the sulfur atom from the activated cysteine residue without affecting its linkage to tRNA.

Thus, a *hybrid aminoacyl-tRNA* was produced in which alanine was covalently attached to a tRNA specific for cysteine. Does this hybrid tRNA recognize the codon for alanine or for cysteine? The answer was provided by using this hybrid tRNA in a cell-free protein-synthesizing system. The template was a random copolymer of U and G in the ratio of (5:1), which normally leads to the incorporation of cysteine (UGU) but not of alanine (GCX). However, alanine was incorporated into a polypeptide when Ala-tRNA^{Cys} was added to the incubation mixture. In other words, alanine was incorporated as though it were cysteine because it was attached to the tRNA specific for cysteine. Thus, *codon recognition does not depend on the amino acid that is attached to tRNA*.

The same result was obtained when mRNA for hemoglobin served as the template. ^{14}C alanyl-tRNA^{Cys} was used as the hybrid aminoacyl-tRNA. The only radioactive tryptic peptide (i.e., containing ^{14}C-alanine) was one that normally contained cysteine but not alanine. On the other hand, no radioactivity was found in any of the peptides that normally contained alanine but not cysteine. This experiment and the preceding one unequivocally demonstrated that the amino acid in aminoacyl-tRNA does not play a role in selecting a codon.

A TRANSFER RNA MOLECULE MAY RECOGNIZE MORE THAN ONE CODON BECAUSE OF WOBBLE

What are the rules that govern the recognition of a codon by the anticodon of a tRNA? A simple hypothesis is that each of the bases of the codon forms a Watson-Crick type of base pair with a complementary base on the anticodon. The codon and anticodon would then be lined up in an antiparallel fashion. In the accompanying diagram, the prime denotes the complementary base. Thus X and X' would be either A and U (or U and A), or G and C (or C and G). A specific prediction of this model is that a particular anticodon can recognize only one codon. The facts are otherwise. *Some pure tRNA molecules can recognize more than one codon.* For example, the yeast alanine tRNA studied by Holley binds to *three* codons: GCU, GCC, and GCA. The first two bases of these codons are the same, whereas the third is different. Could it be that the recognition of the third base of a codon is sometimes less discriminating than of the other two? The pattern of degeneracy of the genetic code indicates that this might be so. XYU and XYC always code for the same amino

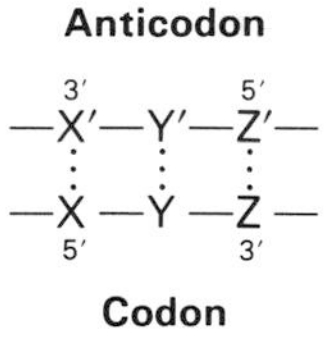

Anticodon

$$-X'-Y'-Z'-$$
$$-X-Y-Z-$$

Codon

Table 26-2

Allowed pairings at the third codon base according to the wobble hypothesis

First anticodon base	Third codon base
C	G
A	U
U	A or G
G	U or C
I	U, C, or A

acid, whereas XYA and XYG usually do. Crick surmised from these data that the steric criteria for pairing of the third base might be less stringent than for the other two. Models of various base pairs were built to determine which ones are similar to the standard A-U and G-C base pairs in terms of the distance and angle between the glycosidic bonds. Inosine was included in this study because it appeared in several anticodons. Assuming some steric freedom ("wobble") in the pairing of the third base of the codon, the combinations shown in Table 26-2 seemed plausible.

The wobble hypothesis is now firmly established. The anticodons of tRNAs of known sequence bind to the codons predicted by this hypothesis. For example, the anticodon of yeast alanine tRNA is IGC. This tRNA recognizes the codons GCU, GCC, and GCA.

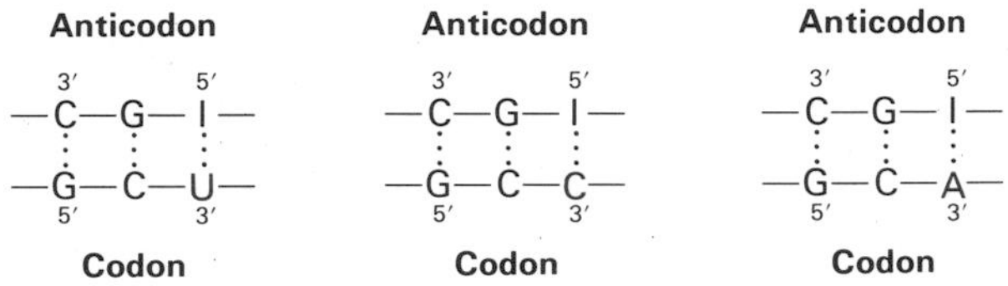

Thus, I pairs with U, C, or A, as predicted. Phenylalanine tRNA, which has the anticodon GAA, recognizes the codons UUU and UUC but not UUA and UUG.

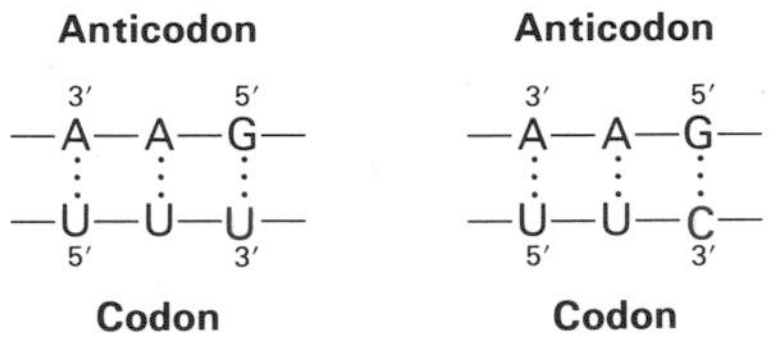

Thus, G pairs with either U or C in the third position of the codon, as predicted by the wobble hypothesis.

Two generalizations concerning the codon-anticodon interaction can be made:

1. The first two bases of a codon pair in the standard way. Recognition is precise. Hence, *codons that differ in either of their first two bases must be recognized by different tRNAs.* For example, UUA and CUA both code for leucine, but are read by different tRNAs.

2. A particular tRNA molecule reads one, two, or three kinds of codons, depending on whether the first base of its anticodon is C or A (1 codon), U or G (2 codons), or I (3 codons). Thus, *part of the degeneracy of the genetic code arises from imprecision (wobble) in the pairing of the third base of the codon.* We see here a strong reason for the frequent appearance of inosine, one of the unusual nucleosides, in anticodons. Inosine maximizes the number of codons that can be read by a particular tRNA molecule (Figure 26-7).

MUTANT TRANSFER RNA MOLECULES CAN SUPPRESS OTHER MUTATIONS

Further insight into the process of codon recognition has come from studies of mutant tRNAs in which a single base in the anticodon is changed. In effect, these spontaneous events in nature are the converse of the ones in which the amino acid attached to tRNA was chemically modified in vitro. These mutant tRNAs were discovered in an interesting way. Geneticists had known for many years that the deleterious effects of some mutations can be reversed by a second mutation. Suppose that an enzyme becomes inactive because of a mutation from GCU (alanine) to GAU (aspartate). What kind of mutations might restore the activity of this enzyme?

1. A reverse mutation of the same base, A $\longrightarrow$ C, would give the original code word.

2. A different mutation, A $\longrightarrow$ U, would give valine (GUU), which might result in a partially or fully active enzyme.

3. A mutation at a different site in the same gene might reverse the effects of the first mutation. Such an altered protein would differ from the wild-type by two amino acids.

4. A mutation in a *different gene* might overcome the deleterious effects of the first mutation. Such a mutation is called an *intergenic suppressor.*

The mode of action of intergenic suppressors was a puzzle for a long time. We now know from genetic and biochemical studies

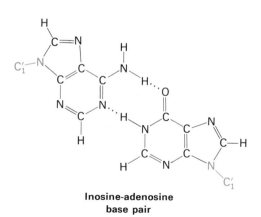

Inosine-uridine base pair

Inosine-cytidine base pair

Inosine-adenosine base pair

Figure 26-7
Inosine can base-pair with cytosine, adenine, or uracil because of wobble.

658

that *most of these suppressors act by altering the reading of mRNA*. For example, consider a mutation that leads to the appearance of the codon UAG, which is a termination signal. The result is an incomplete polypeptide chain. Such mutations are called *nonsense mutations* because incomplete polypeptides are usually inactive. The UAG nonsense mutation can be suppressed by mutations in several different genes. One of these suppressors reads UAG as a codon for tyrosine, which may lead to the synthesis of a functional protein rather than to an incomplete polypeptide chain (Figure 26-8). Why does this mutant tRNA insert tyrosine in response to the UAG codon? Tyrosine tRNA normally recognizes the codons UAC and UAU. This mutant tRNA is identical to the normal tyrosine tRNA except for a single base change in its anticodon: GUA $\longrightarrow$ CUA. The mutation G $\longrightarrow$ C in the first base of its anticodon changes its recognition properties. The altered anticodon can only recognize UAG, as predicted by the wobble hypothesis.

Anticodon

—A—U—C—
  3'   5'

—U—A—G—
  5'   3'

Mutated base in the anticodon of a tyrosine tRNA (normally 3' AUG 5')

This chain-termination codon is read as tyrosine.

Codon

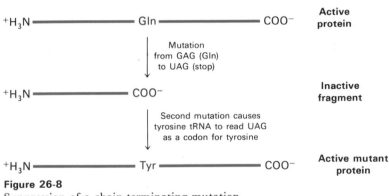

Figure 26-8
Suppression of a chain-terminating mutation by a second mutation in a tRNA molecule.

This kind of suppressor mutation is more likely to be selected if the tRNA undergoing mutation is not essential. In other words, there must be another tRNA species that recognizes the same codons as the tRNA undergoing mutation. In fact, in *E. coli,* there are two different tRNAs that normally recognize both UAC and UAU. The species present in a small amount is the one that becomes altered. The normal role of this minor species of tyrosine tRNA is uncertain. The occurrence of this suppressor mutation raises a second question. If the mutation to UAG is read as tyrosine rather than as a stop signal, what happens to normal chain termination? Surprisingly, most polypeptide chains terminate normally in the suppressor mutant, possibly because the termination signal may be more than just UAG. Also, suppression is not completely efficient.

Other kinds of mutant tRNAs have been identified. *Missense suppressors* alter the reading of mRNA so that one amino acid is inserted instead of another at certain codons (e.g., glycine instead of arginine). *Frameshift suppressors* are especially interesting. One of them has an extra base in its anticodon loop. Consequently, it reads *four* nucleotides as a codon rather than three. In particular, UUUC rather than UUU is read as the codon for phenylalanine. A mutation caused by the insertion of an extra base can be suppressed by this altered tRNA.

RIBOSOMES, THE ORGANELLES OF PROTEIN SYNTHESIS, CONSIST OF A LARGE AND A SMALL SUBUNIT

We turn now to the mechanism of protein synthesis. This complex process occurs in ribosomes, which can be regarded as the organelles of protein synthesis, just as mitochondria are the organelles of oxidative phosphorylation. The ribosome is a highly specialized and complex structure, approximately 200 Å in diameter. The best-characterized ribosomes are those of *E. coli,* which have a molecular weight of about 2.5 million and a sedimentation coefficient of 70S. The *E. coli* ribosome can be dissociated into a *large subunit (50S)* and a *small subunit (30S)*. These subunits can be further split into their constituent proteins and RNAs. The 30S subunit contains twenty-one proteins and a 16S RNA molecule. The 50S subunit contains about thirty-five proteins and two RNA molecules, a 23S species and a 5S species. About two-thirds of the weight of an *E. coli* ribosome is RNA, whereas the other third is protein.

The ribosomes in the cytoplasm of eucaryotic cells are somewhat larger than those of bacteria. These intact eucaryotic ribosomes have a sedimentation coefficient of 80S. Like bacterial ribosomes, they dissociate into a large subunit (60S) and a small subunit (40S).

Figure 26-9
Ribosomes can be dissociated into about fifty-five proteins and three RNA molecules.

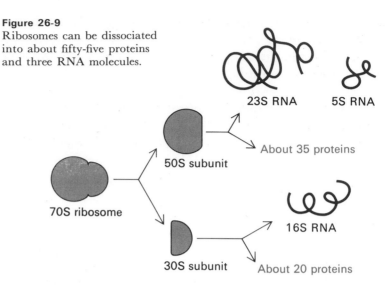

23S RNA   5S RNA

50S subunit   About 35 proteins

70S ribosome

30S subunit   16S RNA

About 20 proteins

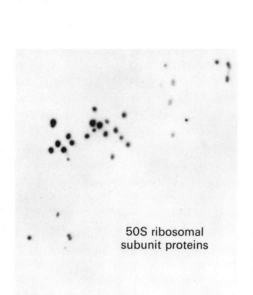

50S ribosomal
subunit proteins

Figure 26-10
Separation of the proteins of the 50S ribosomal subunit by two-dimensional electrophoresis on a polyacrylamide gel. [Courtesy of Dr. Charles Cantor.]

The small subunit contains a single 18S RNA molecule, whereas the large subunit contains three RNA molecules (28S, 7S, and 5S). The ribosomes in mitochondria and chloroplasts are different from those in the cytoplasm of eucaryotes. They are more like 70S than 80S particles. In fact, there are many similarities between protein synthesis in mitochondria, chloroplasts, and bacteria.

RIBOSOMES CAN BE RECONSTRUCTED FROM THEIR CONSTITUENT PROTEINS AND RNA MOLECULES

The 30S ribosomal subunit can be reconstituted from a mixture of 16S RNA and its twenty-one constituent proteins. The spontaneous reassembly of these components to form a fully functional 30S subunit was first achieved by Masayasu Nomura in 1968. The 50S subunit was reconstituted several years later. The significance of these experiments is twofold. First, they demonstrate that all of the information needed for the correct assembly of this organelle is contained in the structure of its components. Nonribosomal factors are not needed. Thus, *the formation of a ribosome in vitro, and probably in vivo, is a self-assembly process.* Second, reconstitution can be used to ascertain whether a particular component is essential for the assembly of the ribosome or for a specific function. For example, the component responsible for sensitivity to the antibiotic streptomycin was identified in this way (p. 670).

The following conclusions have emerged from studies of the reconstitution of the 30S subunit:

1. The 16S RNA is essential for its assembly and function. The requirement is quite specific since 16S RNA from yeast cannot substitute for 16S RNA from *E. coli*. The role of ribosomal RNA molecules is a challenging area of inquiry.

2. Reconstitution was performed with one protein omitted at a time to see whether an intact 30S particle can be formed from the other twenty proteins, in addition to 16S RNA. At least six proteins were found to be essential for the assembly of a 30S particle.

3. Most of the twenty-one proteins were needed (in addition to 16S RNA) for the reassembly of a functionally active 30S particle. Thus, *the 30S subunit is a cooperative functional entity*.

PROTEINS ARE SYNTHESIZED IN THE AMINO-TO-CARBOXYL DIRECTION

One of the first questions asked about the mechanism of protein synthesis was whether proteins are synthesized in the amino-to-carboxyl direction or the reverse direction. The pulse-labeling studies of Howard Dintzis provided a clear-cut answer. Reticulocytes that were actively synthesizing hemoglobin were exposed to ³H-leucine for periods shorter than required to synthesize a complete chain. The released hemoglobin was separated into α and β chains and then treated with trypsin. The specific radioactivity of these peptides—that is, the amount of radioactivity in them relative to their total content of leucine—was determined. There was more radioactivity in the peptides near the carboxyl end than in those near the amino end. In fact, there was a *gradient of increasing radioactivity in going from the amino to the carboxyl end of each chain* (Figure

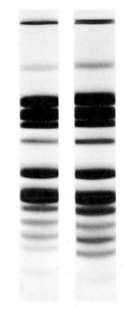

Figure 26-11
Polyacrylamide-gel electrophoresis patterns of the native 30S ribosomal subunit (left) and a reconstituted 30S subunit (right). All of the proteins present in the native subunit are present in the reconstituted one. [Courtesy of Dr. Masayasu Nomura.]

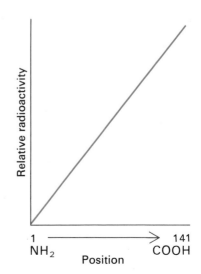

Figure 26-12
Distribution of ³H-leucine in the α chain of hemoglobin synthesized following exposure of reticulocytes to a pulse of tritiated leucine. The higher radioactivity of the carboxyl end relative to the amino end indicates that the carboxyl end was synthesized last.

26-12). The carboxyl end was the most heavily labeled because it was the last to be synthesized. Hence, *the direction of chain growth is from the amino to the carboxyl end.*

MESSENGER RNA IS TRANSLATED IN THE 5′ ⟶ 3′ DIRECTION

The direction of reading of mRNA was determined by using the synthetic polynucleotide

$$\overset{5'}{A}—A—A—(A—A—A)_n—A—A—\overset{3'}{C}$$

as the template in a cell-free protein-synthesizing system. AAA codes for lysine, whereas AAC codes for aspargine. The polypeptide product was

$$^{+}H_3N—Lys—(Lys)_n—Asn—C\overset{\displaystyle O}{\underset{\displaystyle O^-}{\big\langle}}$$

Since asparagine was the carboxyl-terminal residue, the codon AAC was the last to be read. Hence, *the direction of translation is 5′ ⟶ 3′.* Recall that mRNA is also synthesized in the 5′ ⟶ 3′ direction (p. 610). This means that mRNA can be translated as it is being synthesized. In fact, in *E. coli,* the 5′ end of mRNA interacts with ribosomes very soon after it is synthesized (Figure 26-13).

Figure 26-13
Transcription of a section of the DNA of *E. coli* and translation of the nascent mRNA. The lower part of the chromosome is active, whereas the upper part is inactive. [From O. L. Miller, Jr., Barbara A. Hamkalo, and C. A. Thomas, Jr. Visualization of bacterial genes in action. *Science* 169(1970):392.]

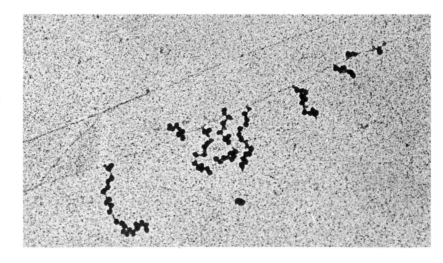

PROTEIN SYNTHESIS IS INITIATED BY
FORMYLMETHIONINE TRANSFER RNA IN BACTERIA

There are three stages in the synthesis of proteins, just as there are in the synthesis of DNA and RNA: *initiation, elongation,* and *termination.* We will consider the mechanism of protein synthesis in *E. coli,* where these processes are best characterized. A clue to the mechanism of chain initiation was provided by the finding that nearly half of the amino-terminal residues of proteins in *E. coli* are methionine, yet this amino acid is uncommon at other positions of the polypeptide chain. The amino terminus of nascent proteins is usually blocked. This amino acid derivative proved to be *formylmethionine* (fMet).

In fact, *protein synthesis in bacteria starts with formylmethionine.* There is a special tRNA that brings formylmethionine to the ribosome to initiate protein synthesis. This initiator tRNA (abbreviated as $tRNA_f$) is different from the one that inserts methionine in internal positions (abbreviated as $tRNA_m$). The subscript f indicates that methionine attached to the initiator tRNA can be formylated, whereas it cannot be formylated when attached to $tRNA_m$.

The same methionyl-tRNA synthetase adds methionine to these different tRNAs. Formylation occurs after methionyl-tRNA is synthesized. Formylation is catalyzed by a specific transformylase, which uses N^{10}-formyltetrahydrofolate as the activated donor. The product is called *formylmethionyl tRNA* (fMet-tRNA).

Formylation of the methionine on the initiator tRNA is essential for the initiation of protein synthesis. The translation of added viral RNA in bacterial extracts is inhibited by the addition of trimethoprim, a potent inhibitor of dihydrofolate reductase. This inhibition is relieved by the addition of formyltetrahydrofolate or of fMet-tRNA.

AUG (OR GUG) IS PART OF THE INITIATION SIGNAL

The signal for initiation is not just a single codon. AUG is recognized both by $tRNA_f$ and $tRNA_m$, since both of these tRNA species have the same anticodon—namely, CAU. However, $tRNA_f$ does not bind to AUG triplets that code for internal methionines, and $tRNA_m$ does not bind to AUG triplets that signify the start of polypeptide chains. Much less frequently, the initiation codon is GUG, which usually signifies an internal valine.

What is the structural basis for this discrimination between AUG codons? One possibility a priori is that AUG signifies initiation only if it is located at the 5′ end of mRNA. However, this explanation was ruled out when base sequences of the ribosome-binding sites of RNA molecules from phages such as R17 and Qβ became known (Figure 26-14). The initiating AUG codon can be identified in these sequences because the amino acid sequences of the proteins encoded by these viral RNAs are known. In Qβ RNA, the first initiating AUG codon is located starting at residue 62. Thus, proximity of AUG to the 5′ end cannot be the distinguishing feature. Is the initiation signal a distinctive base sequence next to an AUG codon? All known base sequences of viral RNAs contain a purine-rich region centered on approximately ten nucleotides on the 5′ side of the initiation codon (Figure 26-15). An intriguing possibility is that this purine-rich region base pairs with a complementary sequence near the 3′ terminus of the 16S RNA. Such an interaction between mRNA and 16S RNA may be critical for marking the site of initiation of protein synthesis.

—CCUAGGAGGUUUGACCU AUG CGA GCU UUU AGU—   **R17 A-protein**

                  fMet—Arg——Ala——Phe——Ser—

—AACAUGAGGAUUACCC AUG UCG AAG ACA ACA—   **R17 replicase**

                  fMet——Ser——Lys——Thr——Thr—

—ACUAAGGAUGAAAUGC AUG UCU AAG ACA GCA—   **Qβ replicase**

                  fMet——Ser——Lys—— Thr——Ala—

Figure 26-14
Ribosome-binding sites of some viral RNAs.

INITIATION OF PROTEIN SYNTHESIS

Protein synthesis starts with the formation of a *30S initiation complex* (Figure 26-15). The interacting components are a 30S ribosomal subunit, mRNA, formylmethionyl-tRNA$_f$, and GTP. The 30S initiation complex is held together by cooperative interactions between these species. Three protein factors, identified symbolically as IF 1, IF 2, and IF 3, are required for the formation of the 30S initiation complex. One of these initiation factors, IF 3, prevents the reassociation of dissociated 50S and 30S subunits, which would interfere with the formation of the 30S initiation complex. It also appears to be essential for the binding of mRNA to the 30S subunit, whereas IF 2 probably enhances the binding of initiator tRNA to the mRNA-30S subunit complex. The sequence of assembly of the initiation complex is being investigated.

A 50S ribosomal subunit then joins a 30S initiation complex to form a *70S initiation complex*. GTP is hydrolyzed in this step. The 70S initiation complex (Figure 26-15) is ready for the elongation phase of protein synthesis. The fMet-tRNA$_f$ molecule occupies the P (peptidyl) site on the ribosome. The other site for a tRNA molecule on the ribosome, called the A (aminoacyl) site, is empty. There is no direct structural evidence for the existence of the P and A sites. Rather, it has been inferred from studies involving the antibiotic puromycin, as will be discussed later (p. 672). The important point now is that fMet-tRNA$_f$ is positioned so that its anticodon pairs with the initiating AUG (or GUG) codon on mRNA. Thus, the reading frame is defined by specific interactions of the ribosome and fMet-tRNA$_f$ with mRNA. Nonspecific initiation may occur when the concentration of Mg^{2+} is high. In fact, poly U and other synthetic polynucleotides devoid of initiation sites were translated because the concentration of Mg^{2+} in the incubation medium was higher than it would be physiologically.

THE ELONGATION CYCLE IN PROTEIN SYNTHESIS

The elongation cycle consists of three steps: (1) *binding of aminoacyl-tRNA (codon recognition)*, (2) *peptide bond formation*, and (3) *translocation*. The elongation cycle starts with the insertion of an aminoacyl-tRNA into the empty A site on the ribosome (Figure 26-16). The particular species inserted depends on the codon of mRNA that

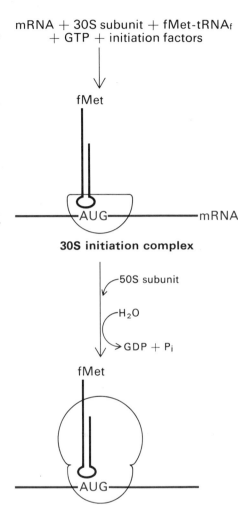

Figure 26-15
Initiation phase of protein synthesis: formation of a 30S initiation complex, followed by a 70S initiation complex.

is positioned in the A site. *GTP* and a protein *elongation factor* referred to as *EF Tu* are required for the binding of aminoacyl-tRNA to the ribosome. GTP is hydrolyzed, and the complex of EF Tu and GDP dissociates from the ribosome. This complex then reacts with EF Ts, another elongation factor, to form a complex of EF Tu and EF Ts. In turn, this complex reacts with GTP to form a complex of EF Tu and GTP, which is ready for another round of protein synthesis. In short, *EF Tu brings aminoacyl-tRNAs to the A site of the ribosome.* An important finding is that *EF Tu does not interact with fMet-tRNA$_f$. Hence, this initiator tRNA cannot read internal AUG codons.*

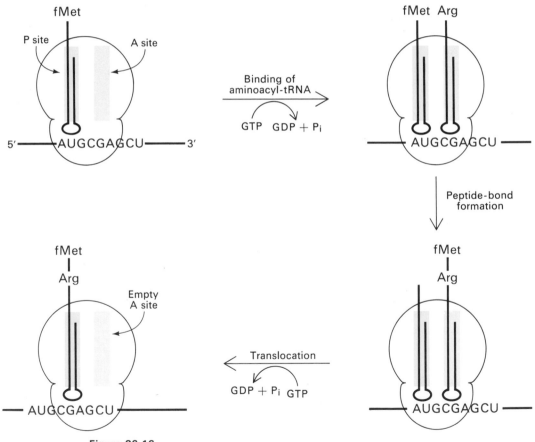

Figure 26-16
Elongation phase of protein synthesis: binding of aminoacyl-tRNA, peptide-bond formation, and translocation.

We now have a complex in which aminoacyl-tRNA occupies the A site and fMet-tRNA occupies the P site. The stage is set for the *formation of a peptide bond* (Figure 26-17). This reaction is catalyzed by *peptidyl transferase,* an enzyme that is an integral part of the 50S subunit. The activated formylmethionine unit of fMet-tRNA$_f$ (in the P site) is transferred to the amino group of the aminoacyl-tRNA (in the A site) to form a dipeptidyl-tRNA.

An uncharged tRNA occupies the P site, whereas a dipeptidyl-tRNA occupies the A site, following the formation of a peptide bond. The next phase of the elongation cycle is *translocation*. Three movements occur: the uncharged tRNA leaves the P site, the peptidyl-tRNA moves from the A site to the P site, and mRNA moves a distance of three nucleotides. The result is that the next codon is positioned for reading by the incoming aminoacyl-tRNA. Translocation requires a third elongation factor—namely, EF G (or the translocase). In addition, GTP must be hydrolyzed for translocation to occur. After translocation, the A site is empty, ready to bind an aminoacyl-tRNA to start another round of elongation.

PROTEIN SYNTHESIS IS TERMINATED BY RELEASE FACTORS

Aminoacyl-tRNA does not normally bind to the A site of a ribosome if the codon is UAA, UGA, or UAG. Normal cells do not contain tRNAs with anticodons complementary to these stop signals. Instead, these stop signals are recognized by release factors, which are proteins. One of these release factors, RF 1, recognizes UAA or UAG. The other release factor, RF 2, recognizes UAA or UGA. Thus, proteins can recognize trinucleotide sequences with high specificity.

The binding of a release factor to a termination codon in the A site somehow activates peptidyl transferase so that it hydrolyzes the bond between the polypeptide and the tRNA in the P site. The specificity of peptidyl transferase is altered by the release factor so that water rather than an amino group is the acceptor of the activated peptidyl moiety. The polypeptide chain then leaves the ribosome. The 70S ribosome then dissociates into 30S and 50S subunits as the prelude to the synthesis of another protein molecule.

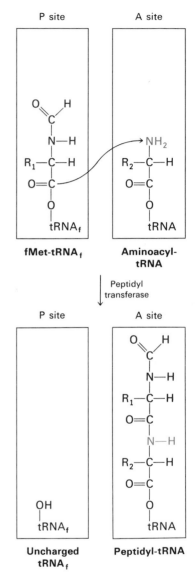

Figure 26-17
Peptide-bond formation.

SOME PROTEINS ARE MODIFIED AFTER THEY ARE SYNTHESIZED

Many of the polypeptides formed by the translation of mRNA are not the final products. Proteins may be subsequently modified in a variety of ways:

1. The formyl group at the amino terminus of bacterial proteins is hydrolyzed by a *deformylase*. One or more amino-terminal residues may be removed by *aminopeptidases*. In both procaryotes and eucaryotes, the terminal methionine is sometimes hydrolyzed while the rest of the polypeptide chain is being synthesized.

2. *Disulfide bonds* may be formed by the oxidation of two cysteine residues.

3. Certain amino acid side chains may be specifically modified. For example, some proline and lysine residues in collagen are *hydroxylated*. Glycoproteins are formed by the *attachment of sugars* to the side chains of asparagine, serine, and threonine. Some proteins are *phosphorylated*. *Prosthetic groups,* such as lipoate, are covalently attached to some enzymes.

4. Polypeptide chains may be specifically *cleaved,* as in the conversion of procollagen to collagen and of proinsulin to insulin. Translation of the mRNA of polio virus yields a very long polypeptide chain, which is hydrolyzed to form several proteins.

SEVERAL RIBOSOMES SIMULTANEOUSLY TRANSLATE A MESSENGER RNA MOLECULE

Many ribosomes can simultaneously translate an mRNA molecule. This markedly increases the efficiency of utilization of the mRNA. The group of ribosomes bound to an mRNA molecule is called a *polyribosome* or a *polysome*. The ribosomes in this unit operate independently, each synthesizing a complete polypeptide chain. The maximum density of ribosomes on mRNA is about one ribosome per eighty nucleotides, for reasons not yet evident. Polyribosomes synthesizing hemoglobin (which contains about 145 amino acids per chain, or 500 nucleotides per mRNA) typically consist of five ribosomes. Ribosomes closest to the 5' end of the messenger have the shortest polypeptide chains, whereas those nearest the 3' end have almost finished chains. Ribosomes dissociate into 30S and 50S subunits after the polypeptide product is released.

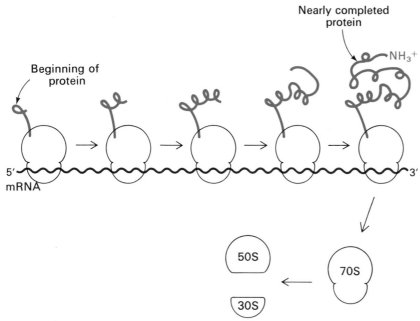

Figure 26-18
Diagram of a polyribosome (polysome). Ribosomes move along the mRNA in the 5' ⟶ 3' direction. They function independently of each other.

SOME RIBOSOMES ARE BOUND TO MEMBRANES
IN EUCARYOTIC CELLS

In eucaryotic cells, some ribosomes are located in the cytosol, whereas others are bound to a membrane system called the rough *endoplasmic reticulum* (rough ER). The proportions of free and membrane-bound ribosomes depend on the cell type. In general, secretory cells (such as those in the pancreas and salivary glands) have

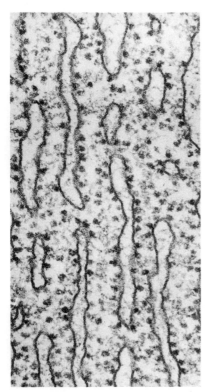

Figure 26-19
Electron micrograph of the rough endoplasmic reticulum. [Courtesy of Dr. George Palade.]

a highly developed rough ER (Figure 26-19). Newly synthesized proteins are transferred from the bound ribosomes into the vesicles of the rough ER. They then migrate to the Golgi complex, where they become incorporated into secretory granules. These granules migrate to the periphery of the cell, where they fuse with the plasma membrane, thereby discharging their contents into the extracellular space (Figure 8-2).

STREPTOMYCIN INHIBITS INITIATION AND CAUSES MISREADING OF MESSENGER RNA

We have already seen that antibiotics are very useful tools in biochemistry because many of them are highly specific in their action. For example, rifampicin is a potent inhibitor of the initiation of RNA synthesis (p. 614). Many antibiotic inhibitors of protein synthesis are known, and the mechanism of action of some of them has been determined. *Streptomycin,* a highly basic trisaccharide, interferes with the binding of formylmethionyl tRNA to ribosomes, and thereby prevents the correct initiation of protein synthesis. Streptomycin also leads to a misreading of mRNA. If poly U is the template, isoleucine (AUU) is incorporated in addition to phenylalanine (UUU). The site of action of streptomycin in the ribosome has been determined by the reconstruction of ribosomal components from streptomycin-sensitive and streptomycin-resistant bacteria. These bacterial strains are related by mutation of a single gene. How do their ribosomes differ? Because ribosomes can be dissociated and assembled in vitro, it was feasible to determine whether the determinant of streptomycin sensitivity resides in the 50S or 30S ribosomal subunit. Hybrid ribosomes consisting of 50S subunits from resistant bacteria and 30S subunits from sensitive bacteria were found to be sensitive, whereas ribosomes prepared from the reverse combination were resistant. This experiment showed that the determinant of streptomycin sensitivity is located in the 30S subunit. The next experiment demonstrated that a protein in the 30S subunit rather than its 16S RNA molecule specifies streptomycin sensitivity. Finally, an extensive series of experiments revealed that a single protein in the 30S subunit, namely protein S12, is the determinant of streptomycin sensitivity.

Table 26-3
Antibiotic inhibitors of protein synthesis

Antibiotic	Action
Streptomycin	Inhibits initiation and causes misreading of mRNA
Tetracyclines	Binds to the 30S subunit and inhibits binding of aminoacyl-tRNAs
Chloramphenicol	Inhibits the peptidyl transferase activity of the 50S ribosomal subunit (procaryotes)
Cycloheximide	Inhibits the peptidyl transferase activity of the 60S ribosomal subunit (eucaryotes)
Erythromycin	Binds to the 50S subunit and inhibits translocation
Puromycin	Causes premature chain termination by acting as an analog of aminoacyl-tRNA

DIPHTHERIA TOXIN BLOCKS PROTEIN SYNTHESIS IN EUCARYOTES BY INHIBITING TRANSLOCATION

Diphtheria was a major cause of death in childhood prior to the advent of effective immunization. The lethal effects of this disease are due mainly to a toxin formed by *Corynebacterium diphtheriae,* a bacterium that grows in the upper respiratory tract. A few micrograms of this 65,000-dalton protein is usually lethal in an unimmunized individual bacause it inhibits protein synthesis. Specifically, diphtheria toxin blocks the elongation phase of protein synthesis in eucaryotes (but not in procaryotes) by inactivating the elongation factor necessary for translocation. This factor in eucaryotes, called aminoacyl transferase II or *translocase,* has a role analogous to that of EF G in bacteria. The translocase is required for the GTP-driven movement of peptidyl tRNA from the A site to the P site and for the associated movement of messenger RNA immediately following the formation of a peptide bond. The mechanism of inactivation of the translocase by diphtheria toxin is especially interesting. *The toxin catalyzes the covalent modification of the translocase.* NAD^+ donates its adenosine diphosphate ribose moiety (ADPR) to the translocase, and nicotinamide is released.

$$NAD^+ + \text{translocase} \underset{\text{(active)}}{\rightleftharpoons} \text{ADPR-translocase} + \text{nicotinamide}$$
$$\text{(active)} \qquad\qquad \text{(inactive)}$$

Streptomycin

PUROMYCIN CAUSES PREMATURE CHAIN TERMINATION BECAUSE IT MIMICS AMINOACYL-TRANSFER RNA

The antibiotic puromycin inhibits protein synthesis by releasing nascent polypeptide chains before their synthesis is completed. Puromycin is an analog of the terminal aminoacyl-adenosine portion of aminoacyl-tRNA (Figure 26-20). It binds to the A site on the ribosome and inhibits the entry of aminoacyl-tRNA. Furthermore, puromycin contains an α-amino group. This amino group, like the one on aminoacyl-tRNA, forms a peptide bond with the carboxyl group of the growing peptide chain in a reaction that is catalyzed by peptidyl transferase. The product is a peptide having a covalently attached puromycin residue at its carboxyl end. Peptidyl puromycin then dissociates from the ribosome. Puromycin has been used to ascertain the functional state of ribosomes. In fact, the concept of A and P sites resulted from the use of puromycin to ascertain the location of peptidyl tRNA. When peptidyl tRNA is in the A site (before translocation), it cannot react with puromycin.

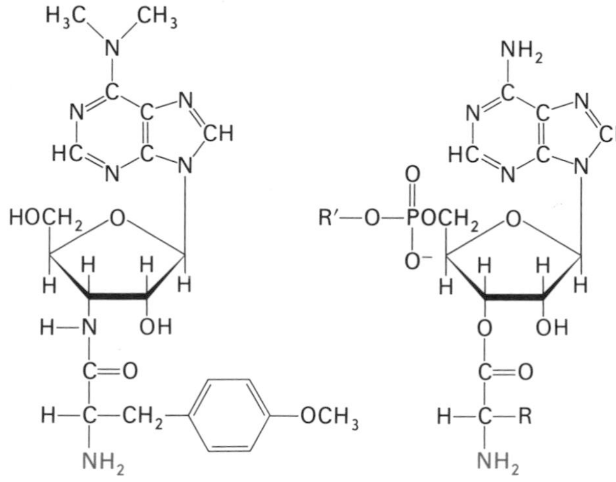

Puromycin   Aminoacyl-tRNA

Figure 26-20
Puromycin resembles the aminoacyl terminus of an aminoacyl-tRNA.

SHORT PEPTIDES ARE NOT SYNTHESIZED BY RIBOSOMES

We turn now to a different mechanism of peptide-bond formation in biological systems. As an example, let us consider the biosynthesis of gramicidin S, a cyclic peptide antibiotic made up of two identical pentapeptides joined head-to-tail. This antibiotic is produced by certain strains of *Bacillus brevis,* a spore-forming bacterium.

The biosynthesis of gramicidin S does not depend on the presence of ribosomes or mRNA. A much simpler synthetic apparatus consisting only of two enzymes, E_I and E_{II}, is required. D-Phenylalanine is activated by E_{II} (100,000 daltons), whereas the other four amino acids of the pentapeptide unit are activated by E_I (180,000 daltons). Both enzymes also participate in peptide-bond formation.

In this system, amino acids are activated by the formation of enzyme-bound *thioesters*. The activated amino acid is attached to a sulfhydryl group of E_I or E_{II} rather than to the 3′ terminal hydroxyl group of a tRNA.

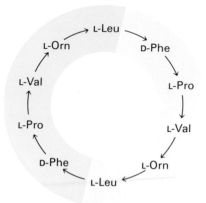

Gramicidin S

The arrows denote the polarity of the polypeptide chain (amino to carboxyl direction).

$$Amino\ acid + ATP \rightleftharpoons Aminoacyl\text{-}AMP + PP_i$$

$$Aminoacyl\text{-}AMP + E\text{—}SH \rightleftharpoons E\text{—}S\sim\overset{\overset{\displaystyle O}{\|}}{C}\text{—}\overset{\overset{\displaystyle H}{|}}{\underset{\underset{\displaystyle R}{|}}{C}}\text{—}NH_3^+ + AMP$$

Thioester

L-Proline, L-valine, L-ornithine and L-leucine form thioester linkages with specific sulfhydryls of E_I when they are incubated in the presence of ATP. Likewise, D-phenylalanine forms a thioester bond with E_{II} in the presence of ATP.

Peptide synthesis in this system is initiated by the interaction of E_I and E_{II}. The D-phenylalanine residue on E_{II} is transferred to the imino group of the L-proline residue on E_I to form a dipeptide.

$$E_{II}\text{—}S\sim\overset{\overset{\displaystyle O}{\|}}{C}\text{—}Phe\text{—}NH_2 + E_I\text{—}S\sim\overset{\overset{\displaystyle O}{\|}}{C}\text{—}Pro\text{—}NH \rightleftharpoons E_{II}\text{—}SH + E_I\text{—}S\sim\overset{\overset{\displaystyle O}{\|}}{C}\text{—}Pro\text{—}Phe\text{—}NH_2$$

Activated dipeptide

The subsequent reactions involve only E_I. The activated carbonyl group of the prolyl residue in the dipeptide reacts with the amino group of the valine residue on the same enzyme to yield a tripeptide. This process occurs again with ornithine and then with leucine to give an enzyme-bound pentapeptide. The growing peptide is transferred to a different sulfhydryl each time a peptide bond is formed.

Finally, the activated pentapeptides attached to two different E_I molecules react with each other to form cyclic gramicidin S.

Two features of this biosynthetic pathway are noteworthy:

1. The amino acid sequence of gramicidin S is determined by the spatial arrangement and specificity of the enzymes in E_I and E_{II}. There is at least one protein subunit for each peptide bond formed. Hence, *this mode of synthesis is uneconomical compared with the ribosomal mechanism. Consequently, peptides containing more than about fifteen residues are not synthesized by this mechanism.*

2. *The synthesis of gramicidin S resembles fatty acid synthesis,* since the activated intermediates in both processes are thioesters. Furthermore, E_I contains a covalently attached phosphopantetheine residue. This thiol probably carries the growing peptide chain from

one site to the next in E_I. Recall that the central thiol in fatty acid synthesis has an analogous function (p. 425). Fritz Lipmann has suggested that *antibiotic polypeptide synthesis may be a surviving relic of a primitive mechanism of protein synthesis used early in evolution.* Ribosomal protein synthesis may have evolved from fatty acid synthesis.

SUMMARY

Protein synthesis (translation) requires the coordinated interplay of more than a hundred macromolecules, which include mRNA, tRNAs, activating enzymes, and protein factors, in addition to ribosomes. Protein synthesis starts with the ATP-driven activation of amino acids by aminoacyl-tRNA synthetases (activating enzymes), which link the carboxyl group of an amino acid to the 2'- or 3'-hydroxyl group of the adenosine unit at the 3' end of tRNA. There is at least one specific activating enzyme for each amino acid. Also, there is at least one specific tRNA for each amino acid. Transfer RNAs of different specificity have a common structural design. They are single RNA chains of about eighty nucleotides, containing some modified (e.g., methylated) derivatives of the standard bases. The base sequences of all known tRNAs can be written in a cloverleaf pattern in which about half of the nucleotides are base-paired. X-ray crystallographic studies have shown that tRNA is L-shaped. The amino acid attachment site at the 3' CCA terminus is at one end of the L, whereas the anticodon is 80 Å away at the other end. Messenger RNA recognizes the anticodon of a tRNA rather than the attached amino acid. A codon on mRNA forms a base pair with the anticodon on tRNA. Some tRNAs recognize more than one codon because pairing of the third base of a codon is less discriminating than for the other two (the wobble hypothesis). Protein synthesis takes place on ribosomes, which consist of a large and a small subunit, each about two-thirds RNA and one-third protein by weight. In *E. coli,* the 70S ribosome (2.5 million daltons) is made up of 30S and 50S subunits.

There are three phases of protein synthesis: initiation, elongation, and termination. Messenger RNA, formylmethionyl-tRNA$_f$, and a 30S ribosomal subunit come together to form a 30S initiation complex. Formylmethionyl-tRNA$_f$ recognizes a special initiation sequence on mRNA. A 50S ribosomal subunit then joins this complex to form a 70S initiation complex, which is ready for the next phase. The elongation cycle consists of the binding of aminoacyl-tRNA (codon recognition), peptide-bond formation, and translocation. The direction of chain growth is from the amino to the carboxyl end. Protein synthesis is terminated by release factors, which recognize the termination codons UAA, UGA, and UAG and lead to the hydrolysis of the bond between the polypeptide and tRNA. GTP is hydrolyzed in the formation of the 70S initiation complex, in the binding of aminoacyl-tRNA to the ribosome, and in the translocation step. Various steps in protein synthesis are specifically inhibited by toxins and antibiotics. Peptide antibiotics and other short polypeptides are synthesized by a nonribosomal pathway that resembles fatty acid synthesis.

SELECTED READINGS

Where to start:

Miller, O. L., Jr., 1973. The visualization of genes in action. *Sci. Amer.* 228(3):34–42. [Available as *Sci. Amer.* Offprint 1267.]

Holley, R. W., 1966. The nucleotide sequence of a nucleic acid. *Sci. Amer.* 214(2):30–39. [Offprint 1033.]

Nomura, M., 1973. Assembly of bacterial ribosomes. *Science* 179:864–873.

Clark, B. F. C., and Marcker, K. A., 1968. How proteins start. *Sci. Amer.* 218(1):36–42. [Offprint 1092.]

Books and reviews on protein synthesis:

Cold Spring Harbor Laboratory, 1969. *Mechanisms of Protein Biosynthesis* (Cold Spring Harbor Symposium on Quantitative Biology, vol. 34). [Contains many valuable articles on all aspects of protein biosynthesis, including an excellent summary by P. Lengyel.]

Nomura, M., Tissières, A., and Lengyel, P., (eds.), 1975. *Ribosomes*. Cold Spring Harbor Laboratory. [Contains excellent reviews on the structure, function, and assembly of ribosomes.]

Haselkorn, R., and Rothman-Denes, L. B., 1973. Protein synthesis. *Annu. Rev. Biochem.* 42:397–438.

Lucas-Lenard, J., and Lipmann, F., 1971. Protein biosynthesis. *Annu. Rev. Biochem.* 40:409–448.

Transfer RNA:

Suddath, F. L., Quigley, G. J., McPherson, A., Sneden, D., Kim, J. J., Kim, S. H., and Rich, A., 1974. Three-dimensional structure of yeast phenylalanine transfer RNA at 3.0 Å resolution. *Science* 248:20–24.

Robertus, J. D., Ladner, J. E., Finch, J. T., Rhodes, D., Brown, R. S., Clark, B. F. C., and Klug, A., 1974. Structure of yeast phenylalanine tRNA at 3 Å resolution. *Nature* 250:546–551.

von Ehrenstein, G., 1970. Transfer RNA and amino acid activation. *In* Anfinsen, C. B., Jr., *Aspects of Protein Biosynthesis,* pp. 139–214. Academic Press.

Smith, J. D., 1973. Genetic and structural analysis of transfer RNA. *Brit. Med. Bull.* 29:220–225.

Söll, D., 1971. Enzymatic modification of transfer RNA. *Science* 173:293–299.

Initiation, elongation, and termination:

Dintzis, H. M., 1961. Assembly of the peptide chains of hemoglobin. *Proc. Nat. Acad. Sci.* 47:247–261. [These important pulse-labeling experiments showed that proteins are synthesized from the amino to the carboxyl end.]

Steitz, J. A., 1969. Polypeptide chain initiation: nucleotide sequence of the three ribosomal binding sites in bacteriophage R17 RNA. *Nature* 224:957–964.

Leder, P., 1973. The elongation reactions in protein synthesis. *Advan. Protein Chem.* 27:213–242.

Gupta, S. L., Waterson, J., Sopori, M., Weissman, S. M., and Lengyel, P., 1971. Movement of the ribosome along the messenger ribonucleic acid during protein synthesis. *Biochemistry* 10:4410–4421.

Caskey, C. T., 1973. Peptide chain termination. *Advan. Protein Chem.* 27:243–276.

Shine, J., and Dalgarno, L., 1974. The 3′-terminal sequence of *Escherichia coli* 16S ribosomal RNA: complementarity to nonsense triplets and ribosome binding sites. *Proc. Nat. Acad. Sci.* 71:1342–1346.

Antibiotics and toxins:

Pestka, S. (1971). Inhibitors of ribosome function. *Annu. Rev. Microbiol.* 25:487–562.

Nonribosomal peptide synthesis:

Lipmann, F., 1971. Attempts to map a process evolution of peptide biosynthesis. *Science* 173:875–884.

Perlman, D., and Bodanszky, M., 1971. Biosynthesis of peptide antibiotics. *Annu. Rev. Biochem.* 40:449–464.

Lipmann, F., 1973. Nonribosomal polypeptide synthesis on polyenzyme templates. *Accounts Chem. Res.* 6:361–367.

PROBLEMS

1. The formation of Ile-tRNA proceeds through an enzyme-bound Ile-AMP intermediate. Predict whether ^{32}P-labeled ATP is formed from $^{32}PP_i$ when each of the following sets of components is incubated with the specific activating enzyme:

   (a) ATP and $^{32}PP_i$.

   (b) tRNA, ATP, and $^{32}PP_i$.

   (c) Isoleucine, ATP, and $^{32}PP_i$.

2. Ribosomes were isolated from bacteria grown in a "heavy" medium (^{13}C and ^{15}N) and also from bacteria grown in a "light" medium (^{12}C and ^{14}N). These 70S ribosomes were added to an in vitro system actively engaged in protein synthesis. An aliquot removed several hours later was analyzed by density-gradient centrifugation. How many bands of 70S ribosomes would you expect to see in the density gradient?

3. How many high-energy phosphate bonds are consumed in the synthesis of a 200-residue protein, starting from amino acids?

4. There are two basic mechanisms for the elongation of biomolecules (Figure 26-21). In type 1, the activating group (labeled X) is released from the growing chain. In type 2, the activating group is released from the incoming unit as it is added to the growing chain. Indicate whether each of the following biosyntheses occurs by means of a type 1 or a type 2 mechanism:

    (a) Glycogen synthesis.

    (b) Fatty acid synthesis.

    (c) $C_5 \longrightarrow C_{10} \longrightarrow C_{15}$ in cholesterol biosynthesis.

    (d) DNA synthesis.

    (e) RNA synthesis.

    (f) Protein synthesis.

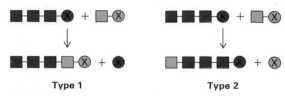

Type 1   **Type 2**

Figure 26-21
Two modes of elongation.

5. Mutations that produce termination codons are called nonsense mutations. These mutations can be suppressed by altered tRNAs. For example, the UGA codon is translated into tryptophan by a mutant tRNA. What is the most likely base change in this mutant tRNA?

6. Design an affinity-labeling reagent for one of the tRNA binding sites in *E. coli* ribosomes. How would you synthesize such a reagent?

For additional problems, see W. B. Wood, J. H. Wilson, R. M. Benbow, and L. E. Hood, *Biochemistry: A Problems Approach* (Benjamin, 1974), Chapter 18.

CHAPTER 27

CONTROL OF GENE EXPRESSION

We have already seen that the activity of many proteins is regulated by a variety of mechanisms, such as proteolytic activation, allosteric interaction, and covalent modification. This chapter deals with the *control of the rate of protein synthesis,* which is also critical in determining the pattern of cellular processes. In bacteria, gene activity is regulated primarily at the level of *transcription* rather than translation. We will focus on the lactose operon of *E. coli* and on regulatory aspects of the bacteriophage lambda, because much is known about the molecular basis of the control of these systems. Furthermore, the study of these systems has revealed some general principles of gene regulation in procaryotes and viruses.

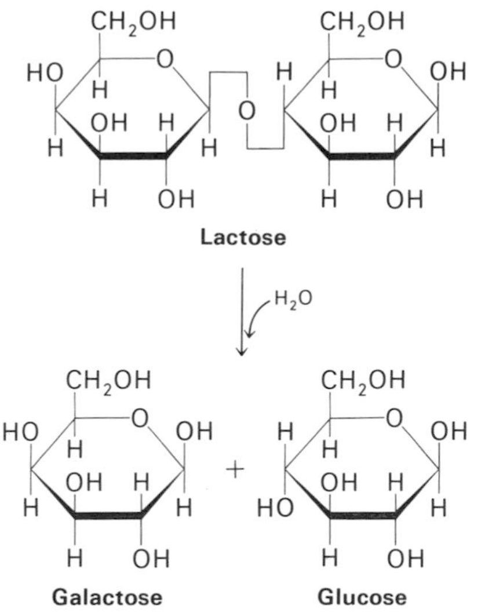

BETA-GALACTOSIDASE IS AN INDUCIBLE ENZYME

E. coli can use lactose as its sole source of carbon. An essential enzyme in the metabolism of this sugar is *β-galactosidase,* which hydrolyzes lactose to galactose and glucose (Figure 27-1). An *E. coli* cell growing on lactose contains several thousand molecules of *β*-galactosidase. In contrast, the number of *β*-galactosidase molecules per cell is fewer than ten if *E. coli* is grown on other sources

Figure 27-1
Lactose is hydrolyzed by *β*-galactosidase.

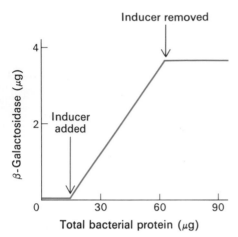

Figure 27-2
The increase in the amount of
β-galactosidase parallels the increase in
the number of cells in a growing culture
of *E. coli*. The slope of this plot indicates
that 6.6% of the protein synthesized is
β-galactosidase.

of carbon, such as glucose or glycerol. Lactose induces a large increase in the amount of β-galactosidase in *E. coli* by eliciting the synthesis of new enzyme molecules rather than by activating a proenzyme (Figure 27-2). Hence, β-galactosidase is an inducible enzyme. Two other proteins are synthesized in concert with β-galactosidase—namely, *galactoside permease* and *thiogalactoside transacetylase*. The permease is required for the transport of lactose across the bacterial cell membrane, whereas the transacetylase is not essential for lactose metabolism. The physiologic role of the transacetylase is not yet known. In vitro, it catalyzes the transfer of an acetyl group from acetyl CoA to the C-6 hydroxyl group of a thiogalactoside.

The physiologic inducer is *allolactose,* which is formed from lactose by transglycosylation. The synthesis of allolactose is catalyzed by the few β-galactosidase molecules that are present prior to induction. A study of the nature of inducers showed that some β-galactosides are inducers without being substrates of β-galactosidase, whereas other compounds are substrates without being inducers. For example, *isopropylthiogalactoside* (IPTG) is a nonmetabolizable inducer (also called a gratuitous inducer).

1,6-Allolactose   IPTG

DISCOVERY OF A REGULATORY GENE

An important clue concerning the nature of the induction process was provided by the finding that the amount of the permease and of the transacetylase increased in direct proportion to β-galactosidase, for all inducers tested. Further insight came from studies of mutants, which showed that β-galactosidase, the permease, and the transacetylase are encoded by three contiguous genes, called *z, y,* and *a*, respectively. Mutants defective in only one of these proteins were isolated. For example, $z^-y^+a^+$ denotes a mutant lacking β-galactosidase but having normal amounts of the permease and the transacetylase. A most interesting class of mutants affecting all three proteins was then isolated. These *constitutive mutants* synthesize large amounts of β-galactosidase, the permease, and the transacetylase whether or not inducer is present. Francois Jacob and Jacques Monod deduced that *the rate of synthesis of these three proteins is governed by a common element that is different from the genes that specify their structure.* The gene for this common regulatory element was named i. Wild-type inducible bacteria have the genotype $i^+z^+y^+a^+$, whereas the constitutive lactose mutants have the genotype $i^-z^+y^+a^+$.

How does the i^+ gene affect the rate of synthesis of the proteins encoded by the z, y, and a genes? The simplest hypothesis was that the i^+ gene determines the synthesis of a cytoplasmic substance called a *repressor,* which is missing or inactive in i^-. This idea was tested in an ingenious series of genetic experiments involving partially diploid bacteria that contained two sets of genes for the lactose region. One set was on the bacterial chromosome whereas the other was on an F′ sex factor, introduced by conjugation. For example, an i^+z^-/Fi^-z^+ diploid was isolated. In this diploid, i^+z^- is on the chromosome, whereas i^-z^+ is on the episome. Is this diploid inducible or constitutive for β-galactosidase? In other words, does i^+ on the bacterial chromosome repress the expression of z^+ on the episome? The experimental result was clearcut: *the diploid was inducible rather than constitutive, which strongly supported the repressor hypothesis.* The same result was obtained for the diploid i^-z^+/Fi^+z^-.

AN OPERON IS A COORDINATED UNIT
OF GENETIC EXPRESSION

These experiments led Jacob and Monod to propose the *operon model* for the regulation of protein synthesis. The genetic elements of this model are a *regulator gene,* an *operator gene,* and a set of *structural genes* (Figure 27-3). The regulator gene produces a *repressor* that can interact with the operator gene. Subsequent work revealed that the repressor is a protein. The operator gene is adjacent to the structural genes it controls. The binding of the repressor to the operator gene prevents the transcription of the structural genes. The operator gene and its associated structural genes are called an *operon.* For the

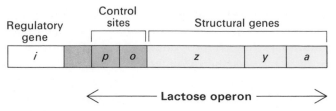

Figure 27-3
Map of the lactose operon and its associated regulatory gene. (The proportions in this diagram are not accurate: the *p* and *o* sites are much smaller than the other genes.)

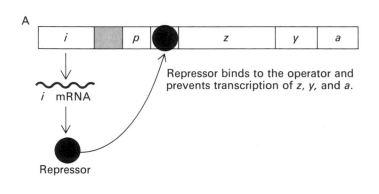

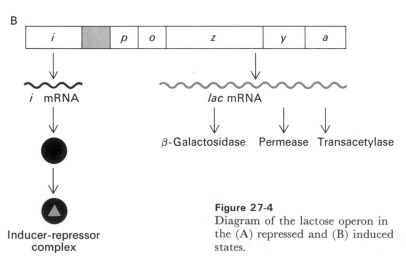

Figure 27-4
Diagram of the lactose operon in the (A) repressed and (B) induced states.

lactose operon, the i gene is the regulator gene, the o gene is the operator gene, and the z, y, and a genes are the structural genes. In addition, there is a *promoter site* (denoted by p) for the binding of RNA polymerase. This site for the initiation of transcription is next to the operator gene. An *inducer* such as IPTG binds to the repressor, which prevents it from interacting with the operator gene. The z, y, and a genes can then be transcribed.

THE *LAC* REPRESSOR IS A TETRAMERIC PROTEIN

The repressor of the lactose operon (called the *lac* repressor) was isolated on the basis of its binding of IPTG. Walter Gilbert and Benno Müller-Hill showed that the *lac* repressor is a protein that binds to DNA carrying the *lac* operon but not to other DNA molecules. As expected, IPTG prevents the binding of *lac* repressor to *lac* operator DNA. A wild-type *E. coli* cell contains only about ten molecules of the *lac* repressor. It is difficult to purify the repressor from these cells because it constitutes only .01% of the total protein. However, much larger amounts of the *lac* repressor are

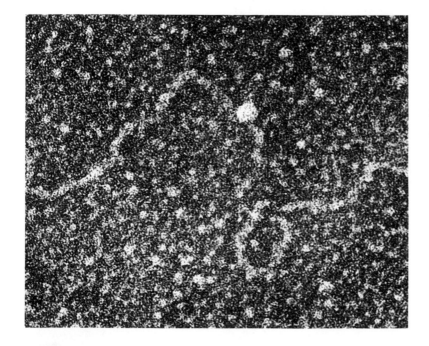

Figure 27-5
Electron micrograph of the *lac* repressor bound to DNA containing the *lac* operator. [Courtesy of Dr. Jack Griffith.]

made in i^{sq} mutants, which probably have a more efficient promoter for the i gene. The amount of *lac* repressor is increased further by using transducing phages that carry the *lac* region. Such an infected *E. coli* cell contains about 20,000 repressors (about 20% of the total protein), which makes it a choice starting material for the purification of *lac* repressor. The repressor is a tetramer consisting of identical 37,000-dalton subunits; the amino acid sequence has recently been determined. Each subunit has one binding site for an inducer. The binding constant for IPTG is about 10^{-6} M. The repressor binds very tightly and rapidly to the operator. The equilibrium constant of the complex is about 10^{-13} M. The rate constants for association and dissociation are 7×10^{9} M^{-1} sec^{-1} and 6×10^{-4} sec^{-1}, respectively. The rate constant for association is strikingly high.

THE *LAC* OPERATOR HAS A SYMMETRIC BASE SEQUENCE

The availability of pure *lac* repressor has made it feasible to isolate the *lac* operator and determine its base sequence. Gilbert and his associates sonicated the DNA of a phage carrying the *lac* region into fragments that were approximately 1000 base pairs long. The *lac* repressor was added to the mixture of fragments, which was then filtered through a cellulose nitrate membrane. DNA fragments without bound *lac* repressor passed through this filter, whereas DNA-repressor complexes bound tightly to it. The bound DNA was released by adding IPTG. These released DNA fragments were then treated with pancreatic deoxyribonuclease in the presence of repressor. The rationale of this step is that the operator region is protected from digestion by deoxyribonuclease when it is complexed to the repressor (Figure 27-6). The base sequence of these operator fragments was then elucidated by determining the sequence of its RNA transcript. *The protected operator region, which is a double-helical fragment containing 27 base pairs, has a very interesting base sequence* (Figure 27-7). *A total of 16 base pairs lie in a symmetrical pattern.* Both strands contain the sequence 5'-AATTGT-3'. Six base pairs on the left side of the fragment are symmetrically related to six on the extreme right side of the fragment.

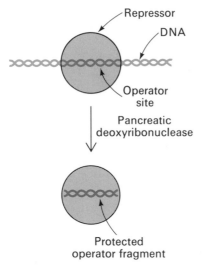

Figure 27-6
The *lac* repressor protects the *lac* operator from digestion by pancreatic deoxyribonuclease.

Figure 27-7
Nucleotide sequence of the *lac* operator. The symmetrically related regions are shown in color.

The sequence of the 5′ terminal region of mRNA transcribed in vitro from the lactose operon has been determined (Figure 27-8). This mRNA starts with the *lac* operator sequence. Thus, the *lac* operator is transcribed. The AUG initiator codon for β-galactosidase starts at position 39 of this transcript. Some of the eighteen bases between the end of the operator region and this AUG probably play a role in the initiation of translation.

SYMMETRY IN THE REPRESSOR-OPERATOR INTERACTION

How does the *lac* repressor select one of six million nucleotide sequences in the *E. coli* chromosome? It seems likely that symmetry is important in this highly specific recognition process. The three-dimensional structure of the *lac* repressor is not known. However, electron micrographs and x-ray diffraction patterns of small crystals of the *lac* repressor suggest that it has an elongated shape whose cross section looks like a dumbbell (Figure 27-9). It has been proposed that the repressor binds to the operator so that its long axis is aligned with the axis of the DNA helix. The twofold symmetry axis of the repressor would then coincide with that of the operator. In other words, *the symmetry of the repressor probably matches that of the operator.* We have already encountered this principle of recognition in the binding of actinomycin D, a peptide antibiotic, to DNA (p. 616).

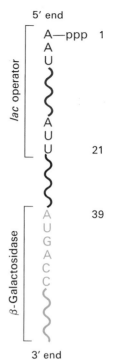

Figure 27-8
Nucleotide sequence of the 5′ end of *lac* mRNA transcribed in vitro.

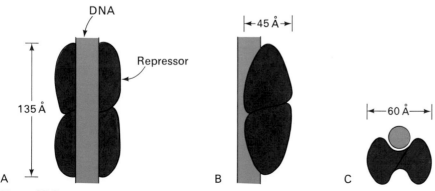

Figure 27-9
Proposed mode of interaction of the *lac* repressor and operator: (A) front view; (B) side view; and (C) top view. [Based on T. A. Steitz, T. J. Richmond, D. Wise, and D. Engelman. *Proc. Nat. Acad. Sci.* 71(1974):596.]

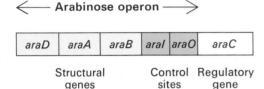

Arabinose

← ——— **Arabinose operon** ——— →

araD	araA	araB	araI	araO	araC

Structural   Control Regulatory
genes   sites   gene

Figure 27-10
Map of the arabinose operon and
its associated regulatory gene.

VARIATIONS ON THE *LAC* OPERON THEME

Several other bacterial operons have been charcterized, such as the
operons for the biosynthesis of histidine, tryptophan, and arginine
and the operons for the degradation of galactose, arabinose, and
glycerol. The *arabinose operon* is interesting because it displays *positive
control* by the regulator protein in addition to negative control.
Arabinose is converted to xylulose 5-phosphate by the sequential
action of arabinose isomerase, ribulokinase, and ribulose 5-phos-
phate epimerase, which are encoded by the *araA, araB,* and *araD*
genes, respectively. These structural genes, together with a promoter
(*araI*) and an operator (*araO*), constitute the arabinose operon
(Figure 27-10). This operon is regulated by *araC,* which is adjacent
to the operator site. The product of *araC* is a protein that can occur
in two functionally different conformations, called P1 and P2. In
the absence of arabinose, this protein acts as a repressor (P1). The
P1 form binds to the operator, which prevents transcription of the
operon. Arabinose removes P1 from the operator and shifts the
conformational equilibrium to P2, the activator form. The P2 form
then binds to the promoter site, which is necessary for transcription.
Thus, *the same regulator protein serves a positive and a negative control
function.*

The operon for the synthesis of tryptophan (the *trp* operon)
illustrates a different variation on the *lac* operon theme. Tryptophan
is synthesized from chorismate in five steps (p. 513). The synthesis
of the enzymes catalyzing these reactions is repressed by tryptophan. In proposing the operon model, Jacob and Monod suggested
that repression and induction occur by similar mechanisms. This
inference has proven to be correct. The *trp* repressor, like the *lac*
repressor, inhibits transcription by binding to a specific operator
site on DNA. However, these operons differ in that the *trp* repressor
binds to the operator only if tryptophan (or a derivative) is *present,*
whereas the *lac* repressor binds to the operator only if an inducer
is *absent.* Thus, *tryptophan is a corepressor of the trp operon.*

One of the largest and most complex operons is the one for the
biosynthesis of histidine in *Salmonella typhimurium.* The nine structural genes of this *his* operon code for the enzymes that convert
phosphoribosylpyrophosphate (PRPP) to histidine (p. 515). These
enzymes are synthesized when there is little histidine in the bacterial cell. Transcription of this operon gives a single mRNA molecule

(containing about 10,000 nucleotides) that is the template for the synthesis of nine proteins. The control of this operon is now being investigated. There are at least six regulatory genes for the *his* operon, which suggests that its control is considerably more complex than that of the *lac* operon.

CYCLIC AMP STIMULATES THE TRANSCRIPTION OF INDUCIBLE OPERONS

It has long been known that *E. coli* grown on glucose have very low levels of catabolic enzymes, such as β-galactosidase, galacto-kinase, arabinose isomerase, and tryptophanase. Clearly, it would be wasteful to synthesize these enzymes when glucose is abundant. The molecular basis of this inhibitory effect of glucose, called *catabolite repression,* has recently been elucidated. A key clue was provided by the observation that glucose lowers the concentration of cyclic AMP in *E. coli.* It was then found that exogenous cyclic AMP can relieve the repression exerted by glucose. Subsequent biochemical and genetic studies revealed that *cyclic AMP stimulates the initiation of transcription of many inducible operons.* Cyclic AMP is synthesized by adenyl cyclase (p. 810). Glucose and some of its derivatives somehow inhibit the synthesis of cyclic AMP. The cyclic AMP synthesized in the absence of glucose binds to a specific receptor protein (called CAP), which is a 45,000-dalton dimer. The complex of cyclic AMP and CAP binds to the promoter site of the *lac* operon and thereby stimulates the initiation of transcription. It seems likely that the cyclic AMP-CAP complex acts similarly at other inducible operons. Thus, *inducible operons are controlled by complementary mechanisms that use cyclic AMP and specific inducers as the signal molecules.*

REPRESSORS AND ACTIVATORS DETERMINE THE DEVELOPMENT OF TEMPERATE PHAGE

We turn now to the role of repressors and activators of transcription in regulating the life cycle of *lambda* (λ) *bacteriophage.* The mature virus particle consists of a linear double-helical DNA molecule (31,000,000 daltons) surrounded by a protein coat. Two developmental pathways are open to this virus: it can destroy its host or

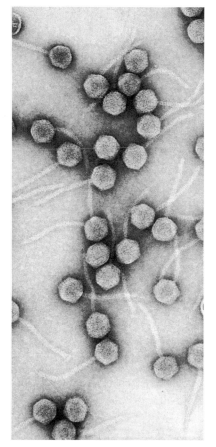

Figure 27-11
Electron micrograph of λ phages. [Courtesy of Dr. A. Dale Kaiser.]

it can become part of its host (hence the name *temperate*). In the *lytic pathway,* the viral functions are fully expressed, leading to the lysis of the bacterium and the production of a burst of about 100 progeny virus particles. Alternatively, λ can enter the *lysogenic pathway,* in which its DNA becomes covalently joined to the host-cell DNA at a specific site. This recombination process involving a circular λ DNA molecule will be discussed later (p. 720). Most of the phage functions are switched off when its DNA is integrated in the host DNA. The viral DNA in this state is called the *prophage*; a host cell containing a prophage is called a *lysogenic bacterium.* The prophage replicates as part of the host chromosome in a lysogenic bacterium, usually for many generations. The lytic functions of λ are dormant but not lost in the lysogenic state. A variety of agents that interfere with DNA replication in the host induce the prophage to undergo lytic development.

Let us first consider the pattern of gene expression of λ in the lytic pathway. The goal of producing a large number of progeny is achieved by the *sequential transcription of viral genes.* Proteins needed for DNA replication and recombination are made first, followed by the synthesis of the head and tail proteins of the virus particle and of the proteins required for lysis of the host cell. Timing is critical; premature destruction of the host would of course be disadvantageous to the virus. There are three stages in the lytic pathway, called immediate-early, delayed-early, and late. In the *immediate-early stage,* RNA synthesis starts at two promoter sites, P_L and P_R. One of these transcripts is the template for the N protein, which has a critical regulatory role. The N protein antagonizes the termination of transcription and thereby permits further expression of λ genes. The N protein turns on the *delayed-early stage*. Proteins necessary for the replication of λ DNA and for recombination are made at this time. In addition, the Q gene is transcribed in the

Figure 27-12
Genetic map of λ phage. Only some of the genes are shown here.

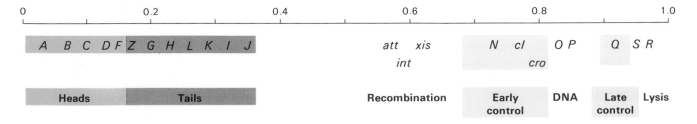

delayed-early stage. The Q protein is another critical regulator of gene expression in λ. The Q protein is required for the *late stage*. The genes for proteins needed for phage head and tail formation and also for the lysis of the host are transcribed in the late stage. *In short, the sequential regulation of lytic development is effected by two positive regulatory proteins, encoded by genes N and Q, which act at the level of transcription.*

There are three stages in the lysogenic cycle: *establishment, mainte-nance,* and *release.* The establishment of the prophage state requires the integration of the viral DNA into the host DNA and the inactivation of the lytic functions of the virus. These processes are complex and not fully understood. In contrast, the maintenance of the prophage state is relatively simple. A. Dale Kaiser showed that *only the cI gene is expressed in the prophage.* This gene codes for the λ repressor, which binds to two operator regions, O_L and O_R. *The λ repressor directly prevents the transcription of the immediate-early genes.* In particular, the N protein is not synthesized and so the lytic pathway is blocked. In addition, the λ repressor stimulates the transcription of the *cI* gene. In other words, it stimulates its own synthesis. Inactivation of the λ repressor leads to the release stage, in which the prophage is excised from the host chromosome and the lytic functions are expressed.

TWO OPERATORS IN LAMBDA CONTAIN A SERIES OF BINDING SITES FOR THE REPRESSOR

The λ repressor has been isolated and studied in detail by Mark Ptashne. The 27,000-dalton monomer is in equilibrium with oli-gomers, which are the forms that bind to DNA. Two distinct operators, O_L and O_R, are recognized by the same λ repressor. Repressor bound to O_L blocks leftward transcription, whereas repressor bound to O_R blocks rightward transcription (Figure 27-13). The *cI* gene that codes for the repressor is located between O_L and O_R. O_L is the operator for the operon that includes the N gene, which codes for a critical regulatory protein as mentioned above.

The operators O_L and O_R are more complex than the *lac* opera-tor. It came as a surprise to find that *O_L and O_R each contain a series of nonidentical binding sites for the λ repressor* (Figure 27-13). The base sequences of O_L and O_R are different, but they exhibit certain

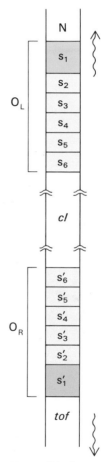

Figure 27-13
Diagram of the operators O_L and O_R and the adjacent genes. The s_1 and s_1' sites have the highest affinity for the λ repressor. The repressor gene is *cI*. One transcript starts with the N gene, whereas the other starts with the *tof* gene.

similarities. The strongest binding site for repressor in both O_L and O_R is the one closest to the start of the first structural gene in the operon. Furthermore, both O_L and O_R contain a binding site for RNA polymerase. Thus, *the promoter site is located inside the operator site in these operons,* in contrast to the *lac* operon, where the promoter and operator only partly overlap. Repressor bound to either of these operators of λ prevents RNA polymerase from binding to the corresponding promoter.

The base sequence of the strongest repressor binding site (s_1) of O_L has recently been determined. *Several groups of base pairs in this operator are symmetrically related,* as in the *lac* repressor. Furthermore, the promoter site in this operator includes this symmetric sequence:

$$5'\quad GTCGAC\quad 3'$$
$$3'\quad CAGCTG\quad 5'$$

These findings reinforce the proposal that symmetry plays a key role in the recognition of specific sites on DNA by proteins.

SUMMARY

Cells regulate the amounts of proteins that are synthesized in a variety of ways. In *E. coli,* gene expression is regulated primarily at the level of transcription rather than translation. Many genes are organized into operons, which are coordinated units of genetic expression. An operon consists of control sites (an operator and a promoter) and a set of structural genes. Outside the operon, there is a regulator gene coding for a protein that can interact with the operator site. The lactose operon is induced by β-galactosides such as allolactose and IPTG. Binding of an inducer to the *lac* repressor results in its displacement from the operator. RNA polymerase can then move through the operator to transcribe the *lac* operon. The tryptophan operon is repressed by tryptophan, which binds to the specific repressor and thereby enables it to interact with the operator. The effect is to switch off the transcription of the genes that code for the biosynthesis of tryptophan. Another variation of the *lac* operon theme is displayed by the arabinose operon, in which the same regulator protein serves a positive and a negative control function. Broad control is exerted by cyclic AMP, which binds to a specific receptor protein. This complex interacts with the promoter sites of several inducible operons and stimulates the initiation of transcription. The level of cyclic AMP is high only if glucose is scarce. Thus, glucose indirectly represses the synthesis of a variety of catabolic enzymes.

The λ bacteriophage can multiply and destroy its host (the lytic pathway) or its DNA can become integrated with that of its host (the lysogenic pathway). Three groups of genes are sequentially transcribed in the lytic pathway. The N protein formed in the immediate-early stage activates the transcription of the delayed-early genes, which in turn yields the Q protein, the activator for the late stage of transcription. The lysogenic pathway is maintained by the λ repressor, which is encoded by the *cI* gene. The repressor binds to the O_L and O_R operators and prevents the transcription of the immediate-early genes.

SELECTED READINGS

Where to start:

Jacob, F., and Monod, J., 1961. Genetic regulatory mechanisms in the synthesis of proteins. *J. Mol. Biol.* 3:318–356. [The operon model and the concept of messenger RNA were proposed in this superb paper.]

Ptashne, M., and Gilbert, W., 1970. Genetic repressors. *Sci. Amer.* 222 (6): 36–44. [Available as *Sci. Amer.* Offprint 1179.]

Lac operon:

Gilbert, W., and Müller-Hill, B., 1966. Isolation of the *lac* repressor. *Proc. Nat. Acad. Sci.* 56:1891–1898.

Müller-Hill, B., 1971. Lac repressor. *Angew. Chem. Int. Ed. Engl.* 10:160–172. [A review, emphasizing *lac* system mutants, the isolation of the repressor, and repressor-operator interactions.]

Bourgeois, S., 1971. The *lac* repressor. *Curr. Top. Cell Regul.* 4:39–75. [A review, emphasizing inducer-repressor and repressor-operator interactions in vitro.]

Beckwith, J. R., and Zipser, D., (ed.), 1970. *The Lactose Operon.* Cold Spring Harbor Laboratory.

Beyreuther, K., Adler, K., Geisler, N., and Klemm, A., 1973. The amino-acid sequence of *lac* repressor. *Proc. Nat. Acad. Sci.* 70:3576–3580.

Gilbert, W., and Maxam, A., 1973. The nucleotide sequence of the *lac* operator. *Proc. Nat. Acad. Sci.* 70:3581–3584.

Maizels, N. M., 1973. The nucleotide sequence of the lactose messenger ribonucleic acid transcribed from the UV5 promoter mutant of *Escherichia coli. Proc. Nat. Acad. Sci.* 70:3585–3589.

Other bacterial operons:

Vogel, H. J., (ed.), 1971. *Metabolic Regulation.* Volume 5 of *Metabolic Pathways* (3rd ed.). Academic Press. [Includes reviews of the arabinose system (by E. Englesberg), the tryptophan pathway (by P. Margolin), the histidine operon (by M. Brenner and B. N. Ames), and the *lac* operon (by G. Zubay and D. A. Chambers).]

Cyclic AMP and catabolite repression:

Pastan, I., and Perlman, R., 1970. Cyclic adenosine monophosphate in bacteria. *Science* 169:339–344.

Control of transcription in λ phage:

Echols, H. E., 1972. Developmental pathways for the temperate phage: lysis versus lysogeny. *Annu. Rev. Genet.* 6:157–190.

Davidson, J., 1973. Positive and negative control of transcription in bacteriophage λ. *Brit. Med. Bull.* 29:208–213.

Hershey, A. D., (ed.), 1971. *The Bacteriophage Lambda.* Cold Spring Harbor Laboratory. [Contains a wealth of information about λ.]

Reichardt, L., and Kaiser, A. D., 1971. Control of λ repressor synthesis. *Proc. Nat. Acad. Sci.* 68:2185–2189.

Maniatis, T., Ptashne, M., and Maurer, R., 1973. Control elements in the DNA of bacteriophage λ. *Cold Spring Harbor Sym. Quant. Biol.* 38:857–868.

Maurer, R., Maniatis, T., and Ptashne, M., 1974. Promoters are in the operators in phage lambda. *Nature* 249:217–221.

PROBLEMS

1. What is the effect of each of these mutations?
   (a) Deletion of the *lac* regulator gene.
   (b) Deletion of the *trp* regulator gene.
   (c) Deletion of the arabinose regulator gene.
   (d) Deletion of the *cI* gene of λ.
   (e) Deletion of the N gene of λ.

2. Superrepressed mutants (i^s) of the *lac* operon behave as noninducible mutants. The i^s gene is dominant over i^+ in partial diploids. What is the molecular nature of this mutation likely to be?

3. A mutant of *E. coli* synthesizes large amounts of β-galactosidase whether or not inducer is present. A partial diploid formed from this mutant and $Fi^+o^+z^-$ also synthesizes large amounts of β-galactosidase whether or not inducer is present. What kind of mutation might give these results?

4. A mutant unable to grow on galactose, lactose, arabinose, and several other carbon sources is isolated from a wild-type culture. The cyclic AMP level in this mutant is normal. What kind of mutation might give these results?

5. An *E. coli* cell bearing a λ prophage is immune to lytic infection by λ. Why?

For additional problems, see L. E. Hood, J. H. Wilson, and W. B. Wood, *Molecular Biology of Eucaryotic Cells* (Benjamin, 1975), Chapter 1.

CHAPTER 28

EUCARYOTIC CHROMOSOMES

Much more is known about the storage and expression of genetic information in procaryotes and viruses than in eucaryotes because they are much simpler. *A human cell contains more than one thousand times as much DNA as an* E. coli *cell and about one hundred thousand times as much DNA as a* λ *phage.* A second major difference is that the DNA of higher forms is associated with basic proteins called *histones,* whereas the DNA of lower forms is not. The distinctive morphology of eucaryotic chromosomes, which is evident under the light microscope, indicates that they have a much higher degree of structural organization than procaryotic genomes. Moreover, the shape of eucaryotic chromosomes changes markedly during the cell cycle. Another major difference—indeed, the one that sets eucaryotes apart from procaryotes—is that eucaryotic chromosomes are bounded by a *nuclear membrane.* This membrane and other internal ones are absent in procaryotes. Most striking and interesting, the cells of higher forms are *highly differentiated.* Clearly, eucaryotic chromosomes are a challenging and fascinating area of inquiry. There is much current interest in them, and the field is moving rapidly.

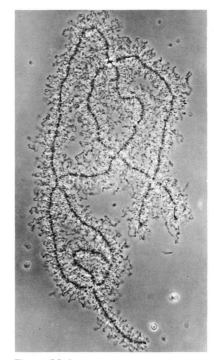

Figure 28-1
Phase-contrast light micrograph of a lampbrush chromosome from an oocyte. [Courtesy of Dr. Joseph Gall.]

A EUCARYOTIC CHROMOSOME CONTAINS A SINGLE MOLECULE OF DOUBLE-HELICAL DNA

Does a chromosome contain one long molecule of DNA? Until recently, this important question was difficult to answer because very large DNA molecules are exquisitely sensitive to degradation by shearing forces. Bruno Zimm has circumvented this problem by using the *viscoelastic technique,* which provides a measure of the size of the *largest DNA molecules in a mixture.* DNA molecules are stretched out by flow and then allowed to recoil to their normal form. The half time for recoiling depends on molecular weight. Cells from fruit flies were lysed in the measuring chamber to avoid the breakage of DNA that would be caused by transferring samples. Nucleases were inactivated by incubating the sample in detergent at 65°C. In addition, pronase was added to digest proteins bound to DNA.

The observed molecular weight of the largest DNA molecules in this mixture was 41×10^9, which agreed closely with the known value of the DNA content of the largest chromosome in *Drosophila melanogaster,* which is 43×10^9 daltons. Excellent agreement was also obtained for a translocation mutant of the largest chromosome, which has an extra piece of DNA, giving it a DNA content of 59×10^9 daltons. The measured molecular weight of this DNA molecule was 58×10^9. Radioautographs of *D. melanogaster* DNA (Figure 28-2) provide confirming evidence for the existence of very long DNA molecules. These studies reveal that *a Drosophila chromosome contains a single uninterrupted molecule of DNA.* Furthermore, this DNA molecule is *linear and unbranched.*

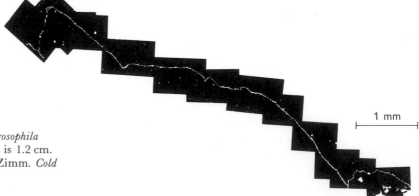

1 mm

Figure 28-2
Radioautograph of a DNA molecule from *Drosophila melanogaster.* The contour length of this DNA is 1.2 cm. [From R. Kavenoff, L. C. Klotz, and B. H. Zimm. *Cold Spring Harbor Symp. Quant. Biol.* 38(1974):4.]

EUCARYOTIC DNA CONTAINS MANY REPEATED BASE SEQUENCES

Studies of the kinetics of reassociation of thermally denatured DNA by Roy Britten and his associates have revealed that eucaryotic DNA, in contrast to procaryotic DNA, contains many repeated base sequences. In these experiments, DNA was sheared to small fragments and then denatured by heating the solution above the melting temperature (T_m) of the DNA. This solution of single-stranded DNA was then cooled to about 25°C below the T_m, which is optimal for the reassociation of complementary strands to form double-helical DNA. The kinetics of reassociation can be measured in a variety of ways. One technique is to follow the absorbance of the solution at 260 nm. At this wavelength, the extinction coefficient of double-stranded DNA is about 40% less than that of single-stranded DNA; this decrease is called *hypochromicity*. Another experimental approach is based on the fact that *double-stranded DNA binds to hydroxyapatite (calcium phosphate) columns, whereas single-stranded DNA passes through*. An attractive feature of this technique is that large amounts of DNA can be fractionated on the basis of their rates of reassociation following thermal denaturation.

The observed kinetics of reassociation of DNA from *E. coli* or T4 phage followed the time course expected for the bimolecular reaction

$$S + S' \xrightarrow{k} D$$

in which S and S′ are complementary single-stranded molecules, D is the reassociated double helix, and k is the rate constant for association. For such a reaction, the fraction f of single-stranded molecules decreases with time according to the expression

$$f = \frac{1}{1 + kC_0 t}$$

in which C_0 is the total concentration of DNA (expressed in moles per liter of nucleotides) and t is time (sec). For a particular DNA and a particular set of experimental conditions (e.g., ionic strength, temperature, size of the DNA fragments), f depends only on $C_0 t$, the product of DNA concentration and time. It is convenient to depict the kinetics of reassociation by plotting f versus the logarithm

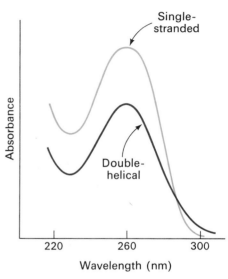

Figure 28-3
The absorption spectrum of DNA depends on its conformation. Double-helical DNA has a lower extinction coefficient than single-stranded DNA.

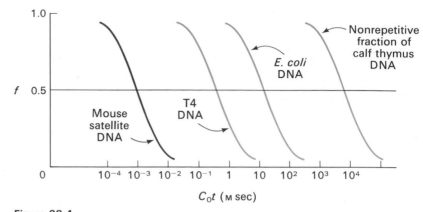

Figure 28-4

These C_0t curves depict the kinetics of reassociation of several kinds of thermally denatured DNA. The fraction of single-stranded molecules is plotted as a function of C_0t. The rapid reassociation of mouse satellite DNA shows that it contains very many repeated sequences. [Based on R. J. Britten and D. E. Kohne. *Science* 161(1968):530.]

of C_0t. Such a C_0t *curve* has a sigmoidal shape (Figure 28-4). An informative index of this plot is $C_0t_{0.5}$, the value of C_0t corresponding to reassociation of half of the DNA ($f = 0.5$). For *E. coli* DNA, $C_0t_{0.5}$ is about 9 M sec, whereas for T4 phage, $C_0t_{0.5}$ is 0.3 M sec. These values indicate that *E. coli* DNA reassociates about thirty times as slowly as T4 phage DNA. The reason is that *E. coli* DNA is larger and so the number of kinds of fragments in a sheared sample is greater than for T4 DNA. Hence, *the concentration of complementary fragments is lower in a solution of sheared* E. coli *DNA compared with one of T4 DNA* (containing the same concentration of nucleotides) *and thus its rate of reassociation is slower.* Studies of a variety of procaryotic DNA molecules showed that $C_0t_{0.5}$ is directly related to the size of the genome.

An unexpected result was obtained when mouse DNA was studied in this way. Since mammalian genomes are about three orders of magnitude larger than those of *E. coli,* a $C_0t_{0.5}$ value of the order of 10^4 M sec seemed likely. A 10^{-4} M solution of DNA having such a $C_0t_{0.5}$ value would take 10^8 sec (about 3 years) to reassociate halfway. It came as a surprise to find that 10% of the mouse DNA was half-reassociated in a few seconds. Indeed, *this fraction of mouse DNA reassociates more rapidly than even the smallest viral DNAs, which revealed that it contains many repeated sequences.* An analysis of the C_0t curve suggested that this fraction of mouse DNA contains *of the order of a million copies of a repeating sequence of about 300 base pairs.* About 20% of the mouse DNA renatured at an intermediate rate, which was interpreted to indicate that this fraction contains between 10^3 and 10^4 copies of certain sequences. The other 70% of mouse DNA renatures very slowly. The $C_0t_{0.5}$ value of this fraction shows that it consists of unique or nearly unique base sequences.

All eucaryotes studied thus far, perhaps except yeast, contain repetitive DNA, whereas procaryotes do not. For example, 30% of human DNA consists of sequences repeated at least twenty times. The proportion of highly repetitive, moderately repetitive, and single-copy DNA varies from one species to another.

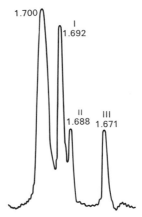

Figure 28-5
Three prominent bands of satellite DNA (ρ = 1.692, 1.688, and 1.671) are evident in this sedimentation pattern. DNA from *Drosophila virilis* was centrifuged to equilibrium in neutral CsCl. [Based on J. G. Gall, E. H. Cohen, and D. D. Atherton. *Cold Spring Harbor Symp. Quant. Biol.* 38(1974):417.]

HIGHLY REPETITIVE DNA (SATELLITE DNA) IS LOCALIZED AT CENTROMERES

Many highly repetitive DNAs can be isolated by density-gradient centrifugation because they have distinctive buoyant densities. For example, *Drosophila virilis* DNA exhibits a main band and three less-dense satellite bands. These satellite bands consist exclusively of highly repetitive DNA. In fact, each consists of a repeating heptanucleotide sequence:

```
5′—ACAAACT—3′        5′—ATAAACT—3′        5′—ACAAATT—3′
   : : : : : : :          : : : : : : :          : : : : : : :
3′—TGTTTGA—5′        3′—TATTTGA—5′        3′—TGTTTAA—5′
   Satellite I           Satellite II           Satellite III
```

The role of highly repetitive DNA is not known. The chromosomal localization of this fraction has been determined by the technique of *in situ hybridization,* which was devised by Joseph Gall and Mary Lou Pardue. Cells were immobilized under a thin layer of agar and treated with alkali to denature the DNA. This preparation was then incubated with tritium-labeled RNA transcribed in vitro using purified mouse satellite DNA as the template. Hybrids of this radioactive RNA and chromosomal regions containing satellite DNA were detected by autoradiography. A clear-cut result was obtained: *mouse satellite DNA is confined to centromeric regions.*

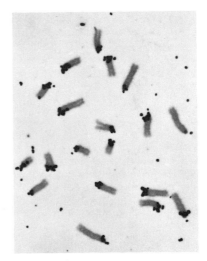

Figure 28-6
Autoradiograph of a mouse cell showing the localization of the satellite DNA. [Courtesy of Dr. Joseph Gall.]

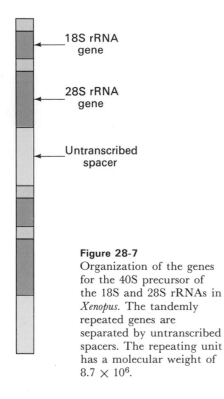

Figure 28-7
Organization of the genes for the 40S precursor of the 18S and 28S rRNAs in *Xenopus*. The tandemly repeated genes are separated by untranscribed spacers. The repeating unit has a molecular weight of 8.7×10^6.

(labels on figure: 18S rRNA gene; 28S rRNA gene; Untranscribed spacer)

GENES FOR RIBOSOMAL RNA ARE TANDEMLY REPEATED SEVERAL HUNDRED TIMES

Much is known about the genes coding for ribosomal RNA molecules, largely because of the efforts of Max Birnstiel, Donald Brown, Oscar Miller, and their associates. The genes for the 18S and 28S rRNAs and also for the 5S rRNA have been purified from the DNA of the African clawed toads, *Xenopus laevis* and *Xenopus mulleri*. These creatures were chosen because their oocytes contain a very large amount of the genes for 18S and 28S rRNA as a result of gene amplification.

The genes for 18S and 28S rRNA are tandemly repeated about 450 times. In situ hybridization studies showed that these genes are located in the *nucleolus*. The repeating unit consists of a *gene for a 40S precursor* and a *spacer* (Figure 28-7). The 40S precursor is cleaved to give an 18S and a 28S rRNA (p. 612). The 5000 base-pair spacer region is not transcribed (Figure 28-8). Its role is an enigma.

The genes for 5S rRNA are also tandemly repeated. Clusters containing between 100 and 1000 repeats are present at the ends

Figure 28-8
The tandem array of genes for rRNA is evident in this electron micrograph of nucleolar DNA. The dense axial fiber is DNA. The fine lateral fibers are newly synthesized RNA molecules with bound proteins. The tip of each arrowhead of the RNA molecules corresponds to an initiation point for transcription. The base regions between arrowheads are the untranscribed spacers. [From O. L. Miller, Jr., and B. R. Beatty. Portrait of a gene. *J. Cell Physiol.* 74 (Supp. 1, 1969):225.]

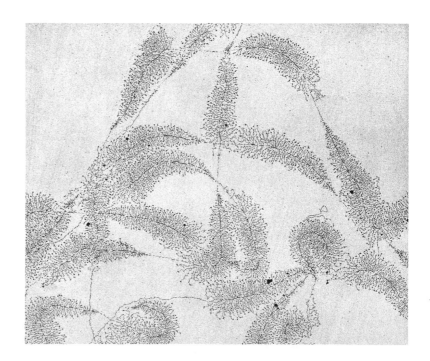

of most of the 18 chromosomes of these toads. Again, there is an untranscribed spacer region between the genes for this rRNA. Indeed, the spacer is several times as long as the gene (Figure 28-9).

Oogenesis in *Xenopus* is accompanied by a thousandfold increase in the number of nucleoli and a selective replication of the genes for the 18S and 28S rRNAs. *The genes for these rRNAs comprise nearly 75% of the total DNA in the oocyte.* This *gene amplification* enables the oocyte to amass the 10^{12} ribosomes needed for very rapid protein synthesis that occurs during cell cleavages. In the absence of gene amplification, it would take several centuries to assemble 10^{12} ribosomes!

THERE IS ONLY ONE GENE FOR SILK FIBROIN IN A HAPLOID GENOME

The genes for ribosomal RNA are tandemly repeated several hundred times in all the eucaryotes that have been studied and they are selectively amplified a thousand times in amphibian oocytes. How general are tandem repetition and selective amplification in eucaryotes? The genes for silk fibroin in the silkworm *Bombyx mori* were studied by Brown to answer this question. This system was chosen because the mRNA for silk fibroin has distinctive chemical characteristics since it is the template for repeating amino acid sequences that are rich in glycine. Furthermore, very large amounts of this protein are synthesized by a single type of giant cell at a particular stage of larval development. Fibroin mRNA was purified on the basis of its large size (6×10^6 daltons) and its high content of G compared with that of rRNA and the total DNA of the silkworm. The gene for silk fibroin was then partially purified by taking advantage of its unusual buoyant density. Hybridization of purified fibroin mRNA with DNA from silkworm revealed that there are between one and three fibroin genes per haploid genome, whether or not a silkworm cell is actively engaged in the synthesis of fibroin. Further studies have indicated that *there is probably only one gene for silk fibroin per haploid genome.*

This finding has important implications concerning gene expression and differentiation in eucaryotes. It shows that large amounts of a single protein can be synthesized without tandem repetition and selective amplification of its gene. *The single gene for silk fibroin*

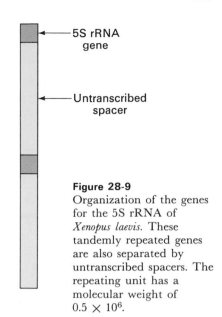

5S rRNA gene

Untranscribed spacer

Figure 28-9
Organization of the genes for the 5S rRNA of *Xenopus laevis*. These tandemly repeated genes are also separated by untranscribed spacers. The repeating unit has a molecular weight of 0.5×10^6.

is the template for the synthesis of 10^4 mRNA molecules, which are stable for days. Each mRNA is the template for the synthesis of 10^5 proteins. Thus, a single gene is sufficient for the synthesis of 10^9 protein molecules in four days. Tandem repetition and selective gene amplification may be special mechanisms for obtaining large amounts of RNAs that are final products. In these cases, RNA is not used repetitively as in protein synthesis, and so a different stage of amplification is needed.

MOST NONREPETITIVE BASE SEQUENCES ARE SEPARATED BY REPETITIVE SEQUENCES

We have seen that eucaryotic DNA contains highly repetitive, moderately repetitive, and single-copy base sequences. It has recently been found, using a variety of experimental approaches, that repetitive and nonrepetitive sequences are generally interspersed in eucaryotic chromosomes. In at least half of the total DNA, nonrepetitive sequences having an average length of 800 base pairs seem to be separated from one another by moderately repetitive sequences averaging 300 base pairs in length. The function of the interspersed moderately repetitive DNA is unknown. They might serve as sites for the binding of specific regulatory macromolecules, which could thereby control the transcription of adjacent single-copy structural genes.

BASIC PROTEINS CALLED HISTONES ARE BOUND TO EUCARYOTIC DNA

We turn now to the protein components of eucaryotic chromosomes. The chromosomes of all animal and plant somatic cells contain several kinds of small basic proteins called *histones*, which can be extracted by salt or dilute acid. They are single polypeptide chains having molecular weights ranging from about 11,000 to 21,000. Their content of lysine plus arginine is about 25%. There are five types of histones (Table 28-1), which can be separated by ion-exchange chromatography.

The amino acid sequences of several histones are known. One striking feature of these sequences is that *basic residues tend to be*

Table 28-1
Types of histones

Class	Lys/Arg ratio	Number of residues	Molecular weight
I (f1)	20	215	21,000
IIbl (f2a2)	1.25	129	14,500
IIb2 (f2b)	2.5	125	13,774
III (f3)	0.72	135	15,324
IV (f2a1)	0.79	102	11,282

Source: R. J. DeLange and E. L. Smith. *Accounts Chem. Res.* 5(1972):396.

clustered. For example, the amino-terminal half of histone IV has a net charge of $+16$, whereas the carboxyl-terminal half has a net charge of only $+3$ and contains most of the hydrophobic residues. It seems likely that the primary binding sites for the negatively charged phosphodiester bridges of DNA are located in the amino-terminal half of this histone.

Each histone exists in a variety of forms because of *posttranscriptional modifications* of certain side chains. For example, a specific lysine (residue 16) is acetylated in histone IV. Histones are also phosphorylated by phosphokinases and dephosphorylated by phosphatases. It has been suggested that these covalent modifications are important in regulating the compactness of chromosomes.

THE AMINO ACID SEQUENCE OF HISTONE IV IS NEARLY THE SAME IN ALL PLANTS AND ANIMALS

Emil Smith and Robert DeLange have shown that the amino acid sequences of histone IV from pea seedlings and calf thymus differ at only two sites out of 102 residues. The changes at these two positions are quite small: valine in place of isoleucine, and lysine in place of arginine. Histone IV from rats and pigs is identical to that from calves. It is interesting to compare the rate of change of histone IV in the course of evolution with those of other proteins. A useful index is the *unit evolutionary period,* which is the time required for the amino acid sequence to change by 1% after two evolutionary lines have diverged. For histone IV, 2% of the sequence has changed in the 1,200 million years following the divergence

```
Ser-Gly-Arg-Gly -Lys -Gly-Gly-Lys -Gly-Leu-     10
Gly -Lys-Gly-Gly -Ala -Lys-Arg-His -Arg-Lys-     20
Val -Leu-Arg-Asp-Asn-Ile -Gln-Gly-Ile -Thr-      30
Lys -Pro-Ala -Ile  -Arg-Arg-Leu-Ala -Arg-Arg-    40
Gly -Gly-Val -Lys -Arg-Ile  -Ser-Gly-Leu-Ile  -  50
Tyr -Glu-Glu-Thr -Arg-Gly-Val -Leu-Lys -Val -    60
Phe-Leu-Glu-Asn-Val -Ile  -Arg-Asp-Ala -Val -    70
Thr -Tyr -Thr-Glu -His -Ala -Lys-Arg-Lys-Thr-    80
Val -Thr-Ala -Met-Asp-Val -Val -Tyr -Ala -Leu-   90
Lys -Arg-Gln-Gly -Arg-Thr-Leu-Tyr -Gly-Phe-     100
Gly -Gly                                         102
```

Figure 28-10
Amino acid sequence of histone IV from calf thymus. Several residues are modified. The α-amino group is acetylated, as is the ε-amino group of Lys 16. The ε-amino group of Lys 20 is methylated or dimethylated. Histone IV from pea seedlings has the same amino acid sequence except that residue 60 is isoleucine and residue 77 is arginine.

of plants and animals, and so its unit evolutionary period is 600 million years. Histone III also displays a remarkable degree of evolutionary conservation. The proteins from calf and carp differ in only one residue, a cysteine in place of a serine. The unit evolutionary periods of three other proteins are known: 20 million years for cytochrome *c*, 6 million years for hemoglobin, and 1 million years for fibrinopeptides.

HISTONES PROBABLY PARTICIPATE IN THE PACKING OF DNA

The high degree of conservation of the amino acid sequences of histones III and IV in evolution implies that these proteins serve critical functions and that most of their structure is essential for these functions, which are not yet known. It seems likely that *histones play a role in the packing of DNA molecules, rendering them more compact.* A striking example of the condensation of DNA is provided by human metaphase chromosomes. A total of 1.2×10^{10} base pairs (corresponding to a length of 400 cm) is packed into forty-six paired cylinders whose total length is 200 μm. The packing ratio of this DNA is about 10^4.

The packing ratio for DNA in eucaryotic chromosomes changes greatly during the cell cycle. Chromosomes are most compact during metaphase. In interphase, some parts of chromosomes are

Figure 28-11
Fluorescence micrograph of human male metaphase chromosomes from cells that replicated twice in a medium containing the thymidine analogue 5-bromodeoxyuridine (BrdU). Chromosomes were first stained with quinacrine (A), destained, and then restained with the bisbenzimidazole dye 33258 Hoechst (B). Since the fluorescence of this latter dye is quenched by BrdU, sister chromatids containing DNA with only one polynucleotide chain substituted with BrdU fluoresce more brightly than those with BrdU in both polynucleotide chains. (The bar is 10 μm long.) [Courtesy of Dr. Samuel Latt.]

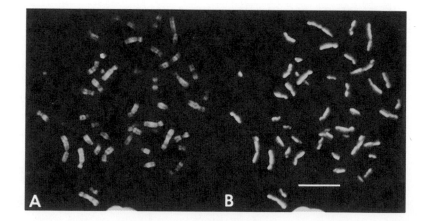

highly dispersed (called *euchromatin*) and may have a packing ratio as low as three. Other parts (called *heterochromatin*) remain highly condensed except during DNA synthesis. *Heterochromatin is transcribed to a much lesser extent than euchromatin and is replicated later in the cell cycle.* In mammalian females, one of the two X chromosomes remains condensed during interphase and appears under the light microscope as a dense clump called the *Barr body*. The genes in this condensed X chromosome are not expressed. The random inactivation of one of the X chromosomes (called the *Lyon effect*) accounts for the finding that females heterozygous for glucose 6-phosphate dehydrogenase deficiency have two populations of red cells: one has normal amounts of this enzyme, whereas the other is deficient (p. 367).

Histones probably do not specifically control gene expression because they are quite uniform in different kinds of cells and also in very different species. Histones are synthesized only during DNA synthesis, which is consistent with their proposed role in the packing of DNA. *Some of the nonhistone proteins associated with chromosomes are more likely participants in the specific control of gene function.* They differ from histones in several ways. First, many of them are acidic. Second, they are highly heterogeneous. The nonhistone chromosomal proteins vary markedly in going from one kind of cell to another in a species. Third, particular sets of these proteins are synthesized in concert with the activation of groups of genes, such as those stimulated by a particular hormone. Fourth, they stimulate transcription in cell-free systems. Much remains to be learned about this interesting group of chromosomal proteins.

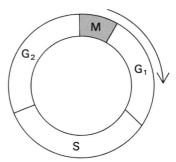

Figure 28-12
Phases in the life cycle of a eucaryotic cell: M (mitosis); G_1 (gap 1, prior to DNA synthesis); S (period of DNA synthesis); and G_2 (gap 2, between DNA synthesis and mitosis). The length of these periods depends on the cell type and conditions of growth. Mitosis is usually the shortest phase.

PROPOSED MODEL FOR CHROMATIN

There is much current interest in how histones interact with eucaryotic DNA to form the compact fiber called chromatin. Recent work indicates that histones interact specifically with each other and with DNA. In solution, two molecules of histone III and two of histone IV interact to form a tetramer. Histones IIb1 and IIb2 form oligomers in solution. It is interesting to note that histones III and IV are richest in arginine and also most conserved in evolution, whereas IIb1 and IIb2 are rich in lysine and somewhat less conserved. Histone I, which has the highest amount of lysine and varies the most in going from one species to another, does not appear to bind strongly to the other histones. These complexes of histones may also be present in chromatin. In fact, a complex formed from $(III)_2(IV)_2$ tetramers, IIb1-IIb2 oligomers, and DNA gives nearly the same low-angle x-ray diffraction pattern as chromatin. These findings have led to a tentative model in which chromatin consists of a repeating unit of two each of histones III, IV, IIb1, and IIb2 and about 200 base pairs of DNA. Many such units would give a flexibly jointed chain, which could be extensively coiled or folded.

EUCARYOTIC DNA IS REPLICATED BIDIRECTIONALLY FROM MANY ORIGINS

Eucaryotic DNA, like all other DNA molecules, is *replicated semi-conservatively*. Newly replicated DNA is also distributed semiconservatively between sister chromatids of metaphase chromosomes (Figure 28-11). Electron microscopic studies have also shown that eucaryotic DNA is replicated *bidirectionally from many origins*. The use of many initiation points is necessary for rapid replication because of the great length of eucaryotic DNAs. For example, the largest chromosome in *Drosophila* has a length of about 2.1 cm or 62,000 kb. A kb (kilobase) is a unit of length equal to 1000 base pairs in double-helical DNA. A DNA replication fork in *Drosophila* moves at the rate of 2.6 kb/min, compared with about 16 kb/min in *E. coli*. The replication of the largest *Drosophila* chromosome would take more than sixteen days if there were only one origin for replication. The actual replication time of less than three minutes is achieved by *the cooperative action of more than 6000 replication forks per DNA molecule*. A DNA molecule from the cleavage nuclei of *Drosophila* exhibits a serial array of replicated regions or "eye forms" (Figure 28-13). The activation of each initiation point generates two diverging replication forks. The array of eye forms expanding in both directions merge to form the two daughter DNA molecules (Figure 28-14). An eye form within an eye form has not been observed, which indicates that an origin cannot be reactivated until after the entire DNA molecule is replicated.

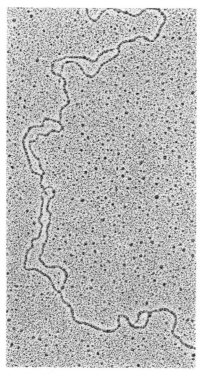

Figure 28-13
Electron micrograph of replicating chromosomal DNA from cleavage nuclei of *Drosophila*. The eye forms are the newly replicated regions. [Courtesy of Dr. Jack Griffith.]

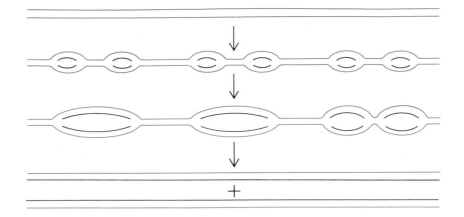

Figure 28-14
Schematic diagram of the replication of a eucaryotic chromosome. The parental DNA is shown in blue, and the newly replicated DNA in red.

CYTOPLASMIC MESSENGER RNA IS DERIVED FROM VERY LARGE PRECURSORS IN THE NUCLEUS

The nucleus of eucaryotic cells contains very large RNA molecules that have a lifetime of only a few minutes, in contrast to many eucaryotic mRNAs. No distinct species have been detected among these molecules, and so they are called *heterogeneous RNA of the nucleus (hnRNA)*. This class of molecules has a distribution of molecular weights up to about 2×10^7, which is an order of magnitude larger than most cytoplasmic mRNAs. The hnRNA molecules are located in the extranucleolar portion of the nucleus. Most of them contain a *long tail of poly A at the 3′ end*. About 150 to 200 residues of A are added stepwise after transcription has been completed.

Messenger RNAs are formed by cleavage of hnRNAs. Only about 10% of the nucleotides in hnRNAs appear in cytoplasmic mRNAs. *Most eucaryotic mRNAs also have a poly A tail at their 3′ end.* The presence of a poly A tail facilitates the isolation of hnRNA and mRNA from eucaryotic cells. These molecules bind strongly to columns containing covalently attached poly U or poly T. It appears that mRNAs arise from the 3′ end of hnRNAs because they retain the poly A tail of their precursor. In contrast, the 5′ terminus and some repeated sequences of hnRNA do not emerge in cytoplasmic mRNA. The poly A tail may play a role in the processing of hnRNA. However, it is not required for the production of all mRNAs in eucaryotes; for example, mRNAs for histones are devoid of poly A. *The eucaryotic mRNAs studied thus far are monocistronic*—that is, they are templates for the synthesis of a single polypeptide chain. In contrast, procaryotic mRNAs are frequently polycistronic (e.g., *lac* mRNA is the template for three polypeptide chains).

SUMMARY

Eucaryotic chromosomes contain a single molecule of double-helical DNA, which is at least one hundred times as large as DNA molecules in procaryotes. The kinetics of reassociation of thermally denatured DNA revealed that eucaryotic DNA, unlike procaryotic DNA, contains many repeated base sequences. Species of *highly repetitive DNA* (satellite DNA) are readily isolated because of their distinctive buoyant density. In situ hybridization studies have shown that it is usually localized near centromeres. Some satellite DNAs have a heptanucleotide sequence repeated more than ten thousand times. Another category is *moderately repetitive DNA,* such as the gene for the precursor of the 18S and 28S rRNAs, which is tandemly repeated several hundred times. These genes are separated by untranscribed spacer regions, about 5000 base pairs in length. A third category is *single-copy DNA,* such as the gene for silk fibroin. Hybridization studies using purified fibroin mRNA showed that there is only one gene for silk fibroin per haploid genome. It has been proposed that nonrepetitive sequences are usually separated from each other by moderately repetitive sequences, which may serve as control sites.

Five types of histones are bound to eucaryotic DNA. These small proteins have a high content of lysine and arginine. Histones III and IV have remained nearly constant throughout a billion years of evolution. The precise role of histones is not known, but it seems likely that they are important for the packing of DNA. Histones probably do not specifically control gene expression because they are quite uniform in different types of cells and in different species. Some of the acidic proteins associated with chromosomes may have regulatory functions. Eucaryotic DNA is replicated semiconservatively from several thousand origins. A large number of initiation points is necessary because of the great length of eucaryotic DNA compared with procaryotic DNA.

SELECTED READINGS

Where to start:

Britten, R., and Kohn, D., 1968. Repeated segments of DNA. *Sci. Amer.* 222(4):24–31. [Available as *Sci. Amer.* Offprint 1173.]

Brown, D. D., 1973. The isolation of genes. *Sci. Amer.* 229(2):20–29. [Offprint 1278.]

Cold Spring Harbor Laboratory, 1974. *Chromosome Structure and Function* (Cold Spring Harbor Symposia on Quantitative Biology, vol. 38). [A most valuable collection of more than ninety articles, presented at a symposium in 1973. This volume provides an excellent view of research on eucaryotic DNA.]

Histones and other chromosomal proteins:

DeLange, R. J., and Smith, E. L., 1972. The structure of histones. *Accounts Chem. Res.* 5:368–373.

Stein, G. S., Spelsberg, T. C., and Kleinsmith, L. J., 1974. Non-histone chromosomal proteins and gene regulation. *Science* 183:817–824.

Phillips, D. M. P., (ed.), 1971. *Histones and Nucleohistones*. Plenum.

Birnstiel, M., Telford, J., Weinberg, E., and Stafford, D., 1974. Isolation and some properties of the genes coding for histone proteins. *Proc. Nat. Acad. Sci.* 71:2900–2904.

Structure of chromosomes:

Huberman, J. A., 1973. Structure of chromosome fibers and chromosomes. *Annu. Rev. Biochem.* 42:355–375.

Simpson, R. T., 1973. Structure and function of chromatin. *Advan. Enzymol.* 38:41–108.

Kornberg, R. D., 1974. Chromatin structure: a repeating unit of histones and DNA. *Science* 184:868–871.

Crick, F., 1971. General model for the chromosomes of higher organisms. *Nature* 234:25–27.

Unique and repetitive sequences in eucaryotic DNA:

Britten, R., and Kohne, D., 1968. Repeated sequences in DNA. *Science* 161:529–540.

Lee, C. S., and Thomas, C. A., Jr., 1973. Formation of rings from *Drosophila* DNA fragments. *J. Mol. Biol.* 77:25–55.

Suzuki, Y., Gage, L. P., and Brown, D. D., 1972. The genes for silk fibroin in *Bombyx mori. J. Mol. Biol.* 70:637–649.

Davidson, E. H., Hough, B. R., Amenson, C. S., and Britten, R. J., 1973. General interspersion of repetitive with nonrepetitive sequence elements in the DNA of *Xenopus. J. Mol. Biol.* 77:1–23.

Replication of eucaryotic DNA:

Prescott, D. M., 1970. The structure and replication of eukaryotic chromosomes. *Advan. Cell Biol.* 1:57–117.

Kriegstein, H. J., and Hogness, D. S., 1974. The mechanism of DNA replication in *Drosophila* chromosomes: structure of replication forks and evidence for bidirectionality. *Proc. Nat. Acad. Sci.* 71:135–139.

Eucaryotic messenger RNA:

Brawerman, G., 1974. Eukaryotic messenger RNA. *Annu. Rev. Biochem.* 43:621–642.

Weinberg, R. A., 1973. Nuclear RNA metabolism. *Annu. Rev. Biochem.* 42:329–354.

Problem sets:

Hood, L. E., Wilson, J. H., and Wood, W. B., 1975. *Molecular Biology of Eucaryotic Cells,* chapters 2–5. Benjamin.

CHAPTER 29

VIRUSES

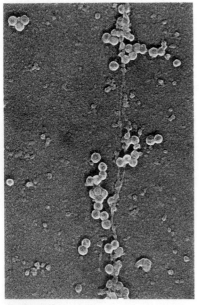

Figure 29-1
Electron micrograph of SV40 tumor virus particles developing in association with the host chromosome. [Courtesy of Dr. Jack Griffith.]

Viruses are packets of infectious nucleic acid surrounded by a protective coat. They are the most efficient of the self-reproducing intracellular parasites. Viruses are unable to generate metabolic energy or to synthesize proteins. They also differ from cells in having either DNA or RNA, but not both. The nucleic acid is single-stranded in some viruses, and double-stranded in others (Table 29-1). Viruses range in complexity from Qβ, an RNA phage having only 3 genes, to the poxviruses, which have about 250 genes. The completed extracellular product of virus multiplication is called a *virion* (or virus particle). In a virion, the viral nucleic acid is covered by a protein *capsid,* which protects it from enzymatic attack and mechanical breakage and also delivers it to susceptible hosts. In some of the more complex animal viruses, the capsid is surrounded by an *envelope* that contains lipid and glycoprotein.

We have already seen that the study of viruses has had a profound impact on the development of molecular biology, as in the discovery of messenger RNA. There is much current interest in viruses for several reasons. First, viral multiplication is a *model for cellular development* because it involves the sequential expression of genes and the assembly of macromolecules to form highly ordered structures. The relatively small number of viral genes, the rapid rate of replication, and the feasibility of genetic analysis contribute to

Table 29-1
Types of viruses

Nucleic acid	Representative virus	Approximate number of genes
Single-stranded DNA	φX174 phage	5
Double-stranded DNA	Polyoma virus	5
	Adenovirus 2	30
	T4 phage	165
	Vaccinia poxvirus	240
Single-stranded RNA	Qβ phage	3
	Tobacco mosaic virus	6
	Polio virus	8
	Influenza virus	12
	Rous sarcoma virus	40
Double-stranded RNA	Reovirus	22

Source: B. D. Davis, R. Dulbecco, H. N. Eisen, H. S. Ginsberg, and W. B. Wood, Jr. *Microbiology* (Harper and Row, 1973).

the value of viruses as models. Second, viruses provide insight into *evolutionary processes* and molecular aspects of *host-parasite relationships*. Third, *some viruses cause cancers in experimental animals.* The possible role of viruses in human cancers is now being intensively studied.

THE COAT OF A SMALL VIRUS CONSISTS OF MANY IDENTICAL PROTEIN SUBUNITS

The total number of amino acids in the coat of a virus always exceeds the number of nucleotides in its genome. For example, the protein coat of a tobacco mosaic virus (TMV) particle contains about 340,000 amino acid residues, whereas its RNA contains only about 6,400 nucleotides. In 1957, Francis Crick and James Watson noted that the protein coat of a virus cannot be one large molecule or an assembly of very many different small proteins because the amount of viral nucleic acid is much too small to code for such a large number of amino acid residues. On the other hand, the protein coat cannot be reduced in size if all the nucleic acid is to be covered. Viruses circumvent their genetic poverty by having coats made up of a *large number of one or a few kinds of protein subunits.*

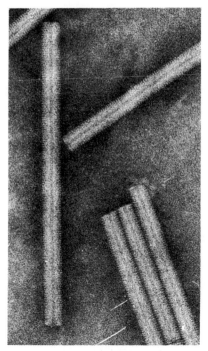

Figure 29-2
Electron micrograph of tobacco mosaic virus (TMV) particles. [Courtesy of Dr. Aaron Klug.]

Figure 29-3
Model of an icosahedron.

For example, the coat of TMV consists of 2,130 identical protein subunits (each 158 residues long).

A viral coat can be built from identical subunits in only a limited number of ways. Stability is achieved by forming a maximum number of bonds and by using the same kinds of contacts again and again. The resulting structure must be symmetric. Two kinds of arrangements of the protein coat are most likely: *a cylindrical shell having helical symmetry and a spherical shell having icosahedral symmetry* (Figure 29-3). In fact, *all small viruses are either rods or spheres* (or a combination of these shapes). The rules governing the structure of spherical viruses were deduced by Donald Caspar and Aaron Klug, who received some of their inspiration from the architectural designs of Buckminster Fuller.

SELF-ASSEMBLY OF TOBACCO MOSAIC VIRUS

The simplest and best understood viral assembly process is that of TMV. This rod-shaped virus is 3,000 Å long and 180 Å in diameter (Figure 29-4), and has a mass of about 40 million daltons. The 2,130 identical subunits in the protein coat are closely packed in a helical array around a single-stranded RNA molecule consisting of 6,390 nucleotides. The RNA is deeply buried in the protein, rendering it insusceptible to attack by ribonucleases. Each protein subunit interacts with three nucleotides. The dissociated RNA is very labile, whereas intact TMV remains infective for decades.

The subunits of TMV are not covalently bonded to one another. TMV can be dissociated into protein and RNA by agents such as concentrated acetic acid. In 1955, Heinz Fraenkel-Conrat and Robley Williams showed that the *dissociated coat subunits and RNA of TMV spontaneously reassemble under suitable conditions into virus particles that are indistinguishable from the original TMV in structure and infectivity.* This was the first example of the self-assembly of an active biological structure. *A self-assembly process is one in which the components associate spontaneously under appropriate environmental conditions to form a specific structure.*

The coat subunits can associate in the absence of RNA to form several kinds of structures, depending on pH, ionic strength, and temperature. At acid pH, they form a helix like that of the intact virus. However, the lengths of the helices formed in the absence

←——100 Å——→

Figure 29-4
Model of a part of TMV, showing the helical array of protein subunits around a single-stranded RNA molecule. [Based on A. Klug and D. L. D. Caspar, *Advan. Virus Res.* 7(1960):274.]

of RNA are highly variable. In contrast, TMV particles have a
uniform length because the coat subunits associate to form a structure just long enough to cover the entire RNA molecule. Thus, *the
length of the RNA determines the length of the protein coat*. A different
structure is formed at pH 7 and low ionic strength in the absence
of RNA. Discs consisting of two layers of subunits are formed under
these conditions. Each layer has 17 protein subunits, which is nearly
the same as the number of subunits ($16\frac{1}{3}$) in a turn of the TMV
helix. At high ionic strength, these discs stack on top of each other
to form a rod.

Klug has proposed that the *two-layer disc is a key intermediate in
the assembly of TMV*. This structure is stable until it interacts with
a specific sequence of about fifty nucleotides at the 5′ end of TMV
RNA. This interaction is thought to convert the disc form into two
turns of helix (Figure 29-5). The structure grows by the addition
of two-layered discs, which unroll into a helical form. This process
is repeated until the entire RNA molecule is coated. What prevents
the discs from forming a helix devoid of RNA? Two carboxyl groups
in each subunit appear to play a critical role in this regard. At
neutral pH, both carboxyl groups are ionized in the helix form but
only one is ionized in the disc. The disc form is favored because
of electrostatic repulsion between these closely spaced carboxylates
in the helix form. The binding of RNA to the helix form provides
enough free energy to overcome the electrostatic repulsion of the
carboxylates. Thus, *the carboxylates are a negative switch to prevent the
formation of a helix devoid of RNA. Only TMV RNA is coated by TMV
coat subunits during viral multiplication because it contains a specific sequence
that favors initiation of a helical structure.*

Figure 29-5
Proposed mechanism of assembly of
TMV. [Based on A. Klug, *Fed. Proc.*
31(1972):40.]

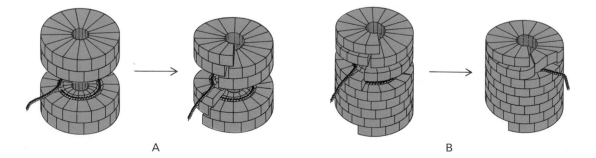

A   B

INFECTION BY T4 PHAGE PROFOUNDLY ALTERS
MACROMOLECULAR SYNTHESIS IN *E. COLI*

The T4 bacteriophage is a much more complex virus than TMV. Its double-stranded DNA contains about 165 genes, compared with 6 in TMV. However, much is known about the structure, multiplication, and assembly of T4 because it has been subjected to intensive genetic and biochemical analysis. The T4 virion consists of a *head,* a *tail,* and six *tail fibers* (Figure 29-6). The DNA molecule is tightly packed inside an icosahedral protein coat to form the head of the virus. The tail is made up of two coaxial hollow tubes that are connected to the head by a short neck. In the tail, a contractile sheath surrounds a central core through which the DNA is injected into the bacterial host. The tail terminates in a baseplate, which has six short spikes and gives off six long, slender fibers.

The tips of the tail fibers bind to a specific site on *E. coli*. An ATP-driven contraction of the tail sheath pulls the phage head toward the end plate and tail fibers, which causes the central core to penetrate the cell wall but not the cell membrane. The naked T4 DNA then penetrates the cell membrane. *A few minutes later, all cellular, DNA, RNA, and protein synthesis stops, and the synthesis of viral macromolecules begins.* In other words, the infecting virus takes over the synthesizing machinery of the cell and substitutes its genes for the bacterial genes.

The T4 DNA contains three sets of genes that are transcribed at different times after infection: *immediate-early, delayed-early,* and *late.* The early genes are transcribed and translated before T4 DNA

Figure 29-6
Electron micrograph of T4 virions from an infected cell. [Courtesy of Dr. Jonathan King, Dr. Yoshiko Kikuchi, and Dr. Elaine Lenk.]

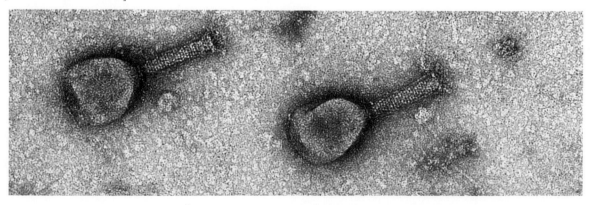

is synthesized. Some proteins coded by these genes are responsible for switching off the synthesis of cellular macromolecules. *Soon after infection, the host cell DNA is degraded by a deoxyribonuclease that is specified by a T4 early gene. This enzyme does not hydrolyze T4 DNA because it lacks clusters of cytosines.* T4 DNA contains *5-hydroxymethylcytosine* (HMC) instead of cytosine. Furthermore, some HMC residues in T4 DNA are *glucosylated*.

The DNA of T4 bacteriophage contains these derivatives of cytosine because of the action of several phage-specified enzymes that are synthesized in the early phase of infection. One of them hydrolyzes dCTP to dCMP to prevent the incorporation of dCTP into T4 DNA. Then, dCMP is hydroxymethylated by a second enzyme to form 5-hydroxymethylcytidylate (dHMCMP). A third enzyme converts dHMCMP to the triphosphate, which is a substrate for DNA polymerases. Finally, some HMC bases in DNA are glucosylated by a fourth enzyme.

The synthesis of the *late proteins* is coupled to the replication of T4 DNA. The capsid proteins and a lysozyme are formed at this stage. The lysozyme digests the bacterial cell wall and causes its rupture when the assembly of progeny virions has been completed. About two hundred new virus particles emerge approximately twenty minutes after infection.

5-Hydroxymethylcytosine

α-Glucosylated derivative of 5-hydroxymethylcytosine

ENZYMES PARTICIPATE IN THE ASSEMBLY OF THE T4 PHAGE

The assembly of T4 is a far more intricate process than that of TMV. The T4 capsid has a much higher degree of structural organization and contains about twenty kinds of proteins. An additional thirty proteins participate in the construction of this virus. Insight into the mechanism of assembly of T4 has come from the combined use of genetic, biochemical, and electron microscopic techniques. The studies of William Wood and Robert Edgar on mutants of T4 that are defective in assembly have revealed that:

1. *There are three major pathways in the construction of the virus.* They lead independently to the head, the tail, and tail fibers (Figure 29-7). A block in the formation of one of these components does not affect the formation of the other two.

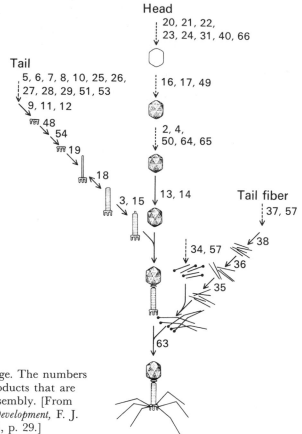

Figure 29-7
Morphogenetic pathway of T4 phage. The numbers next to the arrows refer to gene products that are required for a particular step in assembly. [From W. B. Wood, *Genetic Mechanism of Development,* F. J. Ruddle, ed. (Academic Press, 1973), p. 29.]

2. The head and the tail must be completed before they can combine. Then, completed tail fibers attach to the baseplate.

3. *There is a stringent sequential order in the assembly process.* Because the components are synthesized at the same time, it is evident that the temporal sequence is imposed by the structural features of the intermediates.

The construction of T4 virions is not a simple self-assembly process. The major protein of the head (called P23*), which has a molecular weight of 45,000, is derived from a 55,000-dalton precursor (called P23). The P23 subunits are split when the head is partly assembled, which suggests that it may be the trigger for the packing of DNA. Several other head proteins are also cleaved

at this stage. An enzyme has also been implicated in the attachment
of tail fibers to the otherwise finished phage particle. Thus, *T4 is
constructed by a combination of self-assembly and enzyme-directed assembly.*

BACTERIAL ENDONUCLEASES (RESTRICTION ENZYMES) CLEAVE FOREIGN DNA MOLECULES

We have seen that the T4 phage has enzymatic devices for specifically degrading the DNA of its host. Likewise, bacteria contain
enzymes called *restriction endonucleases* that cleave foreign DNA
molecules. The restriction enzymes of *E. coli* and *Hemophilus influenza*
recognize nucleotide sequences four to eight residues long. *A striking
feature of these target sequences is their symmetry* (Figure 29-8), which
is like that of the *lac* operator (p. 684). The restriction endonucleases
hydrolyze two phosphodiester bonds in the target sequence, one
in each polynucleotide strand. The cleavage of both strands pre-
cludes their repair by DNA ligase. Restriction enzymes are very
useful tools in the analysis of large DNA molecules because of their
very high specificity.

   How does the bacterial DNA evade the action of its own restriction enzyme? *The bacterial DNA is not degraded because its target sequences for the restriction enzyme are methylated* (Figure 29-9). A bacterial methylase (which uses *S*-adenosylmethionine as the methyl
donor) recognizes the same sequence as the restriction endonuclease.

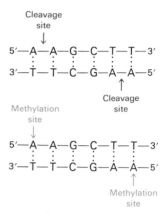

Figure 29-8
Specificity of the *Hind III* restriction
endonuclease from *Hemophilus
influenza* and of the corresponding
methylase.

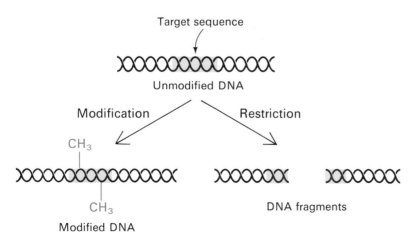

Figure 29-9
Methylation of the target site
prevents cleavage by the
corresponding restriction
endonuclease.

716

Part IV / INFORMATION

Methylation of a particular base in the target sequence prevents hydrolysis by the restriction enzyme. Thus, timing is critical: bacterial DNA is not degraded because it is first methylated. A newly replicated bacterial chromosome, which is methylated only on the parent strand, is resistant to the action of the restriction enzyme. This half-modified DNA becomes fully methylated before the subsequent round of replication.

STRATEGIES FOR THE REPLICATION OF RNA VIRUSES

A special problem arises in the replication of RNA viruses because uninfected host cells lack enzymes to synthesize RNA according to instructions given by an RNA template. Consequently, RNA viruses must contain genetic information for the synthesis of an *RNA-directed RNA polymerase* (also called an *RNA replicase* or an *RNA synthetase*) or for an *RNA-directed DNA polymerase* (also called a *reverse transcriptase*). Four types of mechanisms for the expression of genetic information in RNA viruses are known (Figure 29-10). The first type is exemplified by *poliovirus,* which contains a single-stranded RNA molecule. This RNA molecule, called the (+) strand, acts as a messenger on entering the cytoplasm of its host cell. It is translated by host ribosomes to give capsid proteins and a special RNA polymerase. This (+) RNA also serves as a template for the synthesis of a complementary (−) RNA, which is carried out by the polio RNA polymerase. This enzyme then uses the (−) RNA as the template for the synthesis of many (+) RNA strands (Figure 29-12), which bind capsid proteins to form new virions. An interesting feature of polio (+) RNA is that it is the messenger for the synthesis of a continuous polypeptide chain of two thousand amino acid residues. David Baltimore showed that this giant polypeptide

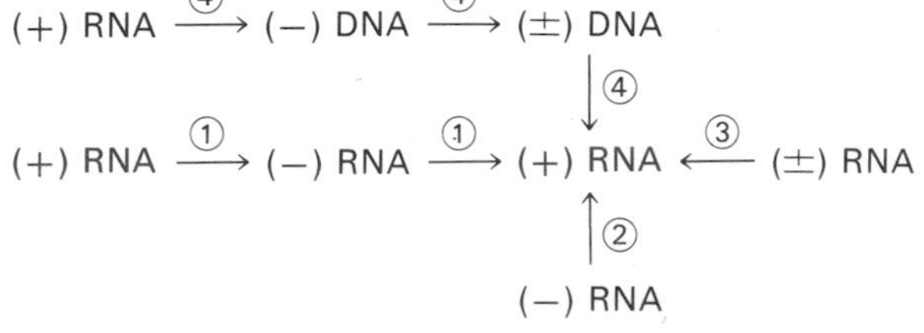

Figure 29-10
Modes of gene expression of RNA viruses. Messenger RNA is shown in red. Examples: 1, poliovirus and Qβ phage; 2, rabies virus; 3, reovirus; 4, *Rous sarcoma* virus. [Based on D. Baltimore, *Bacteriol. Rev.* 35(1971):326.]

$$(+)\ \text{RNA} \xrightarrow{④} (-)\ \text{DNA} \xrightarrow{④} (\pm)\ \text{DNA}$$
$$\Big\downarrow ④$$
$$(+)\ \text{RNA} \xrightarrow{①} (-)\ \text{RNA} \xrightarrow{①} (+)\ \text{RNA} \xleftarrow{③} (\pm)\ \text{RNA}$$
$$\Big\uparrow ②$$
$$(-)\ \text{RNA}$$

is cleaved by host proteases to give six proteins: an RNA polymerase, four coat proteins, and one protein of unknown function.

Rabies virus and *vesicular stomatitis virus* display a second mode of gene expression. The single-stranded RNA molecules in these viruses do not serve as mRNA; hence, they are designated as (−) RNA. The virions of these viruses carry an RNA polymerase that synthesizes several short (+) RNAs following infection. Each of these (+) mRNAs is translated to give a single polypeptide chain. Replication of viral RNA starts after viral protein has been made. A (+) strand having the full length of the (−) strand is synthesized by the viral polymerase. In turn, new (−) strands are made from this (+) template.

Reovirus, a double-stranded RNA virus that infects mammalian cells, multiplies by a third mechanism. The virion contains *ten different double-stranded (±) RNA molecules.* The virion loses its outer shell made of three kinds of proteins on entering its host. Removal of this shell activates an RNA polymerase that is carried in the core of the virion. This polymerase transcribes the ten (±) RNA

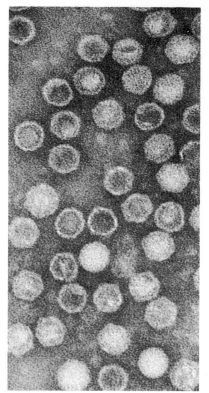

Figure 29-11
Electron micrograph of poliovirus particles. [Courtesy of Dr. John Finch.]

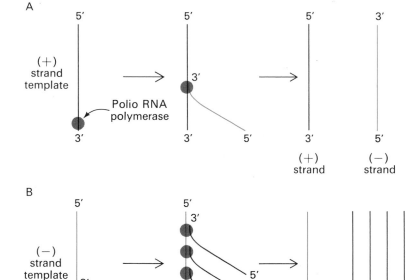

Figure 29-12
Replication of poliovirus RNA.

molecules in entirety, so that the (+) mRNA formed has exactly the same length as the genome segment. The (±) RNA template is transcribed *asymmetrically* and *conservatively*—that is, only (+) RNA is formed, and the original (±) RNA duplex is not disrupted. Each of the ten mRNAs is translated into one protein. A complete set of ten (+) RNA molecules is assembled with capsid proteins and ten (−) strands are then synthesized.

Why is the genome of reoviruses segmented? *In eucaryotic cells, an RNA molecule is translated into only one polypeptide chain,* in contrast to procaryotic mRNA, which is often polycistronic (e.g., *lac* mRNA). Animal viruses circumvent this limitation in several ways. Reovirus does so by having a separate chromosome for each kind of protein synthesized. The device of vesicular stomatitis virus is to transcribe its virion RNA into short mRNAs, each corresponding to one protein. Yet another strategem is used by poliovirus, which uses host proteases to cleave the giant polypeptide formed by translation of its single mRNA.

A fourth mechanism for the expression of genetic information in RNA viruses will be discussed later in this chapter. Most interesting, in this mode, information flows from viral RNA to DNA and then back to RNA.

SOME RNA PHAGES HAVE ONLY THREE GENES

RNA phages such as R17 and Qβ are very simple viruses. They have a regular polyhedral shape and a diameter of about 200 Å. The capsid of these phages contains 180 molecules of the coat protein (14,000 daltons) and one molecule of the A protein (38,000 daltons). *The single-stranded (+) RNA contains three genes, which code for the A protein, the coat protein, and a subunit of the replicase* (Figure 29-13). The (+) RNA molecule of the virion serves both as a messenger for the synthesis of these three proteins and as a template for the synthesis of (−) RNA. Then, many copies of (+) RNA are made, using (−) RNA as the template.

The replicase that synthesizes the (+) and (−) strands of phage RNA is a very interesting enzyme. It is highly specific for the homologous phage RNA. Hence, host RNA molecules do not compete with phage RNA for replication. *The Qβ replicase consists of four subunits, of which only one is encoded by Qβ RNA. The other three*

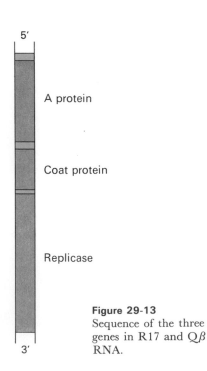

5′

A protein

Coat protein

Replicase

3′

Figure 29-13
Sequence of the three genes in R17 and Qβ RNA.

subunits of the replicase are host proteins that have been diverted by the phage for its own ends. Two of them are elongation factors for protein synthesis, EF Tu and EF Ts, and the third is a component of the 30S ribosomal subunit. Thus, Qβ creates a highly specific enzyme in a very economical way.

The three phage proteins are synthesized in different amounts. The coat protein, which is synthesized throughout infection, is the major product of translation. One reason is that ribosomes bind much more strongly to the initiation site for the coat cistron than to the other two initiation sites on the (+) RNA. Furthermore, the coat protein inhibits translation of the replicase cistron by blocking its initiation site. *Other regulatory mechanisms assure that replication and translation are appropriately timed.* The (+) RNA is both a messenger for protein synthesis and a template for the synthesis of (−) RNA. It would be undesirable for both processes to occur simultaneously on the same (+) RNA since ribosomes travelling in the 5′ to 3′ direction would collide with the replicase moving in the 3′ to 5′ direction. This does not occur because Qβ replicase strongly inhibits the binding of ribosomes to (+) RNA until an adequate number of (−) RNA are synthesized.

DARWINIAN EVOLUTION OF PHAGE RNA OUTSIDE A CELL

The purification of Qβ RNA and Qβ replicase free from contaminating nucleases enabled Sol Spiegelman to examine evolutionary events outside a living cell. One question asked was: *What happens to RNA molecules if the only demand made on them is that they multiply as rapidly as possible?* Molecules of Qβ RNA, Qβ replicase, and ribonucleoside triphosphates were incubated for twenty minutes, an incubation period that facilitates the selection of mutant RNA molecules that are rapidly replicated. An aliquot was diluted into a standard reaction mixture containing Qβ replicase and ribonucleoside triphosphates. A series of seventy-five transfers was carried out, and the RNA products were analyzed. The length of incubation was reduced periodically because the RNA molecules replicated more rapidly as the experiment proceeded. *The striking outcome was that the RNA molecules after seventy-five transfers ("generations") were only 12% as long as the original Qβ RNA.* Nucleotides not essential for replication were lost because the resulting shortened

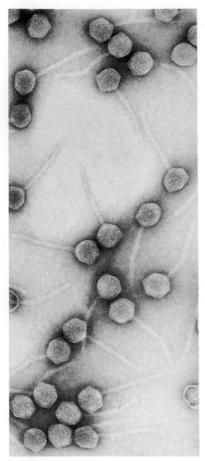

Figure 29-14
Electron micrograph of λ phage.
[Courtesy of Dr. A. Dale Kaiser.]

molecule was replicated more rapidly. The main constraint in this experiment was that the mutants had to retain the initiation sequence recognized by the $Q\beta$ replicase.

LYSOGENIC PHAGES CAN INSERT THEIR DNA INTO THE HOST CELL DNA

Some bacteriophages have a choice of life styles: they can multiply and lyse an infected cell (*lytic pathway*) or their DNA can join the infected cell, retaining the capacity for multiplication and lysis (*lysogenic pathway*). Viruses that do not always kill their hosts are called *temperate* or *moderate*. The best understood temperate virus is *lambda* (λ), which was discussed previously regarding its control of transcription (p. 688).

In the lysogenic pathway, the DNA molecule of λ becomes covalently joined to the DNA molecule of *E. coli*. The DNA in the λ virion is a linear, double-stranded molecule of molecular weight 30.8×10^6, which corresponds to about 46,500 base pairs. The 5′ end of each strand of this linear DNA molecule contains a single-stranded sequence of twelve nucleotides. These sequences are called *cohesive ends* because they are mutually complementary and can base-pair to each other. In fact, the cohesive ends come together following infection. The 5′ phosphate terminus of each strand is then adjacent to its own 3′ hydroxyl end. A host DNA ligase seals these two gaps, yielding a *circular λ DNA molecule* (Figure 29-15).

Figure 29-15
Conversion of linear λ DNA
to the circular form.

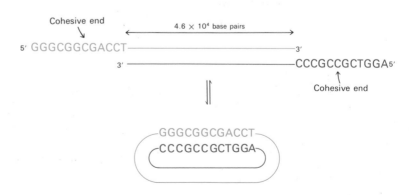

The circular λ DNA molecule then inserts into the bacterial chromosome by a single reciprocal recombination at specific loci of both the λ and *E. coli* DNA. The attachment site for λ in the *E. coli* DNA, called *att*λ, is located between *galE* and *bioA*, genes in the galactose and biotin operons. The base sequence of *att*λ is symbolized by B-B′ (B for bacterial). The specific attachment site on λ, called *att*, is located next to the genes *int* (for integrate) and *xis* (for excise). The base sequence of *att* is symbolized by P-P′ (P for phage). The *int* protein recognizes the P-P′ sequence in the phage DNA and the B-B′ sequence in *E. coli* DNA. A reciprocal transfer then occurs: P joins B′ and B joins P′. This scheme (Figures 29-16 and 29-17) was originally proposed by Allan Campbell on the basis of genetic evidence.

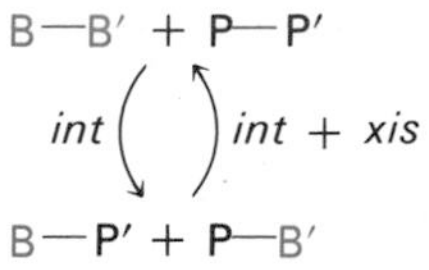

$$\text{B—B′ + P—P′}$$

$$int \left(\begin{array}{c} \\ \end{array}\right) int + xis$$

$$\text{B—P′ + P—B′}$$

Figure 29-16
Insertion and excision of λ DNA. This equation is incomplete because some of the reactants have not yet been identified.

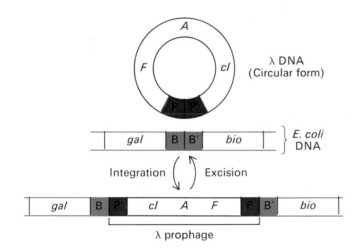

Figure 29-17
Schematic diagram of reciprocal recombination of λ DNA and *E. coli* DNA.

The λ DNA is now part of the *E. coli* DNA molecule. This form of λ is called the *prophage,* and the *E. coli* cell containing the prophage is called a *lysogenic bacterium.* The prophage is stable in the absence of the *xis* protein. Transcription of the *xis* gene is blocked by the λ repressor (p. 688). When repression is released, *xis* and *int* together catalyze the breaking of the B-P′ and P-B′ sequences and again a reciprocal transfer occurs (Figure 29-17): P rejoins P′ and B rejoins B′, which recreates a circular molecule of λ DNA and a nonlysogenic *E. coli* chromosome. A key feature of this recombination system is that the *int* protein alone cannot recognize the two new sequences at the ends of the prophage (B-P′ and P-B′), which makes it stable. Thus, insertion occurs when only the *int* protein is present, whereas excision occurs when both the *int* and *xis* proteins are present.

VIRUSES SUCH AS POLYOMA AND SV40 INDUCE CANCER IN SUSCEPTIBLE HOSTS

A variety of viruses can induce cancer when injected into susceptible animals. They are called *oncogenic viruses* or *tumor viruses*. The simplest oncogenic viruses are *polyoma virus, simian virus 40* (SV40), and *papilloma virus*. These members of the papova group contain a double-stranded circular DNA molecule, which has about 5100 base pairs (Figure 29-18). *These viruses contain only about five genes.* Investigators working on them hope that only one or two of their genes are directly responsible for producing cancer and that it will be feasible to elucidate their mode of action.

Tissue-culture systems have been developed for the study of cancer at the molecular level. Appropriate animal cells are permanently *transformed*—that is, rendered cancerlike—by infection with an oncogenic virus. Purified viral DNA from SV40 or polyoma can transform cells, whereas capsids devoid of DNA do not. *Hence, transformation is caused by viral DNA.* Transformed cells differ from normal cells in their *growth patterns* and in the nature of their *cell surface*. These differences are most pronounced when they are grown in dense cultures (Table 29-2). In addition, transformed cells exhibit signs of harboring *viral DNA,* which accounts for the fact that transformation is a *heritable alteration.* The abnormal characteristics of transformed cells are perpetuated in cultures derived from transformed colonies. Furthermore, some *transformed cells from tissue culture grow into a cancer when they are injected in sufficient number into an appropriate host.*

Polyoma virus and SV40, like λ phage, have two developmental pathways. In some cells (called *permissive hosts*), they go through a *lytic cycle* in which the major events are (1) adsorption, (2) penetration, (3) uncoating of the viral DNA, (4) stimulation of host DNA synthesis, and synthesis of early viral mRNA and early proteins, (5) viral DNA replication, (6) synthesis of late viral mRNA and late proteins, followed by packaging of viral DNA, and (7) cell lysis.

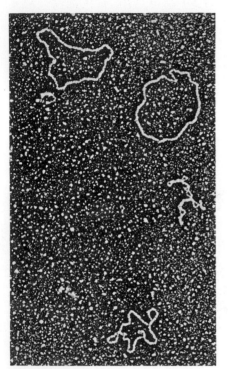

Figure 29-18
Electron micrograph of SV40 DNA molecules in different states of coiling. [Courtesy of Dr. Jack Griffith.]

Table 29-2
Properties of cells transformed by SV40 or polyoma virus

Growth patterns
    High saturation density
    Reduced requirement for serum
    Forms tumors following injection into susceptible animals
    Relatively unoriented growth

Surface properties
    New viral-specific antigen
    Increased rate of transport of nutrients
    Exposed fetal antigens
    Increased agglutinability by plant lectins
    Changes in composition of glycoproteins and glycolipids

Signs of the presence of tumor virus
    Virus DNA sequences are present
    Virus mRNAs are present
    Virus-specific antigens are detectable

Based on J. Tooze, ed. *The Molecular Biology of Tumour Viruses* (Cold Spring Harbor Laboratory, 1973), p. 351.

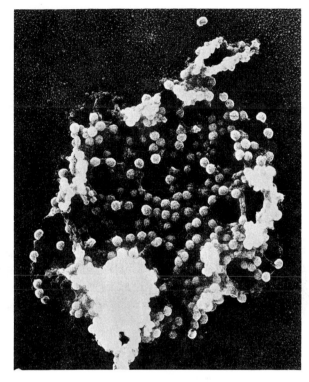

Figure 29-19
Electron micrograph of a fragment of the nuclear membrane of an SV40-infected cell showing nuclear pores and many imbedded virions. [Courtesy of Dr. Jack Griffith.]

In other types of cells (called *nonpermissive hosts*) steps 5, 6, and 7 are blocked for unknown reasons. *These nonpermissive cells become transformed on infection.*

As in lysogeny by λ phage, *the DNA of these oncogenic viruses becomes integrated with the host DNA as a provirus during transformation.* Hybridization studies using radioactive viral RNA have shown that the host DNA of a transformed cell contains several viral genomes that are covalently attached. Integration of polyoma DNA, which probably occurs by reciprocal recombination, requires the function of an early viral gene called *tsa.* Virus-specific RNA has been found in transformed cells. Some transcripts are longer than the entire viral genome, and they contain a tail of poly A. The viral DNA may be integrated so that the functions necessary to maintain transformation are preserved intact, whereas there may be a break in the late genes. The viral RNA molecules are probably translated since virus-specific proteins are present in transformed cells. One of them, called the tumor or *T antigen,* appears in the nuclei of transformed cells. SV40 also induces a *U antigen,* which is present on the nuclear membrane. It is not yet known which gene or genes cause transformation and result in its maintenance.

A DNA MOLECULE PLAYS A KEY ROLE IN THE ONCOGENIC ACTION OF RNA TUMOR VIRUSES

In 1911, Peyton Rous prepared a cell-free filtrate from a connective-tissue tumor that appeared spontaneously in a hen and injected it into normal chickens. The striking result was that the recipients developed tumors of the same kind, called *sarcomas.* The tumor-causing agent in the filtrate, now known as *Rous sarcoma virus* (RSV), could be propagated by serial passage through chickens. Rous sarcoma virus is a member of a group of *oncogenic RNA viruses,* which are also called *oncorna viruses* or *leukoviruses.*

The oncogenic RNA viruses are more complex than SV40 and polyoma virus. They are 1000 Å in diameter and contain an *envelope,* which is rich in lipid. This envelope comes from the host cell when the virus buds from its surface. Most oncogenic RNA viruses, in contrast to the oncogenic DNA viruses, do not destroy their hosts. Inside the envelope, there is a core shell consisting of an outer capsid having icosahedral symmetry. In the center of the particle, there is a ribonucleoprotein nucleoid surrounded by a membrane. The viral genome is a 10^7-dalton RNA molecule, which has a sedimentation coefficient of 60 to 70 S. Smaller amounts of other RNAs are also present. The virions contain more than ten kinds of proteins.

In 1964, Howard Temin observed that infection by RNA tumor viruses such as RSV is blocked by inhibitors of DNA synthesis. Inhibitors such as amethopterin, 5-fluorodeoxyuridine, and cytosine arabinoside are effective if they are present during the first twelve hours following the introduction of viruses. This finding suggested that *DNA synthesis is required for the growth of RNA tumor viruses.* Furthermore, the production of progeny virus particles is inhibited

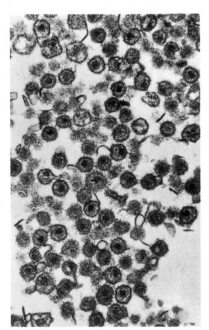

Figure 29-20
Electron micrograph of particles of avian myeloblastosis virus. Injection of this RNA tumor virus into newborn chicks produces a fatal leukemia within three weeks. [Courtesy of Dr. Ursula Heine, National Cancer Institute.]

by actinomycin D. This antibiotic is known to inhibit the synthesis of RNA from DNA templates (p. 614). Hence, *transcription of DNA seemed to be essential for the multiplication of RNA tumor viruses.* These unexpected results led Temin to propose that *a DNA provirus is an intermediate in the replication and oncogenic action of RNA tumor viruses.*

$$\text{RNA tumor} \longrightarrow \text{DNA} \longrightarrow \text{RNA tumor}$$
$$\qquad \text{virus} \qquad\qquad \text{provirus} \qquad\qquad \text{virus}$$

Temin's bold hypothesis that genetic information can flow from RNA to DNA was initially greeted with little enthusiasm by most investigators. It required the existence of an as yet unknown enzyme—one that synthesizes DNA according to instructions given by an RNA template. In 1970, Temin and Baltimore independently discovered such an enzyme, which is called an *RNA-directed DNA polymerase* or a *reverse transcriptase,* in the virions of some RNA tumor viruses. All RNA tumor viruses subsequently studied have a reverse transcriptase. The reverse transcriptase from Rauscher mouse leukemia virus is a single polypeptide chain of 70,000 daltons. Reverse transcriptases, like DNA polymerases, synthesize DNA in the 5' to 3' direction and use short RNAs as primers.

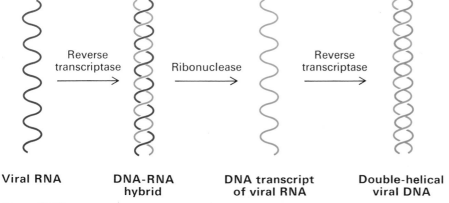

| Viral RNA | DNA-RNA hybrid | DNA transcript of viral RNA | Double-helical viral DNA |

Figure 29-21
Synthesis of DNA from an RNA template by the reverse transcriptase. The primer is not shown here.

The life cycle of a typical RNA tumor virus starts when infecting virions bind to specific receptors on the surface of the host cell and enter the cell. The viral (+) RNA is uncoated and moves to the cell nucleus. The reverse transcriptase from the virion then forms a (−) DNA strand taking instructions from the (+) RNA. Next, a (+) DNA strand is synthesized using the (−) DNA as the template. *The (±) DNA is then integrated into the host genome.* Following cell division, the provirus in the host genome is transcribed. Viral proteins are formed in the cytoplasm by translation of the viral transcripts. Viral (+) RNA molecules interact with capsid proteins to form partial virions, which bind to the cell surface. Mature virus particles are formed as an envelope is acquired. *The infected cell is permanently transformed because its genome contains the provirus.*

Do viruses cause human cancers? The answer to this question is not yet known. However, several lines of evidence suggest a *probable* role for viruses in *some* human cancers. First, hybridization studies have revealed that human leukemias, sarcomas, lymphomas, and breast adenocarcinomas contain 70 S RNAs, which are homologous to RNAs of tumor viruses that cause these malignancies in mice. Second, these human cancer cells contain particles with reverse transcriptase activity. Third, the DNA of some malignant human cells have virus-related sequences that are not found in the DNA of normal cells. The possible viral etiology of some human cancers is being intensively studied in many laboratories. It is hoped that definitive answers will soon emerge.

SUMMARY

Viruses are packets of infectious nucleic acid surrounded by a protective coat. They contain either DNA or RNA, which can be single-stranded or double-stranded. The simplest viruses have only 3 genes, whereas the most complex ones have about 250 genes. Most viral coats consist of a large number of one or a few kinds of protein subunits because viruses have a very limited amount of genetic information. The two principal arrangements are a cylindrical shell with helical symmetry and a spherical shell with icosahedral symmetry. Tobacco mosaic virus, which consists of a helical array of 2,130 identical subunits around a single-stranded RNA molecule, is formed by a self-assembly process in which discs of coat subunits are key intermediates. The assembly of T4 phage is a much more complex process that involves approximately fifty proteins. Three major pathways lead independently to the head, the tail, and the tail fibers. There is a stringent sequential order in the formation of T4, which occurs by a combination of self-assembly and enzyme-directed assembly.

Uninfected cells cannot synthesize RNA from an RNA template. Hence, a common feature of RNA viruses (except for the RNA tumor viruses) is that they specify an RNA-directed RNA polymerase. The RNA in the virion of some RNA viruses (e.g., Qβ phage) serves as a messenger on infection. For other RNA viruses (e.g., reovirus), the genome RNA must first be transcribed following infection. This is accomplished with an enzyme that is carried in the virion.

Not all viruses destroy their hosts. The temperate phage λ can assume a dormant prophage form in which its DNA is integrated into the bacterial chromosome by reciprocal recombination. This prophage retains the capacity for multiplication and lysis, which occur when it is excised from the host DNA. Likewise, the DNA tumor viruses SV40 and polyoma can multiply and lyse their hosts, or their DNA can become integrated into the host cell DNA. Such transformed cells carrying a provirus have abnormal growth patterns and cell surfaces. They cause tumors when injected into susceptible animals. SV40 and polyoma have only about five genes, and so they are attractive models for the study of the molecular basis of cancer. The RNA tumor viruses, such as Rous sarcoma virus, are more complex. The RNA of these viruses is transcribed by a virus-specified RNA-dependent DNA polymerase (reverse transcriptase). The viral DNA formed in this way becomes integrated into the host DNA, which transforms the host cell.

SELECTED READINGS

Where to start:

Wood, W. B., and Edgar, R. S., 1967. Building a bacterial virus. *Sci. Amer.* 217(1):60–74. [Available as *Sci. Amer.* Offprint 1079.]

Luria, S. E., 1970. The recognition of DNA in bacteria. *Sci. Amer.* 222(1):88–92. [Offprint 1167].

Temin, H. M., 1972. RNA-directed DNA synthesis. *Sci. Amer.* 226(1):24–33. [Offprint 1239.]

Dulbecco, R., and Ginsberg, H. S., 1973. Virology. *In* Davis, B. D., Dulbecco, R., Eisen, H. N., Ginsberg, H. S., and Wood, W. B., Jr., *Microbiology,* pp. 1007–1449. Harper and Row. [A very readable introduction to virology that emphasizes animal viruses.]

Assembly of viruses:

Klug, A., 1972. Assembly of tobacco mosaic virus. *Fed. Proc.* 31:30–42.

Wood, W. B., Edgar, R. S., King, J., Lielausis, I., and Henninger, M., 1968. Bacteriophage assembly. *Fed. Proc.* 27:1160–1166.

Eiserling, F. A., and Dickson, R. C., 1972. Assembly of viruses. *Annu. Rev. Biochem.* 41:467–502.

Crick, F. H. C., and Watson, J. D., 1957. Virus structure: general principles. *In* Wolstenholme, G. E. W., (ed.), *Ciba Foundation Symposium on the Nature of Viruses,* pp. 5–13.

Caspar, D. L. D., and Klug, A., 1962. Physical principles in the construction of regular viruses. *Cold Spring Harbor Symp. Quant. Biol.* 27:1–24. [The basic theory of the structure of spherical viruses is presented in this classic paper.]

Restriction and modification:

Linn, S., Lautenberger, J. A., Eskin, B., and Lackey, D., 1974. Host-controlled restriction and modification enzymes of *Escherichia coli* B. *Fed. Proc.* 33:1128–1134.

Meselson, M., Yuan, R., and Heywood, J., 1972. Restriction and modification of DNA. *Annu. Rev. Biochem.* 41:447–466.

Extracellular evolution of Qβ RNA:

Mills, D. R., Peterson, R. L., and Spiegelman, S., 1967. An extracellular Darwinian experiment with a self-duplicating nucleic acid molecule. *Proc. Nat. Acad. Sci.* 58:217–224.

Lysogeny:

Hershey, A. D., (ed.), 1971. *The Bacteriophage Lambda.* Cold Spring Harbor Laboratory. [This valuable reference volume includes several articles on lysogeny.]

Lwoff, A., 1966. The prophage and I. *In* Cairns, J., Stent, G. S., and Watson, J. D., (eds.), *Phage and the Origins of Molecular Biology,* pp. 88–99. Cold Spring Harbor Laboratory. [A delightful account of the discovery of the basis of lysogeny.]

Oncogenic viruses:

Tooze, J., (ed.), 1973. *The Molecular Biology of Tumour Viruses.* Cold Spring Harbor Laboratory.

Eckart, W., 1972. Oncogenic viruses. *Annu. Rev. Biochem.* 41:503–516.

Temin, H. M., 1971. Mechanism of cell transformation by RNA tumour viruses. *Annu. Rev. Microbiol.* 25:609–648.

Dulbecco, R., 1974. Mechanisms of virus-induced oncogenesis. *In* Schmitt, F. O., and Worden, F. G., (eds.), *The Neurosciences: Third Study Program,* pp. 1057–1065. MIT Press.

Sweet, R. W., Goodman, N. C., Cho, J.-R., Ruprecht, R. M., Redfield, R. R., and Spiegelman, S., 1974. The presence of unique DNA sequences after viral induction of leukemia in mice. *Proc. Nat. Acad. Sci.* 71:1705–1709.

Hynes, R. O., 1974. Role of surface alterations in cell transformation: the importance of proteases and surface proteins. *Cell* 1:147–156.

On the facing page: Scanning electron micrograph of retinal rod cells. A rod cell can be excited by a single photon. [Courtesy of Dr. Deric Bownds.]

PART V

MOLECULAR PHYSIOLOGY

interaction of information, conformation, and metabolism in physiological processes

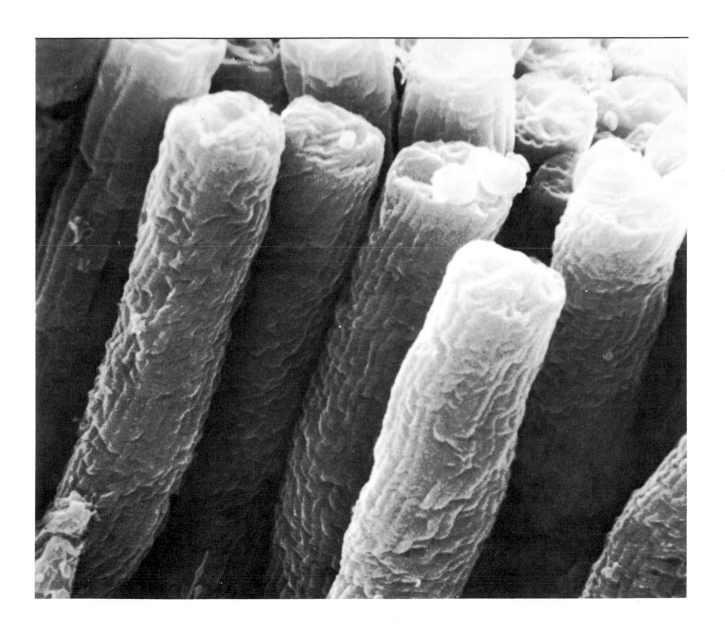

CHAPTER 30

IMMUNOGLOBULINS

By the turn of the century, many important features of the immune response had been discovered. An awareness of the protection conferred was fully appreciated much earlier, as evidenced by Thucydides' account of the plague that struck Athens in 430 B.C.:

> All speculation as to its origins and its causes, if causes can be found adequate to produce so great a disturbance, I leave to other writers . . . for myself, I shall simply set down its nature, and explain the symptoms by which it may be recognized by the student, if it should ever break out again. This I can the better do, as I had the disease myself, and watched its operation in the case of others. . . .
>
> Yet it was with those who had recovered from the disease that the sick and the dying found most compassion. These knew what it was from experience, and had now no fear for themselves; *for the same man was never attacked twice*—never at least fatally. And such persons not only received the congratulations of others, but themselves also, in the elation of the moment, *half entertained the vain hope that they were for the future safe from any disease whatsoever.* . . .

Electron micrograph of a crystal of antibody molecules. [From L. W. Labaw and D. R. Davies. *J. Ultrastruct. Res.* 40(1972):349.]

In characterizing the hope of protection against all disease a vain one, Thucydides clearly appreciated the *specificity* of the protection afforded by the immune system.

BASIC DEFINITIONS

An *antibody* is a protein synthesized by an animal in response to the presence of a foreign substance. The antibody has specific affinity for the foreign material that elicited its synthesis. An *antigen* (or *immunogen*) is a foreign macromolecule capable of eliciting antibody formation. Proteins, polysaccharides, and nucleic acids are usually effective antigens. The specificity of an antibody is directed against a particular site on an antigen called the *antigenic determinant*.

Small foreign molecules do not stimulate antibody formation. However, they can elicit the formation of specific antibody if they are attached to macromolecules. The macromolecule is then the *carrier* of the attached chemical group, which is called a *haptenic determinant*. The small foreign molecule by itself is called a *hapten*.

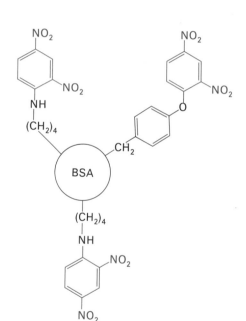

Figure 30-1
Dinitrophenylated bovine serum albumin (DNP-BSA) is an effective immunogen.

SYNTHESIS OF A SPECIFIC ANTIBODY FOLLOWING EXPOSURE TO AN ANTIGEN

Animals can make specific antibodies against virtually any foreign chemical group. The dinitrophenyl group (DNP) is particularly effective in eliciting antibody formation and therefore has been used extensively as a haptenic determinant. Anti-DNP antibody can be obtained in the following way:

1. DNP groups are covalently attached to a carrier protein such as bovine serum albumin (BSA) by reaction of fluorodinitrobenzene with lysine side chains and other nucleophiles of this protein (Figure 30-1).

2. DNP-BSA, the immunogen (antigen), is injected into a rabbit. The level of anti-DNP antibody in the serum of the rabbit starts to rise a few days later (Figure 30-2). These *early antibody molecules* belong to the *immunoglobulin M* (IgM) class and have a molecular weight of nearly 10^6.

3. Approximately ten days after the injection of immunogen, the amount of immunoglobulin M decreases and there is a concurrent increase in the amount of anti-DNP antibody of a different class, called *immunoglobulin G* (IgG), which has a molecular weight of 150,000.

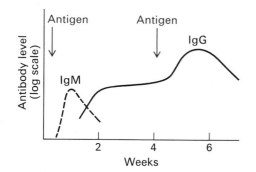

Figure 30-2
Kinetics of the appearance of immunoglobulins M and G (IgM and IgG) in the serum following immunization.

4. The level of anti-DNP antibody of the immunoglobulin G class reaches a plateau approximately three weeks after the injection of immunogen. A booster dose of DNP-BSA given at that time produces a further increase in the level of anti-DNP antibody in the rabbit's serum.

5. Blood is drawn from the immunized rabbit. The resulting serum (called an *antiserum* because it is obtained after immunization) may contain as much as 1 mg of anti-DNP antibody per milliliter. The DNP group is particularly effective in eliciting the formation of large amounts of specific antibody. Nearly all of this antibody is of the immunoglobulin G class, the principal one in serum.

6. The problem now is to separate anti-DNP antibody from antibodies of other specificities and from other serum proteins. The anti-DNP antibody differs from other proteins in the antiserum in its very high affinity for the DNP group. Consequently, it can be separated by *affinity chromatography*. A column consisting of dinitrophenyl groups covalently attached to an insoluble carbohydrate matrix is formed. The antiserum is applied to this DNP column, which is then washed with buffer. Most proteins in the antiserum are eluted because they have little or no affinity for the DNP group or for the carbohydrate support. In contrast, anti-DNP antibody binds tightly to the column. It is released by adding a high concentration of dinitrophenol, which binds to the antibody and thereby displaces it from the DNP groups of the insoluble carbohydrate matrix. A soluble complex of dinitrophenol and anti-DNP antibody emerges from the column. The dinitrophenol is removed by dialysis or by ion-exchange chromatography, yielding a preparation of purified antibody.

THE COMBINING SITES OF ANTIBODIES ARE LIKE THE ACTIVE SITES OF ENZYMES

The *combining sites* of antibodies (i.e., their antigen-binding sites) resemble the active sites of enzymes in several ways:

1. The binding constants for haptens have been determined by equilibrium dialysis and spectroscopic techniques. For example, the binding of a colored hapten such as a DNP derivative quenches the fluorescence of the tryptophan residues of the antibody. The extent of quenching is a measure of the saturation of the antibody combining sites. *Binding constants for most haptens range from $K = 10^{-4}$ to 10^{-10} M.* Thus the standard free energies of binding typically fall between about -6 and -15 kcal/mol, which is the same range as that for enzyme-substrate and enzyme-coenzyme complexes. Furthermore, *the binding forces in hapten-antibody complexes are like those in enzyme-substrate complexes.* Weak, noncovalent interactions of the electrostatic, hydrogen bond, and van der Waals types combine to give strong and specific binding.

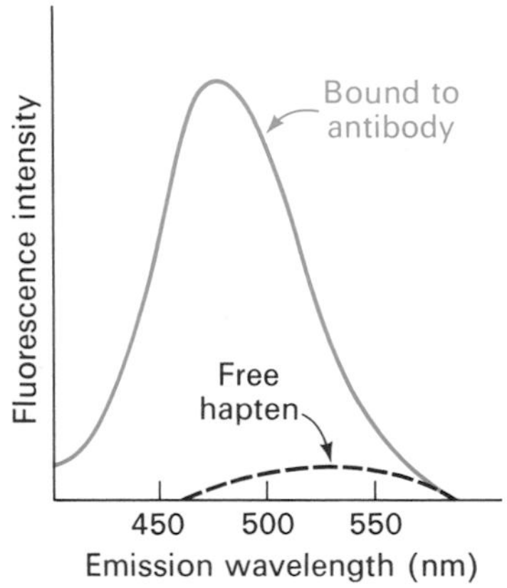

2. An antibody specific for dextran, a polysaccharide of glucose residues, was tested for its binding of oligomers. The full binding affinity was obtained when the oligomer contained six glucose residues. An interesting comparison here is with lysozyme, which also has space for six sugar residues in its active-site cleft. Thus, *the binding site of antidextran antibody is about 25 Å long.*

3. The spectroscopic properties of some haptens provide information about the polarity of the antibody combining site. For example, certain naphthalenes exhibit a weak yellow fluorescence when they are in a highly polar environment (such as water), and an intense blue fluorescence when they are in a markedly nonpolar environment (such as hexane). An intense blue fluorescence is observed when such a naphthalene hapten is bound to specific antibody, indicating that water is largely excluded from the combining site. In general, antibody combining sites are *nonpolar niches,* which enhances the binding of antigens for the reasons discussed previously in regard to enzyme-substrate interactions (p. 147).

4. The fit of hapten is rather precise. *Specificity is high but certainly not absolute.* The hapten is firmly held to the combining site, so that

Figure 30-3
The fluorescence properties of ε-dansyllysine change markedly when this hapten binds to antidansyl antibody. The decrease in the wavelength of the emission maximum and the increase in quantum yield indicate that the hapten is bound to a nonpolar site.

it has little rotational freedom. The rate constant for the binding of a variety of haptens is about 10^8 M^{-1} sec^{-1}. This high rate constant indicates that the binding step is a diffusion-controlled reaction. Large-scale conformational rearrangements do not seem to accompany the binding of hapten.

AN ANTIBODY OF A GIVEN SPECIFICITY IS USUALLY HETEROGENEOUS

Antibodies differ from enzymes in a very important way. *Most normal antibodies of a given specificity, such as anti-DNP antibody, are not a single molecular species.* Analyses of the binding of DNP haptens to a preparation of anti-DNP antibody reveal a wide range of binding affinities. Some antibody molecules bind DNP with a K of 10^{-6} M, whereas others have a K of 10^{-10} M. In contrast, most enzymes have a single binding constant for a particular substrate or coenzyme. Furthermore, a large number of bands are evident on electrophoresis of anti-DNP antibody or other antibodies of a particular specificity. On the other hand, an enzyme displays a single electrophoretic band or a few discrete bands (e.g., the isoenzymes of lactate dehydrogenase).

The heterogeneity of antibodies of a given specificity is an intrinsic feature of the immune response. What is the source of this heterogeneity? An important finding is that *antibodies produced by a single cell are homogeneous.* However, different cells produce different kinds of antibodies. The anti-DNP antibody in serum is the product of many different antibody-producing cells, which accounts for its heterogeneity.

IMMUNOGLOBULIN G CAN BE ENZYMATICALLY CLEAVED INTO ACTIVE FRAGMENTS

Immunoglobulin G has a molecular weight of 150,000. One approach in studying such a large protein is to split it into fragments that retain activity. In 1959, Rodney Porter showed that immunoglobulin G can be cleaved into three active fragments whose molecular weights are 50,000 by the limited proteolytic action of the enzyme papain. Two of these fragments bind antigen. They are

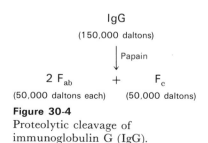

IgG
(150,000 daltons)

Papain

2 F$_{ab}$ + F$_c$
(50,000 daltons each) (50,000 daltons)

Figure 30-4
Proteolytic cleavage of immunoglobulin G (IgG).

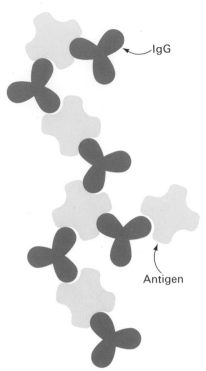

Figure 30-5
Schematic diagram of a lattice formed by the cross-linking of IgG (shown in blue) and an antigen (shown in yellow).

called F_{ab} (ab stands for antigen-binding, F for fragment). *Each F_{ab} contains one combining site for hapten or antigen,* which has the same binding affinity for hapten as does the whole molecule. However, *F_{ab} does not form a precipitate with antigen* because it is univalent (i.e., contains only one combining site), in contrast to an intact immunoglobulin G antibody molecule, which contains two antigen-binding sites. Immunoglobulin G can form a precipitate with an antigen that contains more than one antigenic determinant. An extended lattice is then formed, in which each antibody molecule cross-links two or more antigens, and vice versa (Figure 30-5). The amount of precipitate formed is largest when the antibody and antigen are present in equivalent amounts.

The other fragment having a molecular weight of 50,000, called F_c, does not bind antigen, but it has other important biological activities. This fragment can cross the placental membrane, whereas F_{ab} cannot. Thus, the capacity of immunoglobulin G to traverse the placental membrane and reach the fetal circulation depends on the presence of a *specific placental transmission site on F_c.* There is also a *site for the binding of complement* on F_c. The binding of an antibody to an antigenic determinant on the surface of a foreign cell often leads to the lysis of that cell. This sequel of the combination of antibody and antigen is mediated by a group of proteins known collectively as complement. The binding of one of these proteins to the antigen-antibody complex triggers the reaction sequence that lyses the foreign cell. The F_c fragment derived its name from the fact that it is readily *crystallizable*. This implies that it is homogeneous, in contrast to F_{ab}, which is usually heterogeneous and hence not crystallizable.

IMMUNOGLOBULIN G IS MADE UP OF L AND H CHAINS

Another breakthrough occurred in 1959, when Gerald Edelman showed that immunoglobulin G consists of two kinds of polypeptide chains. The disulfide bonds of the molecule were reduced by mercaptoethanol, and noncovalent interactions between the constituent chains were disrupted by 6 M urea. Chromatographic analysis of the resulting mixture revealed that immunoglobulin G consists of polypeptide chains having molecular weights of 25,000 and 50,000, which are called the *light* (L) and *heavy* (H) chains, respectively.

Porter then found milder conditions, which made it possible to reconstitute immunoglobulin G from two H and two L chains. These studies led him to propose a model (Figure 30-6) for the subunit structure of immunoglobulin G in which each L chain is bonded to an H chain by a disulfide bond. The H chains are bonded to each other by at least one disulfide bond. Thus, *the subunit structure of immunoglobulin G is* L_2H_2.

The fragments produced by papain digestion were then related to this model. Papain cleaves the H chains on the carboxyl-terminal side of the disulfide bond that links the L and H chains. Thus, F_{ab} consists of the entire L chain and the amino-terminal half of the H chain, whereas F_c consists of the carboxy-terminal halves of both H chains. The portion of the H chain contained in F_{ab} is called F_d.

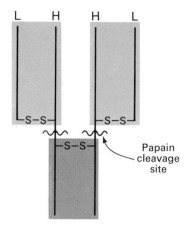

Figure 30-6
The subunit structure of IgG is L_2H_2. The F_{ab} units obtained by digestion with papain are shaded blue, whereas the F_c unit is shaded red.

IMMUNOGLOBULIN G IS A FLEXIBLE Y-SHAPED MOLECULE

Electron microscopic studies by Robin Valentine and Michael Green have shown that immunoglobulin G has the shape of the letter Y, and that the hapten binds near the ends of the F_{ab} units. Furthermore, the F_c and the two F_{ab} units of the intact antibody are joined by a hinge that allows rapid variation in the angle between the F_{ab} units through a wide angular range (Figure 30-7). This kind of flexibility facilitates the formation of antibody-antigen complexes, since the distance between the antigen-binding sites on the antibody can be adjusted to match the distance between specific determinants on the antigen (e.g., a virus with many repeating subunits).

BOTH L AND H CHAINS CONTRIBUTE TO THE ANTIGEN-BINDING SITE

The antigen-binding fragment F_{ab} consists of the entire L chain and part of the H chain. What are the relative contributions of the two chains to binding activity? An answer has been provided by forming hybrid antibody and measuring its binding activity. Specific antibody ($L_2^*H_2^*$, in which the asterisk denotes that it is

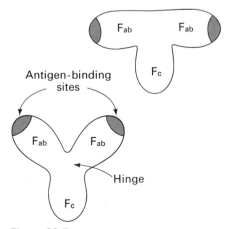

Figure 30-7
Immunoglobulin G has the shape of the letter Y. It contains a hinge that gives it segmental flexibility.

specific for DNP hapten) and nonspecific antibody (L_2H_2, a collection of molecules with no detectable affinity for hapten) were dissociated into their constituent chains. The dissociated chains were then combined to form the original $L_2^*H_2^*$ and L_2H_2, and the hybrids $L_2^*H_2$ and $L_2H_2^*$. As expected, the reconstituted L_2H_2 has no anti-DNP activity. The binding affinity of the other reconstituted antibodies for DNP hapten increased in the order $L_2^*H_2$, $L_2H_2^*$, $L_2^*H_2^*$. Thus, optimal activity was obtained when both chains came from the original specific antibody. Since $L_2H_2^*$ binds hapten more strongly than $L_2^*H_2$, it seems that the H chain makes a larger contribution than the L chain to the binding affinity for DNP. For a different kind of antibody, such as an antibody directed against diphtheria toxin, the L chain makes a larger contribution than the H chain. The generalization that emerges is that *both the L and H chains are important in forming the antigen-binding site,* their relative contributions varying from one antibody to another.

This picture is supported by *affinity-labeling* studies. Haptens containing reactive groups can be readily synthesized. The reactive hapten binds specifically to the antibody combining site and then forms a covalent bond with a suitable neighboring residue on the protein. For example, anti-DNP antibody can be labeled at its combining sites by a nitrophenyldiazonium hapten. The diazonium group is highly reactive and can form a diazo bond with tyrosine, histidine, and lysine residues. Antibody reacted in this way contains the label in both its L and H chains. Thus, *both the L and H chains contribute residues to the binding site.*

p-Nitrophenyldiazonium hapten

ARE ANTIBODIES FORMED BY SELECTION OR INSTRUCTION?

Enzymes acquired their specificities in millions of years of evolution. A specific antibody appears in the serum of an animal only a few weeks after it is exposed to any foreign determinant. How is a specific antibody formed in such a short time?

1. The *instructive theory,* which was proposed by Linus Pauling in 1940, postulated that the *antigen acts as a template that directs the folding of the nascent antibody chain.* In this model, antibody molecules of a given amino acid sequence have the potential for forming

combining sites of many different specificities. The particular one formed depends on the antigen present at the time of folding. The instructive theory predicts that a specific antibody cannot be formed in the absence of the corresponding antigen.

2. The *selective theory,* which was originated by Macfarlane Burnet and Niels Jerne in the 1950s, postulates that the antigen controls only the amount of specific antibody produced. In this model, *the combining site of the specific antibody is completely determined before it ever encounters antigen.*

DEMISE OF THE INSTRUCTIVE THEORY

The instructive theory predicted than an antibody will lose its specificity if it is unfolded and then refolded in the absence of antigen. This prediction contrasted with the experimental results obtained in Christian Anfinsen's laboratory on the refolding of ribonuclease. Recall that unfolded ribonuclease spontaneously assumes the three-dimensional structure, specificity, and catalytic activity of the native enzyme upon removal of the denaturing agent (p. 36). A substrate need not be present during the refolding of ribonuclease. This important experiment on ribonuclease provided the impetus for devising a similar test for antibodies. An F_{ab} fragment from anti-DNP antibody rather than an intact antibody molecule was used because its smaller size simplified the experiment. The F_{ab} fragment was unfolded in 7 M guanidine hydrochloride, a highly effective denaturing agent. This denatured F_{ab} molecule had no detectable affinity for DNP haptens. Hydrodynamic and optical techniques indicated that the conformation of the denatured F_{ab} approached that of a random coil. The guanidine hydrochloride was then removed by dialysis, and the denatured F_{ab} refolded in *the absence of DNP hapten.* The striking result was that the refolded F_{ab} had a high affinity for DNP haptens. Since hapten was absent during refolding, it was evident that *the specificity of the combining site is inherent in the amino acid sequence of the antibody.* This experimental result is consistent with the selective theory but contradicts a critical prediction of the instructive theory. Furthermore, it was found that *antibody-producing cells can synthesize large amounts of antibody in the absence of antigen.* The essence of the selective theory—that the *antigen influences the amount of specific antibody produced but not its amino acid sequence or three-dimensional structure*—is now firmly established.

MYELOMA IMMUNOGLOBULINS ARE HOMOGENEOUS

How is a specific antibody made before the arrival of an antigen? It is evident that detailed structural information about antibodies is needed to solve this problem. However, it is difficult to analyze a heterogeneous population of antibody molecules. This hurdle has been overcome by obtaining sufficient amounts of antibody produced by one kind of cell, since these molecules are homogeneous. This sometimes happens spontaneously in man and other animals. Patients who have a malignant disease called *multiple myeloma* synthesize large amounts of *immunoglobulins of a single kind*. The disorder in multiple myeloma is one of control. An antibody-producing cell divides many times, as do its daughter cells, yielding a very large number of cells of a single kind. They are a *clone,* since they are descended from the same cell and have identical properties. *Myeloma immunoglobulins have a normal structure and are typical of normal immunoglobulins, each being a homogeneous example of the large number of different antibodies present in an individual.* Myelomas in mice can be transplanted to other mice, where they proliferate. Furthermore, these antibody-producing tumors synthesize the same kind of homogeneous antibody generation after generation. In fact, homogeneous myeloma immunoglobulins have contributed to a major breakthrough in molecular immunology.

THE L AND H CHAINS CONSIST OF A VARIABLE REGION AND A CONSTANT REGION

An abnormal protein appears in large amounts in the urine of many patients who have multiple myeloma. This protein attracted the attention of Henry Bence-Jones in 1847 because of its unusual solubility properties: it precipitates on being heated to 50°C and becomes soluble again on boiling. The *Bence-Jones protein* is now known to be a *dimer of the L chains* of a patient's myeloma immunoglobulin, which makes it a very suitable starting material for amino acid sequence studies.

The first complete amino acid sequences of myeloma L chains, which contain about 214 amino acids, were determined in 1965. These studies showed that Bence-Jones proteins from different patients have different amino acid sequences. Most striking, these differences in sequence are confined to the amino-terminal half of the polypeptide chain. Each Bence-Jones protein studied has a unique amino acid sequence from positions 1 to 108. In contrast, the sequences of a number of Bence-Jones proteins are the same starting at position 109. Thus, *the L chain consists of a variable region (residues 1 to 108) and a constant region (residues 109 to 214)* (Figure 30-8). This remarkable finding had no precedent in protein chemistry.

Not all L chains have the same common region. In fact, there are two types, called kappa (κ) and lambda (λ), which exhibit many similarities. For example, each consists of 214 residues and contains an amino-terminal variable half and a carboxyl-terminal constant half. Each half of the κ and λ chains contains an intrachain disulfide loop. Both types have a carboxyl-terminal cysteine residue, which forms a disulfide bond with the H chain. Furthermore, the κ and λ chains have identical amino acid residues at 40% of the positions, which is nearly the same degree of homology exhibited by the α and β chains of human hemoglobin.

All κ chains have the same common region except for residue 191, which is either leucine or valine. This *allotypic* difference is inherited in a classical Mendelian manner. The λ chains also exhibit variability at position 191, which is either lysine or arginine.

The H chain consists of 446 amino acid residues. A comparison of the sequences of H chains from different myeloma immunoglobulins shows that all of the differences in sequence are located in the 108 residues at the amino-terminal end. Thus, *the heavy chain, like the light chain, consists of a variable region and a constant region.* The variable region of the heavy chain has the same length as that of the light chain, whereas the constant region is about three times as long (Figure 30-9).

THE VARIABLE AND CONSTANT REGIONS HAVE DIFFERENT ROLES

The amino-terminal parts of the light and heavy chains contain the variable regions, which are responsible for the binding of

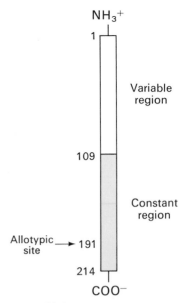

Figure 30-8
The light chain of an immunoglobulin consists of a variable region (V) and a constant region (C).

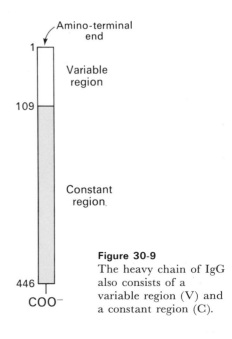

Figure 30-9
The heavy chain of IgG also consists of a variable region (V) and a constant region (C).

antigen. Differences in amino acid sequence yield combining sites that have different specificities. The rest of each chain contains the constant region, which mediates effector functions such as the binding of complement and the transfer of antibodies across the placental membrane. The same effector functions are mediated by antibodies of very different specificity. Thus, *antibodies consist of regions of unique sequence that confer antigen-binding specificity and regions of constant sequence that mediate common effector functions.* A striking structural feature of immunoglobulins is that the variable and constant regions are sharply demarcated in the amino acid sequences of the light and heavy chains.

ANTIBODY MOLECULES EVOLVED BY GENE DUPLICATION AND DIVERSIFICATION

The complete amino acid sequence of an immunoglobulin molecule was elucidated by Gerald Edelman in 1968. A most interesting feature of this sequence is the periodic location of intrachain disulfide bonds in both the light and heavy chains (Figure 30-10). Furthermore, there are many similarities in the amino acid sequences of different parts of the immunoglobulin molecule. The variable region of the light chain (V_L) is homologous with the variable region of the heavy chain (V_H). In addition, the constant

Figure 30-10
Structure of IgG, showing the pattern of disulfide bonds and the regions of homologous sequence. [Based on G. M. Edelman. The structure and function of antibodies. Copyright © 1970 Scientific American, Inc. All rights reserved.]

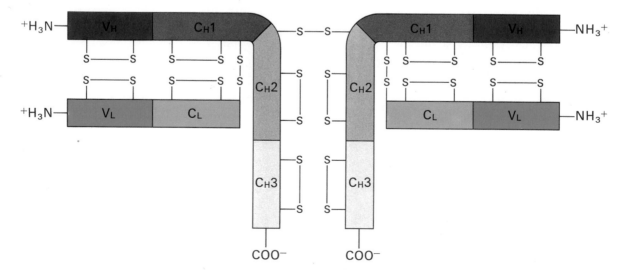

region of the heavy chain (C_H) consists of equal thirds (C_H1, C_H2, and C_H3) that are similar in sequence. Furthermore, the constant region of the light chain (C_L) is homologous with the three domains of the constant region of the heavy chain.

These structural homologies have important evolutionary implications. The similarities in the amino acid sequence of chymotrypsin, trypsin, elastase, and thrombin indicate that they evolved from a common ancestor (p. 191). Likewise, *the internal homologies in the immunoglobulin molecule suggest that antibodies evolved by a process of gene duplication and subsequent diversification.* An ancestral gene may have coded for a primitive antibody having a length of about 108 residues. Such an ancestral gene probably duplicated and then diverged to give rise to different kinds of variable and constant regions.

EACH REGION OF HOMOLOGOUS SEQUENCE FOLDS INTO A COMPACT DOMAIN

The homologies evident in the amino acid sequences suggested that immunoglobulins are folded into compact domains, each consisting of a homologous region containing at least one active site that serves a distinctive molecular function (Figure 30-11). It also seemed

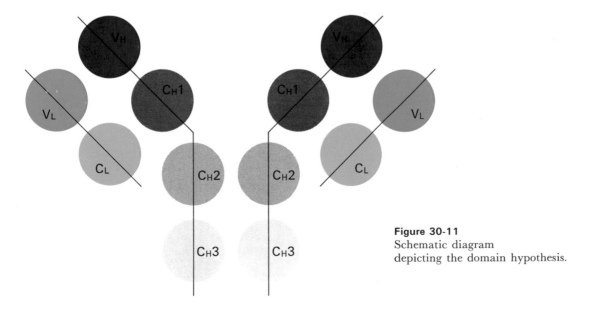

Figure 30-11
Schematic diagram
depicting the domain hypothesis.

likely that these domains would resemble each other in tertiary structure because they have similar amino acid sequences. Recent x-ray crystallographic studies have verified the domain hypothesis. The three-dimensional structure of the $F_{ab'}$ fragment (which is nearly the same as the F_{ab} fragment) of a human immunoglobulin has been solved at 2.0-Å resolution by Roberto Poljak. This fragment consists of a *tetrahedral array of four globular subunits—V_L, V_H, C_L, and $C_H 1$—which are strikingly similar in three-dimensional structure* (Figures 30-12 and 30-13). The binding site for antigen contains residues from both the V_L and V_H subunits. Most of the residues in the antigen-binding site come from the *hypervariable regions* of

Figure 30-12
An α-carbon diagram of the $F_{ab'}$ unit. The L chain is shown in blue and the H chain in red. [From R. L. Poljak, L. M. Amzel, H. P. Avey, B. L. Chen, R. P. Phizackerley, and F. Saul. *Proc. Nat. Acad. Sci.* 70(1973):3306.]

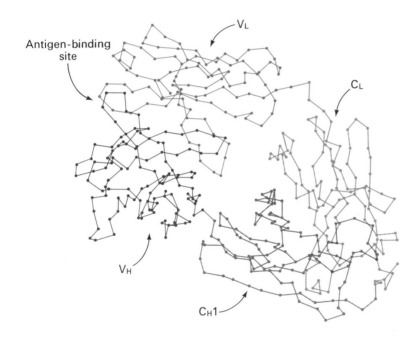

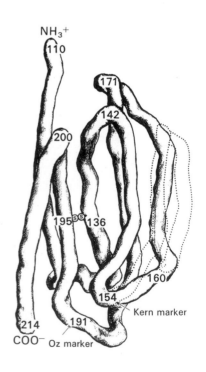

Figure 30-13
Diagram of the basic immunoglubulin fold. The domain shown here is C_L, which starts at residue 110 and extends to 214, the carboxyl terminus. The $C_H 1$ domain has a very similar structure. The V_L and V_H domains contain an additional loop of polypeptide (dashed lines). [From R. L. Poljak, L. M. Amzel, H. P. Avey, B. L. Chen, R. P. Phizackerley, and F. Saul. *Proc. Nat. Acad. Sci.* 70(1973):3306]

these domains—that is, from regions that show greatest variation in going from one immunoglobulin to another.

DIFFERENT CLASSES OF IMMUNOGLOBULINS HAVE DISTINCT BIOLOGICAL ACTIVITIES

There are five classes of immunoglobulins (Table 30-1), each consisting of heavy and light chains. The heavy chains differ from one class to another, whereas the light chains are the same, either λ, or κ. The heavy chains in immunoglobulin G are called γ chains, whereas those in immunoglobulin A, M, D and E are called α, μ, δ, and ϵ, respectively. These distinctive heavy chains give the five classes of immunoglobulins different biological characteristics. As mentioned previously, *immunoglobulin M* (IgM) is the first class of antibodies to appear in the serum after injection of an antigen, whereas *immunoglobulin G* (IgG) is the principal antibody in the serum. For many years, IgG was called γ-globulin, because of its electrophoretic and solubility characteristics. *Immunoglobulin A* (IgA) is the major class of antibodies in external secretions, such as saliva, tears, bronchial mucus, and intestinal mucus. Thus, IgA serves as a first line of defense against bacterial and viral antigens. The benefits conferred by immunoglobulin D (IgD) and E (IgE) are not yet known. However, the harmful effects of *immunoglobulin E* in mediating allergic reactions are clearly established.

Table 30-1
Properties of immunoglobulin classes

Class	Serum concentration (mg/ml)	Molecular weight	Sedimentation coefficient(s)	Light chains	Heavy chains	Chain structure
IgG	12	150,000	7	κ or λ	γ	$\kappa_2\gamma_2$ or $\lambda_2\gamma_2$
IgA	3	180,000–500,000	7, 10, 13	κ or λ	α	$(\kappa_2\alpha_2)_n$ or $(\lambda_2\alpha_2)_n$
IgM	1	950,000	18–20	κ or λ	μ	$(\kappa_2\mu_2)_5$ or $(\lambda_2\mu_2)_5$
IgD	0.1	175,000	7	κ or λ	δ	$\kappa_2\delta_2$ or $\lambda_2\delta_2$
IgE	0.001	200,000	8	κ or λ	ϵ	$\kappa_2\epsilon_2$ or $\lambda_2\epsilon_2$

Note: $n = 1$, 2, or 3.

VARIABLE AND CONSTANT REGIONS ARE PROBABLY ENCODED BY DISTINCT GENES

We turn now to the synthesis of antibody molecules. A comparison of the amino acid sequences of the variable (V) regions of light chains of the κ subclass has shown that they are encoded by at least three genes. In contrast, the constant (C) region is encoded by a single gene. As noted earlier, the C regions of κ chains have the same sequence except at position 191, which is either leucine or valine. Pedigree analyses have shown that these two variants are inherited in a classical Mendelian manner, which indicates that there is probably only one gene for the C region of these chains. Since the V region is encoded by at least three genes and the C region by only one, it seems very likely that the *V and C genes are separate and distinct in the germline*. This conclusion also pertains to λ light chains and to heavy chains.

VARIABLE AND CONSTANT GENES ARE JOINED IN ANTIBODY-PRODUCING CELLS

The proposed presence of distinct V and C genes in the germ line poses a special problem: how do they give rise to a single polypeptide chain? In principle, three mechanisms are possible:

1. The V gene specifies only one polypeptide chain, the C gene another. These two polypeptide chains then become linked by a peptide bond to form a single chain. The splicing of polypeptide chains has recently been observed in *E. coli.*

2. The joining occurs at the level of RNA.

3. The joining occurs at the level of DNA.

The first of these joining mechanisms has been tested by pulse-labeling experiments of the kind described previously for hemoglobin (p. 661). Should the V and C regions of a light chain join after they were separately synthesized, two growing points (one starting at residue 1, the other at residue 109) would be observed. On the other hand, if the L chain were synthesized as a single unit, there would be only one growing point (Figure 30-14).

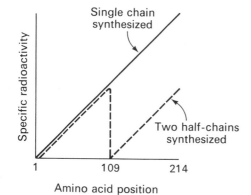

Figure 30-14
Expected gradient of radioactivity in a pulse-labeling experiment for the synthesis of a whole chain and for the synthesis of two half-chains that become fused.

The experimental data seem to be more compatible with synthesis from a single growing point. Thus, V and C probably do not come together at the polypeptide level.

There is no precedent for the joining of two pieces of messenger RNA, in contrast to the insertion of a piece of DNA into another piece (as in lysogeny, p. 720). An attractive hypothesis is that a *V gene is translocated and becomes integrated with the C gene in an antibody-producing cell*. In this model, the V gene is excised from a chromosome by an intrachromosomal recombination event and then diffuses to a C gene on the same chromosome. The V gene becomes joined to the C gene by a highly specific integrating enzyme (Figure 30-15). The resulting VC gene is then transcribed into a single messenger RNA, which is translated into a single polypeptide chain containing both the variable and constant regions.

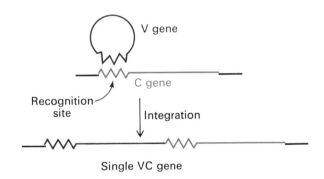

Figure 30-15
Proposed integration of V and C genes in an antibody-producing cell to form a single VC gene.

HOW ARE DIVERSE ANTIBODY SPECIFICITIES GENERATED?

An animal can synthesize large amounts of specific antibody against virtually any foreign determinant within a few weeks after being exposed to it. The number of different antibodies that can be made by an animal is very large, at least ten thousand. We have seen that the structural basis of antibody specificity resides in the amino acid sequences of the variable regions of the light and heavy chains. This brings us to the key question: *How are different variable-region sequences generated?*

This central enigma of immunology has not been solved. The

hypotheses presented in the next section offer very different answers to the following questions:

1. *When* was the diversity generated? During the life of an individual (somatic) or in the course of evolution (genetic)?

2. *How* was the diversity generated? By a random evolutionary process, by somatic recombination, or by somatic hypermutation?

GERM-LINE, SOMATIC-RECOMBINATION, AND SOMATIC-HYPERMUTATION HYPOTHESES

1. *Germ-line hypothesis.* In this model, diversity is already present in the germ line, in which there are a very large number ($>10^4$) of genes for the V regions of antibodies. For each unique amino acid sequence in the V regions, there is a corresponding piece of DNA in the germ line. Thus, this hypothesis proposes that diversity arose in the course of evolution by random mutation and selection.

Is there enough DNA in an organism to code for the very large number of different antibodies that can be made? It seems that there is. The amount of DNA in a human cell may be sufficient to code for 10^7 different proteins having an average length of 200 residues. How many genes are needed to provide for, say, 10^8 different antibody specificities? Since both the light and heavy chains contribute to antibody specificity, the combination of 10^4 light chains with 10^4 heavy chains might give 10^8 different specificities. This corresponds to only 0.2% of the genome, which would be a small cost for a biological function as protective as the formation of specific antibody.

2. *Somatic-recombination hypothesis.* In this model, there are only a small number of genes, of the order of 100, for the V regions of antibodies. These genes, which are similar but not identical, undergo extensive recombination in the antibody-producing cell during the lifetime of an individual. Crossing over is assumed to be intrachromosomal. A plausible calculation suggests that recombination among ten genes could generate 10^6 different amino acid sequences.

3. *Somatic-hypermutation hypothesis.* In this model, diversity is created by point mutations in a single gene for the V region of a

particular subclass. An extraordinarily high mutation rate would be needed to produce the required diversity during the lifetime of an individual. One proposal is that this is achieved in the antibody-producing cell (or its precursor) by a DNA repair mechanism that is designed to make mistakes. In this scheme, a specific nuclease cleaves one of the DNA strands of the V gene, which is then trimmed by an exonuclease. The gaps are subsequently filled by a special DNA repair enzyme that makes mistakes with high frequency, yielding new amino acid sequences for the V region.

Experiments are in progress to distinguish between these proposed mechanisms for the generation of diversity. Messenger RNAs for immunoglobulins are being purified, and their base sequences are being elucidated. An mRNA for an L chain contains 1,250 bases, including a poly A tract of 200 residues. The other 1,050 bases in this mRNA are more than sufficient to serve as the template for the synthesis of the entire L chain, which contains 230 amino acid residues. Another interesting recent result comes from studies of the hybridization of κ-chain mRNA with DNA derived from mouse liver cells. The kinetics of hybridization are biphasic, indicating that there are unique and reiterated sequences. These data suggest that there are fewer than 200 κ-chain V genes.

CLONAL SELECTION THEORY OF ANTIBODY FORMATION

A unifying model of the immune response is provided by the *clonal selection theory,* which was developed by Niels Jerne, Macfarlane Burnet, David Talmage, and Joshua Lederberg in the 1950s. The essential features of this theory, now firmly established, are:

1. An antibody-producing cell has a unique base sequence in its DNA that determines the amino acid sequence of its immunoglobulin chains. The specificity of an antibody is determined by its amino acid sequence.

2. A single cell produces antibody of a single kind. *The commitment to synthesize a particular kind of antibody is made before the cell ever encounters antigen.*

3. Each cell, as it begins to mature, produces small amounts of specific antibody. Some of this antibody becomes bound to the surface of the cell. An immature cell is killed if it encounters the antigen corresponding to its antibody. Hence, an animal does not usually make antibody against its own macromolecules—it is *self-tolerant*. Such antibody-producing cells were eliminated during fetal life. In contrast, *a mature cell is stimulated if it encounters antigen. Large amounts of antibody are then synthesized and the cell is triggered to divide.* The descendants of this cell are a *clone*. They have the same genetic makeup as the cell that was initially triggered by its encounter with antigen, and therefore they all make antibody of the same specificity.

4. A clone tends to persist after the disappearance of the antigen. These cells retain their capacity to be stimulated by antigen if it reappears, thereby providing for *immunological memory*.

AN ANTIBODY-PRODUCING CELL CONTAINS RECEPTORS FOR ANTIGENS ON ITS SURFACE

Specific antibody molecules are located on the surface of antibody-producing cells. The precursor cells are *lymphocytes*, which are very slow to divide unless stimulated by antigen. They then become *plasma cells*, which are very active in the synthesis and secretion of antibodies. If a population of lymphocytes is passed through a column containing a covalently bound dinitrophenyl (DNP) group, a small proportion remains bound. These bound lymphocytes synthesize anti-DNP antibody and contain antibody molecules of this specificity on their surfaces. The other lymphocytes do not make anti-DNP antibody.

How are lymphocytes triggered by exposure to specific antigen? Recent studies have shown that antibody molecules on lymphocyte surfaces are randomly distributed in the absence of antigen. The addition of antigen has a striking effect: the surface antibody molecules with their bound antigens come together to form a cap. Following this redistribution, the antibody molecules on the surface are engulfed into the interior of the cell. *Cap formation* is an energy-requiring process that involves contractile activity. The formation of a surface antibody-antigen lattice seems to be critical for cap

Figure 30-16
Scanning electron micrograph of sheep erythrocytes bound to a lymphocyte (center) from the spleen of a mouse. The surface of this lymphocyte contains immunoglobulins that are specific for a component of the sheep erythrocyte. [Courtesy of Dr. Jerry Thornthwaite and Dr. Robert Leif.]

formation and for stimulation of the lymphocyte. The antigen must be multivalent (i.e., contain more than one specific determinant), so that it can cross-link antibody molecules on the lymphocyte surface. The link between these events on the cell surface and mitotic activity in the cell nucleus is a challenging area of inquiry.

BIOLOGICAL SIGNIFICANCE OF CLONAL SELECTION

The essence of the clonal selection theory, as expressed by Burnet, is that:

> No combining site is in any evolutionary sense adapted to a particular antigenic determinant. The pattern of the combining site is there and if it happens to fit, in the sense that the affinity of adsorption to a given antigenic determinant is above a certain value, immunologically significant reaction will be initiated.

The selection of a preexisting pattern that is randomly generated is not a new theme in biology. Indeed, it is the heart of Darwin's theory of evolution. It is interesting to note that instructive theories have preceded selective theories: Lamarck's before Darwin's, antigen-directed folding before clonal selection.

Jerne has suggested that we should consider the possibility that selective mechanisms play a role in the operations of the nervous system, such as memory. In fact, the instructive and selective theories of learning have been stated long ago. Locke held that the brain could be likened to white paper, void of all characteristics, on which experience paints with almost endless variety. In contrast, Socrates asserted that "all learning consists of being reminded of what is preexisting in the brain." I hope this chapter has seemed familiar to you!

SUMMARY

An antibody is a protein synthesized by an animal in response to a foreign macromolecule, called an antigen or immunogen, for which it has high affinity. Small foreign molecules (haptens) elicit the formation of specific antibody if they are attached to a macromolecule. Antibody synthesis occurs by selection rather than by instruction. An antigen binds to the surface of lymphocytes already committed to make antibodies specific for that antigen. The combination of antigen and surface receptors triggers cell division and the synthesis of large amounts of specific antibody. Antibodies directed against a specific determinant are usually heterogeneous, because they are the products of many antibody-producing cells. The antibodies produced by a single cell or by a clone are homogeneous. There are five classes of antibodies in the serum, of which immunoglobulin G is the principal one. Immunoglobulin M is the first class to appear in the serum following exposure to an antigen, whereas immunoglobulin A is the major class in external secretions. The roles of immunoglobulins D and E are not known.

Antibodies consist of light (L) and heavy (H) chains. Immunoglobulin G, which has the subunit structure L_2H_2, contains two binding sites for antigen. Immunoglobulin G can be enzymatically cleaved into two F_{ab} fragments, which bind antigen but do not form a precipitate, and one F_c fragment, which mediates effector functions such as the binding of complement. A comparison of the amino acid sequences of myeloma immunoglobulins has revealed that L and H chains consist of a variable (V) region (an amino-terminal sequence of about 110 residues) and a constant (C) region. The different specificities of antibodies arise from differences in the amino acid sequences of the variable regions of their L and H chains. The V and C regions of the L chain and also of the H chain are encoded by distinct genes. The mechanism for the generation of diverse antibody specificities is not yet known. Germ-line, somatic-recombination, and somatic-hypermutation hypotheses have been proposed.

SELECTED READINGS

Where to start:

Edelman, G. M., 1970. The structure and function of antibodies. *Sci. Amer.* 223(2):34–42. [Available as *Sci. Amer.* Offprint 1185.]

Porter, R. R., 1973. Structural studies of immunoglobulins. *Science* 180:713–716.

Edelman, G. M., 1973. Antibody structure and molecular immunology. *Science* 180:830–840.

Jerne, N. K., 1973. The immune system. *Sci. Amer.* 229(1):52–60. [Offprint 1276.]

Books:

Eisen, H. M., 1973. Immunology. *In* Davis, B. D., Dulbecco, R., Eisen, H. N., Ginsberg, H. S., and Wood, W. B., Jr., *Microbiology,* pp. 349–624. Harper and Row. [An excellent introduction to molecular and cellular immunology in a comprehensive textbook of microbiology.]

Edelman, G. M., (ed.), 1974. *Cellular Selection and Regulation in the Immune Response.* Raven Press. [A collection of articles on clonal selection, immunogenetics, cell surface of lymphocytes, and regulatory mechanisms.]

Structure of immunoglobulins:

Edelman, G. M., Cunningham, B. A., Gall, W. E., Gottlieb, P. D., Rutishauser, U., and Waxdal, M. J., 1969. The covalent structure of an entire γG immunoglobulin molecule. *Proc. Nat. Acad. Sci.* 63:78–85.

Poljak, R. J., Amzel, L. M., Chen, B. L., Phizackerley, R. P., and Saul, F., 1973. Three-dimensional structure of the $F_{ab'}$ fragment of a human myeloma immunoglobulin at 2.0 Å resolution. *Proc. Nat. Acad. Sci.* 71:3440–3444.

Valentine, R. C., and Green, N. M., 1967. Electron microscopy of an antibody-hapten complex. *J. Mol. Biol.* 27:615–617.

Sarma, V. R., Silverton, E. W., Davies, D. R., and Terry, W. D., 1971. The three-dimensional structure at 6 Å resolution of a human γG1 immunoglobulin molecule. *J. Biol. Chem.* 246:3753–5759.

Synthesis of antibodies:

Haber, E., 1964. Recovery of antigenic specificity after denaturation and complete reduction of disulfides in a papain fragment of antibody. *Proc. Nat. Acad. Sci.* 52:1099–1106. [Provides evidence that antibody specificity is inherent in the amino acid sequence of the immunoglobulin.]

Lennox, E. S., Knopf, P. M., Munro, A. J., and Parkhouse, R. M. E., 1969. A search for biosynthetic subunits of light and heavy chains of immunoglobulins. *Cold Spring Harbor Symp. Quant. Biol.* 32:249–254.

Brownlee, G. G., Cartwright, E. M., Cowan, M. J., Jarvis, J. M., and Milstein, C., 1973. Purification and sequence of messenger RNA for immunoglobulin light chains. *Nature New Biol.* 244:236–239.

Generation of diversity:

Smith, G. P., Hood, L., and Fitch, W. M., 1971. Antibody diversity. *Annu. Rev. Biochem.* 40:969–1012. [A review of some hypotheses for the generation of diversity.]

Honjo, T., Packman, S., Swan, D., Nau, M., and Leder, P., 1974. Organization of immunoglobulin genes: reiteration frequency of the mouse κ chain constant region gene. *Proc. Nat. Acad. Sci.* 71:3659–3663.

Tonegawa, S., Steinberg, C., Dube, S., and Bernardini, A., 1974. Evidence for somatic generation of antibody diversity. *Proc. Nat. Acad. Sci.* 71:4027–4031. [These hybridization experiments suggest that the number of germ-line genes is too small to account for antibody diversity.]

Clonal selection theory:

Burnet, F. M., 1959. *The Clonal Selection Theory of Acquired Immunity.* Cambridge University Press.

Jerne, N. K., 1967. Antibodies and learning: selection versus instruction. *In* Quarton, G. C., Melnechuk, T., and Schmitt, F. O. (eds), *The Neurosciences: A Study Program,* Rockefeller University Press.

CHAPTER 31

BACTERIAL CELL WALLS

Bacterial cells, in contrast to animal cells, are surrounded by a cell wall, which confers mechanical support and gives them their distinctive shape. The cell membrane in a bacterium cannot by itself withstand the high osmotic pressure inside the cell, which may reach 20 atmospheres because of the high internal concentration of metabolites. Bacterial cells lyse in ordinary media unless they are surrounded by a cell wall.

Bacterial cell walls are of considerable medical importance. The cell wall and associated substances are largely responsible for the *virulence of bacteria*. Indeed, the symptoms of numerous bacterial diseases can be produced by injecting isolated bacterial cell walls into experimental animals. Furthermore, the *characteristic antigens* of bacteria are located on their cell walls. Animals acquire immunity to some bacteria following the injection of extracts of bacterial cell walls. Finally, bacteria can be killed by preventing the synthesis of their cell walls. Indeed, this is the mechanism of action of *penicillin* and of some other antibiotics.

For more than a half-century, bacteria have been classified as

Figure 31-1
Electron micrograph of the isolated cell wall of
Bacillus licheniformis. [Courtesy of Dr. Nathan Sharon.]

Figure 31-2
Electron micrograph of the isolated cell wall of
Micrococcus lysodeikticus. [Courtesy of Dr. Nathan
Sharon.]

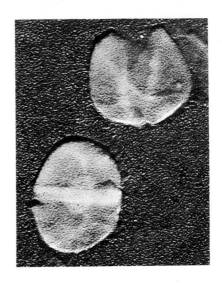

being either gram-negative or gram-positive, depending on their
reaction with the Gram stain. The basis of this empirical distinction
is now understood. These two classes of bacteria have different kinds
of cell walls. *Gram-positive bacteria* are surrounded by a thick cell
wall, typically 500 Å wide, composed of *peptidoglycan* and *teichoic
acid.* The surface of *gram-negative bacteria* is more complex, since it
contains a mosaic of protein, lipid, and lipopolysaccharide beyond
the peptidoglycan layer.

THE CELL WALL IS AN ENORMOUS BAG-SHAPED MACROMOLECULE

Let us consider the structure and biosynthesis of the cell wall of
Staphylococcus aureus, a gram-positive bacterium that causes pus-
producing infections of tissues such as skin, bone, and lung. The
cell-wall macromolecule is called a *peptidoglycan* because it consists
of peptide and sugar units. In the peptidoglycan, *linear polysaccharide
chains are cross-linked by short peptides.* The peptidoglycan is extensively
cross-linked, giving rise to a single enormous bag-shaped macromolecule. The isolated cell wall has the same shape as the bacterium it had formerly enveloped (Figure 31-2).

The peptidoglycan consists of three repeating units (Figure 31-3):
(1) a *disaccharide* of *N*-acetylglucosamine (NAG) and *N*-acetyl-
muramic acid (NAM) in β-1,4-glycosidic linkage; (2) a *tetrapeptide*
of L-alanine, D-glutamine, L-lysine, and D-alanine; and (3) a *pentaglycine bridge peptide.* The tetrapeptide is unusual in two respects.
It contains D-amino acids, which are never found in proteins.
Moreover, the D-glutamine residue forms a peptide linkage via its
side chain γ-carboxyl group.

In the intact peptidoglycan, NAG and NAM alternate in sequence to form a linear polysaccharide chain. The pentaglycine
peptide cross-links NAM residues on different polysaccharide
strands. The amino group of (Gly)₅ forms a peptide bond with the
carboxyl group of D-alanine, whereas the carboxyl group of (Gly)₅
forms a peptide bond with the side-chain ε-amino group of L-lysine.

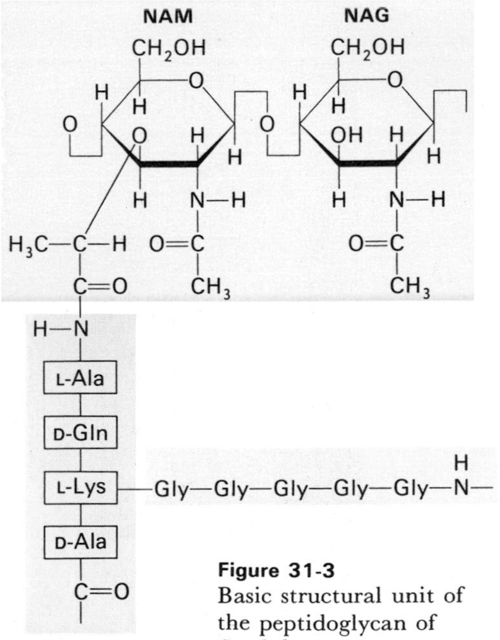

Figure 31-3
Basic structural unit of
the peptidoglycan of
Staphylococcus aureus.

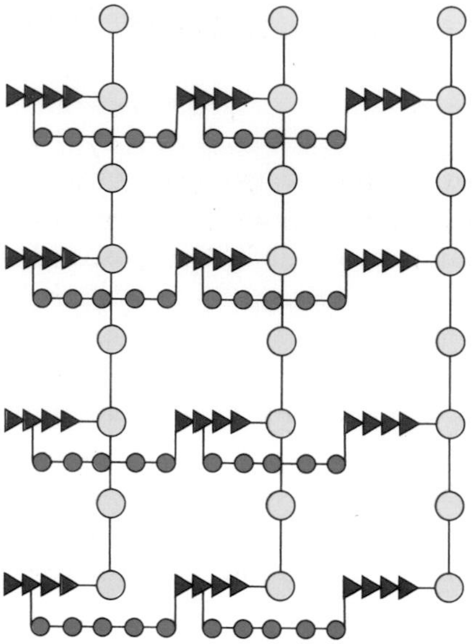

Figure 31-4
Schematic diagram of the peptidoglycan. The sugars are shown in yellow, the tetrapeptides in red, and the pentaglycine bridges in blue. The cell wall is a single enormous bag-shaped macromolecule because of extensive cross-linking.

STAGES IN THE SYNTHESIS OF THE PEPTIDOGLYCAN

There are five stages in the synthesis of the peptidoglycan:

1. *A peptide unit is built on NAM while the sugar is attached to uridine diphosphate.*

2. *The NAM-peptide unit is transferred to a carrier lipid.* A special problem is encountered in cell-wall synthesis. The final polymeric product lies outside the permeability barrier of the cell, whereas the precursors are formed inside the cell. UDP is too polar to get across the membrane. In contrast, the carrier lipid can function as a transmembrane shuttle because it is highly nonpolar.

3. *NAG and the pentaglycine bridge are added to the NAM-peptide unit* while it is attached to the carrier lipid.

4. *The disaccharide peptide unit is transferred from the carrier lipid to a growing polysaccharide chain.*

5. *Different polysaccharide strands are cross-linked by a transpeptidation reaction involving the pentaglycine bridges.*

SYNTHESIS OF A UDP-SUGAR-PEPTIDE UNIT

The biosynthesis of the peptidoglycan starts with the formation of activated sugars. *Uridine diphosphate-N-acetylglucosamine* (UDP-NAG) is synthesized from *N*-acetylglucosamine 1-phosphate and UTP in a reaction that is driven forward by the hydrolysis of PP_i (Figure 31-5). *Uridine diphosphate-N-acetylmuramic acid* (UDP-NAM) is formed from UDP-NAG and phosphoenolpyruvate (Figure 31-6).

Figure 31-5
Synthesis of UDP-NAG.

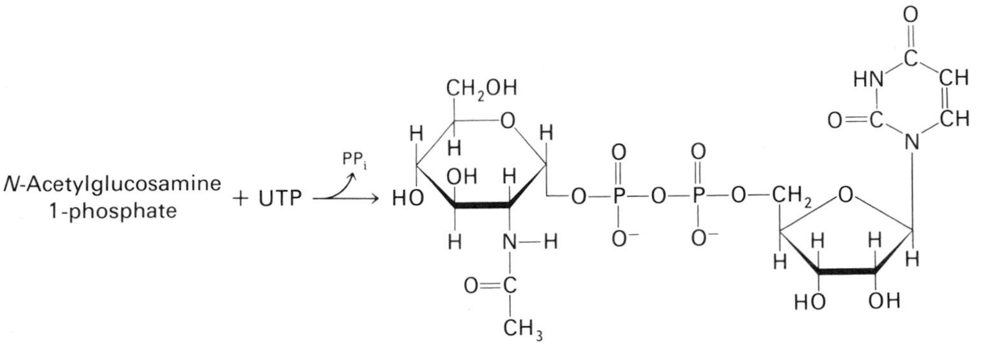

UDP-NAG

Figure 31-6
Synthesis of UDP-NAM.

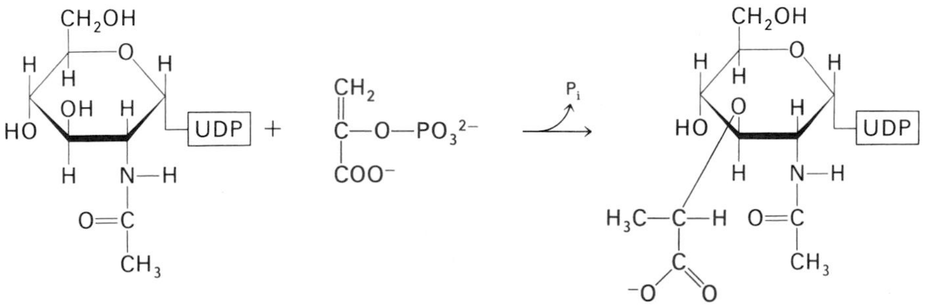

UDP-NAG   Phosphoenolpyruvate   UDP-NAM

Growth of the *peptide chain* starts with the formation of a peptide bond between the amino group of L-alanine and the carboxyl group of the *N*-acetylmuramic acid moiety of UDP-NAM. D-Glutamate, L-lysine, and the dipeptide D-alanyl-D-alanine are successively attached (Figure 31-7). The D-amino acids are formed from their

L-isomer by the action of *racemases* that contain pyridoxal phosphate prosthetic groups. The formation of each of these peptide bonds requires an ATP. *These peptide-bond syntheses are catalyzed by special enzymes, not by the ribosomal pathway for protein synthesis.* The information concerning the amino acid sequence of the peptide is inherent in the specificities of the enzymes rather than in the base sequence of a messenger RNA. A peptide of this kind could not be synthesized by the usual mechanisms of protein synthesis because it contains D-amino acids and a γ-peptide bond.

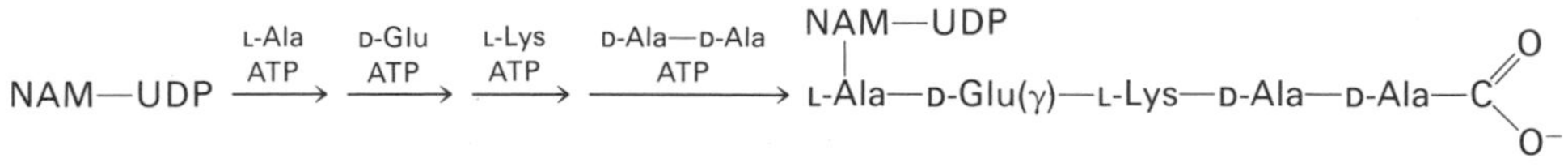

Figure 31-7
Synthesis of a pentapeptide attached to UDP-NAM.

TRANSFER OF THE SUGAR-PEPTIDE UNIT TO A CARRIER LIPID

The activated sugar-peptide unit is then transferred from UDP to a *carrier lipid* (Figure 31-8). This phosphorylated long-chain alcohol contains eleven isoprene units, which make it very hydrophobic, in contrast to UDP. It seems likely that the hydrophobic character of the carrier lipid enables the nascent sugar-peptide unit to traverse the permeability barrier imposed by the cell membrane.

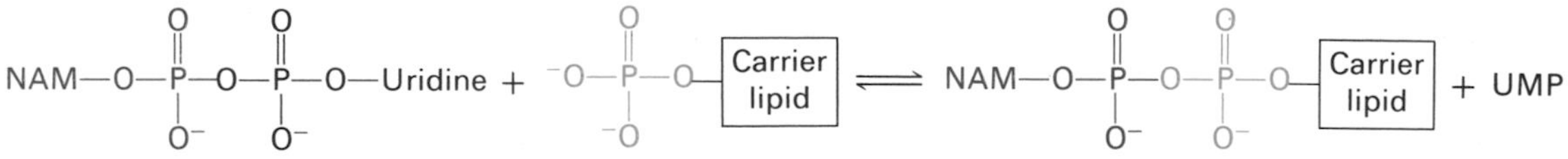

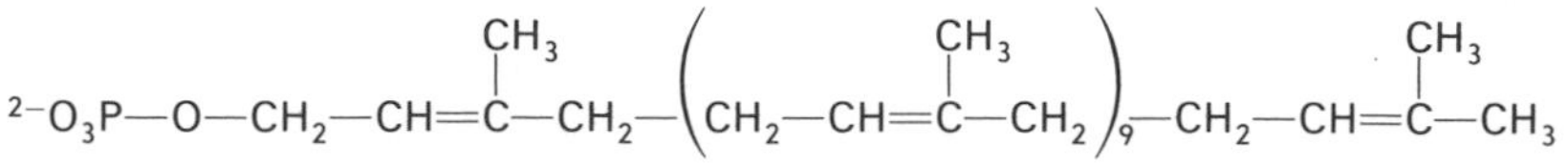

Figure 31-8
Structure of the carrier lipid in cell-wall biosynthesis.

SYNTHESIS OF A DISACCHARIDE-PEPTIDE UNIT ATTACHED TO A CARRIER LIPID

The next step is the addition of NAG to the NAM moiety of the sugar-peptide and carrier-lipid unit. UDP-NAG, the activated sugar donor, reacts with C-4 of the NAM unit to form a β-1,4-glycosidic linkage between these sugars (Figure 31-9). The free α-carboxyl group of D-glutamic acid in the peptide is amidated by NH_4^+ in a reaction that is driven by the hydrolysis of ATP. A *pentaglycine bridge* unit is then added to the ϵ-amino group of the peptide lysine residue. This chain of five glycine residues is formed by the *successive transfer of single glycine residues from glycyl-tRNA.* This is a unique reaction in that tRNA otherwise serves as an amino acid donor only in protein synthesis on ribosomes. The construction of the basic structural unit of the cell wall is now complete.

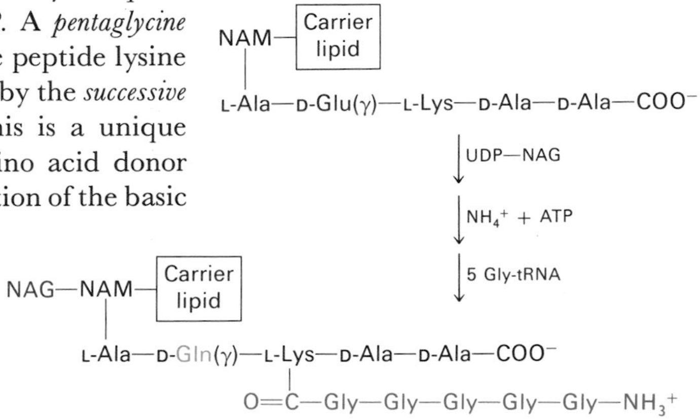

Figure 31-9
Synthesis of a disaccharide-peptide unit attached to the carrier lipid.

TRANSFER OF THE DISACCHARIDE-PEPTIDE UNIT TO A GROWING POLYSACCHARIDE CHAIN

The disaccharide peptide unit is transferred from the carrier lipid to the nonreducing end of a growing polysaccharide chain. The C-1 carbon of the NAM moiety is activated because it is linked to the carrier by a pyrophosphate bond. The C-1 carbon of the activated NAM unit reacts with C-4 of the NAG terminus of a growing polysaccharide chain to form a β-1,4-glycosidic bond (Figure 31-10).

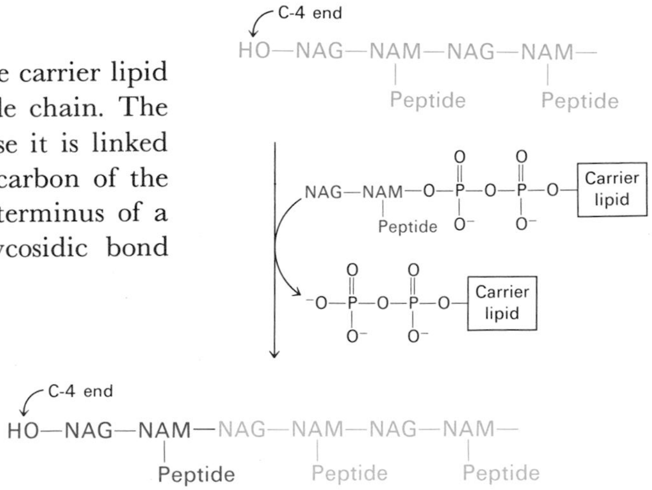

Figure 31-10
Transfer of the disaccharide-peptide unit to a growing polysaccharide chain.

The lipid carrier is released in the pyrophosphate form. It is then hydrolyzed to the monophosphate form by a specific phosphatase. This dephosphorylation step is essential for the regeneration of the species that accepts the sugar-peptide unit from UDP. Bacitracin, a peptide antibiotic, inhibits cell-wall biosynthesis by blocking this step.

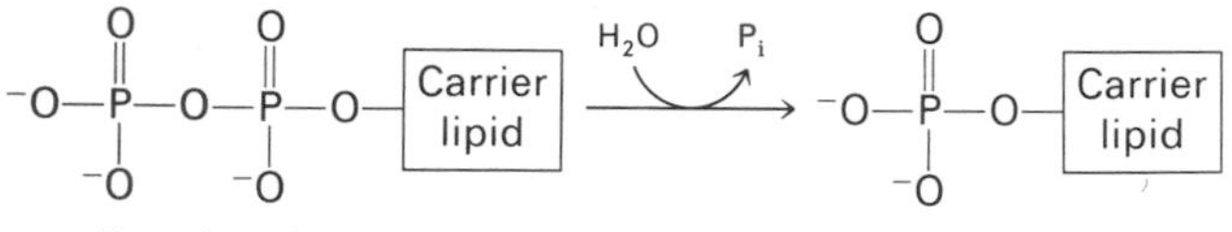

POLYSACCHARIDE CHAINS ARE CROSS-LINKED BY TRANSPEPTIDATION

Different polysaccharide chains are cross-linked by transpeptidation, thereby forming one enormous bag-shaped molecule. *The amino group at the end of one pentaglycine bridge attacks the peptide bond between the D-Ala-D-Ala residues in another peptide unit* (Figure 31-11). A peptide bond is formed between glycine and one of the D-alanine residues, whereas the other D-alanine residue is released. This reaction is catalyzed by *glycopeptide transpeptidase*. Note that ATP is not consumed in the synthesis of this cross-link. This transpeptidation reaction uses the free energy already present in the D-Ala–D-Ala bond. This unusual means of forming a peptide bond may be necessary because the reaction occurs outside the cell membrane, beyond the reach of ATP. Recall that a peptide bond is formed in the cross-linking of fibrin by a transamidation reaction, which does not use ATP (p. 196).

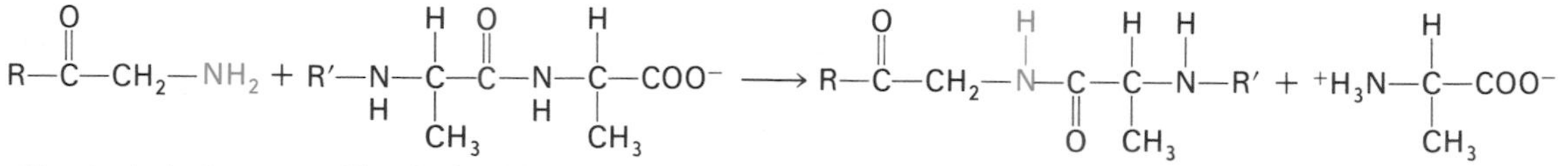

Figure 31-11
The amino group of the pentaglycine bridge attacks the peptide bond between two D-Ala residues to form a cross-link.

TEICHOIC ACID COVERS THE PEPTIDOGLYCAN
IN GRAM-POSITIVE BACTERIA

The surface of gram-positive bacteria consists of teichoic acid, which is a polymer of glycerol (or another sugar such as ribitol) linked by phosphodiester bridges (Figure 31-12). The available hydroxyl groups are esterified to alanine or to sugars such as glucose. Teichoic acid is attached to the NAG-NAM backbone of the peptidoglycan by a phosphodiester bond. The teichoic acid chains are elongated by the transfer of glycerol phosphate from CDP-glycerol to the free terminal hydroxyl group of the chain. Glucose is added to the hydroxyl groups of the teichoic acid backbone by reaction with UDP-glucose.

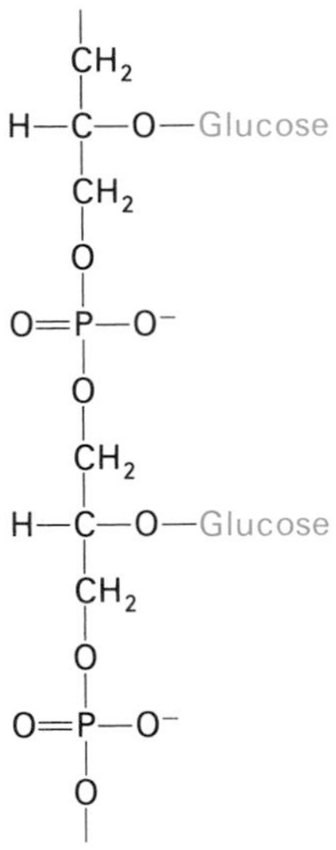

Figure 31-12
Structure of teichoic acid.

PENICILLIN KILLS GROWING BACTERIA BY
INHIBITING CELL-WALL SYNTHESIS

Penicillin was discovered by Alexander Fleming in 1928, the result of a chance observation:

> While working with staphylococcus variants a number of culture-plates were set aside on the laboratory bench and examined from time to time. In the examinations these plates were necessarily exposed to the air and they became contaminated with various micro-organisms. It was noticed that around a large colony of a contaminating mould the staphylococcus colonies became transparent and were obviously undergoing lysis.
>
> Subcultures of this mould were made and experiments conducted with a view to ascertain something of the properties of the bacteriolytic substance which had evidently been formed in the mould culture and which had diffused into the surrounding medium. It was found that broth in which the mould have been grown at room temperature for one or two weeks had acquired marked inhibitory, bactericidal and bacteriolytic properties to many of the more common pathogenic bacteria.

Fleming was further encouraged by the finding that the extract from the *Penicillium* mold had no apparent toxicity when injected into animals. However, he tried to concentrate and purify the active antibiotic but found that "penicillin is easily destroyed, and to all intents and purposes we failed. We were bacteriologists—not chemists—and our relatively simple procedures were unavailing."

Ten years passed before penicillin was studied again. Howard Florey, a pathologist, and Ernst Chain, a biochemist, carried out a series of incisive studies that led to the isolation, chemical characterization, and clinical use of this antibiotic. Penicillin consists of a thiazolidine ring fused to a *β-lactam ring,* to which a variable group (R) is attached by a peptide bond. In *benzyl penicillin,* for example, R is a benzyl group (Figures 31-13 and 31-14). This structure can undergo a variety of rearrangements, thus accounting for the instability first encountered by Fleming. In particular, the β-lactam ring is very labile. Indeed, this property is closely tied to the antibiotic activity of penicillin, as will be discussed shortly.

In 1957, Joshua Lederberg showed that bacteria ordinarily susceptible to penicillin could be grown in its presence if a hypertonic medium were used. The organisms obtained in this way, called *protoplasts,* are devoid of a cell wall, and consequently lyse when transferred to a normal medium. Thus, it was inferred that penicillin interfered with bacterial cell-wall synthesis. In 1965, James Park and Jack Strominger independently deduced that penicillin blocks the last step in bacterial cell-wall biosynthesis, namely the cross-linking reaction.

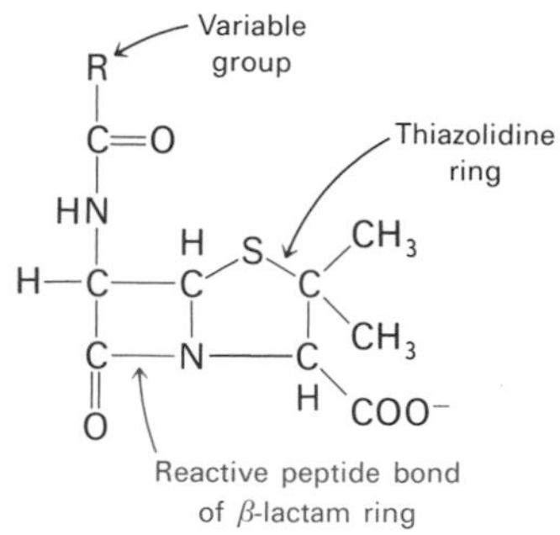

Figure 31-13
The reactive site of penicillin is the peptide bond of its β-lactam ring. The R group is variable. In benzyl penicillin, R is a benzyl group.

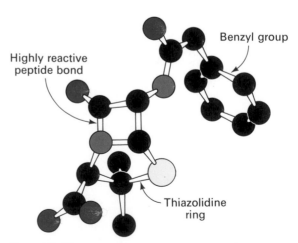

Figure 31-14
Model of benzyl penicillin.

PENICILLIN BLOCKS CELL-WALL SYNTHESIS BY INHIBITING TRANSPEPTIDATION

Penicillin inhibits the transpeptidase that cross-links different peptidoglycan strands.

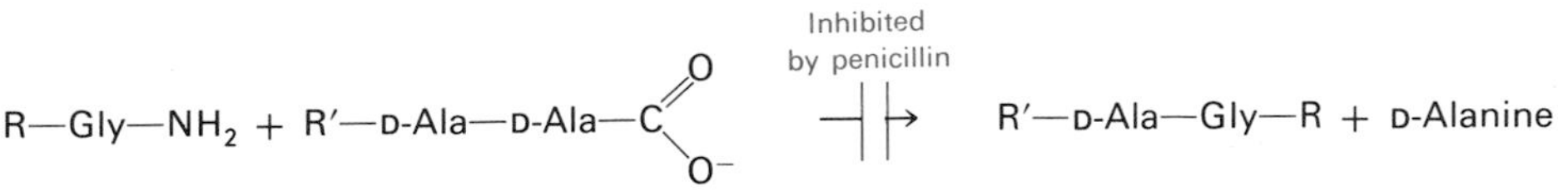

R—Gly—NH$_2$ + R′—D-Ala—D-Ala—C $\begin{matrix} O \\ \diagdown \\ O^- \end{matrix}$ $\xrightarrow[\text{by penicillin}]{\text{Inhibited}}$ R′—D-Ala—Gly—R + D-Alanine

The transpeptidase normally forms an *acyl intermediate* with the penultimate D-alanine residue of the peptide (Figure 31-15). This acyl-enzyme intermediate then reacts with the amino group of the terminal glycine in another peptide. It seems likely that penicillin inhibits the transpeptidase by forming a covalent bond with a residue at the active site. In fact, penicillin forms an acyl-enzyme intermediate (Figure 31-16). *This penicillinoyl-enzyme complex does not undergo deacylation. Hence, the transpeptidase is irreversibly inhibited by penicillin.*

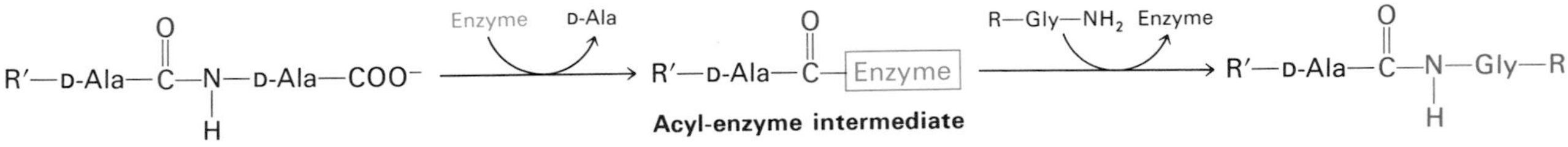

R′—D-Ala—C—N—D-Ala—COO$^-$ $\xrightarrow[\text{}]{\text{Enzyme} \quad \text{D-Ala}}$ R′—D-Ala—C—Enzyme $\xrightarrow[\text{}]{\text{R—Gly—NH}_2 \text{ Enzyme}}$ R′—D-Ala—C—N—Gly—R

Acyl-enzyme intermediate

Figure 31-15
An acyl-enzyme intermediate is formed in the transpeptidation reaction.

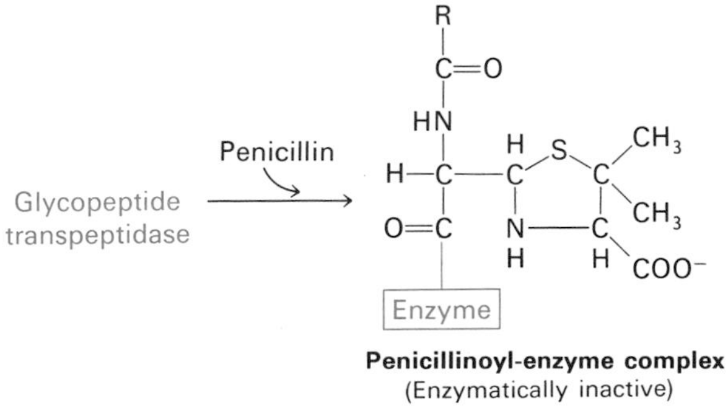

Glycopeptide transpeptidase $\xrightarrow{\text{Penicillin}}$

Figure 31-16
Formation of a penicillinoyl-enzyme complex, which is indefinitely stable.

Penicillinoyl-enzyme complex
(Enzymatically inactive)

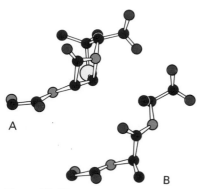

Figure 31-17
The conformation of penicillin in the vicinity of its reactive peptide bond (A) resembles the postulated conformation of the transition state of R—D-Ala—D-Ala (B) in the transpeptidation reaction. [Based on B. Lee. *J. Mol. Biol.* 61(1971):464.]

Why is penicillin such an effective inhibitor of the transpeptidase? Molecular models have revealed that *penicillin resembles acyl-D-Ala-D-Ala, one of the substrates of this enzyme* (Figure 31-17). Moreover, the four-membered β-lactam ring of penicillin is strained, making this peptide bond highly reactive. Indeed, penicillin probably has the structure of the transition state of the normal substrate. In other words, penicillin is a *transition-state analog.*

It is of interest to compare penicillin with other substances that irreversibly inhibit enzymes. The stable, inactive *penicillinoyl-enzyme complex* is analogous to the *phosphoryl-enzyme complex* formed by the reaction of organic fluorophosphates with acetylcholinesterase or the serine proteases. Penicillin is far more specific, however, than the fluorophosphate inhibitors. The reactivity of penicillin is precisely targeted because of its structural resemblance to the D-Ala-D-Ala terminus of nascent strands of peptidoglycan. The low toxicity of penicillin, which is essential for its therapeutic effectiveness, is a consequence of its exquisite specificity. There is no known human enzyme that recognizes D-Ala-D-Ala, and so penicillin does not interfere with our own enzymatic machinery.

SOME BACTERIA ARE RESISTANT TO PENICILLIN BECAUSE THEY SYNTHESIZE A PENICILLIN-DESTROYING ENZYME

Some bacteria secrete *penicillinase,* an enzyme that cleaves the amide bond in the β-lactam ring of penicillin to form penicilloic acid, which is inactive as an antibiotic:

Penicillin → (H₂O, Penicillinase) → **Penicilloic acid**

A variety of related penicillinases having a molecular weight of about 30,000, and a high turnover number (typically 10^3 sec^{-1}) have been isolated. The activity of the enzyme depends markedly on the nature of the R group attached to the β-lactam ring of penicillin. Semisynthetic penicillins having an R group that renders them not susceptible to penicillinase are clinically important.

The gene for penicillinase is located on an extrachromosomal genetic element, which is readily lost or acquired by some bacterial species. The enzyme is inducible in some bacterial strains. *Penicillinase probably evolved as a detoxification mechanism,* since it is only found in organisms that possess a peptidoglycan wall. The enzyme probably does not have a function other than to destroy penicillin, since loss of the gene has no deleterious effect in the absence of the antibiotic. Furthermore, there is a good correlation between the degree of resistance to penicillin and the amount of penicillinase activity.

SUMMARY

Bacterial cells are surrounded by a cell wall that protects them from osmotic lysis. The cell wall in gram-positive bacteria is about 500 Å thick, consisting of peptidoglycan and teichoic acid, whereas the cell surface in gram-negative bacteria is more complex. The cell wall is one enormous bag-shaped macromolecule. The peptidoglycan in *S. aureus* consists of three repeating units: NAG-NAM, a tetrapeptide of D and L residues, and a pentaglycine chain. The pentaglycine unit cross-links different polysaccharide strands. UDP-NAG and UDP-NAM are the activated sugars in the biosynthesis of the peptidoglycan, which takes place in five stages. First, a peptide unit is built on NAM while it is attached to UDP. Second, the NAM-peptide is transferred to a carrier lipid. Third, NAG and the pentaglycine bridge are added to the NAM-peptide unit. Fourth, the NAG-NAM-peptide unit is transferred from the carrier lipid to a growing polysaccharide chain. Fifth, the amino group of a pentaglycine bridge on one strand attacks a terminal D-Ala–D-Ala peptide bond of another strand to form a cross-link.

Penicillin kills bacteria by inhibiting the synthesis of cell walls. Penicillin resembles acyl-D-Ala-D-Ala and contains a highly reactive peptide bond. It irreversibly inhibits glycopeptide transpeptidase, and thereby blocks the cross-linking of the peptidoglycan. Some bacteria are resistant to penicillin because they contain penicillinase, which hydrolyzes penicillin to an inactive form.

SELECTED READINGS

Where to start:

Sharon, N., 1969. The bacterial cell wall. *Sci. Amer.* 220(5):92–98.

Books and reviews:

Leive, L., (ed.), 1973. *Bacterial Membranes and Walls.* Marcel Dekker. [An excellent collection of critical reviews. The article by M. M. Ghuysen and G. D. Shockman on the biosynthesis of peptidoglycan and the one by H. Nikaido on the biosynthesis of the cell envelope of gram-negative organisms are most pertinent to this chapter.]

Salton, M. R. J., 1964. *The Bacterial Cell Wall.* American Elsevier.

Osborn, M. J., 1969. Structure and biosynthesis of the bacterial cell wall. *Annu. Rev. Biochem.* 38:501–538.

Osborn, M. J., 1971. The role of membranes in the synthesis of macromolecules. *In* Rothfield, L. I., (ed.), *Structure and Function of Biological Membranes,* pp. 343–400. Academic Press.

Penicillin and bacitracin:

Fleming, A., 1929. On the antibacterial action of cultures of a penicillinium, with special reference to their use in the isolation of *B. influenzae. Brit. J. Exp. Pathol.* 10:226–236.

Chain, E., 1971. Thirty years of penicillin therapy. *Proc. Roy. Soc. London* (B) 1799:293–319.

Strominger, J. L., Blumberg, P. M., Suginaka, H., Umbreit, J., and Wickus, G. G., 1971. How penicillin kills bacteria: progress and problems. *Proc. Roy. Soc. London* (B) 179:369–383.

Strominger, J. L., Izaki, Matshukashi, M., and Tipper, D. J., 1967. Peptidoglycan transpeptidase and D-alanine carboxypeptidase: penicillin-sensitive enzymatic reactions. *Fed Proc.* 26:9–22.

Pollock, M. R., 1971. The function and evolution of penicillinase. *Proc. Roy. Soc. London* (B) 179:385–401.

Stone, K. J., and Strominger, J. L., 1971. Mechanism of action of bacitracin: complexation with metal ion and C_{55}-isoprenyl pyrophosphate. *Proc. Nat. Acad. Sci.* 68:3223–3227.

CHAPTER 32

MEMBRANE TRANSPORT

Biological membranes are highly selective permeability barriers. The flow of molecules and ions between a cell and its environment is precisely regulated by specific transport systems. Transport processes have several important roles:

1. They *regulate cell volume* and *maintain the intracellular pH and ionic composition within a narrow* range to provide a favorable environment for enzyme activity.

2. They extract and *concentrate metabolic fuels and building blocks* from the environment, and *extrude toxic substances*.

3. They generate ionic gradients that are essential for the *excitability* of nerve and muscle.

The molecular basis of several transport processes is now being elucidated. In this chapter, several bacterial and animal transport systems that carry ions, sugars, and amino acids across biological membranes are considered. The structure and ion-binding properties of some transport antibiotics are also presented because they provide insight into transport mechanisms in general.

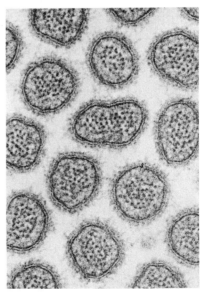

Figure 32-1
Electron micrograph of a cross section of intestinal microvilli, which greatly increase the surface area for transport. [Courtesy of Dr. George Palade.]

DISCOVERY OF SPECIFIC MEMBRANE TRANSPORT SYSTEMS

Genetic and biochemical studies of *E. coli* have shown that membrane transport systems are quite specific. For example, there is a mutant of *E. coli* that grows well on maltose but not on glucose. Since maltose is hydrolyzed to glucose inside the bacterium, it was evident that the defect in this mutant must be in the transport of glucose rather than in its utilization. Furthermore, the transport lesion was specific since maltose traversed the membrane normally. It was deduced that a specific membrane transport protein, called a *permease* or *carrier*, was missing in this mutant. A permease for the transport of lactose and other β-galactosides was subsequently isolated by Eugene Kennedy. This β-galactoside permease is a membrane-bound protein (now termed the M protein) having a subunit molecular weight of about 30,000. It is encoded by the *y* gene of the *lac* operon (p. 682).

FEATURES OF THE CARRIER-MEDIATED TRANSPORT OF MOLECULES

The transport of molecules across membranes by carrier proteins resembles enzyme-substrate interactions in several ways:

1. The transport is *stereospecific*.

2. The transport rate reaches a limiting value at high concentrations of the species transported. In other words, the transport process is *saturable*. This occurs when all the carriers are loaded, just as an enzymatic rate reaches a maximal value at high substrate concentration. Transport kinetics often fit the Michaelis-Menten equation and can then be characterized by V_{max} and K_M. The range of K_M values for transport is like that for enzymatic catalysis.

3. Transport of a species can often be *inhibited by structural analogs* since they compete for the same binding site on the carrier protein.

DISTINCTION BETWEEN PASSIVE AND ACTIVE TRANSPORT

Transport processes are either passive or active, depending on the change in free energy of the transported species. Consider an uncharged solute molecule. The free-energy change in transporting this species from side 1 where it is present at a concentration of c_1 to side 2 where it is present at concentration c_2 is

$$\Delta G = RT \log_e \frac{c_2}{c_1} = 2.303 \; RT \log_{10} \frac{c_2}{c_1}$$

For a charged species, the electrical potential across the membrane must also be considered. The sum of the concentration and electrical terms is called the *electrochemical potential*. The free-energy change is then given by

$$\Delta G = RT \log_e \frac{c_2}{c_1} + ZF \; \Delta V$$

where Z is the electrical charge of the transported species, ΔV is the potential in volts across the membrane, and F is the faraday (23.062 kcal volt^{-1} mol^{-1}).

A transport process must be active when ΔG is positive, whereas it can be passive when ΔG is negative. *Active transport requires a coupled input of free energy, whereas passive transport can occur spontaneously.* For example, consider the transport of an uncharged molecule from $C_1 = 10^{-3}$ M to $C_2 = 10^{-1}$ M.

$$\Delta G = 2.3 \; RT \log_{10} \frac{10^{-1}}{10^{-3}}$$
$$= 2.3 \times 1.98 \times 298 \times 2$$
$$= +2.7 \text{ kcal/mol}$$

At $25°C$ (298 K), ΔG is $+2.7$ kcal/mol, which indicates that this transport process is active, and thus requires an input of free energy. This transport process could be driven, for example, by the hydrolysis of ATP, which yields -7.3 kcal/mol under standard conditions.

DISCOVERY OF THE ACTIVE TRANSPORT SYSTEM FOR SODIUM AND POTASSIUM IONS

Most animal cells have a high concentration of K^+ and a low concentration of Na^+ relative to the external medium. These ionic gradients are generated by a specific transport system that is called the *Na$^+$-K$^+$ pump* because the movement of these ions is linked. The active transport of Na$^+$ and K$^+$ is of great physiological significance. Indeed, more than a third of the ATP consumed by a resting animal is used to pump these ions. The Na$^+$-K$^+$ gradient in animal cells controls cell volume, renders nerve and muscle electrically excitable, and drives the active transport of sugars and amino acids (p. 777).

In 1957, Jens Skou discovered an enzyme that hydrolyzes ATP only if Na$^+$ and K$^+$ are present in addition to Mg^{2+}. This enzyme is called a *Na$^+$-K$^+$ ATPase.*

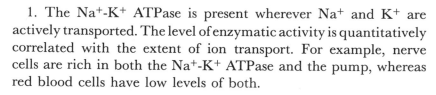

$$ATP + H_2O \xrightarrow{\text{Na}^+, \text{ K}^+, \text{ Mg}^{2+}} ADP + P_i + H^+$$

Skou surmised that the Na$^+$-K$^+$ ATPase might be an integral part of the Na$^+$-K$^+$ pump and that the splitting of ATP could provide the energy needed for the active transport of Na$^+$ and K$^+$. Several lines of evidence have shown that the *Na$^+$-K$^+$ ATPase is in fact part of the Na$^+$-K$^+$ pump:*

1. The Na$^+$-K$^+$ ATPase is present wherever Na$^+$ and K$^+$ are actively transported. The level of enzymatic activity is quantitatively correlated with the extent of ion transport. For example, nerve cells are rich in both the Na$^+$-K$^+$ ATPase and the pump, whereas red blood cells have low levels of both.

2. Both the Na$^+$-K$^+$ ATPase and the pump are tightly associated with the plasma membrane.

3. Both the Na$^+$-K$^+$ ATPase and the pump are oriented in the same way in the plasma membrane.

4. Variations of the concentrations of Na$^+$ and K$^+$ have parallel effect on the ATPase activity and on the rate of transport of these ions.

5. Both the Na$^+$-K$^+$ ATPase and the pump are specifically inhibited by cardiotonic steroids. The concentration of inhibitor that causes half-maximal inhibition is the same for both processes.

6. The pump can be reversed under suitable ionic conditions so that ATP is synthesized from ADP and P$_i$.

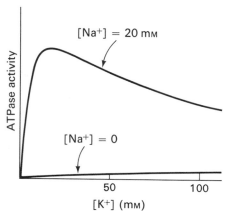

Figure 32-2
The Na$^+$-K$^+$ ATPase, a component of the Na$^+$-K$^+$ pump, hydrolyzes ATP only if both Na$^+$ and K$^+$ are present, in addition to Mg^{2+}.

BOTH THE ENZYME AND THE PUMP ARE ORIENTED IN THE MEMBRANE

Studies of the Na^+-K^+ pump in erythrocyte ghosts have revealed the orientation of this ATPase and pump. In a hypotonic salt solution, an erythrocyte swells and holes appear in its membrane. Hemoglobin diffuses out, leaving a pale cell (hence, a ghost). The inside of the swollen erythrocyte can be equilibrated with the external medium. If the external medium is made isotonic, the membrane again becomes a permeability barrier. Thus, the molecular and ionic composition inside a ghost can be controlled by resealing it in an appropriate medium. Transport and enzymatic studies of such erythrocyte ghosts have shown that the Na^+-K^+ pump is oriented in the following way (Figure 32-3):

1. The Na^+ ion must be inside, whereas K^+ must be outside to activate the ATPase and to be transported across the membrane.

2. ATP is an effective substrate for the ATPase and the pump only when it is located inside the cell.

3. Cardiotonic steroids inhibit the pump and the ATPase only when they are located outside the cell.

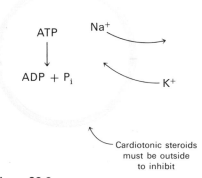

Figure 32-3
The Na^+-K^+ pump is oriented in the plasma membrane.

ATP TRANSIENTLY PHOSPHORYLATES THE SODIUM-POTASSIUM PUMP

How does ATP drive the active transport of Na^+ and K^+? An important clue was provided by the finding that the Na^+-K^+ ATPase is phosphorylated by ATP in the presence of Na^+ and Mg^{2+}. This phosphorylated intermediate (E-P) is then hydrolyzed if K^+ is present. The phosphorylation reaction does not require K^+, whereas the dephosphorylation reaction does not require Na^+ or Mg^{2+}.

$$E + ATP \xrightleftharpoons{Na^+, Mg^{2+}} E{-}P + ADP$$

$$E{-}P + H_2O \xrightarrow{K^+} E + P_i$$

A phosphorylated enzyme intermediate has been following incubation of the enzyme with ATP, Na$^+$, and Mg^{2+} but without K$^+$. The phosphoryl group in this intermediate is attached to the carboxyl side chain of a glutamate residue of the enzyme.

$$\begin{array}{c} H-N \\ | \\ H-C-CH_2-CH_2-C-O-P-O^- \\ | \quad\quad\quad\quad\quad \| \quad\quad \| \\ O=C \quad\quad\quad\quad O \quad\quad O^- \\ | \quad\quad\quad\quad\quad\quad\quad\quad O^- \end{array}$$

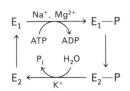

Figure 32-4
Conformational cycle of the Na$^+$-K$^+$ ATPase.

The Na$^+$-dependent phosphorylation and K$^+$-dependent dephosphorylation are not the only critical reactions of the enzyme. The ATPase also interconverts between at least two different conformations (denoted by E$_1$ and E$_2$). There are at least four steps in the hydrolysis of ATP and concomitant transport of K$^+$ and Na$^+$ by the pump (Figure 32-4).

ION TRANSPORT AND ATP HYDROLYSIS ARE TIGHTLY COUPLED

The turnover number of the ATPase is 100 sec^{-1}. The stoichiometry of the pump is *3 Na$^+$ and 2 K$^+$ transported per ATP hydrolyzed.* An important feature of the pump is that *ATP is not hydrolyzed unless Na$^+$ and K$^+$ are transported.* In other words, the system is coupled so that the energy stored in ATP is not dissipated. Tight coupling is a general characteristic of biological assemblies that mediate energy conversion. Recall that electrons do not normally flow through the mitochondrial electron-transport chain unless ATP is concomitantly generated (p. 345). The obligatory coupling of ATP hydrolysis and muscle contraction is another example of this principle.

The Na$^+$-K$^+$ pump can be reversed so that it synthesizes ATP. Net synthesis of ATP from ADP and P$_i$ occurs when the ionic gradients are very steep. This is achieved by incubating red cells in a medium containing a much higher than normal concentration of Na$^+$ and a much lower than normal concentration of K$^+$.

A MODEL FOR THE MECHANISM OF THE SODIUM-POTASSIUM PUMP

How does the phosphorylation and dephosphorylation of the ATPase result in the transport of Na^+ and K^+ across the membrane? The Na^+-K^+ ATPase has only recently been purified. It consists of a catalytic subunit (100,000 daltons) and an associated glycoprotein (50,000 daltons). Little is yet known about the conformation of this transport system. Nevertheless, it is stimulating to consider a simple model for a pump that was proposed by Oleg Jardetzky. In this model, a protein must fulfill three structural conditions to function as a pump:

1. It must contain a cavity large enough to admit a small molecule or ion.

2. It must be able to assume two conformations, such that the cavity is open to the inside in one form and to the outside in the other.

3. The affinity for the transported species must be different in the two conformations.

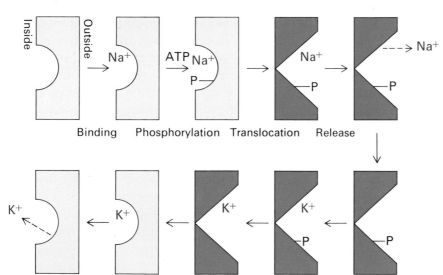

Binding   Phosphorylation Translocation   Release

Release   Translocation Dephosphorylation Binding

Figure 32-5
Schematic diagram of a proposed mechanism for the Na^+-K^+ pump. The upper sequence of reactions depicts the extrusion of Na^+, whereas the lower reactions show the entry of K^+. The E_1 (yellow) and E_2 (blue) forms are shown here as having very different conformations. The actual conformational change may be quite small.

Let us apply this model to Na^+ and K^+ transport (Figure 32-5). The two conformations are the E_1 and E_2 forms described above. It is assumed that (1) the ion-binding cavity in E_1 faces the inside of the cell, whereas in E_2 the cavity faces the outside, and that (2) E_1 has a high affinity for Na^+, whereas E_2 has a high affinity for K^+. The model also makes use of two established facts: (1) Na^+ triggers phosphorylation, whereas K^+ triggers dephosphorylation; and (2) E_2 is stabilized by phosphorylation, and E_1 is stabilized by dephosphorylation. In Figure 32-5, E_1 and E_2 have very different shapes. However, it must be stressed that the conformational differences between these forms need not be large. A shift of a few atoms by a distance of 2 Å might suffice to alter the relative affinity of the cavity for Na^+ and K^+ and change its orientation. There is ample precedent for assuming that phosphorylation can readily produce changes of this magnitude. Recall the effect of phosphorylation on the properties of glycogen phosphorylase and synthetase and also the change in oxygen affinity elicited by the noncovalent binding of diphosphoglycerate to hemoglobin.

CARDIOTONIC STEROIDS ARE SPECIFIC INHIBITORS OF THE Na^+-K^+ ATPase AND PUMP

Certain steroids derived from plants are potent inhibitors of the Na^+-K^+ ATPase and pump. The inhibition of both processes is half-maximal at an inhibitor concentration of the order of 10^{-8} M. Digitoxigenin and ouabain are members of this class of inhibitors, which are known as *cardiotonic steroids* because of their profound effects on the heart (Figure 32-6). The activity of the cardiotonic steroids depends on the presence of a five- or six-membered unsaturated lactone ring in the β configuration at C-17. Also, a hydroxyl group at C-14 and a *cis* fusion of rings C and D are essential. Ouabain and some other cardiotonic steroids contain a sugar residue attached to C-3, but this sugar does not contribute to the inhibition of the ATPase by these compounds.

Cardiotonic steroids inhibit the dephosphorylation reaction of the Na^+-K^+ ATPase.

$$E-P + H_2O \xrightarrow{\quad K^+ \quad} E + P_i$$

Inhibited by
cardiotonic steroids

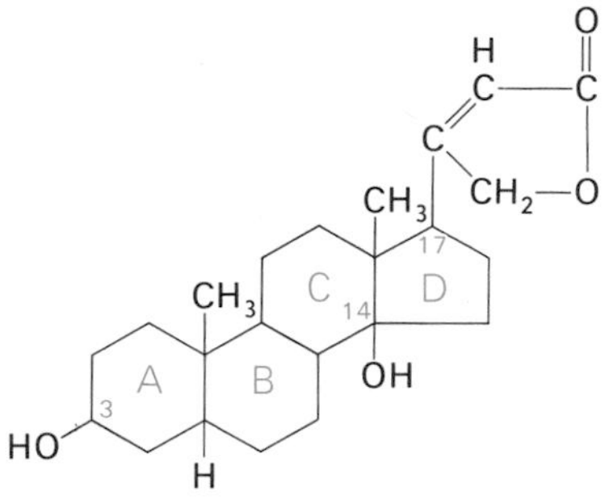

Digitoxigenin

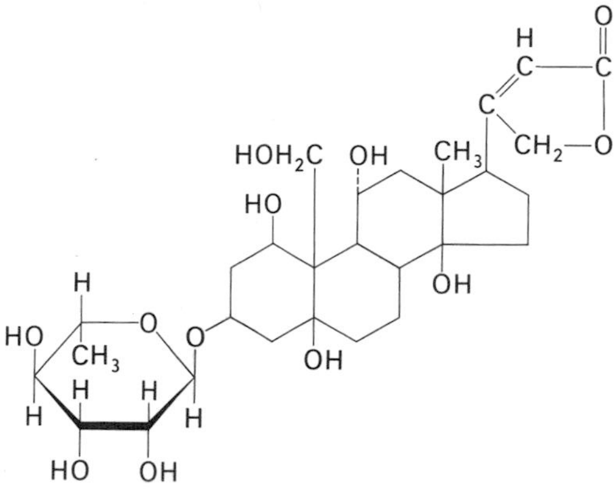

Figure 32-6
Cardiotonic steroids such as digitoxigenin and ouabain inhibit the Na^+-K^+ pump.

Ouabain

Inhibition occurs only if the cardiotonic steroid is located on the *outside* face of the membrane. Thus, inhibition of dephosphorylation by these steroids has the same spatial asymmetry as the activation of dephosphorylation by K^+.

Cardiotonic steroids such as *digitalis* are of great clinical significance. Digitalis increases the force of contraction of heart muscle, which makes it a choice drug in the treatment of congestive heart failure. The molecular basis of this action of digitalis has not been fully elucidated, but there is indirect evidence to support the notion that inhibition of a Na^+-K^+ pump is central to its action. It is interesting to note that digitalis was effectively used long before the discovery of the Na^+-K^+ ATPase. In 1785, William Withering, a physician and botanist, published "An Account of the Foxglove and Some of its Medicinal Uses." He described how he first learned about the use of digitalis to cure congestive heart failure:

> In the year 1775, my opinion was asked concerning a family receipt for the cure of the dropsy. I was told that it had long been kept a secret by an old woman in Shropshire, who had sometimes made cures after the more regular practitioners had failed. . . . This medicine was composed of 20 or more different herbs; but it was not very difficult for one conversant in these subjects to perceive that the active herb could be no other than the foxglove. . . . It has a power over the motion of the heart to a degree yet unobserved in any other medicine, and this power may be converted to salutary ends.

Figure 32-7
Photograph of a foxglove plant. [From A. Krochmal and C. Krochmal. *A Guide to the Medicinal Plants of the United States* (Quadrangle Books, 1973), p. 243. Courtesy of Dr. James Hardin.]

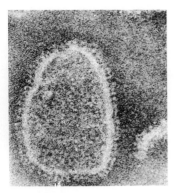

Figure 32-8
Membrane vesicles formed from purified Ca^{2+} ATPase. The globular particles on the surface of the membrane are part of the ATPase molecule, which may extend across the membrane. [From P. S. Stewart and D. H. MacLennan. *J. Biol. Chem.* 249(1974):987.]

CALCIUM IS TRANSPORTED BY A DIFFERENT ATPase

Calcium ion plays an important role in the regulation of muscle contraction (p. 836). Skeletal muscle contains an intricate network of membrane-bound tubules and vesicles. This membrane system, called the *sarcoplasmic reticulum,* regulates the Ca^{2+} concentration surrounding the contractile fibers of the muscle. At rest, Ca^{2+} is pumped into the sarcoplasmic reticulum so that the Ca^{2+} concentration around the muscle fibers is very low. Excitation of the sarcoplasmic reticulum membrane by a nerve impulse leads to a sudden release of large amounts of Ca^{2+}, which triggers muscle contraction. In other words, *Ca^{2+} is the intermediary between the nerve impulse and muscle contraction.*

The transport of Ca^{2+} by the sarcoplasmic reticulum is driven by the hydrolysis of ATP. There is an ATPase in the sarcoplasmic reticulum that is activated by Ca^{2+}. This Ca^{2+} ATPase is an integral part of the Ca^{2+} pump, just as the Na^+-K^+ ATPase is part of the Na^+-K^+ pump. The Ca^{2+} ATPase is also transiently phosphorylated by ATP.

$$E + ATP \xrightleftharpoons{Ca^{2+},\ Mg^{2+}} E{-}P + ADP$$

$$E{-}P + H_2O \longrightarrow E + P_i$$

The very high affinity of this ATPase for Ca^{2+} ($K_M \sim 10^{-7}$ M) enables it to effectively transport Ca^{2+} from the cytosol (where $[Ca^{2+}] < 10^{-5}$ M) into the sarcoplasmic reticulum (where $[Ca^{2+}] \sim 10^{-2}$ M). Two calcium ions are transported for each ATP hydrolyzed. The maximal pump rate is about $10\ sec^{-1}$. The Ca^{2+} ATPase is a large protein, about 100,000 daltons, that is tightly associated with the sarcoplasmic reticulum membrane. The density of Ca^{2+} pumps in this membrane is very high, constituting about a third of the surface area.

A functional Ca^{2+} pump has been reconstituted from purified Ca^{2+} ATPase and phospholipids. The Ca^{2+} ATPase was isolated from sarcoplasmic reticulum membranes following solubilization by cholate, a detergent. This solubilized enzyme was added to phospholipids obtained from soybeans. Membrane vesicles formed when the detergent was removed by dialysis. *These reconstituted vesicles pumped Ca^{2+} at a rapid rate when provided with ATP and Mg^{2+}.*

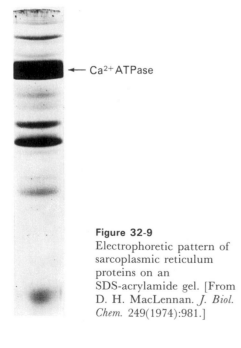

←— Ca^{2+} ATPase

Figure 32-9
Electrophoretic pattern of sarcoplasmic reticulum proteins on an SDS-acrylamide gel. [From D. H. MacLennan. *J. Biol. Chem.* 249(1974):981.]

SOME TRANSPORT PROCESSES ARE DRIVEN BY IONIC GRADIENTS

Some transport processes are not directly driven by the hydrolysis of ATP. For example, the active transport of some sugars and amino acids into animal cells depends on a Na^+ concentration gradient across the plasma membrane. Glucose can be pumped in if the concentration of Na^+ is higher outside than inside. Sodium ion and glucose bind to a specific transport protein and enter together. Such a concerted movement of two species is called *cotransport*. The rate and extent of glucose transport depends on the Na^+ concentration gradient across the membrane (Figure 32-10). The Na^+ that enters the cell with glucose is then extruded by the Na^+-K^+ pump, which uses ATP as its energy source, as described above. Thus, *one membrane transport system can drive another.*

Studies of active transport by vesicles derived from bacterial membranes have revealed a new mechanism for energy coupling. These vesicles are attractive for transport studies because they are much simpler in composition than whole bacteria. Furthermore, they can be supplemented with different energy sources to determine which ones are effective for transport. For example, these vesicles can pump proline so that the internal concentration of this amino acid is fifty times higher than in the surrounding medium. D-Lactate is the most effective energy source in driving this pump. ATP, NADH, glucose, and phosphoenolpyruvate are without effect. D-Lactate is converted to pyruvate by these vesicles, but pyruvate does not elicit transport. This surprising result strongly suggested that a membrane-bound lactate dehydrogenase plays a critical role in transport. This inference is supported by the finding that inhibitors of electron transport (e.g., cyanide) block active transport. In contrast, concomitant ATP synthesis is unnecessary, since inhibitors of oxidative phosphorylation (e.g., oligomycin) do not inhibit active transport. The coupling site for active transport is probably between the membrane-bound lactate dehydrogenase and cytochrome b_1, the first cytochrome in the respiratory chain of *E. coli* (Figure 32-11). Electron flow generates a membrane potential, perhaps by translocating protons. In turn, *the membrane potential drives the active transport of some amino acids and sugars in this bacterial system.*

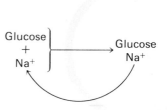

Figure 32-10
Cotransport of glucose and Na^+, followed by the extrusion of Na^+.

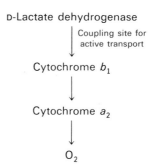

Figure 32-11
Proposed energy coupling site for the active transport of some sugars and amino acids into bacteria.

PHOSPHOENOLPYRUVATE DRIVES THE ACTIVE TRANSPORT OF SOME SUGARS BY PHOSPHORYLATING THEM

In some bacteria, the active transport of sugars is coupled to their phosphorylation. For example, glucose is converted to glucose 6-phosphate during its transport into bacterial cells. The distinctive feature of this kind of transport, called *group translocation,* is that the *solute is changed* during transport. The cell membrane is impermeable to sugar phosphates, and so they accumulate inside the cell.

The transport of some sugars in bacteria such as *E. coli* and *S. aureus* is carried out by a *phosphotransferase system*, which was discovered by Saul Roseman. An unusual feature of this system is that *phosphoenolpyruvate is the phosphoryl donor*, rather than ATP or another nucleoside triphosphate. The overall reaction carried out by the phosphotransferase system is:

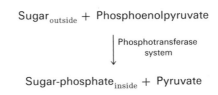

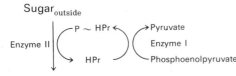

Figure 32-12
Phosphotransferase system for the active transport of some sugars into bacteria.

The phosphorylation of a sugar by this transport system requires four proteins (Figure 32-12). The phosphoryl group of phosphoenolpyruvate is not transferred directly to the sugar. Rather, it is transferred to a specific histidine residue of a small protein called HPr in a reaction catalyzed by enzyme I. This phosphohistidine intermediate has a high group-transfer potential, intermediate between that of phosphoenolpyruvate and ATP. The phosphorylated form of HPr then transfers its phosphoryl group to a sugar in a reaction that is catalyzed by enzyme II. *HPr thus serves as a phosphate carrier protein.* Recall two other carrier proteins: the acyl carrier protein (in fatty acid biosynthesis) and thioredoxin (a hydrogen carrier in ribonucleotide reduction).

Some of the proteins in the phosphotransferase system are general, whereas others are specific. *HPr, enzyme I, and part of the enzyme II complex* (called II-B) *participate in the transport of many sugars.* In contrast, the other part of the enzyme II complex, called II-A, is much more specific. There are different II-A proteins for the transport of glucose, lactose, and fructose, for example. Genetic studies

Phosphohistidine

have confirmed these conclusions. Mutants defective in HPr or enzyme I are unable to transport many different sugars, whereas mutants defective in a II-A protein are unable to transport a particular sugar.

HPr and enzyme I are soluble in aqueous media. In contrast, the enzyme II complex is an integral component of the bacterial membrane. It can be solubilized by detergents such as deoxycholate. The enzyme II complex is enzymatically active only if it is surrounded by appropriate lipids. In fact, a functional enzyme complex has been reconstituted from purified protein II-A, protein II-B, and phosphatidyl glycerol. *This reconstituted assembly transfers phosphate from P ∼ HPr to one of three sugars, depending on which protein II-A is present.*

TRANSPORT ANTIBIOTICS INCREASE THE IONIC PERMEABILITY OF MEMBRANES

Some microorganisms synthesize compounds that make membranes permeable to certain ions. These small molecules are called *transport antibiotics*. For example, *valinomycin* interferes with oxidative phosphorylation in mitochondria by making them permeable to K^+. The result is that mitochondria use the energy generated by electron transport to accumulate K^+ rather than to make ATP. Valinomycin is a repeating cyclic molecule made of four kinds of residues (A, B, C, and D) taken three times (Figure 32-13). The four kinds of residues are alternately joined by ester and peptide bonds.

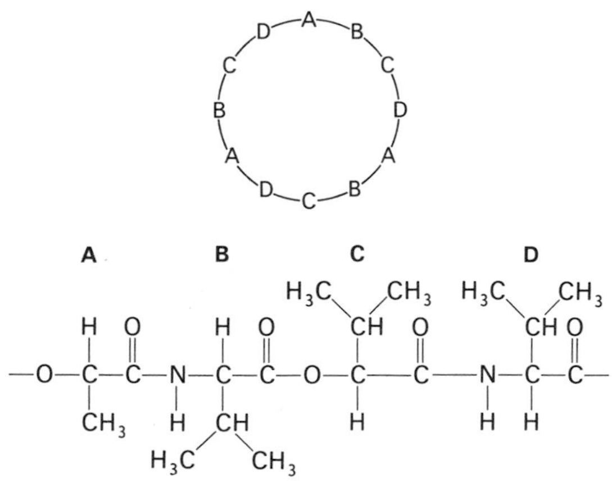

Figure 32-13
Valinomycin is a repeating cyclic molecule made up of L-lactate (A), L-valine (B), D-hydroxyisovalerate (C), and D-valine (D) residues.

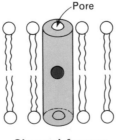

O   H
 \ /
  C
  | L   L   D   L
HN-Val-Gly-Ala-Leu-Ala-5

   D   L   D   L
  Val-Val-Val-Trp-Leu-10
                          O
   L   D   L   D   L ‖
  Trp-Leu-Trp-Leu-Trp-C
                       |
                      N—H
                       |
                      CH₂
                       |
                      CH₂
                       |
                      OH

Figure 32-14
Structure of gramicidin A.

Gramicidin A is another transport antibiotic that has been studied extensively (Figure 32-14). As will be discussed shortly, gramicidin A and valinomycin transport ions in quite different ways. Gramicidin is an open-chain polypeptide consisting of fifteen amino acid residues. A noteworthy feature of the structure is that L-and D-amino acids alternate. Also, the amino and carboxyl termini of this polypeptide are modified.

Ion transport by these antibiotics is readily studied in well-defined model systems such as lipid bilayer vesicles (p. 237) and planar bilayer membranes (p. 238). Vesicles containing a radioactive ion such as $^{42}K^+$ are prepared. The rate of efflux of this radioactive ion is then measured in the presence and absence of a transport antibiotic. Another way to obtain information about the ionic permeability of a bilayer membrane is to measure electrical properties such as membrane resistance and potential. For example, the resistance of a bilayer membrane to K^+ drops by a factor of more than 10,000 in the presence of 10^{-7} M valinomycin or 10^{-9} M gramicidin, when there is a gradient of 0.02 M KCl across the membrane. This large change in conductance is due to the fact that transport antibiotics act *catalytically*.

TRANSPORT ANTIBIOTICS ARE CARRIERS OR CHANNEL FORMERS

Transport antibiotics make membranes permeable to ions in two contrasting ways (Figure 32-15). Some of these antibiotics (e.g., gramicidin A) form channels that traverse the membrane. Ions enter such a channel at one surface of the membrane and diffuse

Figure 32-15
Schematic diagram comparing a channel-forming and a carrier transport antibiotic.

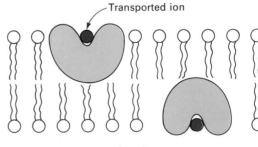

Channel former   Carrier

through it to the other side of the membrane. The channel former itself need not move for ion transport to occur. The other group of antibiotics (e.g., valinomycin) functions by carrying ions through the hydrocarbon region of the membrane. Diffusion of these transport antibiotics is essential for their activity.

Carriers and *channel formers* can be experimentally distinguished in the following way. The ionic conductance of a synthetic lipid bilayer membrane containing the transport antibiotic is measured as a function of temperature through the transition temperature of the lipid. The hydrocarbon interior of the membrane changes from an essentially frozen state to a highly fluid one in going through this thermal transition. A channel former need not diffuse in order to mediate ion transport across the membrane. Hence, freezing of the hydrocarbon core should have little effect on its ion-transport activity. In contrast, freezing should markedly diminish the efficiency of a carrier because a carrier must diffuse through the hydrocarbon region to be active. The experimental data for valinomycin and gramicidin A distinguish between these mechanisms (Figure 32-16). The conductance of a bilayer membrane containing valinomycin increases more than a thousandfold as the membrane is melted. In contrast, the transport activity of gramicidin A is relatively unaffected by the melting of the membrane.

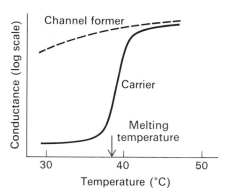

Figure 32-16
Comparison of the temperature-dependence of the conductance of a lipid bilayer membrane containing a channel-forming and a carrier transport antibiotic.

DONUT-SHAPED CARRIER ANTIBIOTICS BIND IONS IN THEIR CENTRAL CAVITIES

Ion-carrying transport antibiotics have some common structural features, as shown by x-ray crystallographic and spectroscopic studies. The carriers studied thus far have a shape like that of a donut. *A single metal ion is coordinated to several oxygen atoms that surround a central cavity.* The number of oxygen atoms that bind the metal ion is typically six or eight. *The periphery of the carrier consists of hydrocarbon groups* (Figure 32-17).

The roles of the central oxygen atoms of the hydrocarbon exterior are evident. In an aqueous medium, a metal ion such as K^+ binds several water molecules through their oxygen atoms. A carrier competes with water for binding the ion by chelating it to several appropriately arranged oxygen atoms in its central cavity. The

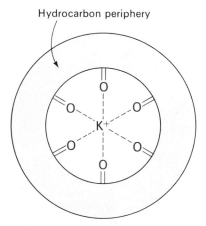

Figure 32-17
Schematic diagram of the chelation of K^+ by oxygen atoms of a carrier transport antibiotic.

hydrocarbon periphery makes the ion-carrier complex soluble in the lipid interior of the membranes. In essence, *these antibiotics catalyze the transport of ions across membranes by making them soluble in lipid.*

The structures of valinomycin and its complex with K^+ are shown in Figure 32-18. The K^+ ion is coordinated to six oxygen atoms, which are arranged octahedrally around the center of the molecule. These oxygen atoms come from the carbonyl groups of the six valine residues in the antibiotic. The methyl and isopropyl side chains constitute the hydrocarbon periphery of valinomycin.

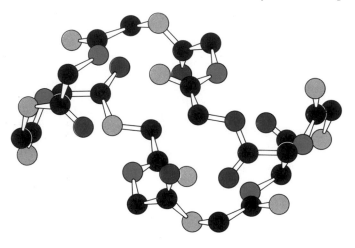

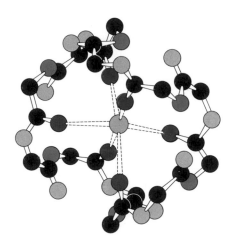

Figure 32-18
Model of valinomycin (above) and its complex with K^+ (at right). The conformation of the antibiotic changes upon binding K^+. [Based on atomic coordinates kindly provided by Dr. William Duax (for valinomycin) and by Dr. Larry Steinrauf (for the valinomycin-K^+ complex.]

The structure of the K^+ complex of nonactin, another ion-carrying antibiotic, is also of interest (Figure 32-19). The ion is bound to eight oxygen atoms in the center of the molecule. Four of these oxygens come from carboxyl groups, whereas the other four are ether oxygens.

Valinomycin and nonactin bind K^+ more strongly than Na^+ by factors of 1000 and 50, respectively. What is the basis of the *ion selectivity* of these antibiotics? A priori, one possibility is that K^+ fits better than Na^+ in the antibiotic-ion complex. The ionic radius of an unhydrated K^+ is 1.34 Å, whereas that of Na^+ is 0.95 Å.

However, nuclear magnetic resonance spectra indicate that the geometry of the K^+-nonactin complex is very similar to that of the Na^+-nonactin complex. A more likely reason for the ionic selectivity of nonactin is that Na^+ binds water more tightly than does K^+. The solvation energies of Na^+ and K^+ differ by about 20 kcal/mol. This difference is not maintained in the nonactin-ion complex. Thus, *nonactin binds K^+ in preference to Na^+ because water is a less effective competitor for the binding of K^+ than for Na^+.*

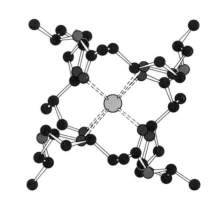

THE FLOW OF IONS THROUGH A SINGLE CHANNEL IN A MEMBRANE CAN BE DETECTED

The movement of ions through a single channel in a membrane has been detected. A planar bilayer membrane was formed in the presence of a very small amount of gramicidin A. The conductance of this membrane to Na^+ fluctuates with time in a step-by-step manner (Figure 32-20). *The first step in conductance arises from ion flow through a single channel formed by gramicidin A.* The next size step corresponds to two open channels. A single channel stays open for about a second.

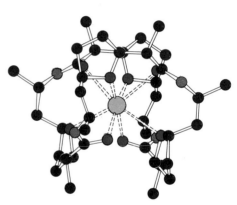

Figure 32-19
Two views of a model of the K^+ complex of nonactin.

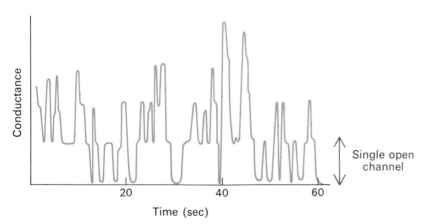

Figure 32-20
The conductance of a lipid bilayer membrane containing a few molecules of gramicidin A fluctuates in a step-by-step manner. The smallest step in conductance arises from a single open channel. [Courtesy of Dr. William Veatch.]

The opening of a transmembrane channel may be the result of a specific association of two gramicidin A molecules, whereas closing may correspond to the dissociation of such a dimer. The conformation of gramicidin A is not yet known. However, model-building studies suggest that gramicidin A may be a cylinder with a hole 4 Å in diameter along its long axis. In this model, the polar groups of gramicidin A line this hole, whereas its hydrophobic side chains are the periphery of the cylinder. Two of these cylinders lined end-to-end could form a transmembrane channel. The channel formed by gramicidin A is highly permeable to cations but not to anions. The maximal transport rate through a single channel is about 2×10^7 cations per second. This high rate is consistent with the postulated presence of a water-filled channel that is 4 Å in diameter, which would permit water and partially hydrated ions to flow through. As might be expected for this kind of pore, gramicidin A transports Na^+ nearly as effectively as K^+. *In general, channel formers are less selective in transporting ions than are carriers. Another difference between these two types of transport antibiotics is that a thousand times as many ions flow through an open channel than can be transported by a single ion-carrying molecule in the same interval.*

SUMMARY

Molecules and ions are transported across biological membranes by oriented transport systems that contain specific carrier proteins. A transport process is passive if ΔG for the transported species is negative, whereas it is active if ΔG is positive. Active transport requires an input of free energy. The most ubiquitous transport system in animal cells is the Na^+-K^+ pump, which hydrolyzes ATP to pump K^+ in and Na^+ out. In the presence of Na^+, ATP phosphorylates a glutamate side chain of the ATPase component of the pump. This phosphorylated intermediate is hydrolyzed in the presence of K^+. The latter reaction is inhibited by cardiotonic steroids, such as digitalis, which are highly specific inhibitors of this pump. It is not yet known how phosphorylation and dephosphorylation lead to active transport. Calcium ion, which plays an important role in the control of muscle contraction, is transported by a different ATPase system in the sarcoplasmic reticulum membrane. A functional Ca^{2+} transport system has been reconstituted from purified Ca^{2+} ATPase and phospholipids.

Some transport systems are directly driven by ionic gradients rather than by the hydrolysis of ATP. For example, the active transport of glucose and amino acids into some animal cells requires the simultaneous entry of Na^+, a process called cotransport. In turn, the Na^+ gradient is maintained by the Na^+-K^+ pump. A different mechanism of energy coupling is responsible for the active transport of some sugars and amino acids into bacteria. The oxidation of lactate by a membrane-bound dehydrogenase generates a membrane potential that drives the pump. Finally, the transported group may be chemically modified (e.g., glucose to glucose 6-phosphate), as by the phosphotransferase system of *E. coli,* which uses phosphoenolpyruvate as the phosphoryl donor.

Transport antibiotics make membranes permeable to certain ions by serving as carriers (e.g., valinomycin) or as channel formers (e.g., gramicidin A). A carrier antibiotic is a donut-shaped molecule that binds a single metal ion in its central cavity. The hydrocarbon periphery enables the complex to traverse the hydrocarbon interior of the membrane. Channel-forming antibiotics create an aqueous pore that traverses the membrane.

SELECTED READINGS

Where to start:

Fox, C. F., 1972. The structure of cell membranes. *Sci. Amer.* 226(2):32–38. [Available as *Sci. Amer.* Offprint 1241.]

Books and general reviews:

Hokin, L. E., (ed.), 1972. *Metabolic Transport.* Academic Press. [A collection of valuable articles on the transport of ions, carbohydrates, amino acids, neurotransmitters, and drugs.]

Fox, C. F., and Keith, A., (eds.), 1972. *Membrane Molecular Biology.* Sinauer.

Oxender, D. L., 1972. Membrane transport. *Annu. Rev. Biochem.* 41:777–814.

Bacterial transport systems:

Kennedy, E. P., 1970. The lactose permease system of *E. coli. In* Beckwith, J. R., and Zipser, D., (eds.), *The Lactose Operon,* pp. 49–109. Cold Spring Harbor Laboratory.

Roseman, S., 1972. Carbohydrate transport in bacterial cells. *In* Hokin, L. E., (ed.), *Metabolic Transport,* pp. 41–89.

Boos, W., 1974. Bacterial transport. *Annu. Rev. Biochem.* 43:123–146.

Kaback, H. R., 1972. Transport mechanisms in isolated bacterial cytoplasmic membrane vesicles. *In* Fox., C. F., (ed.), *Membrane Research,* pp. 473–501. Academic Press.

Hirata, H., Altendorf, K., and Harold, F. M., 1973. Role of an electrical potential in the coupling of metabolic energy to active transport by membrane vesicles of *Escherichia coli*. *Proc. Nat. Acad. Sci.* 70:1804–1808.

Na⁺-K⁺ and Ca²⁺ transport:

Kyte, J., 1972. Properties of the two polypeptides of sodium- and potassium-dependent adenosine triphosphatase. *J. Biol. Chem.* 247:7642–7649.

Hokin, L. E., Dahl, J. L., Deupree, J. D., Dixon, J. F. Hackney, J. F., and Perdue, J. F., 1973. Studies on the characterization of the sodium-potassium transport adenosine triphosphatase. *J. Biol. Chem.* 248:2593–2605.

Jardetzky, O., 1966. Simple allosteric model for membrane pumps. *Nature* 211:969.

Inesi, G., 1972. Active transport of calcium ion in sarcoplasmic reticulum membranes. *Annu. Rev. Biophys. Bioeng.* 1:191–210.

MacLennan, D. H., Yip, C. C., Iles, G. H., and Seeman, P., 1972. Isolation of sarcoplasmic reticulum proteins. *Cold Spring Harbor Symp. Quant. Biol.* 37:469–477.

Racker, E., 1972. Reconstitution of a calcium pump with phospholipids and a purified Ca²⁺-adenosine triphosphatase from sarcoplasmic reticulum. *J. Biol. Chem.* 247:8198–8200.

Transport antibiotics:

Laüger, P., 1972. Carrier-mediated ion transport. *Science* 178:24–30.

Hladky, S. B., and Haydon, D. A., 1972. Ion transfer across lipid membranes in the presence of gramicidin A. *Biochim. Biophys. Acta* 274:294–312.

Krasne, S., Eisenman, G., and Szabo, G., 1971. Freezing and melting of lipid bilayers and the mode of action of nonactin, valinomycin and gramicidin. *Science* 174:412–415.

CHAPTER **33**

EXCITABLE MEMBRANES

Many cell membranes can be excited by specific chemical or physical stimuli. The responses of a nerve axon membrane to an electrical stimulus, of a synapse to a transmitter substance, and of a retinal rod cell to light are mediated by excitable assemblies in membranes. The common features of these processes and others carried out by excitable assemblies are:

1. The stimulus is detected by a highly specific protein receptor, which is an *integral component of the excitable membrane.*

2. The specific stimulus elicits a *conformational change* in the receptor. As a result, the permeability of the membrane or the activity of a membrane-bound enzyme changes. Many of the responses are *highly amplified.*

3. The conformational change in the receptor and the resulting alterations in function are *reversible.* There are mechanisms that take the receptor back to its resting state and restore its excitability.

This chapter deals mainly with the acetylcholine receptor system, which participates in the transmission of nerve impulses at certain synapses, and with the retinal rod cell, which is an exquisitely sensitive receptor for light. These two excitable assemblies have been chosen for discussion because much progress has recently been made in the elucidation of their structures and dynamics.

ACETYLCHOLINE IS A NEUROTRANSMITTER

Nerve cells interact with other nerve cells at junctions called *synapses* (Figure 33-1). Nerve impulses are communicated across most synapses by chemical transmitters, which are small, diffusible molecules such as acetylcholine and norepinephrine. Acetylcholine is also the transmitter at motor end plates (neuromuscular junctions), which are the junctions between nerve and striated muscle.

The *presynaptic membrane* of a *cholinergic synapse*—that is, one that uses acetylcholine as the neurotransmitter—is separated from the *postsynaptic membrane* by a gap of about 500 Å, called the *synaptic cleft*. The end of the presynaptic axon is filled with *synaptic vesicles* containing acetylcholine. The arrival of a nerve impulse leads to the release of acetylcholine into the cleft. The acetylcholine molecules then diffuse to the postsynaptic membrane, where they combine with specific receptor molecules. This produces a *depolarization of the postsynaptic membrane,* which is propagated along the electrically excitable membrane of the second nerve cell. Acetylcholine is *hydrolyzed by acetylcholinesterase* and the polarization of the postsynaptic membrane is restored.

Acetylcholine is synthesized near the presynaptic end of axons by the transfer of an acetyl group from acetyl CoA to choline. This reaction is catalyzed by *choline acetyltransferase* (choline acetylase).

Some of the acetylcholine is then taken up by synaptic vesicles, whereas the remainder stays in the cytosol. There are of the order of *10,000 acetylcholine molecules in a cholinergic synaptic vesicle,* which is typically 400 Å in diameter. The study of synaptic function has been facilitated by the isolation of *synaptosomes* from homogenates of nerve tissue. Synaptosomes are presynaptic endings that have resealed after being pinched off. They consist of mitochondria, cytosol, and synaptic vesicles surrounded by an intact presynaptic

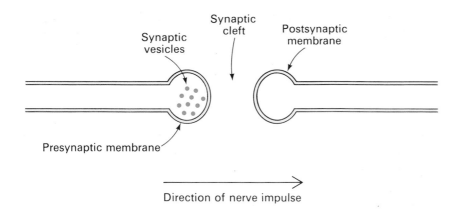

Figure 33-1
Schematic diagram
of a synapse.

membrane. Some synaptosomes also have adhering pieces of postsynaptic membrane. Purified synaptic vesicles can be obtained from a preparation of synaptosomes by density-gradient centrifugation.

ACETYLCHOLINE OPENS CATION GATES IN THE POSTSYNAPTIC MEMBRANE

The resting potential of a postsynaptic membrane or of a motor endplate is about -75 mV. The interaction of acetylcholine with specific receptors produces a large change in the permeability properties of these membranes (Figure 33-2). *The conductance of both Na^+ and K^+ increases markedly within 0.1 msec, and there is a large inward current of Na^+ and a smaller outward current of K^+.* The inward Na^+ current depolarizes the postsynaptic membrane and triggers an action potential in the adjacent axon or muscle membrane. *Acetylcholine opens a single kind of cation channel, which is nearly equally permeable to Na^+ and K^+.* The flow of Na^+ is larger than the flow of K^+ because the electrochemical gradient across the membrane is steeper for Na^+.

ACETYLCHOLINE IS RELEASED IN PACKETS

Studies of nerve-impulse transmission at neuromuscular junctions by Bernard Katz have revealed that acetylcholine is released from the presynaptic membrane in the form of packets containing of the order of 10^4 molecules. The evidence for *quantal release of acetylcholine*

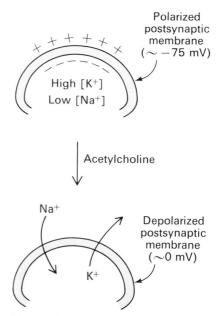

Figure 33-2
Acetylcholine depolarizes the postsynaptic membrane by increasing the conductance of Na^+ and K^+.

comes from an analysis of the membrane potential of the motor end plate, which shows spontaneous electrical activity even when the associated nerve is not stimulated. Depolarizing pulses that are 0.5 mV in amplitude and last about 20 msec occur intermittently. These *miniature end-plate potentials* occur randomly with a probability that stays constant for long periods. A miniature end-plate potential is caused by the spontaneous release of a single packet of acetylcholine, perhaps a single synaptic vesicle. The full depolarization of the end plate elicited by an action potential results from the synchronous release of about a thousand packets of acetylcholine in less than a millisecond. *The number of packets released depends on the potential of the presynaptic membrane.* In other words, *the release of acetylcholine is an electrically controlled form of secretion.* Release of acetylcholine depends on the presence of Ca^{2+} in the extracellular fluid. It has been suggested that the depolarization of the presynaptic membrane leads to the entry of Ca^{2+}, which may promote a transient fusion of the synaptic vesicle membrane and the presynaptic membrane.

ISOLATION OF THE ACETYLCHOLINE RECEPTOR

How does acetylcholine alter the ionic conductance of the postsynaptic membrane? It is evident that detailed structural information is needed to answer this question. Considerable progress has recently been made in the isolation and purification of the acetylcholine receptor. A choice starting material is the *electric organ* of electric fish or electric eel, which is very rich in cholinergic postsynaptic membranes. Electric organs are made of columns of cells called *electroplaxes.* One side of the cell (called the innervated face) receives a nerve ending and is electrically excitable. The other side, called the noninnervated face, is highly folded and not electrically excitable. The voltage produced by a stimulated electroplax results from the asymmetry of the responses of its two faces. The membrane potential of the innervated side changes from -90 to $+60$ mV on excitation, whereas the noninnervated side stays at -90 mV. Consequently, *the potential between the outside of the two faces is 150 mV at the peak of the action potential* (Figure 33-4). In an electric organ, electroplaxes are arranged in series so that their voltages are additive. An organ containing 5,000 rows of electroplaxes can there-

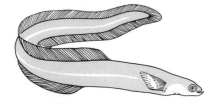

Figure 33-3
An electric eel.

fore generate a 750-volt discharge. It is interesting to note that
electroplaxes have evolved from muscle cells. They retain the
electrically excitable outer membrane of muscle but have lost the
contractile apparatus.

Another exotic biological material has been invaluable in the
isolation of acetylcholine receptors. The receptor must be *specifically
labeled* so that it can be unequivocally identified in a mixture of
macromolecules. This has been accomplished by using neurotoxins
from snakes, such as *α-bungarotoxin*, which comes from the venom
of a Formosan snake, or *cobratoxin*. These neurotoxins block neuromuscular transmission by binding to acetylcholine receptors on
motor end plates, or on the innervated face of the electroplax, in
the electric organ. They are small basic proteins having molecular
weights of about 7,000. The neurotoxins can be made highly radioactive by iodinating them with ^{125}I or by forming a Schiff base
with pyridoxal phosphate and then reducing it with ^3H-borohy-
dride. Labeled cobratoxin binds tightly to the acetylcholine receptor with a dissociation constant of 10^{-9} M. Most important, it does
not bind appreciably to other macromolecules in the postsynaptic
membrane. Thus, the acetylcholine receptor can be specifically
marked with a radioactive label.

The acetylcholine receptor of the electric organ has been solubilized by adding a nonionic detergent (such as Tween 80, a derivative
of polyoxyethylene) to membrane fragments. The soluble extract
was fractionated by gel-filtration chromatography, and by chromatography on hydroxyapatite and ion-exchange columns. In addition,
affinity chromatography on a column containing covalently bound
neurotoxin was used to achieve an 8250-fold purification of the
receptor. Electrophoresis of this preparation on a polyacrylamide
gel showed a single band of protein. The purified receptor binds
one molecule of neurotoxin per 90,000 daltons of protein.

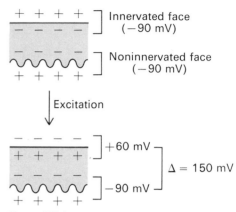

Figure 33-4
Generation of a voltage
by an electroplax cell.

ACETYLCHOLINE IS RAPIDLY HYDROLYZED AND THE END PLATE REPOLARIZES

The depolarizing signal must be switched off to restore the excitability of the postsynaptic membrane. Acetylcholine is hydrolyzed to acetate and choline by *acetylcholinesterase,* an enzyme discovered by David Nachmansohn in 1938. The permeability

properties of the postsynaptic membrane are restored, and the membrane becomes repolarized.

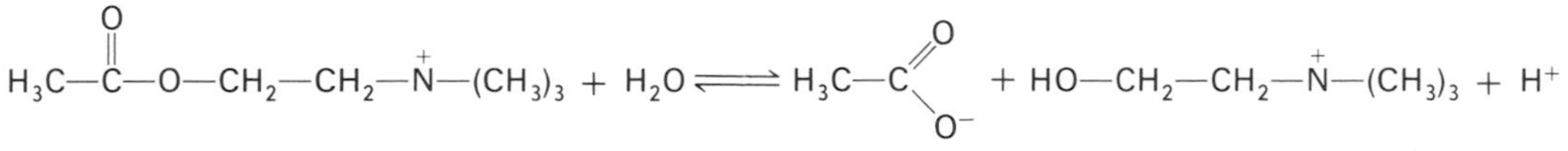

$$H_3C-\overset{\overset{\displaystyle O}{\|}}{C}-O-CH_2-CH_2-\overset{+}{N}-(CH_3)_3 + H_2O \rightleftharpoons H_3C-\overset{\overset{\displaystyle O}{\|}}{C}_{O^-} + HO-CH_2-CH_2-\overset{+}{N}-(CH_3)_3 + H^+$$

Acetylcholine   Acetate   Choline

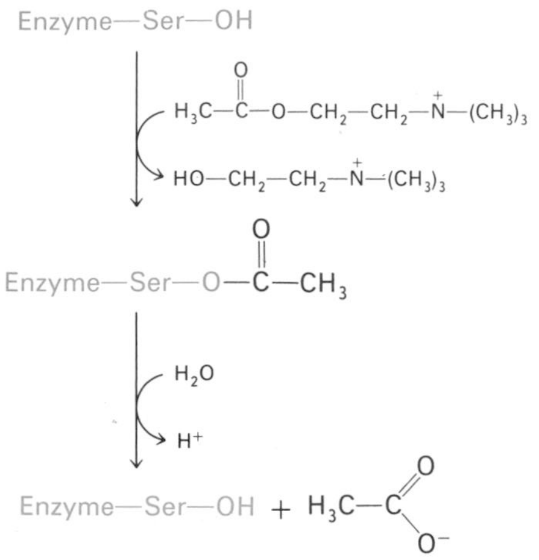

Figure 33-5
Proposed catalytic mechanism of acetylcholinesterase.

Acetylcholinesterase is located in the postsynaptic membrane. It is different from the acetylcholine receptor and can be readily separated from it. Acetylcholinesterase does not bind neurotoxins. It has a molecular weight of 260,000 and an $\alpha_2\beta_2$ subunit structure. A striking feature of this enzyme is its very high turnover number of 25,000 sec^{-1}, which means that it cleaves an acetylcholine molecule in 40 μsec. The high turnover number of the enzyme is essential for the rapid restoration of the polarized state of the postsynaptic membrane. Synapses can transmit 1,000 impulses per second only if the postsynaptic membrane recovers its polarization within a fraction of a millisecond.

The catalytic mechanism of acetylcholinesterase resembles that of chymotrypsin. Acetylcholine reacts with a specific serine residue at the active site of acetylcholinesterase to form a covalent acetyl-enzyme intermediate, and choline is released. The acetyl-enzyme intermediate then reacts with water to form acetate and regenerate the free enzyme (Figure 33-5).

ACETYLCHOLINESTERASE INHIBITORS ARE USED AS DRUGS AND POISONS

The therapeutic and toxic properties of acetylcholinesterase inhibitors are of considerable practical importance. *Physostigmine* (also called eserine) is an alkaloid derived from the Calabar or ordeal bean, which was once used as an ordeal poison in witchcraft trials. Physostigmine and related inhibitors such as neostigmine are *carbamoyl esters* (Figure 33-6). They inhibit acetylcholinesterase by forming a covalent intermediate that is hydrolyzed very slowly. Neostigmine binds to acetylcholinesterase so that its positively charged trimethylammonium group is at the anionic site of the enzyme, whereas its carbamoyl group is adjacent to the reactive serine

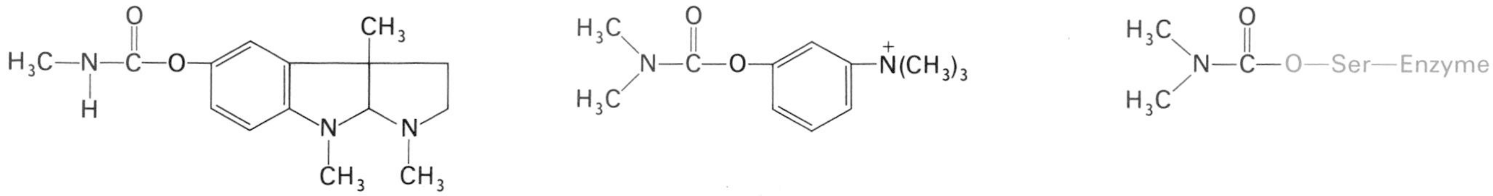

Physostigmine

Neostigmine

Carbamoyl-enzyme complex

Figure 33-6
Physostigmine and neostigmine inhibit acetylcholinesterase by carbamoylating the active-site serine.

residue at the esteratic site. The enzyme is then carbamoylated and the alcohol moiety is released. *The carbamoyl-enzyme intermediate is subsequently hydrolyzed at a very slow rate, in contrast to the acetyl-enzyme intermediate.* Thus, the active site of the enzyme is effectively blocked. Neostigmine is used to treat glaucoma, an eye disease characterized by abnormally high intraocular pressure. The therapeutic rationale is that neostigmine inhibits acetylcholinesterase and thereby enhances the effects of acetylcholine.

Even more potent inhibitors are the *organic fluorophosphates,* such as diisopropyl phosphofluoridate (DIPF). These compounds react with acetylcholinesterase to form *very stable covalent phosphoryl-enzyme complexes* (Figure 33-7). The phosphoryl group becomes bonded to the active-site serine, as in serine proteases that have reacted with DIPF.

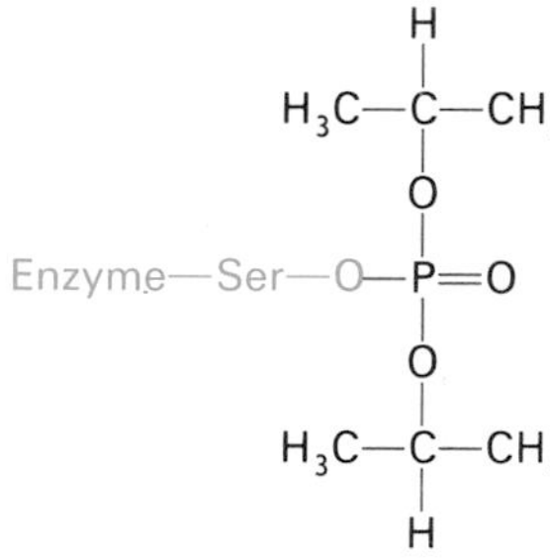

DIPF-inhibited enzyme

Figure 33-7
DIPF-inhibited acetylcholinesterase.

Many organic phosphate compounds have been synthesized for use as *agricultural insecticides* or as *nerve gases* for chemical warfare (Figure 33-8). These compounds can be lethal by causing respiratory paralysis. Tabun and sarin are among the most toxic of them. Parathion has been widely used as an agricultural insecticide.

The number of acetylcholinesterase molecules in the end plate of mouse diaphragm muscle has been counted using radioactive DIPF as a label. The density is 12,000 μm^{-2}, which is nearly the same as was found for acetylcholine receptors using radioactive neurotoxins. Thus, *the postsynaptic membrane is very densely packed with both acetylcholinesterase and the acetylcholine receptor.* Only a small fraction of the acetylcholinesterase is required for the transmission of nerve impulses at low frequencies. However, most of the enzyme molecules must be active to sustain transmission at high rates of firing.

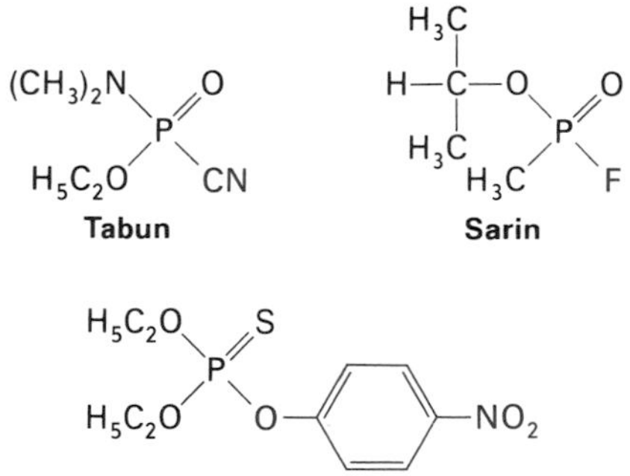

Tabun

Sarin

Parathion

Figure 33-8
Structures of some organic phosphate inhibitors of acetylcholinesterase.

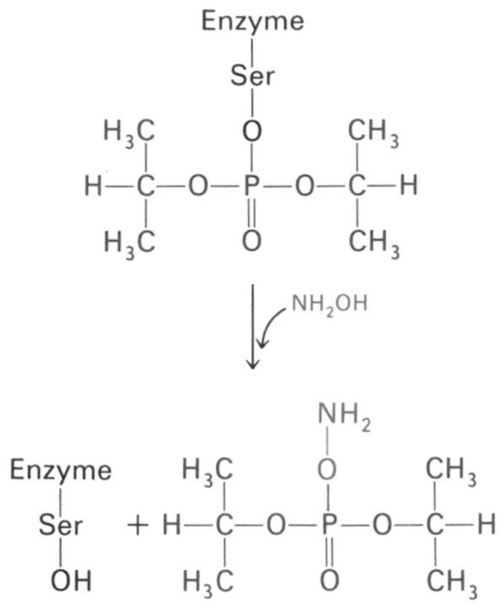

Figure 33-9
Reactivation of DIPF-inhibited
acetylcholinesterase by hydroxylamine.

**Pyridine aldoxime methiodide
(PAM)**

A DESIGNED ANTIDOTE FOR ORGANOPHOSPHORUS POISONING

Acetylcholinesterase inhibited by organic phosphates such as DIPF can be reactivated by derivatives of hydroxylamine (NH_2OH). The starting point in the development of an antidote was the finding by Irwin Wilson that hydroxylamine releases the phosphoryl group that is attached to the serine residue in the inhibited enzyme, and thereby reactivates it (Figure 33-9). The challenge was to make this reaction therapeutically useful. Hydroxylamine could not be used in vivo because it is toxic at the high concentrations required to reactivate DIPF-inhibited acetylcholinesterase. Wilson's approach was to search for a compound that retains the chemical reactivity of hydroxylamine but has specificity and very high affinity for acetylcholinesterase. Since the enzyme was known to have an anionic site for the positively charged choline moiety of acetylcholine, it seemed worthwhile to synthesize a hydroxylamine derivative containing a quaternary ammonium group. Should the substituent be on the oxygen or nitrogen atom of hydroxylamine? It was found that O-methylhydroxylamine was inactive as a reactivator, whereas N-methylhydroxylamine was active. The hydroxyl group of the reactivator cannot be substituted because it is the anionic form that attacks the phosphorus atom in the inhibited enzyme.

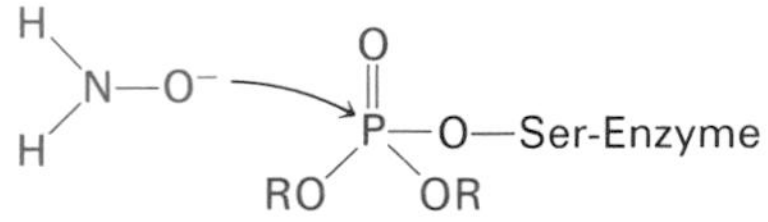

The next step, a crucial one, was to place the quaternary ammonium group at an appropriate distance from the nucleophilic oxygen atom. Furthermore, the orientation of these groups had to be complementary to that of the anionic site and the phosphorus atom of the inhibited enzyme.

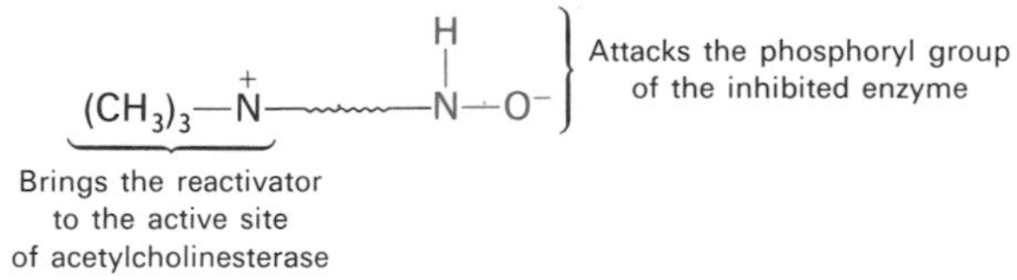

Numerous compounds containing a quaternary ammonium group and a hydroxylamine function were tested as reactivators. The most effective one was *2-pyridine aldoxime methiodide (PAM)*. This compound has a well-defined geometry because of the double bond in the oxime, which tends to fix the oxygen atom in the plane of the ring.

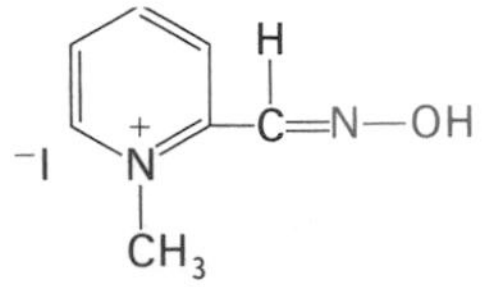

PAM reactivates DIPF-inhibited acetylcholinesterase in much the same way as does hydroxylamine. However, 10^{-6} M PAM is as effective as 1-M hydroxylamine. *This millionfold enhancement makes PAM therapeutically useful in the treatment of organophosphorus poisoning.* The remarkable effectiveness of PAM arises from the optimal positioning of the quaternary ammonium group in relation to the nucleophilic oxygen atom. The development of PAM is a landmark in the rational design of drugs.

INHIBITORS OF THE ACETYLCHOLINE RECEPTOR

Neuromuscular transmission can also be impaired by *compounds that act directly on the acetylcholine receptor. Curare* has been used for centuries by South American Indians. Soon after Columbus returned, d'Anghera wrote *De Orbe Novo* in which he noted that "the natives poisoned their arrows with the juice of a death-dealing herb. . . ." One of the active components of curare is d-*tubocurarine* (Figure 33-10). *Tubocurarine inhibits the depolarization of the end plate by competing with acetylcholine for binding to the acetylcholine receptor.* α-Bungarotoxin and cobratoxin have a similar action. In contrast, compounds such as *decamethonium* bind to the receptor and cause a *persistent depolarization* of the end plate.

$$(CH_3)_3 - \overset{+}{N} - (CH_2)_{10} - \overset{+}{N} - (CH_3)_3$$
Decamethonium

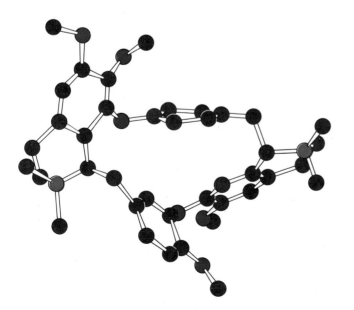

Figure 33-10
Formula and model of *d*-tubocurarine.

Succinylcholine, an analog of acetylcholine, is used to produce muscular relaxation in surgical procedures.

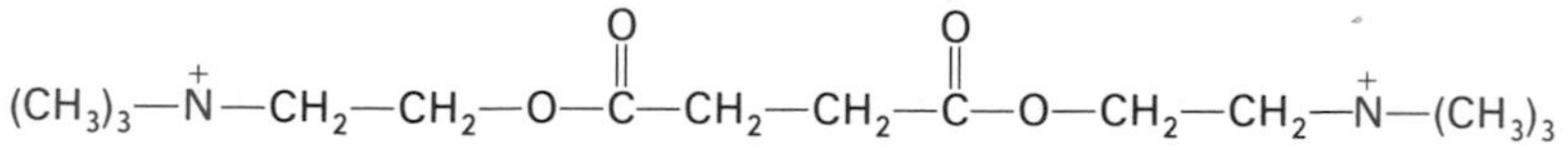

$$(CH_3)_3—\overset{+}{N}—CH_2—CH_2—O—\overset{\overset{O}{\|}}{C}—CH_2—CH_2—\overset{\overset{O}{\|}}{C}—O—CH_2—CH_2—\overset{+}{N}—(CH_3)_3$$

Succinylcholine

Succinylcholine is hydrolyzed very slowly by acetylcholinesterase in the postsynaptic membrane. Hence, it causes a persistent depolarization of the end plate. Succinylcholine is hydrolyzed by less specific cholinesterases in the plasma and in the liver, which are called *plasma cholinesterase* or *pseudocholinesterase* to distinguish them from acetylcholinesterase of the postsynaptic membrane. An attractive feature of succinylcholine is that neuromuscular transmission resumes soon after the infusion of this drug is stopped. In a small proportion of patients, however, muscular relaxation and respiratory paralysis persist for many hours. In such patients, succinylcholine is hydrolyzed at a very slow rate because their plasma cholinesterase has a markedly reduced affinity for succinylcholine. Sensitivity to succinylcholine, like sensitivity to primaquine (p. 366), is an example of a genetically determined drug idiosyncrasy.

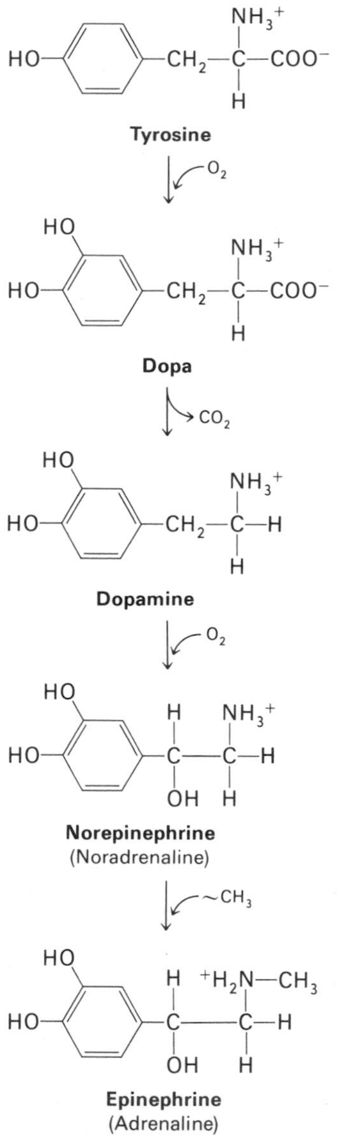

CATECHOLAMINES ALSO SERVE AS NEUROTRANSMITTERS

Several catecholamines act as neurotransmitters. For example, norepinephrine is the transmitter at smooth-muscle junctions that are innervated by sympathetic nerve fibers, in contrast to parasympathetic junctions in which acetylcholine is the transmitter. Epinephrine and dopamine are two other catecholamine transmitters. In fact, these catecholamines are synthesized from tyrosine in sympathetic-nerve terminals and in the adrenal gland (Figure 33-11). The first step, which is rate-limiting, is the hydroxylation of tyrosine to form 3,4-dihydroxyphenylalanine (*dopa*). This reaction is catalyzed by tyrosine hydroxylase, which is like phenylalanine hydroxylase. Molecular oxygen is activated by tetrahydrobiopterin, the cofactor. The second step is the decarboxylation of dopa by dopa decarboxylase, a pyridoxal phosphate enzyme, to yield 3,4-dihydroxyphenylethylamine (*dopamine*). Dopamine is then

Figure 33-11
Pathway for the synthesis of catecholamine neurotransmitters.

hydroxylated to *norepinephrine* by a copper-containing hydroxylase. Finally, *epinephrine* is formed by the methylation of norepinephrine by a transmethylase that utilizes *S*-adenosylmethionine. Catecholamine neurotransmitters are inactivated by methylation of the 3-hydroxyl group of the catechol ring. This reaction is catalyzed by *catechol-O-methyltransferase,* which uses *S*-adenosylmethionine as the methyl donor. Alternatively, these neurotransmitters can be inactivated by oxidative removal of their amino group by *monoamine oxidase* (Figure 33-12).

Figure 33-12
Inactivation of norepinephrine.

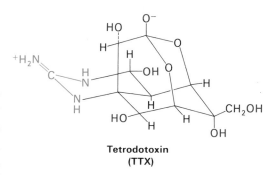

3-*O*-Methylepinephrine   Norepinephrine   3,4-Dihydroxyphenylglycolaldehyde

NERVE AXON MEMBRANES CONTAIN TWO KINDS OF ION GATES

Alan Hodgkin and Andrew Huxley have shown that action potentials in nerves arise from large, transient changes in the permeability of the axon membrane to Na^+ and K^+ ions. When the voltage across a membrane is changed from -65 mV to 0 mV, there is an inward current of Na^+ in the first millisecond, followed by an outward current of K^+. The permeability of the axon membrane to Na^+ and K^+ depends on the membrane potential. What is the molecular basis of these voltage-controlled ion gates? A valuable probe in this regard is *tetrodotoxin* (TTX), a poison from the puffer (or fugu) fish. TTX is a potent poison because it blocks the conduction of nerve impulses along axons and in excitable membrane of muscle fibers, leading to respiratory paralysis. The lethal dose for a mouse is about 0.01 µg. The inhibitory action of TTX is highly specific. *It blocks the early-conductance channels for Na$^+$,* but has no effect on the late-conductance channel for K^+. TTX binds very tightly to nerve axon membranes ($K = 10^{-9}$ M), which has made it feasible to determine the density of Na^+ channels. The interesting result is that *there are only about twenty Na$^+$ channels per square micron of axon membrane.* This

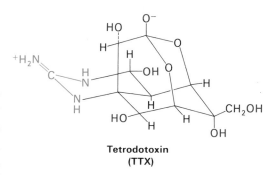

Tetrodotoxin
(TTX)

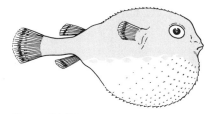

Figure 33-13
A puffer fish, regarded as a culinary delicacy in Japan. [From F. A. Fuhrman. Tetrodotoxin. Copyright © 1967 by Scientific American, Inc. All rights reserved.]

means that the Na$^+$ channels are about 2200 Å apart if they are evenly distributed. It is interesting to note that the density of acetylcholine receptors in postsynaptic membranes is of the order of a thousand times higher.

A RETINAL ROD CELL CAN BE EXCITED BY A SINGLE PHOTON

We now turn to excitable receptors that are activated by *light*. There are two kinds of photoreceptor cells in humans, called rods and cones because of their distinctive shapes. *Cones* function in bright light and are responsible for color vision, whereas *rods* function in dim light but do not perceive color. There are three million cones and a billion rods in a human retina. These photoreceptor cells *convert light into atomic motion and then into a nerve impulse.* Rods and cones form synapses with bipolar cells, which in turn interact with other nerve cells in the retina. The electrical signals generated by the photoreceptors are processed by an intricate array of nerve cells within the retina, and then transmitted to the brain by the fibers

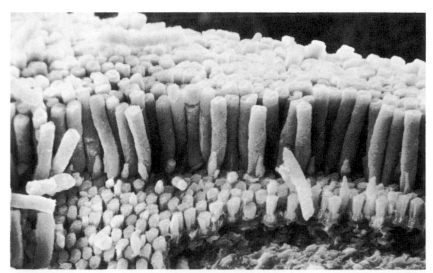

Figure 33-14
Scanning electron micrograph of retinal rod cells.
[Courtesy of Dr. Deric Bownds.]

of the optic nerve. Thus, the retina has a dual function: to transform light into nerve impulses and to integrate visual information.

In 1938, Selig Hecht discovered that *a human rod cell can be excited by a single photon.* Let us explore the molecular basis of the exquisite sensitivity of these cells. Rods are slender, elongated structures, typically 1 μm in diameter and 40 μm in length. The major functions of a rod cell are compartmentalized in a very distinctive way (Figure 33-15). *The outer segment of a rod is specialized for photoreception.* It contains a stack of about 1,000 *discs* (Figure 33-16), which are closed, flattened sacs about 160 Å thick. These membranous structures are densely packed with photoreceptor molecules. The disc membranes are separate from the plasma membrane of the outer segment. A slender cilium joins the outer segment and the *inner segment,* which is rich in mitochondria and ribosomes. The inner segment generates ATP at a very rapid rate, and also is highly active in synthesizing proteins. The discs in the outer segment have a life of only ten days and are continually renewed. The inner segment is contiguous with the nucleus, which is next to the *synaptic body.* Many synaptic vesicles are present in the synaptic body, which forms a synapse with a bipolar cell.

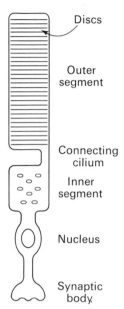

Figure 33-15
Schematic diagram of a retinal rod cell.

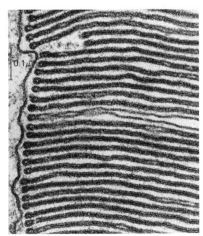

Figure 33-16
Electron micrograph of a rod outer segment, showing the stack of discs. [From John E. Dowling. The organization of vertebrate visual receptors. In *Molecular Organization and Biological Function,* John M. Allen, ed. (Harper and Row, 1967).]

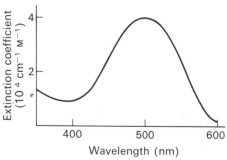

11-*cis*-Retinal

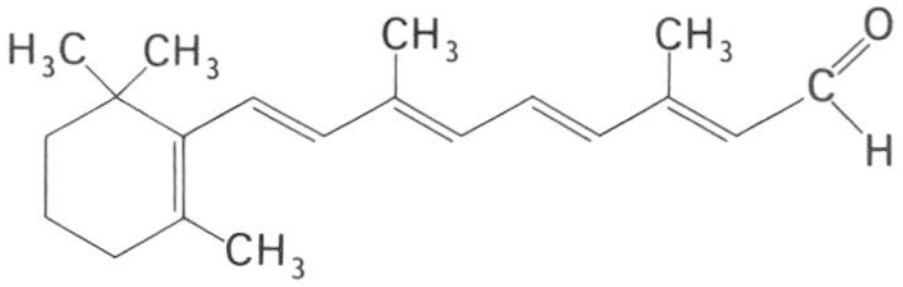

All-*trans*-retinal

All-*trans*-retinol
(Vitamin A)

Figure 33-17
Structure of 11-*cis*-retinal, all-*trans*-retinal, and all-*trans*-retinol (Vitamin A).

RHODOPSIN IS THE PHOTORECEPTOR PROTEIN OF RODS

Light must be absorbed to stimulate a photoreceptor cell. In addition, the light-absorbing group (called a *chromophore*) must undergo a conformational change after it has absorbed a photon. The photosensitive molecule in rods is *rhodopsin,* which consists of *opsin,* a protein, and *11*-cis-*retinal,* a prosthetic group (Figure 33-17). Rhodopsin is an integral membrane protein having a molecular weight of about 38,000. It is tenaciously bound to the lipid bilayer of disc membranes. Opsin, like other proteins devoid of prosthetic group, does not absorb visible light. The color of rhodopsin and its responsiveness to light depend on the presence of 11-*cis*-retinal, which is a very effective chromophore. 11-*cis*-Retinal gives rhodopsin a broad absorption band in the visible region of the spectrum with a peak at 500 nm, which nicely matches the solar output. The intensity of the visible absorption band of rhodopsin is also noteworthy. The extinction coefficient of rhodopsin at 500 nm is $40{,}000 \text{ cm}^{-1} \text{ M}^{-1}$, a high value (Figure 33-18). The integrated absorption strength of the visible absorption band of rhodopsin approaches the maximum value attainable by organic compounds. 11-*cis*-Retinal has these favorable chromophoric properties because it is a *polyene.* Its six alternating single and double bonds constitute a long, unsaturated electron network.

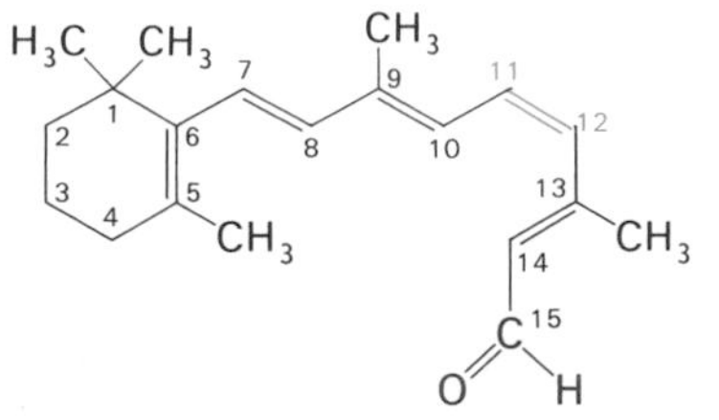

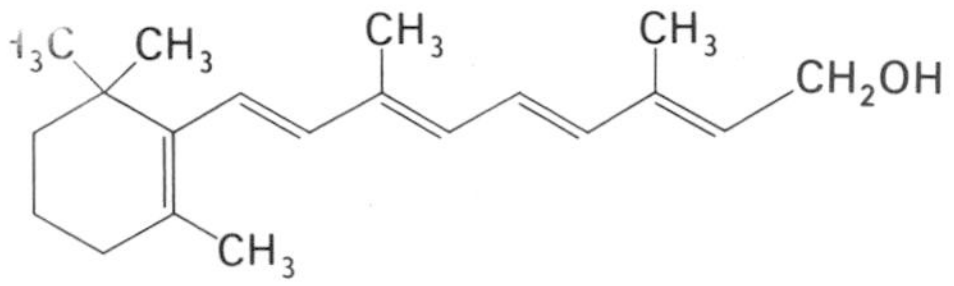

11-*cis*-Retinal is attached to rhodopsin by a *Schiff-base linkage.* The aldehyde group of 11-*cis*-retinal is linked to the ε-amino group of a specific lysine residue of opsin. The spectral properties of rhodopsin indicate that the Schiff base is protonated.

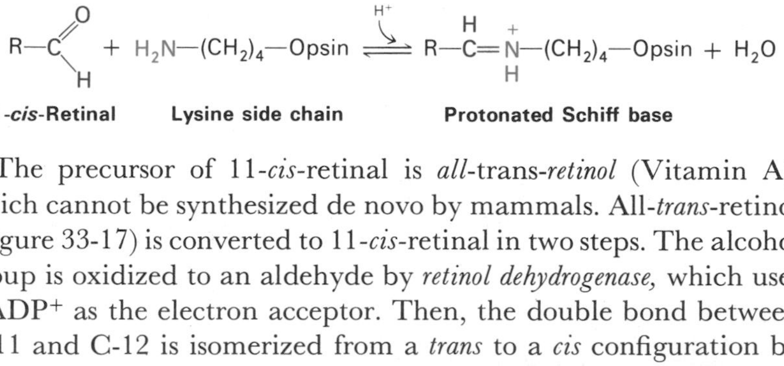

$$R-\underset{\underset{H}{|}}{\overset{\overset{O}{||}}{C}} \; + \; H_2N-(CH_2)_4-Opsin \; \underset{H^+}{\rightleftharpoons} \; R-\underset{\underset{H}{|}}{\overset{\overset{H}{|}}{C}}=\overset{+}{N}-(CH_2)_4-Opsin \; + \; H_2O$$

11-*cis*-Retinal   **Lysine side chain**   **Protonated Schiff base**

The precursor of 11-*cis*-retinal is *all*-trans-*retinol* (Vitamin A), which cannot be synthesized de novo by mammals. All-*trans*-retinol (Figure 33-17) is converted to 11-*cis*-retinal in two steps. The alcohol group is oxidized to an aldehyde by *retinol dehydrogenase,* which uses $NADP^+$ as the electron acceptor. Then, the double bond between C-11 and C-12 is isomerized from a *trans* to a *cis* configuration by *retinal isomerase.* A deficiency of Vitamin A leads to *night blindness* and eventually to the deterioration of the outer segments of rods.

Figure 33-18
Absorption spectrum of rhodopsin.

LIGHT ISOMERIZES 11-*CIS*-RETINAL

The primary event in visual excitation is known. George Wald showed that *light isomerizes the 11-cis-retinal group of rhodopsin to all-trans-retinal*. This isomerization markedly alters the geometry of retinal (Figure 33-19). The Schiff-base linkage of retinal moves approximately 7 Å in relation to the ring portion of the chromophore. In essence, *a photon has been converted into atomic motion.*

All-*trans*-retinal does not fit into the binding site for the 11-*cis* isomer. Consequently, the Schiff base in illuminated rhodopsin is

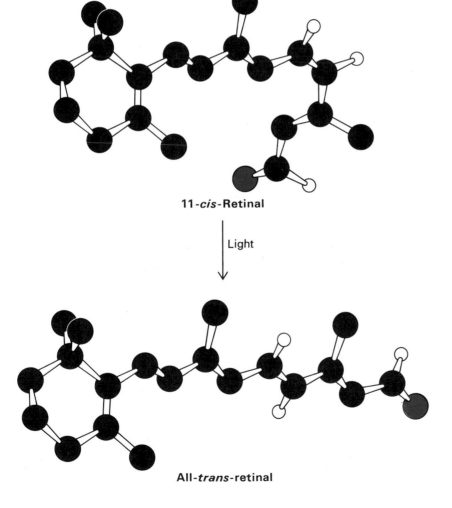

11-*cis*-Retinal

Light

All-*trans*-retinal

Figure 33-19
Molecular models
of 11-*cis*-retinal and
all-*trans*-retinal.

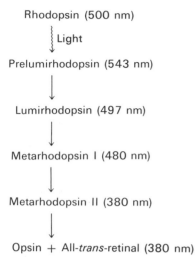

Figure 33-20
Intermediates in the
bleaching of rhodopsin.
Metarhodopsin I appears
about 10^{-5} sec after
illumination and
metarhodopsin II in about
10^{-3} sec.

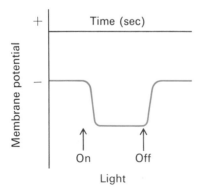

Figure 33-21
Light hyperpolarizes the plasma
membrane of a retinal rod cell.

not stable. *A series of conformational changes occur, leading to the hydrolysis of the Schiff-base linkage.* Several transient species have been identified by their distinctive spectral properties (Figure 33-20). Rhodopsin is rapidly converted to metarhodopsin I and then to metarhodopsin II, which is yellow rather than red. Hence, light bleaches rhodopsin. *A nerve impulse is initiated during this sequence of reactions, which occur in about a millisecond. The hydrolysis of the Schiff base in metarhodopsin II takes about a minute, which is much too slow to be pertinent to the generation of a nerve impulse.* All-*trans*-retinal released from meta-rhodopsin II may be reduced to all-*trans*-retinol. 11-*cis*-Retinal is regenerated by the action of retinol dehydrogenase and retinal isomerase.

LIGHT HYPERPOLARIZES THE PLASMA MEMBRANE OF THE OUTER SEGMENT

The *cis-trans* isomerization of retinal and the consequent conformational changes in rhodopsin are the primary events in visual excitation. An important later step in the generation of a nerve impulse has been elucidated by electrophysiological studies of intact retinas. The plasma membrane of the outer segment becomes hyperpolarized approximately ten milliseconds after the onset of illumination. The membrane potential returns to its initial dark value soon after the light is switched off (Figure 33-21).

Light hyperpolarizes the plasma membrane by decreasing its permeability to Na^+. The plasma membrane of the rod outer segment is unusually permeable to Na^+ in the dark. There is a large inward flow of Na^+ into the outer segment in the dark because there is a Na^+ concentration gradient across the plasma membrane. This gradient is maintained by a Na^+-K^+ active transport system located in the plasma membrane of the inner segment. Thus, in the dark, sodium ion flows into the outer segment, diffuses to the inner segment, and is then extruded by an ATP-driven pump. Light blocks the Na^+ channels in the plasma membrane of the outer segment, the inward flow of Na^+ decreases, and the membrane consequently becomes more negative on the inside. This hyperpolarization is then passively transmitted by the plasma membrane to the synaptic body.

The permeability change and hyperpolarization are *highly amplified responses* of the outer segment. The ratio of the photocurrent to the number of absorbed photons is between 10^5 and 10^6. These responses are *half-maximal when only 30 photons* are absorbed by an outer segment containing forty million rhodopsin molecules. Thus, the properties of the plasma membrane are markedly altered when *just one rhodopsin molecule out of a million is excited by light.*

MISSING LINKS IN VISUAL EXCITATION

How does the *cis-trans* isomerization of a single retinal chromophore in a *disc* membrane produce a change in Na^+ permeability of the *plasma membrane?* The answer is not yet known. The distance between these sites may be several thousand angstroms, and so a direct conformational coupling seems unlikely. The signal from an excited disc membrane is probably conveyed to the plasma membrane by a *diffusible transmitter,* perhaps Ca^{2+}. In one hypothesis, Ca^{2+} is pumped inside discs in the dark. On illumination, many Ca^{2+} ions are released into the medium between the discs for each photon absorbed. The released Ca^{2+} ions diffuse to the plasma membrane, where they close Na^+ gates (Figure 33-22). How might the isomerization of a single 11-*cis*-retinal lead to the release of many Ca^{2+} ions? Recent studies suggest that rhodopsin is at least 75 Å long, which means that it could traverse the disc membrane if suitably oriented. An intriguing possibility is that rhodopsin serves as a light-activated gate. Many Ca^{2+} ions could flow through a single channel. A quite different hypothesis is that rhodopsin is a pro-enzyme that becomes activated by light. In this model, amplification is achieved by the generation of many transmitters by a single activated enzyme. It is evident that detailed information concerning the structure and enzymatic activities of rod outer segments is needed to solve this challenging problem.

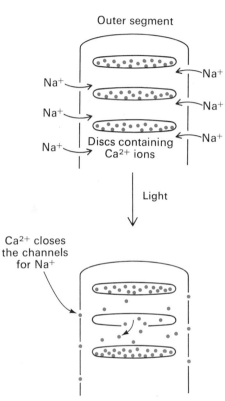

Figure 33-22
A hypothetical mechanism of visual excitation, in which Ca^{2+} is the transmitter.

COLOR VISION IS MEDIATED BY THREE KINDS OF PHOTORECEPTORS

In 1802, Thomas Young proposed that color vision is mediated by *three fundamental receptors.* Spectrophotometric studies of intact retinas more than a century and a half later have revealed that there are in fact *three types of cone cells: blue-, green-, and red-absorbing.* The

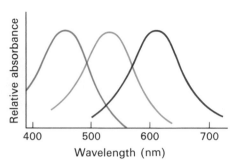

Figure 33-23
Absorption spectra of the three color receptors.

absorption spectra of these three photoreceptor pigments have been obtained by illuminating cones with a beam of light having a diameter of 1 μm (Figure 33-23). Furthermore, microelectrodes have been inserted into many cone cells. The action spectra for the hyperpolarization of their plasma membranes fall into three groups with maxima in the blue, green, and red regions of the visible spectrum. In goldfish, the absorption maxima of the three color receptors are at 455, 530, and 625 nm, whereas that of rhodopsin is at 500 nm.

The chromophore in all three kinds of cones is 11-cis-*retinal.* The protonated Schiff base of 11-*cis*-retinal in the absence of a protein has an absorption maximum at 380 nm. Thus, groups on opsin have a large effect on the chromophoric properties of the bound 11-*cis*-retinal. The dependence of the absorption properties of this chromophore on its protein environment exemplifies a general principle: *the properties of a prosthetic group are modulated by its interaction with the protein.* Another example is that the heme group functions as an oxygen carrier in hemoglobin, an electron carrier in cytochrome *c*, and a catalyst in peroxidase.

Most forms of color blindness are caused by a sex-linked recessive mutation. About 1% of men are red-blind, and 2% are green-blind. Spectral measurements of the intact eye have shown that these individuals lack either the red-absorbing or the green-absorbing photoreceptor molecule. Thus, *color blindness is caused by the absence of one kind of cone opsin.*

11-*CIS*-RETINAL IS THE CHROMOPHORE IN ALL KNOWN VISUAL SYSTEMS

Image-resolving eyes are found in only three phyla: molluscs, arthropods, and vertebrates. The three kinds of eyes are anatomically quite different and are thought to have evolved independently. However, all three use 11-*cis*-retinal as the chromophore in their photoreceptor molecules. *This is a striking example of convergent evolution.* What is so special about 11-*cis*-retinal? First, it has an intense absorption band that can be readily shifted into the visible region of the spectrum. Second, light isomerizes 11-*cis*-retinal with high efficiency. Moreover, its rate of isomerization in the dark is very low. Third, the structural change produced by isomerization is large. Light is converted into atomic motion of sufficient magnitude to trigger the generation of a nerve impulse. Finally, the ultimate precursors of 11-*cis*-retinal are the carotenes, which have a very broad biological distribution.

SUMMARY

Nerve impulses are communicated across most synapses by chemical transmitters. Acetylcholine, the neurotransmitter at many synapses and at motor end plates, is synthesized from acetyl CoA and choline. Acetylcholine molecules are packaged in synaptic vesicles, which fuse with the presynaptic membrane upon arrival of a nerve impulse. The released acetylcholines diffuse across the synaptic cleft and combine with specific protein receptors on the postsynaptic membrane. The consequent increase in the permeability of Na^+ and K^+ leads to the depolarization of this membrane. The acetylcholine receptor has high affinity for α-bungarotoxin and other neurotoxins, a characteristic that has facilitated the purification of this integral membrane protein. The postsynaptic membrane repolarizes following the hydrolysis of acetylcholine by acetylcholinesterase. This enzyme is inhibited by organic phosphates such as DIPF, which can be lethal by causing respiratory paralysis. PAM, a derivative of hydroxylamine, is an antidote for organophosphorus poisoning.

A retinal rod cell can be excited by a single photon. The outer segment of a rod contains a stack of about 1,000 discs, which are closed bilayer membranes densely packed with rhodopsin molecules. The chromophore of rhodopsin, the photoreceptor protein, is 11-*cis*-retinal, which is derived from all-*trans*-retinol (Vitamin A). 11-*cis*-Retinal forms a Schiff-base linkage with a specific lysine residue of opsin. The primary event in visual excitation is the isomerization of the 11-*cis*-retinal group to all-*trans*-retinal. The next defined step in visual excitation is a hyperpolarization of the plasma membrane of the outer segment, which results from a decrease in its permeability to Na^+. The links between the *cis-trans* isomerization in a single rhodopsin molecule (the primary event) and the closing of Na^+ channels in the plasma membrane (the amplified response) are not yet known. Color vision is mediated by three kinds of photoreceptors, each containing 11-*cis*-retinal. In fact, 11-*cis*-retinal is the chromophore in all known visual systems, which is a striking example of convergent evolution.

SELECTED READINGS

Where to start:

Axelrod, J., 1974. Neurotransmitters. *Sci. Amer.* 230(6):59–71. [Available as *Sci. Amer.* Offprint 1297.]

Wald, G., 1968. The molecular basis of visual excitation. *Nature* 219:800–807.

Books and reviews:

Katz, B., 1966. *Nerve, Muscle and Synapse.* McGraw-Hill. [A superb concise introduction.]

Aidley, D. J., 1971. *The Physiology of Excitable Cells.* Liverpool University Press.

Langer, H., (ed.), 1973. *Biochemistry and Physiology of Visual Pigments.* Springer-Verlag.

Hille, B., 1971. Ionic channels in nerve membranes. *Progr. Biophys. Mol. Biol.* 21:1–32.

Hall, Z., 1972. Release of neurotransmitters and their interaction with receptors. *Annu. Rev. Biochem.* 41:925–952.

Barondes, S. H., 1974. Synaptic macromolecules: identification and metabolism. *Annu. Rev. Biochem.* 43:147–168.

Hagins, W. A., 1972. The visual process. *Annu. Rev. Biophys. Bioeng.* 1:131–158.

Daemen, F. J. M., 1973. Vertebrate rod outer segment membranes. *Biochim. Biophys. Acta* 30:255–288.

Acetylcholine receptor:

Klett, R. P., Fulpius, B. W., Cooper, D., Smith, M., Reich, E., and Possani, L. D., 1973. The acetylcholine receptor. I. Purification and characterization of a macromolecule isolated from *Electrophoros electricus. J. Biol. Chem.* 248:6841–6853.

Hazelbauer, G. L., and Changeux, J.-P., 1974. Reconstruction of a chemically excitable membrane. *Proc. Nat. Acad. Sci.* 71:1479–1483.

Photoreceptors:

Tomita, T., 1970. Electrical activity of vertebrate photoreceptors. *Quart. Rev. Biophys.* 3:179–222.

Young, R. W., 1970. Visual cells. *Sci. Amer.* 223(4):80–91. [Offprint 1201.]

Dowling, J. E., 1966. Night blindness. *Sci. Amer.* 215(4):78–84. [Offprint 1053.]

MacNichol, E. F., Jr., 1964. Three-pigment color vision. *Sci. Amer.* 211(6):48–56. [Offprint 197.]

Wu, C.-W., and Stryer, L., 1972. Proximity relationships in rhodopsin. *Proc. Nat. Acad. Sci.* 69:1104–1108.

Miki, N., Keirns, J. J., Marcus, F., Freeman, J., and Bitensky, M. W., 1973. Regulation of cyclic nucleotide concentrations in photoreceptors: an ATP-dependent stimulation of cyclic nucleotide phosphodiesterase by light. *Proc. Nat. Acad. Sci.* 70:3820–3824.

CHAPTER 34

HORMONE ACTION

Hormones are chemical messengers that coordinate the activities of different cells in multicellular organisms. The term hormone was first used in 1904 by William Bayliss and Ernest Starling to describe the action of secretin, a molecule secreted by the duodenum that stimulates the subsequent flow of pancreatic juice. A very fruitful concept emerged from this work—namely, that (1) hormones are molecules synthesized by specific tissues (*glands*); (2) they are secreted directly into the blood, which carries them to their sites of action; and (3) they specifically alter the activities of certain responsive tissues (target organs or *target cells*).

Hormones are chemically diverse. Some hormones, such as epinephrine and thyroxine, are *small molecules derived from amino acids.* Others, such as oxytocin, insulin, and thyroid-stimulating hormone, are *polypeptides* or *proteins.* A third group of hormones consists of the *steroids,* which are derived from cholesterol. The molecular basis of the action of some hormones is now being elucidated. *Hormones can exert their specific effects in three ways: (1) by influencing the rate of synthesis of enzymes and other proteins; (2) by affecting the rate of enzymatic catalysis; and (3) by altering the permeability of cell membranes.* It is interesting to note that no known hormone is an enzyme or co-enzyme. Rather, *hormones act by regulating preexisting processes.*

DISCOVERY OF CYCLIC AMP, A MEDIATOR OF THE ACTION OF MANY HORMONES

A major breakthrough in the elucidation of the mechanism of action of hormones was made by Earl Sutherland as an outgrowth of studies started in the 1950s. The initial aim was to determine how epinephrine and glucagon elicit the breakdown of glycogen and the production of glucose by the liver. Sutherland chose this system because, first, the effects of these hormones on glycogen breakdown are large and reproducible. Second, they occur within a few minutes. Third, liver slices are easy to prepare in large quantities. Fourth, much was known about the biochemistry of glycogen breakdown (Chapter 16). In fact, Sutherland started these studies on the mechanism of action of epinephrine and glucagon in the laboratory of Carl Cori and Gerty Cori.

One of the first steps was to identify the enzymatic reaction in the conversion of glycogen to glucose that was enhanced by these hormones. Measurements of the labeled intermediates following incubation of liver slices in the presence of $^{32}P_i$ showed that phosphorylase, rather than phosphoglucomutase or glucose 6-phosphatase, was rate-limiting. Furthermore, epinephrine and glucagon led to an increase in the activity of phosphorylase. However, the nature of the activation process was not yet evident. Sutherland then found an enzyme that catalyzed the inactivation of active phosphorylase. This inactivating enzyme proved to be a phosphatase, which suggested that phosphorylase might be activated by being phosphorylated. This hypothesis was tested by incubating liver slices in the presence of $^{32}P_i$. Indeed, the rate of incorporation of ^{32}P into phosphorylase was found to be increased by glucagon and epinephrine, in direct proportion to their effects on glycogen breakdown. These studies revealed that *phosphorylase is activated by phosphorylation and inactivated by dephosphorylation. This was the first example of the regulation of an enzyme by covalent modification.*

The activation of phosphorylase by hormones in a preparation of broken liver cells was then studied. The exciting finding was that the addition of epinephrine and glucagon resulted in the activation of phosphorylase, as in liver slices. *This observation of hormone action in a cell-free homogenate is a landmark in biochemistry.* Specific hormone effects had not previously been observed in cell-free systems. Some biologists had felt that hormones could only act on intact target cells. It is interesting to note that a similar attitude had been shattered a half-century earlier when the Buchners showed that fermentation can occur in cell-free extracts of yeast. However, in contrast to the glycolytic enzymes, not all of the components of the epinephrine and glucagon response system are soluble. Experiments showed that the response to these hormones is lost if the liver-cell homogenate is centrifuged. Hence, an essential part of this hormone response system must be located in a particulate fraction. In fact, the hormonal response was restored by adding back the particulate fraction to the original supernatant.

The role of the particulate fraction, which contains the plasma membrane, became evident from the following experiment. When the particulate fraction was incubated with epinephrine and glucagon, a heat-stable factor was produced. This factor activated phosphorylase when added to the supernatant fraction. Thus, *the hormonal response was separated into two parts: the interaction of the hormones with the particulate fraction to produce a heat-stable factor, followed by its effect on the supernatant to activate phosphorylase.* The next challenge was to identify this heat-stable intermediate of which only small amounts were available. Chemical analysis showed that it was an adenine ribonucleotide, but its properties were unusual. Sutherland wrote Leon Heppel about this molecule in the hope that he might be able to help in elucidating its structure. At the same time, David Lipkin wrote Heppel describing a new nucleotide that was produced by treating ATP with barium hydroxide. Heppel surmised that Lipkin and Sutherland were studying the same molecule, and he put them in touch with each other. Indeed, both investigators were studying the same molecule, which turned out to be adenosine 3', 5'-monophosphate, now commonly referred to as *cyclic AMP* or *cAMP*. Another dividend of this chance encounter was that large amounts of cyclic AMP could be prepared for biochemical studies. Furthermore, the laboratory synthesis of cyclic AMP from ATP and barium hydroxide suggested a plausible route for its biosynthesis.

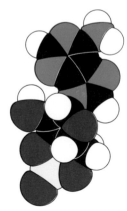

Figure 34-1
Space-filling model of cyclic AMP.

CYCLIC AMP IS SYNTHESIZED BY ADENYL CYCLASE
AND DEGRADED BY A PHOSPHODIESTERASE

Cyclic AMP is formed from ATP by the action of *adenyl cyclase*, a membrane-bound enzyme.

$$ATP \xrightarrow{Mg^{2+}} \text{Cyclic AMP} + PP_i + H^+$$

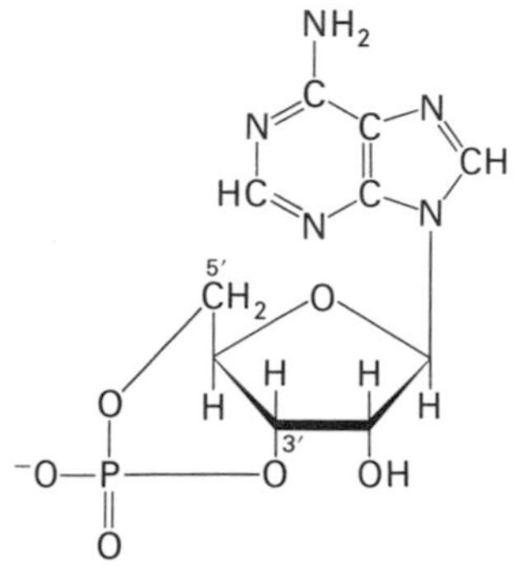

This reaction is slightly endergonic, having a $\Delta G^{\circ\prime}$ of about 1.6 kcal/mol. The synthesis of cyclic AMP is driven by the subsequent hydrolysis of pyrophosphate. Cyclic AMP is destroyed by a specific *phosphodiesterase,* which hydrolyzes it to AMP.

$$\text{Cyclic AMP} + H_2O \xrightarrow{Mg^{2+}} AMP + H^+$$

This reaction is highly exergonic, having a $\Delta G^{\circ\prime}$ of about -12 kcal/mol. Cyclic AMP is a very stable compound in the absence of the phosphodiesterase.

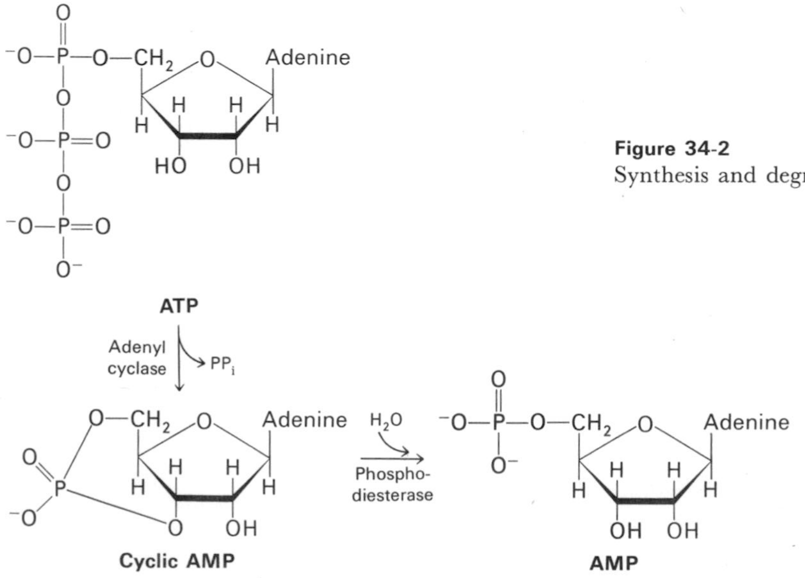

**Adenosine 3′,5′-monophosphate
(Cyclic AMP)**

Figure 34-2
Synthesis and degradation of cyclic AMP.

ATP

Adenyl
cyclase

PP_i

Cyclic AMP

H₂O

Phospho-
diesterase

AMP

CYCLIC AMP IS A SECOND MESSENGER IN THE ACTION OF MANY HORMONES

Sutherland's work led to the concept that cyclic AMP is a *second messenger* in the action of some hormones. The first messenger is the hormone itself. The essential features of this concept are:

1. Cells contain receptors for hormones in the plasma membrane.

2. The combination of a hormone with its specific receptor in the plasma membrane leads to the stimulation of adenyl cyclase, which is also bound to the plasma membrane.

3. The increased activity of adenyl cyclase increases the amount of cyclic AMP inside the cell.

4. Cyclic AMP then acts inside the cell to alter the rate of one or more processes.

An important feature of the second-messenger model is that the hormone need not enter the cell. Its impact is made at the cell membrane. The biological effects of the hormone are mediated inside the cell by cyclic AMP rather than by the hormone itself. A number of experimental criteria have been used to test the validity of this concept:

1. Adenyl cyclase in a target cell should be stimulated by hormones that affect that cell. Hormones that do not elicit a characteristic biological response in such a cell should not elevate its cyclase level.

2. The concentration of cyclic AMP in target cells should change in direct proportion to their biological response to hormonal stimulation, in terms of both kinetics and dependence on hormone concentration.

3. Inhibitors of the phosphodiesterase, such as theophylline and caffeine, should act synergistically with hormones that use cyclic AMP as a second messenger.

4. The biological effects of the hormone should be mimicked by the addition of cyclic AMP or a related compound to the target cells. (In practice, cyclic AMP cannot readily be used in this way

because it penetrates cells poorly. However, less polar derivatives of cyclic AMP, such as dibutyryl cyclic AMP, do enter cells and are active.)

Experiments based on these criteria have shown that cyclic AMP is a second messenger for many hormones in addition to epinephrine and glucagon. Other hormones that act by means of cyclic AMP are norepinephrine, vasopressin, thyroxine, parathyroid and luteinizing hormones, and thyroid-stimulating and melanocyte-stimulating hormones. Furthermore, cyclic AMP affects an exceptionally wide range of cellular processes (Table 34-1).

Table 34-1
Effects of cyclic AMP

Process	Change in rate
Lipolysis	Increase
Glycogen degradation	Increase
Gluconeogenesis	Increase
Ketogenesis	Increase
Ion permeability of epithelia	Increase
Renin production by kidney	Increase
Contraction of cardiac muscle	Increase
HCl secretion by the gastric mucosa	Increase
Amylase release by the parotid gland	Increase
Dispersion of melanin granules	Increase
Insulin release by the pancreas	Increase
Aggregation of platelets	Decrease
Growth of tumor cells in tissue culture	Decrease

CYCLIC AMP STIMULATES A PROTEIN KINASE

How does cyclic AMP influence such a wide range of cellular processes? Is there a common denominator for many of these effects? Indeed there is, and again the answer has come from studies of the control of glycogen metabolism, which has been described in a lighter vein as the metabolic birthplace of cyclic AMP. Edwin Krebs and Donald Walsh discovered that *cyclic AMP activates a protein kinase* in skeletal muscle. This protein kinase phosphorylates both glycogen synthetase (rendering it inactive) and phosphorylase kinase (rendering it active). In this way, cyclic AMP stimulates glycogen breakdown and stops glycogen synthesis in muscle (p. 392). A similar mechanism exists in liver. In fact, *many of the effects elicited by cyclic AMP result from the activation of protein kinases.* Nearly all cells studied thus far contain protein kinases that are activated by cyclic AMP at concentrations of the order of 10^{-8} M. These kinases modulate the activities of different proteins in different cells by phosphorylating them.

The mechanism of activation of the protein kinase in muscle by cyclic AMP is an interesting one. This enzyme consists of two subunits: a regulatory subunit, which can bind cyclic AMP, and a catalytic subunit. In the absence of cyclic AMP, the regulatory and catalytic subunit form a complex that is enzymatically inactive. In the presence of cyclic AMP, the complex dissociates. The regulatory subunit binds cyclic AMP. The free catalytic subunit is then enzymatically active. Thus, *the binding of cyclic AMP to the regulatory subunit relieves its inhibition of the catalytic subunit. Cyclic AMP acts as an allosteric effector.* The presence of distinct regulatory and catalytic subunits in the protein kinase is reminiscent of the subunit structure of aspartate transcarbamoylase.

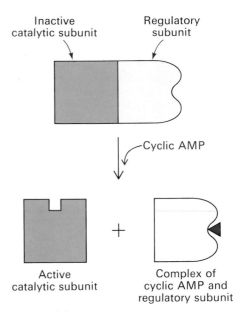

Figure 34-3
Cyclic AMP activates protein kinases by dissociating the complex of the regulatory and catalytic subunits.

CYCLIC AMP IS AN ANCIENT HUNGER SIGNAL

As discussed previously, cyclic AMP has a regulatory role in bacteria, where it stimulates the transcription of certain genes (p. 687). It is evident that *cyclic AMP has a long evolutionary history as a regulatory molecule.* In bacteria, cyclic AMP is a hunger signal. It signifies an absence of glucose and leads to a synthesis of enzymes that can exploit other energy sources. In some mammalian cells, such as in liver and muscle, cyclic AMP retains its ancient role as a hunger signal. However, it acts by stimulating a protein kinase rather than by enhancing the transcription of certain genes. Another important difference is that cyclic AMP has become a second messenger in higher organisms, responding to hormones produced by other cells.

Why was cyclic AMP chosen in the course of evolution to be a second messenger? Three factors seem important:

1. Cyclic AMP is derived from ATP, a ubiquitous molecule, in a simple reaction driven forward by the subsequent hydrolysis of pyrophosphate.

2. Though derived from a molecule that is at the center of metabolic transformations, cyclic AMP itself is not on a major metabolic pathway. It is used only as an integrator of metabolism, not as a biosynthetic precursor or intermediate in energy production. Hence, its concentration can be independently controlled. Furthermore, it is stable unless hydrolyzed by a specific phosphodiesterase.

3. Cyclic AMP has a sufficient number of functional groups so that it can bind tightly and specifically to receptor proteins, such as the regulatory subunit of the protein kinase in muscle, and elicit allosteric effects.

It is important to note that the *hormonal signal is greatly amplified by the use of cyclic AMP as a second messenger.* The concentration of many hormones in the blood is of the order of 10^{-10} M. In a stimulated target cell, the concentration of cyclic AMP is typically 10^{-6} M. Thus, the synthesis of cyclic AMP by an activated adenyl cyclase molecule in the plasma membrane leads to a *10^4-fold enhancement* of the hormonal signal. Activation of the protein kinase by cyclic AMP produces a further amplification since the protein kinase has an appreciable turnover number. The enzymatic cascade in the control of glycogen metabolism has one or two more steps, resulting in an even higher amplification of the original hormonal stimulus.

CHOLERA TOXIN STIMULATES ADENYL CYCLASE

The direct participation of cyclic AMP in a disease process has been clearly established in cholera. This potentially lethal disease is caused by *Vibrio cholerae,* a gram-negative bacterium that is propelled by a single flagellum. The striking clinical feature of this disease is a massive diarrhea. Several liters of body fluid may be lost within a few hours, which may lead to shock and death if the fluids are not replaced. The diarrhea is caused by a bacterial toxin rather than by a direct action of the bacteria themselves. *Cholera toxin (a 90,000-dalton protein) increases the adenyl cyclase activity of the mucosa of the small intestine, which in turn raises the level of cyclic AMP in these cells.* The abnormally high level of cyclic AMP affects the active transport of ions by these intestinal epithelial cells, leading to a very large efflux of Na^+ and water into the gut.

INSULIN STIMULATES ANABOLIC PROCESSES AND INHIBITS CATABOLIC PROCESSES

We now turn to insulin, a polypeptide hormone that has profound effects on a wide range of metabolic processes. A unifying characteristic is that insulin promotes anabolic processes and inhibits catabolic ones in muscle, liver, and adipose tissue. Specifically, insulin increases the rate of synthesis of glycogen, fatty acids, and proteins, and also stimulates glycolysis. An important action of the hormone is that *it promotes the entry of glucose, some other sugars, and amino acids into muscle and fat cells.* Hence, the level of glucose in the blood is lowered by insulin (called the *hypoglycemic effect*). Insulin inhibits catabolic processes such as the breakdown of glycogen and fat. It also decreases gluconeogenesis by lowering the level of enzymes such as pyruvate carboxylase and fructose 1,6-diphosphatase.

It is evident that many of the effects of insulin are contrary to those elicited by epinephrine and glucagon. In essence, *epinephrine and glucagon signal that glucose is scarce, whereas insulin signals that glucose is abundant.*

THE PRIMARY SITE OF INSULIN ACTION IS ON THE PLASMA MEMBRANE

The large effect of insulin in promoting the transport of sugars across the plasma membrane of fat and muscle cells has suggested that the surface of these cells may be the principal site of action of this hormone. This hypothesis was tested by preparing an insulin derivative that does not penetrate the plasma membrane. Insulin was covalently attached through its amino groups to carbohydrate beads (made of agarose, a polymer of galactose) that are comparable in size to fat cells (50–100 μm in diameter). The hormonal activity of insulin attached to these large, impermeable beads was then assayed by measuring the response of isolated fat cells. Two criteria of insulin action were used: the increase in the utilization of glucose and the decrease in the breakdown of fats. The striking finding was that the *hormonal activity of insulin attached to the agarose beads was nearly as high as that of native insulin.* This result strongly suggested that the plasma membrane contains receptors for insulin and that *the interaction of insulin with structures on the surface of the membrane may suffice to elicit the transport and metabolic effects of this hormone.*

PROINSULIN IS THE PRECURSOR OF THE ACTIVE HORMONE

Frederick Sanger showed that bovine insulin consists of two chains—an A chain of 21 residues and a B chain of 30 residues—which are covalently joined by two disulfide links (p. 21). The same pattern exists in insulin molecules from many different species, including humans. How is this two-chain protein synthesized? A priori, one possibility is that the A and B chains are synthesized separately and that correct disulfide pairings are formed after the two chains associate in a specific way because of noncovalent interactions. This proposed mechanism was tested by determining whether active hormones having the correct disulfide pairings can be formed in vitro from reduced A and B chains. Anfinsen's demonstration that ribonuclease, a one-chain protein, can efficiently refold into an active enzyme after it is reduced and unfolded (p. 36) provided the stimulus for this experiment. The outcome with insulin was very different: less that 4% of the molecules regained the correct disulfide pairings. This low yield for the refolding of insulin was like that obtained for chymotrypsin, a three-chain enzyme connected by two interchain disulfide bonds (p. 180). In contrast, chymotrypsinogen, the one-chain precursor, refolded to a much greater degree. The contrasting behavior of chymotrypsin and chymotrypsinogen suggested that insulin did not refold properly because part of the necessary information was missing. These studies raised the possibility that insulin, like chymotrypsin, is derived from a single polypeptide chain.

A breakthrough came from a study of an islet-cell adenoma of the pancreas. This rare human tumor produces large amounts of insulin. Slices of this pancreatic tumor were incubated with tritiated leucine and then analyzed. A new, highly radioactive protein was found (Figure 34-4). Subsequent studies revealed that this new protein, called *proinsulin,* is the *biosynthetic precursor of insulin.*

Proinsulin is a single polypeptide chain containing a sequence of about thirty amino acid residues that is absent from insulin. This *connecting peptide* joins the carboxyl end of the B chain and the amino terminus of the A chain of the future insulin molecule. As expected, *proinsulin forms correct disulfide pairings when its disulfide bonds are reduced and then reoxidized.*

The removal of the connecting peptide by enzymatic digestion converts proinsulin, which is not active as a hormone, into active insulin. The connecting peptides in proinsulins from different species have common structural features. In particular, the connecting peptide in all proinsulins studied thus far contains Arg-Arg at the amino end and a Lys-Arg at the carboxyl end. A proteolytic enzyme similar to trypsin hydrolyzes the polypeptide chain at these positively charged residues. *Proinsulin is converted in the storage granules of the β cells of the pancreas.* The insulin molecules in mature storage granules are secreted when the membrane of the granule fuses with the plasma membrane of the cell.

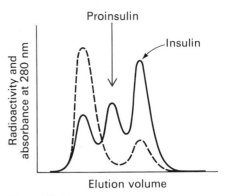

Figure 34-4
The appearance of a highly radioactive protein following pulse-labeling of an islet-cell adenoma led to the discovery of proinsulin. (The solid line refers to radioactivity; the dashed line to absorbance at 280 nm.)

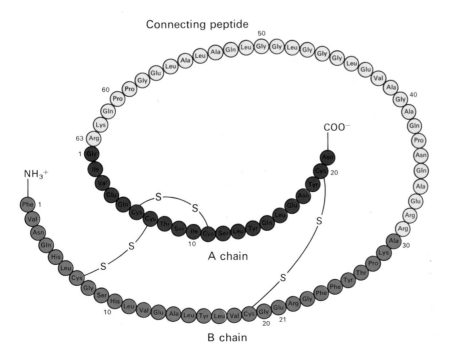

Figure 34-5
Amino acid sequence of porcine proinsulin: the A chain of insulin is shown in red, the B chain in blue, and the connecting peptide in yellow. [Based on R. E. Chance, R. M. Ellis, and W. W. Bromer. *Science* 161(1968):165. Copyright 1968 by the American Association for the Advancement of Science.]

THREE-DIMENSIONAL STRUCTURE OF INSULIN

X-ray crystallographic studies done in the laboratory of Dorothy Crowfoot Hodgkin have revealed the three-dimensional structure of porcine insulin at a resolution of 1.9 Å. This accomplishment was completed approximately thirty-six years after Hodgkin obtained the first x-ray diffraction pattern of a protein crystal (pepsin) when she was a graduate student in the laboratory of John Bernal. In the intervening years, Hodgkin has solved the structures of such biologically important molecules as cholesterol, penicillin, and Vitamin B_{12}. *Insulin has a compact three-dimensional structure* (Figure 34-6). Only the amino and carboxyl termini of the B chain are extended away from the rest of the protein. The A chain is nestled between these extended arms of the B chain. There is a nonpolar core, consisting of buried aliphatic side chains from both the A and B chains. Insulin is also stabilized by several salt links and by hydrogen bonds between groups on the A and B chains, in addition to its two interchain disulfide bonds. The conformation of proinsulin is not yet known. However, spectroscopic studies suggest that proinsulin closely resembles insulin in its three-dimensional structure. Furthermore, proinsulin cocrystallizes with insulin.

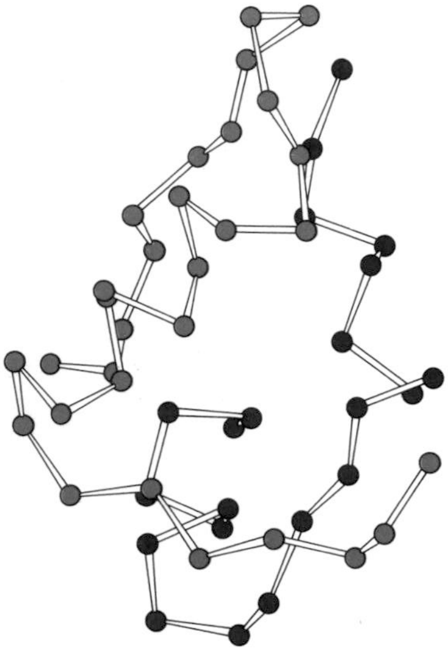

Figure 34-6
Alpha-carbon diagram of insulin: the A chain is shown in red, the B chain in blue.

ISOLATION OF THE INSULIN RECEPTOR

How does insulin act at the molecular level? The structure of insulin is known in magnificent detail, but not that of its reaction partner, the specific receptor for insulin in the plasma membrane. However, significant progress has recently been made by Pedro Cuatrecasas in the isolation of this receptor. An essential first step was to devise a specific assay. The insulin receptor has at least two functions: it must specifically recognize insulin and then convey this information to other molecules. An assay based on the recognition property of the receptor was devised since it was thought that this function might be retained when the plasma membrane is solubilized in detergent. In contrast, the transmission of this information to other membrane components would almost certainly be disrupted by solubilization of the plasma membrane. The binding of insulin made highly radioactive by covalent insertion of ^{125}I was used as the assay for the insulin receptor. There was a good correlation between the binding of labeled insulin to isolated fat cells and its capacity to elicit the characteristic responses. The specificity of binding was demonstrated by the finding that radioactive insulin was displaced from the fat cell by the addition of unlabeled insulin but not by the addition of reduced insulin or other peptide hormones. After homogenization of the isolated fat cells, the binding activity was present in the nonnuclear particulate fraction, specifically in the plasma membrane.

There are very few insulin receptors on the plasma membrane of a fat cell. Titration studies showed that only 10,000 insulin molecules are bound to a cell at saturation, which corresponds to a density of insulin receptors of about one per square micron of plasma membrane. This means that there are about 10^6 phospholipids per insulin receptor and that the distance from a receptor to its nearest neighbor is of the order of 10,000 Å (1 μm).

The binding of insulin to the receptor is very tight. The dissociation constant for the reaction

$$\text{Insulin + receptor} \underset{k_2}{\overset{k_1}{\rightleftharpoons}} \text{insulin-receptor}$$

is 5×10^{-11} M. Such tight binding is necessary since the insulin level in the blood is of the order of 10^{-10} M. The rate constant for association is very high ($k_1 = 1.5 \times 10^7$ M^{-1} sec^{-1}), approaching the value of a diffusion-controlled reaction. The dissociation rate constant is low ($k_2 = 7.4 \times 10^{-4}$ sec), which means that the complex has a half-life of about 16 minutes. *The hormone-receptor complex is reversible.* Neither the receptor nor the hormone is inactivated by binding.

The insulin receptor has been extracted from the plasma membrane by the addition of nonionic detergents, such as Triton X-100. The solubilized receptor has the same binding properties as the receptor in the intact plasma membrane. The solubilized receptor has been purified by affinity chromatography, a separation procedure that exploits the specific binding properties of the receptor. The insulin receptor binds tightly to an affinity column consisting of insulin covalently attached to agarose, whereas other molecules in the solubilized membrane preparation do not. The insulin receptor is then eluted by an acidic urea solution, which denatures the receptor and dissociates it from the insulin-agarose column.

Fortunately, the purified receptor can then be renatured by removing the urea and raising the pH of the solution. A 250,000-fold purification has been achieved by these procedures.

The insulin receptor comprises only about $4 \times 10^{-3}\%$ of the membrane protein of a liver homogenate. This means that five hundred grams of protein present in the homogenate derived from the livers of two hundred rats has to be fractionated to obtain one milligram of pure receptor protein. Obtaining enough receptor for amino acid sequence and x-ray crystallographic studies, and for reconstitution experiments, will be a formidable but not impossible task.

INSULIN DEFICIENCY CAUSES DIABETES

Diabetes mellitus is a complex disease that affects several hundred million people. Diabetes is characterized by an elevated level of glucose in the blood and in the urine. Glucose is excreted in the urine when the blood glucose level exceeds the reabsorptive capacity of the renal tubules. Water accompanies the excreted glucose, and so an untreated diabetic in the acute phase of the disease is hungry and thirsty. The loss of glucose depletes the carbohydrate stores, thereby leading to the breakdown of fat and protein. The mobilization of fats results in the formation of large amounts of acetyl CoA. Ketone bodies (acetoacetate, acetone, and hydroxybutyrate) are formed when acetyl CoA cannot enter the citric acid cycle because there is insufficient oxaloacetate. Recall that animals cannot synthesize oxaloacetate from acetyl CoA. Rather, oxaloacetate is derived from glucose and some amino acids (p. 369). The excretion of ketone bodies impairs the acid-base balance and causes further dehydration, which may lead to coma and death in the acute phase of the disease in an untreated diabetic.

Diabetes mellitus probably arises from a deficiency of insulin, although the underlying causes are unknown. Diabetes can be experimentally produced in an animal by surgical removal of much of the pancreas or by chemical destruction of the β cells. These insulin-producing cells can be selectively destroyed by administration of alloxan. Further evidence that insulin deficiency plays an important role in diabetes is provided by the fact that acute diabetes is rapidly reversed by the administration of insulin.

Alloxan
(2,4,5,6-Tetraoxypyrimidine)

Why is a diabetic deficient in insulin relative to the needs of his tissues? *There may be many kinds of underlying molecular defects that produce the clinical picture called diabetes mellitus.* Possible defects include abnormalities in the mechanism of insulin synthesis, storage, and release. Alternatively, the defect may reside in the utilization of insulin. For example, the binding of insulin to the specific receptor in the plasma membrane may be impaired. Another possibility is that the interactions between the insulin-receptor complex and other membrane constituents may be abnormal. There is every reason to expect that an understanding of these processes in molecular detail will shed light on the causes of diabetes and also lead to more effective therapy.

PROSTAGLANDINS ARE MODULATORS OF HORMONE ACTION

We now turn to the prostaglandins, a group of fatty acids that affect a wide variety of physiological processes. These compounds were discovered in the 1930s but have become prominent only recently, largely because of the pioneering work of Sune Bergstrom. *Prostaglandins are twenty-carbon fatty acids that contain a five-carbon ring.* The major classes are designated PGA, PGB, PGE, and PGF, followed by a subscript, which denotes the number of carbon–carbon double bonds outside the ring. *Prostaglandins seem to modulate the action of hormones rather than act as hormones themselves.* They modulate the activities of many of the cells in which they are synthesized. The nature of these effects may vary from one type of cell to another, in contrast to the uniformity in the action of a hormone.

The mechanism of the effect of prostaglandin PGE_1 on the breakdown of fat in adipose tissue has been studied in detail. Lipolysis is stimulated by hormones such as epinephrine, glucagon, adrenocorticotrophic hormone, and thyroid-stimulating hormone. PGE_1 at a concentration of 10^{-7} M strongly inhibits the lipolytic effects of these hormones. A related finding is that PGE_1 prevents the rise of the intracellular level of cyclic AMP that is elicited by these hormones. However, PGE_1 does not inhibit lipolysis caused by the addition of dibutyryl cyclic AMP. Hence, *PGE_1 probably inhibits adenyl cyclase in fat cells.* In other cells, prostaglandins may have the opposite effect on cyclic AMP level.

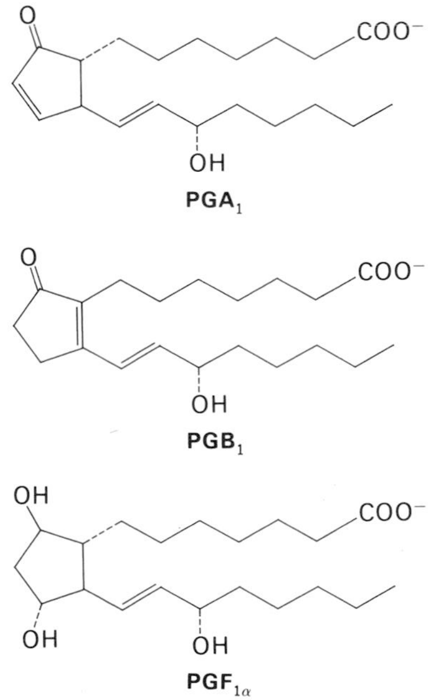

Figure 34-7
Structures of some prostaglandins.

The molecular basis of many of the major actions of prostaglandins is not yet known. Some of the effects of prostaglandins are the stimulation of inflammation, the regulation of blood flow to particular organs, the control of ion transport across some membranes, and the modulation of synaptic transmission. There is much clinical interest in prostaglandins. For example, prostaglandins may be involved in parturition. Infusion of PGE_2 induces delivery within a few hours. Also, prostaglandins are being evaluated for potential use as contraceptives. $PGF_{2\alpha}$ reduces the secretion of progesterone, a hormone needed for implantation of a fertilized ovum in the uterus.

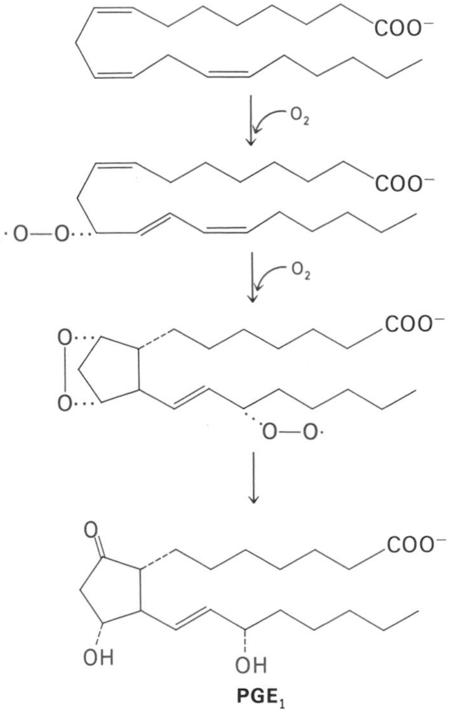

PROSTAGLANDINS ARE SYNTHESIZED FROM UNSATURATED FATTY ACIDS

Prostaglandins are synthesized in membranes from C_{20} fatty acids that contain at least three double bonds. These polyunsaturated fatty acids are required in the diet of mammals (p. 428). The precursors of prostaglandins are released from membrane phospholipids by phospholipases. For example, the biosynthesis of PGE_1 starts from $\Delta^8,\Delta^{11},\Delta^{14}$-eicosatrienoate. A cyclopentane ring is formed and three oxygen atoms are introduced by a microsomal synthetase system. As expected, all three oxygen atoms come from molecular oxygen. An interesting recent finding is that *aspirin inhibits the biosynthesis of prostaglandins by interfering with the enzymatic insertion of oxygen.* Prostaglandins enhance inflammatory effects, whereas aspirin diminishes them. This pharmacologic activity of aspirin may be due to its inhibition of the biosynthesis of prostaglandins.

Figure 34-8
Synthesis of PGE_1 starting from cis-$\Delta^8,\Delta^{11},\Delta^{14}$-eicosotrienoate.

STEROID HORMONES ACTIVATE SPECIFIC GENES

The *primary effect* of steroid hormones such as estradiol, progesterone, and cortisone is on *gene expression* rather than on enzyme activities or transport processes. In contrast to insulin, these hormones must enter their target cells to exert their effects. Furthermore, their major site of action is in the *cell nucleus* rather than on the plasma membrane. The full impact of these steroids is achieved in hours rather than minutes because their biological effects depend on the *synthesis of new proteins.* Actinomycin D inhibits the action of these steroid hormones, which implies that the *synthesis of new messenger RNA is required for their action.*

The first step in the stimulation of uterine growth by 17β-estradiol is the binding of this hormone to a specific receptor in the cytoplasm of the uterine cell. The binding is very tight ($K \sim 10^{-9}$ M). This hormone-receptor complex then migrates to the nucleus of the cell. *The binding of estradiol to the receptor markedly enhances its affinity for DNA.* Furthermore, the receptor acquires a second subunit after binding estradiol. This is reflected in an increase in its sedimentation coefficient from 4 S to 5 S. It is not yet known whether the binding sites for the estradiol receptor on DNA are specified by the base sequence or by chromosomal proteins. The interaction of the hormone-receptor complex with DNA is highly specific. Furthermore, the activation of the receptor by the binding of hormone is also quite specific. Estrone binds to the receptor, but this complex does not interact with DNA. The failure of the estrone-receptor complex to bind to DNA agrees with the fact that estrone does not stimulate uterine growth.

The interactions of estradiol probably exemplify a general pattern for steroid hormones. For example, dexamethasone, a glucocorticoid steroid hormone, also binds to a specific receptor in the cytoplasm of hepatoma cells in tissue culture. The receptor undergoes a conformational change on binding the hormone and then migrates to the cell nucleus, where it complexes to specific sites on DNA. The number of binding sites on DNA for this receptor is estimated to be 1 per 10^6 base pairs. This finding is consistent with the observation that this hormone influences the transcription of a rather small number of genes. The mechanism of this highly selective control of transcription by steroid hormones is an intriguing problem.

SUMMARY

Hormones are a group of diverse molecules that integrate the activities of different cells in multicellular organisms. Cyclic AMP plays a critical role in the action of many hormones (e.g., epinephrine, glucagon, and thyroxine) by serving as a second messenger inside the target cell. Such hormones bind to specific receptors on the plasma membrane of the target cell and stimulate adenyl

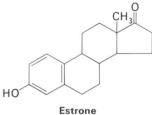
17β-Estradiol

Estrone

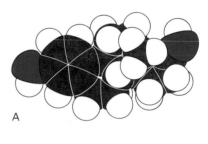
A

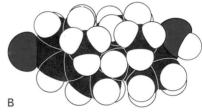
B

Figure 34-9
Space-filling models of (A) 17β-estradiol and (B) testosterone.

cyclase. This membrane-bound enzyme catalyzes the synthesis of cyclic AMP from ATP. The activation of adenyl cyclase leads to an increased amount of cyclic AMP inside the cell. Cyclic AMP then activates a protein kinase, which phosphorylates one or more proteins. For example, the phosphorylation of glycogen synthetase and phosphorylase kinase in muscle and liver results in decreased synthesis and enhanced degradation of glycogen. This reaction cascade amplifies the initial hormonal stimulus. Cyclic AMP has a long evolutionary history as a regulatory molecule, starting as a hunger signal.

Insulin promotes anabolic processes (e.g., the synthesis of glycogen, fatty acids, and proteins) and inhibits catabolic ones (e.g., the breakdown of glycogen and fat). Insulin has a striking effect on membrane transport. Specifically, this hormone stimulates the entry of glucose and amino acids into muscle and fat cells. The hormonal activity of insulin attached to agarose beads is nearly as high as that of the native molecule, which suggests that insulin need not enter a cell to exert its action. Insulin is bound tightly by noncovalent interactions to specific receptors on the plasma membrane. The purification of the insulin receptor rendered soluble by detergent is in progress. The biosynthetic precursor of insulin is proinsulin, which contains a connecting peptide between the prospective A and B chains.

Steroid hormones such as estradiol, progesterone, and cortisone act by a different mechanism. Their primary effect is on gene expression rather than on enzymatic activities or transport processes. Estradiol enters its target cell and binds to a specific receptor in the cytosol. A second subunit joins this complex, which then enters the nucleus and binds to specific sites on DNA.

SELECTED READINGS

Where to start:

Pastan, I., 1972. Cyclic AMP. *Sci. Amer.* 227(2):97–105. [Available as *Sci. Amer.* Offprint 1256.]

Sutherland, E. W., 1972. Studies on the mechanism of hormone action. *Science* 177:401–408.

Schauer, R., 1972. The mode of action of hormones. *Angew. Chem. Int. Ed. Engl.* 11:7–16.

Cyclic AMP:

Robinson, G. A., Butcher, R. W., and Suther-land, E. W., 1971. *Cyclic AMP*. Academic Press. [A valuable reference volume.]

Bitensky, M. W., and Gorman, R. E., 1973. Cellular responses to cyclic AMP. *Progr. Biophys. Mol. Biol.* 26:409–461.

Gill, G. N., and Garren, L. D., 1971. Role of the receptor in the mechanism of action of adenosine 3′:5′-cyclic monophosphate. *Proc. Nat. Acad. Sci.* 68:786–790.

Hirschorn, N., and Greenough, III, W. B., 1971. Cholera. *Sci. Amer.* 225(2):15–21.

Prostaglandins:

Pike, J. E., 1971. Prostaglandins. *Sci. Amer.* 225(5):84–92. [Offprint 1235.]

Bergstrom, S., 1967. Prostaglandins: members of a new hormonal system. *Science* 157:382–191.

Hinman, J. W., 1972. Prostaglandins. *Annu. Rev. Biochem.* 41:161–178.

Vane, J. R., 1971. Inhibition of prostaglandin synthesis as a mechanism of action for aspirin-like drugs. *Nature New Biol.* 231:232–235.

Insulin and diabetes mellitus:

Tager, H. S., and Steiner, D. F., 1974. Peptide hormones. *Annu. Rev. Biochem.* 43:509–538.

Blundell, T., Dodson, G., Hodgkin, D., and Mercola, D., 1972. Insulin: the structure in the crystal and its reflection in chemistry and biology. *Advan. Protein Chem.* 26:280–402.

Steiner, D. F., 1969. Proinsulin and the biosynthesis of insulin. *N. Eng. J. Med.* 280:1106–1107.

Cuatrecasas, P., 1969. Interaction with the cell membrane: the primary action of insulin. *Proc. Nat. Acad. Sci.* 63:450–457.

Cuatrecasas, P., 1972. Affinity chromatography and purification of the insulin receptor of liver cell membranes. *Proc. Nat. Acad. Sci.* 69:1277–1281.

Renold, A. E., Stauffacher, W., and Cahill, G. F., Jr., 1972. Diabetes mellitus. *In* Stanbury, J. B., Wyngaarden, J. B., and Fredrickson, D. S., (eds.), *The Metabolic Basis of Inherited Disease* (3rd ed.). McGraw-Hill.

Steroid hormones:

Jensen, E. V., and DeSombre, E. R., 1972. Mechanism of action of the female sex hormones. *Annu. Rev. Biochem.* 41:203–230.

Yamamoto, K. R., and Alberts, B. M., 1972. In vitro conversion of estradiol-receptor protein to its nuclear form: dependence on hormone and DNA. *Proc. Nat. Acad. Sci.* 69:2105–2109.

Baxter, J. D., Rousseau, G. G., Benson, M. C., Garcea, R. L., Ito, J., and Tomkins, G. M., 1972. Role of DNA and specific cytoplasmic receptors in glucocorticoid action. *Proc. Nat. Acad. Sci.* 69:1892–1896.

Problem sets:

Hood, L. E., Wilson, J. H., and Wood, W. B., 1975. *Molecular Biology of Eucaryotic Cells*, chapter 6. Benjamin.

CHAPTER 35

MUSCLE CONTRACTION AND CELL MOTILITY

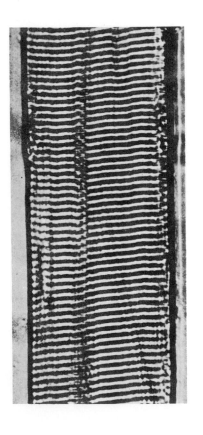

How is chemical-bond energy transformed into coordinated motion? This is one of the most challenging questions in molecular biology today. Directed motion occurs in the movement of chromosomes in cell division, in the injection of bacteriophage DNA into a bacterium, in the beating of cilia and flagella, in the active transport of molecules, in the movement of mRNA in protein synthesis, and, most obviously, in the contraction of muscle. This chapter deals mainly with the structural basis of contraction in vertebrate striated muscle, which is the best understood of these processes. This contractile system consists of interdigitating protein filaments that slide past each other. The energy for contraction comes from the hydrolysis of ATP. The contraction of striated muscle is controlled by the concentration of Ca^{2+}, which is regulated by the sarcoplasmic reticulum, a specialized membrane system that sequesters Ca^{2+} in the resting state and releases it upon the arrival of a nerve impulse.

Figure 35-1
Phase-contrast light micrograph of a skeletal muscle fiber about 50 μm in diameter. The A-bands are dark and the I-bands are light. [Courtesy of Dr. Hugh Huxley.]

MUSCLE IS MADE UP OF INTERACTING THICK AND THIN PROTEIN FILAMENTS

Vertebrate muscle that is under voluntary control has a *striated* appearance when examined under the light microscope (Figure 35-1). It consists of *fibers* that are surrounded by an electrically excitable membrane called the *sarcolemma*. A fiber consists of many parallel *myofibrils,* each about 1 μm in diameter. The myofibrils are immersed in the *sarcoplasm,* which is the intracellular fluid. The sarcoplasm contains glycogen, ATP, phosphocreatine, and glycolytic enzymes. Many regularly spaced mitochondria are found in the more active muscles.

An electron micrograph of a longitudinal section of a myofibril displays a wealth of structural detail (Figures 35-2 and 35-3). The functional unit, called a *sarcomere,* repeats every 2.3 μm (23,000 Å) along the fibril axis. The dark *A band* and the light *I band* alternate regularly. The central region of the A band, termed the *H zone,* is less dense than the rest of the band. A dark *M line* is found in the middle of the H zone. The I band is bisected by a very dense, narrow *Z line.*

The underlying molecular plan of a sarcomere is revealed by electron micrographs of cross sections of a myofibril, which show that there are *two kinds of interacting protein filaments*. The *thick filaments*

Figure 35-2
Electron micrograph of a longitudinal section of a skeletal muscle fiber. [Courtesy of Dr. Hugh Huxley.]

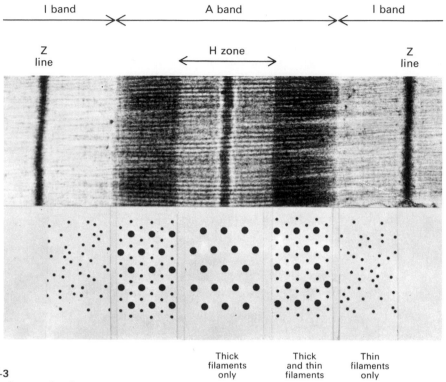

I band   A band   I band

Z line   H zone   Z line

Thick filaments only   Thick and thin filaments   Thin filaments only

Figure 35-3
Electron micrograph of a longitudinal section of a skeletal muscle fiber with diagrams of cross sections. [Courtesy of Dr. Hugh Huxley.]

have a diameter of about 150 Å, whereas the *thin filaments* have a diameter of about 70 Å. The thick filament primarily contains *myosin*. The thin filament contains *actin, tropomyosin,* and *troponin.* Alpha-actinin is present in the Z line, whereas an M-protein is located in the M line. A small amount of β-actinin is found in the thin filament.

The I band consists of only thin filaments, whereas only thick filaments are found in the H zone of the A band. Both types of filament are present in the other parts of the A band. A regular hexagonal array is evident in cross sections, which show that each thin filament has three neighboring thick filaments, and each thick filament is encircled by six thin filaments. The thick and thin filaments interact by *cross-bridges,* which emerge at regular intervals from the thick filaments. They bridge a gap of 130 Å between the surfaces of the thick and thin filaments. In fact, the cross-bridges are the sites at which the contractile force is generated.

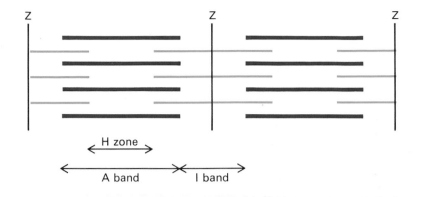

Figure 35-4
Schematic diagram of the structure of striated muscle, showing overlapping arrays of thick and thin filaments, and electron micrograph of a longitudinal section, showing the cross-bridges between thick and thin filaments. [Courtesy of Dr. Hugh Huxley.]

THICK AND THIN FILAMENTS SLIDE PAST EACH OTHER AS MUSCLE CONTRACTS

Muscle shortens by as much as a third of its original length as it contracts. What causes this shortening? In the 1950s, two groups of investigators independently proposed a *sliding-filament model* on the basis of x-ray, light microscopic, and electron microscopic studies. The essential features of this model, which was proposed by Andrew Huxley and R. Niedergerke and by Hugh Huxley and Jean Hanson, are:

1. The lengths of the thick and thin filaments do not change during muscle contraction.

2. Instead, the length of the sarcomere decreases during contraction because the two types of filaments overlap more. *The thick and thin filaments slide past each other in contraction.*

3. The force of contraction is generated by a process that actively moves one type of filament past the neighboring filaments of the other type.

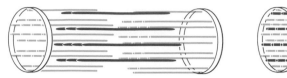

Figure 35-5
Sliding filament model. [From H. E. Huxley. The mechanism of muscular contraction. Copyright © 1965 by Scientific American, Inc. All rights reserved.]

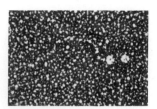

Figure 35-6
Electron micrograph of a myosin molecule. [Courtesy of Dr. Susan Lowey.]

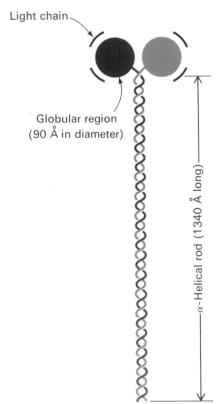

Figure 35-7
Schematic diagram of a myosin molecule.

This model is supported by measurements of the lengths of the A and I bands and the H zone in stretched, resting, and contracted muscle (Figure 35-5). The length of the A band is constant, which means that the thick filaments do not change size. The distance between the Z line and the adjacent edge of the H zone is also constant, which indicates that the thin filaments do not change size. In contrast, the size of the H zone and also of the I band decreases on contraction, because the thick and thin filaments overlap more.

MYOSIN FORMS THICK FILAMENTS, HYDROLYZES ATP, AND BINDS ACTIN

Myosin has three important biological activities. First, myosin molecules spontaneously assemble into filaments in solutions of physiologic ionic strength and pH. In fact, the *thick filament* consists mainly of myosin molecules. Second, myosin is an enzyme. Vladimir Engelhardt and Militsa Lyubimova discovered in 1939 that myosin is an *ATPase*.

$$ATP + H_2O \rightleftharpoons ADP + P_i + H^+$$

This reaction is the immediate source of the free energy that drives muscle contraction. Third, myosin binds to the polymerized form of *actin,* the major constituent of the thin filament. Indeed, this interaction is critical for the generation of the force that translates the thick and thin filaments relative to each other.

Myosin is a very large molecule of *molecular weight 500,000.* It contains two identical major chains (200,000 daltons each) and *four light chains* (about 20,000 daltons each). Electron micrographs show that myosin consists of a double-headed globular region joined to a rod (Figure 35-6). The rod is a double-stranded α-helical cable that is 1340 Å long, and the globular regions have a diameter of about 90 Å (Figure 35-7).

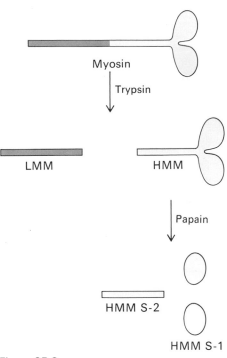

MYOSIN CAN BE CLEAVED INTO ACTIVE FRAGMENTS

Myosin can be enzymatically split into fragments that retain some of the activities of the intact molecule. The fruitfulness of such an experimental approach was evident in the discussion of immunoglobulins (p. 735). In fact, the cleavage of myosin into active fragments came first. In 1953, Andrew Szent-Györgyi showed that myosin is split by trypsin into two fragments, called *light meromyosin* and *heavy meromyosin* (Figure 35-8).

Light meromyosin (LMM), like myosin, forms filaments. However, it lacks ATPase activity and does not combine with actin. Electron micrographs show that LMM is a rod, which accounts for the high viscosity of solution of LMM. The optical rotatory properties of LMM point to an α-helix content of 90%. Furthermore, x-ray diffraction patterns of LMM fibers display a strong reflection at 5.1 Å, which is characteristic of an α-helical coiled coil. The composite picture is that *LMM is a two-stranded α-helical rod for its entire length of 850* Å.

Heavy meromyosin (HMM) has entirely different properties. HMM catalyzes the hydrolysis of ATP and binds to actin, but it does not form filaments. HMM consists of a rod attached to a double-headed globular region (Figure 35-8). It can be split further into two globular subfragments (called HMM S-1) and one rod-shaped subfragment (called HMM S-2). *Each HMM S-1 fragment contains an ATPase active site and a binding site for actin.* Furthermore, the light chains of myosin are bound to the HMM S-1 fragments. *Recent studies suggest that the light chains may modulate the ATPase activities of myosin.* Muscles that vary in the speed of contraction have different kinds of light chains.

Figure 35-8
Enzymatic cleavage of myosin. The four light chains are not shown in this diagram.

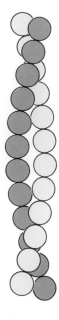

Figure 35-9
F-actin is a
double-stranded helix of
actin monomers.

ACTIN FORMS FILAMENTS THAT BIND MYOSIN

Actin is the major constituent of thin filaments. In solutions of low ionic strength, actin is a 42,000-dalton monomer called *G-actin* because of its globular shape. As the ionic strength is increased to the physiologic level, G-actin polymerizes into a fibrous form called *F-actin*, which closely resembles the thin filaments. An F-actin fiber looks in electron micrographs like two strings of beads wound around each other. X-ray diffraction patterns show that *F-actin is a double-stranded helix of actin monomers.* The helix has a diameter of about 70 Å. The structure repeats at intervals of 355 Å along the helix axis (Figure 35-9).

A complex called *actomyosin* is formed when a solution of actin is added to a solution of myosin. The formation of this complex is accompanied by a large increase in the viscosity of the solution. In the 1940s, Albert Szent-Györgyi showed that this increase in viscosity is reversed by the addition of ATP. This observation revealed that *ATP dissociates actomyosin* to actin and myosin. Szent-Györgyi also prepared threads of actomyosin in which the molecules were oriented by flow. A striking result was obtained when the threads were immersed in a solution containing ATP, K^+, and Mg^{2+}. The *actomyosin threads contracted,* whereas threads formed from myosin alone did not. *These incisive experiments suggested that the force of muscle contraction arises from an interaction of myosin, actin, and ATP.*

ACTIN ENHANCES THE ADENOSINE TRIPHOSPHATASE ACTIVITY OF MYOSIN

The ATPase activity of myosin is markedly enhanced by stoichiometric amounts of F-actin. The turnover number increases 200 times, from 0.05 sec^{-1} to 10 sec^{-1}. The hydrolysis of ATP by myosin alone is rapid but the release of the products ADP and P_i is slow. Actin increases the turnover number of myosin by binding to the myosin-ADP-P_i complex and accelerating the release of ADP and P_i (Figure 35-10). Actomyosin then binds ATP, which leads to the dissociation of actin and myosin. The resulting ATP-myosin complex is then ready for another round of catalysis. These reactions require Mg^{2+}. An important feature of this reaction cycle is that actin has high affinity for myosin and for the myosin-ADP-P_i

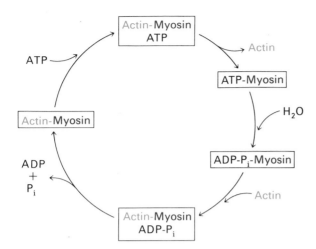

Figure 35-10
The hydrolysis of ATP drives
the cyclic association and
dissociation of actin and
myosin.

complex but it has low affinity for the myosin-ATP complex.
*Consequently, actin alternately binds to myosin and is released from myosin
as ATP is hydrolyzed.*

THE THICK AND THIN FILAMENTS HAVE DIRECTION

The cyclic formation and dissociation of the complex between
myosin and actin produces coherent motion because these molecules
are components of highly ordered assemblies. The organization of
myosin molecules in the thick filament has been elucidated by
Hugh Huxley through studies of intact filaments dissociated from
muscle and of synthetic filaments formed from purified myosin
molecules. *Dissociated thick filaments* have a diameter of 160 Å and
a length of 1.5 μm (15,000 Å). *Cross-bridges* emerge from this filament
in a regular helical array, except for a 1500-Å *bare region* midway
along its length (Figure 35-11).

The same structural features are evident in *synthetic thick filaments,*
which are produced by lowering the ionic strength of a solution
of myosin. The shortest synthetic filaments are about 3000 Å long
and contain a central bare region 1500 Å long. Since the size of
the bare zone is the same in longer synthetic filaments, it is evident
that the thick filament grows by the addition of molecules parallel
to those already assembled. *The myosin molecules on one side of the bare*

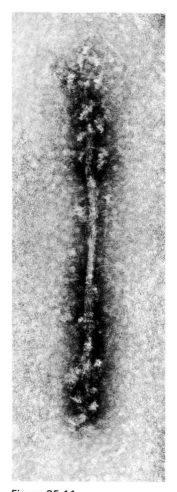

Figure 35-11
Electron micrograph of a reconstituted
thick filament. The projections on either
side of the bare zone are the cross-bridges.
[Courtesy of Dr. Hugh Huxley.]

Figure 35-12
Electron micrograph of filaments of F-actin decorated with HMM. The "arrowheads" all point the same way. [Courtesy of Dr. Hugh Huxley.]

zone point one way, whereas those on the other side point in the opposite direction. Thus, *a thick filament is inherently bipolar.*

LMM, like myosin, aggregates in solutions of low ionic strength to form filaments having an axial period of 430 Å, which is the same as that of the cross-bridges in the intact thick filament. However, filaments formed from LMM are smooth, indicating that the *cross-bridges arise from the HMM portion of myosin, whereas the LMM unit of myosin forms the backbone of the thick filament.*

Thin filaments also have direction. If myosin (or HMM, or HMM S-1) is added to thin filaments or to F-actin, an arrowhead pattern appears in electron micrographs (Figure 35-12). These structures are picturesquely called *decorated filaments.* The arrows on both strands of a decorated filament always point the same way for its entire length. Thus, *thin filaments have inherent directionality, with both strands pointing the same way.* A number of preparations of thin filaments from homogenized muscle contain many filaments still joined to the Z line. Decoration of these filaments with heavy meromyosin reveals that the arrows on all filaments point away from the Z line. Thus, *thin filaments on one side of a Z line all have the same orientation, whereas those on the opposite side have the reverse polarity.*

GENERATION OF THE CONTRACTILE FORCE

The structural polarity of both the thick and thin filaments is crucial for coherent motion. The sliding force developed by the interaction of individual actin and myosin units adds up because the interacting sites have the same relative orientation. Furthermore, *the absolute direction of the sites reverses half-way between Z lines* (Figure 35-13). Consequently, the two thin filaments that interact with a thick filament are drawn toward each other, which shortens the distance between Z lines.

How does the cyclic formation and dissociation of the actin-myosin complex generate the contractile force? A significant clue comes from x-ray diffraction studies showing that myosin cross-bridges move back and forth during contraction. In resting muscle, the cross-bridges are in an ordered helical array and do not make contact with the actin filaments. In the contracted state, the cross-bridges are bound to the actin filaments, but they are no longer in an ordered helical array. The conformational changes that cause myosin cross-bridges to hook onto consecutive sites on the active filament are now being elucidated.

TROPONIN AND TROPOMYOSIN MEDIATE CALCIUM ION REGULATION OF MUSCLE CONTRACTION

The physiologic regulator of muscle contraction is Ca^{2+}. Setsuro Ebashi discovered that the effect of Ca^{2+} on the interaction of actin and myosin is mediated by *tropomyosin* and the *troponin complex,* which are located in the thin filament and comprise about a third of its mass. Tropomyosin is a two-stranded α-helical rod having a molecular weight of 70,000. It is 400 Å long and is located in the groove between the two helical strands of actin. Troponin is a complex of three polypeptide chains, called TpC (18,000 daltons), TpI (24,000 daltons) and TpT (37,000 daltons). The troponin complex is located in the thin filaments at intervals of 385 Å, a period set by the length of tropomyosin. A troponin complex bound to a molecule of tropomyosin regulates the activity of about seven actin monomers.

The interaction of actin and myosin is inhibited by troponin and tropomyosin in the absence of Ca^{2+}. Nerve excitation triggers the release of Ca^{2+} by the sarcoplasmic reticulum, as will be discussed shortly. The released Ca^{2+} binds to the TpC component of troponin and causes conformational changes that are transmitted to tropomyosin and then to actin. These structural changes enable actin to interact with myosin, resulting in muscle contraction and ATP hydrolysis, which persist until Ca^{2+} is removed. Thus, *Ca^{2+} controls muscle contraction by an allosteric mechanism in which the flow of information is:*

$$Ca^{2+} \longrightarrow \text{troponin} \longrightarrow \text{tropomyosin} \longrightarrow \text{actin} \longrightarrow \text{myosin}$$

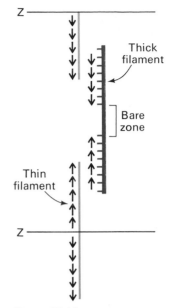

Figure 35-13
The polarity of the thick and thin filaments reverses half-way between the Z lines.

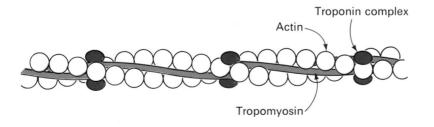

Figure 35-14
Proposed model of a thin filament. [Based on S. Ebashi, M. Endo, and I. Ohtsuki. Control of muscle contraction. *Quart. Rev. Biophys.* 2(1969):164. Published by Cambridge University Press.]

THE FLOW OF CALCIUM IONS IS CONTROLLED BY THE SARCOPLASMIC RETICULUM

The outer membrane of a muscle fiber becomes depolarized following the arrival of a nerve impulse at the end plate, which is the junction between nerve and muscle. The depolarization of the outer membrane is transmitted to the interior of the muscle fiber by the T channels (T stands for their transverse orientation). The T channels are in close proximity to a network of extremely fine channels called the *sarcoplasmic reticulum*, which is a reservoir of Ca^{2+}.

In the resting state, Ca^{2+} is sequestered in the sarcoplasmic reticulum by a Ca^{2+} active transport system (p. 776). This ATP-driven pump lowers the concentration of Ca^{2+} in the cytoplasm of resting muscle to less than 10^{-6} M, and increases the Ca^{2+} level inside the sarcoplasmic reticulum to more than 10^{-3} M. A second protein, called *calsequestrin*, binds Ca^{2+} inside the reticulum. This highly acidic protein (44,000 daltons) has more than forty sites for Ca^{2+}. *Depolarization of the T channel membranes causes a sudden release of Ca^{2+} from the sacs of the sarcoplasmic reticulum.* The released Ca^{2+} then stimulates muscle contraction by the troponin-tropomyosin system described above.

PHOSPHOCREATINE IS A RESERVOIR OF ~P

The amount of ATP in muscle suffices to sustain contractile activity for only a fraction of a second. Vertebrate muscle contains a reservoir of high-potential phosphoryl groups in the form of *phosphocreatine*. This compound has a higher phosphate group-transfer potential than ATP (p. 266). *Creatine kinase* catalyzes the transfer of a phosphoryl group from phosphocreatine to ADP to form ATP.

$$\text{Phosphocreatine} + \text{ADP} \rightleftharpoons \text{ATP} + \text{creatine}$$

Some invertebrates use *phosphoarginine* to store high-potential phosphoryl groups in muscle. Phosphocreatine and phosphoarginine are *phosphagens*.

In active muscle, the supply of phosphocreatine is rapidly depleted and so the level of ATP drops. The concentration of ADP

Phosphocreatine

Phosphoarginine

and P_i rises, as does the level of AMP by the action of adenylate kinase (myokinase).

$$2\,ADP \rightleftharpoons ATP + AMP$$

The reduced energy charge of active muscle stimulates glycolysis, the citric acid cycle, and oxidative phosphorylation (p. 273). The relative contributions of these processes to the generation of ATP depends on the type of muscle. Red muscle, which derives its color from myoglobin and the cytochromes of the respiratory chain, has a much more aerobic metabolism than white muscle.

ACTIN AND MYOSIN ARE PRESENT IN MANY KINDS OF CELLS

Actin and myosin are present not only in muscle. An early indication of the widespread occurrence of actomyosinlike proteins came from the work of Ariel Loewy on *Physarum polycephalum,* a plasmodial slime mold, which contains a streaming mass of cytoplasm. Extracts made at high ionic strength had properties like that of actomyosin from striated muscle. The addition of ATP produced a rapid drop in viscosity, which was followed by a slow rise accompanying the hydrolysis of ATP. In fact, a protein very similar to muscle actin has been isolated from this slime mold. Most interesting, *slime mold actin* reacts with HMM S-1 from *vertebrate striated muscle* to give decorated filaments like those formed from vertebrate actin and myosin. This slime mold also contains a protein that is similar to vertebrate muscle myosin. It forms bipolar filaments and interacts with actin.

Many types of eucaryotic cells contain filaments having a diameter of 50 Å to 70 Å, which are called *microfilaments.* They mediate a wide range of cellular movements: elongation of nerve cells, migration of cells in morphogenesis, retraction of clots by blood platelets, resorption of tadpole tail in metamorphosis, movement of melanin granules in melanocytes, and the beating of microvilli in intestinal epithelium. Microfilaments contain a protein that is nearly the same as muscle actin, and can be decorated by HMM S-1 from striated muscle. Cell movements carried out by microfilaments can be reversibly inhibited by cytochalasin B, an alkaloid derived from fungi. *Myosin has also been isolated from nonmuscle cells of higher organisms.* Actin and myosin in blood platelets form a contractile assembly called thrombosthenin, which is very like actomyosin. The contraction of blood platelets is important in the retraction of clots. It is evident that *actin and myosin are ancient proteins.* It also seems likely that muscle contraction evolved from primitive forms of cellular movement that are mediated by microfilaments.

CILIA AND FLAGELLA CONTAIN MICROTUBULES

A quite different contractile apparatus is present in the cilia and flagella of higher organisms. These hairlike organelles are made up of a bundle of fibers, called an *axoneme,* enclosed by a membrane that is continuous with the plasma membrane. An axoneme contains *nine pairs of outer fibers and two central fibers,* which is often called a *9 + 2 array.* The two central fibers are tubules with diameters of 240 Å. Each of the nine outer doublets looks like a figure of eight, 370 Å × 250 Å. One of the pair, *subfiber A,* is smaller and is joined to the central sheath of the cilium by radial spokes. Two *arms* emerge from each subfiber A. All the arms point in the same direction in a given cilium. The wall of each fiber is made up of *protofilaments* consisting of several globular proteins called *tubulins,* which are different from actin. Treatment of cilia or flagella with detergent removes their outer membrane and solubilizes an ATPase called *dynein.* The outer fibers retain their ninefold cylindrical arrangement but are devoid of arms. The arms can be restored by the addition of dynein under suitable ionic conditions. Thus, *the arms of subfibers A contain the ATPase activity.* ATP induces shearing forces between outer tubules, which result in sliding motions.

Microtubules have been found in many kinds of eucaryotic cells. They are cylinders 240 Å in diameter that participate in the movement of chromosomes in cell division, in the ordered movement of material within cells (e.g., nerve axons), and in the development and maintenance of cell shape. The alkaloid *colchicine* disrupts microtubules, and thereby blocks mitosis and the development of cell structures.

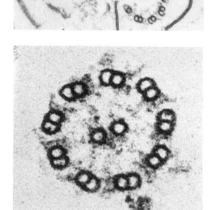

Figure 35-15
Electron micrographs of cross sections of flagellar axonemes. [Courtesy of Dr. Joel Rosenbaum.]

SUMMARY

Vertebrate striated muscle consists of two kinds of interacting protein filaments. The thick filaments contain myosin, whereas the thin filaments contain actin, tropomyosin, and troponin. The hydrolysis of ATP by actomyosin drives the sliding of these filaments past each other. Myosin is a very large protein (500,000 daltons) composed of two major chains and four light chains. The major chains are folded so that myosin contains two globular regions attached to a long α-helical rod. The globular regions and part of the rod form the cross-bridges that interact with actin to generate the contractile force. The rest of the myosin molecule forms the backbone of the thick filament. Actin, the major component of thin filaments, is a globular protein (42,000 daltons) that polymerizes to form double-stranded filaments. The thin and thick filaments have direction, which reverses half-way between the Z lines. ATP drives the cyclic formation and dissociation of complexes between the myosin cross-bridges of thick filaments and the actin units of thin filaments, thereby shortening the distance between Z lines. Muscle contraction is regulated by Ca^{2+}, whose effect is mediated by troponin and tropomyosin. The interaction of actin and myosin is inhibited by these proteins when the level of Ca^{2+} is low. Nerve excitation triggers the release of Ca^{2+} by the sarcoplasmic reticulum. Calcium ion then binds to troponin, which initiates a series of conformational changes that allow actin and myosin to interact.

Actin and myosin are ancient proteins, as shown by their presence in slime molds. They are also present in many kinds of cells of higher organisms. For example, actin and myosin form a contractile assembly in blood platelets that is responsible for clot retraction. Most eucaryotic cells contain microfilaments (50 Å–70 Å in diameter) composed of actin. They also contain microtubules (240 Å in diameter), which are a different type of contractile system composed of tubulins rather than actin. Microfilaments and microtubules mediate a wide range of cellular movements.

SELECTED READINGS

Where to start:

Huxley, H. E., 1965. The mechanism of muscular contraction. *Sci. Amer.* 213(6):18–27. [Available as *Sci. Amer.* Offprint 1026.]

Murray, J. M., and Weber, A., 1974. The cooperative action of muscle proteins. *Sci. Amer.* 230(2):59–69. [Offprint 1290.]

Hoyle, G., 1970. How is muscle turned on and off? *Sci. Amer.* 224(4):85–93. [Offprint 1175.]

Satir, P., 1974. How cilia move. *Sci. Amer.* 231(4):44–52. [Offprint 1304.]

Books and reviews:

Cold Spring Harbor Laboratory, 1972. *The Mechanism of Muscle Contraction* (Cold Spring Harbor Symposia on Quantitative Biology, vol. 37). [An outstanding volume that contains articles on the structure of muscle proteins, the generation of the contractile force, regulatory systems, the sarcoplasmic reticulum, myogenesis, and contractile proteins in nonmuscle tissue.]

Bendall, J. R., 1969. *Muscles, Molecules, and Movement.* Heinemann.

Huxley, H. E., 1971. The structural basis of muscle contraction. *Proc. Roy. Soc. London* (B)178:131–149.

Taylor, E. W., 1972. Chemistry of muscle contraction. *Annu. Rev. Biochem.* 41:577–616.

Tonomura, Y., and Oosawa, F., 1972. Molecular mechanism of contraction. *Annu. Rev. Biophys. Bioeng.* 1:159–190.

Wessells, N. K., Spooner, B. S., Ash, J. F., Bradley, M. O., Luduena, M. A., Taylor, E. L., Wrenn, J. T., and Yamada, K. M., 1971. Microfilaments in cellular and developmental processes. *Science* 171:135–142.

Olmstead, J. B., and Borisy, G. G., 1973. Microtubules. *Annu. Rev. Biochem.* 42:507–540.

APPENDIXES

ANSWERS TO PROBLEMS

INDEX

PHYSICAL CONSTANTS AND CONVERSION OF UNITS

Values of physical constants

Physical constant	Symbol	Value
Atomic mass unit (dalton)	amu	1.661×10^{-24} g
Avogadro's number	N	6.022×10^{23} mol^{-1}
Boltzmann's constant	k	1.381×10^{-23} J deg^{-1}
		3.298×10^{-24} cal deg^{-1}
Electron volt	eV	1.602×10^{-19} J
		3.828×10^{-20} cal
Faraday constant	F	9.649×10^4 C mol^{-1}
		2.306×10^4 cal volt^{-1} eq^{-1}
Curie	Ci	3.70×10^{10} disintegrations sec^{-1}
Gas constant	R	8.314 J mol^{-1} deg^{-1}
		1.987 cal mol^{-1} deg^{-1}
Planck's constant	h	6.626×10^{-34} J sec
		1.584×10^{-34} cal sec
Speed of light in a vacuum	c	2.998×10^{10} cm sec^{-1}

Abbreviations: C, coulomb; cal, calorie; cm, centimeter; deg, degree Kelvin; eq, equivalent; g, gram; J, joule; mol, mole; sec, second.

Mathematical constants

$\pi = 3.14159$

$e = 2.71828$

$\log_e x = 2.303 \log_{10} x$

Standard prefixes

Prefix	Symbol	Factor
kilo	k	10^3
hecto	h	10^2
deca	da	10^1
deci	d	10^{-1}
centi	c	10^{-2}
milli	m	10^{-3}
micro	μ	10^{-6}
nano	n	10^{-9}
pico	p	10^{-12}

Conversion factors

Physical quantity	Equivalent
Length	1 cm $= 10^{-2}$ m $= 10$ mm $= 10^4$ μm $= 10^7$ nm
	1 cm $= 10^8$ Å $= 0.3937$ inch
Mass	1 g $= 10^{-3}$ kg $= 10^3$ mg $= 10^6$ μg
	1 g $= 3.527 \times 10^{-2}$ ounce (avoirdupoir)
Volume	1 cm^3 $= 10^{-6}$ m^3 $= 10^3$ mm^3
	1 ml $= 1$ cm^3 $= 10^{-3}$ l $= 10^3$ μl
	1 cm^3 $= 6.1 \times 10^{-2}$ in^3 $= 3.53 \times 10^{-5}$ ft^3
Temperature	K $= °C + 273.15$
	$°C = 5/9$ ($°F - 32$)
Energy	1 J $= 10^7$ erg $= 0.239$ cal $= 1$ watt sec
Pressure	1 torr $= 1$ mm Hg ($0°C$)
	$= 1.333 \times 10^2$ newton/m^2
	$= 1.333 \times 10^2$ pascal
	$= 1.316 \times 10^{-3}$ atmospheres

ATOMIC NUMBERS AND WEIGHTS OF THE ELEMENTS

Element	Symbol	Atomic number	Atomic weight
Actinium	Ac	89	227.03
Aluminum	Al	13	26.98
Americium	Am	95	243.06
Antimony	Sb	51	121.75
Argon	Ar	18	39.95
Arsenic	As	33	74.92
Astatine	At	85	210.99
Barium	Ba	56	137.34
Berkelium	Bk	97	247.07
Beryllium	Be	4	9.01
Bismuth	Bi	83	208.98
Boron	B	5	10.81
Bromine	Br	35	79.90
Cadmium	Cd	48	112.40
Calcium	Ca	20	40.08
Californium	Cf	98	249.07
Carbon	C	6	12.01
Cerium	Ce	58	140.12
Cesium	Cs	55	132.91
Chlorine	Cl	17	35.45
Chromium	Cr	24	52.00
Cobalt	Co	27	58.93
Copper	Cu	29	63.55
Curium	Cm	96	245.07
Dysprosium	Dy	66	162.50
Einsteinium	Es	99	254.09
Erbium	Er	68	167.26
Europium	Eu	63	151.96
Fermium	Fm	100	252.08
Fluorine	F	9	18.99
Francium	Fr	87	223.02
Gadolinium	Gd	64	157.25
Gallium	Ga	31	69.72
Germanium	Ge	32	72.59
Gold	Au	79	196.97
Hafnium	Hf	72	178.49
Helium	He	2	4.00
Holmium	Ho	67	164.93
Hydrogen	H	1	1.01
Indium	In	49	114.82
Iodine	I	53	126.90
Iridium	Ir	77	192.22
Iron	Fe	26	55.85
Khurchatovium	Kh	104	260
Krypton	Kr	36	83.80
Lanthanum	La	57	138.91
Lawrencium	Lr	103	256
Lead	Pb	82	207.20
Lithium	Li	3	6.94
Lutetium	Lu	71	174.97
Magnesium	Mg	12	24.31
Manganese	Mn	25	54.94
Mendelevium	Md	101	255.09
Mercury	Hg	80	200.59
Molybdenum	Mo	42	95.94
Neodymium	Nd	60	144.24
Neon	Ne	10	20.18
Neptunium	Np	93	237.05
Nickel	Ni	28	58.71
Niobium	Nb	41	92.91
Nitrogen	N	7	14.01
Nobelium	No	102	255
Osmium	Os	76	190.20
Oxygen	O	8	16.00
Palladium	Pd	46	106.40
Phosphorus	P	15	30.97
Platinum	Pt	78	195.09
Plutonium	Pu	94	242.06
Polonium	Po	84	208.98
Potassium	K	19	39.10
Praseodymium	Pr	59	140.91
Promethium	Pm	61	145
Protactinium	Pa	91	231.04
Radium	Ra	88	226.03
Radon	Rn	86	222.02
Rhenium	Re	75	186.20
Rhodium	Rh	45	102.91
Rubidium	Rb	37	85.47
Ruthenium	Ru	44	101.07
Samarium	Sm	62	150.40
Scandium	Sc	21	44.96
Selenium	Se	34	78.96
Silicon	Si	14	28.09
Silver	Ag	47	107.87
Sodium	Na	11	22.99
Strontium	Sr	38	87.62
Sulfur	S	16	32.06
Tantalum	Ta	73	180.95
Technetium	Tc	43	98.91
Tellurium	Te	52	127.60
Terbium	Tb	65	158.93
Thallium	Tl	81	204.37
Thorium	Th	90	232.04
Thulium	Tm	69	168.93
Tin	Sn	50	118.69
Titanium	Ti	22	47.90
Tungsten	W	74	183.85
Uranium	U	92	238.03
Vandadium	V	23	50.94
Xenon	Xe	54	131.30
Ytterbium	Yb	70	173.04
Yttrium	Y	39	88.91
Zinc	Zn	30	65.37
Zirconium	Zr	40	91.22

APPENDIX C

pK' VALUES OF SOME ACIDS

Acid	pK' (at 25°C)
Acetic acid	4.76
Acetoacetic acid	3.58
Ammonium ion	9.25
Ascorbic acid, pK'_1	4.10
pK'_2	11.79
Benzoic acid	4.20
n-Butyric acid	4.81
Cacodylic acid	6.19
Carbonic acid, pK'_1	6.35
pK'_2	10.33
Citric acid, pK'_1	3.14
pK'_2	4.77
pK'_3	6.39
Ethylammonium ion	10.81
Formic acid	3.75
Glycine, pK'_1	2.35
pK'_2	9.78
Imidazolium ion	6.95

Acid	pK' (at 25°C)
Lactic acid	3.86
Maleic acid, pK'_1	1.83
pK'_2	6.07
Malic acid, pK'_1	3.40
pK'_2	5.11
Phenol	9.89
Phosphoric acid, pK'_1	2.12
pK'_2	7.21
pK'_3	12.67
Pyridinium ion	5.25
Pyrophosphoric acid, pK'_1	0.85
pK'_2	1.49
pK'_3	5.77
pK'_4	8.22
Succinic acid, pK'_1	4.21
pK'_2	5.64
Trimethylammonium ion	9.79
Tris (hydroxymethyl) aminomethane	8.08
Water	14.0

STANDARD BOND LENGTHS

Bond	Structure	Length (Å)
C—H	R_2CH_2	1.07
	Aromatic	1.08
	RCH_3	1.10
C—C	Hydrocarbon	1.54
	Aromatic	1.40
C=C	Ethylene	1.33
C≡C	Acetylene	1.20
C—N	RNH_2	1.47
	O=C—N	1.34
C—O	Alcohol	1.43
	Ester	1.36
C=O	Aldehyde	1.22
	Amide	1.24
C—S	R_2S	1.82
N—H	Amide	0.99
O—H	Alcohol	0.97
O—O	O_2	1.21
P—O	Ester	1.56
S—H	Thiol	1.33
S—S	Disulfide	2.05

ANSWERS TO PROBLEMS

CHAPTER 2

1. (a) Phenyl isothiocyanate.
   (b) Dansyl chloride.
   (c) Urea; β-mercaptoethanol to reduce disulfides.
   (d) Chymotrypsin.
   (e) CNBr.
   (f) Trypsin.
2. (a) 3, (b) 12, (c) 4.28, (d) 9.8, and (e) 2.4.
3. .01, .1, 1, 10, and 100.
4. 31,536.
5. 477 Å (318 residues per strand, 1.5 Å per residue).
6. The side-chain carboxylates are ionized at pH 7, whereas they are unionized at pH 4. Electrostatic repulsion between the carboxylates is minimized in the random coil, which has a larger molecular volume than the α helix.
7. Each amino acid residue, except the carboxyl-terminal one, gives rise to a hydrazide on reaction with hydrazine. The carboxyl-terminal residue can be identified because it yields a free amino acid.
8. (a) Approximately $+1$.
   (b) Two peptides.
9. The S-aminoethylcysteine side chain resembles that of lysine. The only difference is a sulfur atom in place of a methylene group.
10. The native conformation of insulin is not the thermodynamically most stable form. See page 816 for further discussion.

CHAPTER 3

1. (a) 2.96×10^{-11} g.
   (b) 2.71×10^8 molecules.
   (c) No. There would be 3.22×10^8 hemoglobin molecules in a red cell if they were packed in a cubic crystalline array. Hence, the actual packing density is about 84% of the maximal possible.
2. 2.65 g (or 4.75×10^{-2} moles) of Fe.

3. (a) In man, 1.44×10^{-2} g (4.49×10^{-4} moles) of O_2 per kg of muscle. In sperm whale, 0.144 g (4.49×10^{-3} moles) of O_2 per kg.
   (b) 128.
4. (a) Approximately $+36$ at pH 2, $+4$ at pH 7, and $+2$ at pH 9.
   (b) Approximately 10. However, there are interactions between titratable groups. The actual isoelectric point is 8.2. For an excellent treatment of this problem, see S. J. Shire, G. I. H. Hanania, and F. R. N. Gurd, *Biochemistry* 13(1974):2967.
5. (a) Three peptides, consisting of residues 1 to 55, 56 to 131, and 132 to 153.
   (b) Short α helices are marginally stable in aqueous solution. They are stabilized by tertiary interactions in myoglobin.

6. $$\text{Log} \frac{[\text{MbO}_2]}{[\text{Mb}]} = \log pO_2 - \log P_{50}$$

7. (a) $k_{\text{off}} = 20 \text{ sec}^{-1}$.

$$K = \frac{[\text{Mb}][O_2]}{[\text{MbO}_2]} = \frac{k_{\text{off}}}{k_{\text{on}}}$$

   (b) Mean duration is 0.05 sec (the reciprocal of k_{off}).

CHAPTER 4

1. (a) Increased, (b) decreased, (c) decreased, and (d) increased oxygen affinity.
2. (a and b) Fewer H^+ bound.
3. Inositol hexaphosphate.
4. (a) $\Delta Y = .977 - .323 = .654$.
   (b) $\Delta Y = .793 - .434 = .359$.
5. The pK is (a) lowered, (b) raised, and (c) raised.
6. The increased concentration of DPG increases ΔY from 0.61 to 0.687.
7. (a) Yes. $K_{\text{AB}} = K_{\text{BA}} (K_{\text{B}}/K_{\text{A}}) = 2 \times 10^{-5}$ M.
   (b) The presence of A enhances the binding of B, and so the presence of B enhances the binding of A.

8. Carbon monoxide bound to one heme alters the oxygen affinity of the other hemes in the same hemoglobin molecule. Specifically, CO increases the oxygen affinity of hemoglobin and thereby decreases the amount of O_2 released in actively metabolizing tissues. Carbon monoxide stabilizes the quaternary structure characteristic of oxyhemoglobin. In other words, CO mimics O_2 as an allosteric effector.

9. (a) For maximal transport, $K = 10^{-5}$ M. In general, maximal transport is achieved when $K = \sqrt{L_A L_B}$.
   (b) For maximal transport, $P_{50} = 44.7$ torr, which is considerably higher than the physiological value of 26 torr. However, it must be stressed that this calculation ignores cooperative binding and the Bohr effect.

CHAPTER 5

1. (a) Lysine or arginine at position 6.
   (b) This hemoglobin migrates least rapidly toward the anode at pH 8 because it has the smallest net negative charge of the three hemoglobins.
2. (a) Hb C, (b) Hb D, (c) Hb J, and (d) Hb N.
3. Mutations in the α gene affect all three hemoglobins because their subunit structures are $\alpha_2\beta_2$, $\alpha_2\delta_2$, and $\alpha_2\gamma_2$, respectively. Mutations in the β, δ, or γ genes affect only one of them.
4. The reaction

$$2(\alpha_2\beta\beta^S) \rightleftharpoons \alpha_2\beta_2 + \alpha_2\beta_2^S$$

is rapid compared with the duration of an electrophoresis experiment. The separation of $\alpha_2\beta_2$ and $\alpha_2\beta_2^S$ by the electric field shifts the equilibrium to the right.
5. Deoxy Hb A contains a complementary site and so it can add on to a fiber of deoxy Hb S. The fiber cannot then grow further because the terminal deoxy Hb A molecule lacks a sticky patch.

CHAPTER 6

1. (a) 31.1×10^{-6} moles.
   (b) 5×10^{-8} moles.
   (c) 622 sec^{-1}.
2. (a) Yes. $K_M = 5.2 \times 10^{-6}$ M.
   (b) $V_{max} = 6.84 \times 10^{-10}$ moles.
   (c) 337 sec^{-1}.
3. (a) In the absence of inhibitor, V_{max} is 47.6 μmole/min and K_M is 1.1×10^{-5} M. In the presence of inhibitor, V_{max} is the same, and the apparent K_M is 3.1×10^{-5} M.
   (b) Competitive.

(c) 1.1×10^{-3} M.
(d) f_{ES} is 0.243 and f_{EI} is 0.488.
(e) f_{ES} is 0.73 in the absence of inhibitor and 0.49 in the presence of 2×10^{-3} M inhibitor. The ratio of these values, 1.49, is the same as the ratio of the reaction velocities under these conditions.
4. (a) V_{max} is 9.5 μmole/min. K_M is 1.1×10^{-5} M, the same as without inhibitor.
   (b) Noncompetitive.
   (c) 2.5×10^{-5} M.
   (d) 0.73, in the presence or absence of this noncompetitive inhibitor.
5. (a) $V = V_{max} - (V/[S]) K_M$.
   (b) Slope $= -K_M$, y-intercept $= V_{max}$, x-intercept $= V_{max}/K_M$.
   (c)

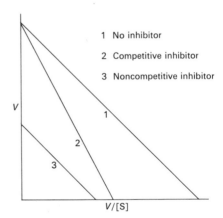

1	No inhibitor
2	Competitive inhibitor
3	Noncompetitive inhibitor

6. The binding of a competitive inhibitor to one active site of an allosteric enzyme may increase the substrate-binding affinity of the other sites on the same enzyme molecule. In the concerted model, such a competitive inhibitor acts in this way by promoting the T $\longrightarrow$ R transition.
7. Potential hydrogen-bond donors at pH 7 are the side chains of the following residues: arginine, asparagine, glutamine, histidine, lysine, serine, threonine, tryptophan, and tyrosine.

CHAPTER 7

1. The fastest is (b) and the slowest is (a).
2. (a) B-C, (b) A-B or E-F, and (c) A-B-C (one sugar

residue does not interact with the enzyme, thus avoiding site D, which is energetically unfavorable).
3. The ^{18}O emerges in the C-4 hydroxyl of di-NAG (residues E-F).
4. (a) In oxymyoglobin, Fe is bonded to five nitrogens and one oxygen. In carboxypeptidase A, Zn is bonded to two nitrogens and two oxygens.
   (b) In oxymyoglobin, one of the nitrogens bonded to Fe comes from the proximal histidine residue, whereas the other four come from the heme. The oxygen atom linked to Fe is that of O_2. In carboxypeptidase A, the two nitrogen atoms coordinated to Zn come from histidine residues. One of the oxygen ligands is from a glutamate side chain, the other from a water molecule.
   (c) Aspartate, cysteine, and methionine.

CHAPTER 8

1. (a) Carboxypeptidase A.
   (b) Lysozyme and carboxypeptidase A.
   (c) Chymotrypsin.
   (d) Carboxypeptidase A and chymotrypsin.
2. (a) Yes.
   (b) Histidine 57 in chymotrypsin, glutamic 35 in lysozyme, and tyrosine 248 in carboxypeptidase A.
3. Serine 195 in chymotrypsin and glutamate 270 (or a hydroxyl ion) in carboxypeptidase A.
4. Precise positioning of catalytic residues and substrates, geometrical strain (distortion of the substrate), electronic strain, and desolvation of the substrate.
5. (a) Tosyl-L-lysine chloromethyl ketone (TLCK).
   (b) First, determine whether substrates protect trypsin from inactivation by TLCK and, second, ascertain whether the D-isomer of TLCK inactivates trypsin.
   (c) Thrombin.
6. (a) Serine.
   (b) A hemiacetal between the aldehyde of the inhibitor and the hydroxyl group of the active site serine.
7. The boron atom becomes bonded to the oxygen atom of the active-site serine. This tetrahedral intermediate has a geometry similar to that of the transition state.

CHAPTER 9

1. (a) Every third residue in each strand of a collagen triple helix must be glycine because there is insufficient space for a larger residue.
   (b) Poly (Gly-Pro-Gly) melts at a lower temperature than poly (Gly-Pro-Pro).

(c) No. Glycine does not occupy every third position.
2. (a and b)

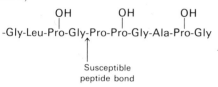

-Gly-Leu-Pro-Gly-Pro-Pro-Gly-Ala-Pro-Gly

Susceptible
peptide bond

3. (a) Disulfides.
   (b) None.
   (c) Peptide bonds between specific glutamine and lysine side chains.
   (d) Aldol cross-link, histidine–aldol cross-link, and lysinonorleucine.
   (e) Aldol cross-link, lysinonorleucine, and desmosine.

CHAPTER 10

1. 2.86×10^6 molecules.
2. Cyclopropane rings interfere with the orderly packing of hydrocarbon chains, and so they increase membrane fluidity.
3. 2×10^{-7} cm, 6.32×10^{-6} cm, and 2×10^{-4} cm.
4. The radius of this molecule is 3.08×10^{-7} cm and its diffusion coefficient is 7.37×10^{-9} cm^2/sec. The average distances traversed are 1.72×10^{-7} cm in 1 μsec, 5.42×10^{-6} in 1 msec, and 1.72×10^{-4} cm in 1 sec.
5. The initial decrease in the amplitude of the paramagnetic resonance spectrum results from the reduction of spin-labeled phosphatidyl cholines in the outer leaflet of the bilayer. Ascorbate does not traverse the membrane under these experimental conditions, and so it does not reduce the phospholipids in the inner leaflet. The slow decay of the residual spectrum is due to the reduction of phospholipids that have flipped over to the outer leaflet of the bilayer.

CHAPTER 11

1. Reactions (a) and (c), to the left; reactions (b) and (d), to the right.
2. None whatsoever.
3. (a) $\Delta G^{\circ\prime} = +7.5$ kcal/mol and $K'_{eq} = 3.16 \times 10^{-6}$.
   (b) 3.16×10^4.
4. $\Delta G^{\circ\prime} = 1.7$ kcal/mol. The equilibrium ratio is 17.8.
5. (a) $+0.2$ kcal/mol.
   (b) -7.8 kcal/mol. The hydrolysis of PP$_i$ drives the reaction toward the formation of acetyl CoA.
6. (a) $\Delta G^{\circ} = 2.303\ RT\ pK$.
   (b) -6.53 kcal/mol at 25°C.

7. An ADP unit (or a closely related derivative, in the case of CoA).
8. The activated form of sulfate in most organisms is 3′-phosphoadenosine 5′-phosphosulfate. See P. W. Robbins and F. Lipmann, *J. Biol. Chem.* 229 (1957):837.

CHAPTER 12

1. (a) Aldose-ketose, (b) epimers, (c) aldose-ketose, (d) anomers, (e) aldose-ketose, and (f) epimers.

2. The methyl carbon of pyruvate is ^{14}C-labeled.

3. There is no net formation of ATP.

$$\text{Glucose} + 2\,NAD^+ \longrightarrow 2\,\text{pyruvate} + 2\,NADH + 2H^+$$

4. (a) $\Delta G°'$ is -29.5 kcal/mol for the reaction

$$\text{Glucose} + 2\,P_i + 2\,ADP \longrightarrow 2\,\text{lactate} + 2\,ATP$$

   (b) $\Delta G = -27.2$ kcal/mol.
5. 3.06×10^{-5}.
6. The equilibrium concentrations of fructose 1,6-diphosphate, dihydroxyacetone phosphate, and glyceraldehyde 3-phosphate are 7.76×10^{-4} M, 2.24×10^{-4} M, and 2.24×10^{-4} M, respectively.
7. All three carbon atoms of 2,3-DPG are ^{14}C-labeled. The phosphorus atom attached to the C-2 hydroxyl is ^{32}P-labeled.

CHAPTER 13

1. (a) After one round of the citric acid cycle, the label emerges in C-2 and C-3 of oxaloacetate.
   (b) After one round of the citric acid cycle, the label emerges in C-1 and C-4 of oxaloacetate.
   (c) The label emerges in CO_2 in the formation of acetyl CoA from pyruvate.
   (d and e) Same fate as in (a).
2. No, because two carbon atoms are lost in the two decarboxylation steps of the cycle. Hence, there is no *net* synthesis of oxaloacetate.
3. 0.90, 0.03, and 0.07.
4. -9.8 kcal/mol.
5. The coenzyme stereospecificity of glyceraldehyde 3-phosphate dehydrogenase is the opposite of that of alcohol dehydrogenase (type B versus type A, respectively.)

CHAPTER 14

1. (a) 15, (b) 2, (c) 38, (d) 16, (e) 36, and (f) 19.
2. (a) $\Delta E_0'$ is $+1.05$ volts and $\Delta G°'$ is -48.4 kcal/mol for the reaction

$$2\,G—SH + \tfrac{1}{2}\,O_2 \rightleftharpoons G—S—S—G + H_2O$$

   (b) $\Delta E_0' = +0.09$ volts and $\Delta G^{0'}$ is -4.15 kcal/mol.
3. (a) Blocks electron transport and ATP synthesis at site 3.
   (b) Blocks electron transport and ATP synthesis by inhibiting the exchange of ATP and ADP across the inner mitochondrial membrane.
   (c) Blocks electron transport and ATP synthesis at site 1.
   (d) Blocks ATP synthesis without inhibiting electron transport.
   (e) Blocks electron transport and ATP synthesis at site 3.
   (f) Blocks electron transport and ATP synthesis at site 2.
4. Oligomycin inhibits ATP formation without blocking electron transport.
5. $\Delta G°'$ is $+16.1$ kcal/mol for oxidation by NAD^+, and $+1.4$ kcal/mol for oxidation by FAD. The reduction of succinate by NAD^+ is not thermodynamically feasible.
6. Cyanide can be lethal because it binds to the ferric form of cytochrome $(a + a_3)$ and thereby inhibits oxidative phosphorylation. Nitrite converts ferrohemoglobin to ferrihemoglobin, which also binds cyanide. Thus, ferrihemoglobin competes with cytochrome $(a + a_3)$ for cyanide. This competition is therapeutically effective because the amount of ferrihemoglobin that can be formed without impairing oxygen transport is much greater than the amount of cytochrome $(a + a_3)$.

CHAPTER 15

1. (a) 5 Glucose 6-phosphate $+$ ATP $\longrightarrow$ 6 ribose 5-phosphate $+$ ADP $+$ H^+

   (b) Glucose 6-phosphate $+$ 12 $NADP^+$ $+$ 7 H_2O $\longrightarrow$ 6 CO_2 $+$ 12 NADPH $+$ 12 H^+ $+$ P_i

2. The label emerges at C-5 of ribulose 5-phosphate.
3. Oxidative decarboxylation of isocitrate to α-ketoglutarate. A β-keto acid intermediate is formed in both reactions.

4. C-1 and C-3 of fructose 6-phosphate are labeled, whereas erythrose 4-phosphate is not labeled.
5. (a, c, and f) No; (b and e) Yes.
6. Form a Schiff base between a ketose substrate and transaldolase, reduce it with tritiated $NaBH_4$, and fingerprint the labeled enzyme.

CHAPTER 16

1. Galactose + ATP + UTP + H_2O + glycogen$_n$ $\longrightarrow$
   glycogen$_{n+1}$ + ADP + UDP + 2 P_i + H^+

2. Fructose + 2 ATP + 2 H_2O $\longrightarrow$
   glucose + 2 ADP + 2 P_i

3. There is a deficiency of the branching enzyme.
4. The concentration of glucose 6-phosphate is elevated in von Gierke's disease. Consequently, the phosphorylated D form of glycogen synthetase is active.
5. (a) High levels of galactose 1-phosphate are more toxic than those of galactose.
   (b) A deficiency of fructose 1-phosphate aldolase has more severe consequences than a deficiency of fructokinase.

CHAPTER 17

1. (a) Glycerol + 2 NAD^+ + P_i + ADP $\longrightarrow$
       pyruvate + ATP + H_2O + 2 NADH + H^+

   (b) Glycerol kinase and glycerol phosphate dehydrogenase.

2. Stearate + ATP + 13½ H_2O + 8 FAD +
   8 NAD^+ $\longrightarrow$ 4½ acetoacetate + 12½ H^+ +
   8 $FADH_2$ + 8 NADH + AMP + 2 P_i

3. (a) Oxidation in mitochondria, synthesis in the cytosol.
   (b) Acetyl CoA in oxidation, acyl carrier protein for synthesis.
   (c) FAD and NAD^+ in oxidation, NADPH for synthesis.
   (d) L-isomer of 3-hydroxyacyl CoA in oxidation, D-isomer in synthesis.
   (e) Carboxyl to methyl in oxidation, methyl to carboxyl in synthesis.
   (f) The enzymes of fatty acid synthesis, but not those of oxidation, are organized in a multienzyme complex.

4. (a) Palmitoleate, (b) linoleate, (c) linoleate, (d) oleate, (e) oleate, and (f) linolenate.

5. C-1 is more radioactive (see p. 661 for a discussion of this experimental approach in the elucidation of the direction of synthesis of a polypeptide chain).
6. (a) Yes.

2 Acetyl CoA + 3 H_2O + FAD + 2 NAD^+ $\longrightarrow$
   oxaloacetate + 2 CoA + $FADH_2$ +
   2 NADH + 4 H^+

   (b) Yes, because glucose can be synthesized from oxaloacetate by the gluconeogenic pathway.

CHAPTER 18

1. (a) Pyruvate, (b) oxaloacetate, (c) α-ketoglutarate, (d) α-ketoisocaproate, (e) phenylpyruvate, and (f) hydroxyphenylpyruvate.

2. Aspartate + α-ketoglutarate + GTP + ATP +
   2 H_2O + NADH + H^+ $\longrightarrow$
   ½ glucose + glutamate + CO_2 + ADP +
   GDP + NAD^+ + 2 P_i

3. Aspartate + CO_2 + NH_4^+ + 3 ATP + NAD^+ +
   4 H_2O $\longrightarrow$ oxaloacetate + urea +
   2 ADP + 4 P_i + AMP + NADH + H^+

4. (a) Label the methyl carbon atom of L-methylmalonyl CoA with ^{14}C. Determine the location of ^{14}C in succinyl CoA. The group transferred is the one bonded to the labeled carbon atom.
   (b) Label the methyl carbon atom of L-methylmalonyl CoA with ^{14}C and the sulfur atom of its CoA moiety with ^{35}S. The transfer of the —CO—S—CoA group is intramolecular if succinyl CoA contains both ^{14}C and ^{35}S.
   (c) The proton that is abstracted from the methyl group of L-methylmalonyl CoA is directly transferred to the adjacent carbon atom.

CHAPTER 19

1. $\Delta E_0' = +0.28$ volt and $\Delta G^{\circ'} = -12.9$ kcal/mol.
2. Aldolase participates in the Calvin cycle, whereas transaldolase participates in the pentose phosphate pathway.
3. The concentration of 3-phosphoglycerate would increase, whereas that of ribulose 1,5-diphosphate would decrease.
4. The concentration of 3-phosphoglycerate would decrease, whereas that of ribulose 1,5-diphosphate would increase.

5. Phycoerythrin and phycocyanin serve as antenna molecules. They absorb light in the spectral region in which chlorophyll a has little absorbance and then transfer their electronic excitation energy to chlorophyll a.

6. (a) It expresses a key aspect to photosynthesis—namely, that water is split by light. The evolved oxygen in photosynthesis comes from water.

   (b) The Van Niel equation for respiration expresses the fact that the combustion of glucose requires the input of six molecules of H_2O. For an interesting discussion, see G. Wald, On the nature of cellular respiration, *Current Aspects of Biochemical Energetics*, N. O. Kaplan and E. P. Kennedy, eds., pp. 27–37 (Academic Press, 1966).

CHAPTER 20

1. Glycerol + 4 ATP + 3 fatty acids + 4 H_2O $\longrightarrow$ triacylglycerol + ADP + 3 AMP + 7 P_i + 4 H^+

2. Glycerol + 3 ATP + 2 fatty acids + 2 H_2O + CTP + serine $\longrightarrow$ phosphatidyl serine + CMP + ADP + 2 AMP + 6 P_i + 3 H^+

3. (a) CDP-diacylglycerol, (b) CDP-ethanolamine, (c) acyl CoA, (d) CDP-choline, (e) UDP-glucose or UDP-galactose, (f) UDP-galactose, and (g) geranyl pyrophosphate.

4. (a and b) None, because the label is lost as CO_2.

CHAPTER 21

1. Glucose + 2 ADP + 2 P_i + 2 NAD^+ + 2 glutamate $\longrightarrow$ 2 alanine + 2 α-ketoglutarate + 2 ATP + 2 NADH + H^+

2. N_2 $\longrightarrow$ NH_4^+ $\longrightarrow$ glutamate $\longrightarrow$ serine $\longrightarrow$ glycine $\longrightarrow$ δ-aminolevulinate $\longrightarrow$ porphobilinogen $\longrightarrow$ heme

3. (a) Tetrahydrofolate, (b) tetrahydrofolate, and (c) N^5-methyltetrahydrofolate

4. γ-Glutamyl phosphate may be a reaction intermediate.

5. The Schiff base of PLP and aminoacrylate is a likely reaction intermediate.

$$\boxed{PLP}-\overset{H}{\underset{H}{C}}=\overset{+}{N}-\overset{COO^-}{\underset{CH_2}{C}}$$

CHAPTER 22

1. Glucose + 2 ATP + 2 $NADP^+$ + H_2O $\longrightarrow$ PRPP + CO_2 + ADP + AMP + 2 NADPH + H^+

2. Glutamine + aspartate + CO_2 + 2 ATP + NAD^+ $\longrightarrow$ orotate + 2 ADP + 2 P_i + glutamate + NADH + H^+

3. (a, c, d, and e) PRPP; (b) carbamoyl phosphate.

4. PRPP and formylglycinamide ribonucleotide.

5. dUMP + serine + NADPH + H^+ $\longrightarrow$ dTMP + $NADP^+$ + glycine

6. There is a deficiency of N^{10}-formyltetrahydrofolate. Sulfanilamide inhibits the synthesis of folate by acting as an analog of p-aminobenzoate, one of the precursors of folate.

CHAPTER 23

1. (a) TTGATC, (b) GTTCGA, (c) ACGCGT, and (d) ATGGTA.

2. (a) [T] + [C] = 0.46.
   (b) [T] = 0.30, [C] = 0.24, and [A] + [G] = 0.46.

3. 5.88×10^3 base pairs.

4. [CT], [GC], [TA], [AC], [TG], and [TC].

5. A G-C base pair has three hydrogen bonds, whereas an A-T base pair has only two.

6. After 1.0 generation, one-half of the molecules would be ^{15}N-^{15}N, the other half ^{14}N-^{14}N. After 2.0 generations, one-quarter of the molecules would be ^{15}N-^{15}N, the other three-quarters ^{14}N-^{14}N. Hybrid ^{14}N-^{15}N molecules would not be observed in conservative replication.

7. Two.

8. FAD, CoA, $NADP^+$.

CHAPTER 24

1. (a) DNA polymerase I is a single chain, whereas RNA polymerase has the subunit structure $\alpha_2\beta\beta'\sigma$.

   (b) Deoxyribonucleoside triphosphates versus ribonucleoside triphosphates.

   (c) $5' \longrightarrow 3'$ for both.

   (d) DNA polymerase I has $5' \longrightarrow 3'$ and $3' \longrightarrow 5'$ nuclease activities, whereas RNA polymerase has none.

   (e) Semiconserved for DNA polymerase I, conserved for RNA polymerase.

   (f) DNA polymerase I needs a primer, whereas RNA polymerase does not.

(g) Both are driven by the hydrolysis of pyrophosphate.
2. 5′ UAACGGUACGAU 3′.
3. (a) AU, (b) GC, (c) UG, and (d) CA.
4. The 2′-OH group in RNA acts as an intramolecular catalyst. A 2′-3′ cyclic intermediate is formed in the alkaline hydrolysis of RNA.
5. Cordycepin terminates RNA synthesis. An RNA chain containing cordycepin lacks a 3′-OH group.
6. (a) pGCp, AGUp, ACp, Up, GUp, and C.
(b) pGp, CAGp, UACUGp, and UC.
(c) pGp, CAp, Gp, UAp, CUGp, and UC.
7. UAGCCUGAAUp.

CHAPTER 25

1. Leu-Pro-Ser-Asp-Trp-Met-
2. Poly (Leu-Leu-Thr-Tyr).
3. (a) Pro (CCC), Ser (UCC), Leu (CUC), and Phe (UUC). Alternatively, the last base of each of these codons could be U.
(b) These C $\longrightarrow$ U mutations were produced by nitrous acid.
4. (a) It required two base changes.
(b) Arg, Asn, Gln, Glu, Ile, Met, or Thr.
5. (a) No, it was produced by the deletion of the first base in the sequence shown below and the insertion of another base at the end of this sequence.
(b) -AGUCCAUCACUUAAU-

CHAPTER 26

1. (a) No, (b) no, and (c) yes.
2. Four bands: light, heavy, a hybrid of light 30S and heavy 50S, and a hybrid of heavy 30S and light 50S.
3. About 799 high-energy phosphate bonds are consumed—400 to activate the 200 amino acids, 1 for initiation, and 398 to form 199 peptide bonds.
4. (b, c, and f) Type 1; (a, d, and e) type 2.
5. The simplest hypothesis is that the CCA anticodon of a tryptophan tRNA has mutated to UCA, which is complementary to UGA. However, analysis of this altered tRNA produces a surprise. Its anticodon is unaltered. Rather, there is a substitution of A for G at position 24. Thus, a residue far from the anticodon in the linear base sequence can influence the fidelity of codon recognition.
6. Bromoacetyl-phenylalanyl-tRNA is an affinity-labeling reagent for the P site of E. coli ribosomes. See H. Oen, M. Pellegrini, D. Eilat, and C. R. Cantor, Proc. Nat. Acad. Sci. 70(1973):2799.

CHAPTER 27

1. (a) The lac repressor is missing in an i^- mutant. Hence, this mutant is constitutive for the proteins of the lac operon.
(b) This mutant is constitutive for the proteins of the trp operon because the trp repressor is missing.
(c) The arabinose operon is not expressed in this mutant because the P2 form of the araC protein is needed to activate transcription.
(d) This mutant is lytic but not lysogenic because it cannot synthesize the λ repressor.
(e) This mutant is lysogenic but not lytic because it cannot synthesize the N protein, a positive control factor in transcription.
2. One possibility is that an i^s mutant produces an altered lac repressor that has almost no affinity for inducer but normal affinity for the operator. Such a lac repressor would bind to the operator and block transcription even in the presence of inducer.
3. This mutant has an altered lac operator that fails to bind the repressor. Such a mutator is called O^c (operator constitutive).
4. The cyclic AMP binding protein (CAP) is probably defective or absent in this mutant.
5. An E. coli cell bearing a λ prophage contains λ repressor molecules, which also block the transcription of the immediate-early genes of other λ viruses.

References to structural formulas are given in **boldface** type.

A. *See* adenine
A band, 827, 828
absolute configuration
   of amino acids, 14
   of D-glyceraldehyde, 279
   of sugars, 281, 304, 305
absorption spectrum
   of chlorophylls *a* and *b*, 460
   of color receptors, 807
   of myoglobin and hemoglobin, 57
   of rhodopsin, 800
accelerin, 201
acetaldehyde, **292**
acetate, **3**
   from acetylcholine, 792
   model of, 3
acetoacetate, **417**
   in amino acid oxidation, 449, 451
   in oxidation of leucine, 449
acetoacetyl CoA, **417**
acetoacetyl-ACP, **423**
acetone, **417**
acetyl CoA. *See* acetyl coenzyme A
acetyl CoA carboxylase, 419
acetyl coenzyme A (acetyl CoA),
      268–269, **269**
   in fatty acid oxidation, 412
   in fatty acid synthesis, 419
   free energy of hydrolysis, 268
   from leucine, 449
   model of, 255, 307
   from pyruvate, 307
   role of, 268
   in synthesis of acetylcholine, 788
acetyl group-transfer potential, 268
acetyl transacylase, 421
acetylcholine, **788,** 788–796
acetylcholine receptor, 790–791
   inhibitors of, 795–796

acetylcholinesterase, 185, 788,
      791–795
   inhibitors of, 792–793
   mechanism of, 792
acetyl-enzyme intermediate, **183**
   in acetylcholinesterase, 792
N-acetylgalactosamine, **244**
N-acetylglucosamine (NAG), **154, 244**
   in bacterial cell walls, 755
tri-*N*-acetylglucosamine, **158**
   binding to lysozyme, 159
N-acetylglucosamine
      l-phosphate, 757
β-*N*-acetylhexosaminidase, 487
acetyllipoamide, **318**
N-acetylmuramic acid (NAM), **154**
   in bacterial cell walls, 755
N-acetylneuraminate, **486**
acetylphosphate, **265**
   free energy of hydrolysis, 266
acid-base equilibria, 42
acids, 42
aconitase, 309
cis-aconitate, 309
ACP. *See* acyl carrier protein
acridine dyes, 642
ACTH. *See* adrenocorticotrophic
      hormone
actin, 828
   binding to myosin, 832
   G and F forms of, 832
   in slime molds, 837
   thin filaments from, 832
   ubiquity of, 837
actinin, 828
actinomycin D, 614–617, **615**
actinomycin-deoxyguanosine
      complex, 616
activated carriers, 269

activated CO_2, 370–371
activated dihydroxyacetone, **365**
activated glycoaldehyde, **364**
activated methyl group, 508–509
activating enzymes, in protein
      synthesis, 647
activation energy, 120
activation of enzymes, 117
active site, 122–124
   of carboxypeptidase A, 171
   of chymotrypsin, 187
   of lysozyme, 163
   of ribonuclease S, 9
active transport, 768–769
actomyosin, 832
acute intermittent porphyria, 523
acute pancreatitis, 193
acyl adenylate, **409**
   in amino acid activation, 647–648
   in fatty acid activation, 409
acyl carnitine, **410**
acyl carrier protein (ACP), 420, **421**
acyl CoA. *See* acyl coenzyme A
acyl CoA dehydrogenase, 411
acyl CoA synthetase, 409
acyl coenzyme A (acyl CoA), 268,
      409, 412
N-acyl sphingosine, **485**
acyl-AMP, **409**
acylation, 183, 187
acyl-enzyme intermediate, 184
   in acetylcholinesterase, 792
   in chymotrypsin, 187–188
   in penicillin, 763
acyl-malonyl-ACP condensing
      enzyme, 422
adaptor hypothesis, 624
adenine, **528, 558**
   rare tautomer of, 641

adenine phosphoribosyl
transferase, **536**
adenine-thymine base pair, **565**
adenosine, **530**
adenosine diphosphate
(ADP), 263–264
adenosine monophosphate (AMP),
263–264, **531**
synthesis of, 535
adenosine 5'-phosphate. *See* adenosine
monophosphate
adenosine triphosphatase (ATPase)
in Ca^{2+} transport, 776, 836
in mitochondria, 349
in muscle, 831
in Na^+-K^+ transport, 770
adenosine triphosphate (ATP),
263–266, **263**
in cilia and flagella, 838
from fatty acid oxidation, 414
free energy of hydrolysis, 264–266
from glycolysis, 287–290
group-transfer potential of, 264
in membrane transport,
771–772, 776
models of, 3
in muscle contraction, 832–833,
836–837
from oxidative phosphorylation,
336, 340–341
from photosynthesis, 467–468
S-adenosylhomocysteine, **509**
S-adenosylmethionine, **508**
in methylation of DNA, 715
in methylation of RNA, 613
molecular model of, 476, 508
in synthesis of phosphatidyl
choline, 482
adenyl cyclase, 810–811
in glycogen metabolism, 390
inhibition by PGE_1, 821
stimulation by cholera toxin,
814–815
adenylate. *See* adenosine
monophosphate
adenylate deaminase, 547
adenylate kinase, 264, 539, 837
adenylosuccinate, **535**
adenylyl transferase, 519
adenylylated enzyme, **515**
adenylylation, 519
ADP. *See* adenosine diphosphate

adrenalin. *See* epinephrine
adrenocorticotrophic hormone
(ACTH), 45, 497
adrenocorticotrophin, 45
adrenodoxin, 496
adrenogenital syndrome, 498–499
affinity chromatography, 733
affinity labeling, 185, 738
alanine, **15**
oxidation of, 441
synthesis of, 504
alanine transaminase, 433
alcaptonuria, 451
alcohol dehydrogenase, 292
alcoholic fermentation, 292
aldehyde cross-links, 218–220, 223
aldimines, **435**
aldol condensation, 219
aldol cross-link, **219**
aldolase, 286
catalytic mechanism of, 294–295
in photosynthesis, 471, 473
aldoses, 279, **304**
aldosterone, **497**
alkaline phosphatase, 618
alkenyl glycerol ether, **483**
allantoicase, 548
allantoinase, 548
allolactose, **680**
allopurinol, **550**
D-allose, **304**
allosteric enzymes, 134
concerted model, 135–137
sequential model, 138–139
allosteric interaction, 118, 541
allosteric proteins
aspartate transcarbamoylase,
540–542
enzymes, 134–139
hemoglobin, 71–91
allotypic site, 741
alloxan, 820
alpha helix, **32**, 32–33
coiled coil of, 33, 831
D-altrose, **304**
amethopterin, **544**, 724
amide bond, 18
amino acid sequence
of adrenocorticotrophin, 45
determination of, 22–30
determines conformation, 36–39
of hemoglobin, 62

of histone IV, 702
of insulin, 21
of myoglobin, 54
of ribonuclease, 37
significance of, 22
amino acids, **13–17**
abbreviations, 17
absolute configuration, **14**
acidic, **16**
aliphatic, **15**
amide, **17**
analysis, 23
aromatic, **16**
basic, **16**
as biosynthetic precursors, 520–521
classification, 14
deamination, 432–437
degradation, 432–453
essential, 503, 511–520
fate of carbon skeleton, 453
hydrogen-bonding potentiality, 143
hydroxyl, **16**
D-isomers, 755
nonessential, 503
optical activity, 14
oxidation, 432–453
p*K* values, 44
structure, 13, 15
sulfur-containing, **17**
synthesis, 503–520
aminoacrylate, **437**
aminoacyl site (A site), 665
aminoacyl transferase II, 671
aminoacyl-adenylate, **647**
aminoacyl-AMP, **647**
aminoacyl-tRNA, **646**
synthetic hybrid, 654
aminoacyl-tRNA synthetases, **646,**
647–649
synthetic hybrid, 654
5-aminoimidazole-4-carboxamide
ribonucleotide, **515**, 534
5-aminoimidazole 4-carboxylate
ribonucleotide, **534**
5-aminoimidazole
ribonucleotide, **533**
5-aminoimidazole-4-*N*-
succinocarboxamide
ribonucleotide, **534**
β-aminoisobutyrate, **549**
δ-aminolevulinate, **521**
δ-aminolevulinate dehydrase, 521

δ-aminolevulinate synthetase, 521–523
aminopeptidase, 668
aminopterin, **544**
2-aminopurine, 641
amino-terminal analysis, 24, **25, 27**
ammonium ion
    in amino acid oxidation, 432–437
    conversion to urea, 437–440
    inherited metabolic defects, 440–441
    from N_2, 504
    in synthesis of amino acids, 504
ammonotelic organisms, 437
amniocentesis, 487
AMP. *See* adenosine monophosphate
amphipathic molecules, 233
amplification
    in blood clotting, 193
    in glycogen metabolism, 394
    in hormone action, 814
    in vision, 802
α-amylase, 397
β-amylase, 397
amylase release, 812
amylopectin, 396
amylose, 396
amytal, 340
anabolic pathways, 275
anabolism, 479
Andersen's disease, 396
androgens, 495
    synthesis of, 498
androstenedione, **498**
anemias, 95–107, 366, 444, 447
Anfinsen, C., 36, 39, 739
angstrom, 30
anthranilate, **513**
antibiotics
    actinomycin D, 614–617, **615**
    gramicidin A, **780**, 783–784
    gramicidin S, **673**, 675
    nonactin, 783
    penicillin, 761–765, **762**
    rifampicin, 614
    rifamycin B, **614**
    valinomycin, **779**, 782
antibodies, 731–752
    active fragments, 735–736
    clonal selection, 749–750
    combining sites, 734
    conformation, 744–745

definition, 732
electron micrograph, 731
evolution, 742–743
flexibility, 737
heterogeneity, 735
purification, 733
specificity, 734
subunit structure, 736–737
synthesis, 732–733, 738–740, 746–751
variable and constant regions, 740–742
See also immunoglobulins
anticodon, **624,** 625
antidextran antibody, 734
antifolate drugs, 544
antifreeze poisoning, 134
antigen, 732
antigen-antibody lattice, 736
antigenic determinant, 732
antihemophilic factor, 200–201
antimycin A, 340–341
antiparallel β pleated sheet, 34
antiserum, 733
apomyoglobin, 61
apoprotein, 47
D-arabinose, **304,** 686
arabinose operon, 686
arachidate, **406**
arginase, 438
arginine, **16**
    oxidation of, 442
    in urea cycle, **438**
argininosuccinase, 439
argininosuccinate, **439**
argininosuccinate synthetase, 438
Arnold, W., 461
Arnon, D., 465
ascorbic acid, **210**
    in hydroxylation of collagen, 208
    in oxidation of hydroxyphenylpyruvate, 451
    reduction of nitroxides, 253
asparagine, **17**
    linkage to sugars, 244
    oxidation of, 448
    synthesis of, 505
aspartate, **16**
    oxidation of, 448
    in pyrimidine biosynthesis, 538
    synthesis of, 504
    in urea cycle, 439

aspartate transcarbamoylase, 540–542
aspirin, 822
asymmetric reactions, 322
asymmetry
    of biological membranes, 244
    of Na^+-K^+ transport system, 244
Atkinson, D., 273
ATP. *See* adenosine triphosphate
ATPase. *See* adenosine triphosphatase
atractyloside, 343
autoradiography, 582–584
avian myeloblastosis virus, 724
Avery, O., 560
avidin, 377
axial substituent, 281
axoneme, 838
azaserine, 554
azide, 340–341

B_1 and B_2 enzymes, 543
Bach, J. S., 217
Bacillus brevis, 637
bacitracin, 760
bacterial cell walls, 754–765, **755**
    action on penicillin, 761–764
    cleavage by lysozyme, 154
    cross-linking, 760
    electron micrographs, 754, 755
    function, 754
    structure, **755**
    synthesis, 756
bacteriophage
    λ, 571
    φX174, 571–572
    RNA phages, 718–720
    T2, 560–562
    T4, 712–715
ball-and-stick models, 2
Baltimore, D., 716, 725
Barcroft, J., 77
Barr body, 703
base analogs, 641
base-pairing, 564–566, **565**
bases, proton acceptors, 42
bases, in nucleotides, 529
Bayliss, W., 807
behenate, **406**
Bence-Jones, H., 740
Bence-Jones protein, 740–741
Benesch, Reinhold, 77
Benesch, Ruth, 77
Benzer, S., 639

benzoate, **408**
benzyl penicillin, **762**
Berg, P., 409
Bergstrom, S., 821
beriberi, 321
Bernal, J., 63, 818
beta pleated sheet, **34**
betaine, **509**
bicarbonate. *See* carbon dioxide
bilayers, 235–236
bile salts, 492–493
bilirubin, **525**
bilirubin diglucuronide, **525**
biliverdin, **525**
bimolecular sheet, 235
biological membranes, 227–250
    asymmetry, 244–245
    common features, 228
    dynamics and fluidity, 245–249
    freeze-etch electron
        microscopy, 243
    fusion, 246–247
    lipid bilayers, 235–240
    membrane lipids, 229–234
    membrane proteins, 240–243
    permeability properties, 237–240
    roles, 227–228
biotin, **370**
    in acetyl CoA carboxylase, 419
    inhibition by avidin, 377
    mechanism of, 371
    in pyruvate carboxylase, 370–371
biotin carboxyl carrier protein, 420
biotin carboxylase, 420
Birnstiel, M., 698
black lipid films, 239
Bloch, K., 487
Blow, D., 181
Bohr effect, 76–77
    mechanism of, 86–88
bone, 217
Bonitus, J., 321
boron acids, 205
bovine spleen phosphodiesterase, 618
Brachet, J., 595
Bragg, L., 63
branching enzyme, 388
Braunstein, A., 435
Brenner, S., 598, 625
Britten, R., 695
5-bromouracil, **641**
Brown, D., 698

Buchanan, J., 532
Buchner, E., 277
Buchner, H., 277
buffer, 43
α-bungarotoxin, 791
Burnet, M., 739–749
butyryl-ACP, **423**

C. *See* cytosine
Ca^{2+} ATPase, 776, 836
$Ca_{10}(PO_4)_6(OH)_2$, 217
caffeine, 811
Cairns, J., 581
calciferol, 270
calcium ion
    effect on phosphorylase kinase, 391
    in muscle contraction, 835–836
    in release of acetylcholine, 790
    in visual excitation, 803
calcium transport, 776
calsequestrin, 836
Calvin, M., 469
Calvin cycle, 471
cAMP. *See* cyclic adenosine
        monophosphate
Campbell, A., 721
cancer
    chemotherapy, 544
    transformed cells, 723
    viral etiology, 726
cap formation, 750
CAP protein, 687
capsid, 708
capsular polysaccharide, 559
carbamoyl esters, 792–793
carbamoyl phosphate, **538**
    free energy of hydrolysis, 26
    synthesis of, 439
    in urea cycle, **438**
carbamoyl phosphate synthetase, 439
N-carbamoylaspartate, **538**
N-carbamoylisobutyrate, **549**
carbanion, 364–366
carbon dioxide
    binding to hemoglobin, 88–89
    in citric acid cycle, 315
    in fatty acid synthesis, 419
    in gluconeogenesis, 370–372
    in pentose phosphate pathway, 357
    in photosynthesis, 469–471
    in purine biosynthesis, 532
carbon monoxide, 94, 340–341

carbon monoxide hemoglobin, 94
carbonic anhydrase, 115
carbonium ion, in action of
        lysozyme, **164, 167**
carboxypeptidase A, 168–174
    active site, 170–174
    catalytic mechanism, 173–174
    conformation, 169
    electric strain, 173–174
    induced fit, 170, 172, 174
    specificity, 169
1-(*o*-carboxyphenylamino)-
        1-deoxyribulose 5-phosphate, **513**
cardiotonic steroids, 774, **775**
carnitine, **410,** 411
carrier in immune response, 732
carrier in membrane transport,
        768, 780
carrier lipid, **758**
Cartier, J., 210
Caspar, D., 710
Caspersson, T., 594
catabolic pathways, 275
catabolism, 479
catabolite repression, 687
catalase, 548
catalytic groups, 122
catalytic mechanisms
    of carboxypeptidase A, 168–174
    of lysozyme, 153–168
catalytic power, 115
catalytic subunit, 813
catecholamines, 796–797
catechol-*O*-methyltransferase, 797
CDP. *See* cytidine diphosphate
CDP-choline, **483**
CDP-diacylglycerol, **481**
    molecular model of, 5, 481
    in phosphoglyceride synthesis,
        480–481
cell cycle, 703
cell motility, 837
cell walls. *See* bacterial cell walls
cellulase, 397
cellulose, 397
ceramide, **485**
    as precursor of gangliosides, 485
    as precursor of sphingomyelin, 485
    synthesis of, 484–485
cerebrosides, 485
Chain, E., 762
Chance, B., 341

Changeux, J.-P., 135
channel former, 780
Chargaff, E., 566
charge relay network, 185–186, **186**
Chase, M., 562
chemical mutagens, 641–642
chemical-coupling hypothesis, 346–347
chitin, 155, 397
chloramphenicol, 671
chlorophylls, 459–460
   chlorophyll *a,* **459**
   chlorophyll *b,* **459**
chloroplast, 460–461
   diagram, 461
   electron micrograph, 456
cholera toxin, 814–815
cholesterol, 229, **233**
   control of synthesis, 492
   molecular model, 487
   numbering system, **494**
   origin of carbon atoms, 487–488
   precursor of bile salts, 493
   precursor of steroid hormones, 494–496
   synthesis, 487–492
choline, **483, 788**
   acetylation of, 788
   in phosphoglycerides, **230**
choline acetylase, 788
choline acetyltransferase, 788
cholyl CoA, **493**
choriocarcinoma therapy, 544
chorismate, **511**
   as precursor of aromatic amino acids, 511–513
Christmas factor, 200–201
chromatin, 703–704
chromophore, 800
chromosomes. *See* eucaryotic chromosomes
chromatography
   gel-filtration, 20
   ion-exchange, 21
   of peptides, 101
chymotrypsin, 179–192
   α and π forms, 180
   catalytic mechanism, 182–188
   charge relay network, 185–186
   conformation, 180–181
   homologous enzymes, 190–192
   specificity, 181–182

   substrate binding, 187
   tetrahedral transition state, 187–188
chymotrypsinogen, 179–180
   activation mechanism, 188–190
cI gene, 689
cilia, 838
cilium, of retinal rod cell, 799
cis double bonds, **243**
cis-trans isomerization, 801
citrate, **309**
   asymmetric fate of, 322
   as carrier of acetyl groups, 426
citrate lyase, 426
citrate synthetase, 309
citric acid cycle, 307–329
   cofactors in, 315
   control of, 326
   discovery of, 327
   free energy changes in, 315
   link to urea cycle, 440
   reactions of, **314,** 315
   stoichiometry of, 313
   types of reactions, 315
citrulline, **438**
citryl CoA, 309
clonal selection theory, 749
Clostridium histolyticum, 221
clotting, 193–203
   amplifying cascade, 193
   extrinsic pathway, 194, 201
   factors, 201
   fibrin clot, 194–196
   hemophilia, 198
   intrinsic pathway, 194, 200–201
   role of Ca^{2+} and phospholipids, 198
CMP. *See* cytidine monophosphate
CMP-*N*-acetylneuraminate, 485
CNBr, 26
CoA. *See* coenzyme A
coat, viral, 709
cobalamin, 444–447
cobalt, 445
cobratoxin, 791
codon-anticodon pairings, 655–657
codons, 625
   table of, 636
coenzyme A (CoA), **269**
   biosynthesis of, 546
coenzyme Q (CoQ), **337**
cohesive ends, 720

coiled coil, 831
colchicine, 838
colinearity, 639
collagen, 206–222
   amino acid sequence, 208
   carbohydrate unit, **209**
   conformation, 211–215
   cross-linking, 218–220
   electron micrograph, 12, 207
   fiber formation, 216–218
   hydroxylation and glycosylation, 208–209
   models, 211–213
   from procollagen, 215–216
   in scurvy, 210
   in tadpole metamorphosis, 221
   thermal stability, 214–215
collagenases, 221–222
color blindness, 804
color vision, 803–804
competitive inhibition, 130–134
complement, 736
complementation test, 200
concanavalin A, 244
concerted feedback control, 518
cone opsins, 804
cones, 798
conformation
   of membranes, 249–250
   of nucleic acids, 563–566, 595, 653
   of polypeptides, 30–34
   of proteins, 30–39
conformational hypothesis, 347
congenital adrenal hyperplasia, 498–499
congenital erythropoietic porphyria, 523
congestive heart failure, 775
connective-tissue proteins, 206–223
constant (C) region, 741
constitutive mutants, 681
contractile force, 834–835
convergent evolution, 191–192, 804
cooperative interactions
   in collagen, 214
   in DNA, 695
   in hemoglobin, 75, 89–90
   in lipid bilayers, 236
copper deficiency, 222
coproporphyrinogen III, **522**
CoQ. *See* coenzyme Q
Corey, R., 30

Cori, C., 279, 380, 808
Cori, G., 279, 380, 808
Cori cycle, 374
Cori's disease, 396
corrin ring system, **445**
corticosterone, **497**
cortisol, **497**
Corynebacterium diphtheriae, 671
cotransport, 777
Coulomb's law, 140
countercurrent distribution, 649–650
coupling factors, 349
covalent modification, 117
cozymase, 278
creatine phosphate, **265**
    free energy of hydrolysis, 266
creatine kinase, 836
Crick, F., 563, 623, 625, 709
cristae, 332
cross-bridges, 828, 834–835
    electron micrograph, 829
crotonyl-ACP, **423**
crystallization, 49
CTP. *See* cytidine triphosphate
Cuatrecasas, P., 818
cumulative feedback control, 518
cyanate, **87**
    effect on hemoglobin, 87
cyanide, 340–341
cyanocobalamin, 445
cyanogen bromide, 26, 27
cyclic adenosine monophosphate
        (cyclic AMP or cAMP),
        390, 810
    in control of transcription, 687
    degradation of, 394, 810
    discovery of, 808–809
    effects of, 812–813
    in glycogen metabolism, 389–394
    as hunger signal, 813
    model of, 909
    as second messenger, 811–812
    synthesis of, 390, 810
cyclic AMP. *See* cyclic adenosine
        monophosphate
cyclic photophosphorylation, 468–469
cycloheximide, 671
cyclopropane fatty acids, **253**
cystathioninase, 510
cystathionine, **510**
cystathionine synthetase, 510
cysteic acid, **30**

cysteine, **17**
    model of, 3
    oxidation of, 441
    synthesis of, 510
cystine, **19**
    oxidation to cysteic acid, **30**
cytidine, **530**
cytidine diphosphate, 531
cytidine diphosphodiacylglycerol.
        See CDP-diacylglycerol
cytidine monophosphate, 531
cytidine triphosphate, **540**
cytochalasin B, 837
cytochrome $(a + a_3)$, 338
    heme group in, **339**
cytochrome b, 338
cytochrome c, 338, 350–352
    evolution of, 351
    heme group in, **339**
    structure of, 350
cytochrome c_1, 338
    heme group in, **339**
cytochrome oxidase, 340
cytochrome P_{450}, 496
cytochromes, 338–340
cytosine, **558**
cytosine arabinoside, 724

dalton, 19
dansyl chloride, **24**
ε-dansyllysine, 734
dark reactions of photosynthesis, 461,
        469–473
deacylation, 183, 188
deadenylylating enzyme, 519
debranching enzyme, 383
decamethonium, **795**
decarboxylation
    of isocitrate, 310
    of α-ketoglutarate, 310
    of malonyl CoA, 422
    of oxaloacetate, 372
    of 6-phosphogluconate, 358
    of pyruvate, 307
decorated filaments, 834
deformylase, 668
degeneracy, 626, 636
    wobble hypothesis, 655–657
dehydrogenases
    in amino acid degradation, 433
    in amino acid synthesis, 504
    in citric acid cycle, 310–313

    in fatty acid oxidation, 414
    in fatty acid synthesis, 422
    in glycolysis, 296–297
    in pentose phosphate pathway,
        357–358
    in photosynthesis, 466
    in oxidation of pyruvate, 316–319
    in oxidative phosphorylation,
        336, 342
dehydratases, 436
5-dehydroquinate, **511**
5-dehydroshikimate, **511**
DeLange, R., 701
delayed-early genes, 688
deletion mutations, 640
DeLucia, P., 581
denaturation, 36
density-gradient equilibrium
        sedimentation, 568–569
deoxyadenosine, **530**
5'-deoxyadenosylcobalamin, **445**
    in methylmalonyl CoA
        mutase, 444
    model of, 446
3-deoxyarabinoheptulosonate-
        7-phosphate, **511**
deoxycytidine, **530**
deoxyguanosine, **530**
deoxyribonuclease, 713
deoxyribonucleic acid (DNA),
        557–591, **558**
    absorption spectrum, 695
    action of nucleases, 578–580
    annealing, 695–696
    autoradiography, 582–584
    base-pairing, 564–566
    base ratios, 565
    bases, 557–558
    calf thymus, 696
    cleavage, 715
    circularity, 571
    C_0t analysis, 695
    electron micrograph, 6
    in eucaryotic chromosomes,
        694–700
    hypochromicity, 695
    ligase, 577–578
    melting, 695
    methylation, 715
    models, 564–568
    mouse satellite, 697
    nearest-neighbor analysis, 576–577

DNA (*continued*)
  nomenclature, 557–559
  packing, 702–704
  polymerase I, 116, 572–577,
    578–580
  polymerase II and III, 580–582
  promoter sites, 607–609
  repair, 578–580, 589
  repetitive sequences, 695–699
  replication, 582–588
  replication in eucaryotes, 704–705
  restriction, 715–716
  schematic diagram, 557
  semiconservative replication,
    568–570
  single-stranded, 571–572
  size, 570–571
  size in *Drosophila,* 694
  synthesis, 572–588
  as templates, 566–567, 603–606
  transformation, 559–561
  untranscribed spacers, 698–699
  Watson-Crick double helix,
    563–567
  x-ray diffraction pattern, 563
deoxyribonucleoside, 530
deoxyribonucleotides, synthesis
  of, 543
β-D-2-deoxyribose, **529**
deoxythymidine, **530**
deoxythymidylate (dTMP), **544**
deoxyuridylate (dUMP), **544**
dephosphocoenzyme A, 546
depolarization, 789
dermatosparaxis, 216
desamido-NAD$^+$, **545**
desmosine, **223**
deuterated NADH, 324
dexamethasone, 823
dextran, 397
α-dextrin, 397
α-dextrinase, 397
DFP. *See*
  diisopropylphosphofluoridate
diabetes mellitus, 820
diacylglycerol 3-phosphate, **230, 479**
dialysis, 19, 20
dibutyryl cyclic AMP, 812
Dickens, F., 357
dielectric constants, 146
difference Fourier method, 158
diffusion coefficient, 247, 253

diffusion in membranes,
  246–249, 253
digestive enzymes, 178–193
digitalis, 775
digitoxigenin, 774, **775**
dihedral angles, **32**
dihydrobiopterin, **450**
dihydrobiopterin reductase, 450
dihydrofolate, **544**
dihydrofolate reductase, 663
dihydrolipoamide, **318**
dihydrolipoyl dehydrogenase, **316**
dihydrolipoyl transacetylase, 316
dihydroorotate, **538**
dihydrosphingosine, **484**
dihydrothymine, **549**
dihydroxyacetone, **279, 305**
dihydroxyacetone phosphate, **286**
  in gluconeogenesis, 372
  in glycolysis, 286
  in photosynthesis, 471
20α, 22-dihydroxycholesterol, **496**
dihydroxyphenylalanine, **796,** 797
dihydroxyphenylethylamine, 796
dihydroxyphenylglycolaldehyde, **797**
diisopropylphosphofluoridate (DIPF),
  129–130, **130, 184**
  antidote for, 794
  as inhibitor of
    acetylcholinesterase, 793
diisopropylphosphoryl-enzyme
  complex, **184**
6-dimethyladenine, **613**
dimethylallyl pyrophosphate, **490**
di-NAG, 166
2, 4-dinitrophenol, **346**
dinitrophenylated bovine serum
  albumin, **732**
Dintzis, H., 661
dipeptide, **18**
DIPF. *See*
  diisopropylphosphofluoridate
diphosphatidyl glycerol, **231**
1, 3-diphosphoglycerate, **287**
  in gluconeogenesis, 372
  in glycolysis, 287
  in photosynthesis, 471
2, 3-diphosphoglycerate, **77, 300**
  abnormal levels, 301
  binding to hemoglobin, 84–86
  degradation, 300
  effect on oxygen affinity, 77–80

  synthesis, 300
diphosphoglycerate mutase, 131, 300
2, 3-diphosphoglycerate
  phosphatase, 300
diphtheria toxin, 671
dipolar ion, 13
disaccharides, 397–399
disaccharide-peptide unit, **759**
discs, 799
distal histidine, 55
distortion in catalysis, 160, 165–167
disulfide bonds, **19**
  cleavage of, 30
  reduction of, 36
divergent evolution, 191
DNA. *See* deoxyribonucleic acid
DNA ligase, 577–578
DNA polymerases
  I, 572–577, 578–580
  II, 580–582
  III, 580–582
  RNA-directed, 725
dnaE mutants, 582
DNP-amino acids, 24
docosanoate, **406**
dodecanoate, **406**
domains, in antibodies, 743
dopa, **796**
dopa decarboxylase, 796
dopamine, **796**
Doty, P., 600
double helix of DNA, 563–567
Dowex-50, 23
DPN$^+$. *See* nicotinamide adenine
  dinucleotide, oxidized form
DPNH. *See* nicotinamide adenine
  dinucleotide, reduced form
Drosophila chromosomes, 570,
  694, 704
Drosophila satellite DNA, 697
drug reactions, 366, 796
dynein, 838

early-conductance channel, 797
Ebashi, S., 835
Edelman, G., 736, 742
Edgar, R., 713
Edman, P., 24
Edman degradation, 24–27, **27**
Ehlers-Danlos syndrome, 216
eicosanoate, **406**
eicosatetraenoate, **406, 822**

elastase, 178, 190
  homology to chymotrypsin,
      190–191
elastin, 222–223
electric eel, 790–791
electric organ, 790
electron carriers, 266–268
electron transport, 331–353, 496
electron-density map, 51
electronic strain, 173–174
electrophoresis, 21, 98
  of hemoglobin, 98–100
electroplaxes, 790
electrostatic bonds, 139–140
elongation factors, 666
  in replication of $Q\beta$, 719
Embden, G., 279
Embden-Meyerhof pathway. *See*
      glycolysis
Emerson, R., 461
endonucleases, 618
  in restriction of DNA, 715–716
energy charge, 273
energy coupling, 262
energy generation, 271–272
energy transfer, 462
Engelhardt, V., 830
enol phosphates, 298
enolate ion, **295**
3-enolpyruvylshikimate-
      5-phosphate, **511**
enoyl CoA, 412
enoyl CoA hydratase, 413
enoyl-ACP reductase, 422
enterokinase, 192
enthalpy, 259
entropy, 258
envelope, viral, 708
enzyme multiplicity, 517
enzyme-directed assembly, 713–715
enzymes, 115–149
  active sites, 122–124
  allosteric, 134–139
  catalytic power, 115
  enzyme-substrate complex, 120
  forces, 139–147
  inhibition, 129–134
  Michaelis-Menten model, 124–129
  regulation, 117, 134–139
  specificity, 116
enzyme-substrate complex, 120
  in carboxypeptidase A, 170–173

  in chymotrypsin, 187
  forces in, 139–147
  in lysozyme, 163
  in ribonuclease, 9, 142
epimerases
  in galactose metabolism, 400
  in pentose phosphate pathway, 360
epinephrine, **389**, 520, 796, 797
episome for penicillin, 765
equatorial substituent, 281
erythrocyte ghosts, 771
erythrocyte membrane, 46
  gel pattern, 241
  freeze-etch electron
      micrograph, 243
erythromycin, 671
D-erythrose, **304**
erythrose 4-phosphate, **360**
  in amino acid synthesis, 511
  in photosynthesis, 471
D-erythrulose, **305**
Escherichia coli chromosome, 588
eserine, 792
essential amino acids, 503
  synthesis of, 511–520
essential fatty acids, 428
ester, hydrolysis of, 116
17β-estradiol, **498**
  model of, 823
estradiol receptor, 823
estrogens, 495
  synthesis of, 498
estrone, **498**
ethanol, **292**
  alcoholic fermentation, 292
  therapeutic use, 134
ethanolamine, **230**
ethylene glycol, **134**
  poisoning, 134
ethyleneimine, 45
eucaryotes, 693
eucaryotic chromosomes, 693–706
  histones, 700–703
  fluorescence micrograph, 702
  light micrograph, 693
  model for chromatin, 703–704
  radioautograph, 694
  repetitive sequences of DNA,
      695–699
  replication, 704–705
euchromatin, 703
excision, 721

excitable membranes, 787–805
5′ → 3′ exonuclease, 579
eye forms, 704
F_1 factor, 349
F_{ab} fragment, 735–736
F_c fragment, 735–736
facilitated exchange diffusion, 343
FAD. *See* flavin adenine dinucleotide,
      oxidized form
$FADH_2$, *See* flavin adenine
      dinucleotide, reduced form
faraday, 335
farnesyl pyrophosphate, **491**
  molecular model of, 491
fat cell, 405
fatty acids
  activation, 409
  essential, 428
  fluidity, 406
  inconvertible to glucose, 418
  influence on fluidity, 247
  in membrane lipids, 229–230
  nomenclature, 404–406
  β-oxidation, 408–418
  packing, 248
  precursors of prostaglandins, 822
  synthesis, 418–428
fatty acid oxidation, 408–418
  ATP yield, 414
  reaction sequence, **412,** 414
fatty acid synthesis, 418–428
  elongation, 427
  energetics, 422
  reactions, 422, **423**
  source of NADPH, 426
  stoichiometry, 424
  synthetase, 421
  unsaturation, 427
fatty acid synthetase, 424–426
fatty acyl CoA: carnitine fatty acid
      transferase, 411
FDNB. *See* fluorodinitrobenzene
feedback inhibition, 118
  in amino acid biosynthesis,
      516–519
Fe-protoporphyrin IX, **47**
ferredoxin, 466
  in nitrogen fixation, 504
  in photosynthesis, 466
ferredoxin-NADP reductase, 466
ferredoxin-reducing substance (FRS),
      466–469

ferricyanide, in Hill reaction, 463
ferrihemoglobin, 48
ferritin, 11
ferritin conjugates, 244
ferrochelatase, 523
ferrocyanide, in Hill reaction, 463
ferrohemoglobin, 48
ferrous ion, in hydroxylation of
    collagen, 208
fibrin, 194–196
fibrinogen, 194–195
fibrinopeptides, 195
fibrin-stabilizing factor, 201
fingerprinting, 101
    of hemoglobin, 100–102
    of 5S RNA, 619
Fischer, E., 122, 390
flagella, 838
flavin adenine dinucleotide, oxidized
    form (FAD), **267**
    biosynthesis of, 545
    reactions of, 267
flavin adenine dinucleotide, reduced
    form ($FADH_2$), 267
flavin mononucleotide (FMN),
    337, 545
Fleming, A., 153, 761
flip-flop, 247, 253
Florey, H., 762
fluid mosaic model, 249–250
fluorescence, 121
    of hapten-antibody complex, 734
    of pyridoxal phosphate, 121
5-fluorodeoxyuridine, 724
fluorodinitrobenzene (FDNB), 24
fMet. *See* formylmethionine
FMN. *See* flavin mononucleotide
$FMNH_2$. *See* flavin mononucleotide
folate derivatives 506–509, 534
formamide, model of, 3
5-formamidoimidazole-4-carboxamide
    ribonucleotide, **534**
N-formiminoglutamate, **443**
N^5-formiminotetrahydrofolate, **507**
formyl-L-tryptophan, **187**
formylglycinamide
    ribonucleotide, **533**
formylglycinamidine
    ribonucleotide, **533**
formylmethionine (fMet), 663
formylmethionyl-$tRNA_f$, 663
N^5-formyltetrahydrofolate, **507**

N^{10}-formyltetrahydrofolate, **507**
foxglove plant, 775
Fraenkel-Conrat, H., 710
frameshift mutation, 626
frameshift suppressors, 659
Franklin, R., 563
free energy, 258–266
    of activation, 120
    ATP as currency of, 263
    criterion of spontaneity, 259
    relationship to K', 261
freeze-etching electron
    microscopy, 243
freeze-fracture electron
    microscopy, 242
FRS. *See* ferredoxin-reducing
    substance
α-D-fructofuranose, **280**
fructose, **279–281, 305**
    α and β anomers, 281
    furanose ring form, 280
    metabolism, 399
    open-chain form, 280
fructose 1,6-diphosphate, **283**
    in gluconeogenesis, 369
    in glycolysis, 283–286
    in photosynthesis, 471
fructose 1-phosphate pathway, 399
fructose 6-phosphate, **360**
    in gluconeogenesis, 369
    in glycolysis, 285
    in photosynthesis, 471
fugu fish, 797–798
Fugue in D Major, 217
Fuller, B., 710
fumarase, 313
fumarate, **312**
    in amino acid oxidation, 448, 451
    in citric acid cycle, 312
    in urea cycle, 439
4-fumarylacetoacetate, **451**
furanose, 280

G. *See* guanine
G_1 phase, 703
G_2 phase, 703
Gaffron, H., 462
galactitol, **401**
galactokinase, 399
galactose, **304, 679**
    metabolism of, 399–401
    as a precursor, 118

galactose 1-phosphate, **400**
galactose 1-phosphate uridyl
    transferase, 400
    deficiency disease, 401
galactosemia, 401
β-galactosidase, 679–683
galactoside permease, 680–682
galactosyl transferase, 398
Gall, J., 697
gangliosides, 485
    defective breakdown, 486
    ganglioside G_{M1}, **486**
    ganglioside G_{M2}, **487**
    ganglioside G_{M3}, **487**
Garrod, A., 451
gas gangrene, 221
GDP. *See* guanosine diphosphate
gel-filtration chromatography, 19, 20
general acid catalysis, 165
genes
    for antibodies, 746–749
    for rRNA, 698–699
    for silk fibroin, 699–700
genetic code, 623–638, **636**
    degeneracy, 626, 636–637
    discovery of, 627–635
    major features, 625–626
    reading frame, 626
    synonyms, 636
    table, 636
    universality, 638
genetic material, 559–563
geranyl pyrophosphate, **491**
Gerhart, J., 540
germ-line hypothesis, 748
Gibbs, J. W., 259
Gilbert, W., 683
glaucoma, 793
glucagon, **389**
glucocorticoids, **495,** 823
    synthesis of, 498
glucogenic amino acids, 441
gluconeogenesis, 367–374
    distinctive reactions, 369
    effect of cyclic AMP on, 812
    effect of insulin, 815
    energetics, 370
    pathway, 372
α-D-glucopyranose, **280**
glucose, **279–281,** 304
    α and β anomers of, 281
    ATP yield from, 343–344

glucose (*continued*)
  chair form of, 281
  formation of, 367
  in glycolysis, 382–384
  inhibition of cyclic AMP
     synthesis, 687
  from lactose, 679
  model of, 3, 281
  open-chain form of, 280
  pyranose ring form of, 280
  transport of, 777
glucose 1,6-diphosphate, **384**
glucose 6-phosphatase
  deficiency disease, 395
  organ distribution, 385
glucose 1-phosphate, **381**
  free energy of hydrolysis, 266
  in glycogen degradation, 380–384
  in glycogen synthesis, 386–388
glucose 6-phosphate, **283**
  free energy of hydrolysis, 266
  in gluconeogenesis, 369
  in glycogen metabolism, 384
  in glycolysis, 283–284
  in pentose phosphate pathway, 362
glucose 6-phosphate
    dehydrogenase, 357–358
  deficiency of, 366–367, 703
α-1,6-glucosidase, 383
glucuronate, 525, **559**
glutamate, **16**, 432
  degradation of, 432–434
  as precursor of other amino acids,
    504–505
  synthesis of, 504
glutamate dehydrogenase, 433, 504
glutamate γ-semialdehyde, **443, 505**
glutamate transaminase, 433
glutaminase, 442
glutamine, **17**
  in cross-linking of fibrin, 196
  as nitrogen donor, 518
  oxidation of, 442
  synthesis of, 505
glutamine synthetase, 505
  control of, 518–520
  electron micrograph of, 518
γ-glutamyl phosphate, **772**
glutathione, **366–367**
D-glyceraldehyde, **279, 304**
  absolute configuration of, 279
glyceraldehyde 3-phosphate, **286, 359**

glyceraldehyde 3-phosphate
    dehydrogenase, 287
  catalytic mechanism of, 296–297
  in photosynthesis, 471
glycerol, **230**
  entry into glycolysis, 408
  in phosphoglycerides, 230
  as precursor of teichoic acid, 761
  from triacylglycerols, 407
glycerol 3-phosphate, **230, 264**
  absolute configuration of, 230
  free energy of hydrolysis, 266
  from glycerol, 408
  as precursor of phosphatidate, 480
glycerol phosphate acyl
    transferase, 480
glycerol phosphate shuttle, 341–343
glycinamide ribonucleotide, **533**
glycine, **15**
  in collagen, 208, 212–213
  oxidation of, 441
  in porphyrin synthesis, 521
  synthesis of, 506
glycocholate, **493**
glycogen, 378–396
  control, 389–394
  degradation, 380–383, 812
  electron micrograph, 379
  granules, 378
  role, 378
  storage diseases, 395–396
  structure, **379**
  synthesis, 385–388
glycogen phosphorylase, 380–381
  control of, 390, 392
  phosphorylase *a*, 390
  phosphorylase *b*, 390
glycogen synthetase, 387
  control of, 391
  D form, 391
  I form, 392
glycogen synthetase phosphatase, 393
glycolipids, 229, 232–233, 244
N-glycolylneuraminate, 485
glycolysis, 277–301
  ATP yield, 290
  control, 298–299
  inherited defects, 301
  reactions, 290–291
  stoichiometry, 291
  thermodynamics, 291
  types of reactions, 291

glycopeptide transpeptidase, 760
glycoproteins, 244–246
  collagen, **209**
glycosidic linkages
  α and β, 154
  α-1,4-, **378**
  α-1,6-, **378**
  $\beta(1 \longrightarrow 4)$, **154**
  N, 529
glycosyl-enzyme intermediate, 168
glycyl-tRNA, in cell wall
    synthesis, 759
glycyltyrosine, 170–174, **171**
glyoxylate, **548**
GMP. *See* guanosine monophosphate
gout, 549–550
Gram stain, 755
gram-negative bacteria, 755
gram-positive bacteria, 755
gramicidin A, **780,** 783–784
gramicidin S, **673,** 675
grana, 456
Green, D., 411
Green, M., 737
green algae, 469
green sulfur bacteria, 462
Greenberg, G. R., 532
Griffith, F., 559
group translocation, 778
Grunberg-Manago, M., 628
GTP. *See* guanosine triphosphate
guanidine hydrochloride, **36**
guanine, **529,** 558
guanine-cytosine base pair, **565**
guanosine, **530**
guanosine diphosphate (GDP), 531
guanosine monophosphate
    (GMP), **535**
guanosine triphosphate (GTP), 531
  in citric acid cycle, 311
  in gluconeogenesis, 372
  in protein synthesis, 665–667
D-gulose, **304**

H zone, 827, 828
Hageman factor, 201
Haldane, J. B. S., 200
half-cell, 333
Hanson, J., 829
hapten, 732
  binding to antibody, 734
haptenic determinant, 732

Harden, A., 278
Haworth projections, 280
Hb. *See* hemoglobin
heavy (H) chains, 736
heavy meromyosin, 831
Hecht, S., 799
helix
  α, **32,** 32–33
  DNA double, 563–566
heme, **47**
  breakdown, 524–525
  coordination positions, **47,** 57
  environment in myoglobin, 58
  model, 48
  synthesis, 521–523
heme A, **339**
heme oxygenase, 524
heme-heme interaction, 75
hemiacetal, 279
hemiketal, 280
hemoglobin (Hb), 62–112
  A, 62
  A_2, 62
  allosteric properties, 71–91
  amino acid sequence, 62, 65
  β-chain structure, 64
  binding of CO, 94
  binding of CO_2, 88
  binding of DPG, 84–86
  binding of H^+, 86–88
  Bohr effect, 76, 86–88
  carbamoylation of, 87
  cooperative binding, 75, 89–90
  critical residues, 66
  δ chain, 62
  effect of CO_2 on, 77
  effect of DPG on, 77
  effect of pH on, 76
  embryonic, 62
  F, 62
  γ chain, 62
  Gun Hill, 110
  Hammersmith, 108
  M, 107–108
  models, 63, 64, 71, 85
  molecular pathology of, 107, 110
  mutants, 107, 638
  quaternary structure, 63
  S, 95–107
  structural changes, 81–84
  subunit interaction, 80
  subunit interfaces, 81

  subunit structure, 62–65
  x-ray analysis, 63
hemophilia, 198–200
Henderson-Hasselbalch equation, 43
Henseleit, K., 437
Heppel, L., 809
heptoses, 279
Herrick J., 95
Herriott, R., 560
Hers' disease, 396
Hershey, A., 562
heterochromatin, 703
heterogeneous RNA of the
    nucleus, 705
heterokaryon, 246
heterotropic effects, 138
hexa-NAG, 161
hexadecanoate, **406**
hexadecenoate, **406**
hexokinase, 11, 283–284
hexokinase deficiency, 301
hexose monophosphate pathway. *See*
    pentose phosphate pathway
hexoses, 279, **304, 305**
high-energy bonds, 266
high-energy phosphate
    compounds, 266
Hill, R., 463
Hill coefficient, 74
Hill plot, 74
Hill reaction, 463
histamine, **520**
histidine, **16**
  oxidation of, 442
  synthesis of, 514–515
histidine–aldol cross-link, **220**
histidinol, **515**
histidinol phosphate, **515**
histones, 700–704
  amino acid sequence of histone
      IV, 702
  evolutionary aspects, 701–702
  packing of DNA, 702–704
  types, 701
Hodgkin, D., 63, 445, 818
homocysteine, **509**
  conversion to cysteine, 510
homocysteine transmethylase, 509
homogentisate, **451**
homogentisate oxidase, 451
homotropic effects, 138
Holley, R., 617, 649

Horecker, B., 357
hormone action, 807–824
  cyclic AMP as mediator, 808–814
hormone-receptor complex, 819
hormones
  effect on enzymes, 118
  effect on lactose synthesis, 399
hnRNA, 705–706
HPr protein, 778–779
Huxley, A., 829
Huxley, H., 829, 833
hybridization, of RNA and DNA,
    600–601
hydrazinolysis, 45
hydride ion, 296
hydrogen bonds, 139, 141–143
  lengths of, 141
hydrolysis, 116
hydrophilic units, 233
hydrophobic interactions, 147
  in lipid bilayers, 236
hydrophobic units, 234
hydroxocobalamin, 446
3-hydroxyacyl-ACP-dehydratase, 422
L-hydroxyacyl CoA, 412
L-3-hydroxyacyl CoA
    dehydrogenase, 413
hydroxyapatite, 217
  columns of, 695
D-3-hydroxybutyrate, **417**
D-3-hydroxybutyryl-ACP, **423**
hydroxyethyl-TPP, **317**
hydroxylamine, 642, 794
21-hydroxylase deficiency, 498
hydroxylation, of collagen,
    208–209, 218
5-hydroxylysine, **208**
hydroxymethylation, 713
5-hydroxymethylcytosine, **713**
  α-glucosylated derivative, **713**
3-hydroxy-3-methylglutaryl CoA, **417**
  in cholesterol synthesis, 489
  in oxidation of leucine, 449
3-hydroxy-3-methylglutaryl CoA
    reductase, 489
p-hydroxyphenylpyruvate, **451, 512**
hydroxyphenylpyruvate oxidase, 451
17α-hydroxyprogesterone, **498**
4-hydroxyproline, **17, 208**
  attachment of sugar, **209**
5-hydroxytryptamine, 521
hyperammonemia, 440

hyperpolarization, 802
hypervariable regions, 744–745
hypochromicity of DNA, 695
hypoglycemic effect, 815
hypoxanthine, **548**
hypoxanthine-guanine phosphoribosyl
    transferase, 536
    deficiency diseases, 550–551

I band, 827, 828
i gene, 681–682
    mutants, 684
ice, 145
icosahedron, 710
D-idose, **304**
IgG. *See* immunoglobulin G
immediate-early genes, 688, 712
imidazole acetol phosphate, **515**
imidazole glycerol phosphate, **515**
4-imidazolone 5-propionate, **443**
imino acids, 15
immune response, 731–733
immunogen, 732
immunoglobulin A, 745
immunoglobulin D, 745
immunoglobulin E, 745
immunoglobulin G, 732
    active fragments, 735–736
    amino acid sequence, 742–743
    antigen-binding site, 738
    conformation of F_{ab}', 744
    domains, 743–744
    flexibility, 737
    L and H chains, 737
    subunit structure, 736–737
    variable and constant regions,
        740–742
immunoglobulin M, 732, 745
immunoglobulins, 731–752
    classes of, 745
    evolution of, 742–743
    in myelomas, 740–741
    variable and constant regions in,
        740–741
    See also antibodies
in situ hybridization, 697
indole-3-glycerol phosphate, **513**
induced fit, 122–123
    in carboxypeptidase A, 170,
        172, 174
inducers, 680, 683
Ingenhousz, J., 458

Ingram, V., 101
inhibition of enzymes, 129
initiation complexes, 665
initiation factors, 665
initiation signals
    in protein synthesis, 637, 664
    in RNA synthesis, 607
inosinate, **534**
inosine-adenine base pair, **657**
inosine-cytosine base pair, **657**
inosine-uracil base pair, **657**
inositol, **230**
insecticides, 185, 793
insertion, 721
    mutations, 640
instructive theory, 738–740
insulin
    action of, 815–816
    amino acid sequence of, 21
    conformation of, 818
    deficiency of, 820–821
    from proinsulin, 816–817
    receptor for, 818–820
insulin-agarose, 815–816
integral membrane protein, 242
intercalation, 614, 642
intercellular recognition, 246
intergenic suppressor, 657
intrinsic factor, 447
intrinsic pathway, 194
iodoacetamide, 129–130, **130**
ion pair, 140
ion-exchange chromatography, 21
    of amino acids, 23
ionic bond, 140
ionophores, 781
IPTG. *See* isopropylthiogalactoside
iron-sulfide cluster, 504
irreversible inhibition, 129–130
islet-cell adenoma, 818
isocitrate, **309**, 310
isocitrate dehydrogenase, 310
isocyanic acid, 106
isoelectric point, 70, 98
isoenzymes, 374
isoleucine, **15**
    oxidation of, 443, 449
    synthesis of, 516
isopentenyl pyrophosphate, **488**
    synthesis of, 490
isoprene, **488**
isoprene units, 758

isopropylthiogalactoside (IPTG), **680**
isovaleryl CoA, **448**
isovaleryl CoA dehydrogenase, 449
isozymes, 374

Jacob, F., 597, 681
Jardetzky, O., 773
Jerne, N., 739, 749

K_i, 133
K_M, 124–129
    definition, 125
    determination, 127
    significance, 127
    table of values, 128
Kaiser, A. D., 689
kappa chains, 741
Katz, B., 789
Keilin, D., 338
Kendrew, J., 48
Kennedy, E., 407, 768
ketimine, **435**
α-keto acids
    in transamination, 436
β-ketoacyl-ACP-reductase, 422
ketoacyl CoA, **412**
α-ketobutyrate, 436
    in cysteine synthesis, **510**
    in isoleucine synthesis, 516
ketogenesis, 812
ketogenic amino acids, 441
α-ketoglutarate, **310, 433**
    in amino acid oxidation,
        433–434, 442
    in hydroxylation of collagen, 208
α-ketoglutarate dehydrogenase
    complex, 311, 320
β-ketoisocaproate, **448**
ketone bodies, 416–418, 820
ketose-aldose isomerization, 358
ketoses, 279, **305**
β-ketothiolase, 413
Khorana, H. G., 631
kinases, 812–813
kinetics
    of enzymes, 124–129
    of reassociation of DNA, 695–697
King George III, 524
Klug, A., 653, 710
Knoop, F., 327, 408
Kornberg, A., 572
Kornberg, R. D., 253

Koshland, D., 173
Kraut, J., 189
Krebs, E., 390, 813
Krebs, H., 271, 327, 437
Krebs cycle. *See* citric acid cycle

lac. See lactose
α-lactalbumin, 398
β-lactam ring, 762
lactase, 398
lactate, **293, 373,** 373–374
lactate dehydrogenase, 293, 373–374
lactonase, 357
lactone analog of tetra-NAG, **166**
lactose, 118, **398,** 679
    model of, 399
    synthesis of, 118, 398
lactose operator, 684–685
lactose operon, 682–683
    mRNA sequence, 685
    role of cyclic AMP, 687
lactose repressor, 681–685
lactose synthetase, 118, 398–399
lambda bacteriophage
    circular DNA of, 720
    control of transcription in, 687–690
    electron micrograph of, 720
    genetic map of, 688
    lysogeny by, 720–722
    reciprocal recombination in, 721
lambda chains, 741
lambda operators, 689–690
lambda repressor, 689–690
lampbrush chromosome, 693
lanosterol, **492**
late genes, 688, 712
late-conductance channel, 797
lateral diffusion, 247
lathyrism, 220
laurate, **406**
Lavoisier, A., 457
lectins, 244
Lederberg, J., 749, 762
Lehninger, A., 409
Leloir, L., 385
Lesch-Nyhan syndrome, 551
leucine, **15**
    oxidation of, 448–450
leukemia therapy, 544
leukoviruses, 724
light
    in photosynthesis, 459–462

in vision, 798, 801–802
light (L) chains, 736
light meromyosin, 831
light reactions of photosynthesis, 461,
    464–469
lignocerate, **406**
Lind, J., 210
linear tetrapyrrole, **522**
linkage, 94
linoleate, **428**
linolenate, **428**
lipases, 193, 407–408
lipid bilayers, 235
    forces in, 236
    self-assembly of, 236
lipid vesicles, 237–238
lipids, 229
    solubilization by bile salts, 492
Lipkin, D., 809
Lipmann, F., 357
lipoamide, **318**
    model of, 318
    in pyruvate dehydrogenase,
        316–319
lipoic acid, 318
lipolysis, 408
    effect of cyclic AMP on, 812
    inhibition by PGE$_1$, 821
liposomes, 237–238
Lipscomb, W., 169, 542
Lister, J., 194
liver, source of cobalamin, 447
lock-and-key model, 123
Loewy, A., 837
lumirhodopsin, 802
Lusitanus, Z., 451
lymphocytes, 750
Lynen, F., 411, 449
Lyon effect, 703
lysine, **16**
lysinonorleucine, **223**
lysogenic bacterium, 688, 721
    stages in life cycle, 689
lysogenic pathway, 688
lysogeny, by lambda phage, 720–722
lysophosphatidate, **473**
lysosomes
    electron micrograph of, 486
    in Tay-Sachs disease, 486
lysozyme, 153–168
    active site, 157–159
    amino acid sequence, 156

binding affinity, 166
catalytic mechanism 160–168
catalytic residues, 163
conformation, 155–157
distortion of substrate, 160,
    165–167
encoded by T4 phage, 713
specificity, 154–155
lysyl oxidase, 219
lytic pathway, 688
Lyubimova, M., 830
D-lyxose, **304**

M line, 827
M phase, 703
M protein, 768
McArdle's disease, 395, 396
McCarty, M., 560
McConnell, H. M., 253
MacLeod, C., 560
magnesium porphyrin, 459
main chain, 18
major groove of DNA, 568
malaria, 104–105, 367
malate, **312**
    in citric acid cycle, 312
malate dehydrogenase, 313
    in cytosol, 426
malate enzyme, 427
4-maleylacetoacetate, **451**
malonate, **130, 327**
malonyl CoA, **419**
malonyl transacylase, 422
malonyl-ACP, 421
maltase, 397
maltose, 397, **398**
    model of, 397
maltotriose, 397
manganese in photosynthesis, 466
D-mannose, **304**
maple syrup urine disease, 450
Marmur, J., 600
Martius, C., 327
Mary, Queen of Scots, 524
matrix, 332
Mayer, J., 458
Mb. *See* myoglobin
melanin, 521
melanin granules, 812, 837
melting
    of collagen, 214
    of DNA, 593

melting temperature, 214
membrane fluidity, 247
membrane lipids, 229–240, **230–232**
  synthesis of, 479–492
membrane potential, 777, 797
membrane proteins, 240–243
membrane transport, 767–785
  active and passive, 768–769
  by antibiotics, 779–784
  of Ca^{2+}, 776
  of Na^+ and K^+, 769–775
  of Na^+ and sugars, 777
  phosphotransferase system of, 778
  roles of, 767
  of sugars, 777–779
membranes, 227–250
  excitable, 787–805
Menten, M., 124
β-mercaptoethanol, **36**
mercaptoethylamine unit, **269**
Meselson, M., 568, 598
messenger RNA (mRNA), 596
  for antibody L chain, 749
  direction of translation, 662
  discovery of, 597–601
  in *E. coli*, 596
metabolic regulation, 273
metabolism, 257
  basic concepts, 257–275
metarhodopsin I, 802
metarhodopsin II, 802
methanol poisoning, 134
MetHb. *See* methemoglobin
methemoglobin (MetHb), 48
methemoglobinemia, 108
N^5N^{10}-methenyltetrahydrofolate, **507**
methionine, **17**
  oxidation of, 443
  synthesis of, 509
methotrexate, **544**
methionyl-tRNA$_f$, **663**
methylcobalamin, 509
β-methylcrotonyl CoA, **449**
β-methylcrotonyl CoA
  carboxylase, 449
N^5N^{10}-methylenetetrahydrofolate,
  507, 544
β-methylglutaconyl CoA, **449**
methylmalonyl CoA
  in amino acid degradation, 443
  in fatty acid oxidation, 444
  inherited metabolic defects, 447

D-isomer, **444**
L-isomer, **444**
  isomerization, 444
methylmalonyl CoA mutase, 444
2'-O-methylribose, **613**
N^5-methyltetrahydrofolate, **507**
  in synthesis of methionine, 509
methylvaline, **615**
mevalonate, **488**
  synthesis of, 489
Meyerhof, O., 279
micelle, 234–235
Michaelis, L., 121
Michaelis-Menten model, 124–129
Micrococcus lysodeikticus, 153
microfilaments, 837
microtubules, 838
microvilli, 767
Miller, O., 698
mineralocorticoids, 495
  synthesis of, 498
miniature end-plate potentials, 790
minor groove of DNA, 568
Minot, G., 444
missense suppressors, 659
mitochondria, 331–332
  electron micrograph, 331
  permeability, 332, 341, 343
  structure, 332
mixed-function oxidases, 496
mm Hg, 72
modifier subunit, 118, 398
molecular disease
  definition, 100
  sickle-cell anemia, 95–112
molecular evolution, 111
molecular fugue, 217
molecular models, 2–3
molecular pathology, 107
molybdenum, 504
monoamine oxidase, 797
Monod, J., 135, 597, 681
mononucleotides. *See* nucleotides
monosaccharides, 279
morphogenesis of T4 phage, 714
motor end plate, 788
mouse satellite DNA, 696–697
mRNA. *See* messenger RNA
Mueller, P., 233
Mueller-Rudin membranes, 239
Müller-Hill, B., 683
multiple myeloma, 740

Murphy, W., 444
muscle, 826–839
  electron micrographs, 12, 827,
    828, 829
  light micrograph, 826
muscle contraction, 826–839
  control, 835–836
  force generation, 834–835
mutations, 638, 640–642
myeloma immunoglobulins, 740
myofibrils, 827
myoglobin (Mb), 46–62
  α-helix content, 54
  absorption spectrum, 57
  amino acid sequence, 54
  apomyoglobin, 61
  conformation, 52–57
  denaturation, 61
  electron-density map, 56
  heme, 47–48
  heme environment, 55–59
  model, 52–53
  oxygen-binding site, 55–57
  renaturation, 61
  x-ray analysis, 48–57
myokinase, 264, 539, 837
myosin, 828, 830–835
  ATPase activity of, 831
  binding of actin to, 832
  cleavage of, 831
  electron micrograph of, 830
  in platelets, 838
  structure of, 830–831
  thick filaments from, 830
  ubiquity of, 837
myristate, **406**

N protein, 688
Nachmansohn, D., 791
NAD$^+$. *See* nicotinamide adenine
  dinucleotide, oxidized form
NADH. *See* nicotinamide adenine
  dinucleotide, reduced form
NADH dehydrogenase, 336
NADP$^+$. *See* nicotinamide adenine
  dinucleotide phosphate,
  oxidized form
NADPH. *See* nicotinamide adenine
  dinucleotide phosphate,
  reduced form
NAG. *See* N-acetylglucosamine

Na⁺-K⁺ ATPase, 770

Na⁺-K⁺ transport, 769–775
  in driving glucose transport, 777
  inhibition of, 774–775
  mechanism of, 771–774
  Na⁺-K⁺ ATPase in, 770
  orientation of, 771

NAM. *See* N-acetylmuramic acid

nearest-neighbor analysis, 576–577

negative nitrogen balance, 503

neostigmine, 792, **793**

nerve gases, 185, 793

nerve growth factor, 13

Neuberg, C., 279

neuromuscular junction, 788

neurotoxins, 791

neurotransmitter, 788

neutral fats. *See* triacylglycerols

niacin, **271**

Nicholson, G., 249

nicotinamide adenine dinucleotide,
  oxidized form (NAD⁺), **266**
  biosynthesis of, 545
  model of, 294
  reactions of, 267

nicotinamide adenine dinucleotide,
  reduced form (NADH), **267**, 268
  from citric acid cycle, 308
  from fatty acid oxidation, 412
  from glycolysis, 290
  oxidation of, 341–343
  stereospecificity of, 324

nicotinamide adenine dinucleotide
  phosphate, oxidized form
  (NADP⁺), **266**

nicotinamide adenine dinucleotide
  phosphate, reduced form
  (NADPH), **356**
  in cholesterol biosynthesis, 492
  in deoxyribonucleotide
    biosynthesis, 543
  formation by malate enzyme, 427
  formation by pentose phosphate
    pathway, 356–358
  formation in photosynthesis,
    465–466
  generation of, 357, 427, 465–466
  in hydroxylation of steroids,
    495–496
  role of, 268, 356

nicotinate, **271, 545**

nicotinate ribonucleotide, **545**

Niedergerke, R., 829

night blindness, 800

ninhydrin, **23**

Nirenberg, M., 627, 631

nitriles, 220

nitrogen fixation, 504

nitrogenase, 504

p-nitrophenol, **183**

p-nitrophenyl acetate, 182–183, **183**

p-nitrophenyldiazonium hapten, **738**

nitrous acid, 641

nitroxide group, 253

Nomura, M., 660

nonactin, 782–783

noncompetitive inhibition, 131–134

noncyclic photophosphorylation, 467

nonessential amino acids, 503
  synthesis of, 503–510

nonheme iron proteins, 312, 337,
  466, 504

nonhistone proteins, 703

nonpermissive hosts, 723

nonsense mutations, 640, 658

noradrenaline, **796**, 797

norepinephrine, **796**, 797

nuclear membrane, 693

3′ ⟶ 5′ nuclease, 579

nucleation sites, 217

nucleolus, 698

nucleophile, 185–186

nucleoside diphosphate kinase, 539

nucleoside diphosphate sugars, 398

nucleoside monophosphate
  kinases, 539

nucleoside nomenclature, 529–531

nucleoside phosphorylases, 547

nucleoside triphosphates, 264

nucleotidases, 547

nucleotides, 529–547, **531**
  biosynthesis, 532–547
  nomenclature, 529–531
  roles, 528

nutritional requirements, 270–271,
  428, 503

O₂, in hydroxylations, 208,
  495–496, 524

Ochoa, S., 411, 628

octadecadienoate, **406**

octadecanoate, **406**

octadecatrienoate, **406**

octadecenoate, **406**

Ogston, A., 323

Okazaki, R., 535

Okazaki fragments, 535

oleate, **230, 428**
  molecular model of, 229

oncogenic viruses, 722–726

oncorna viruses, 724

one-carbon units, 506–509

operator site, 681–685
  in *lac* operon, 684–685
  in λ bacteriophage, 689–690

operons, 681
  arabinose operon, 686
  lac operon, 682–683
  in λ phage, 687–690
  trp operon, 686

opsin, 800

optical transform, 52

organic fluorophosphates, 184, 793

ornithine, **438**

ornithine transcarbamoylase, 438

orotate, **538**

orotidylate, **539**

orotidylate pyrophosphorylase, 539

orthophosphate, resonance forms, 265

Otto, J., 199

ouabain, 774, **775**

outer segment, 799

overlap peptides, 29

oxalic acid, **134**

oxaloacetate, **309**
  in amino acid oxidation, 448
  in citric acid cycle, 309
  in gluconeogenesis, 371–373

oxalosuccinate, **310**

oxidation-reduction potential. *See*
  redox potential

oxidative deamination, 432–434

oxidative phosphorylation, 331–353
  ATP generation, 340–341
  control, 345
  electron carriers, 336–340
  mechanisms, 346–348
  reconstitution, 348
  uncoupling, 345

oxygen evolved in photosynthesis, 463

oxygen carriers. *See* myoglobin;
  hemoglobin

oxygen dissociation curves, 72–77
  of hemoglobin, 73
  of myoglobin, 73

~ P, 266
P:O ratio, 340, 343
P_{450}, 496
P700, 465, 467, 469
Palade, G., 332
palmitate, **230, 406**
  ATP yield from, 415
  molecular model of, 229
  synthesis of, 424
palmitoleate, **428**
palmitoyl CoA, **484**
pamaquine, 366
pancreatic ribonuclease, 618
pancreatic zymogens, 178–179
  activation of, 192
pancreatitis, 192
pantothenate, **271, 546**
  in coenzyme A, 269
pantothenate unit, **269**
papain, 735
paramagnetic analog, 253
parathion, 793
Pardue, M. L., 697
Park, J., 762
Parnus, J., 279
passive transport, 768–769
Pasteur, L., 278
Pauling, L., 30, 98, 111, 167, 738
penicillin, 761–765, **762**
  discovery of, 761
  hydrolysis of, 764
  mechanism of action of, 762–763
  model of, 762
penicillinase, 764–765
penicillinoyl-enzyme complex,
  763, 764
penicillium mold, 761
penicilloic acid, **764**
pentaglycine bridge, 755
pentose phosphate pathway, 356–367
  control, 361–363
  nonoxidative branch, 358–361
  oxidative branch, 358
  table of reactions, 363
  tissue differences, 362–363
pentose shunt. See pentose phosphate
  pathway
pentoses, 279, **304, 305**
pepsin, 178
pepsinogen, 178
peptide bond, **18**
  hydrolysis of, 116

peptide group, 31
peptide synthesis, 673–675
  evolution of, 675
peptides
  chymotryptic, 29
  conformation, 30–32
  hydrolysis, 23
  overlap, 29
  sequence analysis, 22–30
  tryptic, 26, 29
peptidoglycan, 755
peptidyl site (P site), 665
peptidyl transferase, 667
performic acid, **30**
permease, 768
permissive hosts, 722
Perutz, M., 48, 63
peripheral membrane protein, 242
permeability properties
  coefficients, 240
  of lipid bilayers, 237–240
pernicious anemia, 444, 447
pH
  definition of, 42
  optimum for lysozyme, 167
phenoxazone ring, 614
phenylacetate, **408**
phenylalanine, **16**
  hydroxylation, 450
  oxidation, 450–453
  in phenylketonuria, 452
  synthesis, 512
phenylalanine hydroxylase, 450
  deficiency in phenylketonuria, 452
phenylethyl imidazole, **58**
phenylisothiocyanate, **24, 27**
phenylketonuria, 452–453
phenylpropionate, **408**, 452–453
  in synthesis of phenylalanine, 512
phenylthiohydantoin, 26, **27**
φX174 phage, 571–572
Phillips, D., 157
phosphagens, 836
phosphate group-transfer potential,
  264–266
phosphatases
  in cell-wall synthesis, 760
  in gluconeogenesis, 369
  in glycogen metabolism, 393
  in RNA sequencing, 618
  in synthesis of triacylglycerols, 480
phosphatidal choline, **483**

phosphatidate, **230, 480**
  as precursor of
    CDP-diacylglycerol, 481
  as precursor of triacylglycerol, 480
phosphatidyl choline, **231**
  molecular model of, 233
  spin-labeled, 253
  synthesis of, 482, 483
phosphatidyl ethanolamine, **231**
  synthesis of, 482
phosphatidyl inositol, **231**
phosphatidyl serine, **231**
  synthesis of, 481
3-phospho-
  5-pyrophosphomevalonate, **490**
phosphoanhydride bonds, 263
phosphoarginine, **836**
phosphocreatine, **836**
phosphodiester bridges, 557–558
phosphodiesterase, 394, 811
phosphoenolpyruvate, **265, 289**
  in amino acid synthesis, 511
  free energy of hydrolysis, 266
  in glycolysis, 289
  in membrane transport, 778
  in synthesis of UDP-NAM, 757
phosphofructokinase, 285
  control of, 299
phosphoglucomutase, 283
6-phosphogluconate, **358**
6-phosphogluconate
  dehydrogenase, 358
phosphogluconate oxidative pathway.
  See pentose phosphate pathway
6-phosphoglucono-δ-lactone, **358**
phosphoglucose isomerase, 285
2-phosphoglycerate, **289**
3-phosphoglycerate, **289**
  in photosynthesis, 470–471
  in synthesis of serine, 505
phosphoglycerate kinase, 288
phosphoglycerides, 229, **231**
  de novo synthesis of, 480–482
  salvage synthesis of, 482–483
phosphoglyceromutase, 288
phosphohistidine, **778**
3-phosphohydroxypyruvate, **505**
phospholipases, 484
phospholipids, 229
5-phosphomevalonate, **490**
4′-phosphopantothenate, **546**
4′-phosphopantothenyl cysteine, **546**

phosphopentose epimerase, 358
  in pentose phosphate pathway,
    358, 363
  in photosynthesis, 472
phosphoribomutase, 547
N'-5'-phosphoribosyl-AMP, **515**
N'-5'-phosphoribosyl-ATP, **515**
5'-phosphoribosyl-1-amine, **533**
5-phosphoribosyl-1-pyrophosphate
    (PRPP), **532**
  in purine nucleotide
    biosynthesis, 532
  in pyrimidine nucleotide
    biosynthesis, 539
  in synthesis of histidine, 514
  in synthesis of tryptophan, **513**
N-5'phosphoribosylanthranilate, **513**
N'-5'-phosphoribosylformimino-
    5-aminoimidazole-4-carboxamide
    ribonucleotide, **515**
phosphoribulose kinase, 472
N'-5'-phosphoribulosylformimino-
    5-aminoimidazole-4-carboxamide
    ribonucleotide, **515**
phosphorylase. *See* glycogen
    phosphorylase
phosphorylase kinase, 391
phosphorylase kinase phosphatase, 393
phosphorylation
  in active transport, 771, 776, 778
  in ATP synthesis, 288, 289,
    331–353
  in control of enzymes, 325,
    390–393
phosphorylcholine, **483**
phosphoserine, **117, 505**
phosphotransferase system, 778–779
photophosphorylation, 465, 467–469
photoreceptors, 798–805
photosynthesis, 456–476
  Calvin cycle, 469–473
  chloroplasts, 460–461
  dark reactions, 469–473
  discovery of basic equation,
    457–458
  efficiency of energy conversion, 473
  formation of ATP, 467–468
  formation of NADPH, 465–466
  general equation, 463
  light reactions, 464–469
  source of evolved O_2, 463
photosynthetic bacteria, 462

photosynthetic phosphorylation, 465,
    467–469
photosynthetic unit, 461–462
photosystem I, 464–467
photosystem II, 464–467
Physarum polycephalum, 837
physostigmine, 792, **793**
pK
  of amino acids, 44
  definition of, 42
  of residues in proteins, 86, 87
placental transmission site, 736
planar bilayer membranes, 238–240
plasma cells, 750
plasma cholinesterase, 796
plasma membranes, 227
plasma thromboplastin
    antecedent, 201
plasmalogens, **482**
plastocyanin, 468
plastoquinone, **468**
platelets, 812, 837
pneumococci
  polysaccharide coat of, **559**
  R and S forms of, 560
  transformation of, 559–561
pol mutants, 582
polar head group, 234
poliovirus, 716–717
Poljak, R., 744
poly A, 705
poly uridylate (poly U), 627–629
poly-L-lysine, 629
poly-L-phenylalanine, 629
poly-L-proline, 211, 629
polyenes, 460, 800
polynucleotide kinase, 618
polynucleotide phosphorylase, 628
polyoma virus, 570, 722–724
polypeptide chain, **18**
polypyrryl methane, **522**
polyribosomes, 668–669
polysome, 6, 668–669
polystyrene, **58**
Pompe's disease, 396
porphobilinogen, **521**
porphyrias, 523–524
porphyrins
  inherited metabolic defects,
    523–524
  synthesis, 521–523
Porter, R., 735

postsynaptic membrane, 788–789, 793
potassium ion
  binding to carriers, 781–783
  gates, 789, 797
  transport. *See* Na^+-K^+ transport
pregnenolone, **497**
prelumirhodopsin, 802
prephenate, **512**
Priestley, J., 457
primary structure, 34
primer
  in DNA synthesis, 574, 586
  in glycogen synthesis, 387
pro-α1 chain, 216
pro-α2 chain, 216
procarboxypeptidase, 178
procollagen, 215–216, 218
procollagen peptidase
  deficiency disease, 216
  in fiber formation, 218
proconvertin, 201
proelastase, 178
proenzyme, 178
progestagens, 495
progesterone, **497**
  synthesis of, 497
proinsulin, 816–817
  amino acid sequence of, 818
proline, **15**
  oxidation of, 442, 443
  synthesis of, 505
promoter sites, 607–609, 683,
    686, 690
prophage, 688, 721
propionyl CoA, **416**
  in amino acid degradation, **443**
  carboxylation of, 444
propionyl CoA carboxylase, 444
prostaglandins, 821–822
  action of aspirin, 822
  PGA_1, **821**
  PGB_1, **821**
  $PGF_{1\alpha}$, **821**
prosthetic group, 47
protein kinase, 812–813
protein synthesis, 645–676
  activated intermediates, 646–648
  aminoacyl site (A site), 665
  direction, 661–662
  electron micrograph, 662
  elongation, 665–667
  elongation factors, 666

protein synthesis (*continued*)
  inhibitors of, 670–672
  initiation, 663–665
  initiation factors, 665
  peptide-bond formation, 667
  peptidyl site (P site), 665
  posttranslational modifications, 669
  release factors, 667
  ribosomes, 659–661
  start signals, 637
  stop signals, 634–636
  termination, 667–668
  termination factors, 667
  transfer RNAs, 649–657
  translocation, 667
proteins
  amino acid residues, 13–15
  amino acid sequence, 21–22
  conformation, 30–39
  denaturation and renaturation,
    36–39
  function, 11–13
  hydrazinolysis, 45
  levels of structure, 34–36
  peptide bonds, 18
  periodic structures, 32–34
  purification, 19–21
  sequence determination, 22–30
  synthesis of. *See* protein synthesis
  thermodynamic stability, 39
  word origin, 22
proteolytic enzymes, 116
prothrombin, activation by Factor
    X_a, 197
protocollagen hydroxylase, 208
protofilaments, 838
proton pump, 347
protoplasts, 762
protoporphyrin IX, **47, 522**
  molecular model, 522
provirus, 723
proximal histidine, 55
PRPP. *See* 5-phosphoribosyl-
    1-pyrophosphate
pseudocholinesterase, 796
pseudouridine, **613**
D-psicose, **305**
PTH-amino acids, 26–27, **27**
puffer fish, 797–798
purine, **529**
purine biosynthesis, 532–537
  control, 537

de novo pathway, 532
  origin of atoms, 532
  salvage pathway, 536
purine degradation, 547–549
puromycin, 665, 671–672, **672**
pyranose, 280
2-pyridine aldoximine methiodide,
    794, 795
pyridoxal phosphate, **434**
  catalytic versatility of, 436
  fluorescence of, 121
  molecular model of, 435
  in racemases, 758
  in transaminases, 434–436
pyridoxamine phosphate, **434**
pyridoxine, **271, 434**
pyrimidine, **529**
pyrimidine biosynthesis, 538–542
  control, 540
  origin of atoms, 538
pyrimidine degradation, 549
pyrimidine dimers, **580**
  in xeroderma pigmentosum, 589
pyrophosphatase, 410
pyrophosphate
  free energy of hydrolysis, 266
  hydrolysis of, 387, 547, 647
5-pyrophosphomevalonate, **490**
pyrrole, 47
pyrrolidone rings, 213
pyrroline 5-carboxylate, **443, 505**
pyruvate, **288**
  in amino acid degradation, 441–442
  carboxylation of, 371
  conversion to acetyl CoA, 294
  conversion to ethanol, 292
  conversion to lactate, 293
  in glycolysis, 281–288
  oxidative decarboxylation of, 307
  reduction to lactate, 373
  from serine, 436
pyruvate carboxylase, 370–372
pyruvate decarboxylase, 292
pyruvate dehydrogenase complex,
    316–320
  control of, 325
pyruvate kinase deficiency, 301

Q protein, 689
Qβ phage, 718–719
  evolution in vitro, 719–720
quantal release, 789

quaternary structure, 36
Queen Victoria, 198

R17 phage, 718
racemases, 758
Racker, E., 348, 357
Raney nickel, 654
Rasputin, 199
reaction center, 462
reading frame, 626
recombination, 721
reconstitution, of membrane
    systems, 421
red drop, 464
redox couple, 333
redox potential, 333–335
  relation to free energy, 335
  standard values, 335
regulation of gene expression,
    679–690
regulatory gene, 680–683
regulatory subunit, 813
Reichard, P., 543
release factors, 667
releasing factor, 447
renin production, 812
reovirus, 717–718
repair of DNA, 578–580, 589
repetitive DNA sequences, 695–699
replicating forks, 583–584, 704
repressor, 681–683
  in arabinose operon, 686
  electron micrograph, 13
  interaction with operator, 685
  of *lac* operon, 681–685
  in λ bacteriophage, 689–690
  of *trp* operon, 686
resolution, 52
respiratory assembly, 331–332
  sequence of carriers, 336
respiratory control, 345
restriction endonucleases, 715–716
reticulocytes, 661
retinal isomerase, 800
retinal rod cells, 798–799
  electron micrographs, 12, 729, 798
retinal rod membranes, gel
    pattern, 241
retinals, 800
  all-*trans* isomer, **800**
  11-*cis* isomer, 7, **800**

retinals (*continued*)
  isomerization of, 801
retinol, **270**
retinol dehydrogenase, 800
reverse transcriptase, 716, 725
reversible inhibition, 130
Rhizobium bacteria, 504
rho factor, 611
rhodopsin, 800–803
ribitol, precursor of teichoic acid, 761
riboflavin, **271**
riboflavin 5′-phosphate, 545
ribonuclease
  amino acid sequence, **37**
  binding of substrate, 142
  denaturation, 36–37
  renaturation, 38–39
  scrambled, 39
ribonuclease S
  active site of, 9
  model of, 9
ribonuclease T$_1$, 618
ribonuclease U$_2$, 618
ribonucleic acid, **595**
  cleavage, 612–613
  control of synthesis, 679–691
  in eucaryotes, 705–706
  hnRNA, 705
  mRNA. *See* messenger RNA
  modification, 612–613
  as primer in DNA synthesis, 586
  processing, 612–613
  rRNA. *See* ribosomal RNA
  sequencing, 617–620
  structure, 595–596
  synthesis, 602–605, 607–613
  synthetic RNA as mRNA, 627
  tRNA. *See* transfer RNA
  types, 596–597
ribonucleoside, 530
D-ribose, **304**, 529, 595
ribose 5-phosphate, **358**, 359
  in nucleotide biosynthesis, 532
  in photosynthesis, 471
ribosomal RNA (rRNA), 596, 612
  genes for, 698–699
ribosome-binding sites, **664**
ribosomes, 659–661
  aminoacyl (A) site, 665
  in *E. coli,* 659
  electron micrograph, 6
  in eucaryotes, 659, 669

function, 599
  initiation complexes, 665
  peptidyl (P) site, 665
  polysome, 668
  reconstitution, 660–661
  rough endoplasmic reticulum,
    669–670
ribothymidine, **613**
D-ribulose, **305**
ribulose 1,5-diphosphate, **471**
  conversion to fructose
    6-phosphate, 471
  regeneration, 472
ribulose diphosphate
    carboxydismutase, 471
ribulose 5-phosphate, **358,** 360, 473
  formation, 357–358
Rich, A., 653
rif-r mutants, 614
rifampicin, 614
rifamycin B, **614**
rRNA. *See* ribosomal RNA
RNA polymerase, 602–605
  core enzyme, 607
  DNA-dependent, 603
  DNA-directed, 603
  RNA-directed, 716
  holoenzyme, 607
  inhibitors of, 614–617
  subunit structure, 607
RNA replicase, 716
  of Qβ phage, 718–719
RNA synthetase. *See* RNA replicase
RNA tumor viruses, 724–726
RNA viruses, 716–719, 724–726
  modes of replication, 716
  poliovirus, 717
  Qβ phage, 718–719
  R17 phage, 718
  rabies virus, 717
  reovirus, 717–718
  vesicular stomatitis virus, 717
RNase. *See* ribonuclease
rod outer segment, 799
rods, 798–799
Roseman, S., 778
rotenone, 340–341
rough endoplasmic reticulum,
    669–670
Rous, P., 724
Rous sarcoma virus, 724
Rudin, D., 233

S phase, 703
salt bridge, 140
salt linkage, 140
salting-out, 49
salvage pathways
  for nucleotide synthesis, 536
  for phosphoglyceride synthesis,
    482–483
Sanger, F., 21, 618
Sanger's reagent, **25**
sarcolemma, 827
sarcomere, 827
sarcoplasm, 327
sarcoplasmic reticulum, 836
  Ca^{2+} transport by, 776
  gel pattern, 241
sarcosine, **615**
sarin, 793
satellite DNA, 696–697
de Saussure, T., 458
Schachman, H., 451
Schiff bases, **365**
  in aldolase, 294
  in collagen cross-links, 220
  in elastin cross-links, 223
  of pyridoxal phosphate, **435**
  of pyridoxamine phosphate, **435**
  in rhodopsin, 800
  in transaminases, 434–436
scurvy, 210
SDS-acrylamide gel
    electrophoresis, 241
secondary structure, 35
sedoheptulose 1,7-diphosphate,
    471, 472
sedoheptulose 7-phosphate, **359**
segmental flexibility, 737
selective theory, 739–740
self-assembly
  of lipid bilayers, 236
  of tobacco mosaic virus, 710–711
semiconservative replication, 568–570
Senebier, J., 458
Sephadex, 19
sequenator, 29
sequential feedback control, 516
serine, **15**
  deamination of, 436
  linkage to sugars, 244
  oxidation of, 441
  in phosphoglycerides, 230
  synthesis of, 505

serine dehydratase, 436
serine proteases, 185
   homologies, 190–192
serotonin, 521
shikimate, **511**
shrinkage temperature, 214
sialic acids, **485**
sickle-cell anemia, 95–107
   clinical report, 95
   genetics of, 97
   molecular defect, 102–104
   protection against malaria, 104–105
   sickle cells, 95, 97
   therapy, 105–107
sickle-cell hemoglobin (Hb S)
   abnormal solubility, 97
   amino acid substitution, 102–104
   carbamoylation, 106
   electrophoretic mobility, 98
   fibers of, 98
   reaction with cyanate, 106
   sticky patch, 102
sickle-cell trait, 97, 105
side chain, 13, 18
sigma subunit, 607–609
silk fibroin, 699
simian virus 40. *See* SV40 tumor
   virus
Singer, S. J., 249
Sinsheimer, R., 571
Sjostrand, F.,
skeletal models, 2
skeletal muscle, 827
skin cancer, 589
Skou, J., 770
sliding-filament model, 829–830,
   834–835
Smith, E., 701
snake venom phosphodiesterase, 618
Snell, E., 435
sodium dodecyl sulfate, 241
sodium ion, binding to carriers,
   782–783
sodium ion gates, 789, 797
sodium ion transport. *See* Na+-K+
   transport
somatic-hypermutation
   hypothesis, 748
somatic-recombination hypothesis, 748
D-sorbose, **305**
space-filling models, 2, 3

sperm whale myoglobin. *See*
   myoglobin
sphingomyelin, **232**
   molecular model of, 234
sphingosine, **232, 484**
   as precursor of ceramide, 485
   synthesis of, 484
Spiegelman, S., 600, 719
spin-label, as a probe, 253
squalene, **488**
   as precursor of cholesterol, 492
   synthesis of, 490–491
squalene epoxide, **492**
Stahl, F., 568
starch, 396–397
starch-gel electrophoresis, 100
Starling, E., 807
steady state, 125
stearate, **406**
stereospecific hydrogen transfer,
   323–325
steroid hormones, 822–823
   inherited metabolic defects,
      498–499
   synthesis, 494–499
steroids
   α and β substituents, **494**
   nomenclature, 494
stop signals, 634–636
streptomyces, 614
streptomycin, 670–671, **671**
striated muscle, 827
Strominger, J., 762
Stuart factor, 201
submitochondrial particles, 348–349
substitution mutations, 640
substrate-level phosphorylation, 311
substrate specificity, 116
subtilisin, 116, 191–192
subunit, 36
succinate, **311**
succinate dehydrogenase, 312
   competitive inhibition, 130
succinyl CoA, 310
   in amino acid degradation, 443
   in citric acid cycle, 310–311
   in porphyrin synthesis, 521
succinyl CoA synthetase, 311
succinyl choline, **796**
sucrase, 398
sucrose, 397, **398**

sugars, 279, 397–399
sulfanilamide, **554**
sulfhydryl buffer, 366
superhelical cable, 212
supressor mutations, 657–659
Sutherland, E., 808
SV40 tumor virus, 722–724
   electron micrograph of, 7, 708,
      722, 723
symmetry
   in protein–nucleic acid
      interactions, 617
   in repressor-operator interactions,
      685, 690
   in restriction of DNA, 715
   of viruses, 710
synaptic body, 799
synaptic cleft, 788–789
synaptic vesicles, 788–790
synaptosomes, 788
synthetic messengers, 627–635
Szent-Györgyi, Albert, 327, 832
Szent-Györgyi, Andrew, 831

T. *See* thymine
T antigen, 724
T channels, 836
T_m, 214
   of collagen, 214
   of DNA, 593, 695
T_s, 214
T2 bacteriophage, 560–562
T4 bacteriophage, 712–715
   assembly of, 713–715
   infection by, 712–713
tabun, **793**
tadpole metamorphosis, 221
D-tagatose, **305**
tail fibers, 712
Talmage, D., 749
D-talose, **304**
taurine, **493**
taurocholate, 403
tautomers, 640–641
Tay-Sachs disease, 486–487
teichoic acid, 755, **761**
Temin, H., 724, 725
temperate phage, 687
templates, 566
termination signals, 634–636
tertiary structure, 36

testosterone, 498
  model of, 823
tetracosanoate, 406
tetracyclines, 671
tetradecanoate, **406**
tetrahydrobiopterin, **450**
tetrahydrofolate, **506,** 544
  conversions of derivatives, **507**
  one-carbon units carried, 506
tetraiodothyronine, **520**
tetraoxypyrimidine, **820**
tetrapyrrole, 47
tetrodotoxin, **797**
tetroses, 279, **304, 305**
theophylline, 811
thermodynamic efficiency
  of photosynthesis, 472
thermodynamics, 258–262
thermogenesis, 346
theta structures, 583
thiamine deficiency, 321
thiamine pyrophosphate (TPP), **317,**
  363–365
  molecular model of, 364
  in pyruvate dehydrogenase,
    316–318
thiazolidine ring, 762
thick filaments, 827–828, 833–835
thin filaments, 828, 833–835
thioesters, 297
  in peptide synthesis, 673
thiogalactoside transacetylase, 680–682
thiophorase, 418
thioredoxin, 543
thioredoxin reductase, 543
threonine, **16**
  conversion to isoleucine, 516
  deamination of, 436
  linkage to sugars, 244
  oxidative, 443
threonine deaminase, 118, 516
threonine dehydratase, 436
D-threose, **304**
thrombin, 116
  action on fibrinogen, 195
  homology to trypsin, 197
  specificity, 197
thrombosthenin, 838
Thucydides, 731
thykaloid membranes, 460–461
  electron micrograph, 456

thymidine 5′-phosphate, **531**
thymine, **529,** 558
  degradation of, 549
thymine dimer, **580**
thyroxine, **520**
tissue factor, 201
titration curve, 43
tobacco mosaic virus (TMV), 709–711
  model of, 710
  mutations of, 638
  self-assembly of, 710–711
α-tocopherol, **270**
torr, 72
Torricelli, E., 72
tosyl-L-phenylalanine chloromethyl
  ketone, **185**
TpC, 835
TpI, 835
TPN+. *See* nicotinamide adenine
  dinucleotide phosphate, oxidized
  form
TPNH. *See* nicotinamide adenine
  dinucleotide phosphate, reduced
  form
TPP. *See* thiamine pyrophosphate
TpT, 835
transaldolase, 358–360
  mechanism of, 365–366
transamidase, 196
transamidation, 196
transaminases, 433
transamination, 433–436
transcarboxylase, 420
transcription, 594
  control of, 679–691
  direction, 610–611
  electron micrograph, 606
  inhibition of, 614–617
  initiation, 607–610
  RNA polymerase, 602–605, 607
  strand selection, 605–606
  termination, 611–612
transfer RNA (tRNA), **651**
  amino acid attachment site, 624
  anticodon, 652, 654
  common design, 649–652
  conformation, 652–654
  in *E. coli,* 596
  function, 623–625
  initiator tRNA, 663
  sequence determination, 649–651

  sequence of yeast alanine
    tRNA, 651
  suppressor mutants, 657
  template recognition site, 624
  unusual bases, 652
  wobble in recognition, 655–659
transferase, 382
transferrin, 11
transformation by DNA, 559–561
transformation by tumor viruses,
  722–726
transformylase, 663
transglycosylation, 168
transition-state analog, 764
transitions, 640
transketolase, 358–360
  mechanism of, 363–365
  in pentose phosphate pathway,
    358–360
  in photosynthesis, 471, 473
translation. *See* protein synthesis
translocase, 667
translocation in protein synthesis, 667
translocation of ions. *See* membrane
  transport
transmembrane channel, 784
transpeptidation, 760, 763
transport. *See* membrane transport
transport antibiotics, 779–784
transsuccinylase, 320
transverse diffusion, 247, 253
transversions, 640
triacylglycerol synthetase
  complex, 480
triacylglycerols, **404, 480**
  as energy store, 406–408
  synthesis of, 478–480
tricarboxylic acid cycle. *See* citric
  acid cycle
triglycerides. *See* triacylglycerols
trihydroxycoprostanoate, **498**
trimethoprim, 663
tri-NAG, **158**
trinucleotides, 631
triose kinase, 399
triose phosphate isomerase, 286
trioses, 279, **304, 305**
triple helix of collagen, 211–213
triplet, 625
tRNA. *See* transfer RNA
tropocollagen, 206, 211, 218

tropomyosin, 828, 835
troponin complex, 828, 835
trypsin, 116
  homology to chymotrypsin,
      190–191
  in sequence analysis, 26, 28
trypsinogen, 192
tryptophan, **16**
  as corepressor, 686
  synthesis of, 512–514
tryptophan synthetase, 514
  colinearity studies, 639
  fluoresence of, 121
  mutations of, 638
Tsarevich Alexis, 198–200
Tswett, M., 101
TTX, **797**
d-tubocurarine, **795**
tubulins, 838
tumor viruses, 722–726
turnover number, 129
Tween 80, 791
tyrosine, **16**
  oxidation of, 450–451
  from phenylalanine, 450
  as precursor of
      neurotransmitters, 796
  synthesis of, 450, 512
tyrosine-*O*-sulfate, 196
tyrosine hydroxylase, 796

U. *See* uracil
U antigen, 724
ubiquinone, **337**
UDP-galactose, **400**
UDP-galactose-4-epimerase, 400
UDP-glucose. *See* uridine
      diphosphate glucose
UDP-glucuronate, 525
UDP-NAG, **757**
UDP-NAM, **757**
UMP. *See* uridine monophosphate
uncouplers, 345–346
unit evolutionary period, 701–702
unsaturated fatty acids, oxidation
      of, 415
untranscribed spacers, 698
unwinding proteins, 587
uracil, **529, 595**
uracil dimer, 580

urate, **549**
  in gout, 549–550
urea, **36**
  in denaturation, 36, 736
  equilibrium with cyanate, 106
  in purine degradation, 548
  in urea cycle, **438**
urea cycle, 437–441
  inherited defects, 440
  link to citric acid cycle, 440
  reactions, 437
  stoichiometry, 440
urease, 548
ureotelic organisms, 437
uric acid, **548**
uricase, 548
uricotelic organisms, 437
uridine, **530**
uridine diphosphate-*N*-
      acetylglucosamine, **757**
uridine diphosphate galactose, **400**
uridine diphosphate glucose, **386**
  conversion to UDP-galactose, 400
  in glycogen synthesis, 358–387
uridine monophosphate, 531
uridine triphosphate, **540**
uridylate, **539**
urocanate, **443**
uroporphyrinogen I, 522, **523**
uroporphyrinogen III, **522**
uroporphyrinogen III
      cosynthetase, 523
UTP. *See* uridine triphosphate
Utter, M., 370

V_{max}, 124–129
  definition, 126
  determination, 127
  significance, 127
  table of values, 128
cis-vaccenate, 428
Vagelos, P. R., 420
Valentine, R., 737
valine, **15**
  oxidation of, 443, 449
valinomycin, **779**, 782
van der Waals bonds, 139, 144
van der Waals contact distance, 144
van der Waals radius, 144
Van Niel, C., 462

variable (V) region, 741
variegate porphyria, 524
Vennesland, B., 323
vesicles, active transport by, 777
vesicular stomatitis virus, 717
Vibrio cholerae, 814
virion, 708
viruses, 708–724
  assembly of, 709–711, 713–715
  types of, 709
viscoelastic technique, 694
viscosity of membranes, 247
vision, 798–805
Vitamin A, **270, 800**
  efficiency of, 800
Vitamin B$_1$ (thiamine), efficiency
      of, 321
Vitamin B$_2$, **271**
Vitamin B$_6$, **271, 434**
Vitamin B$_{12}$, 444–447
Vitamin D$_2$, **270**
Vitamin E, **270**
Vitamin K$_1$, **270**
vitamins
  as coenzymes, 270–271
  fat-soluble, **270**
  water-soluble, **271**
von Gierke, E., 395
von Gierke's disease, 395, 396

Wakil, S., 419
Wald, G., 801
Walsh, D., 813
Wang, J., 58
Warburg, O., 279, 357
water, 145–147
  effect on nonpolar interactions, 147
  effect on polar interactions, 146
  model of, 3
Watson, J., 563, 709
Watson-Crick double helix, 563–567
Westheimer, F., 323
wheat-germ agglutinin, 244
Wilkins, M., 563
Williams, R., 710
Wilson, I., 794
Withering, W., 775
Wood, W., 713
Wyman, J., 135

x-ray crystallography, 49–50
xanthine, **548**
xanthine oxidase, 547
xanthylate, **535**
Xenopus genes, 698–699
xeroderma pigmentosum, 589
D-xylose, **304**
D-xylulose, **305**

xylulose 5-phosphate, **359, 360**
  in pentose phosphate pathway,
    359–360
  in photosynthesis, 471

Yanofsky, C., 639
Young, W., 278

Z line, 827, 828, 834
Zimm, B., 570, 694
zinc, 169, 170
Zuckerkandl, E., 111
zwitterion, 13
zymase, 278
zymogen granules, 178–179
zymogens, 117, 178–203

COMMON ABBREVIATIONS IN BIOCHEMISTRY

A	adenine
ACP	acyl carrier protein
ACTH	adrenocorticotropic hormone
ADP	adenosine diphosphate
Ala	alanine
AMP	adenosine monophosphate
cAMP	cyclic AMP
Arg	arginine
Asn	asparagine
Asp	aspartate
ATP	adenosine triphosphate
ATPase	adenosine triphosphatase
C	cytosine
CDP	cytidine diphosphate
CoA	coenzyme A
CoQ	coenzyme Q (ubiquinone)
CMP	cytidine monophosphate
CTP	cytidine triphosphate
cyclic AMP	adenosine 3′, 5′-cyclic monophosphate
Cys	cysteine
d	2′-deoxyribo
DIPF	diisopropylphosphofluoridate
DNA	deoxyribonucleic acid
DNP	dinitrophenyl
DPN^+	see NAD^+
DPNH	see NADH
FAD	flavin adenine dinucleotide (oxidized form)
$FADH_2$	flavin adenine dinucleotide (reduced form)
FDNB	fluorodinitrobenzene
fMet	formylmethionine
FMN	flavin mononucleotide (oxidized form)
$FMNH_2$	flavin mononucleotide (reduced form)
G	guanine
Gln	glutamine
Glu	glutamate
Gly	glycine
GDP	guanosine diphosphate
GMP	guanosine monophosphate
GTP	guanosine triphosphate
Hb	hemoglobin
HbCO	carbon monoxide hemoglobin
HbO_2	oxyhemoglobin